T0233889

Lecture Notes in Computer Science 11896

More information about this series at http://www.springer.com/series/7412

Ingela Nyström · Yanio Hernández Heredia ·
Vladimir Milián Núñez (Eds.)

Progress in Pattern Recognition, Image Analysis, Computer Vision, and Applications

24th Iberoamerican Congress, CIARP 2019
Havana, Cuba, October 28–31, 2019
Proceedings

 Springer

Editors
Ingela Nyström
Uppsala University
Uppsala, Sweden

Yanio Hernández Heredia ⓘ
University of Information Science
Havana, Cuba

Vladimir Milián Núñez ⓘ
University of Information Science
Havana, Cuba

ISSN 0302-9743 ISSN 1611-3349 (electronic)
Lecture Notes in Computer Science
ISBN 978-3-030-33903-6 ISBN 978-3-030-33904-3 (eBook)
https://doi.org/10.1007/978-3-030-33904-3

LNCS Sublibrary: SL6 – Image Processing, Computer Vision, Pattern Recognition, and Graphics

This Springer imprint is published by the registered company Springer Nature Switzerland AG
The registered company address is: Gewerbestrasse 11, 6330 Cham, Switzerland

Preface

The Iberoamerican Congress on Pattern Recognition (CIARP), is an annual international conference that publishes original, high-quality papers related to Pattern Recognition, Artificial Intelligence, and related fields, welcoming contributions on any aspect of theory as well as applications. CIARP has become a key research event, and one of the most important in Pattern Recognition for the Iberoamerican community.

As has been the case for previous editions of the conference, CIARP 2019 hosted worldwide participants with the aim of promoting and disseminating ongoing research on mathematical methods and computing techniques for Artificial Intelligence and Pattern Recognition, in particular in Bioinformatics, Cognitive and Humanoid Vision, Computer Vision, Image Analysis and Intelligent Data Analysis, as well as their application in a number of diverse areas such as industry, health, robotics, data mining, opinion mining and sentiment analysis, telecommunications, document analysis, and natural language processing and recognition. Moreover, CIARP 2019 was a forum for the scientific community to exchange research experience, to share new knowledge, and to increase the cooperation among research groups in Artificial Intelligence, Pattern Recognition, and related areas.

CIARP has always been an open international event, and this 24th edition received 128 contributions from 23 countries. The biggest presence was Cuba, Brazil, Colombia, Mexico, Germany, and Chile. We also received contributions from Argentina, Austria, Canada, Ecuador, France, Italy, India, Japan, The Netherlands, Pakistan, Portugal, Romania, Spain, Sweden, Tunisia, UK, and USA.

After a rigorous blind reviewing process, where three highly qualified reviewers reviewed each submission (362 reviews from 146 reviewers who spent significant time and effort in reviewing the papers), 70 papers authored by 239 authors from 14 countries were accepted, which is an acceptance rate of 54.69%. The scientific quality of all the accepted papers was above the overall mean rating. The reviewers were chosen based on their expertise ensuring that they came from different countries and institutions around the world. We would like to thank all the members of the Program Committee for their work, which we are sure contributed to improving the quality of the selected papers.

The conference was held during October 28–31, 2019, at the Hotel Nacional de Cuba, and consisted of four days of papers, tutorials, and keynotes. As has been the case for the most recent editions of the conference, CIARP 2019 was a single-track conference. The program comprised eight oral sessions: Image Analysis and Retrieval; Signals Analysis and Processing; Video Analysis; Mathematical Theory of Pattern Recognition; Applications of Pattern Recognition; Machine Learning and Neural Networks; Data Mining, Natural Language Processing and Text Mining; Speech Recognition. Also, one additional poster sessions included papers on all previous topics.

CIARP 2019 was endorsed by the IAPR, the International Association for Pattern Recognition and therefore the conference gave out the CIARP-IAPR Best Paper Award. This award and the Aurora Pons-Porrata Medal (to an Iberoamerican woman with a prestigious career in Pattern Recognition and related fields) will be published as a Special Section in the *Pattern Recognition Letters* journal. Also, authors of the 20–25 best papers, within the aims and scope of the journals, will be invited to prepare an extended version for a Special Issue of the *Intelligent Data Analysis* (IDA) journal (http://www.iospress.nl/journal/intelligent-data-analysis), with guest editors A. Fazel Famili and José E. Medina Pagolan, and for a Special Issue of the *Pattern Analysis and Applications* (PAAA) journal (https://link.springer.com/journal/10044), with guest editors José Ruiz Shulcloper and Heydi Mendez Vazquez.

The program also included four invited talks by eminent speakers: Petra Perner, Alberto Del Bimbo, Massimo Tistarelli, and Jean-François Bonastre. Additionally, four tutorials, given by Ingela Nyström, Petra Perner, Alberto Del Bimbo, and Massimo Tistarelli completed the program. It was a pleasure to work with this outstanding group of invited speakers.

CIARP 2019 was organized by Universidad de las Ciencias Informáticas (UCI) and the Cuban Association for Pattern Recognition (ACRP). We appreciate their valuable contributions to the success of CIARP 2019. We gratefully acknowledge the help of all members of the Organizing Committee and of the Local Committee for their unflagging work in the organization of CIARP 2019 that allowed us to put together an excellent conference and proceedings.

We are especially grateful to Alfred Hofmann, Anna Kramer, and Volha Shaparava of Springer for their support and advice during the preparation of this LNCS volume.

Special thanks are due to all authors who submitted to CIARP 2019, including those of papers that could not be accepted. Finally, we hope that these proceedings will be a useful reference for the Pattern Recognition research community.

October 2019

Ingela Nyström
Yanio Hernández Heredia
Vladimir Milián Núñez

Organization

General Chairs

Ingela Nyström   Uppsala University, Sweden
Yanio Hernandez   Universidad de las Ciencias Informàticas, Cuba
Heydi Mendez-Vazquez   CENATAV, Cuba
Milton García-Borroto   CUJAE, Cuba

Program Chairs

Jose Ruiz Shulcloper   Universidad de las Ciencias Informàticas, Cuba
José Eladio Medina Pagola   Universidad de las Ciencias Informàticas, Cuba
Vladimir Miliàn Núñez   Universidad de las Ciencias Informàticas, Cuba

CIARP Steering Committee

Bernardete Ribeiro   APRP, Portugal
Eduardo Bayro-Corrochano   MACVNR, Mexico
César Beltrán-Castañón   APeRP, Peru
Julian Fierrez   AERFAI, Spain
José Ruiz Shulcloper   ACRP, Cuba
Marta Mejail   SARP, Argentina
Marcelo Mendoza   AChiRP, Chile
Joao Paulo Papa   SIGPR-BR, Brazil
Álvaro Pardo   APRU, Uruguay

Local Committee

Beatriz Aragón Fernández   Universidad de las Ciencias Informáticas, Cuba
Nadiela Milan Cristo   Universidad de las Ciencias Informáticas, Cuba
Hector Raul Gonzalez Diez   Universidad de las Ciencias Informáticas, Cuba
Edel Garcia-Reyes   CENATAV, Cuba
Javier Lamar   CENATAV, Cuba

CIARP-IAPR Best Paper Award Committee

Jose Ruiz Shulcloper   Universidad de las Ciencias Informáticas, Cuba
Ingela Nyström   Uppsala University, Sweden
Petra Perner   Institute of Computer Vision and Applied Computer Sciences IBaI Leipzig, Germany
Alberto Del Bimbo   Università degli Studi di Firenze, Italy
Massimo Tistarelli   Università di Sassari, Italy

Aurora Pons Porrata Award Committee

Olga Regina Pereira Bellon   Universidade Federal do Parana, Brazil
Maria Vanrell   Universitat Autònoma de Barcelona, Spain
Heydi Méndez Vázquez   CENATAV, Cuba

Program Committee

Sergey Ablameyko   Belarusian State University, Belarus
Niusvel Acosta-Mendoza   CENATAV, Cuba
Luis Alexandre   Universidade da Beira Interior, Portugal
Hector Allende   Universidad Técnica Federico Santa María, Chile
Fernando Alonso-Fernandez   Halmstad University, Sweden
Rene Alquezar   UPC, Spain
Leopoldo Altamirano   INAOE, Mexico
Mauricio Araya   Universidad Técnica Federico Santa María, Chile
Leticia Arco   VUB, Belgium
G. Arroyo-Figueroa   INEEL, Mexico
Akira Asano   Kansai University, Japan
Ali Ismail Awad   Luleå University of Technology, Sweden
Xiang Bai   Huazhong University of Science and Technology, China
Virginia Ballarin   National University of Mar del Plata, Argentina
Antonio Bandera   University of Malaga, Spain
Rafael Bello   UCLV, Cuba
Cesar Beltran Castanon   Pontificia Universidad Católica del Perú, Peru
José Miguel Benedí   Universitat Politécnica de Valéncia, Spain
Josef Bigun   Halmstad University, Sweden
Gunilla Borgefors   Information Technology, Sweden
Maria Elena Buemi   Universidad de Buenos Aires, Argentina
Lázaro Bustio-Martínez   INAOE, Mexico
Leticia Cagnina   Universidad Nacional de San Luis, Argentina
Luis V. Calderita   Universidad de Extremadura, Spain
Jose Ramon Calvo de Lara   CENATAV, Cuba
Sergio Daniel Cano-Ortiz   Universidad de Oriente, Cuba
Jesús Ariel Carrasco-Ochoa   INAOE, Mexico
César Castellanos   Universidad Nacional de Colombia, Colombia
Gerard Chollet   CNRS, France
Eduardo R.   Universidad de Cienfuegos, Cuba
  Concepcion-Morales
Raúl Cruz-Barbosa   Universidad Tecnológica de la Mixteca, Mexico
Mauricio Delbracio   Universidad de la República, Uruguay
Mariella Dimiccoli   Universitat Autónoma de Barcelona, Spain
Marcelo Errecalde   Universidad Nacional de San Luis, Argentina
Alfonso Estudillo-Romero   Universidad Nacional Autónoma de México, Mexico
Jacques Facon   Universidade Federal do Espírito Santo, Brazil

José Kadir Febrer-Hernández	Advanced Technologies Applications Center, Cuba
Gustavo Fernandez Dominguez	AIT Austrian Institute of Technology, Austria
Alicia Fernández	Universidad de la República, Uruguay
Carlos Ferrer	Universidad Central Marta Abreu de Las Villas, Cuba
Francesc J. Ferri	Universitat de Valencia, Spain
Marcelo Fiori	Universidad de la República, Uruguay
Giorgio Fumera	University of Cagliari, Italy
Antonino Furnari	Università di Catania, Italy
Silvio Guimaraes	PUC Minas, Brazil
Edel Garcia-Reyes	CENATAV, Cuba
Jose Garcia-Rodriguez	University of Alicante, Spain
María Matilde García Lorenzo	UCLV, Cuba
Milton García-Borroto	CUJAE, Cuba
Eduardo Garea-Llano	CENATAV, Cuba
Alexander Gelbukh	Instituto Politécnico Nacional, Mexico
Petia Georgieva	University of Aveiro, Portugal
Daniela Godoy	National Council for Scientific and Technological Research, Argentina
Lev Goldfarb	IIS, Canada
Luis Gomez Deniz	University of Las Palmas de Gran Canaria, Spain
Marta Gomez-Barrero	Hochschule Darmstadt, Germany
Pilar Gomez-Gil	National Institute of Astrophysics, Optics an Electronics, Mexico
Hector Raul Gonzalez Diez	UCI, Cuba
Rocio Gonzalez-Diaz	University of Seville, Spain
Antoni Grau	Technical University of Catalonia, Spain
Miguel Angel Guevara Lopez	University of Minho, Portugal
Bilge Gunsel	Istanbul Technical University, Turkey
Laurent Heutte	Université de Rouen, France
Michal Haindl	Institute of Information Theory and Automation ASCR, Czech Republic
Yanio Hernandez	Universidad de las Ciencias Informáticas, UCI
Javier Hernandez-Ortega	Universidad Autonoma de Madrid, Spain
Gabriel Hernandez-Sierra	Advanced Technologies Applications Center, Cuba
Raudel Hernández	CENATAV, Cuba
Xiaoyi Jiang	University of Münster, Germany
Maria-Jose Jimenez	University of Seville, Spain
Vitaly Kober	CICESE, Mexico
Martin Kampel	Vienna University of Technology, Austria
Tomas Krajnik	Czech Technical University, Czech Republic
Walter Kropatsch	Vienna University of Technology, Austria
Manuel S. Lazo-Cortés	INAOE, Mexico

Marcos Levano	Universidad Católica de Temuco, Chile
Aristidis Likas	University of Ioannina, Greece
Hélio Lopes	PUC-Rio, Brazil
Aurelio Lopez-Lopez	Instituto Nacional de Astrofisica, Optica y Electronica, Mexico
Juan Valentín Lorenzo-Ginori	Universidad Central Marta Abreu de Las Villas, Cuba
Octavio Loyola-González	Tecnologico de Monterrey's Puebla Campus, Mexico
Cristian López Del Alamo	Universidad La Salle, Peru
Filip Malmberg	Uppsala University, Sweden
Rebeca Marfil	University of Malaga, Spain
Francesco Marra	University Federico II of Naples, Italy
Jose Francisco Martinez-Trinidad	Instituto Nacional de Astrofísica Óptica y Electrónica, Mexico
Nelson Mascarenhas	UFSCar, Brazil
Rosana Matuk Herrera	Universidad Nacional de Luján, Argentina
José Eladio Medina Pagola	UCI, Cuba
Jiri Mekyska	Brno University of Technology, Czech Republic
Heydi Mendez-Vazquez	CENATAV, Cuba
Ana Maria Mendonça	Universidade do Porto, Portugal
Marcelo Mendoza	Universidad Técnica Federico Santa María, Chile
Vladimir Milián Núñez	University of Informatics Sciences, Cuba
Miguel Moctezuma-Flores	UNAM, Fac Ingenieria, Mexico
Manuel Montes-y-Gómez	INAOE, Mexico
Eduardo Morales	Instituto Nacional de Astrofísica, Óptica y Electrónica (INAOE), Mexico
Aythami Morales	Universidad Autonoma de Madrid, Spain
Annette Morales-González	CENATAV, Cuba
Sebastian Moreno	Universidad Adolfo Ibañez, Chile
Vadim Mottl	Computing Center of the Russian Academy of Sciences, Russia
Pablo Muse	Universidad de la República, Uruguay
Alberto Muñoz	University Carlos III, Spain
Alfredo Muñoz-Briseño	CENATAV, Cuba
Michele Nappi	Università di Salerno, Italy
João Neves	IT - Instituto de Telecomunicações, Portugal
Lawrence O'Gorman	Nokia Bell Labs, USA
Volodymyr Ponomaryov	Instituto Poltecnico Nacional, Mexico
Kalman Palagyi	University of Szeged, Hungary
Joao Papa	Sao Paulo State University, Brazil
Glauco Vitor Pedrosa	University of Brasilia, Brazil
Billy Peralta	Andres Bello University, Chile
Eanes Pereira	UFCG, Brazil
Nicolai Petkov	University of Groningen, The Netherlands
Ignacio Ponzoni	Universidad Nacional del Sur, Argentina
Osvaldo Pérez-García	CENATAV, Cuba

Adrián Pérez-Suay	Universitat de València, Spain
Airel Pérez-Suárez	CENATAV, Cuba
Maria Alejandra Quiros Ramirez	Max Planck Institute for Intelligent Systems, Germany
José Felipe Ramírez Pérez	University of Informatics Sicences, Cuba
Gregory Randall	Universidad de la República, Uruguay
Pedro Real	Universidad de Sevilla, Spain
Carlos A Reyes-Garcia	INAOE, Mexico
Dayana Ribas	University of Zaragoza, Spain
Bernardete Ribeiro	University of Coimbra, Portugal
Edgar Roman-Rangel	ITAM, Mexico
Alejandro Rosales-Perez	Tecnologico de Monterrey, Mexico
Jose Ruiz Shulcloper	University of Informatics Sciences, Cuba
Cesar San Martin	Universidad de la Frontera, Chile
Guillermo Sanchez-Diaz	Universidad Autonoma de San Luis Potosi, Mexico
Carlo Sansone	University of Naples Federico II, Italy
William Schwartz	Universidade Federal de Minas Gerais, Brazil
Juan Humberto Sossa Azuela	Instituto Politecnico Nacional, Mexico
Elaine Sousa	University of São Paulo, Brazil
Enrique Sucar	INAOE, Mexico
Josep Salvador Sánchez Garreta	Universitat Jaume I, Spain
Antonio-José Sánchez-Salmerón	Universitat Politécnica de Valéncia, Spain
Alberto Taboada-Crispi	UCLV, Cuba
Ruben Tolosana	Universidad Autónoma de Madrid, Spain
Esau Villatoro-Tello	Universidad Autonoma Metropolitana, Mexico
Ventzeslav Valev	Bulgarian Academy of Sciences, Bulgaria
Sergio A Velastin	Cortexica Vision Systems Ltd, UK
Ruben Vera-Rodriguez	Universidad Autonoma de Madrid, Spain
Max Viergever	University Medical Center Utrecht, The Netherlands
Gui-Song Xia	Wuhan University, China
Vera Yashina	Federal Research Center Computer Science and Control of the Russian Academy of Sciences, Russia

Additional Reviewers

Carlos Caetano	Universidade Federal de Minas Gerais, Brazil
Delia Irazu Hernandez Farias	INAOE, Mexico
Abbas Khosravani	Telecom SudParis, France
Gabriel Lucas Silva Machado	Universidade Federal de Minas Gerais, Brazil
Cristian Martinez	UNSa, Argentina
Ana Montalvo	CENATAV, Cuba

Thierry Moreira	UNICAMP, Brazil
Antonio Carlos Nazaré	Universidade Federal de Minas Gerais, Brazil
Rosa María Ortega Mendoza	Universidad Politécnica de Tulancingo, Mexico
Dijana Petrovska Delacretaz	Telecom SudParis, France
Raphael Prates	Universidade Federal de Minas Gerais, Brazil
Luciano Arnaldo Romero Calla	Universidad La Salle de Arequipa, Peru
Claudio Santos	UFSCar, Brazil
Jefry Sastre Pérez	Pontifical Catholic University of Rio de Janeiro, Brazil
Guilherme Schardong	Pontifical Catholic University of Rio de Janeiro, Brazil
Luiz José Schirmer Silva	Pontifical Catholic University of Rio de Janeiro, Brazil
Marcos Cleison Silva Santana	São Paulo State University, Brazil
Caio Cesar Viana da Silva	Universidade Federal de Minas Gerais, Brazil
Tomas Vintr	Czech Technical University, Czech Republic

Contents

Image Analysis and Retrieval

Machine Learning and Neural Networks

Mathematical Theory of Pattern Recognition

Pattern Recognition and Applications

Signals Analysis and Processing

Keynote Lecture

Incremental Learning of People Identities

Federico Bartoli, Federico Pernici, Matteo Bruni, and Alberto Del Bimbo[✉]

MICC, Media Integration and Communication Center,
Department of Information Engineering, University of Firenze, Florence, Italy
alberto.delbimbo@unifi.it

Abstract. Face recognition in unconstrained open-world settings is a challenging problem. Differently from the closed-set and open-set face recognition scenarios that assume that the face representations of known subjects have been manually enrolled in a gallery, the open-world scenario requires that the system learns identities incrementally from frame to frame, discriminate between known and unknown identities and automatically enrolls every new identity in the gallery, so to be able to recognize it every time it is observed again in the future. Performance scaling with large number of identities is likely to be needed in real situations. In this paper we discuss the problem and present a system that has been designed to perform effective open-world face recognition in real time at both small-moderate and large scale.

Keywords: Open-world recognition · Incremental learning · Large scale

1 Introduction

Deep face recognition is now believed to surpass human performance in many scenarios of face identity face verification and authentication [1,2] and is widely used in many fields such as military, public security and many other contexts of ordinary daily life. Despite this progress, many fundamental questions are still open and should be answered in order to build robust applications for real world. Different tasks have been identified for the evaluation of face recognition systems: face verification is traditionally relevant in access control systems, re-identification in multi camera systems; closed-set face identification [32] is relevant when searching individuals into a gallery of known subjects, such as for example in forensics applications; open-set identification [4] is relevant to search systems where the system should also be able to reject probes that are not present in the gallery; open-world recognition [6] is finally relevant to those cases where the system should also be able to reject probes that are not present in the gallery and at the same time enroll the rejected subjects as new identities in the gallery. While most of the research has addressed the verification, closed set and open set tasks, very little research has been done on the latter task, the open world task, despite of the highest relevance it has in real contexts, mainly due to the difficulty to solve the many problems that this task implies.

© Springer Nature Switzerland AG 2019
I. Nyström et al. (Eds.): CIARP 2019, LNCS 11896, pp. 3–15, 2019.
https://doi.org/10.1007/978-3-030-33904-3_1

In this paper, we discuss face recognition in unconstrained open-world settings and present a system that has been designed to operate in real time at both small-moderate and large scale. The open-world scenario inherits the basic working principles of face (re-)identification of the closed-set and open-set scenarios in which recognition of identities are performed by considering the distance (or similarity) between meaningful features of the subject. Features of the same identity are expected to be close in the representation space, while for different identities they are expected to be far apart [3,9,10]. However, while both closed-set and open-set scenarios require that the face representations of known subjects have been manually enrolled in a gallery, in the open-world setting the system must learn identities incrementally from frame to frame, discriminate between known and unknown identities and automatically enroll every new identity in the gallery, so to be able to recognize it every time it is observed again in the future. Large scale scenarios are likely to be implied in real situations. Key challenges of open-world recognition are therefore to avoid the possible indefinite fragmentation of identities and performance scaling.

1.1 Main Issues

Deep face recognition is a mature field of research. Network architectures, such as Deepface [11], DeepID [12], VGGFace [3], FaceNet [13], and VGGFace2 [14], have been demonstrated to be able to provide very discriminative face descriptors and effective recognition. Pose invariance is traditionally a critical issue of face recognition. In the real case, we expect that observations of the same subject under changes of pose or illumination or partial occlusions originate different (although correlated) representations. While pose invariance is mandatory in the open-world recognition context, changes in facial appearance by the aging process is not an issue instead (in most real cases there is only a short time lag between the first and last appearance of a subject). A few attempts have been published to obtain pose invariance in the deep face representation [15,16].

However, on-line incremental learning from video streams almost naturally suggests that a complete model of the identity is built as a set a collection of distinct representations, each of which refers to a specific observed pose of the face. Collecting such distinct representations as faces are detected in the video sequence unsupervisedly, so that the models are continuously self-updated by the novel information observed, is not anyway free of complexity.

This feature is not supported by the current Deep face recognition systems that rely on a separate (off-line) training phase. In fact, these systems assume that a training phase is performed off-line exploiting large face datasets. Such architectures are therefore unsuited to unsupervisedly and incrementally learn person identities, since would require continuous retraining of the network as new identities are discovered.

On-line incremental learning from video streams requires therefore **the inclusion of a memory mechanism in learning** [33]. The presence of memory allows to break the temporal correlations of the observations and combine more recent and less recent representations to complete the model of appearance of the observed subject. Different memory mechanisms for incremental learning have been proposed in the literature. In Reinforcement Learning [17,18], memory has been used to store the past experience with some priority, assuming temporal coherence. Mini batches are sampled to perform incremental learning. More recently, deep network architectures named Neural Turing Machine have been proposed in [19,20] and [21] that train an external memory module to quickly encode and retrieve new information. These architectures have the ability to rapidly bind never-before-seen information after a single presentation. Both these solutions are anyway unfit to our problem. In both cases training is provided supervisedly. In Reinforcement Learning, feedbacks are explicitly provided by humans, by assigning weights of relevance to the observations and the network is updated periodically. In the Neural Turing machine, the memory is trained offline in a supervised way. Moreover, both these solutions don't scale with massive video streams.

As in the open-set setting, open-world recognition requires the capability to discriminate between already known and unknown classes [4]. The open-set classification has been modeled as a problem of balancing known space (specialization) and unknown open space (generalization) according to the class rejection option. Solutions have formalized the open space risk as the relative measure of open space compared to the overall space [4,5,7,8]. The underlying assumption is that data is independent and identically distributed, in order to allow sampling the overall space uniformly. However, in our open-world context, since observations come from a continuous video stream, **discrimination between already known and unknown classes observations cannot assume independent and identically distributed data.**

Finally, in a typical open-world recognition context, the system should manage a very large number of different identities possibly in real time. So it should scale with massive data. If identities are represented as sets of distinct representations, **some forgetting mechanism is required in the memory** that discards unuseful representations and only retains the representations useful to build unique identities. Moreover, it is unlikely that such massive data can be retained in the memory. So, **some smart mechanism that allows to switch between main and secondary memories with effective indexing is required.**

2 Principles of Operation

The block diagram of the solution proposed is shown in Fig. 1. We used the state of the art Tiny Face Detector [22] for detection and the VGG-face descriptor [3] to represent faces. A memory module is used to collect the face descriptors [36]. The matching module is a discriminative classifier that associates to each new observation the same identity id of the most similar past observations already in the memory. So, clusters of descriptors are dynamically formed each of which is ideally representative of a single identity. The memory controller has the task of discarding redundant descriptors and implements a forgetting mechanism that attempts to keep descriptors of both the most recent and frequent and

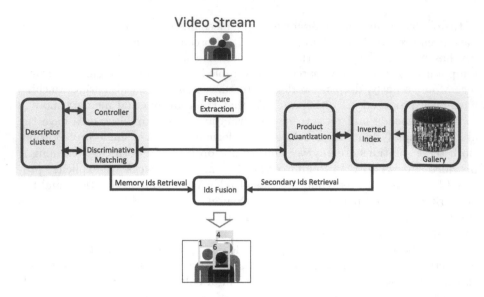

Fig. 1. Block diagram of the incremental identity learning with the main and secondary memories.

the rare observations. Consolidated clusters are periodically transferred into the secondary memory and indexed to guarantee fast access at large scales. Indexing is implemented using the FAISS framework [27] that guarantees high retrieval performance for very large number of instances. At regular time intervals the index is updated with the memory clusters, while the system continues to collect face descriptors in the main memory. As soon as a face is detected, its descriptor is matched first with the gallery in the secondary memory. Then it is either associated to an existing cluster in the memory (with its identity label) or a new cluster is formed based on Euclidean distance and Reverse Nearest Neighbor. Ideally, a new identity should be created whenever a new individual is observed that has not been observed before.

3   Matching Face Representations in Memory

In our open-world scenario, we cannot exploit Nearest Neighbor with distance ratio criterion to assess matching between face descriptors in the frame and descriptors in the memory. In fact, it is likely that faces of the same subject in consecutive frames have little differences. So, similar feature descriptors will rapidly be accumulated in the memory. Due to this, in most cases the distance ratio between the descriptor of a face observation to its nearest and the second nearest descriptor in memory will be close to 1 and the matching will be undecidable. Reverse Nearest Neighbor (ReNN) with the distance ratio criterion [31] is therefore used to assess matching. With ReNN, each descriptor in memory is NN-matched with the descriptors of the faces detected in the frame and distance ratio is used to assess matching. Since the faces detected in the frame are of different persons the distance ratio criterion can be used effectively in this case. ReNN matching could determine ambiguous assignments when distinct face observations match with descriptors of the same identity in memory, or an observation matches with descriptors of different identities. To resolve such ambiguities, we assign no identity id to the observations in the first case, and the id with the largest number of descriptors in memory, in the second case. Duplicated ids assignments to distinct face observations in the same frame are not allowed.

Accumulating matched descriptors in memory allows to dynamically create models of identities with no need of prior information about the identities and their number. At the same time, it allows to disregard the non-iid nature of data. Time of observation is not considered anymore and the descriptors in memory don't maintain the order of occurrence of the observations in the video sequence. However, the temporal coherence of the observation is useful as a form of supervision to decide whether non matched observations should be considered as new identities. Assuming that faces of the same individual have similar descriptors in consecutive frames, non-matched descriptors are assigned a new identity id only if the same identity is assessed also in the following frames (two consecutive assignments of the same id and at least one in the following three frames was verified to provide good results).

This incremental learning mechanism of memory module has two drawbacks. On the one hand a large amount of redundant information is likely will be included for each identity model (consecutive frames have similar face descriptors); on the other hand, the matching mechanism would not scale its performance at very large scales. To solve these drawbacks, we implemented the forgetting mechanism for main memory and the secondary memory indexing, respectively.

4   Forgetting Mechanism

The forgetting mechanism has the goal to avoid redundancy in the identity clusters, so that ideally, they only retain the most useful descriptors to discriminate between distinct identities. To avoid redundancy, we associate to each i-th descriptor-identity pair a dimensionless quantity e_i referred to as *eligibility-to-be-learned* (shortly *eligibility*) that indicates the relevance of the descriptor to be used as representative of the identity. Eligibility is set to 1 when the descriptor is loaded into the memory. At each match eligibility is down-weighted by a proportional amount to the matching distance ratio:

$$e_i(t+1) = \eta_i \, e_i(t) \ \text{ with } \ \eta_i = \left[\frac{1}{\bar{\rho}} \frac{d_i^1}{d_i^2} \right]^{\alpha}. \tag{1}$$

So descriptors in memory that have smaller distance ratio (i.e. are more similar to the observation) will have their eligibility decreased more than the others and therefore will have higher chance to be replaced in the future. In this equation, distance ratio threshold $\bar{\rho}$ is used for normalization and parameter α helps to emphasize the effect of the distance-ratio. As the eligibility of a face descriptor in memory drops below a given threshold (that happens after a number of matches), the descriptor is removed from the memory and will not be used as a representative of the identity. Effects of Eq. 1 can be appreciated in Fig. 2 that simulates a matching condition. Descriptors in memory that are very similar to observation o_1 and dissimilar to o_2 (dark red region) have have low matching distance ratio η; their eligibility is more down-weighted and will have higher chance to be replaced in the future. Descriptors less similar to o_1 and dissimilar to o_2 (light red) have higher η and their eligibility is less down-weighted; so remaining in memory is higher.

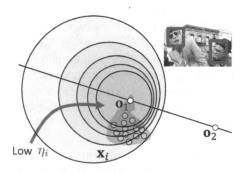

Fig. 2. Matching of descriptors in memory (x_i) with observations in the frame (o_1,o_2) and effects on the eligibility (Color figure online)

According to this, the set of descriptors of a face identity in memory includes both the rare views and the most recent occurrences of frequent. The eligibility-based forgetting mechanism is accompanied by removal by aging, that helps to remove descriptors that did not receive matches for a long time (typically false positives of the detector). Similarly to [26], removal is made according to Least Recently Used Access strategy. This learning schema is well suited for the *open-world* face recognition scenario.

5   Matching Face Representations to the Secondary Memory

If the number of identities increases indefinitely, ReNN matching with forgetting may not be sufficient to maintain high recognition performance and cannot prevent memory overflow in the long term.

So, storage of descriptors in secondary memory must be used also. In this case, the descriptors extracted from the frame are matched against descriptors in both the main and secondary memory. The two outputs are taken into account to assign identity id to face descriptor.

In order to perform efficient similarity search of face descriptors in the secondary memory, we use the FAISS indexing [27] based on Inverted Index [34] and Product Quantization [30]. FAISS supports efficient search and clustering of compressed representations of the vectors at the cost of a less precise search, but allows to scale matching to billions of vectors on a single server. The inverted index groups similar descriptors in the same bucket, and represents each bucket by its centroid. Product Quantization compresses face descriptors of size 4096 into a representation of size 64, before their storage in the secondary memory. Only a limited set of candidates is considered for the nearest neighbour matching, so drastically reducing the computational cost.

6 Experiments

In the following, we will present and discuss performances of our open-world recognition system without and with the secondary memory module, in order to assess the performance in the two scenarios of small-moderate scale and large scale open-world recognition.

6.1 Small-Moderate Scale

As discussed in the introduction, besides its highest relevance in real world contexts open-world recognition has received little attention. For the small-moderate scale, considering the affinity of the problem with the Multiple Object Tracking (MOT) problem, we have compared our system (referred to as IdOL, Identity Online Learning) against a few of the most effective MOT methods published in the literature. For a correct comparative evaluation it is anyway appropriate the main differences between the two tasks. MOT methods perform data associations off-line and build identity shot-level tracklets on the basis of the whole video information (at each time instant they exploit past, present and future information).

We used the publicly available Music [23] and Big Bang Theory [24] datasets. The Music dataset includes short YouTube videos of live vocal concerts with limited number of annotated characters in continuous fast movement. In total, there are 117,598 face detections and 3,845 face tracks. The difficulty of the dataset is mainly due to the presence of frequent shot changes, rapid changes in pose, scale, viewpoint, illumination, camera motion, makeup, occlusions and special effects. The Big Bang Theory dataset collects six episodes of Big Bang Theory TV Sitcom, Season 1, approx 20' each. They include indoor ordinary scenes under a variety of settings and illumination conditions and crowding conditions. In total the dataset contains a much larger number of identities (approximately 100). In total, there are 373,392 face detections and 4,986 face tracks. Faces have large variations of appearance due to rapid changes in pose, scale, makeup, illumination, camera motion and occlusions.

Tables 1 and 2 provide the MOTA and IDS scores [25] for the experiments with the Music [23] and Big Bang Theory [24] datasets, respectively. In the Music dataset our system has lower MOTA in most videos (although almost the same of ADMM and IHTLS) but comparable IDS for HELLOBUBBLE, APINK PUSSYCATSDOLLS and WESTLIFE videos and lower IDS for T-ARA. We obtained similar results for the Big Bang Theory dataset. However, the presence of less frequent cuts and less extreme conditions due to editing effects and camera takes determines sensibly lower Identity Switch and similar MOTA in almost all the videos.

Table 1. MOTA and ID SWITCH scores comparative. *Music* dataset

METHOD	APINK		BRUNOMARS		DARLING		GIRLSALOUD	
	IDS ↓	MOTA ↑	IDS ↓	MOTA ↑	IDS ↓	MOTA ↑	IDS ↓	MOTA ↑
mTLD2*	173	77.4	278	52.6	278	59.8	322	46.7
Siamese*	124	79.0	126	56.7	214	69.5	112	51.6
Triplet*	140	78.9	126	56.6	187	69.2	80	51.7
SymTriplet*	78	80.0	105	56.8	169	70.5	64	51.6
IdOL	191	55.1	420	48.8	449	62.1	339	49.3

METHOD	HELLOBUBBLE		PUSSYCATDOLLS		TARA		WESTLIFE	
	IDS ↓	MOTA ↑	IDS ↓	MOTA ↑	IDS ↓	MOTA ↑	IDS ↓	MOTA ↑
mTLD2*	139	52.6	296	68.3	251	56.0	177	58.1
Siamese*	105	56.3	107	70.3	106	58.4	74	64.1
Triplet*	82	56.2	99	69.9	94	59.0	89	64.5
SymTriplet*	69	56.5	82	70.2	75	59.2	57	68.6
IdOL	88	51.4	83	30.7	270	39.5	76	58.9

* Values reported from [24]

Table 2. MOTA and ID SWITCH scores comparative. *Big Bang Theory* dataset

METHOD	BBT_S01E01		BBT_S01E02		BBT_S01E03		BBT_S01E04		BBT_S01E05		BBT_S01E06	
	IDS ↓	MOTA ↑	IDS ↓	MOTA ↑	IDS ↓	MOTA ↑	IDS ↓	MOTA ↑	IDS ↓	MOTA ↑	IDS ↓	MOTA ↑
mTLD2*	223	58.4	174	43.6	142	38.0	103	11.6	169	46.4	192	37.7
Siamese*	144	69.0	116	60.4	109	52.6	85	23.0	128	60.7	156	46.2
Triplet*	164	69.3	143	60.2	121	50.7	103	18.0	118	60.5	185	45.4
SymTriplet*	156	72.2	102	61.6	126	51.9	77	19.5	90	60.9	196	47.6
IdOL	26	60.37	55	45.2	14	46.1	75	53.9	35	44.7	204	43.0

* Values reported from [24]

Figure 3 shows plots of MOTA computed at each frame for the two cases. Note that MOTA has low score initially, when identity models are largely incomplete and then stabilizes at good asymptotic values as more observations are received.

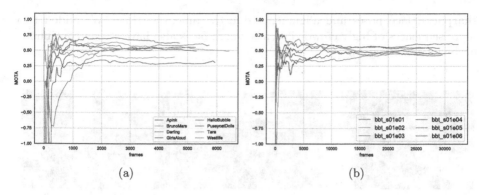

(a)   (b)

Fig. 3. MOTA computed at each frame for the videos in the *Music* (left) and *Big Bang Theory* (right) dataset.

6.2   Large Scale

A realistic scenario for online incremental learning of face identities at large scale is surveillance of public areas, such as railway stations, subway access, airports, malls, or open air crowded places. According to this, we evaluated the system in the large scale scenario with two different datasets, namely the SubwayFaces [35] and the ChokePoint [29], that have different image resolutions and represent different setting conditions.

The SubwayFaces is a dataset that has been used for face tracking. It includes four video sequences of real crowded subway scenes, 25 frames per second with 1920×1080 resolution (full HD). Faces are annotated with their bounding boxes and identity id (see Fig. 4). Most of the faces are of caucasian people. The dataset is available upon request for research purposes only.

Fig. 4. Consecutive frames from the SubwayFaces dataset.

The ChokePoint dataset includes 48 sequences of gates and portals from three different cameras. The sequences are collected at 30 frames per second with 800×600 resolution. The faces have different lighting conditions, image sharpness, face poses and alignments due to camera settings. We added two more sequences of higly crowded scenes where faces also have strong continuous occlusions (see Fig. 5).

Fig. 5. Images from different portals in the ChokePoint dataset.

To simulate the large scale scenario, we populated the secondary memory with the faces of the UMDFaces dataset [28], about 400.000 face images of about 8.000 different subjects. The FAISS index was initialized with this dataset and performance of incremental learning was evaluated separately for the Subway-Faces [35] and ChokePoint [29] datasets. Descriptors removed from the main memory according to the forgetting mechanism are stored in the secondary memory at regular time intervals, and the index is updated. The system performance was evaluated with different distance ratio thresholds $\bar{\rho}$ and different parameters of descriptor quantization and FAISS index. We present here only the best results obtained with $\bar{\rho} = 2.4$.

Figure 6 shows the MOTA plots of the system for the two datasets. For each case we compared the system performance in the large scale scenario with the performance measured in the small-moderate scenario (where incremental learning does not consider the faces in the secondary memory, but applies the descriptor forgetting mechanism to expel redundant descriptors from memory).

In general, it can be noticed that for both datasets, MOTA plots of the two scenarios are very close to each other. This shows that the system is able to scale to large number of identities almost with no loss of performance. Some little improvements can be observed in the large scale scenario that can be explained with the fact that in this case, descriptors that are removed from memory are not discarded but stored in the secondary memory.

MOTA plots show different behavior and performance differences between the two datasets. For the Subwayfaces dataset, plots have similar behavior as those of the Music and Big Bang theory: MOTA stabilizes almost at the same

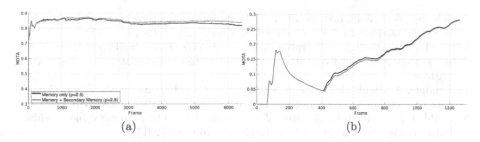

Fig. 6. MOTA computed at each frame for the sequences in the *SubwayFaces* (left) and *ChokePoint* (right) dataset.

values after an initial interval. For the ChokePoint dataset, there are several factors that determine the large performance drop. First the camera setting: the sequences are taken from cameras placed on top of the gate, that determines face crops with smaller size and face images have lower resolution than in the Subwayfaces dataset. Second, in the sequences, people remain at the gate area only for a limited number of frames that makes identity learning largely flawed.

7   Conclusions

We have discussed open-world face recognition for both small-moderate and large scale scenarios. in both cases incremental learning of identities is performed online, at real time pace. We demonstrated how it can be performed with almost no performance drop between the two cases. Most of the good performance of the system is to be ascribed to the smart forgetting mechanism in the main memory that allows to keep both the most recent frequent descriptors and the rare descriptors in memory, and the effective indexing that supports operations in the large scale scenario.

Acknowledgments. This research has been partially supported by Leonardo SpA and TICOM, Consorzio Per Le Tecnologie Dell'informazione E Comunicazione, Italy.

References

1. Deng, W., Hu, J., Zhang, N., Chen, B., Guo, J.: Fine-grained face verification: FGLFW database, baselines, and human-DCMN partnership. Pattern Recogn. **66**, 63–73 (2017)
2. Phillips, P.J., et al.: Face recognition accuracy of forensic examiners, super recognizers, and face recognition algorithms. In: Proceedings of the National Academy of Sciences, p. 201721355 (2018)
3. Parkhi, O.M., Vedaldi, A., Zisserman, A., et al.: Deep face recognition. In: BMVC, vol. 1, p. 6 (2015)
4. Scheirer, W.J., de Rezende Rocha, A., Sapkota, A., Boult, T.E.: Toward open set recognition. IEEE Trans. Pattern Anal. Mach. Intell. **35**(7), 1757–1772 (2013)

5. Scheirer, W.J., Jain, L.P., Boult, T.E.: Probability models for open set recognition. IEEE Trans. Pattern Anal. Mach. Intell. (T-PAMI) **36** (2014)
6. Bendale, A., Boult, T.: Towards open world recognition. In: Proceedings of the IEEE Conference on Computer Vision and Pattern Recognition, pp. 1893–1902 (2015)
7. Bendale, A., Boult, T.E.: Towards open set deep networks. In: The IEEE Conference on Computer Vision and Pattern Recognition (CVPR), June 2016
8. Rudd, E.M., Jain, L.P., Scheirer, W.J., Boult, T.E.: The extreme value machine. IEEE Trans. Pattern Anal. Mach. Intell. **40**, 762–768 (2017)
9. Wen, Y., Zhang, K., Li, Z., Qiao, Y.: A discriminative feature learning approach for deep face recognition. In: Leibe, B., Matas, J., Sebe, N., Welling, M. (eds.) ECCV 2016. LNCS, vol. 9911, pp. 499–515. Springer, Cham (2016). https://doi.org/10.1007/978-3-319-46478-7_31
10. Wang, H., et al.: CosFace: large margin cosine loss for deep face recognition. In: Proceedings of the IEEE Conference on Computer Vision and Pattern Recognition, pp. 5265–5274 (2018)
11. Taigman, Y., Yang, M., Ranzato, M., Wolf, L.: DeepFace: closing the gap to human-level performance in face verification. In: CVPR, pp. 1701–1708 (2014)
12. Sun, Y., Liang, D., Wang, X., Tang, X.: DeepID3: face recognition with very deep neural networks. arXiv preprint arXiv:1502.00873 (2015)
13. Schroff, F., Kalenichenko, D., Philbin, J.: FaceNet: a unified embedding for face recognition and clustering. In: CVPR, pp. 815–823 (2015)
14. Cao, Q., Shen, L., Xie, W., Parkhi, O.M., Zisserman, A.: VGGFace2: a dataset for recognising faces across pose and age. arXiv preprint arXiv:1710.08092 (2017)
15. Chen, G., Shao, Y., Tang, C., Jin, Z., Zhang, J.: Deep transformation learning for face recognition in the unconstrained scene. Mach. Vis. Appl. **29**, 1–11 (2018)
16. Zhao, J., Cheng, Y., et al.: Towards pose invariant face recognition in the wild. In: CVPR, pp. 2207–2216 (2018)
17. Mnih, V., et al.: Human-level control through deep reinforcement learning. Nature **518**(7540), 529–533 (2015)
18. Schaul, T., Quan, J., Antonoglou, I., Silver, D.: Prioritized experience replay. In: International Conference on Learning Representations, Puerto Rico (2016)
19. Graves, A., Wayne, G., Danihelka, I.: Neural turing machines. arXiv preprint arXiv:1410.5401 (2014)
20. Graves, A., et al.: Hybrid computing using a neural network with dynamic external memory. Nature **538**(7626), 471–476 (2016)
21. Santoro, A., Bartunov, S., Botvinick, M., Wierstra, D., Lillicrap, T.: Meta-learning with memory-augmented neural networks. In: International Conference on Machine Learning, pp. 1842–1850 (2016)
22. Hu, P., Ramanan, D.: Finding tiny faces. In: The IEEE Conference on Computer Vision and Pattern Recognition (CVPR), July 2017
23. Zhang, S., et al.: Tracking persons-of-interest via adaptive discriminative features. In: Leibe, B., Matas, J., Sebe, N., Welling, M. (eds.) ECCV 2016. LNCS, vol. 9909, pp. 415–433. Springer, Cham (2016). https://doi.org/10.1007/978-3-319-46454-1_26
24. Bäuml, M., Tapaswi, M., Stiefelhagen, R.: Semi-supervised learning with constraints for person identification in multimedia data. In: IEEE Conference on Computer Vision and Pattern Recognition (CVPR), June 2013
25. Leal-Taixé, L., Milan, A., Reid, I., Roth, S., Schindler, K.: Motchallenge 2015: towards a benchmark for multi-target tracking. arXiv preprint arXiv:1504.01942 (2015)

26. Santoro, A., Bartunov, S., Botvinick, M., Wierstra, D., Lillicrap, T.: One-shot learning with memory-augmented neural networks. arXiv preprint arXiv:1605.06065 (2016)
27. Johnson, J., Douze, M., Jégou, H.: Billion-scale similarity search with GPUs. arXiv preprint arXiv:1702.08734 (2017)
28. Bansal, A., Nanduri, A., Castillo, C.D., Ranjan, R., Chellappa, R.: UMDFaces: an annotated face dataset for training deep networks. arXiv (2016)
29. Wong, Y., Chen, S., Mau, S., Sanderson, C., Lovell, B.C.: Patch-based probabilistic image quality assessment for face selection and improved video-based face recognition. In: Computer Vision and Pattern Recognition (CVPR) Workshops, pp. 81–88 (2011)
30. Jégou, H., Douze, M., Schmid, C.: Product quantization for nearest neighbor search. IEEE Trans. Pattern Anal. Mach. Intell. **33**(1), 117–128 (2011). https://doi.org/10.1109/TPAMI.2010.57. inria-00514462v2
31. Korn, F., Muthukrishnan, S.: Influence sets based on reverse nearest neighbor queries. In: Proceedings of the 2000 ACM SIGMOD International Conference on Management of Data, pp. 201–212. ACM, New York (2000)
32. Liu, W., Wen, Y., Yu, Z., Li, M., Raj, B., Song, L.: SphereFace: deep Hypersphere embedding for face recognition. In: The IEEE Conference on Computer Vision and Pattern Recognition (CVPR), June 2017
33. Kumaran, D., Hassabis, D., McClelland, J.L.: What learning systems do intelligent agents need? Complementary learning systems theory updated. Trends Cogn. Sci. **20**, 512–534 (2016)
34. Sivic, J., Zisserman, A.: The inverted file from "Video Google: a text retrieval approach to object matching in videos." In: ICCV (2003)
35. Wen, L., Lei, Z., Lyu, S., Li, S.Z., Yang, M.H.: Exploiting hierarchical dense structures on hypergraphs for multi-object tracking. IEEE Trans. Pattern Anal. Mach. Intell. (TPAMI) **38**, 1983–1996 (2016)
36. Pernici, F., Bartoli, F., Bruni, M., Del Bimbo, A.: Memory based online learning of deep representations from video streams. In: Proceedings of the IEEE Conference on Computer Vision and Pattern Recognition, pp. 2324–2334 (2018)

Case-Based Reasoning – Methods, Techniques, and Applications

Petra Perner[(⊠)]

Institute of Computer Vision and Applied Computer Sciences,
IBaI, Leipzig, Germany
pperner@ibai-institut.de
http://www.ibai-institut.de

Abstract. Case-based reasoning (CBR) solves problems using the already stored knowledge, and captures new knowledge, making it immediately available for solving the next problem. Therefore, CBR is seen as a method for problem solving and also as a method to capture new experience and make it immediately available for problem solving. The CBR paradigm has been originally introduced by the cognitive science community. The CBR community aims to develop computer models that follow this cognitive process. Up to now many successful computer systems have been established on the CBR paradigm for a wide range of real-world problems. We will review in this paper the CBR process and the main topics within the CBR work. Hereby we try bridging between the concepts developed within the CBR community and the statistics community. The CBR topics we describe are the similarity, memory organization, CBR learning, and case-base maintenance. Then we will review based on applications the open problems that need to be solved. The applications we are focusing on are meta-learning for parameter selection, image interpretation, incremental prototype-based classification and novelty detection and handling. Finally, we summarize our concept on CBR.

Keywords: Case-based reasoning · Incremental learning · Similarity · Memory organization · Signal processing · Image processing · CBR meta-learning

1 Introduction

CBR [1] solves problems using the already stored knowledge, and captures new knowledge, making it immediately available for solving the next problem. Therefore, CBR can be seen as a method for problem solving, and also as a method to capture new experience and make it immediately available for problem solving. It can be seen as an incremental learning and knowledge-discovery approach, since it can capture from new experience general knowledge, such as case classes, prototypes and higher-level concepts.

The CBR paradigm has originally been introduced by the cognitive science community. The CBR community aims at developing computer models that follow this cognitive process. For many application areas computer models have successfully been

I. Nyström et al. (Eds.): CIARP 2019, LNCS 11896, pp. 16–30, 2019.
https://doi.org/10.1007/978-3-030-33904-3_2

developed based on CBR, such as signal/image processing and interpretation tasks, help-desk applications, medical applications and E-commerce-product selling systems.

In this paper we will explain the CBR process scheme in Sect. 2. We will show what kinds of methods are necessary to provide all the necessary functions for such a computer model. Then we will focus on similarity in Sect. 3. Memory organization in a CBR system will be described in Sect. 4. Both similarity and memory organization are concerned in learning in a CBR system. Therefore, in each section an introduction will be given as to what kind of learning can be performed. In Sect. 5 we will describe open topics in CBR research for specific applications. We will focus on meta-learning for parameter selection, image interpretation, incremental prototype-based classification and novelty detection and handling. In Sect. 5.1 we will describe meta-learning for parameter selection for data processing systems. CBR based image interpretation will be described in Sect. 5.2 and incremental prototype-based classification in Sect. 5.3. New concepts on novelty detection and handling will be presented in Sect. 5.4. While reviewing the CBR work, we will try bridging between the concepts developed within the CBR community and the concepts developed in the statistics community. In the conclusion, we will summarize our concept on CBR in Sect. 6.

2   Case-Based Reasoning

CBR is used when generalized knowledge is lacking. The method works on a set of cases formerly processed and stored in a case base. A new case is interpreted by searching for similar cases in the case base. Among this set of similar cases the closest case with its associated result is selected and presented to the output.

In contrast to a symbolic learning system, which represents a learned concept explicitly, e.g. by formulas, rules or decision trees, a CBR learning system describes a concept C implicitly by a pair (CB, sim) where CB is the case base and sim the similarity, and changes the pair (CB, sim) as long as no further change is necessary because it is a correct classifier for the target concept C.

Formal, we like to understand a case as the following:

Definition 1. A case F is a triple (P, E, L) with a problem description P, an explanation of the solution E and a problem solution L.

The problem description summarizes the information about a case in the form of attributes or features. Other case representations such as graphs, images or sequences may also be possible. The case description is given a-priori or needs to be elicited during a knowledge acquisition process. Only the most predictive attributes will guarantee us to find exactly the most similar cases.

Equation 1 and Definition 1 give a hint as to how a case-based learning system can improve its classification ability. The learning performance of a CBR system is of incremental manner and it can also be considered as on-line learning. In general, there are several possibilities to improve the performance of a case-based system. The system can change the vocabulary V (attributes, features), store new cases in the case base CB, change the measure of similarity sim, or change V, CB and sim in combinatorial manner.

That brings us to the notion of knowledge containers introduced by Richter [2]. According to Richter, the four knowledge containers are the underlying vocabulary (or features), the similarity measure, the solution transformation, and the cases. The first three represent compiled knowledge, since this knowledge is more stable. The cases are interpreted knowledge. As a consequence, newly added cases can be used directly. This enables a CBR system to deal with dynamic knowledge. In addition, knowledge can be shifted from one container to another container. For instance, in the beginning a simple vocabulary, a rough similarity measure, and no knowledge on solution transformation are used. However, a large number of cases are collected. Over time, the vocabulary can be refined and the similarity measure defined in higher accordance with the underlying domain. In addition, it may be possible to reduce the number of cases, because the improved knowledge within the other containers now enables the CBR system to better differentiate between the available cases.

The abstraction of cases into a more general case (concepts, prototypes and case classes) or the learning of the higher-order relation between different cases may reduce the size of the case base and speed up the retrieval phase of the system [3]. It can make the system more robust against noise. More abstract cases which are set in relation to each other will give the domain expert a better understanding about his domain. Therefore, beside the incremental improvement of the system performance through learning, CBR can also be seen as a knowledge-acquisition method that can help to get a better understanding about the domain [4, 5] or learn a domain theory.

The main problems with the development of a CBR system are the following: What makes up a case?, What is an appropriate similarity measure for the problem?, How to organize a large number of cases for efficient retrieval?, How to acquire and refine a new case for entry in the case base?, How to generalize specific cases to a case that is applicable to a wide range of situations?

3   Similarity

3.1   Similarity Measures

Although similarity is a concept humans prefer to use when reasoning over problems, they usually do not have a good understanding of how similarity is formally expressed. Similarity seems to be a very incoherent concept.

From the cognitive point of view, similarity can be viewed from different perspectives [8]. A red bicycle and a blue bicycle might be similar in terms of the concept "bicycle", but both bicycles are dissimilar when looking at the colour. It is important to know what kind of similarity is to be considered when reasoning over two objects. Overall similarity, identity, similarity, and partial similarity need to be modelled by the right flexible control strategy in an intelligent reasoning system. It is especially important in image data bases where the image content can be viewed from different perspectives. Image data bases need to have this flexibility and computerized conversational strategies to figure out from what perspective the problem is looked at and what kind of similarity has to be applied to achieve the desired goal. From the mathematical point of view, the Minkowski metric is the most used similarity measure for technical problems:

$$d_{ii'}^{(p)} = \left[\frac{1}{J} \sum_{j=1}^{J} |x_{ij} - x_{i'j}|^p \right]^{1/p} \tag{1}$$

the choice of the parameter p depends on the importance we give to the differences in the summation. Metrical properties such as symmetry, identity and unequality hold for the Minkowski metric.

If we use the Minkowski metric for calculating the similarity between two 1-dimensional curves, such as the 1-dimensional path signal of a real robot axis, and the reconstructed 1-dimensional signal of the same robot axis [9], calculated from the compressed data points stored in a storage device, it might not be preferable to chose $p = 2$ (Euclidean metric), since the measure averages over all data points, but gives more emphasis to big differences. If choosing $p = 1$ (City-Block metric), big and small differences have the same influence (impact) on the similarity measure. In case of the Max-Norm ($p = \infty$) none of the data point differences should exceed a predefined difference. In practice it would mean that the robot axis is performing a smooth movement over the path with a known deviation from the real path and will never come in the worse situation to perform a ramp-like function. In the robot example the domain itself gives us an understanding about the appropriate similarity metric.

Unfortunately, for most of the applications we do not have any a-priori knowledge about the appropriate similarity measure. The method of choice for the selection of the similarity measure is to try different types of similarity and observe their behaviour based on quality criteria while applying them to a particular problem. The error rate is the quality criterion that allows selecting the right similarity measure for classification problems. Otherwise it is possible to measure how well similar objects are grouped together, based on the chosen similarity measure, and at the same time, how well different groups can be distinguished from each other. It changes the problem into a categorization problem for which proper category measures are known from clustering [24] and machine learning [30].

In general, distance measures can be classified based on the data-type dimension. There are measures for numerical data, symbolical data, structural data and mixed-data types. Most of the overviews given for similarity measures in various works are based on this view [10, 12, 16]. A more general view to similarity is given in Richter [11].

Other classifications on similarity measures focus on the application. There are measures for time-series [54], similarity measures for shapes [53], graphs [29], music classification [13], and others.

Translation, size, scale and rotation invariance are another important aspect of similarity as concerns technical systems.

Most real-world applications nowadays are more complex than the robot example given above. They are usually comprised of many attributes that are different in nature. Numerical attributes given by different sensors or technical measurements and categorical attributes that describe meta-knowledge of the application usually make up a case. These n different attribute groups can form partial similarities Sim_1, Sim_2, ..., Sim_n, that can be calculated based on different similarity measures and may have a meaning for itself. The final similarity might be comprised of all the partial similarities. The simplest way to calculate the overall similarity is to sum up over all partial similarities: $Sim = w_1 Sim_1 + w_2 Sim_2 ... + w_n Sim_n$ and model the influence of the particular similarity by different weights w_i. Other schemas for combining similarities are possible as well. The usefulness of such a strategy has been shown for meta-learning of segmentation parameters [14] and for medical diagnosis [15].

The introduction of weights into the similarity measure in Eq. 1 puts a different importance on particular attributes and views similarity not only as global similarity, but also as local similarity. Learning the attribute weights allows building particular similarity metrics for the specific applications. A variety of methods based on linear or stochastic optimization methods [18], heuristics search [17], genetic programming [25], and case-ordering [20] or query ordering in NN-classification, have been proposed for attribute-weight learning.

Learning distance function in response to users' feedback is known as relevance feedback [21, 22] and it is very popular in data base and image retrieval. The optimization criterion is the accuracy or performance of the system rather than the individual problem-case pairs. This approach is biased by the learning approach as well as by the case description.

New directions in CBR research build a bridge between the case and the solution [23]. Cases can be ordered based on their solutions by their preference relations [26] or similarity relation [27] given by the users or a-priori known from application. The derived values can be used to learn the similarity metric and the relevant features. That means that cases having similar solutions should have similar case descriptions. The set of features as well as the feature weights are optimized until they meet this assumption. Learning distance function by linear transformation of features has been introduced by Bobrowski et al. [19].

3.2   Semantic of Similarity

It is preferable to normalize the similarity values between 0 and 1 in order to be able to compare different similarity values based on a scale. A scale between 0 and 1 gives us a symbolic understanding of the meaning of the similarity value. The value of 0 indicates identity of the two cases while the value of 1 indicates the cases are unequal. On the scale of 0 and 1 the value of 0.5 means neutral and values between 0.5 and 0 means more similarity and values between 0.5 and 1 means more dissimilarity

Different normalization procedures are known. The most popular one is the normalization to the upper and lower bounds of a feature value.

The main problem arises when the case base is not yet filled and contains only a small number of cases while the other cases are collected incrementally as soon as they arrive in the system. In this case the upper and lower bounds of a feature value $[x_{min,i,k}, x_{max,i,k}]$ can only be judged based on this limited set of cases at the point in time t_k and must not meet the true values of $x_{min,i}$ and $x_{max,i}$ of feature i. Then the scale of similarity might change over time periods t_k which will lead to different decisions for two cases at the points in time t_k and t_{k+l}. The problem of normalization in an incremental case-based learning system needs to be considered in a different way. A first explanation of this problem is given in [55].

4 Organization of Case Base

The case base plays a central role in a CBR system. All observed relevant cases are stored in the case base. Ideally, CBR systems start reasoning from an empty memory, and their reasoning capabilities stem from their progressive learning from the cases they process [28].

Consequently, the memory organization and structure are in the focus of a CBR system. Since a CBR system should improve its performance over time, imposes on the memory of a CBR system to change constantly.

In contrast to research in data base retrieval and nearest-neighbour classification, CBR focuses on conceptual memory structures. While k-d trees [31] are space-partitioning data structures for organizing points in a k-dimensional space, conceptual memory structures [29, 30] are represented by a directed graph in which the root node represents the set of all input instances and the terminal nodes represent individual instances. Internal nodes stand for sets of instances attached to that node and represent a super-concept. The super-concept can be represented by a generalized representation of the associated set of instances, such as the prototype, the mediod or a user-selected instance. Therefore a concept C, called a class, in the concept hierarchy is represented by an abstract concept description (e.g. the feature names and its values) and a list of pointers to each child concept $M(C) = \{C_1, C_2, ..., C_i, ..., C_n\}$, where C_i is the child concept, called subclass of concept C.

The explicit representation of the concept in each node of the hierarchy is preferred by humans, since it allows understanding the underlying application domain.

While for the construction of a k-d tree only a splitting and deleting operation is needed, conceptual learning methods use more sophisticated operations for the construction of the hierarchy [33]. The most common operations are splitting, merging, adding and deleting. What kind of operation is carried out during the concept hierarchy construction depends on a concept-evaluation function. There are statistical functions known, as well as similarity-based functions.

Because of the variety of construction operators, conceptual hierarchies are not sensitive to the order of the samples. They allow the incremental adding of new examples to the hierarchy by reorganizing the already existing hierarchy. This flexibility is not known for k-d trees, although recent work has led to adaptive k-d trees that allow incorporating new examples.

The concept of generalization and abstraction should make the case base more robust against noise and applicable to a wider range of problems. The concept description, the construction operators as well as the concept evaluation function are in the focus of the research in conceptual memory structure.

The conceptual incremental learning methods for case base organization puts the case base into the dynamic memory view of Schank [32] who required a coherent theory of adaptable memory structures and that we need to understand how new information changes the memory.

Memory structures in CBR research are not only pure conceptual structures, hybrid structures incorporating k-d tree methods are studied also. An overview of recent research in memory organization in CBR is given in [28].

Other work goes into the direction of bridging between implicit and explicit representations of cases [34]. The implicit representations can be based on statistical models and the explicit representation is the case base that keeps the single case as it is. As far as evidence is given, the data are summarized into statistical models based on statistical learning methods such as Minimum Description Length (MDL) or Minimum Message Length (MML) learning. As long as not enough data for a class or a concept have been seen by the system, the data are kept in the case base. The case base controls the learning of the statistical models by hierarchically organizing the samples into groups. It allows dynamically learning and changing the statistical models based on the experience (data) seen so far and prevents the model from overfitting and bad influences by singularities.

This concept follows the idea that humans have built up very effective models for standard repetitive tasks and that these models can easily be used without a complex reasoning process. For rare events the CBR unit takes over the reasoning task and collects experience into its memory. The aspects of case-based maintenance can be found in [6] and the lifecycle of a CBR system is described in [7].

5 Applications

CBR has been successfully applied to a wide range of problems. Among them are signal interpretation tasks [35], medical applications [36], and emerging applications such as geographic information systems, applications in biotechnology and topics in climate research (CBR commentaries) [37]. We are focussing here on hot real-world topics such as meta-learning for parameter selection, image & signal interpretation, prototype-based classification and novelty detection & handling. We first give an overview on CBR-based image interpretation system.

5.1 Meta-Learning for Parameter Selection of Data/Signal Processing Algorithms

Meta learning is a subfield of Machine learning where automatic learning algorithms are applied on meta-data about machine-learning experiments. The main goal is to use such meta-data to understand how automatic learning can become flexible as regards solving different kinds of learning problems, hence to improve the performance of existing learning algorithms. Another important meta-learning task, but not so widely studied yet, is parameter selection for data or signal processing algorithms. Soares et al. [39] have used this approach for selecting the kernel width of a support-vector machine, while Perner and Frucci et al. [14, 40] have studied this approach for image segmentation.

The meta-learning problem for parameter selection can be formalized as follows: For a given signal that is characterized by specific signal properties A and domain properties B find the parameters of the processing algorithm that ensure the best quality of the resulting output signal:

$$f : A \cup B \to P_i \tag{2}$$

with P_i the i-th class of parameters for the given domain.

What kind of meta-data describe classification tasks, has been widely studied within meta-learning in machine learning. Meta-data for images comprised of image-related meta-data (gray-level statistics) and non-image related meta-data (sensor, object data) are given in Perner and Frucci et al. [14, 40]. In general the processing of meta-data from signals and images should not require too much processing and they should allow characterizing the properties of the signals that influence the signal processing algorithm.

The mapping function f can be realized by any classification algorithm, but the incremental behaviour of CBR fits best to many data/signal processing problems where the signals are not available ad-hoc but appear incrementally. The right similarity metric that allows mapping data to parameter groups and in the last consequence to good output results should be more extensively studied. Performance measures that allow to judge the achieved output and to automatically criticize the system performances are another important problem.

Abstraction of cases to learn domain theory are also related to these tasks and would allow to better understand the behaviour of many signal processing algorithms that cannot be described anymore by standard system theory [41].

5.2   Case-Based Image Interpretation

Image interpretation is the process of mapping the numerical representation of an image into a logical representation such as is suitable for scene description. This is a complex process; the image passes through several general processing steps until the final result is obtained. These steps include image preprocessing, image segmentation, image analysis, and image interpretation. Image pre-processing and image segmentation algorithm usually need a lot of parameters to perform well on the specific image. The automatically extracted objects of interest in an image are first described by primitive image features. Depending on the particular objects and focus of interest, these features can be lines, edges, ribbons, etc. Typically, these low-level features have to be mapped to high-level/symbolic features. A symbolic feature such as *fuzzy margin* will be a function of several low-level features.

The image interpretation component identifies an object by finding the object to which it belongs (among the models of the object class). This is done by matching the symbolic description of the object to the model/concept of the object stored in the knowledge base. Most image-interpretation systems run on the basis of a bottom-up control structure. This control structure allows no feedback to preceding processing components if the result of the outcome of the current component is unsatisfactory. A mixture of bottom-up and top-down control would allow the outcome of a component to be refined by returning to the previous component.

CBR is not only applicable as a whole to image interpretation, it is applicable to all the different levels of an image-interpretation system [12, 42] and many of the ideas mentioned in the chapters before apply here. CBR-based meta-learning algorithms for parameter selection are preferable for the image pre-processing and segmentation unit [14, 40]. The mapping of the low-level features to the high-level features is a classification task for which a CBR-based algorithm can be applied. The memory organization [29] of the interpretation unit goes along with problems discussed for the case base organization in Sect. 5. Different organization structures for image interpretation systems are discussed in [12]. The organization structure should allow the incremental updating of the memory and learning from single cases more abstract cases. Ideally the system should start working with only a few samples and during usage of the system new cases should be learnt and the memory should be updated based on these samples. This view at the usage of a system brings in another topic that is called life-time cycle of a CBR system. Work on this topic takes into account that a system is used for a long time, while experience changes over time. The case structure might change by adding new relevant attributes or deleting attributes that have shown not to be important or have been replaced by other ones. Set of cases might not appear anymore, since these kinds of solutions are not relevant anymore. A methodology and software architecture for handling the life-time cycle problem is needed so that this process can easily be carried out without rebuilding the whole system. It seems to be more a software engineering task, but has also something to do with evaluation measures that can come from statistics.

5.3   Incremental Prototype-Based Classification

The usage of prototypical cases is very popular in many applications, among them are medical applications [43], Belazzi et al. [45] and by Nilsson and Funk [44], knowledge management systems [46] and image classification tasks [48]. The simple nearest-neighbour- approach [47] as well as hierarchical indexing and retrieval methods [43] have been applied to the problem. It has been shown that an initial reasoning system could be built up based on these cases. The systems are useful in practice and can acquire new cases for further reasoning during utilization of the system.

There are several problems concerned with prototypical CBR: If a large enough set of cases is available, the prototypical case can automatically be calculated as the generalization from a set of similar cases. In medical applications as well as in applications where image catalogues are the development basis for the system, the prototypical cases have been selected or described by humans. That means when building the system, we are starting from the most abstract level (the prototype) and have to collect more specific information about the classes and objects during the usage of the system.

Since a human has selected the prototypical case, his decision on the importance of the case might be biased and picking only one case might be difficult for a human. As for image catalogue-based applications, he can have stored more than one image as a prototypical image. Therefore we need to check the redundancy of the many prototypes for one class before taking them all into the case base.

According to this consideration, the minimal functions a prototype-based classification system should realize are: classifications based on a proper similarity-measure, prototype selection by a redundancy-reduction algorithm, feature weighting to determine the importance of the features for the prototypes and to learn the similarity metric, and feature-subset selection to select the relevant features from the whole set of features for the respective domain.

Statistical methods focus on adaptive k-NN that adapts the distance metric by feature weighting or kernel methods or the number k of neighbours off-line to the data. Incremental strategies are used for the nearest- neighbour search, but not for updating the weights, distance metric and prototype selection.

A prototype-based classification system for medical image interpretation is described in [48]. It realizes all the functions described above by combining statistical methods with artificial intelligence methods to make the system feasible for real-world applications. A system for handwriting recognition is described in [49] that can incrementally add data and adapt the solutions to different users' writing style. A k-NN realization that can handle data streams by adding data through reorganizing a multi-resolution array data structure and concept drift by realizing a case forgetting strategy is described in [50].

The full incremental behaviour of a system would require an incremental processing schema for all aspects of a prototype-based classifier such as for updating the weights and learning the distance metric, the prototype selection and case generalization.

5.4   Novelty Detection by Case-Based Reasoning

Novelty detection [51], recognizing that an input differs in some respect from previous inputs, can be a useful ability for learning systems.

Novelty detection is particularly useful where an important class is under-represented in the data, so that a classifier cannot be trained to reliably recognize that class. This characteristic is common to numerous problems such as information management, medical diagnosis, fault monitoring and detection, and visual perception.

We propose novelty detection to be regarded as a CBR problem under which we can run the different theoretical methods for detecting the novel events and handling the novel events [34]. The detection of novel events is a common subject in the literature. The handling of the novel events for further reasoning is not treated so much in the literature, although this is a hot topic in open-world applications.

The first model we propose is comprised of statistical models and similarity-based models. For now, we assume an attribute-value based representation. Nonetheless, the general framework we propose for novelty detection can be based on any representation. The heart of our novelty detector is a set of statistical models that have been learnt in an off-line phase from a set of observations. Each model represents a case-class. The probability density function implicitly represents the data and prevents us from storing all the cases of a known case-class. It also allows modelling the uncertainty in the data. This unit acts as a novel-event detector by using the Bayesian decision-criterion with the mixture model. Since this set of observations might be limited, we consider our model as being far from optimal and update it based on new observed examples. This is done based on the Minimum Description Length (MDL) principle or the Minimum Message Length (MML) learning principle [52].

In case our model bank cannot classify an actual event into one of the case-classes, this event is recognized as a novel event. The novel event is given to the similarity-based reasoning unit. This unit incorporates this sample into their case base according to a case-selective registration-procedure that allows learning case-classes as well as the similarity between the cases and case-classes. We propose to use a fuzzy similarity measure to model the uncertainty in the data. By doing that the unit organizes the novel events in such a fashion that is suitable for learning a new statistical model.

The case-base-maintenance unit interacts with the statistical learning unit and gives an advice as to when a new model has to be learnt. The advice is based on the observation that a case-class is represented by a large enough number of samples that are most dissimilar to other classes in the case-base.

The statistical learning unit takes this case class and proves based on the MML-criterion, whether it is suitable to learn the new model or not. In the case that the statistical component recommends to not learn the new model, the case-class is still hosted by the case base maintenance unit and further up-dated based on new observed events that might change the inner-class structure as long as there is new evidence to learn a statistical model.

The use of a combination of statistical reasoning and similarity-based reasoning allows implicit and explicit storage of the samples. It allows handling well-represented events as well as rare events.

6   Conclusion

In this paper we have presented our thoughts and work on Case-Based Reasoning. We presented the methods, techniques, and applications. More work on CBR can be found in [38]. CBR solves problems using already stored knowledge, and captures new knowledge, making it immediately available for solving the next problem. To realize this cognitive model in a computer-based system we need methods known from statistics, pattern recognition, artificial intelligence, machine learning, data base research and other fields. Only the combination of all these methods will give us a system that can efficiently solve practical problems. Consequently, CBR research has shown much success for different application areas, such as medical and technical diagnosis, image interpretation, geographic information systems, text retrieval, e-commerce, user-support systems and so on. CBR systems work efficiently in real-world applications, since the CBR method faces on all aspects of a well-performing and user-friendly system.

We have pointed out that the central aspect of a well-performing system in the real-world is its ability to incrementally collect new experience and reorganize its knowledge based on these new insights. In our opinion the new challenging research aspects should have its focus on incremental methods for prototype-based classification, meta-learning for parameter selection, complex signals understanding tasks and novelty detection. The incremental methods should allow changing the system function based on the newly obtained data.

Recently, we are observing that this incremental aspect is in the special focus of the quality assurance agency for technical and medical application, although this is in opposition to the current quality performance guidelines.

While reviewing the CBR work, we have tried bridging between the concepts developed within the CBR community and the concepts developed in the statistics community. At the first glance, CBR and statistics seem to have big similarities. But when looking closer at it one can see that the paradigms are different. CBR tries to solve real-world problems and likes to deliver systems that have all the functions necessary for an adaptable intelligent system with incremental learning behavior. Such a system should be able to work on a small set of cases and collect experience over time. While doing that it should improve its performance. The solution need not be correct in the statistical sense, rather it should help an expert to solve his tasks and learn more about it over time.

Nonetheless, statistics disposes of a rich variety of methods that can be useful for building intelligent systems. In the case that we can combine and extend these methods under the aspects necessary for intelligent systems, we will further succeed in establishing artificial intelligence systems in the real world.

Our interest is to build intelligent flexible and robust data-interpreting systems that are inspired by the human CBR process and by doing so to model the human reasoning process when interpreting real-world situations.

References

1. Althoff, K.D.: Case-based reasoning. In: Chang, S.K. (ed.) Handbook on Software Engineering and Knowledge Engineering (2001)
2. Richter, M.M.: Introduction to case-based reasoning. In: Lenz, M., Bartsch-Spörl, B., Burkhardt, H.-D., Wess, S. (eds.) Case-Based Reasoning Technology: From Foundations to Applications. LNAI, vol. 1400, pp. 1–16. Springer, Heidelberg (1998)
3. Smith, E.E., Douglas, L.M.: Categories and Concepts. Harvard University Press, Cambridge (1981)
4. Branting, L.K.: Integrating generalizations with exemplar-based reasoning. In: Proceedings of the 11th Annual Conference of Cognitive Science Society, Ann Arbor, MI Lawrence Erlbaum 1989, pp. 129–146 (1989)
5. Bergmann, R., Wilke, W.: On the role of abstraction in case-based reasoning. In: Smith, I., Faltings, B. (eds.) EWCBR 1996. LNCS, vol. 1168, pp. 28–43. Springer, Heidelberg (1996). https://doi.org/10.1007/BFb0020600
6. Iglezakis, I., Reinartz, T., Roth-Berghofer, T.R.: Maintenance memories: beyond concepts and techniques for case base maintenance. In: Funk, P., González Calero, P.A. (eds.) ECCBR 2004. LNCS (LNAI), vol. 3155, pp. 227–241. Springer, Heidelberg (2004). https://doi.org/10.1007/978-3-540-28631-8_18
7. Minor, M., Hanft, A.: The life cycle of test cases in a CBR system. In: Blanzieri, E., Portinale, L. (eds.) EWCBR 2000. LNCS, vol. 1898, pp. 455–466. Springer, Heidelberg (2000). https://doi.org/10.1007/3-540-44527-7_39
8. Smith, L.B.: From global similarities to kinds of similarities: the construction of dimensions in development. In: Smith, L.B. (ed.) Similarity and Analogical Reasoning, pp. 146–178. Cambridge University Press, New York (1989)

9. Fiss, P.: Data Reduction Methods for Industrial Robots with Direct Teach-In Programming, Diss A, Technical University Mittweida (1985)
10. Pekalska, E., Duin, R.: The Dissimilarity Representation for Pattern Recognition. World Scientific (2005)
11. Richter, M.: Similarity. In: Perner, P. (ed.) Case-Based Reasoning on Images and Signals. SCI, pp. 01–21. Springer, Heidelberg (2008)
12. Perner, P.: Why case-based reasoning is attractive for image interpretation. In: Aha, D.W., Watson, I. (eds.) ICCBR 2001. LNCS (LNAI), vol. 2080, pp. 27–43. Springer, Heidelberg (2001). https://doi.org/10.1007/3-540-44593-5_3
13. Weihs, C., Ligges, U., Mörchen, F., Müllensiefen, D.: Classification in music research. J. Adv. Data Anal. Classif. 1(3), 255–291 (2007)
14. Perner, P.: An architecture for a CBR image segmentation system. J. Eng. Appl. Artif. Intell. Eng. Appl. Artif. Intell. 12(6), 749–759 (1999)
15. Song, X., Petrovic, S., Sundar, S.: A case-based reasoning approach to dose planning in radiotherapy. In: Wilson, D.C., Khemani, D. (eds.) The Seventh Intern. Conference on Case-Based Reasoning, Belfast, Northern Ireland, Workshop Proceeding, pp. 348–357 (2007)
16. Wilson, D.R., Martinez, T.R.: Improved heterogeneous distance functions. J. Artif. Intell. Res. 6, 1–34 (1997)
17. Wettschereck, D., Aha, D.W., Mohri, T.: A review and empirical evaluation of feature weighting methods for a class of lazy learning algorithms. Artif. Intell. Rev. 11, 273–314 (1997)
18. Zhang, L., Coenen, F., Leng, P.: Formalising optimal feature weight settings in case-based diagnosis as linear programming problems. Knowl.-Based Syst. 15, 391–398 (2002)
19. Bobrowski, L., Topczewska, M.: Improving the K-NN classification with the euclidean distance through linear data transformations. In: Perner, P. (ed.) ICDM 2004. LNCS (LNAI), vol. 3275, pp. 23–32. Springer, Heidelberg (2004). https://doi.org/10.1007/978-3-540-30185-1_3
20. Stahl, A.: Learning feature weights from case order feedback. In: Aha, D.W., Watson, I. (eds.) ICCBR 2001. LNCS (LNAI), vol. 2080, pp. 502–516. Springer, Heidelberg (2001). https://doi.org/10.1007/3-540-44593-5_35
21. Bhanu, B., Dong, A.: Concepts learning with fuzzy clustering and relevance feedback. In: Perner, P. (ed.) MLDM 2001. LNCS (LNAI), vol. 2123, pp. 102–116. Springer, Heidelberg (2001). https://doi.org/10.1007/3-540-44596-X_9
22. Bagherjeiran, A., Eick, C.F.: Distance function learning for supervised similarity assesment. In: Perner, P. (ed.) Case-Based Reasoning on Images and Signals. SCI, pp. 91–126. Springer, Heidelberg (2008). https://doi.org/10.1007/978-3-540-73180-1_3
23. Bergmann, R., Richter, M., Schmitt, S., Stahl, A., Vollrath, I.: Utility-oriented matching: a new research direction for case-based reasoning. In: Schnurr, H.-P., et al. (eds.) Professionelles Wissensmanagement, pp. 20–30. Shaker Verlag (2001)
24. Jain, A.K., Dubes, R.C.: Algorithms for Clustering Data, 320 p. Prentice Hall, Inc., Upper Saddle River (1988)
25. Craw, S.: Introspective learning to build case-based reasoning (CBR) knowledge containers. In: Perner, P., Rosenfeld, A. (eds.) MLDM 2003. LNCS, vol. 2734, pp. 1–6. Springer, Heidelberg (2003). https://doi.org/10.1007/3-540-45065-3_1
26. Xiong, N., Funk, P.: Building similarity metrics reflecting utility in case-based reasoning. J. Intell. Fuzzy Syst. 17, 407–416 (2006)
27. Perner, P., Perner, H., Müller, B.: Similarity guided learning of the case description and improvement of the system performance in an image classification system. In: Craw, S., Preece, A. (eds.) ECCBR 2002. LNCS (LNAI), vol. 2416, pp. 604–612. Springer, Heidelberg (2002). https://doi.org/10.1007/3-540-46119-1_44

28. Bichindaritz, I.: Memory structures and organization in case-based reasoning. In: Perner, P. (ed.) Case-Based Reasoning on Images and Signals. SCI, pp. 175–194. Springer, Heidelberg (2008). https://doi.org/10.1007/978-3-540-73180-1_6
29. Perner, P.: Case-base maintenance by conceptual clustering of graphs. Eng. Appl. Artif. Intell. **19**(4), 381–395 (2006)
30. Fisher, D.H.: Knowledge acquisition via incremental conceptual clustering. Mach. Learn. **2**(2), 139–172 (1987)
31. Bentley, J.: Multidimensional binary search trees used for associative searching. Commun. ACM **18**(9), 509–517 (1975)
32. Schank, R.C.: Dynamic Memory. A Theory of Reminding and Learning in Computers and People. Cambridge University Press, Cambridge (1982)
33. Jaenichen, S., Perner, P.: Conceptual clustering and case generalization of two dimensional forms. Comput. Intell. **22**(3/4), 177–193 (2006)
34. Perner, P.: Concepts for novelty detection and handling based on a case-based reasoning process scheme. In: Perner, P. (ed.) ICDM 2007. LNCS (LNAI), vol. 4597, pp. 21–33. Springer, Heidelberg (2007). https://doi.org/10.1007/978-3-540-73435-2_3
35. Perner, P., Holt, A., Richter, M.: Image processing in case-based reasoning. Knowl. Eng. Rev. **20**(3), 311–314 (2005)
36. Holt, A., Bichindaritz, I., Schmidt, R., Perner, P.: Medical applications in case-based reasoning. Knowl. Eng. Rev. **20**(3), 289–292 (2005)
37. De Mantaras, R.L., Cunningham, P., Perner, P.: Emergent case-based reasoning applications. Knowl. Eng. Rev. **20**(3), 325–328 (2005)
38. CBR Commentaries: The Knowledge Engineering Review, vol. 20, no. 3
39. Soares, C., Brazdil, P.B.: A meta-learning method to select the kernel width in support vector regression. Mach. Learn. **54**, 195–209 (2004)
40. Frucci, M., Perner, P., di Baja, G.S.: Case-based reasoning for image segmentation by watershed transformation. In: Perner, P. (ed.) Case-Based Reasoning on Images and Signals. SCI, vol. 73, pp. 319–353. Springer, Heidelberg (2008). https://doi.org/10.1007/978-3-540-73180-1_11
41. Wunsch, G.: Systemtheorie der Informationstechnik. Akademische Verlagsgesellschaft, Leipzig (1971)
42. Perner, P.: Using CBR learning for the low-level and high-level unit of an image interpretation system. In: Singh, S. (ed.) International Conference on Advances in Pattern Recognition, pp. 45–54. Springer, London (1999). https://doi.org/10.1007/978-1-4471-0833-7_5
43. Schmidt, R., Gierl, L.: Temporal abstractions and case-based reasoning for medical course data: two prognostic applications. In: Perner, P. (ed.) MLDM 2001. LNCS (LNAI), vol. 2123, pp. 23–34. Springer, Heidelberg (2001). https://doi.org/10.1007/3-540-44596-X_3
44. Nilsson, M., Funk, P.: A case-based classification of respiratory sinus arrhythmia. In: Funk, P., González Calero, P.A. (eds.) ECCBR 2004. LNCS (LNAI), vol. 3155, pp. 673–685. Springer, Heidelberg (2004). https://doi.org/10.1007/978-3-540-28631-8_49
45. Bellazzi, R., Montani, S., Portinale, L.: Retrieval in a prototype-based case library: a case study in diabetes therapy revision. In: Smyth, B., Cunningham, P. (eds.) EWCBR 1998. LNCS, vol. 1488, pp. 64–75. Springer, Heidelberg (1998). https://doi.org/10.1007/BFb0056322
46. Bichindaritz, I., Kansu, E., Sullivan, K.M.: Case-based reasoning in CARE-PARTNER: gathering evidence for evidence-based medical practice. In: Smyth, B., Cunningham, P. (eds.) EWCBR 1998. LNCS, vol. 1488, pp. 334–345. Springer, Heidelberg (1998). https://doi.org/10.1007/BFb0056345

47. Aha, D.W., Kibler, D., Albert, M.K.: Instance-based learning algorithm. Mach. Learn. **6**(1), 37–66 (1991)
48. Perner, P.: Prototype-Based Classification, Applied Intelligence, to appear (online available)
49. Vuori, V., Laaksonen, J., Oja, E., Kangas, J.: Experiments with adaptation strategies for a prototype-based recognition system for isolated handwritten characters. Int. J. Doc. Anal. Recogn. **3**(3), 150–159 (2001)
50. Law, Y.-N., Zaniolo, C.: An adaptive nearest neighbor classification algorithm for data streams. In: Jorge, A.M., Torgo, L., Brazdil, P., Camacho, R., Gama, J. (eds.) PKDD 2005. LNCS (LNAI), vol. 3721, pp. 108–120. Springer, Heidelberg (2005). https://doi.org/10.1007/11564126_15
51. Markou, M., Singh, S.: Novelty detection: a review-part 1: statistical approaches. Sig. Process. **83**(12), 2481–2497 (2003)
52. Wallace, C.S.: Statistical and Inductive Inference by Minimum Message Length, Series: Information Science and Statistics. Springer, Heidelberg (2005). https://doi.org/10.1007/0-387-27656-4
53. Shapiro, L.G., Atmosukarto, I., Cho, H., Lin, H.J., Ruiz-Correa, S., Yuen, J.: Similarity-based retrieval for biomedical applications. In: Perner, P. (ed.) Case-Based Reasoning on Images and Signals, vol. 73, pp. 355–387. Springer, Heidelberg (2008). https://doi.org/10.1007/978-3-540-73180-1_12
54. Sankoff, D., Kruskal, J.B. (eds.): Time Warps, String Edits, and Macromolecules: The Theory and Practice of Sequence Comparison. Addison-Wesley, Reading (1983)
55. Attig, A., Perner, P.: The problem of normalization and a normalized similarity measure by online data. Trans. Case-Based Reason. **4**(1), 3–17 (2011)

Foveated Vision for Deepface Recognition

Souad Khellat-Kihel, Andrea Lagorio, and Massimo Tistarelli$^{(\boxtimes)}$

Computer Vision Laboratory, University of Sassari,
Viale Italia 39, 07100 Sassari, Italy
tista@uniss.it

Abstract. In the last decade deep learning techniques have strongly influenced many aspects of computational vision. Many difficult vision tasks can now be performed by deploying a properly tailored and trained deep network. The enthusiasm for deep learning is unfortunately paired by the present lack of a clear understanding of how they work and why they provide such brilliant performance. The same applies to biometric systems. Deep learning has been successfully applied to several biometric recognition tasks, including face recognition. VGG-face is possibly the first deep convolutional network designed to perform face recognition, obtaining unsurpassed performance at the time it was firstly proposed. Over the last years, several and more complex deep convolutional networks, trained on very large, mainly private, datasets, have been proposed still elevating the performance bar also on quite challenging public databases, such as the Janus IJB-A and IJB-B. Despite of the progress in the development of such networks, and the advance in the learning algorithms, the insight on these networks is still very limited. For this reason, in this paper we analyse a biologically-inspired network based on the HMAX model, not with the aim of pushing the recognition performance further, but to better understand the representation space produced by including the retino-cortical mapping performed by the log-polar image resampling.

Keywords: Face representation space · Hierarchical model 'HMAX' · Biological foveated vision

1 Introduction

The invariance that means the ability to recognize a pattern under various transformations, with the ability to discriminate between different patterns makes the HMAX as a suitable network for pattern recognition. The HMAX [1] is composed of three types of neurons that can be distinguished based on how they respond to visual stimuli that they called: simple cells, complex cells, and hyper complex cells. This architecture is totally based on the functional organization and basic physiology of neurons in V1 however it does not take into consideration the retino-cortical mapping which takes advantage of the space-variant topology of the human retina. We propose to combine between the HMAX and log-polar

© Springer Nature Switzerland AG 2019
I. Nyström et al. (Eds.): CIARP 2019, LNCS 11896, pp. 31–41, 2019.
https://doi.org/10.1007/978-3-030-33904-3_3

mapping to fill the previous gap. The HMAX model is composed of four layers (S1, C1, S2 and C2) each layer represent a response of different type of cells (simple, complex and hyper complex). The number and kind of features change as we go higher in the model. The S1 and C1 represent the responses to the Gabor filters of various orientations. S2 and C2 are the responses to more complex features. The log-polar mapping [2] is an approximation of the retino-cortical mapping that is performed by the early stages of the primate visual system. We add the log-polar mapping to the HMAX for completing the functional vision perception system. Eventually, the log-polar mapping is used to understand the representation of faces based on some specific landmarks such as the eyes and the mouth and also to represent the outer faces based on foveal region. Even if the HMAX is powerful for recognition, the top level of the HMAX is not well understood. Also, it is not known how this representation can be affected using different variations. In this work we highlight the face space representation by visualizing the space based on uniform images and log-polar images. We propose also to understand how the HMAX model represent the faces in space as well as we fuse the advantages of the foveated vision and the peripheral vision using log-polar mapping. By this sequence of processes we can understand how the visual system can turn retinotopic images into features that accomplish the face recognition.

2   Related Works

The Deep Convolutionel networks have been used widely for face recognition. Yaniv Taigman et al. [3], proposed a convolutionel pipeline that consists of four stages: detection, alignment, representation and classification. The face representation is conducted from nine layer deep neural network. The accuracy achieved is 97.35% on the Labeled Faces in the Wild (LFW) dataset. In [4], they proposed a deep convolutional network trained to directly optimize the embedding itself. This approach shows a good representational efficiency as well as a 99.63% of accuracy on the Labeled Faces in the Wild (LFW) dataset. Jun-Cheng Chen et al. [5], proposed an approach based on training and testing stages. The network includes 10 convolutional layers, 5 pooling layers and 1 fully connected layer. They obtain 97.2% as recognition rate for IJB-A dataset. Face representation in Deep Convolutional Neural Networks is introduced in [6], a t-distributed Stochastic Neighbor Embedding (t-SNE) is used to reduce the multidimensional data for visualization. The visualization shows the image information remains in the top-level. The images with high quality is located along periphery such as the frontal view, well-lit and little occlusion. In the other side, the images with low quality are located near the origin such as extreme viewpoints, harshly lit, blurry and heavily occluded faces. Recently Alice O'toole et al. proposed a study [7] for the nature of the code produced by the DCNNs and why is this poorly understood. As main reason the DCNNs designed to model primate visual system with millions of computations between the image and representation in the cortex, also the uncontrolled nature of the datasets used to train the networks can demonstrate the success of the DCNNs. They propose also a visualization of face representations based on t-SNE dimension reduction from 320 dimensions to 2 dimensions.

The HMAX model, which was developed before CNNs started to be extensively used to solve computer vision problems, was developed to demonstrate the feasibility of a biologically plausible architecture for face recognition. The model was tested on several publicly available databases such as LFW, PubFig and SURF-W [8] providing results at the state of the art. In [9], a new C3EFs inspired from ventral and dorsal stream of visual cortex has been used. They proposed a model to extract new view-independent features, using visual attention model and ventral stream model to achieve the goal of view-independent face recognition. A higher layer has been added to the HMAX model. Xiao lin Hu [10] proposed an improved version of the HMAX model, named as sparse HMAX. This model addresses the local-to-global structure gradually along the hierarchy by applying a patch-based learning approach to the output of the previous layer. The major difference between the models is that in the sparse HMAX S2 bases are learned by sparse coding, and therefore the S2 codes are calculated by sparse coding.

3   Foveal and Peripheral Vision

Looking at a face from a short distance should produce a different perception from the same face viewed from a long distance. However, the human visual system is capable of coping for the size change due to distance by capturing a high resolution description of the most salient features of the viewed face. In an artificial system, this can be accomplished either by "foveating", in rapid succession, these parts of the scene or moving an interest window on a high resolution image [11]. Certainly, facial features are important for recognition, but it is not said that the face itself is better characterized by the most prominent features taken in isolation, rather than by the context in which they are located. For this reason, it is not sufficient to scan the face or the image with a fovea, but it is also necessary to provide some information on the area around the fovea. A way to meet both requirements is to adopt a space-variant sampling strategy of the image, where the central part of the visual field is sampled at a higher resolution than the periphery. In this way the peripheral part of the visual field is coded at low resolution, but can be still used to describe the context in which foveal information is located. A great advantage of this approach is the considerable data reduction with respect to adopting a uniform resolution schema, while a wide field of view (i.e. peripheral vision) is preserved [12,13]. The high acuity in the fovea is due to the dense packing of cone photoreceptors. On the other hand, the low acuity in the peripheral area of the retina, is due to the lack of cones and the relative sparsity of rods (Fig. 1). In the proposed system, the fovea is directed to capture the information lying on the eyes and mouth regions, while the periphery captures information on the outer region of the entire face.

The arrangement of the cones in the human retina, and the corresponding variable size of the ganglion cells receptive fields across the visual field, produces

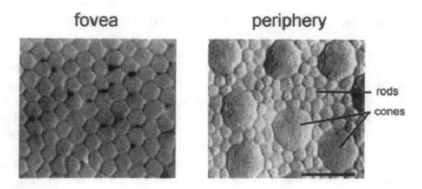

Fig. 1. Microscopic images showing the spatial distribution and size of the cones in fovea and periphery of the human retina.

a space-variant topological transformation of the retinal image into its cortical projection [14]. This transformation can be presented by a log-polar mapping. This retino-cortical mapping can be described through a transformation from the retinal plane onto the cortical plane, which is scale and rotation invariant, as depicted in Fig. 2. If (x,y) are Cartesian coordinates and (ρ, θ) are the polar coordinates, by denoting $z = x + jy = \rho e^{j\theta}$.

A point in the complex plane, the log-polar mapping is

$$w = ln(z). \tag{1}$$

As the resolution in the fovea is almost constant, this transformation is a good approximation of the non-foveal part of the retinal image. Therefore, it is applied to reproduce peripheral vision by re-sampling the outer region of the face.

The transformation has been implemented through the algorithm proposed in [15]. Different parameters can be tuned, including: the number of cells per eccentricity (CP), the number of eccentricities (NE), the cell dimension (CD), the size of the overlapping area along the eccentricity (OE) and the radius (OR). The Caltech database, composed of 450 frontal face images of 27 subjects [16], was employed to select the parameters for the log-polar mapping. Several classification experiments were carried on the re-mapped face images, by assigning different values to the parameters of the log-polar transformation. The resulting scores are reported in Table 1.

4 The Proposed Hierarchical Model

The HMAX model is an hierarchical model for object representation and recognition inspired by the neural architecture of the early stages of the visual cortex in the primates. The general architecture of the HMAX model is represented in Fig. 3. Proceeding to the higher levels of the model, the number and typicality of the extracted features change. Each layer is projected to the next layer by applying template matching or max pooling filters. Proceeding to the higher levels of

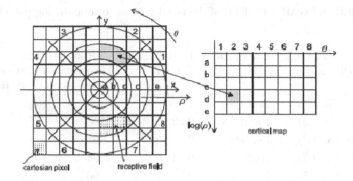

Fig. 2. Graphical representation of the log-polar mapping. Every pixel [x,y] on the Cartesian plane is represented on the basis (ρ, θ) as $[\ln(\rho), \theta]$ on the cortical plane.

the model, the number of (X, Y) pixel positions in a layer is reduced. The input to the model is the gray level image. S1 and C1 represent the responses to a bank of Gabor filters tuned to different orientations. S2 and C2 are the responses to more complex filtering stages.

The first layer S1 in the HMAX network consists of a bank of Gabor filters applied to the full resolution image. The response to a particular filter G, of layer S, at the pixel position (X, Y) is given by:

$$R(X,Y) = \left| \frac{\sum X_i G_i}{\sqrt{\sum X_i^2}} \right| \qquad (2)$$

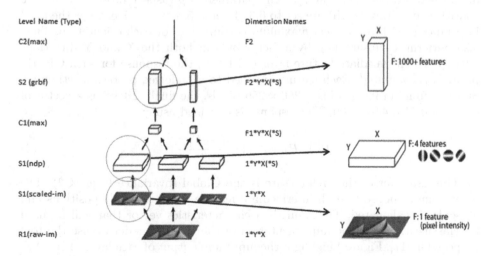

Fig. 3. General architecture of the HMAX model.

Table 1. EER and GAR@ 1% FAR of the facial regions to tune the log-polar mapping parameters

Parameters (CP, NE, CD, OE, OR)	Used regions	EER (Equal Error Rate)	VR at 1% FAR (Verification Rate at 1% False Acceptance Rate)
(50, 50, 50, 1.5, 1.5)	Left eye	2	98
(50, 50, 50, 1.5, 1.5)	Right eye	0.22	100
(50, 50, 50, 1.7, 1.7)	Left eye	2	98
(50, 50, 50, 1.7, 1.7)	Right eye	0.44	98
(70, 50, 50, 1.5, 1.5)	Left eye	2	96
(70, 50, 50, 1.5, 1.5)	Right eye	1.67	98
(70, 50, 50, 1.7, 1.7)	Left eye	4	96
(70, 50, 50, 1.7, 1.7)	Right eye	2	98
(32, 32, 120, 1.7, 1.7)	Left eye	0.22	100
(32, 32, 120, 1.7, 1.7)	Right eye	0	100
(32, 64, 130, 1.5, 1.5)	Face	0.11	100
(32, 64, 130, 1.5, 1.5)	Face	0	100

The size of the Gabor filter is 11×11 and it is formulated as:

$$G(x, y) = exp(\frac{-(x^2\gamma^2Y^2)}{2\sigma^2})cos(\frac{2\pi}{\lambda}X) \quad (3)$$

Where $X=xcos\theta-ysin\theta$ and $Y=xsin\theta+ycos\theta$. x and y vary between -5 and 5, and θ varies between 0 and π. The parameters ρ (aspect ratio), σ (effective width), and λ (wavelengh) are set to 0.3, 4.5 and 5.6, respectively. For the local invariance (C1) layer, a local maximum is computed for each orientation. They also perform a subsampling by a factor of 5 in both the X and Y directions [19]. In the intermediate feature layer (S2 level), the response for each C1 grid position is computed. Each feature is tuned to a preferred pattern as stimulus. Starting from an image of size 256×256 pixels, the final S2 layer is a vector of dimension $44 \times 44 \times 4000$. The response is obtained using:

$$R(X, P) = exp(\frac{||X - P||^2}{\sigma^2}) \quad (4)$$

The last layer of the architecture is the Global Invariance layer (C2). The maxi-mum response to each intermediate feature over all (X, Y) positions and all scales is calculated. The result is a characteristics vector that will be used for classification. For the implementation of the HMAX model we use the tool proposed in [17]. Figure 4 highlight the input and output of each layer of HMAX. Figure 5 shows how the foveal and peripheral images are fused. As local fixation we propose to use the eyes, the region that cover the eye gaze and also the region that cover the mouth and the nose and this due to the obtained results in [18] also recent studies show that the eye and point of gaze measures the motion of the face.

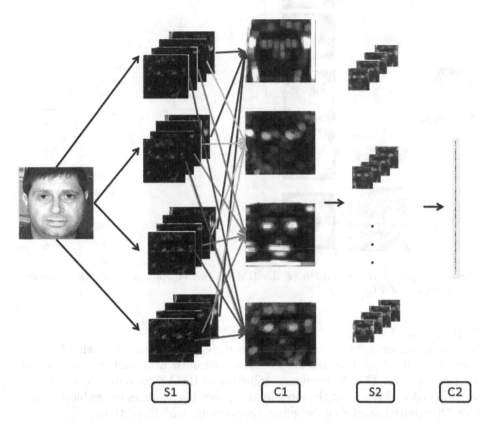

Fig. 4. Graphical representation of the HMAX model applied to a uniformly sampled face image. .

5   Experimental Results

The experiments are conducted with different databases such as PubFig dataset which is a large dataset, real-world face dataset consisting of 58797 images of 200 people collected from the internet. These images are taken in completely uncontrolled situations. This database contains a variation in pose, lighting, expression, camera, imaging conditions. The PubFig dataset is similar to LFW dataset. However, the PubFig dataset has enough examples per each subject. The main purpose is to visualize the face space representation based on uniform images as well as log-polar images.

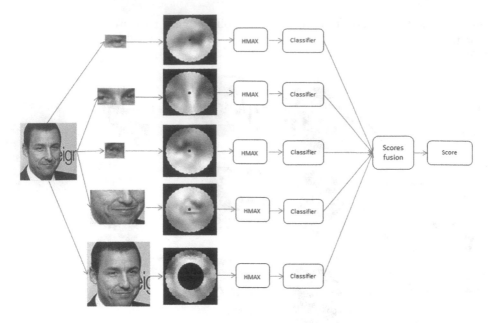

Fig. 5. Graphical representation of the HMAX model applied to several log-polar mapped areas of a face image.

Experiment 1

In the first experiment we visualize the face space representation using Caltech database [16]. Figure 6 represents the space obtained from uniform images and log-polar images. The 256 features obtained from HMAX are reduced to 2 dimensions in order to represent the faces in 2D space. The features are reduced based on t-Distributed Stochastic Neighbor Embedding (t-SNE) method [19].

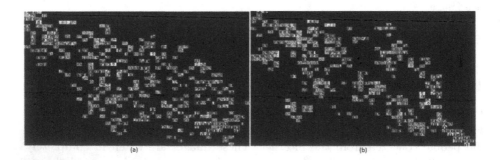

Fig. 6. Two-dimensional representation of the face-space obtained from the HMAX model applied to (a) uniformly sampled images, (b) log-polar mapped images.

From these visualizations we can distinguish that the log-polar outpart faces are more clustered per each subject. We prove this discrimination by calculating the accuracy. Table 2 shows the results obtained on the Caltech database, using the uniform images, the outpart faces and the local regions. These images are the input of an HMAX to extract the features. Then the features are classified using a neural network based on the SoftMax function. The lossfunction for the SoftMax layer is based on the computation of the crossentropy [20]:

$$L_i = -log(\frac{e^{f_{y_i}}}{\sum_j e^{f_j}}) \tag{5}$$

Where f_j is the j-th element of the feature vector representing subject f, while L_i is the full loss over the training examples.

Table 2. Performances using peripheral and foveal regions.

HMAX faces (%)	Peripheral faces (%)	Foveal region1 (%)	Foveal region2 (%)	Foveal region3 (%)	Foveal region4 (%)	Fusion peripheral and Foveal regions
57.22	90.00	57.78	72.78	52.22	77.22	95.56

Experiment 2

In the second experiment we visualize the face space representation based on uniform images as well as the log-polar images. Figure 7 represents the PubFig images represented based on the top level of the proposed hierarchical model.

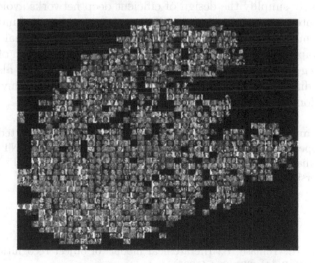

Fig. 7. Two-dimensional representation of the face-space obtained by the HMAX on images from the PubFig dataset resampled with the log-polar mapping.

As it can be observed from the presented results, the log-polar mapped images not only allow a better discrimination of the different identities, but also the representation of each subject is distributed within the face space. For example, while the frontal lighting images are represented in the center of each subject's cluster, the face images with dark lighting are projected far from the center. The log-polar mapping applied to the outer areas of the face also improves the discrimination power of the HMAX model.

6 Conclusion

In the last decade, Deep learning models have demonstrated to be very successful in solving complex computer vision problems, including face recognition. This is not surprising as the human visual system is based on a very deep neural architecture, encompassing several computational processes including convolutions and pooling. In this paper the HMAX model has been exploited to represent the visual pathway from the retinal stage to V1 area in the brain. A log-polar mapping stage has been added to the HMAX model to incorporate also the complex retino-cortical mapping decomposing the visual field into foveal and peripheral regions with a variable resolution mapping. The proposed hierarchical network has shown a remarkable performance in discriminating both the gender and the identities. This work represents a step forward in the process of understanding both the complex neural mechanisms and associated perceptual process subduing face recognition in the visual system of the primates, and the mechanisms behind the recently developed deep learning networks. From the presented results it is clear that using the available knowledge on the neural architecture of the brain can greatly simplify the design of efficient deep networks avoiding a computationally intensive training process. However, the addition of more layers to the simplified model adopted in this paper to account for additional variabilities in the face samples, as well a better analysis of the implementation of the retino-cortical mapping, coupled with a suitable attentional mechanism also involving the temporal dimension, are also required. They will be further investigated in the continuation of this research work .

Acknowledgements. This research work has been partially supported by a grant from the European Commission (H2020 MSCA RISE 690907 "IDENTITY"), by a grant of the Italian Ministry of Research (PRIN 2015) and by a special research grant from the University of Sassari.

References

1. Maximilian, R., Tomaso, P.: Hierarchical models of object recognition in cortex. Nat. Neurosci. **2**(11), 321–354 (1999)
2. Daniilidis, K.: Attentive visual motion processing: computations in the log-polar plan. In: Kropatsch, W., Klette, R., Solina, F., Albrecht, R. (eds.) Theoretical Foundations of Computer Vision. Computing Supplement, vol. 11, pp. 1–20. Springer, Vienna (1996). https://doi.org/10.1007/978-3-7091-6586-7_1

3. Taigman, Y., Yang, M., Ranzato, M.A., Wolf, L.: DeepFace: closing the gap to human-level performance in face verification. In: IEEE Conference on Computer Vision and Pattern Recognition (CVPR), pp. 1701–1708 (2014)
4. Schrott, F., Kalenichenko, D., Philbin, J.: FaceNet: a unified embedding for face recognition and clustering. In: IEEE Conference on Computer Vision and Pattern Recognition (CVPR), pp. 815–823 (2015)
5. Jun-Cheng, C., Patel, V., Chellappa, R.: Unconstrained Face Verification using Deep CNN Features. In: IEEE Winter Conference on Applications of Computer Vision (WACV), pp. 1–9 (2016)
6. Parde, C.J., et al.: Face representations in deep convolutional neural networks. In: IEEE Conference on Automatic Face and Gesture Recognition (FG), Gender and Discourse (2017)
7. O'Toole, A., Castillo, C., Parde, C.J., Hill, M.Q., Chellappa, R.: Face representation in deep convolutional neural networks. Trends Cogn. Sci. **22**(9), 794–809 (2018)
8. Liao, Q., Leibo, J., Poggio, T.: Learning invariant representations and applications to face verification. In: Advances in Neural Information Processing Systems 26, NIPS 2013, pp. 3057–3065 (2013)
9. Esmaili, Ş., Maghooli, K., Nasrabadi, A.: C3 effective features inspired from ventral and dorsal stream of visual cortex for view independent face recognition. Int. J. Adv. Comput. Sci. **5**, 1–9 (2016)
10. Hu, X., Zhang, J., Li, J., Zhang, B.: Sparsity-regularized HMAX for visual recognition. PLoS One, 1–12 (2014)
11. Burt, P.G.: Smart sensing in machine vision. In: Machine Vision: Algorithms, Architectures, and Systems, pp. 1–30. Academic Press (1988)
12. Massone, L., Sandini, G., Tagliasco, V.: Form-invariant topological mapping strategy for 2-D shape recognition. In: CVGIP, vol. 30, no. 2, pp. 169–188 (1985)
13. Sandini, G., Tistarelli, M.: Vision and space-variant sensing. In: Wechsler, H. (ed.) Neural Networks for Perception: Human and Machine Perception. Academic Press (1991)
14. Schwartz, E.: Anatomical and physiological correlates of visual computation from striate to infero-temporal cortex. IEEE Trans. Syst. Man Cybern. **SMC-14**(2), 257–271 (1984)
15. Grosso, E., Lagorio, A., Pulina, L., Tistarelli, M.: Towards practical space-variant based face recognition and authentication. In: 2nd International Workshop on Biometrics and Forensics, pp. 1–6, October 2014
16. Caltech database. http://www.vision.caltech.edu/html-files/archive.html
17. HMAX toolbox. http://maxlab.neuro.georgetown.edu/hmax.html
18. Khellat-Kihel, S., Lagorio, A., Tistarelli, M.: Face recognition 'on the move' combining incomplete information. In: Proceedings of 6th International Workshop on Biometrics and Forensics (IWBF), pp. 1–6, September 2018
19. van der Maaten, L.J.P.: Accelerating t-SNE using tree-based algorithms. J. Mach. Learn. Res. **15**, 3221–3245 (2014)
20. CS231n: Convolutional Neural Networks for Visual Recognition, Spring 2018

Representation Learning for Underdefined Tasks

Jean-François Bonastre$^{(\boxtimes)}$

Avignon University, Avignon, France
jean-francois.bonastre@univ-avignon.fr

Abstract. In the neural network galaxy, the large majority of approaches and research effort is dedicated to defined tasks, like recognize an image of a cat or discriminate noise versus speech records. For these kind of tasks, it is easy to write a labeling reference guide in order to obtain training and evaluation data with a ground truth. But for a large set of high level human tasks, and particularly for tasks related to the artistic field, the task itself is not easy to define, only the result is known, and it is difficult or impossible to write such a labeling book. We name this kind of problem as "Underdefined task". In this presentation, a methodology based on representation learning is proposed to tackle this class of problems and a practical example is shown in the domain of voice casting for voice dubbing.

Keywords: Representation learning · Underdefined task · Knowledge distillation · Transfer learning · Voice casting · Voice dubbing

1 Introduction

Usually, in the definition of *Machine Learning* appears very early a distinction between "supervised" and "unsupervised" classification tasks. "Supervised" indicates if the program has to attribute to a given input a class label between a close set of known labels. "Unsupervised" usually means that you have no prior information on the number of classes and the program has to determine the classes, the number of classes and the tying between an input and a class label. This "Supervised/Unsupervised" classification distinction is usually extended to the training scenario, with a slightly different meaning. At the training level, "supervised" means usually that class-labeled training data are available when "unsupervised" indicates that the labels are not available (several intermediate levels, like "slightly-" or "self-supervised" training were also proposed recently).

More important than this differences is the fact that in both cases knowledge on the classes is available: it is easy to write a labeling reference able to guide the human-based labeling of training or evaluation data. But for a large set of high level human tasks, and particularly for tasks related to the artistic field, even the human operators who are doing the job everyday have difficulties to define precisely the objectives and the details of the targeted tasks.

© Springer Nature Switzerland AG 2019
I. Nyström et al. (Eds.): CIARP 2019, LNCS 11896, pp. 42–47, 2019.
https://doi.org/10.1007/978-3-030-33904-3_4

This presentation is dedicated to this kind of tasks that we denote as *underdefined task*. After some definitions, We will present a methodology based on representation learning [1] able to tackle this class of problems. We will also show a practical example based on our first, and on going, work in this domain.

2   A Better Definition of *Underdefined Tasks*

As said in the previous section, we define a task as *underdefined* when an expert of the task is unable to define precisely "the objective and the details of the task", even if the expert is doing practically the job everyday. This definition is clearly too large and will benefit from precision and restrictions.

First, underdefined tasks correspond usually to an high level intellectual task. Second, we have to restrict to human tasks done at a large scale in order to be able to reduce the human factor (by the number of concerned human operators). Third. Even if in general the decisions are not documented, as we need data for our machine learning processes, we need obviously physical traces of the human decisions.

3   Why to Work on *Underdefined Tasks*?

Looking at the definition proposed in the previous sections, it is clear that underdefined problems are hard problems, involving fine and human-being information and fewly documented. Obviously, this kind of tasks is not easy to automatize. Moreover, it is not clear if there is a direct business interest to automatize it.

But underdefined problems show a large potential in different directions:

- In terms of knowledge gathering. Work on an underdefined task allows to understand better the corresponding high level human-being task. It allows to describe the task, its objective, its difficulties and specificities. And finally, the hope is to highlight the underlined human decision process.
- In terms of knowledge transfer and cultural heritage safegarding. As said before, underdefined tasks are very often related to artistic and cultural aspects. To better understand the creative process allows to document it, to transfer more easily the knowledge to a younger generation and to protect the knowledge.
- In terms of quality of work life. To replace the human expert by a machine, in this area at least, is clearly an heresy, in terms of both feasibility and economic interest. But machine learning area could improve a lot the quality of work life and helps to insure the quality of the work done. The machine learning could check if there is a possibility of a mistake (for example when the expert is trying to associate two specific sounds in a sound-based creation). It could also use the human expert choices in order to recommend some alternative choices or additional choices (still for sound-based creation, propose a given sound based on the first ones selected by the expert).

4   How to Deal with *Underdefined Task*: A General Methodology Based on Representation Learning

Start to work on an underdefined problem is difficult as the only available evidence of the targeted human process is a set of works done by some human experts (like sound ambiances created for specific purposes). Usually, talking with the expert learns to you that the original command order was not very clear, or was not using a generally accepted vocabulary. Talking with the same experts shows also that the experts know what they are doing but are not able to explain it... Finally, in general, only the positive examples are saved and you don't have access to the rejected choices (the expert keeps the final sound ambiance but not the thousands of trials done before).

The methodology we propose starts from this point and goes step by step until an detailed and explainable set of decision criterion. The four main steps of our methodology are presented bellow.

4.1   Check if There Is Evidence of the Task in the Data

First, we have to verify if there is in the available data, or in its projection in the computing world, some traces of the human expert work, of the targeted task. The main idea here is to build a binary recognizer which wishes to decide if an example comes from the expert work or not (in our work, a siamese neural network is used here). If the recognizer is statistically better than a random decision, there is evidence of the human work in the data. If not..., unfortunately you could just conclude that you didn't find this evidence, not that there is no evidence as your recognizer accepts maybe some lacks. The main difficulty is to build a strong evaluation protocol in order to insure the statistical significance of the experiments.

4.2   Build a Data Representation Able to Highlight the Task Information

If you know that there is evidence of the task in the data, it becomes easier to build a well suited representation for the task. Here, some methods inspired from *word embedding* could be easily used (in our case, we use a variant inspired from xvector). This kind of approach uses dicriminative training and allows to build a representation space with the adapted size and structure able to emphasize the in-interest information.

Two remaining problems should be solve at this level. First, in quite all the situations, the size of the available dataset is small and, thanks to the representation build previously, it is interesting to take advantage of *knowledge distillation* and *transfer learning* [2–10] in order to reinforce the model by using other available datasets.

To assess the effectiveness of the representation at this step is not easy. Some measures on the richness of the information present in the representation could be done, in order to verify that the model contains more information on the task than in the initial step. But the only strong criteria is still to use the binary classifier seen previously in order to verify, at least, that your refined representation still contains evidence of the task...

4.3 Find, in This Representation, the Decision Criterion

At this step, the learned model is usually working at the complete example level. In this case, the decisions are made at this level. For example, it will look at a complete sound ambiance of three hours and attributes only one internal label to this example. It will also group together all the examples associated with the same internal label. These characteristics bring a lack in precision and generalization power.

To overcome this limitation, this step of our method will re-clusterize the examples in the representation space build at the previous step. It will also split the examples in homogeneous sub-parts. Then, the same optimization process than in the previous step is use. Thanks to this step, we wish to obtain a representation space with a set of criteria, each of them potentially shared between several training example-level labels. The criterion could be present only on a part of a given training example.

To assess this step, the same difficulties and limitations than for the previous one are present. Another time, the only strong criterion is the binary decision approach and protocol settle at the step 1, but working in this step representation space.

4.4 Make the Representation *Explainable*

This step is dedicated to the feedback of the human experts. Now we have a system able to attribute to an example of the underdefined task several decision criteria, which are (potentially) shared between different examples of the task. In order to understand better the work done by the human experts and in order to facilitate their work, it is mandatory to explain the decision criteria in terms understandable by the experts. This step is dedicated to this a posteriori "explainability" task. It will mix classical human perception experiments and a recommendation approach, dedicated to register the expert feedback.

5 Voice Casting: A Nice Practical Example

The search for professional voice-actors for cultural productions is performed by a human operator and suffers from several difficulties. First, industrialized production has a strong appetite for the mass of new voice-talents but operators generally cannot perform large auditions. Second, the subjectivity bias of artistic directors makes it difficult to automatize the voice casting process.

Voice casting is clearly an underdefined task even if some existing works are present in the literature [11,12]. It is linked to artistic and cultural perception.

There is not precise description of the task or of the decision criterion. The impact of the task, both in artistic and financial terms, could be high. There is thousands of available examples of the expert work (all the already dubbed movies, for example). For all these reasons, we selected voice casting to apply our methodology. The two first steps were already presented in our previous works [12,13]. The third is ongoing and the last step of the methodology is... planed.

Acknowledgment and Credits. The voice casting for voice dubbing work was supported by Avignon University foundation "Pierre Berge" PhD program and by ANR TheVoice project ANR-17-CE23-0025 (DIGITAL VOICE DESIGN FOR THE CREATIVE INDUSTRY).

The main part of the presented work on voice casting was done by Adrien Gresse during his PhD. Some ongoing parts are directly issued from Mathias Quillot's (on going) PhD. Both provided a large part of the figures and tabs of this presentation.

References

1. Variani, E., Lei, X., McDermott, E., Moreno, I.L., Gonzalez-Dominguez, J.: Deep neural networks for small footprint text-dependent speaker verification. In: International Conference on Acoustics, Speech and Signal Processing (ICASSP). IEEE (2014)
2. Lopez-Paz, D., Bottou, L., Schölkopf, B., Vapnik, V.: Unifying distillation and privileged information. In: International Conference on Learning Representations (2016)
3. Vapnik, V., Izmailov, R.: Learning using privileged information: similarity control and knowledge transfer. J. Mach. Learn. Res. **16**, 2023–2049 (2015)
4. Hinton, G., Vinyals, O., Dean, J.: Distilling the knowledge in a neural network (2015)
5. Price, R., Iso, K.-I., Shinoda, K.: Wise teachers train better DNN acoustic models. EURASIP J. Audio Speech Music Process. **2016** (2016)
6. Markov, K., Matsui, T.: Robust speech recognition using generalized distillation framework. In: INTERSPEECH (2016)
7. Li, J., Seltzer, M.L., Wang, X., Zhao, R., Gong, Y.: Large-scale domain adaptation via teacher-student learning (2017)
8. Watanabe, S., Hori, T., Le Roux, J., Hershey, J.R.: Student-teacher network learning with enhanced features. In: Acoustics, Speech and Signal Processing (ICASSP). IEEE (2017)
9. Asami, T., Masumura, R., Yamaguchi, Y., Masataki, H., Aono, Y.: Domain adaptation of DNN acoustic models using knowledge distillation. In: International Conference on Acoustics, Speech and Signal Processing (ICASSP). IEEE (2017)
10. Joy, N.M., Kothinti, S.R., Umesh, S., Abraham, B.: Generalized distillation framework for speaker normalization. In: INTERSPEECH (2017)
11. Obin, N., Roebel, A., Bachman, G.: On automatic voice casting for expressive speech: speaker recognition vs. speech classification. In: International Conference on Acoustics, Speech and Signal Processing (ICASSP). IEEE (2014)

12. Gresse, A., Rouvier, M., Dufour, R., Labatut, V., Bonastre, J.-F.: Acoustic pairing of original and dubbed voices in the context of video game localization. In: INTERSPEECH (2017)
13. Gresse, A., Quillot, M., Dufour, R., Labatut, V., Bonastre, J.-F.: Similarity metric based on Siamese neural networks for voice casting. In: International Conference on Acoustics, Speech and Signal Processing (ICASSP). IEEE (2019)

Data Mining. Natural Language Processing and Text Mining

Applying Self-attention for Stance Classification

Margarita Bugueño and Marcelo Mendoza[✉]

Instituto Milenio Fundamentos de los Datos, Departamento de Informática,
Universidad Técnica Federico Santa María, Santiago, Chile
margarita.bugueno.13@sansano.usm.cl, mmendoza@inf.utfsm.cl

Abstract. Stance classification is the task of automatically identify the user's positions about a specific topic. The classification of stance may help to understand how people react to a piece of target information, a task that is interesting in different areas as advertising campaigns, brand analytics, and fake news detection, among others. The rise of social media has put into the focus of this task the classification of stance in online social networks. A number of methods have been designed for this purpose showing that this problem is hard and challenging. In this work, we explore how to use self-attention models for stance classification. Instead of using attention mechanisms to learn directly from the text we use self-attention to combine different baselines' outputs. For a given post, we use the transformer architecture to encode each baseline output exploiting relationships between baselines and posts. Then, the transformer learns how to combine the outputs of these methods reaching a consistently better classification than the ones provided by the baselines. We conclude that self-attention models are helpful to learn from baselines' outputs in a stance classification task.

Keywords: Stance classification · Self-attention models · Social networks

1 Introduction

Stance classification is the task of identifying the attitude that a person has towards target information. Due to the rise of Web 2.0, many researchers have put their efforts in the classification of stance in online social networks (OSN) [14]. In OSN the dissemination of messages is often convoluted. Currently, the popularity of microblogging websites and social networks makes them essential for information dissemination where millions of users could spontaneously post messages to share opinions about various information every day.

In OSN, and as has been reported in numerous previous works [21], stance classification is focused on the position of different users regarding a conversation in the social network. Each conversation is defined by a post that initiates the thread and a set of nested responses that form a conversational thread. Thus, the

© Springer Nature Switzerland AG 2019
I. Nyström et al. (Eds.): CIARP 2019, LNCS 11896, pp. 51–61, 2019.
https://doi.org/10.1007/978-3-030-33904-3_5

task corresponds to classify each message according to the type of interaction in the conversation. There is consensus in addressing this problem using four stances: supporting, denying, questioning, or commenting. In this context, the stance of the user is always about the original post.

Much research has focused on the analysis of the original post and the post of the user that defined a stance in the conversation. To address this problem, text-based representations has been ingested into standard machine learning algorithms (e.g., support vector machines) and also into deep learning architectures as recurrent or convolutional neural networks. The task has shown to be difficult and challenging. Many of the stance classification methods can detect one or two stances at the cost of deteriorating the performance in the remaining classes [9,17]. For example, there are stance methods able to detect supporting stances but with a high rate of false positives in denying or questioning attitudes. It remains unclear how to provide a model able to work well in the four classes.

In this paper we study how to combine different baseline models for stance classification taking advantages of the strengths of each of them. We use the transformer [18], a successful deep learning architecture used for translation in natural language processing, that can detect which part of the data ingested is useful to solve a given task. Instead of using the transformer to encode a representation of the post, we provide to the transformer the outputs of several baselines that in combination with the text, produce an encoding of the baselines' outcomes and the analyzed post. Then, the encoding is provided to a classifier to infer the stance of the message. Model fitting is driven to minimize the loss produced by the transformer. As a consequence, the encoding learned by the transformer can consistently produce a better prediction of the stance than the ones provided by the baselines. Also, we use the attention matrices proportioned by the transformer to understand which baselines are more important for the model. Since the encoding of the transformer jointly encodes the outputs of the baselines and the text of the post analyzed, in several cases, the forecast differs from the result delivered by a committee of machines outperforming the results of an ensemble of classifiers.

This work is organized as follows. In Sect. 2 we present a literature review. In Sect. 3 we introduce our proposal. Section 4 presents experimental results. We conclude in Sect. 5 providing remarks and outlining future work.

2   Related Work

One of the first works that addressed stance classification pointed out to the recognition of stances in on-line ideological debates [17]. In that work, the authors explored the utility of sentiment opinions, building an arguing lexicon from a manually annotated corpus. Using the entries of the lexicon as features, the authors used supervised learners for stance classification on four different online debate forums. Anand *et al.* [1] examined stance classification on a corpus of 4873 posts across 14 topics gathered from ConvinceMe.net. The authors showed that rebuttal posts were hard to classify using text features. They also demonstrate that methods that take into account the context of the post (in an OSN this refers to the conversational thread) might be helpful for this task. Walker *et al.* [19] coded an extensive collection of posts tagging the level of agreement between consecutive posts. The released corpus, named the Internet Argument Corpus (IAC) comprises posts manually sided for the topic of discussion extracted from 4forums.com, a site devoted to online debate. The authors concluded that rebuttal posts are hard to detect even for human annotators. Main reasons for these difficulties are based on the extensive use of stylistic ambiguity in rebuttal as sarcasm and irony. Stance classification at document level was investigated by Faulkner [8], who proposed a set of text-based features to capture the stance of student's essays (answer) concerning an essay prompt (affirmation). Several discriminative and generative machine learning methods were studied for stance classification in two-sided debates [9]. The authors concluded that there is no clear winner between generative or discriminative models but sequence models as Hidden Markov Models outperforms its competitors in many cases.

The first work on stance classification for news comments was authored by Sobhani *et al.* [16]. The authors used topic modeling for argument tagging, who are subsequently used for stance classification (agree/disagree) of news comments. Stance classification on Twitter was addressed by Lukasic *et al.* [11] who explored the use of temporal dependences along sequences of tweets to improve the performance of a stance classifier based on Hawkes processes. The relation between stance and rumor veracity was studied in the PHEME project [5], where several resources were developed to tag misinformation, disinformation, rumors, and speculations in OSN. As part of this project, several posts were labeled according to the stance towards target information. In this line of research, Zubiaga *et al.* [21] studied how a tree-CRF classifier performed on stance classification modeling conversational threads in Twitter replies. Another contribution of this work is the construction of a fine granularity dataset for stance classification, extending the scenario of stance classification from two classes (agree/disagree) to four categories (supporting/denying/questioning/commenting). There is a consensus that these four classes represent in a better way the complexity of the task.

During the last years, deep learning architectures have become a dominant approach in stance classification. LSTM networks were applied for the first on this task by Zubiaga *et al.* [20]. The authors showed that the use of sequential learning architectures are useful for this task. Bidirectional LSTM-based encodings of tweets were used for stance classification by Augenstein *et al.* [2] on the Twitter Stance Detection corpus, a dataset released by Mohammad *et al.* during the SemEval 2016 challenge [13]. Chen *et al.* [4] used convolutional neural networks (CNN) to jointly address stance classification and rumor detection on the corpus released for the SemEval 2017 challenge [6] while Lozano *et al.* [10] used an ensemble classification approach of combining convolutional neural networks with both automatic rule mining and manually written rules on the same corpus, achieving the first place in that competition. In the same competition, Bahuleyan *et al.* [3] used XGBoost and additive and iterative tree-based supervised machine learning approach where a strong classifier is sequentially constructed from multiple weak learners. XGBoost achieved second place in the competition. Recent efforts point to jointly learn rumor detection and stance classification using two-layered gated recurrent units networks (GRU) [12].

3   Applying Self-attention for Stance Classification

3.1   Architecture

For a given tweet and given target information (e.g., a seed post), each baseline provides a stance label from the set of possible tags, i.e., supporting, denying, questioning or commenting (SDQC). To ingest these outputs into the transformer, we encode each label using a one-hot encoding vector of the class, producing orthogonal vectors for different categories. To ingest the text of the tweet, we use its BERT vector[1]. BERT (Bidirectional Encoder Representations from Transformers) [7] is a new method of pre-training language representations computed using the transformer. The vectors of BERT are computed at word-level. It is possible to produce a single vector at sentence-level using vector compositionality. The compositional function is provided by Google research as a service of the BERT library.

We use five baselines. We concatenate these five vectors with the BERT vector of the tweet. This vector is ingested to the transformer. We use the encoder of the transformer using two encoders, each one with four attention heads. Each encoder has an attention module and a position wise feed forward layer. The position wise layer is a key module that allows to code from which baseline the data is encoded. The outputs of the encoders are concatenated and then by applying a Hadamard product between them; we obtain a state vector that represents what the transformer learned from the baselines and the tweet. This vector is ingested into a softmax layer, who is in charge of producing a stance label. The model is depicted in Fig. 1.

As Fig. 1 shows, we only use the encoder of the transformer. The original architecture [18], proposed as a sequence transduction model based on an encoder-decoder structure is simplified by us making use of the encoder and the attention mechanisms provided by the architecture. The transformer uses stacked self-attention and fully connected layers. The encoder used for stance classification is composed of a stack of two layers. The transformer uses a residual connection between each module and a normalization. The links inside the transformer are produced by inputs and outputs of the same dimension. The attention mechanism of the transformer is wired using a scaled dot-product operator. Then, multi-head attention consists of several attention layers running in parallel. We use four attention heads. After the attention module, a position wise feed forward module is applied to each position, consisting of two linear transformations with ReLU activations in between. The output sequence produced by the encoder gives six vectors, one for each input ingested into the

---

[1] https://github.com/google-research/bert#pre-trained-models.

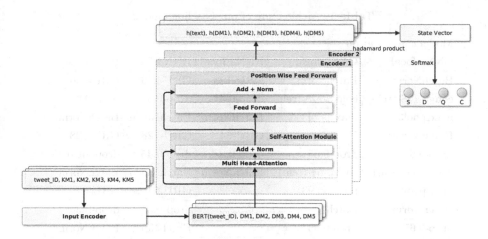

Fig. 1. To apply self-attention for stance classification, each baseline is encoded and concatenated with the BERT vector of the tweet. This encoding is ingested into the transformer using two levels of encodings, each one with a self-attention module of four attention heads. At the output of the encoder, we apply a Hadamard product. The resulting vector is ingested in a softmax layer producing the stance.

encoder. We combine these vectors using the Hadamard product (a.k.a. Schur product), producing a state vector for the transformer.

3.2   Baselines

We used five baselines in our model. Two of them are based on convolutional neural networks (CNN) and the other three on recurrent neural networks (RNN). One CNN was implemented using one convolutional layer, and the other used two layers. For both CNN, we used ReLU activation functions. In the case of RNNs, we used GRU layers. For one RNN, we used two GRU layers, while the other RNNs were implemented using three GRU layers. For the five baselines, the output was produced using a softmax. As a loss function, we used focal loss. For the last baseline, we replaced the focal loss function with a categorical cross entropy loss function. In Table 1 we show the parameters of each architecture.

4   Experiments

4.1   Data Description

We use a publicly available Twitter dataset named RumourEval, used on SemEval 2017 (Task 8) [6], consisting of two subtasks: (a) stance classification, and (b) veracity classification. Subtask A corresponds to the problem of use discourse around claims to verify or disprove them.

RumourEval data has already been annotated for veracity and stance following a published annotation scheme [21]. The labeling process was conducted

Table 1. Architecture of our baseline models

Model configuration				
CNN1_focal	CNN2_focal	RNN2_focal	RNN3_focal	RNN3_cce
BatchNorm	BatchNorm	GRU-128	GRU-256	GRU-256
conv5-128 (relu)	conv5-128 (relu)	drop-0.45	drop-0.65	drop-0.65
max_pool3	max_pool3	BatchNorm	BatchNorm	BatchNorm
BatchNorm	BatchNorm	GRU-64	GRU-128	GRU-128
drop-0.65	drop-0.65	drop-0.2	drop-0.45	dropout-0.45
conv5-64 (relu)	conv5-64 (relu)	BatchNorm	BatchNorm	BatchNorm
max_pool3	max_pool3		GRU-64	GRU-64
BatchNorm	BatchNorm		drop-0.35	drop-0.35
drop-0.65	drop-0.65		BatchNorm	BatchNorm
F-100 (relu)	F-128 (relu)			
BatchNorm	BatchNorm			
	dropout-0.3			

as part of the PHEME project [5] where the relation between rumor veracity and stance was studied. This way, referring to subtask A, each tweet presents a stance with respect to a rumor defined as follows:

- Supporting(S): the tweet supports the veracity of the rumor.
- Denying(D): the tweet denies the veracity of the rumor.
- Questioning(Q): the tweet demands additional evidence.
- Commenting(C): the tweet is related to the rumor but it is not helpful to infer the veracity of the rumor.

The dataset considers three partitions: training, validation, and testing. Training and validation partitions comprise 297 rumourous threads collected for eight events in total, which include 4,519 tweets in total. These events include popular breaking news such as the Charlie Hebdo shooting in Paris, the Ferguson unrest in the US, and the German wings plane crash in the French Alps. The testing partition includes 1,021 tweets in total. These include 20 threads extracted from the same events as the training set and eight threads from two other events. The distribution of tweets per partition is summarized in Table 2.

Table 2. Distribution of tweets in training, validation, and testing data partitions.

	Supporting	Denying	Questioning	Commenting
Training	841	333	330	2734
Validation	69	11	28	173
Testing	68	69	106	778
Total	**978**	**413**	**464**	**3685**

4.2   Data Preprocessing

As social media data sources are unstructured and noisy, we require a careful preprocessing procedure before ingest the data. It is a well-known fact that text preprocessing save space and computational time during the learning stage. In addition, text preprocessing prevents the ingest of noisy data, limiting the effect of artifacts during the learning process. Accordingly, the normalization procedure applied to the dataset of tweets considered stopwords removal, punctuation marks and digits removal, and transform to lowercase. To process jargon, we removed emojis and then we applied the following rules:

- HTML marks were replaced by the term $<url>$
- Replacement of $\#word$ with the term $<hashtag>$
- $@word$ terms were replaced by the term $<user>$
- Numerical terms were replaced with $<number>$

Once each tweet was preprocessed, we used GloVe word embeddings to represent each word of each tweet. GloVe vectors were pre-trained on a Twitter corpus of two billion tweets (27 billion tokens) with a vocabulary of 1.2 million words [15]. These vectors were used in the five baselines and were ingested one-at-a-time as a sequence of word vectors per tweet. We used word vectors with 200 dimensions in the baselines.

Once the baselines' outputs were computed, we encoded each output using class vectors with 768 dimensions, to be consistent with the dimensionality of the BERT vectors. The tweet was encoded using BERT as service[2], a library that is able to map a variable length-sentence to a fixed-length vector with 768 dimensions.

In the transformer, we used gradient descent for parameter update. The size of the hidden units was set to 128 with a dropout of 0.35 using 64 elements per mini batch. We varied the learning rate throughout training, according to recommendations provided by the transformer's authors. We used focal loss as loss function with class weights inversely proportional to each class size. The number of epochs was fixed at 75.

4.3   Results

We provide a description and analysis of the experimental results. Table 3 presents the performance of the studied methods on test data. Results on the validation partition were omitted due to lack of space but we found that those results were similar to the ones showed in Table 3. Using our baseline models, we trained three variants of committee machines, using the output with the highest confidence between the baselines (*best_fit*), using the normalized Hadamard product between baselines' outputs (*norm*) and using a majority voting approach (*voting*). Results achieved by our competitors are shown in the sequel. We show the results achieved using the CNN proposed by Chen *et al.* [4] and Lozano

---

[2] https://github.com/hanxiao/bert-as-service.

et al. [10] and the XGBoost proposed by Bahuleyan *et al.* [3], which achieved the first places in the SemEval 2017 competition in this dataset. We also show the results achieved by Ma *et al.* using a two-layered GRU [12]. The performance achieved by our proposal is shown in the last row of the table. We also trained a variant of our proposal as a baseline using just the BERT vectors on the model (*BERT*).

Table 3. Results reported on test data. Best results per column are shown with bold fonts. F-score results are shown per class and at macro level.

Method	Accuracy	F1 macro	S	D	Q	C
Voting Lozano *et al.* [10]	0.749	0.453	**0.427**	0.022	0.512	0.852
CNN Chen *et al.* [4]	0.392	0.407	0.195	0.114	0.507	0.813
XGBoost [3]	**0.780**	0.453	0.397	0.052	0.494	**0.869**
Two-layered GRU [12]	0.622	0.434	0.314	0.158	0.531	0.739
CNN1 focal	0.724	0.431	0.121	0.161	0.610	0.830
CNN2 focal	0.575	0.410	0.106	0.213	0.611	0.711
RNN2 focal	0.599	0.396	0.188	0.153	0.519	0.726
RNN3 focal	0.738	0.375	0.020	0.157	0.482	0.842
RNN3 cce	0.372	0.307	0.195	0.225	0.356	0.451
Best-fit committee	0.732	0.375	0.020	0.149	0.495	0.838
Norm committee	0.756	0.418	0.021	0.185	0.615	0.851
Voting committee	0.747	0.434	0.057	0.204	0.630	0.844
BERT	0.764	0.216	0.000	0.000	0.000	0.864
Our proposal	0.672	**0.475**	0.232	**0.232**	**0.649**	0.785

Table 3 shows that our model achieves the best performance on D and Q classes among the evaluated methods. Supporting tweets are more difficult to be recognized by the transformer. In terms of F1 score at the macro level, our proposal also achieves the best performance. Note that the methods that won the SemEval 2017 competition achieve good performance in one class, S in the case of the CNN of Lozano *et al.* [10] or C in the case of XGBoost [3] but at the cost of very poor performance on classes with few examples.

4.4   Discussion of Results

Our method achieves the best results on minority classes (D and Q). The transformer was able to learn from categories with few examples, outperforming the results achieved by all the methods in these categories. The proposed architecture allows the identification of patterns among the methods, producing a performance that exceeds in some classes the performance provided by each technique alone. Note, for example, that BERT vectors by themselves over-fit the majority class, but when combined with other methods, produce better-balanced results. This allows the architecture to obtain the best results in F1.

The matrices of attention learned by both encoders are shown in Fig. 2. The first encoder discovered more relations between the inputs than the second encoder, suggesting that our architecture will not take advantage of the inclusion of more than two encoders. The crossing between the third and fifth method was helpful for the transformer, as the first and fourth head show. The same occurs with these baselines and the BERT vector as the fourth head shows. The other methods show interesting crosses evidencing feedback between them as the first and the second head shows. In summary, the transformer takes advantages from combining the inputs, identifying new patterns from data that were hidden before data crossing.

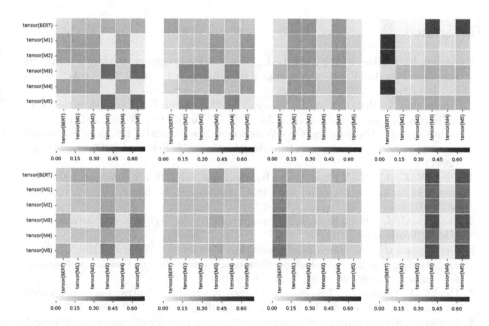

Fig. 2. Multi-head attention matrices learned by the transformer. Each row shows one encoder and each column one of four heads.

5   Conclusions

We have presented a method based on the transformer architecture for stance classification. Experimental results on a benchmark dataset (SemEval 2017) show that our proposal performs well in the hardest classes of this challenging task. Our architecture achieves the best results in terms of F1 macro being also very competitive in terms of accuracy. The main limitation of our approach is its poor performance in the S class.

This work can be extended in many research lines. Currently, we are working with more datasets to study the performance of these models in different scenarios. The question about the performance of stance classification and its relation to specific topics is a promising line. How to improve the performance of the S class using the transformer is a challenging task that we are addressing imitating the method proposed by Lozano *et al.* [10], which is the best method in this class. Finally, the question about the performance of these architectures in foreign languages as Spanish or Portuguese is an issue that it is in the first place of our list of to-do tasks.

Acknowledgements. Mr. Mendoza and Ms. Bugueño acknowledge funding from the Millennium Institute for Foundational Research on Data. Mr. Mendoza was partially funded by the project BASAL FB0821.

References

1. Anand, P., Walker, M.A., Abbott, R., Tree, J.E.F., Bowmani, R., Minor, M.: Cats rule and dogs drool!: classifying stance in online debate. In: WASSA@ACL 2011, pp. 1–9 (2011)
2. Augenstein, I., Rocktäschel, T., Vlachos, A., Bontcheva, K.: Stance detection with bidirectional conditional encoding. In: EMNLP 2016, pp. 876–885 (2016)
3. Bahuleyan, H., Vechtomova, O.: UWaterloo at SemEval-2017 t-8: detecting stance towards rumours with topic independent features. In: SemEval 2017, pp. 461–464 (2017)
4. Chen, Y.-C., Liu, Z.-Y., Kao, H.-Y.: IKM at SemEval-2017 t-8: convolutional neural networks for stance detection and rumor verification. In: SemEval 2017, pp. 465–469 (2017)
5. Derczynski, L., Bontcheva, K.: Pheme: veracity in digital social networks. In: UMAP Workshops (2014)
6. Derczynski, L., Bontcheva, K., Liakata, M., Procter, R., Wong Sak Hoi, G., Zubiaga, A.: SemEval-2017 t-8: RumourEval: determining rumour veracity and support for rumours. In: SemEval 2017, pp. 69–76 (2017)
7. Devlin, J., Chang, M., Lee, K., Toutanova, K.: BERT: pre-training of deep bidirectional transformers for language understanding. CoRR 1810.04805 (2018)
8. Faulkner, A.: Automated classification of stance in student essays: an approach using stance target information and the Wikipedia link-based measure. In: FLAIRS 2014 (2014)
9. Hasan, K.S., Ng, V.: Stance classification of ideological debates: data, models, features, and constraints. In: IJCNLP 2013, pp. 1348–1356 (2013)
10. Lozano, M.G., Lilja, H., Tjörnhammar, E., Karasalo, M.: Mama Edha at SemeVal-2017 t-8: stance classification with CNN and rules. In: SemEval 2017, pp. 481–485 (2017)
11. Lukasik, M., Srijith, P.K., Vu, D., Bontcheva, K., Zubiaga, A., Cohn, T.: Hawkes processes for continuous time sequence classification: an application to rumour stance classification in Twitter. In: ACL 2016 (2016)
12. Ma, J., Gao, W., Wong, K.-F.: Detect rumor and stance jointly by neural multitask learning. In: WWW 2018, pp. 585–593 (2018)
13. Mohammad, S., Kiritchenko, S., Sobhani, P., Zhu, X., Cherry, C.: SemEval-2016 t-6: detecting stance in tweets. In: SemEval 2016, pp. 31–41 (2016)

14. Mohammad, S.M., Sobhani, P., Kiritchenko, S.: Stance and sentiment in tweets. ACM Trans. Internet Technol. **17**(3), 26:1–26:23 (2017)
15. Pennington, J., Socher, R., Manning, C.D.: Glove: global vectors for word representation. In: EMNLP 2014, pp. 1532–1543 (2014)
16. Sobhani, P., Inkpen, D., Matwin, S.: From argumentation mining to stance classification. In: ArgMining@HLT-NAACL 2015, pp. 67–77 (2015)
17. Somasundaran, S., Wiebe, J.: Recognizing stances in ideological on-line debates. In: NAACL CAAGET 2010, pp. 116–124 (2010)
18. Vaswani, A., et al.: Attention is all you need. In: NIPS 2017, pp. 6000–6010 (2017)
19. Walker, M., Tree, J.F., Anand, P., Abbott, R., King, J.: A corpus for research on deliberation and debate. In: LREC 2012, pp. 812–817 (2012)
20. Zubiaga, A., Kochkina, E., Liakata, M., Procter, R., Lukasik, M.: Stance classification in rumours as a sequential task exploiting the tree structure of social media conversations. In: COLING 2016, pp. 2438–2448 (2016)
21. Zubiaga, A., Liakata, M., Procter, R., Hoi, G.W.S., Tolmie, P.: Analysing how people orient to and spread rumours in social media by looking at conversational threads. PloS One **11**(3), e0150989 (2016)

Applying OWA Operator in the Semantic Processing for Automatic Keyphrase Extraction

Manuel Barreiro-Guerrero[1], Alfredo Simón-Cuevas[1(✉)] [iD],
Yamel Pérez-Guadarrama[2], Francisco P. Romero[3] [iD],
and José A. Olivas[3] [iD]

[1] Universidad Tecnológica de La Habana José Antonio Echeverría,
Ave. 114, No. 11901, Marianao, La Habana, Cuba
{mbarreiro,asimon}@ceis.cujae.edu.cu
[2] Centro de Aplicaciones de Tecnologías de Avanzada (CENATAV), 7ma A,
No. 21406, e/214 y 216, Playa, La Habana, Cuba
yperez@cenatav.co.cu
[3] Universidad de Castilla-La Mancha, Paseo de la Universidad, 4,
Ciudad Real, Spain
{FranciscoP.Romero,JoseAngel.Olivas}@uclm.es

Abstract. The automatic keyphrases extraction from texts is a useful task for many computational systems in the natural language processing and text mining fields. Although several solutions to this problem have been developed, the semantic analysis has been one of the linguistic features less exploited in the most reported proposal, causing that the obtained results still show low accuracy and performance rates. This paper presents an unsupervised method for keyphrase extraction, which is based on the use of lexical-syntactic patterns for extracting information from texts and a fuzzy modelling of topics. An OWA operator which combines several semantics measures has been applied in the topic modelling process. This new approach was evaluated with Inspec and 500N-KPCrowd datasets and compared with other reported systems, obtaining promising results.

Keywords: Automatic keyphrase extraction · Linguistic patterns · Topic modelling · Semantic processing · OWA operator

1 Introduction

The exponential growth of textual and unstructured data in digital format have provoked that to distill the most relevant information from the amount of available information constituted a significant challenge to the textual information processing. The development of computational solutions based on the application of natural language processing (NLP) and text mining techniques emerging as the most promising alternatives to deal with this challenge. In this context, a high-level description of a document can be achieved through relevant words or phrases, by its strong relationship with the main topic (s) that are addressed in the documents, so that the automatic

© Springer Nature Switzerland AG 2019
I. Nyström et al. (Eds.): CIARP 2019, LNCS 11896, pp. 62–71, 2019.
https://doi.org/10.1007/978-3-030-33904-3_6

keyphrase extraction constitute an essential task for many text mining solutions [3, 10]. The keyphrase provides a concise understanding of a text, enabling one to grasp the central idea and the main topics discussed in a text document and facilitates the construction of text representation models, as graph-based models.

Several automatic keyphrases extraction solutions have emerged over the last few years, some following a supervised approach [8, 9] and others following unsupervised techniques [1, 6, 11, 14–17]. In this work we focused on unsupervised keyphrase extraction, where a human-annotated training data for applying some machine learning algorithm do not require in this process. The solutions reported still show low rates of accuracy and performance [3, 10], and the semantic constitutes one of the linguistics feature less exploited in the most reported proposal, fundamentally in unsupervised approaches. According to [10], it is essential to focus on semantically and syntactically correct phrase aspects and make sure that the keyphrases are semantically relevant to the document topic and context. Topic modelling for keyphrase extraction from texts has been reported in [1, 14, 15], however the semantic analysis in those proposals has not been considered or not in all its possible dimensions, constituting a weakness. The semantic analysis of textual content, at the level of words meaning or relationships among them, is usually subject to subjectivity, vagueness and imprecision problems, due to the inherent ambiguity of the natural language, which constitutes a challenge for the computational solutions that requires intensive semantic processing. The fuzzy logic offers several techniques for dealing with these problems, such as fuzzy set techniques, fuzzy clustering algorithms, aggregation operators, and others. Despite these advantages, few keyphrase extraction proposals that apply a fuzzy logic approach to carried out some semantic analysis level have been identified [15].

In this paper, an unsupervised method for automatic keyphrase extraction from a single document is proposed. The method was conceived through the combination of the use of lexical-syntactic patterns with a graph-based topic modeling, which is carried out from the fuzzy logic perspective. In this sense, syntactic and semantic measures are combined, applying the aggregation operator OWA (Ordered Weighted Averaging) [18], to increase the semantic processing level of the candidate phrase in the topic identification. The method was evaluated with the Inspec [8] and 500N-KPCrowd [9] datasets, and the performance was measured using the precision, recall, and F-score metrics. The results were compared with those obtained by other state-of-the-art unsupervised proposals, reaching improvement the results respect to those systems included in the comparison. Concretely, the contributions of this paper are the following: (1) we propose a new way for processing the semantic information in topic modelling based keyphrase extraction solutions, applying a fuzzy aggregation operator (OWA), and (2) we prove on two datasets that the fuzzy topic modelling proposed can improve the accuracy in the unsupervised automatic keyphrase extraction process.

The rest of the paper is organized as follows: Sect. 2 summarizes the analysis of the related works; Sect. 3 describes the proposed method; Sect. 4 presents the experimental results and the corresponding analysis. Conclusions and future works are given in Sect. 5.

2 Related Works

The solutions of automatic keyphrase extraction in text documents are usually designed in 4 phases: pre-processing, identification and selection of candidate phrases, keyphrase determination, and evaluation [10]. The unsupervised approach has the advantage of using only the contained information in the input text to determine the keyphrases, and several solutions have been reported [1, 6, 11, 14–17].

In TopicRank [1], a strategy based on the identification and analysis of topics to extract the relevant phrases is proposed. In this method, the longest sequences of nouns and adjectives in the text are extracted as candidate phrases, and the syntactically similar noun phrases are clustered into a theme or topic, using a hierarchical agglomerative clustering (HAC) algorithm [12]. Next, a graph is constructed where each vertex represents a topic and the arcs are labeled with a weight that represents the strength of the contextual relationship in the text among the contained candidate phrase in a topic regarding those that were grouped in another topic to which it relates. Finally, the selection of only one keyphrase from each topic is carried out, which is a weakness because a topic can be represented by more than one keyphrase in the same text. This proposal is improved in [14], through conceiving a more flexible procedure of keyphrase selection from topics and incorporating the definition of a distance between phrases function in the candidate-phrases clustering process, although semantic processing remains limited, as in the case of [1]. Liu et al. [6] also consider the clustering of candidate phrases to represent the document themes, and a cooccurrence-based relatedness measure is applied for computing the semantic relatedness of candidate terms in this process.

In TextRank [11] the candidate terms and their relationships are represented in an unweighted and undirected graph, whose vertexes represent the terms and the arcs represent co-occurrence relationships between them. An algorithm similar to PageRank [2] is applied to the constructed graph for determining the relevance of each vertex. Next, the third part of the vertexes of the whole graph is selected as the most relevant vertexes. Finally, the relevant terms are marked in the text and the sequences of adjacent words are selected as keyphrases. A similar solution is considered in the Salience Rank algorithm [17], but the use of PageRank [2] to obtain a ranking of the words in the document is combined with other word salience measures in the context of an LDA (Latent Dirichlet Allocation) based topic modelling approach. In [15], the co-occurrence graph of the input text's words is created, which is customized for each topic by employing the semantic information obtained from the topic model (built over the Wikipedia's articles) to form the topical graphs. Next, the communities and central nodes of these topical graphs are detected. In this process, the fuzzy modularity criterion for measuring the goodness of overlapped community structures is applied. The co-occurrence graph is also applied in RAKE [16]. In this approach, the graph is constructed with all individual words founded in the candidate keyphrases, and used to calculate the scores of each word and keyphrase. The word score is calculated through the word degree as well as the word frequency. For multiple-word expressions, they calculated the weights by summing the members' weights up.

According to the related works analyzed, the graph-based terms representation and the topic modelling appear as promising alternatives for the unsupervised keyphrase extraction from text. The unsupervised methods offer more significant strengths than supervised ones; nevertheless, have as a weakness that graph-based approach not guarantee that the extracted keyphrases represent all the main topics of the document and fail to reach a reasonable coverage level of the text document [10]. The good keyphrases of a document should are semantically relevant to the document theme or topic and cover the whole document well [6]. In this sense, can be seen in the analyzed works a low use of the semantic analysis in the clustering and topic modelling process carried out or in other task included. This semantic processing has focused on computing the co-occurrence relatedness [6, 11, 15, 16] or distance-based contextual relationship [14]. However, there are other semantic analysis level and measures, such as: semantic similarity and semantic relatedness measures, which have not been explored. Our work is aimed at assessing the benefits of these other semantic measures in the topic modelling from the fuzzy logic perspective to improve the outcomes of the unsupervised keyphrases extraction process.

3 Keyphrase Extraction Using OWA Operator

The proposed method was conceived through the combination of the use of lexical-syntactic patterns with a topic modelling carried out from a fuzzy perspective. This method has four phases: (1) text pre-processing, (2) fuzzy identification of topics, (3) relevance evaluation of topics, and (4) keyphrases selection. The lexical-syntactic patterns were defined for extracting candidate phrases from the text, and a fuzzy clustering of candidate phrases is proposed for identifying the main topics in the texts, to increase the semantic analysis in this process respect other proposals [1, 14, 15]. Also, a more flexible mechanism of keyphrases selection from the relevant topics identified is incorporated, that allows extracting more than one keyphrase and solving the weakness identified in TopicRank [1].

3.1 Text Pre-processing

In this phase, different NLP tasks are carried out for extracting the syntactic information from the text, which is required in the candidate phrases extraction process. Initially, plaintext from the input file is extracted, segmented into paragraphs and sentences, and the set of tokens (e.g., words, numbers, and others) are obtained from each sentence. Subsequently, the deep syntactic analysis using the Freeling parser is performed. The extraction of candidate phrases is based on the identification of conceptual phrases and a set of defined lexical-syntactic patterns are defined for this purpose, such as: [D | P | Z] + [<s-adj>] + NN; [D | P | Z] + [<s-adj>] + NN + NN; [Z] + <sn>; NN + [IN] + NN; JJ + NN + [NN], in a similar way to that reported in [14]. These patterns have been defined according to the grammar labeling used by Freeling, and they combine a set of relevant grammatical categories in the composition of concepts. Most of these patterns have their origins in the most frequent patterns identified in the concepts included in several ontological knowledge resources analyzed, e.g. the ontology of the

DBpedia project [4], which has more than 1000 concepts of different domains from Wikipedia. Through these patterns the coverage of the text in this process is increased, respect to other proposals that only consider noun phrases.

3.2 Fuzzy Identification of Topics

The topics identification process is carried out using a hierarchical agglomerative clustering algorithm [12] of the extracted candidate phrases, which is addressed as a fuzzy logic problem for reinforcing the semantic analyses in the phrases clustering. Although the use of clustering algorithms for topics modelling has also been reported in [1, 14, 15], the semantic analysis in those proposal has not been considered or not in all its possible dimensions. This is a weakness considering the assumption that a topic could be modelled through the cluttering of concepts that frequently appear together as well as concepts with similar meanings or semantically related. To address this weakness, in our new unsupervised approach the phrases clustering process is carried out considering the resultant score of combining the syntactic similarity and distance between phrases measures reported in [14] with other two semantic similarity measures applying a fuzzy aggregation operator. These two semantic similarity measures were conceived according to the sentence-to-sentence similarity metric reported in [5] and using two word-to-word semantic similarity-relatedness metrics from WordNet::Similarity package, specifically the Jiang & Conrath and Leacock & Chodorow metrics [13]. Additionally, the words distance metric reported in [14] was redefined (Eq. 1):

$$D(F_1, F_2) = \begin{cases} 1 & \text{if } F_1 \text{ and } F_2 \text{ appear in the same paragraph} \\ 1 - \frac{ave\_dist(F_1, F_2)}{TW} & \text{in other cases} \end{cases} \quad (1)$$

where $ave\_dist(F1, F2)$ is the average distance [14] in words that exists between the words included in the pair of phrases F_1 and F_2, and TW is the total of words in the text.

In this method, the OWA operator [18] is applied for aggregating the resultant numerical values (a_i) from the four defined measures into a single one similarity-relatedness score (SRS) of a pair of candidate phrases. These measures represent features with different semantic "meaning" for the phrases clustering, as well as different relevance levels for the decision making in this process. OWA operators are very useful for dealing with such problems, modelling the semantics and relevance levels through weights assigned to each measure. To combine these syntactic, distance and semantic measures using an OWA Operator allows to achieve clusters of phrases strongly related among them from different semantic dimensions, and at the same time, to achieve a wide coverage of the whole document in the topic modelling process.

Definition: *An OWA operator of dimension n is a mapping denoted $f_{owa} : \mathbb{R}^n \to \mathbb{R}$ that has associated an n-dimensional weight vector $W = [w_1, w_2, ..., w_n]^T$ such as $w_i \in [0, 1]$ and $\sum_i w_i = 1$. The function f_{owa} is defined according to Eq. 2, with b_j the jth largest element in the collection $a_1 ... a_n$.*

$$f_{owa}(a_i, \ldots, a_n) = \sum_{j=1}^{n} w_j b_j \quad (2)$$

There are different methods for determining the weights to be used in an OWA operator and the use of linguistic quantifiers is one of them [20], e.g. RIM (Regular Increasing Monotone) quantifiers. Yager proposed a method to calculate the weights of an operator OWA by means of RIM quantifiers [19], which is defined in Eq. 3. Specifically, in our proposal, we apply the RIM quantifier "Most (Feng & Dillon)" reported in [7] (see Eq. 4), as the first approach to measure the performance of the OWA operator in the keyphrase extraction problem.

$$w_j = Q\left(\frac{j}{n}\right) - Q\left(\frac{j-1}{n}\right) \tag{3}$$

$$Q(x) = \begin{cases} 0 & si\ 0 \leq x \leq 0.5 \\ (2x-1)^{0.5} & si\ 0.5\ x \leq 1 \end{cases} \tag{4}$$

The hierarchical agglomerative clustering process is carried out by means of creating a square symmetric matrix of size n (total of candidate phrases identified), where each topic identifies a row, and a column and the intersection between each pair of topics contains the SRS (weight value) between a pair of candidate phrases that represent the corresponding topics. Initially, each candidate phrase is considered as a topic. In each iteration, the pair of topics with the highest weight value is clustered. The weight values average is used as a clustering strategy of a pair of topics, according to the reported in TopicRank [1]. The phase concludes generating a graph representation of the text, in which the identified topics are represented as vertices and these are linked by labeled arcs with the weight of the relation between them. Each weight represents the strength of the existing semantic relationship between the pair of topics. The topics A and B have a strong semantic relationship if the candidate phrases that includes those topics which frequently appear closer in the text. The weight W_{ij} is calculated according to Eqs. (5) and (6). Equation (6) refers to the reciprocal distance between the positions of the candidate phrases c_i and c_j in the text, where $pos\ (c_i)$ represents all positions (p_i) of c_i.

$$W_{i,j} = \sum_{c_i \in T_i} \sum_{c_j \in T_j} D(c_i, c_j) \tag{5}$$

$$D(c_i, c_j) = \sum_{p_i \in pos(c_i)} \sum_{p_j \in pos(c_j)} \frac{1}{|p_i - p_j|} \tag{6}$$

3.3 Relevance Evaluation of Topics

In this phase, the relevance of each topic represented in the constructed topics graph is evaluated using the TextRank [11] model. The relevance score computed to each topic T_i is based on the concept of "voting" (inspired in the PageRank algorithm [2]): the adjacent topics of T_i with the highest score contribute more to the relevance evaluation of the topic T_i. The relevance score $S(T_i)$ is obtained through the Eq. 7, where V_i is the set of adjacent topics of T_i in the graph, and λ is a muffled factor that usually is 0,85 [2].

$$S(T_i) = (1 - \lambda) + \lambda * \sum_{T_j \in V_i} \frac{W_{i,j} * S(T_i)}{\sum_{T_k \in V_j} W_{j,k}} \tag{7}$$

3.4 Keyphrases Selection

The selection of keyphrases from the most relevant topics identified in the previous phases is carried out according to the following criteria: (1) candidate phrase that first appears in the text; (2) most frequently used candidate phrase; and (3) candidate phrase that has more relationship with the others of each topic (centroid role). A mechanism that allows combining the three criteria has been implemented in our proposal, offering the possibility of extracting more than one keyphrase from each topic and greater flexibility in its execution, respect to the reported in [1] (only one of the criteria is considered affecting the coverage in the keyphrases extraction process). If more than one candidate phrase (associated with a topic) with the same higher frequency is identified, and the frequency value is higher than 1, then all of them are selected. Otherwise, only the first candidate phrase that appears in the text will be chosen.

4 Experimental Results

The proposed method was evaluated using the Inspec [8] and 500N-KPCrowd [9] datasets, which contain texts collections written in English. Inspec is a collection of 500 paper abstracts of Computer Science & Information Technology journals with manually assigned keyphrases by the authors. 500N-KPCrowd contains 450 broadcast news stories from 10 different categories and is considered to see how the proposal perform on texts of general domain. The performance of the method was measured using the precision (P), recall (R), and F-score (F) metrics and the obtained results were compared with those obtained by others unsupervised methods reported, which have been evaluated with the selected datasets.

As shown in Table 1, the proposed method reached higher values in most metrics and both datasets, respect to those obtained by the state-of-the-art proposals compared. The best results were obtained with Inspec, where the method achieved a better balance between precision and recall, which is a very challenging target to reach and convenient for the increasing the applicability of this type of solutions. In this case, it's evidenced that the proposed fuzzy approach for the semantic processing and topic modelling not only contributed to increase the accuracy in the keyphrase extraction, but also increase the recall, which obtained result was very encouraging (near to the 60%).

The achieved recall with 500N-KPCrowd was the less satisfactory results of our proposal, although the results of precision and F-score were significantly better than those obtained by the other proposals. Although the recall of TSAKE [15] is the highest in the case of 500N-KPCrowd, its precision is approximately 30% lower than our method, and in the same way the F-score (9% lower).

Table 1. Experimental results with Inspec and 500N-KPCrowd datasets

Systems	Inspec			500N-KPCrowd		
	P	R	F	P	R	F
TextRank [11]	31.2	43.1	36.2	26.5	6.3	10.3
TopicRank [1]	36.4	39.0	35.6	26.2	23.9	25.0
TSAKE [15]	40.1	20.3	26.9	14.3	**46.6**	21.9
Salience Rank [17]	26.5	29.8	26.6	25.3	22.2	22.9
RAKE [16]	33.7	41.5	37.2	12.0	3.8	5.8
Method proposed	**42.1**	**59.9**	**47.9**	**45.5**	22.8	**30.8**

The low value obtained of recall can be derived from the presence of a high number of annotated named entities as keyphrase in 500N-KPCrowd. The identification of named entities as candidate phrase from the text was not considered within the defined patterns in the pre-processing phase of the proposed approach, because this type of sentence is not identified often as a keyphrase. On the other hand, the OWA operator applied in the proposed fuzzy modelling of topics includes the aggregation of several semantic measures, which may fail in the case of named entities. This situation suggests specific analysis for this type of phrases in the next approaches of our proposal. Nevertheless, through the experiments carried out, the achieved effectiveness improvement by our method and the fuzzy-based semantic processing proposed in the automatic keyphrase extraction from two types of texts, such as: paper abstracts and news stories, has been demonstrated.

5 Conclusions and Future Works

In this paper, a new unsupervised method for automatic keyphrase extraction from text was presented, in which the use of lexical-syntactic patterns to identify the candidate phrases was combined with a fuzzy modelling of topics. The use of the linguistic patterns allowed to increase the possibilities for identifying the candidate phrases, and the coverage of the text. Several syntactic and semantic measures to modeling the most relevant linguistics features of the candidate phrase were aggregated applying an aggregation operator OWA. The aggregation of these measures through the OWA operator allowed to increase the semantic processing of the candidate phrase in the topic identification, which is a little-considered aspect in most of the reported proposals. The proposed method was evaluated on two datasets with different types of texts, and the obtained results were compared with those obtained by other unsupervised schemes. The most significant results were obtained on Inspec, where a better balance between precision and recall was achieved, at the same time that their values were higher than the obtained by other proposals. These metrics were also improved on 500N-KPCrowd, although the recall must be enhanced. Considering the obtained results, the proposed method reached higher values in most of the metrics, demonstrating the contribution of the applied fuzzy topic modeling for improving the keyphrases extraction process, in paper abstract and in more general domain texts, such as the news stories.

The improvement of the recall results on general domain texts will be one of the challenges to solve in the future, considering specific analysis for the named entities. Additionally, others linguistic quantifiers applied to the OWA operator will be evaluated for measuring their performances in the keyphrase extraction process.

Acknowledgments. This work has been partially supported by FEDER and the State Research Agency (AEI) of the Spanish Ministry of Economy and Competition under grant MERINET: TIN2016-76843-C4-2-R (AEI/FEDER, UE).

References

1. Bougouin, A., Boudin, F., Daille, B.: TopicRank: graph-based topic ranking for keyphrase extraction. In: Proceedings of the 6th International Joint Conference on NLP, pp. 543–551 (2013)
2. Brin, S., Page, L.: The anatomy of a large-scale hypertextual web search engine. Comput. Netw. ISDN Syst. **30**(1–7), 107–117 (1998)
3. Hasan, K.S., Ng, V.: Automatic keyphrase extraction: a survey of the state of the art. In: Proceedings of the 52nd Annual Meeting of the ACL, pp. 1262–1273 (2014)
4. Lehmann, J., et al.: DBpedia - a large-scale multilingual knowledge base extracted from Wikipedia. Semant. Web J. **1**, 1–27 (2012)
5. Li, Y., McLean, D., Bandar, Z.A., O'Shea, J.D., Crockett, K.: Sentence similarity based on semantic nets and corpus statistics. IEEE Trans. Knowl. Data Eng. **18**, 1138–1150 (2006)
6. Liu, Z., Li, P., Zheng, Y., Sun, M.: Clustering to find exemplar terms for keyphrase extraction. In: Proceedings of the 2009 Conference on Empirical Methods in NLP, pp. 257–266 (2009)
7. Liu, X., Han, S.: Orness and parameterized RIM quantifier aggregation with OWA operators: a summary. Int. J. Approx. Reason. **48**(1), 77–97 (2008)
8. Hulth, A.: Improved automatic keyword extraction given more linguistic knowledge. In: Proceedings of the 2003 Conference on Empirical Methods in NLP, pp. 216–223 (2003)
9. Marujo, L., Ribeiro, R., de Matos, D.M., Neto, J.P., Gershman, A., Carbonell, J.: Key phrase extraction of lightly filtered broadcast news. In: Sojka, P., Horák, A., Kopeček, I., Pala, K. (eds.) TSD 2012. LNCS (LNAI), vol. 7499, pp. 290–297. Springer, Heidelberg (2012). https://doi.org/10.1007/978-3-642-32790-2_35
10. Merrouni, Z.A., Frikh, B., Ouhbi, B.: Automatic keyphrase extraction: an overview of the state of the art. In: Proceedings of the 4th IEEE International Colloquium on Information Science and Technology, pp. 306–313 (2016)
11. Mihalcea, R., Tarau, P.: TextRank: bringing order into texts. In: Proceedings of the 2004 Conference on Empirical Methods in NLP, pp. 404–411 (2004)
12. Müllner, D.: Modern hierarchical, agglomerative clustering algorithms. CoRR, abs/1109.2378 (2011)
13. Pedersen, T., Patwardhan, S., Michelizzi, J.: WordNet::Similarity - measuring the relatedness of concepts. In: Proceedings of the 19th National Conference on Artificial Intelligence (AAAI 2004), pp. 1024–1025 (2004)
14. Pérez-Guadarrama, Y., Rodríguez, A., Simón-Cuevas, A., Hojas-Mazo, W., Olivas, J.A.: Combinando patrones léxico-sintácticos y análisis de tópicos para la extracción automática de frases relevantes en textos. Procesamiento del Lenguaje Natural **59**, 39–46 (2017)

15. Rafiei-Asl, J., Nickabadi, A.: TSAKE: a topical and structural automatic keyphrase extractor. Appl. Soft Comput. J. **58**, 620–630 (2017)
16. Rose, S., Engel, D., Cramer, N., Cowley, W.: Automatic keyword extraction from individual documents. Text Mining: Applications and Theory, pp. 1–20 (2010)
17. Teneva, N., Cheng, W.: Salience rank: efficient keyphrase extraction with topic modeling. In: Proceedings of the 55th Annual Meeting of the Association for Computational Linguistics, pp. 530–535 (2017)
18. Yager, R.R.: On ordered weighted averaging operators in multicriteria decisionmaking. IEEE Trans. Syst. Man Cybern. **18**, 183–190 (1988)
19. Yager, R.: Quantifier guided aggregation using OWA operators. Int. J. Intell. Syst. **11**, 49–73 (1996)
20. Zadeh, L.A.: A computational approach to fuzzy quantifiers in natural languages. Comput. Maths. Appl. **9**, 149–184 (1983)

Brazilian Presidential Elections in the Era of Misinformation: A Machine Learning Approach to Analyse Fake News

Jairo L. Alves[1], Leila Weitzel[1(✉)], Paulo Quaresma[2],
Carlos E. Cardoso[1], and Luan Cunha[1]

[1] Fluminense Federal University, Rio de Janeiro, Brazil
jairo.luciano@gmail.com, {leila_weitzel, carloseac,
luanpereiracunha}@id.uff.br
[2] Universidade de Évora, Évora 17, Portugal
pq@uevora.pt

Abstract. As Brazil faced one of its most important elections in recent times, the fact-checking agencies handled the same kind of misinformation that has attacked voting in the US. However, stopping fake content before it goes viral remains an intense challenge. This paper examines a sample database of the 2018 Brazilian election articles shared by Brazilians over social media platforms. We evaluated three different configuration of Long Short-Term Memory. Experiment results indicate that the 3-layer Deep BiLSTMs with trainable word embeddings configuration was the best structure for fake news detection. We noticed that the developments in deep learning could potentially benefit fake news research.

Keywords: Fake news · Machine learning · Long Short-Term Memory · Word embeddings · Deep learning · Recurrent neural network

1 Introduction

Recent political events, notably the Brexit referendum in the U.K., the presidential election of 2016 in the U.S. and the Brazilian's economic and political crisis in 2016 have led to a wave of interest in the phenomenon of fake news. There is already strong concern about the impact that the spread of fake news could cause to the 2018 Brazilian elections, the most important elections in recent times.

Fact-checking is a journalistic method by which it is possible to ascertain whether accurate information has been obtained from reliable sources and then assess whether it is true or false, whether it is sustainable or not. Fact-checking is a costly process, given the large volume of news produced every minute in our post-truth era, making it difficult to check contents in the real time. The rate and the volumes at which false news are produced overturn the possibility to fact-check and verify all items in a rigorous way, i.e. by sending articles to human experts for verification. The process of fact checking requires researching and identifying evidence, understanding the context of information and reasoning about what can be inferred from the evidence. Besides,

© Springer Nature Switzerland AG 2019
I. Nyström et al. (Eds.): CIARP 2019, LNCS 11896, pp. 72–84, 2019.
https://doi.org/10.1007/978-3-030-33904-3_7

not all journalists have the knowledge to investigate the databases that would allow a rigorous verification, and access to the real data is not always possible. The goal of automated fact checking is to reduce the human burden in assessing the veracity of a claim [1, 2].

The Brazilian fact-checking agencies are handling the same kind of misinformation that has attacked voting in the US. It is essential to cite the concern of the fact-checking agencies in order to try to resolve the impact caused by fake news before it goes viral. However, stopping fake content before it goes viral remains an intense challenge. Fake news are deliberately been created to mislead the readers, resulting in an adversarial scenario where it is very hard to distinguish real facts from fakes. While fake news is not a new phenomenon, questions such as why it has emerged as a world topic and why it is attracting increasingly more public attention are particularly relevant at this time. The leading cause is that fake news can be created and published online faster and cheaper when compared to traditional news media such as newspapers and television [3].

Hence, based on the scenario described herein, the main goal of this research is to detect fake news, which is a classic text classification problem with the applications of NLP (Natural Language Processing). We have crawled the labeled articles published by one independent fact-checking agency, "Aos Fatos[1]", for our purpose. The present analysis focuses on the period from May 2018 to the end of September 2018. During this time, we collected 2,996 articles. It must be highlighted that the language analyzed is Brazilian Portuguese. Brazilian Portuguese can be considered different from European Portuguese. There are grammatical peculiarities of Brazilian Portuguese, such as, the syntactic position of sentence subjects; the preferred position of clitic pronouns; the pronominal paradigm and other characteristics, which make it different. To the best of our knowledge, it is the first experiment that uses a dataset in the Brazilian Portuguese language for fake news detection.

2 Fake News Characterization

The term fake news is used in a variety of (sometimes-conflicting) ways, thereby making conceptual analysis more difficult. It must be stressed the difference between satire and fake news. Satire is meant to be comedic in nature. Satire presents stories as news that are factually incorrect, but the intent is not to deceive but rather to call out, ridicule, or expose behavior that is shameful, corrupt, or otherwise "bad". Satirical news as designed specifically to entertain the reader, usually with irony or wit, to critique society or a social figure. Fake news are defined as a news story that is factually incorrect and designed to deceive the consumer into believing it is true [4].

Shu, Sliva, Wang, Tang and Liu [5] argue that there is no agreed definition of the term fake news. They draw attention to the fact that fake news is intentionally written to mislead readers, which makes it nontrivial to detect simply based on news content. Fake news tries to distort truth with diverse linguistic styles. For example, fake news may cite true evidence within the incorrect context to support a non-factual claim.

---

[1] https://aosfatos.org/.

The concepts of fake news have existed since the beginning of civilization. There are several historical examples of its use over the centuries as a strategic resource for winning wars, gaining political support, manipulating public opinion, defame peoples and religions. Years have passed and the Web has turned into the ultimate manifestation of User-Generated Content (UGC) systems [6]. The UGC can be virtually about anything including politics, products, people, events, etc. Hence, the popularization of social networks, the use of fake news to serve the most different purposes has grown dramatically, allowing the free creation and large-scale dissemination of any type of content. Words in news media and political discourse have a considerable power in shaping people's beliefs and opinions [7].

Facebook has begun to mark each news story depending on whether the news is truthful or not. In 2016, "Google News" began to mark news about the USA. Since then, they expanded this practice to the United Kingdom, Germany, France, Mexico, Brazil and Argentina. In 2018, the International Federation of Library Associations and Institutions[2] (IFLA) issued a statement that contains recommendations to governments and libraries regarding fake news marking, and it is accompanied by a toolkit of resources. IFLA is concerned by the risk that this can pose to access to information, where people do not have the skills to spot it, but also by the way it is used by governments to justify potentially repressive policies. The agency posted an infographic on Facebook and Twitter to help combat fake news, which contains eight rules. The set of rules include; (i) Consider the source, (ii) Read beyond the headline, (iii) Check the author, (iv) What is the support? (v) Check the date, (vi) Is this some kind of joke? (vii) Check your biases, and (viii) Consult the experts.

3   LSTM Main Characteristic

In the mid-90s, a variation of RNN called LSTMs was proposed by the German researchers Sepp Hochreiter and Juergen Schmidhuber [8] as a solution to the vanishing gradient problem. RNNs are a family of neural networks well suited to model sequential data like time series, speech, text, financial data, audio, language, and other similar types of data. In general, the recurrence aspect allows RNNs to form a much deeper understanding of a sequence and its context, compared to other algorithms, such as static neural networks or purely statistical approaches. The core concept of LSTM are the memory cell state and its various gates. The gate can regulate the flow of information and it can automatically learn which data in a sequence is important to keep track of and witch data can be thrown away. By doing that, it can pass relevant information down the line in order to make better predictions for long sequences. As there is a direct path to pass the past context along, it suffers much less from vanishing gradients than vanilla RNNs. The gates typically are implemented with sigmoid activation functions and can learn what information is relevant to keep or forget during training with BPTT algorithm.

---

[2] https://www.ifla.org/.

To avoid overfitting, we employ standard regularization techniques, such as, early stopping and dropout. Early stopping monitors and may stop the training process once the validation loss metric starts to increase or when validation accuracy starts to decrease. The main idea of dropout is to randomly remove computing units in a neural network during training to reduce hard memorization of training data. In this case, we want to trade training performance for better generalization on unseen data. Both regularization methods are used when indicated to prevent overfitting [9].

4 Related Work

Fake news is a concern, because they can affect the minds of millions of people every day, these have led to the term post-truth. Fake news detection has attracted the interest of researchers in recent years with several approaches being proposed [4–6, 10–15]. Several previous studies have relied on feature engineering and standard machine learning methods, such as Support Vector Machines (SVM), Stochastic Gradient Descent, Gradient Boosting, Bounded Decision Trees, Random Forests and Naïve Bayes. Several other studies have proposed *Deep Learning Methods* [16–19].

Pfohl [18] organized an analytic study on the language of news media in the context of political fact-checking and fake news detection. Two approaches were applied: lexicon and neural network based. The lexicon approach shows that first-person and second person pronouns are used more in less reliable or deceptive news types. In the second methodology, the models were trained with Max-Entropy classifier with L2 regularization on n-gram TF-IDF feature vectors (up to trigrams). The output layer includes four labels: trusted, satire, hoax and propaganda. The model achieved F1 scores of 65%. The author also trained a Long Short Term Memory (LSTM) network and Naïve Bayes, both used lexicon measurements in order to concatenate to the TF-IDF vectors. Ajao, Bhowmik and Zargari [19] proposed a framework to detect and classify fake news from Twitter posts. The authors use a hybrid approach consisting of CNN - Convolutional Neural Networks and LSTM with dropout regularization and Word Embedding layer. Wang [16] first introduced the Liar dataset. The Liar dataset contains 12.8K manually labeled short statements in various contexts sampled from one decade of content in politifact.com, which provides detailed analysis reports and links to source documents for each case. Wang [16] has evaluated several popular machine-learning methods on this dataset. The baselines include logistic regression classifiers, SVMs, LSTMs and CNN. A model that combines three characteristics for a more accurate and automated prediction was proposed by Ruchansky, Seo and Liu [17]. They incorporated the behavior of both parties, users and articles, and the group behavior of users who propagate fake news. They proposed a model, which is composed of three modules. The first module named *Capture* captures the abstract temporal behavior of user encounters with articles, as well as temporal textual and user features, to measure response as well as the text. To extract temporal representations of articles they use a Recurrent Neural Network (RNN). The second module, *Score*, estimates a source suspiciousness score for every user, which is then combined with the first module by integration to produce a predicted label.

5 Methodology

5.1 Dataset Construction

The data was acquired from the website of the fact-checking agency "Aos Fatos". Fact checking is a task that is normally performed by trained professionals. Depending on the complexity of the specific claim, this process may take from less than one hour to a few days. Daily, journalists of this agency check the statements of politicians and authorities of national expression in order to check if they are speaking the truth. The labeling methodology is based on the following rules: The use of the *True* badge is simple, the statement or the information is consistent with the facts and does not lack contextualization to prove correct. The *Unsustainable* label represents the statements that cannot be refuted or confirmed. i.e., there are no facts, data or any consistent information to support the claim. When the statement receives the *Inaccurate* label, it means that it needs context to be true and sometimes lack contextualization to prove to be true. The label *Exaggerated* means that the fact was exaggerated, for example, "more than 150 million Brazilians live below poverty line". In reality, only 50 million live like that. The *Contradictory* label is when content of the statement is exactly the opposite of the real fact that it happened. Suppose the statement "I have nothing against homosexuals or women, I am not xenophobic". This can be contradictory, if a person has already said homophobic, sexist and xenophobic statements at other times. The *Distorted* label is used only for rumors and news with misleading content. It serves for those texts, images and audios samples that bring information factually correct, but applied with the intention of confusing. If a statement or news or rumor has information without any factual support, they receive the *False* label.

The labeling methodology and other features are available on the website https:// aosfatos.org/. As can be observed, some labels have high semantic similarity, which makes it difficult, even for a human, to perform this analysis.

We performed our data collection by means of an in-house procedure, using Python and some other libraries, such as Beautifulsoup and Regex. Colleting news was not an easy task, since the website "AosFatos" was not in standard formatting, making it difficult to scrape. The dataset has the following attributes: *body-text, *date, *label and *text. The column *body-text* is the original sentence that someone thought or published in a social network. The column *date* is the day of the publication. The column *label* identifies in one of seven labels discussed herein. Finally, column *text* explains why the news was categorized with a certain label. The columns of interest to this research are *label* (as output) and the *body-text* (as input). The data set is available for download at <blind review>.

Following the recommendation of Shu, Sliva, Wang, Tang and Liu [5], due to minimal semantic differences between the many of the labels and in accordance with our goal, we decide to use only the *true* and *fake* labels as binary classification. Hence, after that, only 1,187 news were retained. The following preprocessing phases were performed (not necessarily in the succeeding order): we remove Hashtags, emoticons, punctuation and special characters ($, @, etc.); the text was tokenized (strip white space) and normalized (converting all letters to lower case).

5.2   Model Configuration

We conducted several experiments with different combinations of feature sets. Table 1 shows all the model configurations. The models are implemented using the Keras framework [20] with the TensorFlow backend engine [21]. In the following paragraphs, we describe our models of Regular LSTM (Naïve), Bidirectional LSTM (BiLSTM) and Deep BiLSTM. For all models, a Dense (Fully Connected) 3-neuron output layer is present at the end to generate the final output prediction. Table 1 shows the network configurations.

Table 1. Model configurations.

Model	Model configuration	LSTM inputs	Trainable embeddings?	wmDropout	Embeddings dimension
M20	3 layers of Bidirectional CuDNNLSTMs, 64 internal states on each CuDNNLSTM block	Random embeddings	Yes	Yes (rate = 0.3)	200
M21	3 layers of Bidirectional LSTMs, 32/16/16 internal states	Random embeddings	Yes	Yes (rate = 0.5, recurrent dropout)	50
M22	3 layers of Bidirectional LSTMs, 32/16/16 internal states + 1 additional Dense(8) layer after the LSTMs	Random embeddings	Yes	Yes (rate = 0.5, recurrent dropout)	50
M23	3 layers of Bidirectional CuDNNLSTMs, 32/16/16 internal states + 1 additional Dense(8) layer after the LSTMs	Random embeddings	Yes	Yes (rate = 0.5)	500
M24	**3 layers of Bidirectional CuDNNLSTMs, 64 internal states on each CuDNNLSTM block**	**Random embeddings**	**Yes**	**Yes (rate = 0.5)**	**80**
M25	3 layers of Bidirectional CuDNNLSTMs, 8 internal states on each CuDNNLSTM block	Random embeddings	Yes		50
M26	3 layers of Bidirectional CuDNNLSTMs, 8 internal states on each CuDNNLSTM block	Pre-trained embeddings	Yes	No	50 (embeddings multiplied by 2)

(*continued*)

Table 1. (*continued*)

Model	Model configuration	LSTM inputs	Trainable embeddings?	wmDropout	Embeddings dimension
M27	3 layers of Bidirectional CuDNNLSTMs, 8 internal states on each CuDNNLSTM block	Pre-trained embeddings	Yes	No	50 (embeddings multiplied by 0.25)
M28	3 layers of Bidirectional CuDNNLSTMs, 8 internal states on each CuDNNLSTM block	Pre-trained embeddings	Yes	No	50 (embeddings multiplied by 0.10)
M29	**3 layers of Bidirectional CuDNNLSTMs, 8 internal states on each CuDNNLSTM block**	**Pre-trained embeddings**	**Yes**	**No**	**50 (embeddings multiplied by 0.05)**
M30	3 layers of Bidirectional CuDNNLSTMs, 8 internal states on each CuDNNLSTM block	Pre-trained embeddings	Yes	No	50 (embeddings multiplied by 0.03)
M31	3 layers of Bidirectional CuDNNLSTMs, 8 internal states on each CuDNNLSTM block	Random embeddings	Yes		–
M32	3 layers of Bidirectional CuDNNLSTMs, 8 internal states on each CuDNNLSTM block	Pre-trained embeddings	Fixed	No	50 (embeddings multiplied by 0.03)
M33	1 layer of Bidirectional CuDNNLSTM, 128 internal states	Pre-trained embeddings	Yes	No	50
M34	3 layers of Bidirectional CuDNNLSTMs, 8 internal states on each CuDNNLSTM block	Pre-trained embeddings	Yes		50
M35	3 layers of Bidirectional CuDNNLSTMs, 64/32/32 internal states, +1 additional Dense(8) layer after the LSTMs + maximum input length increased to 250 words	Random embeddings	Yes	Yes (0.5 Recurrent)	50
M36	3 layers of CuDNNLSTMs, 6 internal states on each CuDNNLSTM block + maximum input length increased to 300 words	Pre-trained embeddings	Yes		50 (embeddings multiplied by 0.05)

6 Results

As discussed herein, we consider fact checking as an ordinal classification task. Hence, in theory it would be possible to frame it as a supervised classification task using algorithms that learn from annotated samples with the ground truth labels. We tested four main architectures with different parameters, as seen on Table 1. We tested three-layers of Bidirectional LSTM and Bidirectional CuDNN-LSTM[3] with different internal state sizes. We also tested pre-trained and random embedding as input methods. Word embeddings can be learned from text data and reused among projects. They can also be learned as part of fitting a neural network on the specific textual data. Keras offers an Embedding layer that can be used as part of a deep learning model where the embedding is learned along with the rest of the model (trainable embeddings). Also, this layer can be used to load a pre-trained word-embedding model, a type of transfer learning. We used pre-trained Portuguese word embedding (FastText, 50 dimension) available at http://nilc.icmc.usp.br/embedding.

When indicated, dropout was used and recurrent dropout when the blocks were regular LSTMs (in that case, the dropout varied between 0.3 and 0.5). We also tested models with different batch size (varying between 100 and 50). In most cases, batch size did not affect the result, so most scenarios were run with batch size equal to 100. We noticed that the text processing affected positively the performance of the models, albeit marginally. We also observed that fixed, pre-trained word embedding hurts performance when compared to random, trainable embeddings. The best performance is achieved by pre-trained trainable embeddings, especially when attenuated by a previous 0.05 factor multiplier on the magnitude of the pre-trained embedding vectors.

Naive LSTM – BiLSTM: The first model was a simple layer LSTM - (Regular-Naïve, see Fig. 1). The accuracy achieved was about 60%, which can be considered low. We first tested encoding the words as one-hot vectors and, then, as embeddings in various configurations (refer to Table 1). A one hot encoding is a representation of categorical variables as binary vectors where only one dimension is set to 1, while all others are zero. Traditional approaches to NLP, such as bag-of-words models whilst useful for various machine learning (ML) tasks, tend to not capture enough information about a word's meaning or context. Such models often provide sufficient baselines for simple NLP tasks. Also, it is well known that one-hot encodings, however simpler, do not capture syntactic (structure) and semantic (meaning) relationships across collections of words and, therefore, represent language in a rather limited way.

---

[3] It is high-level deep learning library and it can only run on GPU.

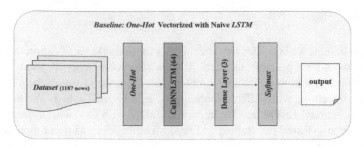

Fig. 1. Baseline model – Naïve LSTM

The input was a single LSTM layer, followed by a Dense (fully-connected) layer and the activation function was Softmax. The non-regularized model is our baseline model. This model had computational issues. It would take more than 500 s to run each epoch of training. In order to tackle this issue, we adopted the cuDNN-LSTM[4] (CUDA is Deep Neural Network library) [22–24] functions on the Google Colab[5] platform instead of the regular TensorFlow LSTM functions. This allows our models to run on Google's Collaboratory GPU hardware. Despite the lack of some parameters (recurrent dropout for example), we noticed that cuDNN-LSTMs were much faster than the "regular" version. Thus, unless indicated otherwise, all remaining references to LSTMs layers are implemented with cuDNN-LSTMs APIs. We also noticed that, the number of trainable parameters in the LSTM layer is strongly affected by its input sentence length and since our samples sequence lengths were rather long (>200 words for some sentences). The number of words in the sentence allows us to set a fixed sequence length, zero pad shorter sentences, and truncate longer sentences to that length as appropriate. Henceforth, we limited the input length to a maximum of 100 words, also padding to zero the shorter sentences. The pad_sequences() function in the Keras library was used. The default padding value is 0.0, which is suitable for most applications, although this can be changed if needed. Thus, every sequence length was exactly 100 words long. The dimensions for data are [Sequence Length (100), Input Dimension (2887)] with 50 epochs, early stopping and 128 hidden states. The batch size limits the number of samples to be shown to the network before a weight update can be performed. The accuracy was about 60%.

LSTM Bidirectional (BiLSTM): The next model was Bidirectional LSTMs (BiLSTMs). Using BiLSTMs is advisable because we can better capture dependencies in both ends of the input. The performance at training phase was about 70% (Fig. 2). Conversely, a validation accuracy started decreasing. Hence, we decided to set 64 the number of hidden states. Over again, we cannot acquire an improvement in performance.

---

[4] cuDNN provides highly tuned implementations for standard routines such as forward and backward convolution, pooling, normalization, and activation layers.

[5] Colab is a research tool for machine learning based on Jupyter notebook.

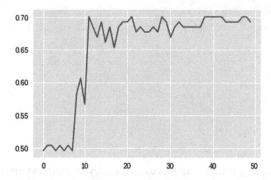

Fig. 2. Visualizing the training process. Accuracy (y-axis) and epoch (x-axis).

Deep BiLSTM: In the next model, we increase the depth of BiLSTM layers, by adding two more layers. We use Dropout regularization between each BiLSTM layer. As more capacity was added when adding layers, we reduced the number of hidden states of each individual BiLSTM to 32. This configuration achieved an average of 75% testing accuracy. We also used in this stage the callback module ModelCheckpoint from Keras. The metric used for triggering the checkpointing callback was the validation accuracy. During training, the best model weights are saved whenever an improvement is observed. The best validation accuracy achieved was about 80% (see Fig. 3).

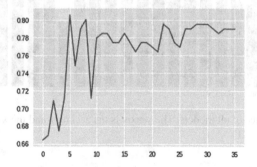

Fig. 3. Visualizing the training process. Accuracy (y-axis) and epoch (x-axis).

Deep BiLSTM and Embedding Layers: The last experiment aimed to take advantage of the similar meanings between correlate words, instead of relying exclusively on the LSTM to infer these relationships. We added an Embedding layer prior to the BiLSTMs. Word embeddings provide a dense representation of words and their relative meanings. The Embedding layer is initialized with random weights and is allowed to learn during the training. The average validation accuracy improved to just above 80%. Figure 4 shows the evolution of validation accuracy with this configuration.

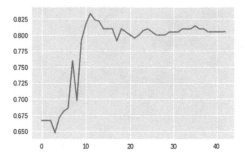

Fig. 4. Visualizing the training process. Accuracy (y-axis) and epoch (x-axis)

Figure 5 shows the evaluation of classification accuracy on the test set of 1,187 news. The accuracy performance ranged between 71% and 81%. The models M24 and M29 achieved the best performance (the accuracy was about 81%). The configurations of these networks are described in Table 1.

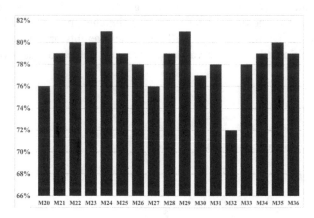

Fig. 5. The performance of all models. The y-axis is the accuracy and the x-axis are the models (the settings are in Table 1).

7   Conclusions

The increasing attention in fake news is motivated by the fact that people are typically not suited to distinguish between good information and fake news, in particular when the source of information is the Internet (and especially social media). Fake news is viewed as one of the greatest threats to journalism and it has weakened public trust in governments. The paper studied various recurrent neural network configurations based on LSTM and embeddings as binary classifiers for detecting Brazilian fake news. We have also built (and exploited) a new dataset collected from the website of Brazilian fact-checking agency AosFatos. Experimental results indicate that, from the models tested, the 3-layer Deep BiLSTMs with trainable word embeddings configuration was the best structure for the task of fake news detection. We noticed that fake news research could potentially benefit from recent developments in deep learning techniques. Several works in NLP tackle this problem in the English language. Nevertheless, in contrast, very little has been done in the (Brazilian) Portuguese language. This research is a small step toward filling this void. Despite being a simple approach (basic LSTMs), it proved rather efficient in classifying fake news with a reasonable performance. For future work, we intend to explore more advanced models for embeddings, such as, BERT or ELMO as well as using more sophisticated structures such as attentional mechanisms.

References

1. Shao, C., Ciampaglia, G.L., Varol, O., Yang, K.-C., Flammini, A., Menczer, F.: The spread of low-credibility content by social bots. Nat. Commun. **9**(1), 4787 (2018)
2. Vlachos, A., Riedel, S.: Fact checking: task definition and dataset construction. In: Fact Checking: Task Definition and Dataset Construction (2014)
3. Allcott, H., Gentzkow, M.: Social media and fake news in the 2016 election. J. Econ. Perspect. **31**(2), 211–236 (2017)
4. Golbeck, J., et al.: Fake news vs satire: a dataset and analysis. In: Proceedings of the 10th ACM Conference on Web Science, Amsterdam, Netherlands (2018)
5. Shu, K., Sliva, A., Wang, S., Tang, J., Liu, H.: Fake news detection on social media: a data mining perspective. SIGKDD Explor. Newsl. **19**(1), 22–36 (2017)
6. Tandoc, E.C., Lim, Z.W., Ling, R.: Defining "Fake News". Digital J. **6**(2), 137–153 (2018)
7. Fallis, D.: A functional analysis of disinformation. In: A Functional Analysis of Disinformation, iSchools edn (2014)
8. Hochreiter, S., Schmidhuber, J.: Long short-term memory. Neural Comput. **9**(8), 1735–1780 (1997)
9. Goodfellow, I., Bengio, Y., Courville, A.: Deep Learning. The MIT Press (2016)
10. Ferrara, E.: Disinformation and social bot operations in the run up to the 2017 French presidential election (2017)
11. Rashkin, H., Choi, E., Jang, J.Y., Volkova, S., Choi, Y.: Truth of varying shades: analyzing language in fake news and political fact-checking. In: Truth of Varying Shades: Analyzing Language in Fake News and Political Fact-Checking (2017)
12. Hanselowski, A., et al.: A retrospective analysis of the fake news challenge stance detection task. arXiv preprint arXiv:1806.05180 (2018)
13. Lazer, D.M., et al.: The science of fake news. Science **359**(6380), 1094–1096 (2018)
14. Vosoughi, S., Roy, D., Ara, S.: The spread of true and false news online. Science **359**(6380), 1146–1151 (2018)
15. Zannettou, S., Caulfield, T., De Cristofaro, E., Sirivianos, M., Stringhini, G., Blackburn, J.: Disinformation warfare: understanding state-sponsored trolls on twitter and their influence on the web. arXiv preprint arXiv:1801.09288 (2018)
16. Wang, W.Y.: "Liar, Liar Pants on Fire": a new benchmark dataset for fake news detection. arXiv preprint arXiv:1705.00648 (2017)
17. Ruchansky, N., Seo, S., Liu, Y.: CSI: a hybrid deep model for fake news detection. In: Proceedings of the 2017 ACM on Conference on Information and Knowledge Management, Singapore, Singapore (2017)

18. Pfohl, S.R.: Stance detection for the fake news challenge with attention and conditional encoding. In: Stance Detection for the Fake News Challenge with Attention and Conditional Encoding (2017)
19. Ajao, O., Bhowmik, D., Zargari, S.: Fake news identification on twitter with hybrid CNN and RNN models. Proceedings of the 9th International Conference on Social Media and Society, Copenhagen, Denmark (2018)
20. Chollet, F.: 'Keras'. In: Book Keras (2015)
21. Abadi, M., et al.: TensorFlow: large-scale machine learning on heterogeneous distributed systems. In: TensorFlow: Large-Scale Machine Learning on Heterogeneous Distributed Systems (2016)
22. Appleyard, J., Kocisky, T., Blunsom, P.: Optimizing performance of recurrent neural networks on GPUs. In: Optimizing Performance of Recurrent Neural Networks on GPUs (2016)
23. Chetlur, S., et al.: cuDNN: efficient primitives for deep learning. arXiv preprint arXiv:1410. 0759 (2014)
24. Lei, T., Zhang, Y., Wang, S.I., Dai, H., Artzi, Y.: Simple recurrent units for highly parallelizable recurrence. In: Simple Recurrent Units for Highly Parallelizable Recurrence, pp. 4470–4481 (2018)

Improved Document Filtering by Multilevel Term Relations

Adrian Fonseca Bruzón[1]([✉]), Aurelio López-López[1],
and José E. Medina Pagola[2]

[1] Instituto Nacional de Astrofísica, Óptica y Electrónica, Puebla, Mexico
{adrian,allopez}@inaoep.mx
[2] University of Informatic Sciences, Havana, Cuba
jmedinap@uci.cu

Abstract. Humans tend to organize information in documents in a logical and intentional way. This organization, which we call textual structure, is commonly in terms of sections, chapters, paragraphs, or sentences. This structure facilitates the understanding of the content that we want to transmit. However, such structure, in which we usually encode the semantic content of information, is not usually exploited by the filtering methods for the construction of user profile. In this work, we propose the use of term relations considering different context levels for enhancing document filtering. We propose methods for obtaining the representation, considering the existence of imbalance between the documents that satisfy the information needs of users, as well as the Cold Start problem (having scarce information) during the initial construction of the user profile. The experiments carried out allowed to assess the impact on the filtering task of the proposed representation.

Keywords: Document Filtering · Term Relations · Cold Start · Document structure

1 Introduction

Information Filtering (IF) is a task that can be a possible solution to face the constant increment in the amount of information generated each day. This task has the objective of classifying documents coming from an information stream, in Relevant or Non Relevant, according to the information needs expressed by a particular user. IF is oriented to satisfy needs that are relatively stable over time. However, if these interests are allowed to change gradually over time, we are talking then of Adaptive Information filtering systems.

One of the key components of a filtering system is the representation of the user's interest, usually called profile. This profile is maintained as long as the information need or interest lasts.

The first author was partially supported by CONACYT, Mexico, under scholarship 635046.

I. Nyström et al. (Eds.): CIARP 2019, LNCS 11896, pp. 85–95, 2019.
https://doi.org/10.1007/978-3-030-33904-3_8

People are very different from each other, and in the same way, a user's information needs are different from those of others. Some users are usually interested in general topics, like certain sport, e.g. baseball. While others can be interested in much more specific content, such as air incidents caused by passengers. This behavior represents a challenge to the applications that implement information filtering. In addition to this problem, we must take into account that each time a user formulates a new information need, the system must address the problem of information shortage, known as Cold Start. Another phenomenon to face is related to the non homogeneity among the documents provided by the user, for a correct modeling of his/her information need. The topic of interest of a user can consist in turn of several subtopics, which do not necessarily have to be equally represented, neither in the starting information provided by the user, nor in the document flow.

Most of the reported approaches that have tackled the IF task use the traditional bag-of-words (BoW) model for document representation, even when this model fails to capture the existing semantic relationships among the terms of a document, commonly referred to as latent semantics. Some document representations, like Latent Dirichlet Allocation (LDA) [2], avoid the BoW limitations, however, in an environment where new documents with new terms frequently emerge, these representations are not convenient. As an alternative, we have pattern based representations. These representations, such as Pattern Taxonomy Mining (PTM) [10], have the advantage that they do not require information available beyond that appearing in the existing documents for profile construction. For that reason, these methods are more attractive to be employed in an online environment. However, none of the above representations takes into account the fact that the terms of a document maintain different relationships depending on the context level in which they are used.

The main contributions of this paper are: a profile representation based on multilevel terms relations, and an algorithm to get them. The algorithm tackles both the phenomenon of Cold Start and the imbalance between the different subtopics underlying the user information need.

The rest of the paper is structured as follow: in Sect. 2 we explore main approach related to this work. Then, Multilevel Term Relations are presented in Sect. 3, followed by the experimental settings, results, and discussion in Sect. 4. Finally, the conclusions of our work are detailed in Sect. 5.

2   Related Work

In the literature, we can find many works that utilizes techniques designed for Information Retrieval for solving the IF task. These approaches usually use the BoW model for representing both the documents and the user profile. Usually, in these works, the profile representation consists of a single vector that condenses the user's need. Commonly, they employ the Rocchio algorithm [11], but other formalist like Logistic Regression [14] are also used. The main differences between the works in this approach are on the way in which a threshold is established and how the terms are selected, and their relevance, which are part of the vector representing the user's profile.

Since the approaches based on IR methods employ a single vector for the user profile, this has the consequence that less represented subtopics in the profile can be affected by those more abundant. These methods also have the drawback of requiring a threshold for determining if a document must be presented to the user or not. But selecting a threshold value that works well in every possible situation is quite complicated.

The use of classifiers designed for the task of Text Categorization is the other great route followed by researchers, to do IF. In this approach, the filtering task is treated as a text categorization, in which a binary classifier is employed. This classifier must assign a label for each new document between two possibilities, Relevant or Not Relevant. If the classifier separates the document as Relevant, this document is reported to the user, otherwise, the document is discarded and the user does not have access to it. In the literature, researchers have explored several classifiers, for instance Winnow [9], k-NN [1], Support Vector Machine (SVM) [5], facets [13], Bayesian Models [12], among other. Methods based on Text Categorization algorithms are seriously affected by the Cold Start, and they usually do not take into account the phenomenon of possible imbalance between different subtopics in the user needs. Like previous methods, they do not work effectively on specific topics.

One the most recent models based on frequent patterns applied to the Information Filtering task is known as Pattern Taxonomy Model (PTM) [10]. This model divides a document into paragraphs and extracts sequences whose support values exceed a preset threshold. Once frequent sequences are extracted from each of the training set documents, the user profile is constructed by condensing the sequences into a single vector. Once the user profile is built, to determine if a new document should be selected and reported to the user, the weights of the patterns that appear in the new document are added and compared against a given threshold value. Several modifications and extensions have been proposed from the PTM model. Among them we find MPBTM (Maximum matched Pattern based Topic Model) [3], where classic PTM is combined with LDA. The idea of combining LDA and frequent patterns was also explored in EFITM algorithm (Enhanced Frequent Itemsets Based on Topic Modeling in Information Filtering) [8].

Methods that use patterns in Information Filtering usually require thresholds for determining the patterns that must be extracted, or to decide which documents must be selected for presenting to the user. The problem with the use of thresholds is that the document availability for the construction of the profile is not the same all the time. The thresholds that are used when the available information is very small may not be adequate when the amount of information available increases.

None of previous models takes into account that the terms maintain different relationships with each other depending on the level of granularity of the context in which they relate, i.e. co-occurrence at document level is different to that at sentence level. Also, existing methods do not consider the imbalance between interest subtopics and usually do not propose any solution to the Cold Start.

3   Multilevel Term Relations

The information needs can be quite different from one user to another. While some users can be interested in very general topics, such as education, football, or politics, other users can have very specific interest, such as traffic accidents involving urban buses and fatalities. In the first case, identifying relations of co-occurrences of frequent sets probably will be enough for obtaining acceptable results in the filtering process. The second interest is more specific and extracting straightforward co-occurrences of terms at the level of document are not enough for expressing the user information needs. Here, we are required to extract relations that involve more specific contexts. For instance, the relation of co-occurrence between *urban* and *bus* which has to be detected in the same sentence or even in the same noun phrase.

The problem is that is not possible to know beforehand which are the right contexts that must be used for satisfying an information need for a particular user. To ameliorate this situation, we do not restrict to consider terms in a single context level. To handle terms, we use a representation that takes into account the co-occurrence relationships between the terms and the level of context in which this relationship occurs.

A multilevel terms relation (symbolically denoted as $\Phi_{\Omega_i}(x_1, x_2, \ldots, x_k)$) is a set of items x_i that co-occur in a determined context level. Items in a relation could be single terms or other relation with a more specific context. Ω_i is the context level employed for relating items x_i. For instance, Ω_i can express Document (D), Paragraph (P), Sentence (S) or noun phrase (N) level. By convention, we are assuming that a term is always related (co-occurs) with itself, at any level. Some examples of multilevel term relations are:

* $\Phi_S (harvest, \Phi_N (organic, fertilizer))$ * $\Phi_S (car, seller)$

* $\Phi_P (\Phi_N (bus, driver), \Phi_S (dealer, square))$ * $\Phi_D (politic)$

In these examples, $(\Phi_S (harvest, \Phi_N (organic, fertilizer)))$ expresses a sentence containing the word *harvest* and a noun phrase with *organic* and *fertilizer*.

For the algorithm that follows, the user profile consists of a set of relations, which we denote as STR. For each document d_j in the stream, d_j is selected as Relevant for the user iff: $\exists R \in STR$, d_j *matches* R.

The process for extracting relations favors more general relations over those more specific. We understand that one relation is more general than another to the extent that a more general context is used in the relation and involves a smaller number of items. The use of more general contexts in relations and with a small number of items allows the matching process in documents to be less restrictive, which implies that more documents are selected for the user. In our algorithm, we start with the use of relationships that involve a single term in a Document context. To these relations, terms are added, or more specific contexts are explored, insofar as they fail to differentiate Relevant from Non Relevant documents in the training set available for the initial construction of the user profile. In this way, relations are selected that allow differentiating the user's interest from the rest of the documents without abruptly affecting the generalization capacity of the extracted relations. Algorithm 1 describes the main steps of our proposed procedure to get the relations.

Algorithm 1. Multilevel Term Relations Extraction Algorithm.

Data: P: Set of Relevant Documents (Positives)

N: Set of No Relevant Documents (Negatives)

T: Set of terms in P

Ω: Context levels to be considered from the most general to the most specific.

Output: STR: Set of Multilevel Term Relations.

1   $STR = \{\}$;

2   $Q \leftarrow \emptyset$ /* Empty Priority Queue */;

3   **foreach** t **in** T **do**

4     $Enqueue\,(Q, \Phi_{\Omega_1}(t))$;

5   **while** $|Q| \neq 0$ **do**

6     $X = Dequeue\,(Q)$;

     /* X is a relation of the form $\Phi_{\Omega_j}(x_1, \ldots, x_n)$ */

7   **if** $not\ Compatible\,(X, P, STR)$ **then**

8   **continue**;

9   **if** $DF\,(X, N) == 0$ **then**

10   Add X to STR ;

11   **else**

12   **foreach** t **in** $T \setminus \{s|s\ \text{in}\ X\}$ **do**

13   **for** $i = j$ **to** $|\Omega|$ **do**

14        $R = \Gamma\,(\Omega_i, X, t)$;

15        $W = \{\}$;

16   **foreach** R_k **in** R **do**

17   **if** $DF\,(R_k, P) \neq \emptyset$ **then**

18   Add R_k a W;

19   **if** $|W| \neq 0$ **then**

20   **foreach** R_k **in** W **do**

21          $Enqueue\,(Q, R_k)$;

$DF\,(R, K)$ is a function that receives a relation R and a set of documents K and return the subset of documents of K that satisfy the relation R. That is: $DF\,(R, K) = \{d_i \mid d_i \in K \wedge R \Mapsto d_i\}$. Γ represents a function that, given a relation R, a context level Ω_i and a term t, returns the next relations to be explored. To determine the new relations to be explored, operators σ and ρ are used, defined as: $\sigma(R, S) = \Phi_{\Omega_i}(r_1, r_2, \ldots, r_k, s_1, s_2, \ldots, s_n)$ and $\rho(R, S) = \Phi_{\Omega_R}(r_i | r_i \notin S)$, where S is a set of items $S = \{s_1, s_2, \ldots, s_n\}$. Function Γ is defined as follows:

$$\Gamma(\Omega_i, R, x) = \begin{cases} \{\sigma(R, \{x\})\} & \Omega_R = \Omega_i \\ \Psi(R, H_1, x, \Omega_i) \bigcup \Delta(R, H_2, x, \Omega_i) & \Omega_R \neq \Omega_i \end{cases} \tag{1}$$

With, $H_1 = \{r_i | r_i\ is\ a\ term\}$ and $H_2 = \{r_i\} \setminus H_1$. Ψ and Δ are functions defined as:

$$\Psi(R, C, x, \Omega_i) = \bigcup_{c_k \in \{\wp(C) \setminus \emptyset\}} \{\sigma(\rho(R, c_k), \Phi_{\Omega_i}(t_j | t_j \in c_k, x))\} \tag{2}$$

$$\Delta(R, C, x, \Omega_i) = \bigcup_{c_k \in \{\wp(C) \setminus \emptyset\}} \bigcup_{V \in \Xi(c_k, x, \Omega_i)} \{\sigma(\rho(R, c_k), V)\} \tag{3}$$

Ξ is defined as $\Xi(C, x, \Omega_i) = \Gamma(C_1, x, \Omega_i) \times \ldots \times \Gamma(C_k, x, \Omega_i)$ and $\wp(C)$ is the power set of C.

One important detail to remark of the algorithm is that there is no threshold to determine if a relationship should be considered. Instead, we employ the concept of compatibility between a relation R and a set of relations S.

Definition: A candidate relation R is compatible with a set of relations $S = \{S_1, \ldots, S_n\}$ if, given $X = DF(R, P)$, $G^1 = \{S_j | |DF(S_j, P)| > |X|\}$, and $G^2 = \{S_j | |DF(S_j, P)| = |X| \wedge |S_j| < |R|\}$, the following condition holds:

$$X \setminus \bigcup_{G_i^1} DF(G_i^1, P) \neq \emptyset \wedge X \setminus \bigcup_{G_i^2} DF(G_i^2, P) \neq \emptyset \tag{4}$$

Here $|R|$ with R a relation expresses the number of different terms in R.

The use of the definition of compatibility, in Algorithm 1, guarantees that we can discard the relations that are subsumed by other relations with a greater presence among the Relevant documents, or by those with equal presence but that involve fewer terms. To the extent that a larger number of relations can be discarded and not explored, the algorithm will be more efficient. So the use of this definition allows us to deal with the problem of imbalance among the subtopics existing in the documents available for the construction of the profile, given that the under-represented subtopics are not discriminated by the number of documents that address them.

In our proposal, Cold Start is attended by two different approaches. First, the algorithm favors the extraction of relationships in the most general contexts and with the least number of terms, which facilitates the matching in new documents. Other approach is by employing externals resources during the indexing process.

The number of relations that can be extracted from a set of documents is huge, however, most of them are noisy or unrepresentative elements. The extraction of a large volume of unrepresentative relations not only influences the

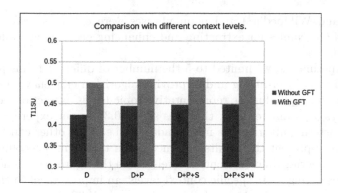

Fig. 1. Results obtained when considering different context levels.

efficiency of the system but also affects the quality obtained, since documents that are not really of interest can be selected.

For addressing this situation, in the algorithm, a relation to be selected for representing the user needs must contain, at least a Global Frequent Term (*GFT*). *GFT*s are terms that frequently appear in the different subtopics associated to a user need. For obtaining these terms, firstly documents in the user profile are structured in subtopics by means of a clustering procedure. Once the set of existing subtopics $L = \{L_1, L_2, \ldots, L_c\}$ has been obtained, *GFT*s are terms satisfying the following property:

$$GFT = \left\{ t_i \mid t_i \in T \wedge \frac{|\{L_k \mid df(t_i, L_k) > 0\}|}{|L|} \geq \gamma \right\} \tag{5}$$

where *df* denotes the number of documents of the subtopic L_k that contain the term t_i, and γ is a parameter.

4   Experiments

For evaluating our proposal, we followed the experimental setting proposed on the TREC Conferences for Adaptive Filtering task [7]. We used the subset of the 50 topics proposed by the organizers (with an associated set of three documents to build the profile). This subset was created from news of the Reuter Corpus Volume 1 (RCV1) [4]. This collection is composed of news and press digest. The length of documents varies from 1 or 2, up to 20 sentences. However, in our method, the length does not have an impact. We employed the $T11SU$ measure, whose range is between 0 and 1, with 1 being the best value.

In the experiments, the documents were previously preprocessed, where tags and stop words were removed and a lemmatization process was applied by means of FreeLing tools[1]. Also Named Entity Recognition was done (using JRC

---

[1] http://nlp.lsi.upc.edu/freeling/index.php.

resources[2], and Wikipedia[3]) and noun phrases were extracted. We also took advantage of GeoNames for extracting and enhancing geographical information when needed.

In the experiments, we limited to 5 the number of different terms present in a relation to avoid over-fitting and improving algorithm efficiency. We selected Compact Clustering [6] for the representative topic terms extraction. This algorithm involves a parameter *beta*, that was set to 0.33 in our experiments. This particular clustering algorithm is not mandatory, i.e. some other can be used.

In Fig. 1, we present the results obtained when taking into account, or not, the *GFT*s considering different context levels. From the figure, we can notice that when are considering more specific context levels, an improvement in the performance of the algorithm is obtained. The most marked difference is found when we go from using only the D level to using D + P, being in the other cases more discreet improvements. At this point, we must highlight two important elements. First, our algorithm determines automatically the levels of context required to correctly differentiate Relevant from Non-Relevant documents. Therefore, even when all the possible context levels are considered, from the most general to the most specific, not all the information needs will be used, in some cases it may be enough with level D, while in others, is required to go as far as employing the most specific. That is why even though the improvement is slight when the levels S and N are used, they are still noticeably. The other relevant point to note is that in the collection the vast majority of paragraphs consist of a single sentence, which obscures the improvement in performance when this context level is considered. In addition, from Fig. 1, we can note that using the strategy of limiting the extracted relations to only those that contain GFT allows increasing the quality of results. Figure 2 presents the results obtained in the filtering process when is considered, in addition to Multilevel Term Relations (MLTR), the GFT and the use of Wikipedia and GeoNames for attacking the Cold Start

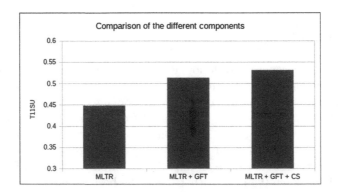

Fig. 2. Comparison of how different components improve the results obtained.

---

[2] https://ec.europa.eu/jrc/en/language-technologies/jrc-names.

[3] https://en.wikipedia.org.

(CS) problem. We can notice that, as mentioned above, when *GFT*s are used to eliminate unrepresentative relations, the results improved. In addition, these results increase if we employ external resources to ameliorate the CS problem.

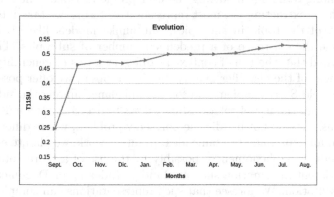

Fig. 3. Evolution of results over time.

An important aspect to analyze in an Adaptive Filtering Algorithm is how it behaves over time. In Fig. 3, we present how our proposed method evolves with time. As we can observe, at the beginning the obtained result is relatively poor but as months passes by and feedback is available, this performance tends to improve.

Finally, we compare our proposal with algorithms of the state of the art. In Table 1, we present a comparison of our approach with some of the most prominent methods reported. The first three rows are for pattern based algorithms, while the other two rows are associated with other techniques, and closing with our proposal. As we can note, our approach obtains the best performance. The results obtained are significantly better than those obtained by the other pattern based methods. Even, our proposal is slightly higher than the best results reported, with the added advantage that with our method, it will be easier to explain to the user the reasons for selecting certain document.

Table 1. Comparison with previous methods

Method	T11SU
PTM	0.32
MPBTM	0.25
EFITM	0.21
Bayesian Model	0.5
Logistic Regression	0.52
Our proposal	**0.531**

5   Conclusions

In this paper, we proposed a new method that analyses in depth the relations that exist among terms for modeling the user profile in Document Filtering. In our approach for extracting the most representative terms, we take into account the structure of the topic in subtopics, but unlike models like MPBTM, we do not assume the existence of a predefined number of subtopics. The results obtained showed that the use of context at different levels of generality leads to an improvement of the classifier quality. In our proposal, we offer possible solutions for the Cold Start problem and for the imbalance among the subtopics of interest. Although the use of relationships allows to overcome the independence between terms, they do not handle the polysemy problem, where different terms could be used for expressing the same idea. In future work, we plan to extend our model to handle this phenomenon, possibly by the use of a representation that incorporates latent information such as Random Indexing or Document Occurrence Representation. We foresee that such representations can complement our approach.

References

1. Ault, T., Yang, Y.: KNN, Rocchio and metrics for information filtering at TREC-10. In: Proceeding of the Tenth Text REtrieval Conference, pp. 84–93. National Institute of Standards and Technology (2001)
2. Blei, D.M., Ng, A.Y., Jordan, M.I.: Latent Dirichlet allocation. J. Mach. Learn. Res. **3**(Jan), 993–1022 (2003)
3. Gao, Y., Xu, Y., Li, Y.: Pattern-based topics for document modelling in information filtering. IEEE Trans. Knowl. Data Eng. **27**(6), 1629–1642 (2015)
4. Lewis, D.D., Yang, Y., Rose, T.G., Li, F.: RCV1: a new benchmark collection for text categorization research. J. Mach. Learn. Res. **5**(Apr), 361–397 (2004)
5. Montejo-Ráez, A., Perea-Ortega, J.M., Díaz-Galiano, M.C., Ureña-López, L.A.: Experiments with Google news for filtering newswire articles. In: Peters, C., et al. (eds.) CLEF 2009. LNCS, vol. 6241, pp. 381–384. Springer, Heidelberg (2010). https://doi.org/10.1007/978-3-642-15754-7_46
6. Pons-Porrata, A., Berlanga-Llavori, R., Ruiz-Shulcloper, J.: On-line event and topic detection by using the compact sets clustering algorithm. J. Intell. Fuzzy Syst. **12**(3, 4), 185–194 (2002)
7. Soboroff, I., Ounis, I., Macdonald, C., Lin, J.J.: Overview of the TREC-2012 microblog track. In: TREC, vol. 2012, p. 20 (2012)
8. Wai, T.T., Aung, S.S.: Enhanced frequent itemsets based on topic modeling in information filtering. In: 2017 IEEE/ACIS 16th International Conference on Computer and Information Science (ICIS), pp. 155–160. IEEE (2017)
9. Wu, L., Huang, X., Niu, J.: FDU at TREC 2002: filtering, Q&A, web and video tasks. In: Proceeding of the Eleventh Text REtrieval Conference, pp. 232–247 (2002)
10. Wu, S.T., Li, Y., Xu, Y., Pham, B., Chen, P.: Automatic pattern-taxonomy extraction for web mining. In: Proceedings of the 2004 IEEE/WIC/ACM International Conference on Web Intelligence, pp. 242–248. IEEE Computer Society (2004)

11. Xu, H., et al.: TREC-11 experiments at CAS-ICT: filtering and web. In: Proceeding of the Eleventh Text REtrieval Conference (TREC 2011), pp. 141–151 (2002)
12. Zhang, L., Zhang, Y.: Hierarchical Bayesian models with factorization for content-based recommendation. arXiv preprint arXiv:1412.8118 (2014)
13. Zhang, L., Zhang, Y., Xing, Q.: Learning from labeled features for document filtering. CoRR abs/1412.8125 (2014)
14. Zhang, Y.: Using Bayesian priors to combine classifiers for adaptive filtering. In: Proceedings of the 27th Annual International ACM SIGIR Conference on Research and Development in Information Retrieval, SIGIR 2004, pp. 345–352. ACM, New York (2004)

TSTM: A Model to Predict Topics' Popularity in News Providers

Mario Alfonso Prado-Romero[1](✉)[ID], Alessandro Celi[2], Giovanni Stilo[2], and Alex Coto-Santiesteban[1][ID]

[1] University of Havana, Havana, Cuba
{mario.prado,a.coto}@matcom.uh.cu
[2] University of L'Aquila, L'Aquila, Italy
{alessandro.celi,giovanni.stilo}@univaq.it

Abstract. The volume of news increases everyday, triggering competition for users' attention. Predicting which topics will become trendy has many applications in domains like marketing or politics, where it is crucial to anticipate how much interest a product or a person will attract. We propose a model for representing topic popularity behavior across time and to predict if a topic will become trendy in the future. Furthermore, we tested our proposal on a real data set from Yahoo News and analysed the performance of various classifiers for the topic popularity prediction task. Experiments confirmed the validity of the proposed model.

Keywords: Collective attention · Predicting model · Topics' Model · Temporal data · Users' interactions · Machine learning

1 Introduction

In the last decades, web and social media have changed the way the public consumes and produces information. In particular, users, by providing content, have facilitated the study of patterns for the emergence, production, and consumption of information [5,11]. The process where people collectively contribute to the creation of shared knowledge relies on the notion of *collective intelligence* [16] and it is involved in what is called the *wisdom of the crowds* [19].

In addition, news media have increased from newspapers and news channels to web blogs and social networks, democratizing the news sharing process, but also increasing the volume of information available. This explosive growth in the amount of information has intensified the online competition for the *collective attention*, since only a small number of items become popular while the rest remains unknown [20].

Predicting news popularity has been the focus of many works, but it is a challenging and still open problem. Findings in [2] suggest that popularity is disconnected from the inherent structural characteristics of news content and

© Springer Nature Switzerland AG 2019
I. Nyström et al. (Eds.): CIARP 2019, LNCS 11896, pp. 96–106, 2019.
https://doi.org/10.1007/978-3-030-33904-3_9

that the problem of predicting popularity at cold start cannot be easily modeled. Instead, news popularity may be more accurately predicted if early-stage popularity measurements are incorporated into prediction models as features. The main drawback of this approach is that the life of news items is very short and the traffic of most articles monotonically decreases after 12 h [7], thus reducing the usefulness of the prediction.

Tracking new topics, ideas, and "memes" across the Web has been an issue of considerable interest [15]. Topics can remain in collective attention far beyond the news that first introduced them. Also, analyzing and predicting topic popularity is a problem of great interest in domains like marketing, where companies spend 12% of their revenues in advertising [18] and where identifying the next rising *star* can be an advantage. Also, this is useful in politics, where there is a need to understand how much interest a public figure will attract and which topic would be discussed during a political campaign. Furthermore, knowing the expected popularity of concepts could be useful for journalist in helping them to choose the topic for their next articles.

To address the aforementioned scenarios, we perform an analysis of topic popularity on news. We propose the conceptual model of topics' popularity: *Time Sensitive Topics' Model* (TSTM). Our model allows to capture effectively the temporal behaviour of a topic. We show that the proposed model can be employed to efficiently predict topics' popularity using simple and well established predictive models. To recap the main contributions of this work are:

- **Understanding the evolution of topics' popularity**: We focus on the analysis of topic behavior, contrary to most works focusing on news. Furthermore, we study the number of clicks on news containing the topic, which is a feature rarely disclosed by news platforms [20].
- **Conceptual Model of Topics' Popularity**: We present TSTM a model that effectively captures the temporal behavior of a topic. The first use of TSTM is to predict the popularity of a topic understanding its behavioural pattern in the near past. Moreover, we would like to remark that the measured popularity is obtained using only information contained in the news network and does not require access to external sources like tweets or other social media platforms.
- **Validating our proposal on real data**. We conduct an exhaustive empirical evaluation on a real world dataset R10 (Yahoo News Feed). R10 dataset is a massive collection of news based on Yahoo News' portal, one of the most popular news aggregator services receiving 7.09% of all news site visits. To validate our model we test it on several popular and well established classifiers: Majority Class, Naive Bayes, Logistic Regression, Linear SVM [6] and Random Forest [14].

The remainder of this paper is structured as follows: In Sect. 2, we analyze existing works on popularity prediction. In Sect. 3, we present the TSTM model. In Sect. 4, we discuss the used experimental framework, the adopted methodology (Sect. 4.1) and the used data (Sect. 4.2). Section 5 describes the conducted experiments and the obtained results. Finally, in Sect. 6, we present our conclusions and discuss open challenges and future directions.

2 Related Work

Due to the great increase in web content generation and the advantage of knowing which content will be popular, there is a great amount of research focusing on predicting the popularity of the diverse forms of web content. Tatar et al. [20] provided a comprehensive analysis of the works on this topic. In the news domain most works have focused on the problem of predicting the popularity of news items. The techniques used in these works can be divided in two main classes, the ones predicting item popularity based on its features, and the ones predicting item popularity at time d based on its previous popularity at time h.

Many works have focused on predicting news popularity at cold start. In [4], the authors use the number of times an article is shared in Twitter to measure popularity of news from Feedzilla news aggregator. To accomplish this task the authors used three different kinds of regression. Furthermore, Fernandes et al. [12] proposed an Intelligent Decision Support System that first extracts a broad set of features that are known prior to an article publication, in order to predict its future popularity. Then, it optimizes a subset of the article features in order to enhance its expected popularity. Also, Choudhary et al. [8] use a Genetic Algorithm to select the features that most influence in the popularity of a news article and then use machine learning techniques as Naive Bayes and Neural Networks to predict popularity based on those features. Other approaches to this problem, using machine learning, include the use of Hierarchical Neural Networks [13] and ensembles like AdaBoost [10] focused on improving the precision of the prediction.

Predicting popularity at cold start is a challenging task and Arapakis et al. [2] reached to the conclusion, that popularity is disconnected from the inherent structural characteristics of news content. The authors affirm that the good precision results are in most cases due to the imbalanced nature of the datasets where the majority of items are not popular. Furthermore, they advise that popularity may be more accurately predicted if early-stage popularity measurements are incorporated into the prediction models.

There are a group of studies focusing on predicting news popularity using early stage popularity measures. Ahmed et al. [1] predicted popularity of content in Youtube, Vimeo and Digg. They perform a clustering of the users' popularity behavior in each time step obtaining clusters representing different behaviors, then they calculate the probability of an item to make a transition to a particular popularity pattern in the future. They achieve better performance than linear regression models and noted that popularity behaviour varies among the 3 data sets. Tatar et al. [21] formulate the prediction task as a ranking problem and tried to predict the rank of an article in the future, based on its popularity. To achieve this task they use a linear regression model on a logarithmic scale, and a constant scaling model, obtaining better accuracy than traditional ranking methods. Van Canneyt et al. [22], proposed an hybrid approach using features from the articles and information about its popularity in the first day after its publication. This technique outperforms previous approaches to news popularity prediction. In general due to the short life span of news, the usefulness of the predictions decreases as the time used for making the prediction increases. On the other hand, lifetime of topics can last far beyond the news that first introduced them.

Some works have focused on the problem of analyzing and predicting topic popularity, most of them focused on social media. Leskovek et al. [15] track the behavior of memes and topics through news sites and blogs. They determined that the behavior of topics is very difficult to describe using a single function due to the singular behavior during the 8 h around the popularity peak. Furthermore, Ardon et al. [3] perform an analysis of topic popularity on Twitter. They take inspiration from disease propagation models and analyzed how the geographic location and the community where a topic is originated could affect its popularity. Zaho et al. [23] approached this task as an information propagation problem and proposed a model to predict topic spreading speed on the Sina Weibo micro-blogging service. They used early data to calculate current spreading speed and then, using a differential equations model, they performed a short-term prediction of the future spreading speed of the topic among users.

Our work, is focused on understanding and predicting topic popularity in news. We present a conceptual model of topics' popularity. The proposed model allows to understand and predict topics behavior over time. Furthermore, our analysis is validated on a massive set of news using the most popular machine learning models. The dataset provides information on users views of the news, which is information rarely provided by news sources: this allows us to analyze topic popularity from a less studied point of view.

3 Conceptual Model of Topics' Popularity

Popularity is a complex phenomenon that is not simple to model. Different measures try to capture different aspects of this phenomenon. Moreover, to cope with the extremely volatile aspect of popularity we consider also the time dimension of this problem. For this reason the proposed model (TSTM) is *Topic* oriented, with a user-defined time-slot as a parameter. We formally define TSTM a *Time Sensitive Topics' Model* as a conceptual model that allows to effectively predict topics' popularity in highly time sensitive scenarios. First, we define user related features and then we will define the features that are topic oriented:

– Let U be the set of *Users*, N be the set of *News*, T be the set of *Topics* and let $\mathcal{T}$ be the set of all possible *time intervals* defined by a chosen *grain* e.g. day, hour, week or month. Each news item $n \in N$ is also described by a set of Topics $T_n \subseteq T$: let $N_t \subseteq N$ be the set of News where the topic t appears. We also introduce N_t^d, which extends the notion of N_t introducing the temporal aspect $d \in \mathcal{T}$.

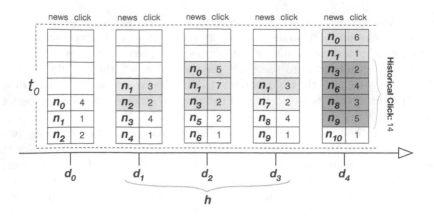

Fig. 1. Example of historical clicks $H_t^{d,h}$ where $t = t_0$, $d = d_4$ and $h = \{d_1, d_2, d_3\}$

- The function $I(u, n, d) : U \times N \times T \longrightarrow \mathbb{N}$ collects all interactions between Users and News in a certain time interval (e.g. hour, day, week) $d \in T$. In general, there are several kinds of interactions (click, view, edit, comment, like, re-post, etc.). In this work, we will focus on arguably the most important interaction, the Clicks C. We remark that, we are interested in modelling Topics' Popularity, for this reason, from now on we will omit the concept of user and aggregate all users' interactions together, considering only the point of view of Topics.
- We refer to C_n^d as the number of *Clicks* that a news item $n \in N$ receives in time interval $d \in T$. Similarly we refer to C_t^d as the number of *Clicks* that a topic $t \in T$ inherits from all the news N_t^d in time interval $d \in T$.
- One of the most important measures we defined is $\widehat{N}_t^d$ which captures the number of news that were *published* in a specific time interval $d \in T$; similarly $\widehat{C}_t^d$ captures the number of *Clicks* obtained by only the *new* news $\widehat{N}_t^d$.
- Moreover, a topic can be popular (receive a high number of *Clicks*) from a global perspective, but this can happen because internet is saturated with news that discuss about it. To capture the real importance of a topic we introduce the normalization function $norm(C) : \mathbb{N} \longrightarrow \mathbb{R}$, $norm(C_t^d) = C_t^d / |N_t^d|$ obtained by dividing the number of *Clicks* C_t^d by the number of involved *News* N_t^d in the same time interval $d \in T$. The normalization function can be applied also to $\widehat{C}_t^d$ as follows: $norm(\widehat{C}_t^d) = \widehat{C}_t^d / |\widehat{N}_t^d|$.
- One central concept introduced by this work is the one that captures the *Historical Clicks*. The measure $H_t^{d,h}$ captures the clicks obtained by the topic t in the time interval d considering the news $\widehat{N}_t^h$ published in time interval h; note that h, generally speaking, can be an aggregation of several contiguous time interval (e.g. let be $h = d_0, d_1, d_2$) and then can be expressed as:

$$H_t^{d,h} = sum_{n \in \widehat{N}_t^h} \widehat{C}_n^d \qquad (1)$$

Figure 1 shows an example of the information that must be consider in order to obtain the value of $H_t^{d,h}$. As it is possible to see, the figure depicts the detailed time series for topic t_0 over the five days d_0, d_1, d_2, d_3, d_4. For each day d_i the list of News $N_{t_0}^{d_i}$ and their clicks are reported. News inside white boxes are those (also identified by the set $\widehat{N}_{t_0}^{d_i}$ previously defined), published in day d_i (e.g. n_5 and n_6 in d_2). The light-gray boxes contain news that receive clicks on that day but had already been published. News that are in the dark-grey boxes are those that produce the *Historical Clicks*. In Fig. 1 are marked only the news that generate *Historical Clicks* $H_{t=t_0}^{d=d_4,h=[d_1,d_2,d_3]}$ of d_4 where $t = t_0$, $d = d_4$ and $h = \{d_1, d_2, d_3\}$. Note that in d_4, the news n_0, n_1 and n_{10} do not increase the value of historical clicks. The news n_0 and n_1 are outside the chosen time interval $h = [d_1, d_2, d_3]$ of 3 days and news n_{10} was published in d_4.

- For each time interval $d \in \mathcal{T}$ we defined two possible rankings R^d and $\overline{R^d}$, based respectively on the amount of *Clicks* C^d and their normalized version $norm(C_t^d)$ of every topic. We can access the rankings from the topics' perspective, where we refer to R_t^d and $\overline{R_t^d}$ as the position of topic t in ranking R^d and $\overline{R^d}$. We can also access the rankings from the positions' perspective, where we refer to $R^d(r)$ and $\overline{R^d}(r)$ as the topic at position r.
- The scope of this work is to define and validate a model useful to predict topics that capture the *Collective Attention* at an early stage. According to [20] and inspired by [17] we choose to predict if a certain topic t falls under the attention of the crowd. For this reason we introduced a parameter $b \in \mathbb{N}^+$ related to the prediction of the popularity of a topic t. If in time interval d, $R_t^d \leq b$ (or $\overline{R_t^d} \leq b$), then we say that the topic t captures the collective attention. From a psychological perspective the number b is equivalent to the number 9 in [17]. However, in this work we will not study how to determine which is the best value of b ($b = 7 \pm 2$ was found to be the best value to represent the attention's capacities of a single person), leaving this aspect for future explorations. For simplicity, we summarize the condition $R_t^d \leq b$ into a binary value (True or False) captured by the notation $A_b^d(t)$.

4 Experimental Framework

In this section we are going to present the experimental framework used to validate the TSTM conceptual model. First, we present the methodology that we follow and then we discuss the used dataset and the preprocessing steps needed to apply the proposed model.

4.1 Methodology

Our goal is to predict, according to its early popularity pattern, if a topic will capture the collective attention. In order to solve this problem, we modeled it (as previously discussed) as a classification task, and we built a predictor based on temporal behavior captured by TSTM, assuming statistical independence between topics. The tuple $B_t^{d,h}$, collects the behaviour of topic t at time d considering also the historical clicks occurred from time h. A behavioral tuple $B_t^{d,h}$, contains the following features:

$$B_t^{d,h} = \{|N_t^d|, |\widehat{N}_t^d|, \widehat{C}_t^d, norm(C_t^d), H_t^{d,h}, \overline{R_t^d}\}$$

In order to conduct our experiments, we built a time-series (one for every topic $t \in T$), that contain all the tuples $B_t^{d,h}$ for every time interval in T. We slice all the time-series using a sliding window with a horizon of size $z \in \mathbb{N}$. In the end, a meaningful data set DS^z, composed by all the windows $w \in DS^z$, is produced. Every window w is also decorated with the binary variable $A_b^d(t)$ that measure if the collective attention was captured by topic t considering a budget b, in the time interval d that immediately follows w. We refer to w_t^d as the window w of topic t that start at time d. Then each record/window w_t^d of the constructed dataset DS^z is as follows:

$$w_t^d = \{B_t^{d,h}, B_t^{(d+1),h}, \ldots, B_t^{(d+z),h}, A_b^{(d+z+1)}(t)\}$$

Clearly $A_b^{(d+z+1)}(t)$ represents the value that we like to predict, and the prediction made at time $d + z + 1$, is solely based on the behavior of topic t captured by TSTM between time d and time $d + z$. It is important to mention that our model does not need to be trained with a particular topic to predict its popularity. We train it using popularity patterns coming from all training topics combined. This way, classifiers learn to predict popularity for new topics based on its behavior without the need to have been trained using those topics.

4.2   Used Data and Preprocessing

In this section, first we present an overview of the characteristics of the used data (Yahoo News Feed - R10). Secondly, we present the preprocessing strategies applied to the data in order to build the experimental dataset DS^z (described in Sect. 4.1) needed for our experiments.

Real World Data: R10 is a massive collection of data, which rounds about 13.5 TB of information. It is based on a sample of user's interaction with the news stream on the Yahoo Homepage, Yahoo Sports, Yahoo Finance, Yahoo Entertainment, Yahoo News, and Yahoo Real Estate. To build the dataset, a random sample of 17 million active Yahoo users was selected in June 2015. Then their news item interaction data from February 2015 through May 2015 was obtained and processed to remove all identifiable properties. This results in a set of over **6 million** news items, and **101 billion** user-news interaction examples.

In its original format the dataset is composed by three sub-collections:

– *Users collection*, contains one entry per user. Note that, in compliance with the G.D.P.R. [9] Users are represented as meaningless, anonymous ids, so that no identifying information is revealed.

- *Items collection*, contains one entry per news item, with two fields: *item-id* and *content* field, which is a JSON encoded object storing metadata about the corresponding news (the *category* of the item, list of *topics* contained in the item, a *summary* and the *title*);
- *Event collection* table links together news items and users. Every entry corresponds to a possible interaction of a user with a news item. It contains the following fields: *user-id, event-id, timestamp, item-id, city, state, property* and other fields that describe the user's surfing features in respect to the specific triplet item-event-user.

Preprocessing: In order to validate the *Time Sensitive Topics' Model*, we need to process the real world data to build a detailed topic oriented time-series that contains all features previously defined by TSTM and required by the experimental methodology described in Sect. 4.1. The first step consists on aggregating all the users-news' clicks on the same time intervals d. Then, for every news item $n \in N$, we build a time-series, where on the x axis we have the time interval d and on the y we collect all the received clicks by n at time interval d. Considering all the topics T that are already available for every news of R10, we expand the news' time-series into the topics' time-series where on the x axis we have the time interval d and on the y axis we have all the clicks that a certain topic t collects on time interval d. Similarly, we can have the normalized version of the topics' time-series were we use $norm(C_t^d)$ instead of C_t^d. Processing several times the time-series, backward and forward, we were able to compute all the features defined by TSTM at every time interval. Table 1 shows an extract of the constructed time-series for topics t_1, t_2, t_3, t_4 in two consecutive days d_i and d_{i+1} considering a $|h|$ of 4 days and $b = 5000$.

Table 1. Example of the values for $[|N_t^d|,|\widehat{N}_t^d|,\widehat{C}_t^d,norm(C_t^d),H_t^{d,h},\overline{R^d},A_b^d]$ with $|d| = 1$ day, $|h| = 4$ days and $b = 5000$.

	d_i							d_{i+1}						
	$|N_t^d|$	$|\widehat{N}_t^d|$	$\widehat{C}_t^d$	$norm(C_t^d)$	$H_t^{d,h}$	R^d	A_b^d	$|N_t^d|$	$|\widehat{N}_t^d|$	$\widehat{C}_t^d$	$norm(C_t^d)$	$H_t^{d,h}$	R^d	A_b^d
t_1	258	110	71508	2829.25	658439	1024	True	286	105	8165	1531.79	429929	1700	True
t_2	511	203	78695	431.30	141704	3970	True	538	272	81116	674.60	281821	3049	True
t_3	740	419	112995	403.40	185527	4119	True	830	415	58525	333.55	218327	4630	True
t_4	601	309	67284	586.15	284995	3345	True	661	329	38288	265.02	0	5150	False

5   Experiments and Results

To test the proposed model we followed the methodology described in Sect. 4.1 and we applied the preprocessing shown in Sect. 4.2 on May 2015 of Yahoo News Feed dataset R10. In our first experiment we want to measure the impact of the size z of the considered window on the obtained prediction. To do so we built several datasets DS^z varying the parameter z from 1 to 7. We obtained

280,914 windows for DS^1, **274,340** for DS^2, **267,666** for DS^3, **260,362** for DS^4, **252,754** for DS^5, **244,730** for DS^6, **236,476** for DS^7. In order to conduct a fair experimentation we follow the *cross validation* approach where we used a Stratified Shuffle Split strategy with 10 splits. Each split is divided into a 90% of the windows to learn the classifier and 10% to test the learned predictor. We used the Area Under ROC Curve (ROC AUC) to measure the quality of the prediction. ROC curve captures the relation between True Positive Rate and False Positive Rate at various thresholds. Our model is tested using several popular and well established classifiers: Majority Class, Naive Bayes, Logistic Regression, Linear SVM [6] and Random Forest [14]. In this work we would like not only to validate the proposed model, but also to establish baseline results for future works in this research area. Table 2 shows the averaged ROC AUC (across 10 splits), achieved by each classifier using different sizes z.

Table 2. Area Under ROC Score using severals ML algorithm using TSTM

$|z|$	Majority	Naive Bayes	Log. Regr.	Linear SVM	Rand. Forest
1	0.5	0.7768	0.8212	**0.8217**	0.7971
2	0.5	0.7371	0.8220	0.8232	**0.8287**
3	0.5	0.7191	0.8209	0.8233	**0.8332**
4	0.5	0.7107	0.8182	0.8256	**0.8397**
5	0.5	0.7068	0.8195	0.8265	**0.8451**
6	0.5	0.7020	0.8221	0.8267	**0.8504**
7	0.5	0.6982	0.8221	0.8303	**0.8569**

Table 2 clearly shows that Random Forest is the best (marked in boldface) and most robust model among the tested algorithms. Linear SVM was confirmed to be a strong predictor as well. Even if the window size does not dramatically affects the quality of the performance, it is possible to notice a significant increment in performances (about 5%) for most classifiers. The obtained results also show the power of TSTM to capture enough information for being effectively used to predict topics' behavior over time.

6 Conclusions

In this work we proposed to approach the problem of topic popularity prediction on News. Furthermore, we proposed a model for representing topic popularity behavior on time sensitive scenarios and an experimental framework for learning popularity patterns and predict when a topic will be in collective attention. Our experiments demonstrate that predicting topic popularity can be effectively performed using our model. Moreover we tested various machine learning algorithms for popularity prediction, providing a good baseline for future works on this subject.

There are some challenges we would like to focus on future work. First, in addition to predict if the topic would be popular in the following day, it would be convenient to rank how popular it will be and for how long it will remain in collective attention. Second, we will test more powerful machine learning models (based on Deep Learning) to predict topic popularity.

References

1. Ahmed, M., Spagna, S., Huici, F., Niccolini, S.: A peek into the future: predicting the evolution of popularity in user generated content. In: Proceedings of the Sixth ACM International Conference on Web Search and Data Mining, pp. 607–616 (2013)
2. Arapakis, I., Cambazoglu, B.B., Lalmas, M.: On the feasibility of predicting popular news at cold start. J. Assoc. Inf. Sci. Technol. **68**, 1149–1164 (2017)
3. Ardon, S., Bagchi, A., Mahanti, A., Ruhela, A., Seth, A.: Spatio-temporal and events based analysis of topic popularity in twitter. In: Proceedings of ACM International Conference on Information & Knowledge Management (2013)
4. Bandari, R., Asur, S., Huberman, B.A.: The pulse of news in social media: forecasting popularity. In: ICWSM, vol. 12, pp. 26–33 (2012)
5. Bermejo, F.: The economics of attention: style and substance in the age of information - by Richard A. Lanham. J. Commun. **57** (2007)
6. Boser, B.E., Guyon, I.M., Vapnik, V.N.: A training algorithm for optimal margin classifiers. In: Proceedings of the Fifth Annual Workshop on Computational Learning Theory, COLT 1992, pp. 144–152 (1992)
7. Castillo, C., El-Haddad, M., Pfeffer, J., Stempeck, M.: Characterizing the life cycle of online news stories using social media reactions. In: 17th ACM Conference on Computer Supported Cooperative Work Social Computing, pp. 211–223. ACM (2014)
8. Choudhary, S., Sandhu, A.S., Pradhan, T.: Genetic algorithm based correlation enhanced prediction of online news popularity. In: Behera, H.S., Mohapatra, D.P. (eds.) Computational Intelligence in Data Mining. AISC, vol. 556, pp. 133–144. Springer, Singapore (2017). https://doi.org/10.1007/978-981-10-3874-7_13
9. Council of European Union: Council regulation (EU) no 679/2016 (2016)
10. Deshpande, D.: Prediction & evaluation of online news popularity using machine intelligence. In: 2017 International Conference on Computing, Communication, Control and Automation (ICCUBEA), pp. 1–6 (2017)
11. Dow, P.A., Adamic, L.A., Friggeri, A.: The anatomy of large Facebook cascades. In: ICWSM (2013)
12. Fernandes, K., Vinagre, P., Cortez, P.: A proactive intelligent decision support system for predicting the popularity of online news. In: Pereira, F., Machado, P., Costa, E., Cardoso, A. (eds.) EPIA 2015. LNCS (LNAI), vol. 9273, pp. 535–546. Springer, Cham (2015). https://doi.org/10.1007/978-3-319-23485-4_53
13. Guan, X., Peng, Q., Li, Y., Zhu, Z.: Hierarchical neural network for online news popularity prediction. In: Chinese Automation Congress 2017, pp. 3005–3009 (2017)
14. Ho, T.K.: Random decision forests. In: Proceedings of the Third International Conference on Document Analysis and Recognition, ICDAR, p. 278 (1995)
15. Leskovec, J., Backstrom, L., Kleinberg, J.: Meme-tracking and the dynamics of the news cycle. In: Proceedings of the 15th ACM SIGKDD International Conference on Knowledge Discovery and Data Mining, pp. 497–506 (2009)

16. Levy, P.: Collective Intelligence: Mankind's Emerging World in Cyberspace (1997)
17. Miller, G.A.: The magical number seven, plus or minus two: some limits on our capacity for processing information. Psychol. Rev. **63**(2), 81 (1956)
18. Pemberton, C.: Gartner CMO spend survey 2016–2017 shows marketing budgets continue to climb (2017)
19. Surowiecki, J.: The wisdom of crowds: why the many are smarter than the few and how collective wisdom shapes business. Little, Brown (2004)
20. Tatar, A., de Amorim, M.D., Fdida, S., Antoniadis, P.: A survey on predicting the popularity of web content. J. Internet Serv. Apps **5**(1), 8 (2014)
21. Tatar, A., Antoniadis, P., de Amorim, M.D., Fdida, S.: From popularity prediction to ranking online news. Soc. Netw. Anal. Min. **4**(1), 174 (2014)
22. Van Canneyt, S., Leroux, P., Dhoedt, B., Demeester, T.: Modeling and predicting the popularity of online news based on temporal and content-related features. Multimed. Tools Appl. **77**, 1409–1436 (2018)
23. Zhao, J., Wu, W., Zhang, X., Qiang, Y., Liu, T., Wu, L.: A short-term prediction model of topic popularity on microblogs. In: Du, D.-Z., Zhang, G. (eds.) COCOON 2013. LNCS, vol. 7936, pp. 759–769. Springer, Heidelberg (2013). https://doi.org/10.1007/978-3-642-38768-5_69

Meta-learning of Text Classification Tasks

Jorge G. Madrid[1(✉)] and Hugo Jair Escalante[1,2]

[1] Computer Science Department, Instituto Nacional de Astrofísica,
Óptica y Electrónica, Tonantzintla, 72840 San Andrés Cholula, PUE, Mexico
{jgmadrid,hugojair}@inaoep.mx
[2] Computer Science Department, Centro de Investigación y Estudios Avanzados del
IPN, Zacatenco, 07360 Mexico city, Mexico

Abstract. A text mining characterization is proposed consisting of a set of meta-features, unlike previous meta-learning approaches, some of them are extracted directly from raw text. Such novel description is useful for comparing text mining tasks and study their differences. The problem of determining the task associated to a text classification dataset is introduced and approached with our characterization. Experimental results on a set of 81 corpora show that the proposed meta-features indeed allow to recognize tasks with acceptable performance using only a few meta-features.

Keywords: Meta-learning · Text classification · Meta-features

1 Introduction

For humans, experiences from the past are usually helpful when learning a new skill or solving a new problem. Equivalently, in the context of machine learning, meta-learning takes advantage of prior experience acquired when solving related tasks for approaching new problems [12]. The main goals are to speed up the learning process and to improve the quantitative performance of models. Meta-learning has had an impact into several machine learning problems such as learning to design optimization algorithms [1], automatically suggesting supervised learning pipelines [4], learning architectures for deep neural networks [3] and few-shot learning [10].

Text classification is one of the most studied tasks in NLP, this is because of the number of problems and applications that can be approached as text classification tasks. Many techniques for pre-processing, feature extraction, feature selection and document representation have been developed over the last decades. Each of these being appropriate for different scenarios and types of tasks. However, despite the progress achieved by the NLP community, nowadays

This work was partially supported by CONACyT under grant A1-S-26314 *Integración de Visión y Lenguaje mediante Representaciones Multimodales Aprendidas para Clasificación y Recuperación de Imágenes y Videos.*

I. Nyström et al. (Eds.): CIARP 2019, LNCS 11896, pp. 107–119, 2019.
https://doi.org/10.1007/978-3-030-33904-3_10

it is still an NLP expert who determines the pipeline of text classification systems, including preprocessing methods, representation and classification models together with their hyperparameters.

This paper takes a first step towards the characterization of text classification problems with the ultimate goal of suggesting text classification pipelines for any type of problem, that is *Meta-learning of text classification tasks*. Earlier work in this direction (see Sect. 2) has defined straightforward meta-features and worked over a small number of datasets. What is more, previous work has focused exclusively on tabular data (i.e., they have extracted meta-features from a document-text matrix). Since natural language presents different characteristics from those of generic tabular data, herein we define a set of meta-features that are derived from the analysis of raw text and combine them with traditional meta-features. To the best of our knowledge this the first work on meta-learning extracting information from raw text directly.

As a first approximation, we approach the problem of learning to determine the type of task (e.g., topic-based vs. sentiment analysis) using the meta-features as predictive variables. We provide empirical evidence on the suitability of the proposed meta-features for characterizing text classification tasks. Additionally, we perform an analysis of the most important features for the approached meta-learning problem. Experimental results are encouraging and show that meta-learning of text classification is a promising research venue for NLP.

Our contributions are threefold: (1) introduction of the task-type prediction problem; (2) introduction of novel and effective meta-features that can be used for other meta-learning tasks; (3) experiments of larger scale than previous work (we proposed 73 meta-features, compared to 11 from previous references and report experiments on 81 corpora, compared to 9 from related work).

2   Background and Related Work

Meta-learning aims to learn from prior learning-experience in order to speed up the learning process when approaching a new task. A common way to learn from/across tasks is by characterizing them with a set of *meta-features* [13]. These attempt to describe a task (i.e., a dataset) by information readily available at a task/dataset level. In this way, each task is usually represented by a vector where dimensions are associated to meta-features. Meta-features can be as simple as the number of instances and features in a dataset and as complex as statistical measures from the data distribution. [11] provide a comprehensive description of the most commonly used meta-features.

In the machine learning context, meta-learning has been studied for a while [12,13]. But it is only recently that it has become a mainstream topic, this mainly because of its successes in several tasks. For instance, Feurer et al. [5] successfully used a set of meta-features to warm-start a hyper-parameter optimization technique in the popular state-of-the-art AutoML solution *Autosklearn*. Likewise, the success of deep learning together with the difficulty in defining appropriate architectures and hyperparameters for users, has motivated a boom on neural architecture search, where meta-learning is common [3].

2.1 Meta-learning in Text Classification

In the context of text mining, meta-features from clustering text documents have been used directly for classification [2]. In the context of meta-learning these features have been used only in very specific domains [8]. Efforts dealing with generic datasets and closely related to the proposed research are reviewed in the remainder of this section.

Lam and Lai [7] introduced a meta-learning approach for text classification, they characterized subsets of the Reuters corpus with 8 *document-feature meta-characteristics* that were extracted from the document-term matrix representation. These consisted of simple meta-data such as the average document length or simple term statistics. These meta-features were later used to estimate the classification error of 6 classifiers and recommended a model depending on the prediction. Similarly, Gomez, et al. [6] proposed 11 meta-features which were also collected from a matrix representation of the documents, 9 different corpus were characterized with them. This method learned a set of rules that determine a suitable algorithm depending on the meta-feature values of the corpus.

Unlike previous approaches we do not assume a predefined representation of the documents, instead we derive meta-features from the raw text and combine these with traditional ones. This allows us to capture more language-relevant information. Also, we perform experiments of larger scale than previous work, considering 81 datasets (previous work used 6–9 collections) that have been characterized by 73 meta-features (in the past 8–11 meta-features have been considered).

3   Meta-learning Text Classification Tasks

We propose in this work a set of 73 meta-features with the aim of characterizing tasks (i.e., datasets), where the proposed meta-features comprise both, traditional and NLP-based ones. The ultimate goal of our work is to automatically suggest pipelines for solving text classification problems. As a first step in such direction, we show in this work that the proposed meta-features can be used as predictive variables to learn models able to recognize the type of task associated to a dataset. Different text classification tasks can be derived given the same dataset, our set of meta-features also acknowledges this since some of the proposed measures provide statistical information about the classes.

In NLP it is empirically known that certain methods work better according to the type of task that is aimed, for example, character-based n-grams are known to perform better than other representations in authorship attribution tasks because they determine better an author's style. Identifying correctly the type of task that is tackled is a fundamental step when modeling a text classification *pipeline*, thus we propose to automate this in pursuit of an automated recommendation system. In this work, we limit ourselves to learn to discriminate among types of tasks, and postpone to future work the problem of pipeline recommendation.

3.1 Proposed Meta-features

A common form of characterizing tasks are meta-features. Some sets of meta-features have proven to be useful for supervised machine learning problems, however we consider that these are not enough to characterize tasks in text classification; extracting them usually requires a tabular representation of the data, in the case of text documents some representation such as Bag-of-Words would be necessary. When a representation is selected some fundamental characteristics of language are lost, extracting *traditional* meta-features from it would result in a limited characterization of the task. We propose a set of 73 meta-features combining meta-learning traditional features with NLP ones. Below we organized them in groups.

- **General meta-features.** The *number of documents* and the *number of categories*.
- **Corpus hardness.** Most of these originally used in [9] to determine the hardness of short text-corpora.
  Domain broadness. Measures related to the thematic broadness/narrowness of words in documents. We included measures based on the vocabulary length and overlap: *Supervised Vocabulary Based (SVB)*, *Unsupervised Vocabulary Based (UVD)* and *Macro-averaged Relative Hardness (MRH)*.
  Class imbalance. Class Imbalance (CI) ratio.
  Stylometry. Stylometric Evaluation Measure (SEM)
  Shortness. Vocabulary Length (VL), Vocabulary Document Ratio (VDR) and average *word length*.
- **Statistical and information theoretic.** We derive meta-features from a document-term matrix representation of the corpus.
  min, max, average, standard deviation, skewness, kurtosis, ratio average-standard deviation, and entropy of: vocabulary distribution, documents-per-category and words-per-document:
  Landmarking. 70% of the documents are used to train 4 simple classifiers and their performance on the remaining 30% was used based on the intuition that some aspects of the dataset can be inferred: *data sparsity - 1NN, data separability - Decision Tree, linear separability - Linear Discriminant Analysis, feature independence Naïve Bayes*. The *percentage of zeros* in the matrix was also added as a measure for sparsity.
  Principal Components (PC) statistics. Statistics derived from a PC analysis: *pcac* from Gomez, et al. [6]; for the first 100 components, the same statistics from documents per category and their *singular values sum, explained ratio and explained variance*, and for the first component its *explained variance*.
- **Lexical features.** We incorporated the distribution of parts of speech tags. We intuitively believe that the frequency of some lexical items will be higher depending on the task associated to a corpus, for instance a corpus for sentiment analysis may have more adjectives while a news corpus may have less. We tagged the words in the document and computed the average number of *adjectives, adpositions, adverbs, conjunctions, articles, nouns, numerals, particles, pronouns, verbs, punctuation marks* and *untagged words* in the corpus.

- **Corpus readability.** Statistics from text that determine readability, complexity and grade from textstat library[1]: *Flesch reading ease:*

$$206.835 - 1.015 \left(\frac{total\_words}{total\_sentences} \right) - 84.6 \left(\frac{total\_syllables}{total\_words} \right)$$

SMOG grade:

$$1.043\sqrt{polysyllables \times \frac{30}{total\_sentences}} + 3.1291$$

Flesch-Kincaid grade level:

$$0.39 \left(\frac{total\_words}{total\_sentences} \right) + 11.8 \left(\frac{total\_syllables}{total\_words} \right) - 15.59$$

Coleman-Liau index:

$$0.0588L - 0.296S - 15.8$$

where L is the average number of letters per 100 words and S the average number of sentences per 100 words, *automated readability index:*

$$4.71 \left(\frac{total\_chars}{total\_words} \right) + 0.5 \left(\frac{total\_words}{total\_sentences} \right) - 21.43$$

Dale-Chall readability score:

$$0.1579 \left(\frac{difficult\_words}{total\_words} \right) + 0.0496 \left(\frac{total\_words}{total\_sentences} \right)$$

the number of difficult words, Linsear Write formula:

$$\frac{3(complex\_words) + (non\_complex\_words)}{2(total\_sentences)}$$

where complex words are those with more than 3 syllables *Gunning fog scale:*

$$0.4 \left(\frac{total\_words}{total\_sentences} \right) + 40 \left(\frac{complex\_words}{total\_words} \right)$$

and the *estimated school level to understand the text* that considers all the above tests.

Apart from general, statistical and PC based, the rest of the listed features have not been used in a meta-learning context.

---

[1] https://github.com/shivam5992/textstat.

Table 1. Meta-features identified as relevant after feature selection. We show the ranked features for each problem, in bold we show the features used for obtaining the results from Table 5.

Hate	Irony	Sentiment	Topics	Author	All 5 TASKS
number of categories	number of categories	dpc min	adverbs	dpc min	number of categories
dpc min	**dpc kurtosis**	numerals	**MRH**	**dpc min**	**dpc kurtosis**
Flesch reading ease	**adpositions**	SMOG	pronouns	**dpc max**	**dpc min**
dpc kurtosis	**wpd average**	unmarked	nouns	feature independence NB	**dpc entropy**
zeros in matrix	**Flesch reading ease**	**pca singular sum**	punctuation marks	number of documents	**MRH**
voc skewness	**zeros in matrix**	**pca explained variance**	dpc entropy	pca kurtosis	**adverbs**
dpc entropy	**readability index**	**adpositions**	number of categories	pca explained ratio	**adjectives**
pca explained variance	**Kincaid grade**	**pca max**	scholar grade	dpc entropy	**wpd average**
imbalance degree	**dpc min**	**number of categories**	SMOG	pcac	**Flesch reading ease**
voc kurtosis	**dpc skewness**	**wpd average**	data separability DT	pca explained variance	**pca explained variance**
	Linsear	**Gunning**			**zeros in matrix**
	MRH	**Linsear**			**SMOG**
	pca singular sum	**zeros in matrix**			
	voc kurtosis	**Articles**			
	pca min	**Flesch reading ease**			
		dpc skewness			
		MRH			
		adverbs			
		Coleman Liau			
		number of documents			
		nouns			
		wpd entropy			
		pca explained variance			
		conjuctions			
		dpc entropy			

3.2 Datasets

For the extraction of the meta-features and the experimental evaluation we collected 81 text corpora associated to different problems. We associated each corpus with a task-type-label according to the associated classification problem, where the considered labels were: *authorship analysis, sentiment analysis, topic/thematic tasks, irony* and *hate speech detection.* Table 2 illustrates the distribution of the datasets as labeled by their task.

Table 2. Tasks by their type.

Type-task	Frequency	Avg. documents	Avg. classes
Topic	16	93,797($\pm$191,833)	15.81($\pm$20)
Author	13	10,490($\pm$15,790)	12.31($\pm$14)
Irony	7	13,579($\pm$10,372)	2.00($\pm$0)
Sentiment	39	362,660($\pm$905156)	2.95($\pm$2)
Hate	6	14,969($\pm$9134)	2.33($\pm$1)

The full list of datasets is available in the appendix material. Some of the considered datasets are well known benchmarks (e.g. Yelp) while the rest can be found in competition sites like Kaggle and SemEval. After pre-processing each corpus to share the same format and codification, we extracted the 73 meta-features for each of the 81 collections and we assigned a task type label to each dataset according to the associated classification problem. To accelerate the feature extraction process we limited the number of documents to 90,000 for each collection, where these were randomly sampled from the categories of the corpus. The resultant matrix of size 81×73 comprises our *knowledge base* characterizing multiple corpora.

3.3 Meta-learning of Task Labels

We approached the problem of recognizing the classification task of a dataset by using the proposed meta-features. We studied the prediction problem as both multiclass (predicting one of the 5 task labels) and binary (distinguishing one label from the rest at a time) classification problems. The following classifiers were considered for the evaluation: Random Forest (RF), XGBoost (XG), Support Vector Machines and 1NN.

4 Experiments

For the evaluation we adopted a leave-one-out cross-validation: 80 tasks were used for training and 1 for testing, repeating this process 81 times, each time changing the test task; the average performance over the 81 folds is reported. As evaluation measures we report accuracy and f_1 measure for the positive class; in the case of the multiclass problem average accuracy and Macro-f_1 are reported.

Table 3. Task prediction results with 73 meta-features

	Accuracy		f_1	
Task/model	XG	RF	XG	RF
Hate	0.94	0.94	0.29	0.29
Irony	**0.95**	0.93	**0.67**	0.25
Sentiment	**0.89**	0.85	**0.83**	0.77
Topics	0.86	**0.89**	0.62	**0.64**
Author	**0.90**	0.89	**0.60**	0.52
All 5-tasks	**0.77**	0.75	**0.64**	0.59

Table 3 shows the results obtained by the 2-best performing classifiers (XG and RF). Table 4 shows the results of experiments with different classifiers. It can be seen that performance for all of the tasks is greater than random guessing. The high accuracy contrasted by moderate f_1 values reveals the models are favouring the majority class. In fact, high imbalance makes prediction quite difficult, specially for the *hate* and *irony detection* tasks where there are 6 and 7 positive examples, respectively.

Table 4. Task prediction f1 score for different classification models.

	f_1			
Task/model	XG	1NN	SVM	RF
Hate	0.29	**0.36**	0.23	0.29
Irony	**0.67**	0.50	0.35	0.25
Sentiment	**0.83**	0.60	0.39	0.77
Topics	0.62	0.30	0.53	**0.64**
Author	**0.60**	0.38	0.33	0.53
All 5-tasks	**0.64**	0.43	0.32	0.59

An additional experiment involved a feature selection process prior to the classification stage. Mutual information was used to select the top K features and used for training and predicting. Table 5 shows the best performance obtained when performing feature selection together with the number of meta-features selected. It can be observed that there is a performance improvement after the selection of meta-features in all binary cases. Improvements are dramatic in terms of the f_1 measure in some cases (e.g., *Hate, Topics, Author*). Surprisingly, for some problems only few meta-features were required to achieve better performance, see, e.g., *Hate*. For the multiclass problem meta-feature selection did not improve the initial results on either evaluation measures.

Table 5. Results with meta-feature selection

Task	Model	K	Accuracy	f_1
Hate	RF	2	0.94 (+0%)	0.55 (+89.6%)
Irony	XG	15	0.96 (+1%)	0.73 (+8.9%)
Sentiment	XG	24	0.90 (+1%)	0.85 (+2.4%)
Topics	RF	3	0.90 (+1%)	0.75 (+17.1%)
Author	RF	3	0.91 (+1%)	0.70 (+16.6%)
5 tasks	XG	12	0.70 (−7%)	0.64 (+0%)

Table 1 shows the complete subsets of features considered for obtaining the results from Table 5. Meta-features are ordered by their mutual information values. It is hard to find a common pattern but we found that some features are part of almost every subset: the *percentage of adverbs,* the *number of categories,* vocabulary overlapping in classes: *MRH,* and some statistic of *documents per category.* Hence showing the importance of the novel meta-features extracted from raw text. For hate detection and authorship analysis simple statistical measures appear to be better to describe the corpora, for the rest of the tasks the subsets that improved the original performance include a wide variety of meta-features from the groups presented in Sect. 3 (Fig. 1).

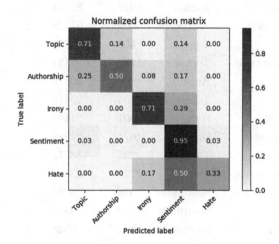

Fig. 1. Normalized confusion matrix of predicting all 5-tasks with XG.

5 Conclusions

We introduced the problem of automatically predicting the type of text classification tasks from meta-features derived from text. A set of 73 meta-features have been proposed and evaluated in 81 data sets associated to 5 types of tasks. Experimental results demonstrate that the proposed meta-features entail discriminative information that could be useful for other meta-learning tasks. Results of a meta-feature selection analysis showed that *traditional* meta-features are not good enough to characterize datasets by themselves, proving the effectiveness of the newly introduced ones. This paper comprises the first steps in trying to meta-learn from raw text directly, we foresee our work will pave the way for the establishment of meta-learning in NLP.

A Appendix

See Tables 6 and 7

Table 6. List of datasets.

Name	Task	# of docs	Voc size	# of classes
20 Newsgroups	Topics	18828	229710	20
Women's reviews	Author	23473	15153	8
Amazon cellphones	Sentiment	999	2241	2
Every song	Author	20779	48752	40
authorship_poetry	Author	200	9141	6
SouthPark episodes	Author	11953	14068	5
Spanish songs	Author	3947	35571	23
Bias Politics	Sentiment	5000	21328	2
Brown	Topics	500	48778	15
Progressive tweets	Topics	1159	5491	4
ccat	Author	1000	20416	10
Classic	Topics	7095	29518	4
Cyber trolls	Hate	20001	21193	2
Davidson hate	Hate	24678	24289	2
BBC News	Topics	2225	33771	5
BBC Summaries	Topics	2225	22921	5
Doctor deception	Sentiment	556	4453	2
Op_spam-	Sentiment	800	8819	2
Op_spam+	Sentiment	800	6548	2
Restauran reviews	Sentiment	400	5353	2
Deflate	Sentiment	11786	25616	5
Gender-microblog	Author	781	2439	2
Gender-twitter	Author	19953	50910	4
Imperium	Hate	6593	28031	2

(*continued*)

Table 6. (*continued*)

Name	Task	# of docs	Voc size	# of classes
Hate tweets	Hate	24783	41639	3
Iro-eduReyes	Irony	20000	32714	2
Iro-humReyes	Irony	19870	30485	2
iro-mohammad	Irony	1929	6040	2
Iro-polReyes	Irony	20000	31882	2
iro-riloff	Irony	2080	6132	2
Iro-semeval18	Irony	4466	10906	2
Kaggle hate	Hate	6594	25646	2
Machado	Topics	246	79461	8
Hate-Malmasi	Hate	7162	14456	3

Table 7. List of datasets.

Name	Task	# of docs	Voc size	# of classes
masc_tagged	Topics	389	43234	20
Medium papers	Topics	185	530	3
Movie reviews	Sentiment	2000	39768	2
polarity	Sentiment	1386	36614	2
Politic	Topics	5000	21328	9
Pros cons	Sentiment	45875	14015	2
Women's clothing	Sentiment	23486	15160	5
rawdata_cric	Author	158	13787	4
rawdata_fin	Author	175	15517	6
rawdata_nfl	Author	97	8940	3
rawdata_travel	Author	172	15560	4
Recommendations	Sentiment	23486	15160	2
Relevance economic news	Sentiment	8000	53162	3
Relevance short news	Sentiment	5007	20111	3
Reuters	Topics	13328	41600	84
Sarcasm Headlines	Irony	26709	25437	2
IMDB short	Sentiment	748	3401	2
Sent-semeval16	Sentiment	30631	36451	3
sent-semevalSA	Sentiment	6999	18042	3
Twitter-airline	Sentiment	14640	18614	3

(*continued*)

Table 7. (*continued*)

Name	Task	# of docs	Voc size	# of classes
Twitter-self-dirve	Sentiment	7156	18017	6
Short yelp	Sentiment	1000	2379	2
Sharktank	Sentiment	706	5175	2
smsspam	Sentiment	310	1610	2
Socialmedia disaster	Sentiment	10860	33768	2
Starter test	Sentiment	10876	33606	3
subjectivity	Sentiment	10000	21001	2
Tripadvisor reviews	Sentiment	17223	32423	5
Sentences polarity	Sentiment	10662	18408	2
Yahoo answers	Sentiment	1459998	180241	10
YouTube	Sentiment	1956	5929	2
Yelp	Sentiment	699998	125757	5
Ag News	Topics	127598	64504	4
Kickstarter	Sentiment	215513	81252	2
News_Categories	Topics	124989	37183	30
Ohsumed	Topics	56984	79479	23
Short Amazon	Sentiment	568454	68831	5
Amazon	Sentiment	3649998	139289	5
sarcasm	Sentiment	1010826	62765	2
Amazon B	Sentiment	3999998	138968	2
Sentiment140	Sentiment	1600000	93115	2
Semeval17	Sentiment	62618	62304	3
Yelp B	Sentiment	597998	113897	2
Sogou news	Topics	509998	42991	5
Dbpedia	Topics	629998	199912	14
Victorian authorship	Author	53678	9977	45
Stanford	Sentiment	25000	95550	2

References

1. Andrychowicz, M., et al.: Learning to learn by gradient descent by gradient descent. In: Advances in Neural Information Processing Systems, pp. 3981–3989 (2016)
2. Canuto, S., Sousa, D.X., Gonçalves, M.A., Rosa, T.C.: A thorough evaluation of distance-based meta-features for automated text classification. IEEE Trans. Knowl. Data Eng. **30**(12), 2242–2256 (2018)
3. Elsken, T., Metzen, J.H., Hutter, F.: Neural architecture search: a survey. arXiv preprint arXiv:1808.05377 (2018)

4. Feurer, M., Klein, A., Eggensperger, K., Springenberg, J., Blum, M., Hutter, F.: Efficient and robust automated machine learning. In: Advances in Neural Information Processing Systems, pp. 2962–2970 (2015)
5. Feurer, M., Springenberg, J.T., Hutter, F.: Initializing bayesian hyperparameter optimization via meta-learning. In: Twenty-Ninth AAAI Conference on Artificial Intelligence (2015)
6. Gomez, J.C., Hoskens, S., Moens, M.F.: Evolutionary learning of meta-rules for text classification. In: Proceedings of the Genetic and Evolutionary Computation Conference Companion, pp. 131–132. ACM (2017)
7. Lam, W., Lai, K.Y.: A meta-learning approach for text categorization. In: Proceedings of the 24th Annual International ACM SIGIR Conference on Research and Development in Information Retrieval, pp. 303–309. ACM (2001)
8. Lee, S.I., Chatalbashev, V., Vickrey, D., Koller, D.: Learning a meta-level prior for feature relevance from multiple related tasks. In: Proceedings of the 24th International Conference on Machine Learning, pp. 489–496. ACM (2007)
9. Pinto, D.: On clustering and evaluation of narrow domain short-text corpora. Ph.D. UPV (2008)
10. Ravi, S., Larochelle, H.: Optimization as a model for few-shot learning. In: 5th International Conference on Learning Representations, ICLR 2017, Toulon, France, 24–26 April 2017, Conference Track Proceedings. OpenReview.net (2017). https://openreview.net/forum?id=rJY0-Kcll
11. Rivolli, A., Garcia, L.P., Soares, C., Vanschoren, J., de Carvalho, A.C.: Towards reproducible empirical research in meta-learning. arXiv preprint arXiv:1808.10406 (2018)
12. Vanschoren, J.: Meta-learning. In: Hutter, F., Kotthoff, L., Vanschoren, J. (eds.) Automated Machine Learning. TSSCML, pp. 35–61. Springer, Cham (2019). https://doi.org/10.1007/978-3-030-05318-5_2. http://automl.org/book
13. Vilalta, R., Drissi, Y.: A perspective view and survey of meta-learning. Artif. Intell. Rev. 18(2), 77–95 (2002)

Portuguese POS Tagging Using BLSTM Without Handcrafted Features

Rômulo César Costa de Sousa and Hélio Lopes[✉]

Pontifícia Universidade Católica do Rio de Janeiro, Rio de Janeiro, RJ, Brazil
{rsousa,lopes}@inf.puc-rio.br

Abstract. Training state-of-the-art Part-of-speech (POS) taggers traditionally requires many handcraft features and external data. In this paper, we propose a neural network architecture for POS tagging task for both contemporary and historical Portuguese texts. The proposed architecture does not use the two traditional requirements cited above. It uses word embeddings and character embeddings representations combined with a BLSTM layer. We apply the architecture on three Portuguese corpora and obtaining state-of-the-art accuracy of 97.87% on the Mac-Morpho corpus, 97.62% accuracy on the revised Mac-Morpho and 97.36% on Tycho Brahe. We also improve the tagging accuracy for Out of Vocabulary (OOV) words in the Mac-Morpho corpus and in the revised Mac-Morpho.

Keywords: Part-of-speech tagging · Deep learning · Word embeddings

1 Introduction

Part-of-speech (POS) tagging is a process of labeling each word in a sentence with a morphosyntactic class (verb, noun, adjective and etc). POS tagging is considered a hard task to perform due to the fact that some words could have more than one class, depending on the context that it is used. POS tagging is a fundamental part of the linguistic pipeline [9], most natural language processing (NLP) applications demand, at some step, part-of-speech information [8]. For example, its use can be found in sentiment analysis [4], in machine translation [15] and question answering [27].

Several works have been done to solve this task and they push the state-of-art accuracy to more than 97%. But still, have room for improvement since any small gain can generate an impact on other NLP tasks. Recently, neural network approaches have become very used to solve NLP problems [12] and word embeddings, also have been applied with a great success [16].

In [18], the authors build an LSTM neural network that extracted character embeddings information for using in POS tagging classification. They tested the model on four different languages. In [24], the authors explore the problem of using a deep neural network that uses a convolutional layer to learn character level representation of words, and achieve great results on three Portuguese corpora.

© Springer Nature Switzerland AG 2019
I. Nyström et al. (Eds.): CIARP 2019, LNCS 11896, pp. 120–130, 2019.
https://doi.org/10.1007/978-3-030-33904-3_11

In this work, we propose the use of a neural network architecture that is effective to solve the Portuguese POS tagging task. More precisely, our proposal combines BLSTM with pre-trained word embeddings and character embeddings. We evaluate our model on three Portuguese corpora: the original Mac-Morpho corpus [2], the revised Mac-Morpho [7] and Tycho Brahe [25] corpus. We outperforms previous state-of-the-art results on these corpora, obtaining 97.87% accuracy on the Mac-Morpho corpus, 97.62% accuracy on the revised Mac-Morpho and 97.36% on Tycho Brahe.

The structure of the paper is as follows. Section 2 discusses related work. Section 3 introduces the neural network architecture proposal. Section 4 presents the experiments and the results. Finally, Sect. 5 gives a conclusion and points out our next steps.

2   Related Work

There has been many efforts to build Portuguese POS tagging. In [24], authors built a system without any handcrafted features, using a neural network. They employ a convolutional layer that allows effective feature extraction from words of any size. They combine this representation along with word embeddings to perform POS tagging. They evaluate on three Portuguese corpora: Mac-Morpho-v1, Mac-Morpho-v2 and Tycho Brahe corpus.

In [5], authors used a Large Margin Structured Perceptron algorithm to solve POS tagging. They used a small set (four) of handcraft features to improve their performance. They empirically evaluate their system on the two versions of the Portuguese Mac-Morpho corpus.

Fonseca et al. [6] used a multilayer perceptron neural network for training a POS tagger. This neural network receives word embedding information and handcraft features such as the presence of capital letters and word endings. The authors report state-of-the-art performance on Portuguese corpora. With 97.57% overall accuracy on the Mac-Morpho-v1 corpus, 97.48% on Mac-Morpho-v2, and 97.33% on Mac-Morpho-v3 presented in their work.

There are also several other neural networks proposed for POS tagging in other languages. Wang et al. [26] proposed to use a BLSTM neural network with word embedding for POS tagging with an English corpus. In [23], authors used a BLSTM neural network for POS tagging with an auxiliary loss function that accounts for rare words. They evaluated the neural network across 22 languages.

In [16], Ma and Hovy propose a neural network to solve POS tagging and named entity recognition (NER). They constructed a model by feeding character embeddings and word embeddings to a BLSTM on top of a CRF (Conditional Random Fields) layer. They used a Convolution Neural Network (CNN) for extracting character embeddings representations of words. And evaluate the model on the English Penn Treebank WSJ corpus. There are four main differences between their model and what we are to propose here. We did not employ CNNs to extracting character embeddings information, instead we use a BLSTM

layer. We did not use CRF layer in the end of the model. We use word embeddings information provided by Wang2vec instead of GloVe. And they apply their
method for English language, while we are focused on Portuguese.

The work of [20] is what most resembles our neural network proposal. The
authors used a BLSTM layer fed by character and word embeddings information
to perform POS Tagging. They used a BLSTM for extracting character embeddings representations of words. Our work mainly differs from the model proposed
since we use Wang2vec instead of FastText and they apply their method to Italian, and we apply it to Portuguese.

3   Neural Network Architecture Proposal

In this section, we describe our proposal for the neural network architecture,
which consists in a BLSTM architecture combined with word embeddings and
character embeddings. Figure 1 illustrates our proposal.

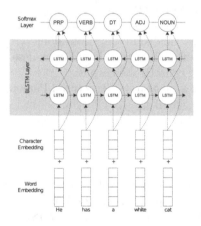

Fig. 1. Neural network architecture for POS tagging

Word-Level Representation. Word embeddings have become commonly used
in modern NLP systems [10], this technique represents each word as a vector with
real numbers in an d-dimensional space. This allows words with similar meaning
to have a similar representation. Word embeddings capture both the semantic
and syntactic information of words [13]. This representation have been applied
with a great success [16] in many NLP taks.

In this work, we used pre-trained word embedding models provided by [13].
They collected a large corpus from several sources in order to obtain a multi-
genre corpus, representative of the Portuguese language. Seventeen different corpora were used, totaling 1,395,926,282 tokens. They training the word embedding
models in algorithms such as Word2vec [21], FastText [3], Wang2vec [14] and

Glove [22]. We tested which algorithm for word embedding is better fit in our neural network.

Character-Level Representation. Morphological information of words can be helpful in POS tagging classification. Suffixes could indicate a word class. For example, the suffix like "ly" in "quietly" indicate an adverb, and a capital letter could suggest that this word is a noun. But, handcrafted this kind of features is costly to develop [17], and make this hard to adapt to other domains or languages. Previous studies have shown that neural networks are a powerful way to extract automatically morphological information. In [24], the authors reached state-of-art results in Portuguese for POS tagging using a convolutional layer to extract this information. And [20] exhibit positive results in Italian for POS tagging by using BLSTM for extract morphological information.

For this subtask, we used a model proposed in [19] that applied a BLSTM layer to produce character embeddings representation. Differently from word embeddings, which are able to capture syntactic and semantic information, character embeddings can capture intra-word morphological and shape information [20]. We define a character's vocabulary that contains all uppercase and lower-case letter as well as numbers and punctuation present on data. Given a word w as an input, decomposed in m characters $\{c_1, c_2, ..., c_m\}$, where m is the length of w. Each c_i is encoded as a one hot vector, with one on the index of c_i in vocabulary. The representation of the word w is obtained by combining the forward and backward states from the BLSTM layer.

Network Architecture. Long short-term memory (LSTM) proposed by [11], are a variant of RNNs. This kind of neural network is well known by their power of capturing long-term dependencies, and have been widely used for sequence labeling tasks. LSTM maintain previous information using memory cells. For example, it takes a sequence of vectors, $(x_1, x_2, ..., x_n)$ as input and produces another sequence $(h_1, h_2, ..., h_n)$ as output [1]. The LSTM captures just previous information and has no knowledge of what comes next, but for many sequence labeling tasks, it is helpful to have access to both past and future contexts [16]. Bi-directional LSTM (BLSTM) is a solution for filling this gap. BLSTM take information in two separate hidden states one to the past and other to the future and then concatenated the two separate hidden as one final state.

Given a decomposed sentence with n words, $w_1, w_2, w_3, ..., w_n$, and n tags, $t_1, t_2, t_3, ..., t_n$, we use the BLSTM main layer to predict the tag probability distribution of each word. As shown in Fig. 1, to perform this prediction, we firstly represent each word as a vector, which is the result of the concatenation of the word embeddings representation and the character embeddings representation. In this way, we can capture semantic and syntactic information (word embeddings) and morphological information (character embeddings) of the words.

4 Experiments

In this section, we provide details about training the neural network and discusses our results. We implemented our model using Keras[1] on top of Tensor-Flow[2]. In the output layer, we use a softmax activation function.

Corpora. We evaluate the model on three different Portuguese corpora: the original Mac-Morpho (v1) corpus [2], a revised version of Mac-Morpho (v2) [7] and Tycho Brahe corpus [25]. The Mac-Morpho is a large manually POS-tagged corpus Portuguese, collected from Brazilian newspaper articles [2]. The original version has 53,374 sentences and 41 morphosyntactic class, as shown in Table 1, and the revised version has 49,900 sentences and 30 morphosyntactic class. The Tycho Brahe Corpus is composed of historical Portuguese with texts written in Portuguese by authors born between 1380 and 1881. We use the same train/development/train split as [24] and [6] in order to directly compare results.

Table 1. Corpus splits

Corpus		Train	Development	Test	Total	Tagset
Mac-Morpho v1	Sentences	42,022	2,211	9,141	53,374	41
	Tokens	957,439	50,232	213,794	1,221,465	
Mac-Morpho v2	Sentences	42,742	2,249	4,999	49,990	30
	Tokens	807,796	43,141	94,991	945,928	
Tycho Brahe	Sentences	29,162	1,534	10,233	40,929	265
	Tokens	734,889	40,673	259,913	1,035,475	

Hyperparameters. We use the development sets to tune the neural network hyperparameters. First, we examine the size of the hidden layer on the main BLSTM. This feature has a limited impact on results. We also analyze word embedding dimensions, character embedding dimensions and dropout rate on the main BLSTM layer. We used the same set of hyperpameters shown on Table 2 for all corpus and experiments.

Table 2. Hyper-parameters for all experiments

Parameter	Value
Hidden Layer (Main BLSTM)	512
Word embedding dimensions	100
Char. embedding dimensions	50
Dropout (Main BLSTM)	0.6

---

[1] https://keras.io/.
[2] https://www.tensorflow.org/.

Results. Firstly, we test the performance of our model on four different methods to pre-train word embedding. We used as metrics overall accuracy (ALL) and Out of Vocabulary (OOV) accuracy. We run experiments with Word2vec, Fasttext, Wang2vec and Glove. As presented in Table 3, all word embeddings generate higher accuracy. However, Wang2vec produce the highest overall accuracy in our model on all corpora. Wang2vec's good performance may be explained by its focus in capture syntactic information [14], this is really useful for POS tagging classification.

Table 3. Performance of our model with different word embeddings on the development set

Corpus	Embedding	Accuracy (%)
Mac-Morpho v1	Word2vec	97.84
	Fasttext	97.78
	Wang2vec	**97.97**
	Glove	97.70
Mac-Morpho v2	Word2vec	97.56
	Fasttext	97.44
	Wang2vec	**97.67**
	Glove	97.53
Tycho Brahe	Word2vec	96.93
	Fasttext	96.94
	Wang2vec	**97.17**
	Glove	97.11

In order to measure the impact of word embeddings information and character embeddings, we compare the performance of our model with two baseline systems. Table 4 shown the accuracy and OOV accuracy for the three systems in the three corpora evaluated. The first baseline (Perceptron-handcrafted-features) uses a multilayer perceptron with handcraft features. We utilize as features the presence of capital letters, word suffixes, word prefixes, previous three tokens, and the next three tokens. This model does not use any word embeddings or character embeddings information. The second baseline (BLSTM-WE) was built similar to our main model, but without any character embeddings information, just word embeddings vectors. And the system BLSTM-WE-CE is our model describe in Sect. 3. Both BLSTM-WE and BLSTM-WE-CE use Wang2vec word embeddings and the same hyperparameters as shown in Table 2.

According to the results shown Table 4, BLSTM-WE performed better than Perceptron-handcrafted-features on Mac-Morpho v1 and Mac-Morpho v2. Since Tycho Brahe is formed by history of Portuguese text, many words are not present in the pre-trained word embeddings, and this corpus could not benefit like the other by using BLSTM-WE. BLSTM-WE-CE system outperforms the BLSTM-WE on the three corpora, especially on OOV accuracy. This result support what already has been demonstrated by [24] and [16], that character embeddings are important for linguistic sequence labeling tasks like POS tagging.

Table 4. Performance of our model and two baseline on the test set

Corpus	System	All (%)	OOV (%)
Mac-Morpho v1	Perceptron-handcrafted features	96.31	88.45
	BLSTM-WE	96.81	90.33
	BLSTM-WE-CE	97.87	95.44
Mac-Morpho v2	Perceptron-handcrafted features	96.28	88.56
	BLSTM-WE	96.58	90.57
	BLSTM-WE-CE	97.62	94.84
Tycho Brahe	Perceptron-handcrafted features	96.04	75.72
	BLSTM-WE	95.88	74.01
	BLSTM-WE-CE	97.36	85.74

Comparisons. We compare our results with two top performance systems. In Table 5, we compare overall accuracy and OOV accuracy on the three corpora. Our system outperforms the previously best systems by improving in 0.23% overall accuracy for the Mac-Morpho v1, 0.31% for Mac-Morpho v2 and 0.19% for Tycho Brahe corpus. We also improve OOV accuracy in 3.18% for the Mac-Morpho v1 and 1.50% in Mac-Morpho v2. These results define a new state-of-the-art of this three corpora and demonstrate effectiveness of using BLSTM in Portuguese POS tagging.

Table 5. Comparison with Portuguese POS taggers

Corpus	Embedding	All (%)	OOV (%)
Mac-Morpho v1	**This work**	**97.87**	**95.44**
	[6]	97.57	93.38
	[24]	97.47	92.49
Mac-Morpho v2	**This work**	**97.62**	**94.84**
	[6]	97.48	94.34
	[24]	97.31	93.43
Tycho Brahe	**This work**	**97.36**	85.74
	[6]	96.91	84.14
	[24]	97.17	**86.58**

Table 6. F_1 per tags in Mac-Morpho v1

Class	Meaning	F_1(%)
CUR	Currency	100.0
ART	Article	99.0
PREP	Preposition	99.0
V	Verb	99.0
KC	Coor. conjunction	98.0
N	Noun	98.0
PROADJ	Adjective pronoun	98.0
PROPESS	Personal pronoun	98.0
NPROP	Proper noun	97.0
NUM	Number	97.0
PCP	Participle	97.0
VAUX	Auxiliary verb	96.0
ADJ	Adjective	95.0
PRO-KS-REL	Sub. c. r. pronoun	94.0
ADV	Adverb	93.0
KS	Sub. conjunction	91.0
PDEN	Denotative word	91.0
PROSUB	Nominal pronoun	89.0
ADV-KS-REL	Sub. rel. adverb	86.0
PRO-KS	Sub. con. pronoun	75.0
ADV-KS	Sub. con. adverb	55.0
IN	Interjection	52.0

Table 7. F_1 per tags in Mac-Morpho v2

Class	Meaning	F_1(%)
CUR	Currency	100.0
PREP+PRO-KS-REL	Prep. sub. c. r. pronoun	100.0
PU	Punctuation	100.0
ART	Article	99.0
PREP+ART	Preposition + article	99.0
PREP+PROADJ	Prep. + adjective pronoun	99.0
PREP+PROPESS	Prep. + personal pronouns	99.0
PROPESS	Personal pronouns	99.0
V	Verb	99.0
KC	Coor. conjunction	98.0
N	Noun	98.0
PREP	Preposition	98.0
NPROP	Proper noun	97.0
NUM	Number	97.0
PCP	Participle	97.0
PROADJ	Adjective pronoun	97.0
VAUX	Auxiliary verb	96.0
ADJ	Adjective	95.0
ADV	Adverb	93.0
PREP+ADV	Preposition + adverb	92.0
PRO-KS-REL	Sub. c. r. pronoun	92.0
KS	Sub. conjunction	91.0
PROSUB	Nominal pronoun	91.0
PDEN	Denotative word	90.0
ADV-KS-REL	Sub. rel. adverb	86.0
PREP+PROSUB		83.0
PRO-KS	Sub. con. pronoun	76.0
IN	Interjection	73.0
ADV-KS	Sub. con. adverb	71.0
PREP+PRO-KS	Prep. + s. c. pronoun	00.0

Performance Per Tags. We used F1-score $F_1 = 2 * Precision * Recall/ Precision + Recall$ to evaluated our best model (BLSTM-WE-CE). We analyzed the tag-wise performance obtained by BLSTM-WE-CE in the test set for the Mac-Morpho v1 and the Mac-Morpho v2.

Table 6 shows F_1 scores per tag on the Mac-Morpho v1, only non punctuation tags were included. Most of the tags reached of more than 90% of F_1. The worst results are for IN (Interjection) with 52% and for ADV-KS (Subordinating connective adverb) with 55%. These two tags have a low representation on Mac-Morpho v1 corpus, IN represents just 0.034% of the tokens and ADV-KS just 0.032%. Due to their poor distributions, the model could not learn to classify these classes properly.

In Table 7 we have the F_1 scores per tag on the Mac-Morpho v2. Most of the tags reached more than 90% of F_1. The model completely failed to predict for PREP+PRO-KS (Preposition + subordinating connective pronoun), the system is not learning anything due to their poor distributions. PREP+PRO-KS have the lowest representation on the dataset just 0.033%.

5   Conclusions

In this study, We have empirically investigated a new approach that does not require handcrafted features to deal with Portuguese POS tagging task. We show that BLSTM with word embeddings and character embeddings gives superior performance for Portuguese POS tagging reaching state-of-art in overall accuracy in Mac-Morpho v1, Mac-Morpho v2, and Tycho Brahe corpus. As future work, we intend to extend this architecture to solve another natural language processing tasks in Portuguese, such as named entity recognition.

References

1. Alam, F., Chowdhury, S.A., Noori, S.R.H.: Bidirectional LSTMs-CRFs networks for bangla POS tagging. In: 2016 19th International Conference on Computer and Information Technology (ICCIT), pp. 377–382. IEEE (2016)
2. Aluísio, S., Pelizzoni, J., Marchi, A.R., de Oliveira, L., Manenti, R., Marquiafável, V.: An account of the challenge of tagging a reference corpus for Brazilian Portuguese. In: Mamede, N.J., Trancoso, I., Baptista, J., das Graças Volpe Nunes, M. (eds.) PROPOR 2003. LNCS (LNAI), vol. 2721, pp. 110–117. Springer, Heidelberg (2003). https://doi.org/10.1007/3-540-45011-4_17
3. Bojanowski, P., Grave, E., Joulin, A., Mikolov, T.: Enriching word vectors with subword information. Trans. Assoc. Comput. Linguist. **5**, 135–146 (2017). http://aclweb.org/anthology/Q17-1010
4. Das, O., Balabantaray, R.C.: Sentiment analysis of movie reviews using POS tags and term frequencies. Int. J. Comput. Appl. **96**(25), 36–41 (2014)
5. Fernandes, E.R., Rodrigues, I.M., Milidiú, R.L.: Portuguese part-of-speech tagging with large margin structure learning. In: 2014 Brazilian Conference on Intelligent Systems (BRACIS), pp. 25–30. IEEE (2014)
6. Fonseca, E.R., Rosa, J.L.G., Aluísio, S.M.: Evaluating word embeddings and a revised corpus for part-of-speech tagging in portuguese. J. Braz. Comput. Soc. **21**(1), 2 (2015)
7. Fonseca, E.R., Rosa, J.L.G.: Mac-Morpho revisited: towards robust part-of-speech tagging. In: Proceedings of the 9th Brazilian Symposium in Information and Human Language Technology (2013)
8. Giménez, J., Marquez, L.: SVMtool: a general POS tagger generator based on support vector machines. In: Proceedings of the 4th International Conference on Language Resources and Evaluation. Citeseer (2004)
9. Gimpel, K., et al.: Part-of-speech tagging for twitter: annotation, features, and experiments. In: Proceedings of the 49th Annual Meeting of the Association for Computational Linguistics: Human Language Technologies: short papers-Volume 2, pp. 42–47. Association for Computational Linguistics (2011)
10. Hartmann, N., Fonseca, E., Shulby, C., Treviso, M., Silva, J., Aluísio, S.: Portuguese word embeddings: evaluating on word analogies and natural language tasks. In: Proceedings of the 11th Brazilian Symposium in Information and Human Language Technology, pp. 122–131 (2017)

11. Hochreiter, S., Schmidhuber, J.: Long short-term memory. Neural Comput. **9**(8), 1735–1780 (1997)
12. Jung, S., Lee, C., Hwang, H.: End-to-end Korean part-of-speech tagging using copying mechanism. ACM Trans. Asian Low-Resour. Lang. Inf. Process. (TALLIP) **17**(3), 19 (2018)
13. Lai, S., Liu, K., He, S., Zhao, J.: How to generate a good word embedding. IEEE Intell. Syst. **31**(6), 5–14 (2016)
14. Ling, W., Dyer, C., Black, A.W., Trancoso, I.: Two/too simple adaptations of word2vec for syntax problems. In: Proceedings of the 2015 Conference of the North American Chapter of the Association for Computational Linguistics: Human Language Technologies, pp. 1299–1304. Association for Computational Linguistics (2015). https://doi.org/10.3115/v1/N15-1142. http://aclweb.org/anthology/N15-1142
15. Ma, J., Liu, H., Huang, D., Sheng, W.: An English part-of-speech tagger for machine translation in business domain. In: 2011 7th International Conference on Natural Language Processing and Knowledge Engineering (NLP-KE), pp. 183–189. IEEE (2011)
16. Ma, X., Hovy, E.: End-to-end sequence labeling via bi-directional LSTM-CNNs-CRF. In: Proceedings of the 54th Annual Meeting of the Association for Computational Linguistics (Volume 1: Long Papers), vol. 1, pp. 1064–1074 (2016)
17. Ma, X., Xia, F.: Unsupervised dependency parsing with transferring distribution via parallel guidance and entropy regularization. In: Proceedings of the 52nd Annual Meeting of the Association for Computational Linguistics (Volume 1: Long Papers), vol. 1, pp. 1337–1348 (2014)
18. Makazhanov, A., Yessenbayev, Z.: Character-based feature extraction with LSTM networks for POS-tagging task. In: 2016 IEEE 10th International Conference on Application of Information and Communication Technologies (AICT), pp. 1–5. IEEE (2016)
19. Marujo, W.L.T.L.L., Astudillo, R.F.: Finding function in form: Compositional character models for open vocabulary word representation (2015)
20. Marulli, F., Pota, M., Esposito, M.: A comparison of character and word embeddings in bidirectional LSTMs for POS tagging in Italian. In: De Pietro, G., Gallo, L., Howlett, R.J., Jain, L.C., Vlacic, L. (eds.) KES-IIMSS-18 2018. SIST, vol. 98, pp. 14–23. Springer, Cham (2019). https://doi.org/10.1007/978-3-319-92231-7_2
21. Mikolov, T., Chen, K., Corrado, G.S., Dean, J.: Efficient estimation of word representations in vector space. CoRR abs/1301.3781 (2013)
22. Pennington, J., Socher, R., Manning, C.: Glove: global vectors for word representation. In: Proceedings of the 2014 Conference on Empirical Methods in Natural Language Processing (EMNLP), pp. 1532–1543 (2014)
23. Plank, B., Søgaard, A., Goldberg, Y.: Multilingual part-of-speech tagging with bidirectional long short-term memory models and auxiliary loss. In: Proceedings of the 54th Annual Meeting of the Association for Computational Linguistics (Volume 2: Short Papers), pp. 412–418 (2016)
24. dos Santos, C.N., Zadrozny, B.: Training state-of-the-art Portuguese POS taggers without handcrafted features. In: Baptista, J., Mamede, N., Candeias, S., Paraboni, I., Pardo, T.A.S., Volpe Nunes, M.G. (eds.) PROPOR 2014. LNCS (LNAI), vol. 8775, pp. 82–93. Springer, Cham (2014). https://doi.org/10.1007/978-3-319-09761-9_8
25. Temponi, C.N., et al.: O corpus anotado do português histórico: um avanço para as pesquisas em lingüística histórica do português. Revista Virtual de Estudos da Linguagem: ReVEL **2**(3), 1 (2004)

26. Wang, P., Qian, Y., Soong, F.K., He, L., Zhao, H.: Part-of-speech tagging with bidirectional long short-term memory recurrent neural network. arXiv preprint arXiv:1510.06168 (2015)
27. Wang, W., Auer, J., Parasuraman, R., Zubarev, I., Brandyberry, D., Harper, M.: A question answering system developed as a project in a natural language processing course. In: Proceedings of the 2000 ANLP/NAACL Workshop on Reading Comprehension Tests as Evaluation for Computer-based Language Understanding Sytems-Volume 6, pp. 28–35. Association for Computational Linguistics (2000)

A Binary Variational Autoencoder for Hashing

Francisco Mena$^{(\boxtimes)}$ and Ricardo Ñanculef$^{(\boxtimes)}$

Federico Santa María University, Santiago, Chile
francisco.mena@alumnos.inf.utfsm.cl, jnancu@inf.utfsm.cl

Abstract. Searching a large dataset to find elements that are similar to a sample object is a fundamental problem in computer science. Hashing algorithms deal with this problem by representing data with similarity-preserving binary codes that can be used as indices into a hash table. Recently, it has been shown that variational autoencoders (VAEs) can be successfully trained to learn such codes in unsupervised and semi-supervised scenarios. In this paper, we show that a variational autoencoder with binary latent variables leads to a more natural and effective hashing algorithm that its continuous counterpart. The model reduces the quantization error introduced by continuous formulations but is still trainable with standard back-propagation. Experiments on text retrieval tasks illustrate the advantages of our model with respect to previous art.

Keywords: Hashing · Variational autoencoders · Deep learning · Gumbel-Softmax distribution · Neural information retrieval

1 Introduction

A wide range of applications in computer science rely on similarity search, i.e., finding elements in a database that are similar to a given sample object [1]. The greater availability of complex data types such as image, audio, and text, has increased the interest for this type of search in the last years and raised the need for methods that can reduce the processing time and storage cost of traditional paradigms. Among these methods, hashing has emerged as a popular approach.

The main idea of hashing methods is to represent the data using binary codes that preserve their semantic content and can be used as addresses into a hash table. Items similar to a query can then be found by accessing all the cells of the table that differ a few bits from the query. As binary codes are storage-efficient, hashing can be performed in main memory even for very large datasets [11].

Hashing algorithms can be broadly categorized into data-independent and data-dependent methods. Data-independent methods exploit properties of some probability distributions to ensure that the similarity function of the original space is approximately preserved by the embedding into the code space [8]. These methods usually require codes much longer than those obtained with data-dependent techniques, that leverage data and machine learning techniques

© Springer Nature Switzerland AG 2019
I. Nyström et al. (Eds.): CIARP 2019, LNCS 11896, pp. 131–141, 2019.
https://doi.org/10.1007/978-3-030-33904-3_12

to explicitly optimize the embedding, at the cost of some training time [12,15]. Supervised, unsupervised and semi-supervised approaches have been studied. Supervised methods rely on explicit annotations, such as topic or similarity labels, to learn the hash codes [9]. Unfortunately, the performance of these methods degrades quickly when there is not enough labelled data for training or it is noisy. Unsupervised methods deal with this issue, providing learning mechanisms that do not require explicit supervisory signals [12] and can thus leverage unlabelled data, which is usually abundant and cheap [14]. Often, these methods can be transformed into semi-supervised models that can also exploit labels if available.

Recently, significant progress has been made in the field of deep generative models. The so-called variational autoencoder (VAE) framework [7], provides algorithms for probabilistic inference and learning that scale to very large datasets and provide state-of-the-art performance in many tasks. A natural question is whether these advances can be exploited to devise novel hashing algorithms. It has been shown indeed that VAEs can be successfully trained to learn hash codes [3], improving on previous techniques when labelled data is scarce. A disadvantage of this approach is that, as conventional VAEs use a Gaussian encoder, the continuous representation learnt by the model needs to be quantized to obtain binary codes. This step introduces an error that is not account for in the learning process and can seriously degrade information retrieval performance.

In this paper, we propose to learn hash codes using a VAE with binary latent variables that directly represent the different bits of the code assigned to an object. The main technical difficulty of this approach, i.e. back-propagation through discrete nodes, can be circumvent by specializing the method proposed in [6] to handle Bernoulli distributions. Experiments on text retrieval tasks demonstrate that this approach works well for hashing, leading to more effective and interpretable binary codes than those produced by a continuous VAE.

The rest of this paper is organized as follows. In the next section, we outline the idea of hashing for similarity search. Related work is discussed in Sect. 3. In Sect. 4, we present the proposed formulation. In Sect. 5, we report experimental results, comparing the codes of our method with those of a continuous VAE. Finally, Sect. 6 summarizes the conclusions of this work.

2   Problem Statement and Background

Similarity Search. Consider a dataset $D = \{x^{(1)}, x^{(2)}, \ldots, x^{(n)}\}$, with $x^{(\ell)} \in \mathbb{X} \ \forall \ell \in [n] = \{1, \ldots, n\}$, and the problem of searching D to find elements that are *similar* to some sample object $q \in \mathbb{X}$ (not necessarily in D) referred to as *query*. If $\mathbb{X}$ is equipped with a similarity function $s : \mathbb{X} \times \mathbb{X} \to \mathbb{R}$, such that the greater the value of s, the more similar are the objects, and n is small, a simple approach to solve this problem is a *linear scan*: compare q with all the elements in D and return $x^{(\ell)}$ if $s(x^{(\ell)}, q)$ is greater than some threshold θ. The value of θ (*search radius*), can be given in advance, computed to return exactly k results or chosen to maximize information retrieval metrics such as precision and recall [1]. If $\mathbb{X} \subset \mathbb{R}^d$, with small d, specialized data structures (e.g. KD-trees) perform efficient scans when n is large. Unfortunately, if d becomes large, as in large-scale collections of images, audio, and text, the performance of these data structures degrades quickly [11] and novel methods are required.

Hashing. Hashing algorithms address similarity-search problems by devising an embedding $h(\boldsymbol{x})$ of the feature space $\mathbb{X}$ into the Hamming space $\mathbb{H}_B = \{0,1\}^B$, and substituting searches in $\mathbb{X}$ by searches in $\mathbb{H}_B$. Since binary codes can be efficiently stored and compared, searches in $\mathbb{H}_B$ can be orders of magnitude faster, even using a simple $\mathcal{O}(n)$ linear scan. Recent data structures however allow to search binary codes in $\mathcal{O}(1)$ time if B is a small constant [11]. Of course, for this approach to make sense, the embedding has to preserve similarity.

Quantization Error. Many hashing approaches obtain $h(\boldsymbol{x})$ by learning a continuous embedding $\phi(\boldsymbol{x}) \in \mathbb{R}^B$ that is then discretized by thresholding, i.e. by computing $h(\boldsymbol{x}) = \mathbf{1}(\phi(\boldsymbol{x}) - b)$, where $\mathbf{1}(\cdot)$ denotes the indicator function. The term $\|h(\boldsymbol{x}) - \phi(\boldsymbol{x})\|$ is called *the quantization error* and can have a significant impact in the quality of the obtained hashes for search applications [5].

Focus. We focus on learning a hash function $h(\cdot)$ using a *deep probabilistic graphical model* that reduces the quantization error. Our final goal is to obtain better codes for similarity search tasks focused on the unsupervised case.

3   Related Work

Up to our knowledge, the use of a deep graphical model to learn hash codes without supervision was first proposed in [12] using a stack of restricted Boltzmann machines (RBM). At training time, the nodes of the deepest layer allowed to identify topics from which the visible nodes had to generate/reconstruct the data. The hash codes were obtained by thresholding the binary nodes of the topic layer. The model can be seen as a stochastic autoencoder where encoder and decoder are tied together in the same neural architecture. Unfortunately, training this model is often computationally hard. Perhaps for this reason, most subsequent research on hashing have adopted simpler models.

In [15], unsupervised hashing is posed as the problem of partitioning a graph where the vertices represent training points and the edges are weighted using similarity scores. In [8], the hash codes are obtained by projections onto random hyperplanes related to the data by means of a kernel function. The method in [5] computes the codes by first projecting the data into the top PCA directions and then learning a rotation matrix that minimizes the quantization error.

The use of deterministic neural architectures for hashing that, in contrast to [12], can be trained using efficient back-propagation, is related to [2]. Here, a shallow autoencoder is trained to minimize the reconstruction error. In [9] a decoder-free approach is proposed where the encoder is a feed-forward neural net trained to maximize the variance of the binary vectors. The method in [4] employs a similar architecture for the encoder but changes the training objective, introducing a linear decoder and minimizing the data reconstruction error.

Recently, [3] have proposed to obtain hash codes by first training a standard VAE [7], i.e., a stochastic autoencoder, and then thresholding the continuous latent representation around the median. This method, called *Variational Deep Semantic Hashing* (VDSH), improve the results of previous unsupervised techniques besides being more scalable and stable than [12]. A discrete VAE is presented in [13] for discovering topics in text documents. In this model only one topic can be active at the same time and thus it cannot be directly used for hashing in a way we can easily conceive.

4   Proposed Method

We propose to learn the hash function $h : \mathbb{X} \to \mathbb{H}_B$ using a variational autoencoder (VAE) framework [7], in which the hash code $\boldsymbol{b} \in \{0,1\}^B$ assigned to a data object $\boldsymbol{x}$ is treated as a random variable and it is generated according to a conditional probability distribution $q_\phi(\boldsymbol{b}|\boldsymbol{x})$, with parameters ϕ. In standard VAE, the distribution $q_\phi(\boldsymbol{b}|\boldsymbol{x})$ is called *the encoder*, and it is typically a Gaussian $\mathcal{N}(\mu(\boldsymbol{x}), \sigma(\boldsymbol{x}))$ where $\mu(\boldsymbol{x}), \sigma(\boldsymbol{x})$ are modeled by a neural net $f(\boldsymbol{x}; \phi)$. Our difference here is that the latent variable $\boldsymbol{b}$ is no longer continuous but binary.

A first advantage of a binary VAE formulation for hashing is interpretability. The latent variables $b_i \in \{0,1\}$, can be directly understood as the bits of the code assigned to $\boldsymbol{x}$. If $\boldsymbol{b}$ is Gaussian, as in [3], the relationship between the hash code and the representation learnt by the model is more ambiguous. A second advantage regards the smaller error introduced by the quantization step required to transform the latent representation into a binary hash code. The method proposed in [3] uses a thresholding operation around the median of the Gaussian that incurs significant quantization error and can seriously degrade the search performance (see Fig. 1 for an illustration). If the latent variables are binary, the quantization step is no longer required and the codes used for hashing are the same codes optimized in the learning process. Unfortunately, the presence of discrete random variables, makes optimization more difficult. Below we explain how our model, called *Binary-VAE* (B-VAE), addresses this problem.

4.1   Model Architecture and Learning Goal

Since $\boldsymbol{b}$ is now binary, we let the encoder $q_\phi(\boldsymbol{b}|\boldsymbol{x})$ be a multi-variate Bernoulli distribution $\text{Ber}(\alpha(\boldsymbol{x}))$, where the probabilities $\alpha(\boldsymbol{x}) = p(\boldsymbol{b}=\boldsymbol{1}|\boldsymbol{x})$ are represented and learnt using a neural net $f(\boldsymbol{x}; \phi)$. We can train this model by defining an auxiliary decoder $p_\theta(\boldsymbol{x}|\boldsymbol{b})$ that reconstructs an input pattern $\boldsymbol{x}$ from the binary code $\boldsymbol{b}$ assigned to it. The form of $p_\theta(\boldsymbol{x}|\boldsymbol{b})$ depends on the type of data. For instance, in text hashing, it can be chosen to be a Multinomial distribution on the words/tokens of a document $\boldsymbol{x}$, $p(\boldsymbol{x}|\boldsymbol{b}) = \prod_{w \in \boldsymbol{x}} p(w|\boldsymbol{b})^{n_w}$, where n_w is the frequency of w. Just like the encoder, the probabilities $p(w|\boldsymbol{b})$ can be learnt using a neural net $g(\boldsymbol{b}; \theta)$.

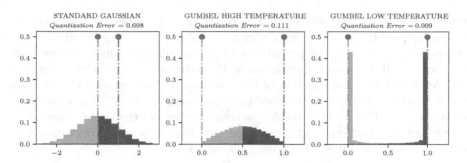

Fig. 1. 1-bit quantization of a Gaussian variable (standard VAE) and two Gumbel-Softmax variables (B-VAE) at different temperatures. In practice, all the yellow/green points are rounded to 0/1 to obtain binary codes. A Gumbel-Softmax distribution at low temperature reduces the quantization error inducing a saturation around 0/1. (Color figure online)

The composition of $p_\theta(\boldsymbol{x}|\boldsymbol{b})$ and $q_\phi(\boldsymbol{b}|\boldsymbol{x})$ leads to a stochastic auto-encoder with parameters ϕ and θ that can be learnt by maximizing the data log-likelihood $\ell(\theta, \phi; D)$. Unfortunately, since $\boldsymbol{b}$ is unobserved, optimizing ℓ is difficult. VAEs are instead trained to maximize a lower bound of $\ell(\theta, \phi; D)$, as for a point $\boldsymbol{x}^{(\ell)}$

$$\ell(\theta, \phi; \boldsymbol{x}^{(\ell)}) \geq \mathcal{L} = \mathbb{E}_{q_\phi(\boldsymbol{b}|\boldsymbol{x}^{(\ell)})} \left[\log p_\theta(\boldsymbol{x}^{(\ell)}, \boldsymbol{b}) - \log q_\phi(\boldsymbol{b}|\boldsymbol{x}^{(\ell)}) \right]$$

$$\mathcal{L} = \mathbb{E}_{q_\phi(\boldsymbol{b}|\boldsymbol{x}^{(\ell)})} \left[\log p_\theta(\boldsymbol{x}^{(\ell)}|\boldsymbol{b}) \right] - D_{\mathrm{KL}} \left(q_\phi(\boldsymbol{b}|\boldsymbol{x}^{(\ell)}) \| p_\theta(\boldsymbol{b}) \right), \quad (1)$$

where the first term of $\mathcal{L}$ corresponds to the expected reconstruction error and the second enforces the consistency between the posterior implemented by the encoder $q_\phi(\boldsymbol{b}|\boldsymbol{x})$ and some prior $p_\theta(\boldsymbol{b})$, using the KL divergence. For common choices of $p_\theta(\boldsymbol{b})$, the KL divergence can be integrated analytically, which leads to expressions easy to differentiate. However, traditional (Monte-Carlo) estimators of the first term in (1), lead to unstable gradients [7]. The framework presented in [7] solves this problem using the so-called *re-parametrization trick*. Unfortunately, this method does not apply to discrete latent distributions and so we need a more specialized method.

4.2   Re-parameterization via Gumbel-Softmax

As shown [10], the so-called Gumbel-Softmax distribution proposed in [6], can be adapted to obtain a continuous approximation of Bernoulli random variables. Indeed, with $\sigma(\xi) = 1/(1 + \exp(-\xi))$, if $\boldsymbol{b}_{i,\ell} \sim \mathrm{Ber}\left(\alpha_i(\boldsymbol{x}^{(\ell)})\right)$, $\epsilon_i \sim \mathcal{U}(0, 1)$ $\forall i \in [B]$, we have that

$$\hat{b}_{i,\ell} = \sigma \left(\left(\log \frac{\alpha_i(\boldsymbol{x}^{(\ell)})}{1 - \alpha_i(\boldsymbol{x}^{(\ell)})} + \log \frac{\epsilon_i}{1 - \epsilon_i} \right) / \lambda \right), \quad (2)$$

converges to $\boldsymbol{b}_{i,\ell}$ in the sense that $P(\lim_{\lambda \to 0} \hat{\boldsymbol{b}}_{i,\ell} = 1) = \alpha_i(\boldsymbol{x})$. Thus, we can take samples of $\hat{\boldsymbol{b}}_{i,\ell}$ to obtain approximate samples of $\boldsymbol{b}_{i,\ell}$. As depicted in Fig. 1, at low temperatures λ, the probability of getting samples which are not 0 or 1 is very small, because (2) saturates at the extremes. Since, in addition, $\hat{\boldsymbol{b}}_{i,\ell}$ is a deterministic transformation of the auxiliary random variable ϵ, that does not depend on the encoder parameters ϕ, we can estimate $\mathbb{E}_{q_\phi} [\log p_\theta(\boldsymbol{x}|\boldsymbol{b})]$ by sampling $p(\epsilon)$. This leads to stable gradients in terms of the model parameters (ϕ, θ), and then back-propagation can be used to train our VAE. According to the experimentation of [6,10] a good value of λ is $2/3$.

4.3 Priors

As in traditional VAEs, we introduce a prior $p_\theta(\boldsymbol{b})$ that helps to regularize the learning process. We propose to adopt the non-informative Bernoulli distribution, $p_\theta(\boldsymbol{b}_i) = \mathrm{Ber}(0.5) \, \forall i \in [B]$. The interpretation of this prior is a preference for balanced hash codes: in average, half of the data points will have bit $\boldsymbol{b}_i$ active and half inactive. With this choice, the KL divergence in (1), for a data point $\boldsymbol{x}$, can be calculate analytically and leads to

$$D_{\mathrm{KL}} \left(q_\phi(\boldsymbol{b}|\boldsymbol{x}) \| p_\theta(\boldsymbol{b}) \right) = \sum_i^B \mathbb{E}_{q_\phi(\boldsymbol{b}_i|\boldsymbol{x})} [\log q_\phi(\boldsymbol{b}_i|\boldsymbol{x})] - \mathbb{E}_{q_\phi(\boldsymbol{b}_i|\boldsymbol{x})} [\log p_\theta(\boldsymbol{b}_i)]$$

$$= B \cdot \log 2 + \sum_i^B \alpha_i(\boldsymbol{x}) \cdot \log \alpha_i(\boldsymbol{x}) + (1 - \alpha_i(\boldsymbol{x})) \cdot \log (1 - \alpha_i(\boldsymbol{x})), \qquad (3)$$

where the second term represent the regularization factor, expressed as the negative binary entropy $(-\mathbb{H}(\alpha_i))$ of the distribution over the binary latent variables.

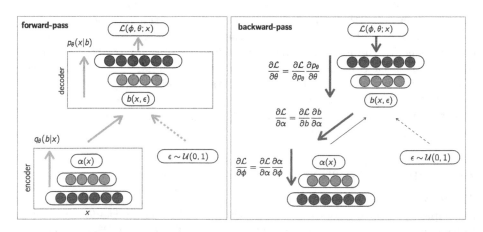

Fig. 2. Illustration of the forward (*orange*) and backward (*red*) pass implementing the proposed method as a deep neural net. The dashed line represents a stochastic layer. (Color figure online)

4.4   Implementation

We illustrate in Fig. 2 the neural net architecture of our method. As other VAEs [7], it can be easily trained with vanilla back-propagation. Only the forward pass requires passing through stochastic layers.

4.5   Hashing

As our encoder is stochastic, we need to sample $q_\phi(b|x)$ to obtain hash codes. Note that as we model $b \sim \text{Ber}(\alpha(x))$, we always obtain binary codes. A discretization is no required. However, in practice one may prefer deterministic codes. In that case, we can take the expected value of the stochastic representation $\alpha(x)$ and compute $b = \mathbf{1}(\alpha(x) - \frac{1}{2})$, where the threshold value $\frac{1}{2}$ is consistent with the model priors. This quantization procedure does not degrade significantly the codes learnt by our model, because in the training procedure, the encoder has learnt probabilities $\alpha(x)$ that are very close to 0 or 1. As shown in Fig. 1 at low temperatures the saturation around 0/1 comes naturally.

5   Experiments

We evaluate our method on text retrieval tasks, previously used to assess hashing algorithms [3,16], and defined on three well-known corpora: *20 Newsgroups*, containing 18000 long documents organized into 20 mutually exclusive classes; *Reuters21578*, containing 11000 news documents annotated with 90 non-exclusive tags (topics); and *Google Search Snippets*, with 12000 short documents organized into 84 mutually exclusive classes (domains). Please check [3,16] for details.

Pre-processing. Documents are pre-processed by removing extra-spaces, stop-words and any character that is not a letter. We then lower-case and lemmatize the text, removing lemmas of length smaller than 3. The 10^4 most frequent lemmas are used to get a term frequency representation tf_d of each document. As shown in [3], a change on this approach does not lead to significant improvements. Early experiments reveled however that the transformation $\log(\text{tf}_d + 1)$ helped to make training more stable and thus was applied from there on.

Evaluation Protocol. As the test set was provided, a split was done on the rest of documents to create training and validation sets (75%/25%). The model was trained on the training set and used to embed the corpus into the Hamming space. Based on this embedding, each test or validation document was then provided to the system as a query and used to retrieve similar documents from the training set. Two items were considered similar if they have at least one label in common. We consider two querying methods: (1) *top-K*: retrieve $K = 100$ documents whose hash codes are the most similar to the hash of the query, and (2) *ball search*: retrieve all the documents at a Hamming distance of at most θ bits. The results are evaluated using precision (P) and recall (R).

Table 1. Precision (P) and recall (R) of alternative architectures on **20 Newsgroups**.

VDSH *Base*	$P = 0.284$	$R = 0.193$
VDSH *Symmetric*	$P = 0.277$	$R = 0.189$

B-VAE *Base*	$P = 0.316$	$R = 0.216$
B-VAE *Symmetric*	$P = 0.353$	$R = 0.241$

Table 2. Precision and recall on the validation set using the first querying mechanism (top-100). As for selecting the bits B, the best results are presented in bold.

Dataset	Method	Precision				Recall			
		4 bits	8 bits	16 bits	32 bits	4 bits	8 bits	16 bits	32 bits
Newsgroup	VDSH	0.213	0.251	0.285	**0.299**	0.147	0.172	0.196	**0.205**
	B-VAE	0.325	0.338	0.340	**0.359**	0.225	0.232	0.232	**0.246**
Reuters	VDSH	0.452	**0.517**	0.496	0.495	0.142	**0.188**	0.178	0.183
	B-VAE	0.587	0.569	0.599	**0.602**	0.193	0.198	0.224	**0.233**
Snippets	VDSH	0.389	**0.426**	0.352	0.341	0.109	**0.119**	0.099	0.096
	B-VAE	**0.475**	0.436	0.401	0.404	**0.138**	0.123	0.113	0.114

Baseline and Architecture. We adopt the VAE recently proposed in [3] as our baseline with the original architecture for encoder and decoder. We adopt the same architecture for our encoder, but, inspired by [12], we define the decoder to obtain a symmetric model. As shown in Table 1, imposing symmetry improves the performance of our method (B-VAE) but slightly worsens the baseline (VDSH).

Results. In Table 2, we investigate the effect of the number of bits B in the validation set. We can see that the proposed method outperforms the baseline in all the cases, with an advantage both in terms of precision and recall. As noted also by [3], the best results are not always obtained with a greater number of bits, probably due to over-fitting. If we reduce the number of bits, our method seems to be more robust in the results compared to the baseline, which, in general, suffers a more clear impact in terms of performance. After these experiments on the validation set, we fix the number of bits to $B = 32$.

Table 3. Precision and recall on the test set using the first querying mechanism (top-100). Best results in bold.

Dataset	Method	Precision	Recall
Newsgroups	VDSH	0.319	0.084
	B-VAE	**0.441**	**0.116**
Reuters	VDSH	0.556	0.174
	B-VAE	**0.698**	**0.246**
Snippets	VDSH	0.297	0.099
	B-VAE	**0.381**	**0.127**

Table 4. Examples of most probable words by activating a bit on the hash code.

Newsgroup	Reuters	Snippets
bit 9	*bit 31*	*bit 25*
Complexity	Device	Interaction
Heterosexual	Recognize	Biogeography
Likelihood	Responsibility	Composer
Inconsistent	Analyze	Radiology
Skeptic	Printing	Gymnastics
Presidential	Undoubtedly	Patient
Homosexuality	Projecting	Strength

In Table 3, we compare the test performance of the methods, using the first querying mechanism (top-K). We can see that the proposed method outperforms the baseline in all the datasets, with a large (absolute) improvement in terms of precision and a more conservative but systematic (absolute) improvement in terms of recall. This demonstrates the practical advantage of using binary latent variables for hashing. In relative terms, the precision improves ∼38% in Newsgroups, ∼26% in Reuters and ∼28% in Snippets, while recall improves ∼38% in Newsgroups, ∼47% in Reuters and ∼28% in Snippets.

In Fig. 3, we show the performance of the different methods using the second querying mechanism, *ball search*, on the test set. The advantage of the proposed method is robust to the choice of the search parameter (radius) θ (which is problem dependent); leading to a better precision and recall in almost all the cases. We can also see the advantage of using the second querying mechanism instead of the first one. For example, using $\theta = 8$ (bits) in Reuters, our method can increase the recall from ∼0.25 to ∼0.55 without significantly reducing the precision. Using $\theta = 6$ (bits) in Snippets, our method can increase the precision from 0.38 to approx ∼0.5, keeping the advantage in terms of recall.

Interpretation of the Hash Codes. To illustrate the interpretability of our model, we sketch in Table 4 results of experiments in which we have activated a bit of the latent representation and ranked the words according to the probabilities predicted by the decoder. In Newsgroups, bit 9 seems to detect political discussions regarding sexuality. In Reuters, bit 31 captures computer-related concepts. In Snippets, bit 25 seems to detect terms associated with health or sport.

Effect of Priors. It is worth mentioning that in all the experiments we have observed that the hash tables produced by our method are well-balanced, i.e., the number of documents colliding into a cell is approximately constant. This is important for computational efficiency [11] and attributed to the model priors.

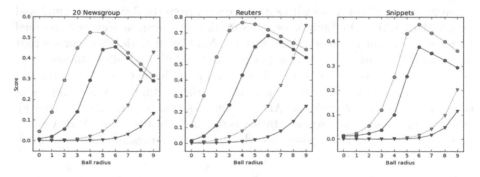

Fig. 3. Precision (circles) and recall (triangles) using the second querying mechanism (ball search). Points are obtained using different values of θ. Green curves is for our method (B-VAE) and blue curves are for the baseline (VDSH). (Color figure online)

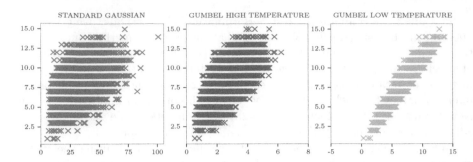

Fig. 4. Similarity before thresholding (x-axis) and after tresholding (y-axis) for 10^5 pairs of samples drawn from different latent distributions (32-bits).

Table 5. Accuracy score on training and testing set, using the representation obtained before and after thresholding is done. Best results on each set are presented in bold.

Dataset	Method	Before Thresholding		After Thresholding	
		Train	Test	Train	Test
Newsgroup	VDSH	**93.46**	**78.15**	**69.61**	62.74
	B-VAE	69.61	67.87	68.11	**63.52**
Reuters	VDSH	**92.63**	**82.51**	63.67	65.80
	B-VAE	69.51	69.48	**71.36**	**70.84**
Snippets	VDSH	**96.79**	**80.72**	70.67	69.42
	B-VAE	85.98	83.46	**86.56**	**82.66**

Effect of Thresholding. In Fig. 4 we compare the distance between codes before and after quantization (Euclidean and Hamming respectively), computed on samples drawn from different distributions. We observe that Gumbel-Softmax samples at low temperature lead to similarities well correlated before and after quantization. This contrasts with samples drawn from the distribution employed by standard VAEs. On Table 5, we measure the classification accuracy obtained by using the latent representations with a KNN classifier. Here we can see that our embedding has quite similar performance before and after thresholding, besides getting a quite low generalization error (difference between train and test). We can also see that the superiority of the continuous VDSH representation is lost after thresholding. All this suggests that a standard VAE has an advantage if a continuous representation is required but the binary VAE we propose is better suited for applications where a binary representation is required, as in hashing.

6   Conclusions

We have investigated the use of a variational autoencoder with binary latent variables to learn hash codes. This formulation is easy to interpret, reduces the quantization error of thresholding continuous codes, and consents the use of back-propagation for training. Experiments on unsupervised text hashing show that the method is more effective for information retrieval than its continuous counterpart, even if the representation of a standard VAE can have an advantage before discretization. In future work, we plan to evaluate the model on image retrieval tasks using convolutional nets and to handle semi-supervised scenarios.

Acknowledgement. F. Mena thanks the *Programa de Iniciación Científica* PIIC-DGIP of the Federico Santa María University for funding this work.

References

1. Baeza-Yates, R., Ribeiro-Neto, B.: Modern Information Retrieval. ACM Press, New York (1999)
2. Carreira-Perpinán, M.A., Raziperchikolaei, R.: Hashing with binary autoencoders. In: Proceedings of the CVPR, pp. 557–566 (2015)
3. Chaidaroon, S., Fang, Y.: Variational deep semantic hashing for text documents. In: Proceedings of the 40th SIGIR, pp. 75–84. ACM (2017)
4. Do, T.-T., Doan, A.-D., Cheung, N.-M.: Learning to hash with binary deep neural network. In: Leibe, B., Matas, J., Sebe, N., Welling, M. (eds.) ECCV 2016. LNCS, vol. 9909, pp. 219–234. Springer, Cham (2016). https://doi.org/10.1007/978-3-319-46454-1_14
5. Gong, Y., Lazebnik, S., Gordo, A., Perronnin, F.: Iterative quantization: a procrustean approach to learning binary codes for large-scale image retrieval. IEEE Trans. Pattern Anal. Mach. Intell. **35**(12), 2916–2929 (2013)
6. Jang, E., Gu, S., Poole, B.: Categorical reparameterization with Gumbel-softmax. In: Proceedings of the ICLR (2017)
7. Kingma, D.P., Welling, M.: Auto-encoding variational Bayes (2013)
8. Kulis, B., Grauman, K.: Kernelized locality-sensitive hashing. IEEE Trans. Pattern Anal. Mach. Intell. **34**(6), 1092–1104 (2012)
9. Liong, V.E., Lu, J., Wang, G., Moulin, P., Zhou, J.: Deep hashing for compact binary codes learning. In: Proceedings of the CVPR, vol. 2015, pp. 2475–2483 (2015)
10. Maddison, C.J., Mnih, A., Teh, Y.W.: The concrete distribution: a continuous relaxation of discrete random variables. arXiv preprint arXiv:1611.00712 (2016)
11. Norouzi, M., Punjani, A., Fleet, D.J.: Fast exact search in hamming space with multi-index hashing. IEEE PAMI **36**(6), 1107–1119 (2014)
12. Salakhutdinov, R., Hinton, G.: Semantic hashing. Int. J. Approximate Reasoning **50**(7), 969–978 (2009)
13. Silveira, D., Carvalho, A., Cristo, M., Moens, M.F.: Topic modeling using variational auto-encoders with Gumbel-softmax and logistic-normal mixture distributions. In: International Joint Conference on Neural Networks (IJCNN). IEEE (2018)
14. Wang, J., Kumar, S., Chang, S.F.: Semi-supervised hashing for large-scale search. IEEE Trans. Pattern Anal. Mach. Intell. **34**(12), 2393–2406 (2012)
15. Weiss, Y., Torralba, A., Fergus, R.: Spectral hashing. In: NIPS (2009)
16. Xu, J., et al.: Convolutional neural networks for text hashing. In: Proceedings of the IJCAI 2015 (2015)

Prototypes Generation from Multi-label Datasets Based on Granular Computing

Marilyn Bello[1,2]($\boxtimes$), Gonzalo Nápoles[2], Koen Vanhoof[2], and Rafael Bello[1]

[1] Computer Science Department,
Universidad Central de Las Villas, Santa Clara, Cuba
mbgarcia@uclv.cu
[2] Faculty of Business Economics, Hasselt University, Hasselt, Belgium

Abstract. Data reduction techniques play a key role in instance-based classification to lower the amount of data to be processed. Prototype generation aims to obtain a reduced training set in order to obtain accurate results with less effort. This translates into a significant reduction in both algorithms' spatial and temporal burden. This issue is particularly relevant in multi-label classification, which is a generalization of multiclass classification that allows objects to belong to several classes simultaneously. Although this field is quite active in terms of learning algorithms, there is a lack of prototype generation methods. In this research, we propose three prototype generation methods from multi-label datasets based on Granular Computing. The experimental results show that these methods reduce the number of examples into a set of prototypes without affecting the overall performance.

Keywords: Multi-label classification · Prototype generation · Granular Computing · Rough Set Theory

1 Introduction

Classification is one of the most popular Data Mining topics. Its aim is to learn from labeled patterns a model able to predict the decision class for future, never seen before, data samples [1]. The best way to solve a classification problem is usually to have as much information as possible. In practice, however, this is not always the case. The performance of learning algorithms may decrease due to the abundance of information, because many examples may be very irrelevant to the resolution of the problem or may provide the same information [14,17].

On the other hand, the abundance of information could increase the computational complexity of the method, particularly in the case of instance-based learners such as the kNN (k Nearest Neighbors) [10] algorithm. However, it is possible to reduce or modify the datasets without affecting the learning process, improving process performance by reducing computational cost.

One approach to doing this is the classification based on the Nearest Prototype (NP) [6,12]. It is an approach in which the decision class of a new object

© Springer Nature Switzerland AG 2019
I. Nyström et al. (Eds.): CIARP 2019, LNCS 11896, pp. 142–151, 2019.
https://doi.org/10.1007/978-3-030-33904-3_13

is calculated by analyzing its proximity to a set of prototypes selected or generated from the initial set of objects. Strategies are needed to reduce the number of examples of input data into a set of representative prototypes. It is possible to say that the data reduction methods with respect to the instances are divided into two categories: *selection of prototypes* [11] and *generation of prototypes* [25]. Prototype selection algorithms select a set of representative objects according to a well-defined criterion, while prototype generation algorithms are capable of generating a set of new objects in the application domain from the initial objects.

On the other, Multi-Label Classification (MLC) is a type of classification where each of the objects in the data has associated a vector of outputs, instead of being associated with a single value [26,31]. ML-kNN is the first learning method that uses the kNN rule in multi-label prediction [30]. This method finds the k nearest neighbours in the datasets using the maximum a posteriori principle in order to determine the label set of the test object. The solution is based on the prior and posterior probabilities of the frequency of each label within the k nearest neighbours. Consequently, this method has the same drawbacks of kNN, because as the datasets increases, so does the computational cost of the algorithm, since for each test object its distance to all existing objects in the training set is calculated.

Despite extensive work on multi-label learning, as far as we know, only in [7] a method of prototypes selection is proposed. In this paper, we develop three methods of prototype generation from MLC datasets. Unlike the method proposed in [7], the three methods proposed are independent of the learning algorithm to be used. By doing so, we rely on Granular Computing [3,4,21], and two different ways of the granularity of the information. Two classical granulations are condition granulation and decision granulation, to name the granularity of the universe according to conditional attributes and decision class, respectively.

In the case of the first two methods proposed, the granulation of a universe is performed using a similarity relation that builds similarity classes (or granules) of objects in the universe from conditional attributes. By using similarity relations, methods can be used in the presence of mixed data, i.e., when there are both numerical and nominal attributes. On the other hand, the third method performs a granulation of the universe from an equivalence relation, and taking into account the different labels existing in the universe of discourse. From this, an equivalence class (or granule) is built for each label, and a prototype is generated for each granule.

The paper is organized as follows. Section 2 motivates our research, while Sect. 3 presents the theoretical background on Granular Computing. Section 4 introduces the three prototype generation methods from MLC datasets, and Sect. 5 is dedicated to evaluating the performance of the ML-kNN algorithm on the set of prototypes generated with our methods. Finally, in Sect. 6 we provide some concluding remarks and research directions.

2 Motivation

Some classification algorithms, such as the ones founded on examples-based learning, use a training set to estimate the class label, which causes the scalability problems when the size of the training set increases. In this case, the number of training objects affects the computational cost of a method [13,15]. The nearest neighbour rule is an example of a high computational cost method when the number of examples is large [2].

The most popular algorithm in this category is kNN. The computational complexity of kNN is $O(nm)$, where n and m are the size of dataset and the dimensionality of embedding space. Thus, these methods are computationally very expensive on large-scale datasets. The purpose of the NP approach is to reduce storage costs and learning technique processes based on examples. In the literature, several papers [2,15,25] on this issue have been proposed in the context of single-label learning.

Nevertheless, to the best of our knowledge, the only relevant work relating to NP in the field of multi-label classification is the $kNNc$ method described in [7]. It works in two stages, by combining prototype selection techniques with example-based classification. First, a reduced set of objects is obtained by prototype selection techniques used in classical classification [18]. The goal of this stage is to determine the set of labels which are nearest to the ones in the object to be classified. Then, the full set of samples is used, but limiting the prediction to the labels inferred in the previous step.

Unlike that study, this research proposes three new proposals for the generation of prototypes using different alternatives to granulate the datasets.

3 Granular Computing

Basic issues of Granular Computing may be studied from two related aspects, the construction of granules and computation with granules. The former deals with the formation, representation, and interpretation of granules, while the latter deals with the utilization of granules in problem solving [22,29].

In the construction of granules, it is necessary to study a criteria for deciding if two elements should be put into the same granule, based on available information. Typically, elements in a granule are drawn together by indistinguishability, similarity, proximity, or functionality [29].

With the granulation of universe, one considers elements within a granule as a whole rather than individually. The loss of information through granulation implies that some subsets of the universe can only be approximately described. The Rough Set Theory (RST) [19,23] is one of the most representative theories within Granular Computing. It deals mainly with the approximation aspect of information granulation. It uses two main components: an information system and an indiscernibility relation. The former is defined as $IS = (U, A)$, where U is a non-empty finite set of objects, and A is a non-empty finite set of attributes that describe each object. A particular case are the decision systems where $DS = (U, A \cup \{d\})$, whereas $d \notin A$ is the decision class.

In the classical RST, the relation of indiscernibility (R) is defined as an equivalence relation [20]. From this point on, $[x]_R$ defines an equivalence class of an element $x \in U$ under R, where $[x]_R = \{y \in U : yRx\}$, i.e. the equivalence class of an element includes all objects in the universe indiscernible from x. Each equivalence class may be viewed as a granule consisting of indistinguishable elements. Two objects are equivalent if they have exactly the same value with respect to a set of attributes. It means that two inseparable objects could incorrectly be labeled as separable, making the relationship excessively strict [28].

This problems can be alleviated in some extent by extending the concept of inseparability relation [24] and replacing the equivalence relation with a weaker binary relation. Equation (1) shows an indiscernibility relation,

$$R : xRy \Longleftrightarrow \delta(x,y) \geq \xi \tag{1}$$

where $0 \leq \delta(x,y) \leq 1$ is a similarity function. This weak binary relation states that objects x and y are inseparable as long as their similarity degree $\delta(x,y)$ exceeds a similarity threshold $0 \leq \xi \leq 1$. This relation actually defines a similarity class $\overline{R}(x) = \{y \in U : yRx\}$ that replaces the equivalence class.

The similarity function could be formulated in a variety of ways, for example, $\delta(x,y) = 1 - \varphi(x,y)$ with $\varphi(x,y)$ being the distance between objects x and y. In reference [27] the authors studied the properties of several distance functions which allow comparing heterogeneous instances, i.e., objects comprising both numerical and nominal attributes.

4   Methods for the Generation of Prototypes from MLC Datasets Based on Granular Computing

As mentioned, in MLC scenarios an object may be associated with multiple labels. Let $mlDS = (U, A \cup L)$ be a multi-label decision system, where the set U is a non-empty finite set of objects, A is a non-empty finite set of attributes that describe each observation, and $L = \{L_1, L_2, \ldots, L_k\}$ is a non-empty finite set of labels such that the label domain is $L_i = \{0, 1\}$.

Prototype-based classification determines the value of the decision class of a new object by analyzing its similarity to a set of prototypes generated from the initial set of objects. By doing so, we must define what is considered to be a decision class in the MLC context. For example,

- Each combination C_i of labels represents a decision value. For example, let $L = \{L_1, L_2, L_3\}$ denote the set of labels, a combination of labels could be "101", pointing out that the object belongs to the labels L_1 and L_3, then "101" defines a decision class, so that all objects associated with labels L_1 and L_3 belong to that decision class.
- Each label (L_i) is considered a decision value, so that all the objects associated that label belong to this decision class. According with this definition, in the example above there are three decision classes.

The basic idea of the *first two proposed algorithms* below is similar. Both are iterative algorithms in which a similarity class is built using the similarity relationship defined in Eq. (1). An object may belong to several similarity classes at the same time. However, when an object is included in a similarity class, it is not taken into account to build a new similarity class from it. Each similarity class consists of a granule that is used to build a prototype.

From this, a prototype or centroid is built for a set of similar objects. Each prototype is composed of both conditional and label attributes. To add their information both by condition (attribute values) and by decision (labels values) an aggregation operator is used. In the case of conditional attributes, the average can be used as the aggregation operator if the attribute value is numeric, or the mode if the attribute value is nominal.

The way in which the part of the prototype related to the labels is built differs between the two algorithms, exactly based on what is considered a decision class. In the case of Algorithm 1 each combination of labels represents a decision value, however Algorithm 2 considers each label independently as a decision value. In this way, the first algorithm builds its decision class from the most common combination of labels in the granule, while the second algorithm does it taking into account the labels independently. The resulting prototype will have as decision values the most common labels of the objects in the granule.

Algorithm 1. GP1mlTS

1: Initialize objects' counter
   Used [i] = 0 i=1,..., n
   PrototypeSet = ∅
2: While $\exists i$: Used [i] = 0
   j = i
   Construct the similarity class $\overline{R}(O_j)$
   Construct a vector $P = [P_{cond}, P_{dec}]$ from all objects in $\overline{R}(O_j)$
      P_{cond} is calculated from the set of values of the attributes (A) of all objects
      in $\overline{R}(O_j)$ and using an aggregation operator
      P_{dec} is calculated from the most common label combination (C) among
      all existing label combinations in the objects in $\overline{R}(O_j)$
   PrototypeSet = PrototypeSet ∪ P
   Used [j] = 1 for all the objects in $\overline{R}(O_j)$
3: Return PrototypeSet

In contrast to the first two algorithms, the *third algorithm* performs a different granulation of the universe. In this case, the decision class of the objects is taking into account to build the granulation of the data. The basic idea is to perform a granulation of the universe taking into account the labels instead of the condition attributes. Therefore, a granule is built for each label, so it will include all objects that are labeled with that decision label.

The partition and the covering of an information space are two common types of granulation of the universe [9]. The granulation obtained by this algorithm is

---

Algorithm 2. GP2mlTS

1: Initialize objects' counter
    Used [i] = 0 i=1,..., n
    PrototypeSet = ∅
2: While $\exists i$: Used [i] = 0
    j = i
    Construct the similarity class $\overline{R}(O_j)$
    Construct a vector $P = [P_{cond}, P_{dec}]$ from all objects in $\overline{R}(O_j)$
        P_{cond} is calculated from the set of values of the attributes (A) of all objects
        in $\overline{R}(O_j)$ and using an aggregation operator
        $P_{dec} = \{L_1, L_2, \ldots, L_k\}$, where $L_k = 1$ if most of the objects in $\overline{R}(O_j)$
        are labeled with that label, otherwise $L_k = 0$
    PrototypeSet = PrototypeSet ∪ P
    Used [j] = 1 for all the objects in $\overline{R}(O_j)$
3: Return PrototypeSet

---

a covering, since any objects could belong to two or more information granules. A prototype is then generated for each granule similar to the Algorithm 2. This procedure is formalized in Algorithm 3.

Algorithm 3. GP3mlTS

1: PrototypeSet = ∅
2: For each $Li \in L$
    Construct the equivalence class $[L_i]_R$
    Construct a vector $P = [P_{cond}, P_{dec}]$ from all objects in $[L_i]_R$
        P_{cond} is calculated from the set of values of the attributes (A) of all objects
        in $[L_i]_R$ and using an aggregation operator
        $P_{dec} = \{L_1, L_2, \ldots, L_k\}$, where $L_k = 1$ if most of the objects in $[L_i]_R$
        are labeled with that label, otherwise $L_k = 0$
    PrototypeSet = PrototypeSet ∪ P
3: Return PrototypeSet

---

5   Results and Discussion

In this section, we explore the performance of our prototype generation methods when coupled with the ML-kNN classification algorithm. To accomplish that, we use *Hamming Loss* (HL) metric which is a well-known performance measure in MLC scenarios [16]. This metric is defined as follows,

$$HL = \frac{1}{n}\frac{1}{k}\sum_{i=1}^{n} |Y_i \Delta Z_i| \tag{2}$$

where Δ operator returns the symmetric difference between Y_i (the real label set of the ith instance) and Z_i (the predicted one).

To perform the simulations, we rely on 12 multi-label datasets taken from the well-known RUMDR [8] repository. Table 1 summarizes the number of instances, attributes, and labels for each dataset. In the adopted datasets, the number of instances ranges from 1,675 to 10,491, the number of attributes from 294 to 1,836, and the number of labels from 6 to 400.

Table 1. Characterization of the MCL datasets used in our study.

	Domain	Instances	Attributes	Labels
bibtex(D1)	Text	7395	1836	159
corel5k(D2)	Images	5000	499	374
enron(D3)	Text	1702	1001	53
scene(D4)	Images	2407	294	6
stackex_chemistry(D5)	Text	6961	540	175
stackex_chess(D6)	Text	1675	585	227
stackex_cooking(D7)	Text	10491	577	400
stackex_cs(D8)	Text	9270	635	274
stackex_philosophy(D9)	Text	3971	842	233

We also studied the reduction coefficient, $Red(.)$ [5]. This measure, in Eq. (3) indicates by how much the number of objects is reduced, that is, the proportion between the size of the set of prototypes (P) and the universe (U),

$$Red(.) = \frac{|U| - |P|}{|U|} * 100 \qquad (3)$$

Figure 1 displays the reduction coefficient achieved once the proposed prototype generation methods are used on each dataset. In this experiment, we have adopted the Heterogeneous Euclidean-Overlap Metric (HEOM), which computes the normalized Euclidean distance between numerical attributes and an overlap metric for nominal attributes [27]. The similarity threshold ξ used in Eq. (1) ranges from 0.85 to 0.95.

It is worth mentioning that, for each datasets, we have estimated the HL value by using a 10-fold cross validation scheme. For each fold, this procedure splits the whole training set into two data pieces, namely, the training set and the test set. It should be highlighted that, while the training set is used to generate the set of prototypes. The test set is never modified so that it only serves to compute the HL associated with the current fold.

From the results in Fig. 1 we can conclude that our methods achieve a reduction rate higher than 20% in most problems. The $GP1mlTS$ and $GP2mlTS$ methods have a similar behaviour. However, the $GP3mlTS$ method reports reduction coefficients even higher than 90%. On the other hand, Fig. 2 displays the HL values achieved by the ML-kNN method with the original multi-label datasets, and the results obtained after using the set of prototypes generated by each of the methods proposed in this paper.

The results show that the prototypes generated for each dataset leads to HL values similar to those obtained with the original dataset. Only in the case of the *scene* dataset there is a significant difference in the LH value, especially when we

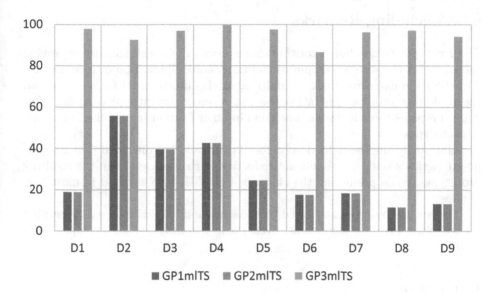

Fig. 1. Reduction percent achieved by each edition method.

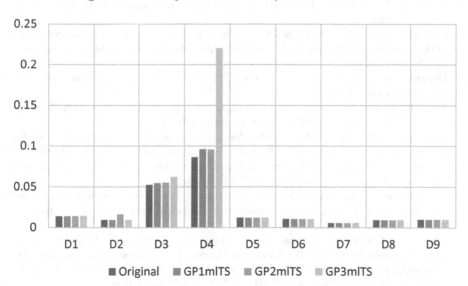

Fig. 2. HL values achieved by the ML-kNN method.

use the set of prototypes generated by the $GP3mlTS$ method with respect to the original dataset. This is due to the fact that this dataset has few labels (exactly 6 labels), thus only a few prototypes are generated. In short, the results showed that our proposal provides a suitable trade-off between algorithm's performance and the number of training examples in the dataset.

6   Concluding Remarks

The *Prototype Generation* algorithms have proved their usefulness by improving some kNN issues such as computational time, noise elimination or memory use. Although the extensive work in multi label classification, as far as we know, the topic of prototype generation has not received any attention so far. This paper proposes three methods based on Granular Computing for the generation of prototypes.

After analyzing the reduction coefficient, it could be concluded that the proposed methods achieve a significant reduction of the datasets from the resulting prototypes, while preserving the efficacy of the ML-kNN method in most case studies.

The set of prototypes generated by these methods could be used as a learning set for other learning algorithms, even those not intended for example-based learning.

References

1. Aggarwal, C.C.: Data Classification: Algorithms and Applications. CRC Press, New York (2014)
2. Barandela, R., Cortés, N., Palacios, A.: The nearest neighbor rule and the reduction of the training sample size. In: Proceedings 9th Symposium on Pattern Recognition and Image Analysis, vol. 1, pp. 103–108 (2001)
3. Bargiela, A., Pedrycz, W.: Granular Computing: An Introduction, vol. 717. Springer, Berlin (2012)
4. Bello, R., Falcón, R., Pedrycz, W.: Granular Computing: At the Junction of Rough Sets and Fuzzy Sets, vol. 224. Springer, Berlin (2007)
5. Bermejo, S., Cabestany, J.: A batch learning vector quantization algorithm for nearest neighbour classification. Neural Process. Lett. **11**(3), 173–184 (2000)
6. Bezdek, J.C., Kuncheva, L.I.: Nearest prototype classifier designs: an experimental study. Int. J. Intell. Syst. **16**(12), 1445–1473 (2001)
7. Calvo-Zaragoza, J., Valero-Mas, J.J., Rico-Juan, J.R.: Improving knn multi-label classification in prototype selection scenarios using class proposals. Pattern Recogn. **48**(5), 1608–1622 (2015)
8. Charte, F., Charte, D., Rivera, A., del Jesus, M.J., Herrera, F.: R ultimate multilabel dataset repository. In: Martínez-Álvarez, F., Troncoso, A., Quintián, H., Corchado, E. (eds.) HAIS 2016. LNCS (LNAI), vol. 9648, pp. 487–499. Springer, Cham (2016). https://doi.org/10.1007/978-3-319-32034-2_41
9. Chen, B., Sun, M., Zhou, M.: Granular rough theory: a representation semantics oriented theory of roughness. Appl. Soft Comput. **9**(2), 786–805 (2009)

10. Cover, T.M., Hart, P.E., et al.: Nearest neighbor pattern classification. IEEE Trans. Inf. Theory **13**(1), 21–27 (1967)
11. García, S., Cano, J.R., Herrera, F.: A memetic algorithm for evolutionary prototype selection: a scaling up approach. Pattern Recogn. **41**(8), 2693–2709 (2008)
12. García, S., Luengo, J., Herrera, F.: Data Preprocessing in Data Mining. Springer, Berlin (2015)
13. García-Durán, R., Fernández, F., Borrajo, D.: A prototype-based method for classification with time constraints: a case study on automated planning. Pattern Anal. Appl. **15**(3), 261–277 (2012)
14. Guan, D., Yuan, W., Lee, Y.K., Lee, S.: Nearest neighbor editing aided by unlabeled data. Inf. Sci. **179**(13), 2273–2282 (2009)
15. Hernández, F., et al.: An approach for prototype generation based on similarity relations for problems of classification. Computación y Sistemas **19**(1), 109–118 (2015)
16. Herrera, F., Charte, F., Rivera, A.J., del Jesus, M.J.: Multilabel classification. Multilabel Classification, pp. 17–31. Springer, Cham (2016). https://doi.org/10.1007/978-3-319-41111-8_2
17. Kim, S.W., Oommen, B.J.: A brief taxonomy and ranking of creative prototype reduction schemes. Pattern Anal. Appl. **6**(3), 232–244 (2003)
18. Nanni, L., Lumini, A.: Prototype reduction techniques: a comparison among different approaches. Expert Syst. Appl. **38**(9), 11820–11828 (2011)
19. Pawlak, Z.: Rough sets. Int. J. Comput. Inf. Sci. **11**(5), 341–356 (1982)
20. Pawlak, Z., Skowron, A.: Rough sets: some extensions. Inf. Sci. **177**(1), 28–40 (2007)
21. Pedrycz, W.: Granular Computing: Analysis and Design of Intelligent Systems. CRC Press, New York (2016)
22. Pedrycz, W., Homenda, W.: Building the fundamentals of granular computing: a principle of justifiable granularity. Appl. Soft Comput. **13**(10), 4209–4218 (2013)
23. Pedrycz, W., Skowron, A., Kreinovich, V.: Handbook of Granular Computing. Wiley, Hoboken (2008)
24. Slowinski, R., Vanderpooten, D.: A generalized definition of rough approximations based on similarity. IEEE Trans. Knowl. Data Eng. **12**(2), 331–336 (2000)
25. Triguero, I., Derrac, J., Garcia, S., Herrera, F.: A taxonomy and experimental study on prototype generation for nearest neighbor classification. IEEE Trans. Syst. Man Cybern. Part C Appl. Rev. **42**(1), 86–100 (2012)
26. Tsoumakas, G., Katakis, I., Vlahavas, I.: Mining multi-label data. In: Maimon, O., Rokach, L. (eds.) Data Mining and Knowledge Discovery Handbook, pp. 667–685. Springer, Boston (2009). https://doi.org/10.1007/978-0-387-09823-4_34
27. Wilson, D.R., Martinez, T.R.: Improved heterogeneous distance functions. J. Artif. Intell. Res. **6**, 1–34 (1997)
28. Yao, Y., Zhong, N.: Granular computing using information tables. In: Lin, T.Y., Yao, Y.Y., Zadeh, L.A. (eds.) Data Mining, Rough Sets and Granular Computing. STUDFUZZ, vol. 95, pp. 102–124. Springer, Heidelberg (2002). https://doi.org/10.1007/978-3-7908-1791-1_5
29. Zadeh, L.A.: Toward a theory of fuzzy information granulation and its centrality in human reasoning and fuzzy logic. Fuzzy Sets Syst. **90**(2), 111–127 (1997)
30. Zhang, M.L., Zhou, Z.H.: Ml-knn: a lazy learning approach to multi-label learning. Pattern Recogn. **40**(7), 2038–2048 (2007)
31. Zhang, M.L., Zhou, Z.H.: A review on multi-label learning algorithms. IEEE Trans. Knowl. Data Eng. **26**(8), 1819–1837 (2014)

A Simple Proposal for Sentiment Analysis on Movies Reviews with Hidden Markov Models

Billy Peralta[1]([envelope]), Victor Tirapegui[2], Christian Pieringer[3], and Luis Caro[2]

[1] Andres Bello University, Santiago, Chile
billy.peralta@unab.cl
[2] Catholic University of Temuco, Temuco, Chile
{vtirapegui,lcaro}@uct.cl
[3] INACAP, Santiago, Chile
cpieringer@inacap.cl

Abstract. Sentiment analysis of texts is the field of study which analyses and studies opinions, sentiments, value judgments, affections and emotions in texts like blogs, news and treating of products, organisations, events and topics. If information on subjective content is required, such as the emotion aroused by an event, computer techniques must be applied to analyse the pattern of public opinion. A common technique for analysing texts is the "Bag of Words", which provides good results assuming that the words are independent of one another. In this work we propose the use of Hidden Markov Chains to determine the polarity of the opinions expressed on movie reviews. We propose a method for simulating hidden states through clustering techniques; we then carry out a sensitivity analysis of the model in which we apply variations to model parameters such as the number of hidden states or the number of words used. The results show that our proposal gives a 3% improvement over the basic model using F-score for real databases of public opinion.

Sentiment analysis of texts is the field of study which analyses and studies opinions, sentiments, value judgments, affections and emotions in texts like blogs, news and treating of products, organisations, events and topics. If information on subjective content is required, such as the emotion aroused by an event, computer techniques must be applied to analyse the pattern of public opinion. A common technique for analysing texts is the "Bag of Words", which provides good results assuming that the words are independent of one another. In this work we propose the use of Hidden Markov Chains to determine the polarity of the opinions expressed on movie reviews. We propose a method for simulating hidden states through clustering techniques; we then carry out a sensitivity analysis of the model in which we apply variations to model parameters such as the number of hidden states or the number of words used. The results show that our proposal gives a 3% improvement over the basic model using F-score for real databases of public opinion.

Keywords: Sentimental analysis · Hidden Markov Models · Clustering

© Springer Nature Switzerland AG 2019
I. Nyström et al. (Eds.): CIARP 2019, LNCS 11896, pp. 152–162, 2019.
https://doi.org/10.1007/978-3-030-33904-3_14

1   Introduction

The growth of Internet-based means of communication like blogs and social networks has promoted interest in sentiment analysis. With the proliferation of opinions, value judgments, recommendations and other forms of expression in the net, on-line opinion has become a sort of virtual currency for companies seeking to sell their products, identify new opportunities and manage their reputations [1]. As companies seek ways to automate processes such as filtering, understanding of conversations, identification of relevant content and appropriate execution, many of them are looking towards sentiment analysis. There are many factors which determine how opinions, value judgments and criticisms are written. Cultural factors, linguistic subtleties and differential contexts make it difficult to interpret a chain of text and obtain subjective information such as a person's emotions or posture with respect to a particular context. Sentiment analysis is an area of study which analyses opinions, sentiments, evaluations, aptitudes and emotions towards entities such as products, services, organisations, individuals, topics, events and their attributes. The problem has applications in a wide range of fields; it is also known as text mining, subjectivity analysis, review mining, emotion analysis, opinion mining and opinion extraction, depending on the use to be given to the information.

During the past ten years, the amount of subjective information posted in the Internet has grown exponentially due to the expansion of Web 2.0. The ability to extract and apply a set of subjective information related with a specific context, using methods such as Hidden Markov Chains, logistic regression, SVM or deep neural networks, make it possible to obtain data to which sentiment analysis can be applied.

Sentiment analysis of texts is acquiring greater importance every year in the Internet; for example, on-line opinion has become an important factor for companies seeking to identify new business opportunities or understand correctly what customers think about the company or its products. Different techniques are used to process the information, for example word filters or identification of relevant content. However these are not entirely appropriate since they are simple filter processes which do not include a search for patterns. Many people are therefore turning to sentiment analysis which offers more appropriate tools for determining text qualities. Some methods of sentiment analysis allow the construction of models which can determine a text's qualities, but as this is a relatively new field of informatics, it is not yet known exactly which methods are best suited to this kind of problem. There are many feasible methods, each with its degree of complexity in implementation. Hidden Markov Chains are a probabilistic model for modelling a Markov process with unknown parameters; more explicitly, we can determine the unknown parameters of the chain through observable parameters. This type of model has many applications, e.g. face [2] and voice [3] recognition. The Bag of Words method on the other hand is used to process natural language and for recovering information in order to represent documents, principally for document classification [4]. The HMM method is based on the relation between two words. This enriches the descriptive power of the model compared to the Bag of Words method, which ignores word order. Although there are many works on HMM, there appear to be none which propose the application of this technique to sentiment analysis. This research proposes an HMM-based methodology for sentiment analysis. We consider the particular problem of predicting opinions about films and different phases in opinion formation, and especially obtaining HMM states, knowing that these are inaccessible.

2   Sentiment Analysis

Among the existing applications in the field of sentiment analysis, [5] present a model capable of identifying and determining opinions and value judgments about products using the comments posted by product users. The method consists in extracting all the characteristics of the opinion, giving greater weight to words with greater significance for each polarity. These are used as the entry parameters of the model, while irrelevant opinions are discarded. The proposed system, which they call "Wikisent", does not require class distinction for training.

Studies of sentiment analysis using social network data as the data source already exist: [6] carry out a sentiment analysis based on information from Twitter. This work takes tweets and classifies them as "positive", "negative" or "neutral" with respect to a specific topic. Word chains are treated as "trees"; in other words a sentence is taken and broken down to identify all the words by type, e.g. noun, pronoun, verb or adjective. This enables the model to filter the sentences and give greater weight to words more strongly oriented towards a polarity, according to the system, and to ignore those of little importance for classification. Twitter users often make use of emoticons, such as ":), :D, :(), :c", and acronyms, like "gr8t", "lol" and "roft". These are also ways of expressing polarity within a sentence and the model presented in the study cited allows expressions of this kind to be converted into value judgments or emotions.

Looking at existing methods of carrying out sentiment analysis, [7] compare the effectiveness of different classifiers in the context of sentiment analysis. The same work also presents a new method based on hybrid use of multiple classifiers to improve sentiment analysis performance; the idea of this method is that if one classifier fails, the system passes automatically to another until the opinion is classified or no further classifiers exist. The methods presented in the paper are: General Inquirer Based Classifier (GIBC), Rule-Based Classifier (RBC), Statistics Based Classifier (SBD), Induction Rule Based Classifier (IRBC), Support Vector Machine (SVM) and the hybrid classification which combines all these methods.

There are various techniques for sentiment analysis. The work of [8] presents a study of the whole corpus called EmotiBlog. This is a collection of blog entries in which the focus is on detecting subjective expressions in new texts given the context of opinions about telephones. It also shows a comparison between the results of the EmotiBlog corpus and those of a bigger corpus known as JRC; EmotiBlog was found to have a better performance.

More recently, Deep Learning techniques have been applied to the sentiment analysis problem in Twitter [9]. These techniques consist in the application of neural networks using a high number of layers and convolution. Unfortunately, these techniques tend to abstract the reasoning used for the decision.

[10] present different problems which arise in sentiment analysis. It explains the different ways of approaching a set of subjective data, such as: "sentiment classification", which focuses on determining moods in a text; "polarity classification", which is aimed at determining the orientation of words as positive or negative; "subjectivity classification" which focuses on determining how subjective the user's opinion is, referring to a particular context; and "text summaries" which seeks to summarise the information in order to clearly understand opinions within long paragraphs.

Finally, Hidden Markov Models (HMM) have also been applied in sentiment analysis. [3] explains in detail the concept of HMM and its application in voice recognition. [11] apply HMM to sentiment analysis by considering the label information as positive, negative or neutral. The hidden state and the position of words are both considered known. Although this is an interesting work it assumes knowledge of the labels, which requires greater human effort. In the present work, we propose an alternative strategy for applying Hidden Markov Models to sentimental analysis without knowing states labels focusing on the prediction of the polarity of opinions, that is, whether they are positive or negative. For such case, we emulate the state labels using the clusters obtained by a clustering algorithm. We detail our method in the next section.

3 Proposed Method

The use of Hidden Markov Models in sentiment analysis is justified because an opinion consists of a limited number of words which together represent what the person is trying to express; to understand it, a person proceeds to read the words sequentially from left to right, since usually each word is related to the previous one to create a meaningful sentence. Thus the words used to write an opinion can be modelled as observations in a Hidden Markov Model. Nonetheless, this method requires to know the HMM states.

In this work we propose an HMM-based method for sentiment analysis where our main idea is that the HMM state can be modelled approximately by considering a hidden variable given by patterns which are independent of the class of text. In our case, we propose the use of word clusters since these may indicate a significant word pattern. Now we describe the steps of our proposed method:

3.1 Construction of a Word Dictionary

In this stage a dictionary of words is constructed for use by the sentiment analysis models. Only words which are neither numbers nor symbols are included. A unique identifier is assigned to each word. The omission of numbers and symbols allows the size of the training data set to be reduced, thus accelerating the training process.

3.2   Filtering the Dictionary of Words

The object of filtering is to select the words which are most strongly polarised as positive or negative opinions. Firstly the database files are represented to show the occurrence of each word in each file of the database. For example, the first word, whose identifier is 1, will have an occurrence value of 1 if it is present in the text file under review and 0 if it is not present. This process is repeated for all the words in the database for all the opinion files. All the values are then totalled in order to obtain a unique vector representing the occurrence of each word in all the files of a training database.

Once the representations are obtained for both classes, these vectors are taken and the occurrence value of the word in the negative class is subtracted from the occurrence value of the word in the positive class; the result is saved in a new vector whose dimension is the number of words in the database. This can be seen in Fig. 1.

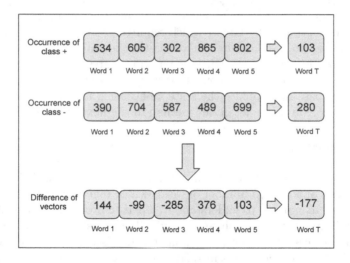

Fig. 1. Result of the count of occurrences in each class. Each word is represented by the difference of occurrences between positive and negative opinions.

The vectors are now ordered such that words with a tendency towards the negative polarity are placed to the left and words with a tendency towards the positive polarity are placed to the right. A set of words taken from either extreme is selected according to a certain criterion and these will be the significant words for our data set. Figure 2 shows a visual example of this step.

The final occurrence vector is used to reduce the number of words; the words from each extreme are selected because they are the words with the greatest difference in occurrence between classifications, and therefore are the most discriminatory.

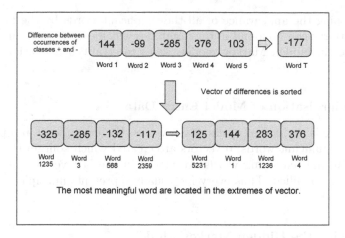

Fig. 2. Ordering the occurrence vectors based on polarity. We expect that a meaningful word for polarity to be placed in the extremes of the list.

3.3 Simulation of the Model's Hidden States

Since the states of the words are unknown, we propose simulating the states using the K-Means clustering algorithm. Using the significant words filtered in the previous step, an occurrence count is done in each file as to establish whether every significant word is present; vectors are created whose length is equal to the number of significant words, and the number of vectors is equal to the number of files per class in the database.

The K-means algorithm is applied to these occurrence vectors, to cluster each opinion file by closeness. This information is used to total the occurrence of each word in each file. The totals are divided by the total number of words in the database in order to calculate the probability of a word in consideration of each centroid, which can be interpreted as the probability of an observation belonging to a state. Figure 3 shows how each word has a certain probability of association with a certain cluster.

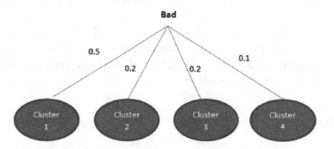

Fig. 3. Simulation of hidden states using the K-Means algorithm. The state is represented by a cluster, which indicates a similar group of words.

By grouping the appearance of all the significant words by the file of each, polarity patterns can be found in the databases which can be used during training as states in hidden Markov models so as to establish a relation with the observations.

3.4   Standardisation of Model Entry Data

Each entry file is divided into sets of ten words. For example, if an opinion file contains 55 significant words, 6 vectors are created which contain the words in groups of 10; the remaining words are placed in the last vector and repeated until the vector is filled. Thus every file usually represents multiple entries into the HMM model.

3.5   Training the Hidden Markov Model

The hidden states of the HMM are represented by the centroids of the clusters, assuming that there are N states. The observations correspond to the significant words in the training set, taking M as the cardinal. The probability distribution of the observations is defined as the probability of the occurrence of a significant word in a cluster. The probability of observations by state is represented stochastically considering the probability of an observation, given a state and considering the frequency of words within clusters. The initial probability of states is random. We apply the Baum-Welch algorithm to train the HMM model.

3.6   Classification of Opinions Using Hidden Markov Models

A Hidden Markov Model is trained for each class, in this case positive and negative. A test text entry S is then sent to each model and the probabilities of occurrence are calculated using the Forward algorithm. Each HMM returns a probability and the model with the highest probability of occurrence will indicate the class of this test text.

$$Class(S) = \arg\max_c p(c)p(S/HMM_c) = \arg\max_c p(c) \prod_{i=1}^{K_S} p(S_i/HMM_c)$$

We divide each file with K_S disjoint parts $(S = \cup_{i=1}^{K_S} S_i)$, where each part has 10 words and assumes that they are conditionally independent given the class. Then, each part is processed by an HMM and multiply the probabilities given by the Forward algorithm for each one.

4   Results

In this section we present and discuss the results obtained by implementing the proposed model. The base method used for comparison is the typical Bag of Words model, which considers that all the words are independent. The databases used in this work are opinions about films; each opinion is pre-defined to a polarity, either positive or negative. Each text file contains one user's opinion about a particular film.

The first database, "Review Movie Dataset", was used in the article "A Sentimental Education: Sentiment Analysis Using Subjectivity Summarization Based on Minimum Cuts" [12]. The second database, "Large Movie Review Dataset", is a data set assembled in "Learning Word Vectors for Sentimental Analysis" [13]. Finally, the third database "Movie reviews sentiment", focuses mainly on polarity of movie reviews [14]. The characteristics of the databases used are summarised in Table 1.

Table 1. Details of Datasets used. All the datasets have two classes in these experiments.

Database	Records	Classes	Sample	Average number of words by sample
DB 1: Review Movie	2000	2	2000	632
DB 2: Large Movie R.	50000	2	2000	281
DB 3: Movie Review S.	10662	2	2000	20

4.1   Experiments

In our experiments we tested the sensitivity of the model for the number of words and the number of hidden states in the Markov model. Each modification resulted in a change in the model's behaviour, since when the number of significant words is modified, the number of symbols accepted by the Hidden Markov Model increases or diminishes; while when the number of hidden states is modified, the observations may be associated differently with the states. The results of these experiments are shown in Table 2.

In all the databases the HMM model obtains better results than the Bag of Words method. In the first database (MR), the best configuration for HMM modelling is with 200 words and 10 hidden states, with a performance of 83.5%; the best configuration for the Bag of Words technique was with 100 words giving a performance of 81.7%. In the second database (LMR), the best configuration for HMM modelling is with 400 words and 10 hidden states, with a performance of 83.4%; the best configuration for the Bag of Words technique was with 400 words giving a performance of 79.4%. In the third database (MRS), the best configuration for HMM modelling is with 400 words and 20 hidden states, with a performance of 74.6%; the best configuration for the Bag of Words technique was with 400 words giving a performance of 66.9%. In the experimental phase we sought to maximise the probability for both techniques, achieving improvements of up to 8% for HMM and 4% for Bag of Words. It is possible to go on modifying the number of entry words for the models since a configuration may exist which gives better results. Finally the two techniques require quite different computing times. Bag of Words requires minutes to carry out cross-validation while HMM

Table 2. Results of RM Database with 100, 200 and 400 observations. In general, the use of HMM overcomes BoW in all the tested settings, where the best result for F-value is considering 400 observations.

Method	Accuracy	Precision	F-Value
Considering 100 observations			
HMM(100, 5)	82.4 (2.1)	82.7 (1.8)	82.3 (2.4)
HMM(100, 10)	82.2 (1.7)	83.6 (1.9)	81.8 (2.0)
HMM(100, 20)	**82.7 (2.2)**	**83.0 (1.8)**	**82.6 (2.5)**
BoW(100)	81.7 (2.2)	81.4 (2.7)	81.7 (2.3)
Considering 200 observations			
HMM(200, 5)	82.9 (2.4)	83.0 (2.5)	82.9 (2.4)
HMM(200, 10)	**83.5 (2.6)**	**83.9 (2.8)**	**83.3 (2.7)**
HMM(200, 20)	83.4 (2.4)	83.6 (2.4)	83.3 (2.5)
Bag-of-Words(200)	80.8 (2.0)	80.9 (2.4)	83.3 (2.5)
Considering 400 observations			
HMM(400, 5)	**83.3 (2.6)**	**82.3 (2.2)**	**83.5 (2.8)**
HMM(400, 10)	83.1 (2.3)	82.8 (2.2)	83.2 (2.4)
HMM(400, 20)	83.0 (2.2)	81.6 (1.8)	83.3 (2.6)
Bag-of-Words(400)	81.5 (2.2)	82.3 (2.6)	81.7 (2.2)

Table 3. Results of LMR dataset with 100, 200 and 400 observations. In general, the use of HMM overcomes BoW in all the tested settings, where the best result for F-value is again considering 400 observations.

Method	Accuracy	Precision	F-Value
Considering 100 observations			
HMM(100, 5)	78.2 (2.1)	78.1 (2.9)	78.2 (1.9)
HMM(100, 10)	**78.5 (2.2)**	**78.9 (3.7)**	**78.4 (1.8)**
HMM(100, 20)	78.2 (2.4)	78.7 (2.8)	78.0 (2.3)
Bag-of-Words(100)	76.2 (1.9)	76.3 (2.7)	76.1 (1.7)
Considering 200 observations			
HMM(200, 5)	79.9 (2.2)	80.4 (3.2)	79.7 (2.1)
HMM(200, 10)	**80.3 (1.6)**	**79.5 (2.8)**	**80.6 (1.3)**
HMM(200, 20)	79.7 (2.0)	80.1 (2.8)	79.6 (2.0)
Bag-of-Words(200)	75.8 (3.2)	75.0 (3.4)	75.5 (3.3)
Considering 400 observations			
HMM(400, 5)	82.7 (2.6)	82.6 (3.1)	82.7 (2.5)
HMM(400, 10)	**83.4 (2.6)**	**83.7 (3.6)**	**83.3 (2.3)**
HMM(400, 20)	83.0 (2.6)	83.9 (2.9)	82.8 (2.6)
Bag-of-Words(400)	79.4 (1.9)	79.0 (2.0)	79.5 (2.1)

Table 4. Results of MRS dataset with 100, 200 and 400 observations. In general, the use of HMM overcomes BoW in all the tested settings, where the F-value best result considers 400 observations.

Method	Accuracy	Precision	F-Value
Considering 100 observations			
HMM(100, 5)	65.5 (3.4)	65.0 (4.0)	66.4 (2.7)
HMM(100, 10)	**66.4 (2.3)**	**67.7 (3.5)**	**66.3 (1.8)**
HMM(100, 20)	66.1 (2.8)	65.2 (3.2)	67.3 (2.3)
Bag-of-Words(100)	63.9 (2.2)	64.0 (2.3)	63.7 (2.7)
Considering 200 observations			
HMM(200, 5)	69.5 (3.4)	68.7 (3.5)	70.1 (3.3)
HMM(200, 10)	**70.0 (3.0)**	**70.3 (2.9)**	**69.7 (3.3)**
HMM(200, 20)	69.0 (3.9)	68.2 (3.9)	69.7 (3.7)
Bag-of-Words(200)	65.7 (2.8)	65.6 (3.5)	66.0 (2.5)
Considering 400 observations			
HMM(400, 5)	73.7 (2.5)	73.6 (3.1)	73.8 (2.3)
HMM(400, 10)	73.7 (2.8)	74.1 (3.6)	73.6 (2.5)
HMM(400, 20)	**74.6 (3.3)**	**74.8 (4.2)**	**74.6 (2.9)**
Bag-of-Words(400)	66.9 (3.6)	66.6 (4.3)	67.3 (2.9)

requires at least 1 h, or even more depending on the number of words used as possible entry values and the number of model entry data. The details of our implementation and the used datasets are available in[1] (Tables 3 and 4).

5   Conclusions

We conclude that the Hidden Markov Models technique is able to model the sentiment analysis, in particular the proposed technique can be used to classify the polarity in opinions about movies. Moreover, we also compared the proposed HMM-based technique with the Bag of Words method, obtaining better results with the proposed technique in all the real databases tested, indicating that the proposed method is competitive for sentiment analysis. Variations of the HMM model were applied to determine whether better performance could be obtained. The improvements varied with different configurations for each database, from which we conclude that it is necessary to experiment with the parameters of the HMM to find the best configuration for each database tested. As a future work, we propose to use a more powerful model as Hidden Semi-Markovian Models and mixture of Hidden Markov Models.

---

[1] https://drive.google.com/open?id=1Ke84q27ovr3bnfxC4BeDDyC1tqHcosfl.

References

1. Mukherjee, S., Bhattacharyya, P.: Sentiment analysis: a literature survey. Technical report, Indian Institute of Technology, Bombay (2013)
2. Nefian, A.V., Hayes III, M.H.: Face detection and recognition using hidden markov models. In: Proceedings of International Conference on Image Processing, vol. 1, pp. 141–145. IEEE (1998)
3. Rabiner, L.: A tutorial on hidden markov models and selected applications in speech recognition. Proc. IEEE **77**(2), 257–286 (1989)
4. Guzella, T.S., Caminhas, W.M.: A review of machine learning approaches to spam filtering. Expert Syst. Appl. **36**(7), 10206–10222 (2009)
5. Mukherjee, S., Bhattacharyya, P.: Feature specific sentiment analysis for product reviews. In: Gelbukh, A. (ed.) CICLing 2012. LNCS, vol. 7181, pp. 475–487. Springer, Heidelberg (2012). https://doi.org/10.1007/978-3-642-28604-9_39
6. Agarwal, A., Xie, B., Vovsha, I., Rambow, O., Passonneau, R.: Sentiment analysis of twitter data. In: Proceedings of the Workshop on Languages in Social Media, Association for Computational Linguistics, pp. 30–38, June 2011
7. Prabowo, R., Thelwall, M.: Sentiment analysis: a combined approach. J. Informetrics **3**(2), 143–157 (2009)
8. Fernández, J., Boldrini, E., Gómez, J.M., Martínez-Barco, P.: Análisis de sentimientos y minería de opiniones: el corpus emotiblog. Procesamiento del lenguaje natural **47**, 179–187 (2011)
9. Tang, D., Wei, F., Qin, B., Liu, T., Zhou, M.: Coooolll: a deep learning system for twitter sentiment classification. In: Proceedings of the 8th International Workshop on Semantic Evaluation (SemEval 2014), pp. 208–212 (2014)
10. Kim, H.D., Ganesan, K., Sondhi, P., Zhai, C.: Comprehensive review of opinion summarization. Technical report, University of Illinois at Urbana-Champaign (2011)
11. Jin, W., Ho, H.H., Srihari, R.K.: Opinionminer: a novel machine learning system for web opinion mining and extraction. In: Proceedings of the 15th ACM SIGKDD International Conference on Knowledge Discovery and Data Mining, pp. 1195–1204. ACM (2009)
12. Pang, B., Lee, L.: A sentimental education: Sentiment analysis using subjectivity summarization based on minimum cuts. In: Proceedings of the 42nd Annual Meeting on Association for Computational Linguistics. ACL 2004, Association for Computational Linguistics (2004)
13. Maas, A.L., Daly, R.E., Pham, P.T., Huang, D., Ng, A.Y., Potts, C.: Learning word vectors for sentiment analysis. In: Proceedings of the 49th Annual Meeting of the Association for Computational Linguistics: Human Language Technologies, Portland, Oregon, USA, Association for Computational Linguistics, pp. 142–150, June 2011
14. Pang, B., Lee, L.: Sentence polarity dataset v1.0 (2017). https://www.kaggle.com/nltkdata/sentence-polarity. Accessed 10 Jan 2017

Optimizing Natural Language Processing Pipelines: Opinion Mining Case Study

Suilan Estevez-Velarde[1](✉), Yoan Gutiérrez[2], Andrés Montoyo[3],
and Yudivián Almeida-Cruz[1]

[1] School of Math and Computer Science,
University of Habana, Habana, Cuba
{sestevez,yudy}@matcom.uh.cu
[2] University Institute for Computing Research (IUII),
University of Alicante, Sant Vicent del Raspeig, Spain
ygutierrez@dlsi.ua.es
[3] Department of Languages and Computing Systems,
University of Alicante, Sant Vicent del Raspeig, Spain
montoyo@dlsi.ua.es

Abstract. This research presents NLP-Opt, an Auto-ML technique for optimizing pipelines of machine learning algorithms that can be applied to different Natural Language Processing tasks. The process of selecting the algorithms and their parameters is modelled as an optimization problem and a technique was proposed to find an optimal combination based on the metaheuristic Population-Based Incremental Learning (PBIL). For validation purposes, this approach is applied to a standard opinion mining problem. NLP-Opt effectively optimizes the algorithms and parameters of pipelines. Additionally, NLP-Opt outputs probabilistic information about the optimization process, revealing the most relevant components of pipelines. The proposed technique can be applied to different Natural Language Processing problems, and the information provided by NLP-Opt can be used by researchers to gain insights on the characteristics of the best-performing pipelines. The source code is made available for other researchers. In contrast with other Auto-ML approaches, NLP-Opt provides a flexible mechanism for designing generic pipelines that can be applied to NLP problems. Furthermore, the use of the probabilistic model provides a more comprehensive approach to the Auto-ML problem that enriches researcher understanding of the possible solutions.

Keywords: Natural Language Processing · Pipeline optimization ·
Metaheuristics · Opinion mining

1 Introduction

Text mining tasks generally require the evaluation of different machine learning and natural language processing algorithms [5]. Selecting the right algorithms

© Springer Nature Switzerland AG 2019
I. Nyström et al. (Eds.): CIARP 2019, LNCS 11896, pp. 163–173, 2019.
https://doi.org/10.1007/978-3-030-33904-3_15

for each task requires either a comprehensive knowledge of the problem domain, previous experience in similar contexts, or trial and error [4]. These algorithms interact with each other creating a complex pipeline [11]. As the possible combinations of algorithms for creating a pipeline could vary widely [7], it is unfeasible to evaluate all of them. Evaluating only a few combinations is risky if the combinations are manually selected because we may miss the best solutions. This is also the case when using grid-search, and similar techniques, which are either too expensive when the grid resolution is very high, or the optimal parameter combination with a lower resolution cannot be found. Consequently, a new research area has emerged recently, namely Auto-ML (Automatic Machine Learning), which designs techniques for automatically finding the optimal configuration of machine learning pipelines through the use of optimization techniques.

Current Auto-ML approaches are often designed as black-box optimization tools, and are thus applicable mainly to machine learning problems in which the input is a feature matrix. However, Natural Language Processing (NLP) problems have to deal with a variety of tasks before obtaining the feature matrix, involving preprocessing (e.g., removing stopwords, performing stemming, etc.) and vectorization (e.g., using TF-IDF or embeddings). For this reason, applying existing Auto-ML techniques to NLP problems still requires a degree of manual feature engineering. An Auto-ML tool suitable for NLP problems needs to consider also the preprocessing and vectorization tasks in order to completely automate the solution of the NLP problem.

This paper proposes a Natural Language Processing Optimizer –hereafter referred to as NLP-Opt– for enhancing pipelines of algorithms for NLP problems where several choices are possible at each stage. NLP-Opt delivers a combination of algorithms and their parameters in a feasible time frame, by performing an intelligent exploration of all possible combinations. In this paper, a pipeline is defined as a sequence of algorithms, each one receiving as input the output of the previous stage. We model the selection of algorithms and features as an optimization problem and propose using the metaheuristic Population-Based Incremental Learning (PBIL) proposed by Baluja [3] to find an optimal combination. To validate the methodology behind NLP-Opt, we focus on a specific NLP task, i.e., Opinion Mining [14] in Spanish language tweets. Opinion Mining is an NLP problem that involves several tasks for preprocessing and extracting relevant features, as well as classification, and posses an interesting challenge in terms of modelling the possible pipelines to solve it.

The following tangible results are highlighted:

- A formal optimization model for a generic pipeline of algorithms, and an optimization technique based on PBIL metaheuristic (Sect. 3) to find the best combination of algorithms (i.e., NLP-Opt).
- NLP-Opt is validated in an NLP problem, in this case opinion mining, using a specific pipeline designed for this task (Sect. 4).
- NLP-Opt permits the discovery of the most relevant parameters for each pipeline component (Sect. 4.1).

– The source code is made available to encourage other authors to further develop this line of research[1].

The rest of the paper is organized as follows: Sect. 2 gives a brief overview of current research in Auto-ML techniques; Sect. 3 explains the core of our proposal, NLP-Opt; Sect. 4 presents an application of NLP-Opt for opinion mining in Twitter, to illustrate the types of problems and domains where NLP-Opt is suitable; Subsect. 4.1 describes the experiments performed and the characteristics of the corpus used in the algorithm's evaluation; Sect. 4.2 presents a general discussion and the analysis of the results obtained; and finally, Sect. 5 presents the main conclusions of the research and outlines the possible future lines of development.

2   Related Work

The rising complexity of modern machine learning techniques for users not specialized in the ML field has motivated research into techniques for the automatic configuration and optimization of machine learning pipelines, i.e., Auto-ML.

The most common approaches for designing Auto-ML techniques are based on Evolutionary Computation (EC), Bayesian Optimization (BO) and Monte Carlo Tree Search (MCTS). Examples of EC are Recipe [17] and TPOT [13], while BO is used in Auto-Sklearn [6], Auto-Weka [19] and Hyperot [8]. ML-Plan [12] is a recent technique that models machine learning pipelines as hierarchical task networks and applies MCTS.

The previous examples provide out-of-the-box machine learning pipelines, which can be automatically configured for a specific dataset, selecting among a large collection of shallow classification, dimensionality reduction, pre-processing and feature selection techniques. Auto-ML tools provide an end-to-end solution for the optimization of machine learning pipelines. However, several techniques have been proposed to deal with specific parts of the machine learning process, rather than whole pipelines [18]. For example, Harmony Search has been used for feature selection in text clustering [1]. Also, several different metaheuristics have been compared for dynamic reduction in text document clustering [2].

The main challenge when applying current Auto-ML tools to NLP problems is related with the feature engineering phase. The existing tools assume an input representation based on feature matrix, hence the researcher still needs to manually select which preprocessing tasks (e.g., stopword removal, stemming, etc.) to perform and how to encode them in features (e.g., using TF-IDF or word embeddings).

3   Proposal for Pipeline Optimization

NLP-Opt, based on the PBIL metaheuristic [3], is designed to find out the optimal configuration of a machine learning pipeline for Natural Language Processing. Pipelines are represented as a sequence of stages for which there are different

---

[1] For blind-review purposes this link is omitted.

options to apply (i.e., there are different algorithms or techniques for solving each stage).

The possible combinations of a generic pipeline with m stages are modeled as a subset of the space N^m. Each vector θ in this subset corresponds to one specific combination of algorithms and vice-versa. The cost function P_θ corresponds to the evaluation of the precision in that combination. The problem of finding the best combination could be modelled as an optimization problem in integers:

$$\max_{\theta}\{P_\theta(X_t, y_t, X_p, y_p) \in [0, 1]\}$$

$$\text{s.t. } \theta \in d_1 \times d_2 \times \ldots \times d_m \subset N^m$$

where:

P_θ is a quality metric of the solution for the combination θ;
X_t, y_t is a list of data points and classes of the training set;
X_p, y_p is a list of data points and classes of the test set; and,
$\quad d_i$ is the domain of the ith component, i.e., the number of options in that stage.

The optimization process of NLP-Opt consists of a generation and evaluation loop that simultaneously searches at each stage for the combination of options that provides the optimal fitness (i.e., the optimal performance when trained on a specific dataset). Pipelines are generated by sampling from a probability distribution that represents which options are more likely to produce the best performance. During the optimization loop, the probability distribution is adjusted according to a selection of the best performing pipelines discovered for each iteration. Figure 1 shows a graphic representation of this process.

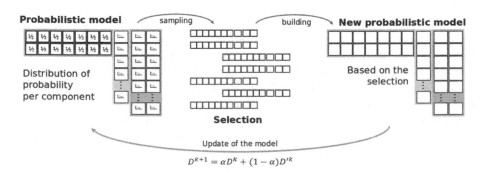

Fig. 1. Representation of process of optimization reusing PBIL.

In the optimization problem the search space is considered as a stochastic cost function. Hence, it is desirable to obtain not only an optimal solution, but also a structure that helps to identify the best combinations, irrespective of the bias introduced by the noisy evaluations. In this sense, we decided to reuse PBIL for providing not only a particular final solution, but also a distribution function for each component of the solution. The analysis of this distribution enables the selection of the most promising algorithms according to their behavior during the entire experimentation period and avoids bias towards a single noisy result.

PBIL [3] is one of the first Estimation of Distribution Algorithms—EDA— created [10]. This is a population-based algorithm with an incremental learning designed for combinatorial problems. Every component of the solution vector is modeled as a gene, which is associated with a probability distribution over its discrete values.

The algorithm NLP-OPT initialize with a uniform distribution for each gene. These distributions change in time, based on the probability distribution of the genes of the best solutions in the population. In every iteration the best n solutions are selected and the probability distribution of each component is modified. The speed of the model adjustment is controlled by a parameter α called "learning factor". Algorithm 1 shows a simplified implementation of this process.

Algorithm 1. Human Language Technology Optimization (NLP-Opt)

—— **Define Problem**
$p \leftarrow$ defines a pipeline (number of stages)
$p_i \leftarrow$ defines for each component of p all the possibles options

—— **Parameters of the optimization algorithm**
$popsize \leftarrow$ number of pipelines to evaluate in each iteration
$n \leftarrow$ number of best pipelines selected each iteration
$\alpha \leftarrow$ learning factor

—— **Initialization**
$D \leftarrow \{D_1, ..., D_m\}$ marginal distribution by p_i component (uniform)
    In this moment all options are equally likely
Best $\leftarrow \Box$

—— **Optimization process**
while generations remain **do**
    $P \leftarrow \{\}$

    for $i = 1 ... popsize$ **do**
        $S_i \leftarrow$ pipeline built by sampling of D
        $f(S_i) \leftarrow$ calculate *fitness* of S_i
        if Best $= \Box$ or $f(S_i) < f(\text{Best})$ **then**
            Best $\leftarrow S_i$
        end if
        $P \leftarrow P \bigcup \{S_i\}$
    end for

    $P^* \leftarrow$ best n pipelines of P

    —— **Promote generation of the best pipelines**
    for each gene j in D **do**
        $N_j \leftarrow$ marginal distribution of the gene j in P^*
        $D_j \leftarrow \alpha N_j + (1 - \alpha)D_j$
    end for

end while

---

return Best

---

4   NLP-Opt Applied to Opinion Mining

This section presents a case study involving a classic NLP problem: opinion mining in Twitter. The process of solving and evaluating the Opinion Mining problem has been tackled using different approaches [9,14,21], of which Supervised classification is the most common [14,16]. Hence, for this case study, we decide to define the process of classifying messages in different stages that can be analyzed independently. The pipeline consists of 4 stages: (I) text preprocessing; (II) dimensionality reduction; (III) classification into objective-subjective; and (IV) classification into positive-neutral-negative. For each stage we select a wide range of possible algorithms to explore. This structure, with several different strategies available for the preprocessing and dimensionality reduction steps, and having two different classification steps, makes Opinion Mining a non-standard learning problem unsuitable for black-box Auto-ML tools.

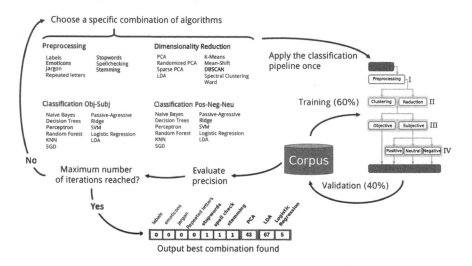

Fig. 2. Graphic representation of the optimization process, from the point of view of the Opinion Mining case study. The represented loop occurs inside the fitness evaluation of the PBIL algorithm, for each of the combinations analyzed during the whole optimization process. The color-highlighted options are an illustrative example of a possible combination.

In our proposal we model the algorithmic pipeline as a vector and apply the optimization process described in Sect. 3 over the space of all possible solutions. Each solution (combination of algorithms) is fitted on a training set, and its performance is measured on a separate test set. This process is repeated with various combinations of algorithms, until convergence is achieved or a predefined number of iterations is reached. At this point, we obtain the combination of algorithms that performs best on the validation set. Figure 2 shows this optimization process in the opinion mining task. The source code is made available to encourage other authors to further develop this line of research (See footnote 1.).

4.1 NLP-Opt Experimental Results

The optimization process described in this section optimizes all the steps of the pipeline *simultaneously*. For description purposes, we will present an analysis independently of each step of the pipeline. However, it is important to keep in mind that in each step, the "optimal" algorithm is determined in correspondence with the "optimal" options for the other steps. The algorithms used are implemented in *sklearn* [15], and highlighted in Fig. 2, Tables 2 and 3.

The parameters selected for the experimentation are 100 individuals per generation, with a selection by truncation of the best 20, and a learning factor $\alpha = 0.1$, following the recommended parameters in the literature. A higher learning rate has been shown to hinder convergence, and 0.1 seems to provide a fairly good performance in several experiments performed by Baluja et al. [3]. The population size is chosen to allow sufficient variation in the pipelines sampled for each generation. We establish a limit of a sufficiently high number of iterations –1,000 generations– and the metaheuristic evolution was monitored until a convergence in most of the components was observed. At this point, the experimentation reached 126 iterations, for a total of 12,600 evaluations of the objective function.[2]

The experimentation was performed in a single commodity computer, with an i5 microprocessor and GB of RAM memory. A total of approximately 720 computing hours were used, with an average 3.4 min per pipeline, although the actual computational cost of each pipeline varies with the complexity of the algorithms involved. The total number of possible combinations is $2^7 \cdot 52 \cdot 67 \cdot 67$, which equates to 29,878,784. At this rate, performing the experiment for all the possible combinations would require more than 1175 years of computing power.

Figure 3 shows the best combination found by NLP-Opt expressed in terms of the corresponding algorithms associated to each solution.

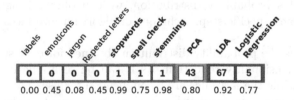

Fig. 3. Best solution found by NLP-Opt.

The experimentation process was performed on the *general* corpus, from the collection of tweets published in TASS 2014 [20]. This corpus has the advantage of being small (6,000 training examples), which decreases the computational cost of the experimentation.

---

[2] As illustrated in Fig. 2, each evaluation consists of a full run of the classification pipeline with a specific combination of preprocessing, reduction and classification.

The corpus is a collection of Spanish language messages from approximately 150 personalities in the field of politics, economics, communication, mass media and culture. The authors are of different nationalities, including Spanish, Mexican, Colombian, Puerto Rican and United States, among other countries, which provides a more diverse linguistic scenario. The Spanish language was chosen given that is a less pervasive language in terms of linguistic resources and thus an interesting challenge. In all cases, a cross-validation was carried out comprising a subset of 60% of the examples for the training set and the remaining 40% for validation.

The mean accuracy obtained is of 0.542 with a standard deviation of 0.0095. Comparatively, the accuracy obtained with a random baseline is of 0.299 with a standard deviation of 0.0078, for a relative improvement of the 44.9%. Table 1 summarizes these results, including a test for a statistical significant (t-Student).

Table 1. Comparative results of the final proposal according to NLP-Opt and a random classification. μ is the mean accuracy and σ is the standard deviation of the mean accuracy in 30 independent evaluations with 60% training and 40% validation.

Algorithm	μ	σ	p-value
Random	0.299	0.0078	-
NLP-Opt (Fig. 3)	0.542	0.0095	4.294e−68

4.2   Results and Discussion

The probability model generated determines, for each stage of the pipeline, a probability distribution of the best options (i.e., algorithms) to use in that stage. By analyzing this probability distribution, we can obtain insights about the influence that each specific step, technique or algorithm produces in the overall process.

Table 2 shows the probability distribution obtained for each stage of preprocessing and dimensionality reduction. As it can be seen, the first 4 tasks are not beneficial, while the last ones are recommended for use. The dimensionality reduction algorithms were modeled as a single component in the input vector. According to Table 2 the best options are two variants of the algorithm Principal Component Analysis (PCA). These two options are also the only ones that display an improvement in the final precision with respect to not using any algorithm for dimensionality reduction.

The process of classification is modeled similar to the dimensionality reduction process. With this aim (i.e. classifying) we defined two components, mentioned in Sect. 4, which were labelled as Objective-Subjective and Positive-Negative-Neutral, respectively.

According to the probabilities measured, see Table 3, in the first classification stage (i.e. III) the algorithm that obtains the highest probability is Linear

Discriminant Analysis. In the second classification stage (i.e IV), the algorithm that obtains the highest probability is Logistic Regression. The probability distribution obtained by the algorithms highlights an advantage for algorithms with a linear decision surface. This fact can be justified since lineal algorithms have less internal parameters to learn and therefore are less likely to overfit a small training set as the one used in this study.

Table 2. Best success probability of preprocessing algorithms (**left**) and dimensionality reduction algorithms (**right**).

Phase	Don't use	Use
tags	**1.000**	0.000
smiles	**0.553**	0.447
jargon	**0.916**	0.084
repetition	**0.548**	0.452
stop words	0.008	**0.992**
spelling	0.255	**0.745**
stemming	0.021	**0.979**

Algorithm	Probability
LDA	0.0000
S-PCA	0.0000
ICA	0.0000
None	0.0026
Mean Shift	0.0034
DBSCAN	0.0036
K-Means	0.0039
Ward	0.0047
Spectral Clustering	0.0066
R-PCA	**0.1724**
PCA	**0.8029**

Table 3. Best success probability of classification algorithms, discovered.

Algorithm	Obj-Subj	Pos-Neg-Neu
Random	0.0000	0.0000
Naive Bayes	0.0000	0.0000
Decision Tree	0.0000	0.0000
Perceptron	0.0000	0.0000
Random Forest	0.0004	0.0000
KNN	0.0006	0.0000
SGD	0.0007	0.0001
Passive-Aggressive	0.0042	0.0000
Ridge	0.0278	0.0079
SVM	0.0062	**0.2245**
Logistic Regression	0.0411	**0.7672**
Linear Discriminant	**0.9189**	0.0002

5   Conclusions and Future Work

This paper presents NLP-Opt, a metaheuristics-based Auto-ML strategy for optimizing NLP pipelines. In contrast with other Auto-ML approaches, NLP-

Opt provides a flexible mechanism for designing generic pipelines that can be applied to NLP problems, where the preprocessing and vectorization tasks are also automated. Furthermore, NLP-Opt not only finds the best pipeline architecture, but also provides probabilistic information about the exploration processes involved. This information can be used by researchers to gain insights on the characteristics of the best performing pipelines.

Improving the representation for including continuous features is a promising direction for future work. This would allow researchers to optimize not only what algorithms to select, but also the specific parameters for those algorithms, without having to define a priori a fixed subset of parameter combinations. Even though the focus of this work is NLP classification problems, the technique proposed can be easily extended to any domain where a pipeline of algorithms is used. Therefore, in future research, the technique will be adapted to more complex pipelines that are not limited to the NLP domain. Specifically, we will explore extending NLP-Opt to the optimization of deep learning architectures.

Acknowledgments. This research has been supported by a Carolina Foundation grant in accordance with the University of Alicante and the University of Havana. This work has also been partially funded by both aforementioned universities, the Generalitat Valenciana and the Spanish Government through the projects SIIA (PROMETEU/2018/089), LIVINGLANG (RTI2018-094653-B-C22) and INTEGER (RTI2018-094649-B-I00).

References

1. Abualigah, L.M., Khader, A.T., Al-Betar, M.A.: Unsupervised feature selection technique based on genetic algorithm for improving the text clustering. In: 2016 7th International Conference on Computer Science and Information Technology (CSIT), pp. 1–6. IEEE (2016)
2. Abualigah, L.M., Khader, A.T., Al-Betar, M.A., Alomari, O.A.: Text feature selection with a robust weight scheme and dynamic dimension reduction to text document clustering. Expert Syst. Appl. **84**, 24–36 (2017). https://doi.org/10.1016/j.eswa.2017.05.002. http://www.sciencedirect.com/science/article/pii/S0957417417303172
3. Baluja, S.: Population-based incremental learning. A method for integrating genetic search based function optimization and competitive learning. Technical report, DTIC Document (1994)
4. Bishop, C.M.: Model-based machine learning. Phil. Trans. R. Soc. A **371**(1984), 20120222 (2013)
5. Dinakar, K., Reichart, R., Lieberman, H.: Modeling the detection of textual cyberbullying. Soc. Mobile Web **11**(02), 11–17 (2011)
6. Feurer, M., Klein, A., Eggensperger, K., Springenberg, J., Blum, M., Hutter, F.: Efficient and robust automated machine learning. In: Advances in Neural Information Processing Systems, pp. 2962–2970 (2015)
7. Jain, S., Shukla, S., Wadhvani, R.: Dynamic selection of normalization techniques using data complexity measures. Expert Syst. Appl. **106**, 252–262 (2018). https://doi.org/10.1016/j.eswa.2018.04.008. http://www.sciencedirect.com/science/article/pii/S095741741830232X

8. Komer, B., Bergstra, J., Eliasmith, C.: Hyperopt-sklearn: automatic hyperparameter configuration for scikit-learn. In: ICML Workshop on AutoML, pp. 2825–2830. Citeseer (2014)
9. Kontopoulos, E., Berberidis, C., Dergiades, T., Bassiliades, N.: Ontology-based sentiment analysis of twitter posts. Expert Syst. Appl. **40**, 4065–4074 (2013)
10. Luke, S.: Essentials of Metaheuristics. Lulu 2009. http://cs.gmu.edu/~sean/book/metaheuristics/ (2011)
11. Manning, C., Surdeanu, M., Bauer, J., Finkel, J., Bethard, S., McClosky, D.: The stanford corenlp natural language processing toolkit. In: Proceedings of 52nd Annual Meeting of the Association for Computational Linguistics: System Demonstrations, pp. 55–60 (2014)
12. Mohr, F., Wever, M., Hüllermeier, E.: ML-Plan: automated machine learning via hierarchical planning. Mach. Learn. **107**(8), 1495–1515 (2018). https://doi.org/10.1007/s10994-018-5735-z
13. Olson, R.S., Moore, J.H.: TPOT: a tree-based pipeline optimization tool for automating machine learning. In: Hutter, F., Kotthoff, L., Vanschoren, J. (eds.) Automated Machine Learning. TSSCML, pp. 151–160. Springer, Cham (2019). https://doi.org/10.1007/978-3-030-05318-5_8
14. Pang, B., Lee, L.: Opinion mining and sentiment analysis. Found. Trends Inf. Retrieval **2**(1–2), 1–135 (2008)
15. Pedregosa, F., et al.: Scikit-learn: machine learning in Python. J. Mach. Learn. Res. **12**, 2825–2830 (2011)
16. Rosenthal, S., Farra, N., Nakov, P.: Semeval-2017 task 4: sentiment analysis in Twitter. In: Proceedings of the 11th International Workshop on Semantic Evaluation (SemEval-2017), pp. 502–518 (2017)
17. de Sá, A.G.C., Pinto, W.J.G.S., Oliveira, L.O.V.B., Pappa, G.L.: RECIPE: a grammar-based framework for automatically evolving classification pipelines. In: McDermott, J., Castelli, M., Sekanina, L., Haasdijk, E., García-Sánchez, P. (eds.) EuroGP 2017. LNCS, vol. 10196, pp. 246–261. Springer, Cham (2017). https://doi.org/10.1007/978-3-319-55696-3_16
18. Salimans, T., Ho, J., Chen, X., Sutskever, I.: Evolution strategies as a scalable alternative to reinforcement learning. arXiv preprint arXiv:1703.03864 (2017)
19. Thornton, C., Hutter, F., Hoos, H.H., Leyton-Brown, K.: Auto-weka: combined selection and hyperparameter optimization of classification algorithms. In: Proceedings of the 19th ACM SIGKDD International Conference on Knowledge Discovery and Data Mining, pp. 847–855. ACM (2013)
20. Villena-román, J., Lana Serrano, S., Martínez Cámara, E., González Cristóbal, J.C.: TASS workshop on sentiment analysis at SEPLN. Procesamiento del Lenguaje Natural (2013)
21. Zhang, L., Ghosh, R., Dekhil, M., Hsu, M., Liu, B.: Combining lexicon based and learning-based methods for twitter sentiment analysis. HP Laboratories, Technical Report HPL-2011 89 (2011)

Predicting Customer Profitability Dynamically over Time: An Experimental Comparative Study

Daqing Chen[1($\boxtimes$)], Kun Guo[1], and Bo Li[2]

[1] School of Engineering, London South Bank University, London SE1 0AA, UK
{chend, guok}@lsbu.ac.uk
[2] School of Electronics and Information, Northwestern Polytechnical University,
Xi'an 710072, China
libo803@nwpu.edu.cn

Abstract. In this paper a comparative study is presented on dynamic prediction of customer profitability over time. Customer profitability is measured by Recency, Frequency, and Monetary (RFM) model. A real transactional data set collected from a UK-based retail is examined in the analysis, and a monthly RFM time series for each customer of the business has been generated accordingly. At each time point, the customers can be segmented by using the k-means clustering into high, medium, or low groups based on their RFM values. Twelve different models of three types have been utilized to predict how a customer's membership in terms of profitability group would evolve over time, including regression, multilayer perceptron, and Naïve Bayesian models in open-loop and closed-loop modes. The experimental results have demonstrated a good, consistent and interpretable predictability of the RFM time series of interest.

Keywords: Time series analysis · RFM model · CRM · Predictive modelling

1 Introduction

Over the last decades marketing has made a significant shift from traditional product/brand based to customer-centric and data-driven by intensively using analytical models and tools. One of the important aspects of applying analytics in marketing is to predict customer profitability over time based on customer purchasing history and a certain profitability measure, such as customer life-time value (CLV), and recency, frequency, and monetary (RFM) values. With regard to modelling techniques, there are mainly two categories of models [1]: probabilistic models and machine learning models. A fundamental question to be asked here is that if a customer's profitability is predictable, and what models can be best suitable a given prediction problem [2]. The primary aim of this research is to provide a case study for such a prediction.

In this paper, a UK-based online retail is examined for customer profitability prediction. A real transactional data set collected from the retail is used for the analysis. The RFM values of each customer are employed as a profitability measure, and an associated monthly RFM time series for every customer can be created accordingly

© Springer Nature Switzerland AG 2019
I. Nyström et al. (Eds.): CIARP 2019, LNCS 11896, pp. 174–183, 2019.
https://doi.org/10.1007/978-3-030-33904-3_16

based on their historical purchasing records. By using the k-means clustering, at any given time point, all the customers are segmented into three groups based on their RFM values: high, medium, or low profitability groups. The prediction problem concerned here is how a customer's membership in terms of profitability group would evolve over time. For comparison purpose, twelve different models of three types are utilized for prediction including both probabilistic and machine learning models in open-loop and closed-loop modes: regression, multilayer perceptron (MLP), and Naïve Bayesian models. Choosing RFM-based measure is because of its simplicity and easy interpretability in practice, and the models selected are classic, simple and widely used in business for marketing purpose.

A comparative analysis with the given data set and the models has demonstrated a good predictability of the chosen measure for the business under consideration in terms of customer profitability. It also shows how to use the certain context of the business to help to interpret the modelling outcomes.

The remainder of this paper is organized as follows. Section 2 gives a brief discussion on the relevant work. Sections 3 describes in detail the methodology adopted in this work including the creation of RFM-based time series, customers grouping and model selection. Detailed experimental settings and the experimental results are provided in Sect. 4, and a discussion on the outcomes is given in Sect. 5. Finally concluding remarks are given in Sect. 6 along with suggested further work.

2 Related Works

In recent years, predicting customer's profitability over time has been an active, yet very challenging research topic. In general, such a prediction mainly involve three interrelated factors:

- The nature of the business under consideration;
- Which measure(s) to be used to indicate a customer's profitability; and
- Which models to be employed to best fit the modelling requirements.

The nature of the business under consideration will be directly linked to what measures could be adopted, for example, an on-line business in the retail industry and a marketing consultancy company in the fashion industry may use a completely different set of measures. On the other hand, depending on the measures to be adopted, a static or a dynamic model could be applied for modelling purpose.

In [3] an RFM score-based time series was created using the k-means clustering analysis and used to measure and describe a customer's profitability for an on-line retail. Furthermore, multilayer feedforward neural network models were trained to identify the dynamics in terms of how customer profitability evolved over time.

Interestingly in [4], RFM was employed to calculate customer loyalty and Apriori algorithm was used to determine the association rules of product bundles. In addition, the work in [5] suggested convolutional neural network structures for predicting the CLV of individual players of video games, and in [6], recurrent neural networks were proposed for customer behavior prediction based on the client loyalty number and RFM values.

Other measures have been also considered, such as Pareto/NBD (negative binomial distribution) [7].

In summary, the main work in this area appears to be subject/domain-specific and has no unified approaches. CLV and RFM are the most popular measures adopted to reflect customer profitability/loyalty. The most diverse aspect of the relevant research is on modelling approaches, and a range of models have been proposed, from very classic regression models to deep learning paradigms.

This research presents a case study for customer profitability prediction in which multiple models are used with a simple yet practically easy-to-implement profitability measure.

3 Methodology

This Section gives in detail the main approaches, models, and procedures adopted in this research.

3.1 Recency, Frequency, and Monetary Model

The RFM model [8] has received much attention and has been widely used in customer relationship management (CRM) and direct marketing due to its simplicity and effectiveness for evaluating a customer's profitability.

Given a set of transactional records of a business over a certain period of time, Recency indicates how recently a customer made a purchase with the business; Frequency shows how often a customer has purchased; and Monetary indicates the total (or average) a customer has spent. Therefore, each customer of a business can be characterized by a set of RFM values, and furthermore all the customers can be grouped into meaningful segments based on their RFM values so that various marketing strategies can be adopted to different customer groups accordingly.

Note that a time series of RFM values can be generated for each customer if they are calculated at consecutive time points, such as at the end of each calendar month over a period of time.

3.2 k-Means Clustering

k-means clustering is one of the most popular algorithms in data mining for grouping samples into a certain number of groups (clusters) based on Euclidean distance measure. Assume $V_1, V_2, \ldots, V_n$ are a set of vectors, for instance, a vector represents a customer's RFM values in the form a vector, and these vectors are to be assigned to k clusters $S_1, S_2, \cdots, S_k$. Then the objective function of the k-means clustering is expressed as

$$f(\mu_1, \mu_2, \cdots, \mu_k) = \sum_{i=1}^{k} \sum_{V_j \in S_i} \left\| V_j - \mu_i \right\|^2 \tag{1}$$

where μ_i represents the centroid of cluster S_i. The k-means clustering algorithm in the form of pseudocode is shown in Table 1.

Table 1. The k-means clustering algorithm.

Step 0: Initialise the centroids of k clusters $\mu_1, \mu_2, \cdots, \mu_k$

Step 1: Assign V_j $(1 \leq j \leq n)$ to cluster S_i, if $\|V_j - \mu_i\|^2 \leq \|V_j - \mu_n\|^2$ $(1 \leq n \leq k)$

Step 2: Update centroids $\mu_1, \mu_2, \cdots, \mu_k$ using $\mu_i = \frac{1}{m}\Sigma_{V_j \in S_i} V_j$, where m represents the number of vectors in a cluster

Step 3: Stop if the centroids $\mu_1, \mu_2, \cdots, \mu_k$ remain unchanged; Otherwise, go back to Step 1.

In this paper, a group of customers are segmented into three segments using the k-means clustering based on their RFM values: low, medium, or high profitability groups.

3.3 Open-Loop Model and Closed-Loop Model for Time Series Prediction

Time series prediction can be in general formalized by open-loop and closed-loop models. Given a time series $\{\theta(t)|t = 1, 2, \cdots n)\}$, a prediction based on an open-loop model is expressed as

$$\hat{\theta}(t) = f(\theta(t-1), \theta(t-2), \cdots, \theta(t-n)) \tag{2}$$

where $f(\cdot)$ donates a mapping, and $\hat{\theta}(t)$ represents the predicted value of variable $\theta(t)$ at time t using the prior n observed values of the variable at time points $t-1$, $t-2$, $\cdots$, $t-n$.

A closed-loop model can be expressed as

$$\hat{\theta}(t) = f\left(\hat{\theta}(t-1), \hat{\theta}(t-2), \cdots, \hat{\theta}(t-n)\right) \tag{3}$$

which uses the previous n predicted values $\hat{\theta}(t-1), \hat{\theta}(t-2), \cdots, \hat{\theta}(t-n)$ to predict to the value of variable $\theta(t)$ at time t.

3.4 Model Selection

The mapping $f(\cdot)$ in an open-loop or a closed model (Eqs. (2) and (3)) can be in different forms. In this paper, three models are considered for comparison purpose: Linear Regression, Multilayer Perceptron (MLP), and Naïve Bayesian.

Linear regression is perhaps the simplest model to be considered. Using this model for prediction, Eqs. (2) and (3) can be re-written, respectively, as

$$\hat{\theta}(t) = w_0 + \sum_{i=1}^{n} w_i \theta(t - i) \tag{4}$$

and

$$\hat{\theta}(t) = w_0 + \sum_{i=1}^{n} w_i \hat{\theta}(t - i) \tag{5}$$

where $\{w_i | i = 0, 1, \ldots n)\}$ are regression coefficients.

A multi-layer perceptron can be thought of as a regression model on a set of derived inputs via layered and successive non-linear transformations. In this paper, an MLP is used with a single hidden layer and a linear transformation for output nodes, which can be expressed as

$$h_j(t) = \frac{1}{1 + e^{-(w_{0j} + \sum_{i=1}^{n} w_{ij}\theta(t-i))}}, j = 1, 2 \cdots, m \tag{6}$$

$$\hat{\theta}_l(t) = w_{0l} + \sum_{j=1}^{m} w_{jl}h_j(t), l = 1, 2 \ldots, k \tag{7}$$

where w_{ij} and w_{jl} are connection weights between the ith input node to the jth hidden node, and the jth hidden node to the lth output node, respectively; w_{0j} and w_{0l} donate the bias to the jth hidden node and the bias to the lth output node, respectively; and $h_j(t)$ and $\hat{\theta}_l(t)$ donate the output of the jth hidden node and the lth output node, respectively. For the closed-loop model the inputs $\{\theta(t - i)\}$ are substituted by $\left\{\hat{\theta}(t - i)\right\}$.

3.5   Naïve Bayesian Model

A Naïve Bayesian model is based on Bayes' theorem as shown below

$$p(A|B) = \frac{p(A, B)}{p(B)} = \frac{p(B|A)p(A)}{P(B)} \tag{8}$$

where $p(\cdot)$ and $p(\cdot|\cdot)$ represent a probability and a conditional probability, respectively. Applying Naïve Bayesian model, Eqs. (2) and (3) can be re-written and simplified as

$$p\left(\hat{\theta}(t)|\theta(t - 1), \ldots, \theta(t - n)\right) = \frac{p\left(\hat{\theta}(t), \theta(t - 1), \ldots, \theta(t - n)\right)}{p(\theta(t - 1), \ldots, \theta(t - n))} \tag{9}$$

$$\propto p\left(\hat{\theta}(t), \theta(t - 1), \ldots, \theta(t - n)\right) \approx p\left(\hat{\theta}(t)\right) \prod_{i=1}^{n} p(\theta(t - i))$$

$$p\left(\hat{\theta}(t)\mid \hat{\theta}(t-1),\ldots,\hat{\theta}(t-n)\right) = \frac{p\left(\hat{\theta}(t),\hat{\theta}(t-1),\ldots,\hat{\theta}(t-n)\right)}{p(\theta(t-1),\ldots,\theta(t-n))} \quad (10)$$

$$\propto p\left(\hat{\theta}(t),\hat{\theta}(t-1),\ldots,\hat{\theta}(t-n)\right) \approx p\left(\hat{\theta}(t)\right)\prod_{i=1}^{n} P\left(\hat{\theta}(t-i)\right)$$

4 Case Study

4.1 Data Set and Data Pre-processing

A UK-based online retail is considered in this study [3, 9]. A data set was collected from the retail which contains all the transactions occurring from December 2010 to November 2011. The data set has 11 variables as described in Table 2. Note that the data set can be found at: https://archive.ics.uci.edu/ml/datasets/online+retail.

It is worth mentioning that, over the years, the business has been functioning as both wholesale and retail, and has maintained a stable and healthy number of customers.

Table 2. Variables in the dataset.

Variable	Data type	Description; Typical values and meanings
Invoice	Nominal	Invoice number
StockCode	Nominal	Product (item) code
Description	Nominal	Product (item) name; CARD I LOVE HAVANA
Quantity	Numeric	Quantity of each product (item) per transaction
Price	Numeric	Product price per unit in sterling; £45.23
InvoiceDate	Numeric	Day and time when a transaction occurred; 23/08/2011 15:59
Address Line 1	Nominal	Delivery address line 1; 103 Borough Road
Address Line 2	Nominal	Delivery address line 2; Elephant and Castle
Address Line 3	Nominal	Delivery address line 3; London
PostCode	Nominal	Delivery address postcode; SE1 0AA
Country	Nominal	Delivery address country; England

Appropriate pre-processing was carried out to address quality issues of the data set. Outliers and extreme values have been removed. The resultant target data set contains 751 valid customers from the UK only.

4.2 Settings for Modelling

To start the analysis, a time series of RFM values for each customer was first calculated at the end of each calendar month successively from December 2010 to November 2011, and therefore each RFM time series consists of 12 data points.

Further at each time point of the monthly-based RFM time series, the customers were grouped using the k-means clustering into three profitability groups as shown in Fig. 1, where Recency is in month and Monetary is in Sterling, and symbols '*', '+', and 'o' indicate high, medium, and low profitability groups, respectively. The subgraphs in Fig. 1 are arranged sequentially by month in ascending order. As such, each customer belongs to a certain profitability group at a given time point of the time series. Before conducting the clustering, the RFM values were normalised by using range normalisation.

Next, the three types of predictive models discussed in the previous section were applied to predict each customer's profitability group using open- and closed- loop models. The three profitability groups were encoded into three orthogonal unit vectors $[1, 0, 0]$, $[0, 1, 0]$ and $[0, 0, 1]$, and these vectors were used as the desired outputs of all

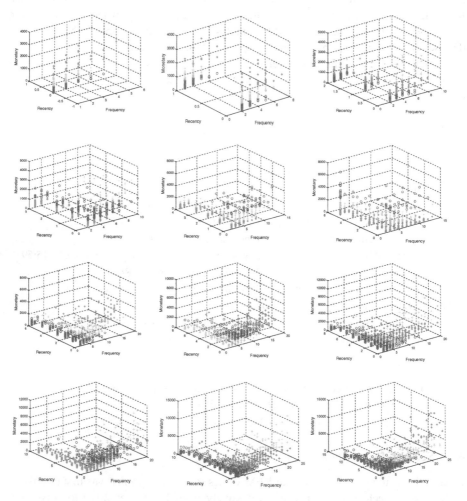

Fig. 1. Customers segmented into three profitability groups: high (*), medium (+), or low (o). Calculations were made at the end of each calendar month from Dec. 2010 to Nov. 2011.

the models for training to represent mutually exclusive three classes. Both the open- and closed- loop linear regression models had two or three terms. The topology of the MLP models were set to: three input nodes, ten hidden nodes and three output nodes. The initial connection weights and biases were generated randomly.

All the models were trained and tested 10 times, and each time 70% of the samples in the data set were randomly selected for training and the remaining 30% for testing. The data in December 2010 and January 2011 was used as the initial inputs for the closed-loop models. Note that, regardless what predictive models to be used, the training procedures for both the open-loop and closed-loop models are the same; However, when applying a trained closed-loop model, the first n observations will be used as the initial inputs to the model, and then the predicted values will be fed back sequentially to the model as inputs to generate further predications in an autonomous manner.

4.3 Experimental Results

With the given settings, the relevant experiments were conducted accordingly to examine how well a customer's membership in terms of profitability groups can be predicted over time. The average prediction accuracies generated by different models are given in Tables 3 and 4.

Table 3. Average prediction accuracy using observations at one previous time point.

Model	Open-loop		Closed-loop	
	Training	Testing	Training	Testing
Linear Regression	84.82	84.34	84.82	74.90
MLP	84.82	84.34	84.82	74.90
Naïve Bayesian	84.82	84.34	84.82	74.90

Table 4. Average prediction accuracy using observations at two previous time points.

Model	Open-loop		Closed-loop	
	Training	Testing	Training	Testing
Linear Regression	83.79	83.75	83.79	73.89
MLP	84.85	84.90	84.85	74.91
Naïve Bayesian	84.85	84.90	84.85	74.91

5 Discussion

From the experiment results obtained, it is evident that the RFM time series under consideration was well predictable, and a customer's profitability group was stable.

Under all the experimental conditions, the prediction models using observations at one previous time point performed well and had a similar performance to those using observations at two previous time points. This can be further interpreted as the transit probability of a customer from one profitability group to another at any two consecutive time points was low.

An examination on the transit probability of the customers from one profitability group to another over time has revealed that, on average, the transit probability was not more than 6%. A summary of the average transit probability is given in Table 5, where the element TP_{ij}, $i,j = 1,2,3$, in the 3×3 matrix indicates the average transit probability from the ith group to the jth group if $i \neq j$, and the average percentage of customers remained the ith group if $i \neq j$.

Since the business has been running as wholesale as well, the prediction results are quite interpretable and understandable. As such, the profitability of a customer in month t only depended on the profitability of the customer in month $t - 1$. Therefore, it's not necessary to use more past time points in the prediction.

Table 5. Average customer transit probability over time.

Group	1	2	3
1	0.05	0.01	0.01
2	0.01	0.19	0.05
3	0.01	0.06	0.61

In addition, the MLP and the Naïve Bayesian models were slightly more stable than the regression models.

The open-loop prediction models could achieve 84% accuracy and those models were useful for a short-term prediction. The closed-loop prediction models have achieved an accuracy of 79% and they could be applied for a long-term prediction.

6   Conclusions and Future Work

In this study, a comparative study has been conducted on predicting customer profitability dynamically based on monthly RFM time series using multiple models. The study shows a good predictability of the time series under consideration. The context of the business of interest has helped to interpret the prediction results.

Further work includes:

- Using real transactional data collected over a longer period of time, such as two or three years, to examine the predictability of the RFM time series;
- To investigate how prediction accuracy might be affected by the frequency at which the RFM values are calculated with a given transactional data; and
- Using other possible profitability measures to conduct comparative research.

References

1. Google Cloud: Predicting customer lifetime value with AI platform: introduction. https://cloud.google.com/solutions/machine-learning/clv-prediction-with-offline-training-intro. Accessed 10 May 2019
2. Malthouse, E.C., Blattberg, R.C.: Can we predict customer lifetime value? J. Interact. Mark. **1** (19), 2–16 (2005)
3. Chen, D., Guo, K., Ubakanma, G.: Predicting customer profitability over time based on RFM time series. Int. J. Bus. Forecasting Mark. Intell. **2**(1), 1–18 (2015)
4. Beheshtian-Ardakani, A., Fathian, M., Gholamian, M.: A novel model for product bundling and direct marketing in e-commerce based on market segmentation. Decis. Sci. Lett. **7**(1), 39–54 (2018)
5. Chen, P.P., Anna Guitart, A., Fernandez, A., Perianz, A.: Customer lifetime value in video games using deep learning and parametric models. In: IEEE International Conference on Big Data, pp. 2134–2140. IEEE, Seattle (2018)
6. Hojjat Salehinejad, H., Rahnamayan, H.: Customer shopping pattern prediction: a recurrent neural network approach. In: 2016 IEEE Symposium Series on Computational Intelligence (SSCI). IEEE, Athene (2016)
7. Gauthier, J.R.: An introduction to predictive customer lifetime value modeling. https://www.datascience.com/blog/intro-to-predictive-modeling-for-customer-lifetime-value. Accessed 02 May 2019
8. Blattberg, R.C., Kim, B.-D., Neslin, S.A.: Database Marketing. ISQM, vol. 18. Springer, New York (2008). https://doi.org/10.1007/978-0-387-72579-6
9. Chen, D., Sain, S.L., Sain, S.L., Guo, K.: Data mining for the online retail industry: a case study of RFM model-based customer segmentation using data mining. J. Database Mark. Customer Strategy Manag. **19**(3), 197–208 (2012)

Image Analysis and Retrieval

Face Spoofing Detection on Low-Power Devices Using Embeddings with Spatial and Frequency-Based Descriptors

Rafael Henrique Vareto[(✉)], Matheus A. Diniz, and William Robson Schwartz

Smart Sense Laboratory, Department of Computer Science,
Universidade Federal de Minas Gerais, Belo Horizonte, Brazil
{rafaelvareto,matheusad,william}@dcc.ufmg.br

Abstract. A face spoofing attack occurs when an intruder attempts to impersonate someone with a desirable authentication clearance. To detect such intrusions, many researchers have dedicated their efforts to study visual liveness detection as the primary indicator to block spoofing violations. In this work, we contemplate low-power devices through the combination of Fourier transforms, different classification methods, and low-level feature descriptors to estimate whether probe samples correspond to spoofing attacks. The proposed method has low-computational cost and, to the best of our knowledge, this is the first approach associating features extracted from both spatial and frequency domains. We conduct experiments with embeddings of Support Vector Machines and Partial Least Squares on recent and well-known datasets under same and cross-database settings. Results show that, even though devised towards resource-limited single-board computers, our approach is able to achieve significant results, outperforming state-of-the-art methods.

Keywords: Face spoofing · Liveness detection · Fourier transform · Machine learning · Biometrics

1 Introduction

Biometric techniques seek for recognizing humans taking into account their intrinsic behavioral or observable aspects, ranging from face and fingerprint to iris and voice. Even though the biometric authentication field has prospered significantly in the recent years, experts claim that new technologies are constantly susceptible to malicious attacks and can be exposed to emerging high-quality spoof mechanisms [18].

Spoofing, also known as copy or presentation attack, is a real threat for biometric systems. More precisely, it occurs when an intruder attempts to impersonate someone who holds a desirable authentication clearance. The criminal usually employs falsified data to bypass the security procedure and gain illegitimate access. As a countermeasure to copy attacks, some researchers dedicate

I. Nyström et al. (Eds.): CIARP 2019, LNCS 11896, pp. 187–197, 2019.
https://doi.org/10.1007/978-3-030-33904-3_17

their efforts to study human liveness detection as the leading indicator to antic-ipate spoofing violations [10, 15, 16, 19, 28].

In general, a spoofing attack involves the display of still or motion pictures of authentic users registered in a set of known individuals present in a face recognition system. These images are easily acquired since the person's face is probably the most typical biometric model due to its noninvasive and availabil-ity characteristics when compared to others, such as fingerprint and iris. With the expansion of surveillance cameras and the increasing number of people dis-tributing personal pictures on social networks, it is practically impossible to keep faces from spreading out [12]. Thus, face spoofing has become an easy approach to deceive biometric-based applications.

This paper is inspired on the works of Pinto et al. [20] and Vareto et al. [26]. However, due to the high demand for low computational-cost algorithms to be embedded on low power devices (e.g., IoT devices), we devise an anti-spoofing algorithm for limited-resource equipments. We propose a spoofing detection app-roach that associates simple handcrafted features extracted from spatial and frequency domains. Classifiers act as bootstrap aggregating meta-algorithms to achieve competitive results on the five most prominent benchmarks, to mention a few, MSU-MFSD [27], OULU-NPU [5] and SIW [14] datasets. We conduct cross-dataset experiments in the interest of assessing the method's generalization and verify how it responds to "unfamiliar" media presentations. This work compares the proposed method with state-of-the-art approaches and investigates how much display devices and image capture quality have an impact on our results.

To the best of our knowledge, this is the first approach associating features extracted from the spatial and frequency domains to tackle the spoofing detec-tion problem. The leading premise is that modeling the association between spatial and frequency domains can be suitable for improving the accuracy and robustness of face anti-spoofing tasks. We assume that authentic and counterfeit biometric data enclose distinct noise signatures derived from the media acqui-sition. In fact, we believe that the combination of different feature descriptors contributes to achieving higher performance considering that they acquire dis-tinctive characteristics, which are capable of enriching the classifier's robustness and generalization potential.

The main contributions of this work are: (1) combination of classification models fitted on randomly generated subsets in a bootstrap aggregating mode; (2) aggregation of features extracted in spatial and temporal domains; (3) effi-cient method for image and video-based copy attack receiving as input high-resolution videos; (4) low complexity and computational cost algorithm, capa-ble of being deployed in embedded systems and computers with small processing capabilities; (5) clear study and experimental evaluation of the proposed app-roach considering fundamental feature descriptors, such as GLCM [11], HOG [8] and LBP [17].

2 Related Works

In the past years, Deep Neural Networks (DNN) have confirmed to be effective in several computer vision and biometric problems. Feng et al. [9] extract deep features from a convolutional neural network to identify real and fake faces. Similarly, Li et al. [13] employ a multiple-input hierarchical neural network combining either *shearlet* or optical-flow-based features. Valle et al. [25] present a transfer learning method using a pre-trained DNN model on static features to recognize photo, video and mask attacks. Liu et al. [14] combine DNN and Recurrent Neural Networks (RNN) to estimate the depth of face images along with rPPG signals to boost the detection of unauthorized access.

Some authors carry on working on long-established traditional approaches, dealing with handcrafted feature extraction and learning design: Pinto et al. [20] explore the spatial domain during the recapture process as it takes over the noise with Fourier transforms followed by visual rhythm algorithms and the extraction of gray-level co-occurrence matrices. Wen et al. [27] come up with an algorithm built on image distortion analysis and low-level feature descriptors. It consists of an embedding of SVM classification algorithms evaluated on cross-dataset scenarios. Pinto et al. [19] extract low-level feature descriptors gathering temporal and spectral information across biometric samples. Boulkenafet et al. [3,4] detect copy attacks using color texture analysis and low-level descriptors via exploring luminance and chrominance information of each image color channel separately.

Even though handcrafted features may end up being restricted to specific datasets domains, they are commonly faster and present lower memory usage than DNN-based methods, especially when it comes to resource-limited equipments. Most neural networks are not invariant to image rotation or scale and may fail to manage scenarios consisting of differing capturing instruments, illumination conditions and shooting angles [2]. In addition, top performing DNNs tend to suffer from either low speed or being too large to fit into single-board computers, preventing their deployment on remote applications. On the contrary of deep neural networks, both traditional features and straightforward classifiers employed in our approach do not require cloud processing services or powerful dedicated servers since embedded devices are capable of running the proposed low-cost standalone algorithm fast enough to be employed in real environments.

3 Proposed Approach

We propose an approach that captures visual noise signatures in both spatial and frequency domains. First, the method extracts low-level features with GLCM [11], HOG [8] and LBP [17]. Then, an ensemble of classifiers is created as we group several identical classifiers to enhance the method's overall efficacy [6]. Figure 1 illustrates the steps that compose the proposed approach.

Different feature descriptors make it possible to combine color, gradient magnitude and texture information, providing complementary evidence for presentation attacks. More precisely, GLCM is a statistical descriptor that analyses spatial

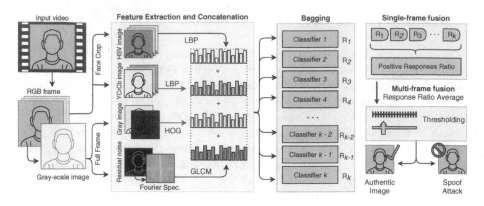

Fig. 1. Overview of the proposed face spoofing detection approach – *Training:* GLCM, HOG and LBP descriptors are extracted from the frames of the videos available for training. These features are concatenated and used for learning several classification models in an embedding fashion. Distinct models are learned containing different video samples in each subset. *Test:* The same features are extracted from the probe video frames and projected to all binary classifiers. Then, it executes a score fusion on the classifiers' responses to determine whether the probe video refers to an authentic presentation.

relationship of pixels and may identify noise artifacts originated from the recapturing process. HOG captures regions of abrupt intensity changes around edges and corners, such as screen frames and picture borders, through the magnitude of gradients. LBP evaluates color and texture patterns in search of *crude* attacks as it compares pixels with their surrounding points in different colorspaces.

3.1   Feature Extraction

The feature extraction process explores distinct spatial colorspaces and frequency domain to gather discriminating spoofing patterns. The procedure starts converting every RGB colorspace video frame into HSV, YC_RC_B and gray-scale images. On the contrary of the RGB color model, which holds high correlation among color components, HSV and YC_RC_B are capable of isolating luminance from chrominance and more robust to illumination variations [21].

As the RGB video frame is converted into HSV and YC_RC_B images, the method locates the region of interest, which is delimited on the subject's face. The approach extracts LBP descriptors from each HSV and YC_RC_B image color channel in an attempt to gather color and texture distinctive information. In fact, it computes local texture representation from all color bands comparing every pixel with its surrounding neighborhood of pixels. Both HSV and YC_RC_B corresponding feature descriptors derive from the integration of each channel's histogram that accounts for the number of times every LBP pattern occurs [4].

Monochromatic video frames go through low-pass filtering techniques (blurring) for artifact and noise reduction. Residual noises are then obtained by subtracting a gray-scale image and its slightly blurred version [20]. A logarithmic-

Fig. 2. Comparison among Fourier spectra extracted from different presentation images. Note that there are some artifacts spread throughout print and replay attacks.

scaled Fourier transform function $\mathcal{F}_{log}(v, u)$ decomposes each residual image $r(x, y)$ of size $M \times N$ into its sine and cosine components where each pixel constitutes a frequency from the spatial domain as

$$\mathcal{F}_{log}(v, u) = log(1 + |\sum_{x=0}^{M-1} \sum_{y=0}^{N-1} r(x, y)e^{-j2\pi[\frac{vx}{M} + \frac{uy}{N}]}|).$$

The employed low-level feature descriptors provide great accuracy vs. speed trade-off due to their fast computation. The gray-scale image and its corresponding spectrum generate HOG and GLCM features, respectively, whereas LBP descriptor receives HSV and YC_RC_B image color bands. HOG carries shape information by counting occurrences of gradient orientation using histograms while GLCM measures the residual image texture with the generation of co-occurring gray-scale values at a determined offset. As shown in Fig. 1, we concatenate HOG and LBP features from the spatial-domain with GLCM information from the *log*-scaled Fourier spectrum to build a robust feature descriptor.

3.2 Classification Methods

Instead of learning a unique binary classifier, we learn a set of models as it seems to be more appropriate to handle contrasting chromatic distortions and to reduce the risk of overfitting. The classification embedding consists either of Support Vector Machines (SVM) [24] or Partial Least Squares (PLS) [22] learning algorithms. While the former chooses the hyperplane that maximizes the distance to the nearest data points, the latter weights features to discriminate throughout different classes and handle high-dimensional data.

During the training stage, the proposed method employs several identical binary learning algorithms trained on random subsets of the training set to create an array of classifiers C. It guarantees a balanced division within each classification model since v genuine live and v presentation attack videos are randomly selected, with replacement, out of all video samples available for training. Then, it fits the learning algorithm on the extracted features where the positive class only contains "authentic" feature vectors and the negative class holds features extracted from copy attacks. This process is repeated k times, where $k = |C|$ is a user-defined parameter that defines the number of classification models.

In the prediction stage, the method projects every single frame onto all classification models as it iterates over the probe video. For each frame, the algorithm computes the ratio of the number of positive responses attained to the total number of classification models k. If most $c \in C$ classifiers return positive responses, it implies that the frame is likely to be a *bona fide* (authentic) sample. Otherwise, if they return negative responses, then the probe sample is likely to belong to a spoofing attack. As the approach examines multiple frames of a probe video, it obtains the numerical mean of all frame ratio scores. A probe video is considered authentic if the averaged ratio score of all frames satisfies a threshold t (t would be chosen according to the biometric system specifications).

4   Experimental Results

This section contains an objective evaluation of the proposed algorithm, which generates many binary classification models combined with a majority voting scheme that determines whether a query image corresponds to a legitimate image or a spoofing attack.

Table 1. Evaluation on different SIW protocols with an increasing number of PLS classification models (PLS approach). Note that the method becomes more discriminative with the addition of classifiers.

Protocol	Metric	50	100	200
1	APCER	0.68 ± 0.00	0.14 ± 2.17	0.00 ± 0.00
	BPCER	4.67 ± 0.00	2.17 ± 0.00	0.67 ± 0.00
2	APCER	11.70 ± 10.73	7.86 ± 6.84	3.93 ± 4.14
	BPCER	3.34 ± 4.74	1.29 ± 1.16	0.66 ± 1.10
3	APCER	17.37 ± 14.53	10.59 ± 7.72	6.99 ± 1.68
	BPCER	4.92 ± 4.09	2.17 ± 1.67	1.17 ± 0.33

Feature Descriptors. Three feature descriptors are employed in this work: The GLCM texture descriptor [11] is computed with directions $\theta \in \{0, 45, 90, 135\}$ degrees, distance $d \in \{1, 2\}$, 16 bins and six texture properties: contrast, dissimilarity, homogeneity, energy, correlation, and angular second moment. The HOG shape descriptor [8] is set with 96×96 cells and holding eight orientations. Lastly, the LBP texture descriptor [17] comprises 256 bins, a radius equal to 1, and eight points arranged in a 3×3 matrix thresholded by its central point. Their low complexity and computational cost endorse our method so that it can be deployed to embedded systems with reduced processing capabilities.

Spoofing Datasets. For a thorough evaluation, we select datasets with distinct protocols, medium characteristics and different lighting conditions. Therefore, experiments are carried out on five benchmarks: CASIA-FASD [29], MSU-MFSD [27], OULU-NPU [5], REPLAY-ATTACK [7] and SIW [14]. CASIA-FASD, MSU-MFSD and REPLAY-ATTACK are traditional benchmark databases made up of genuine live recordings and distinct spoofing attack shots captured by distinct cameras in different scenarios. Both OULU-NPU and SIW are recent datasets containing full high-definition videos of multiethnic individuals and featuring 30-FPS live and presentation attack videos.

Evaluation Metrics. We employ ISO/IEC 30107-3 metrics [1] called Attack Presentation Classification Error Rate, APCER $= \frac{1}{V_{PA}} \sum_{i=1}^{V_{PA}} (1 - Res_i)$; and Bona Fide Presentation Classification Error Rate, BPCER $= \frac{1}{V_{BF}} \sum_{i=1}^{V_{BF}} (Res_i)$. V_{PA} indicates spoofing attacks whereas V_{BF} outlines authentic presentations. Res_i receives 0 when the i-th probe video is considered an *bona fide* presentation and 1 otherwise. On cross-datasets evaluations, it is customary to employ Half Total Error Rate, HTER $= \frac{FAR+FRR}{2}$, which is half the sum of the False Rejection Rate (FRR) and the False Acceptance Rate (FAR) [14,23]. The reader must bear in mind that the closer APCER, BPCER and HTER values get to zero, the more accurate the described methods are.

Evaluation Setup. Experiments were conducted on a Raspberry Pi 3 Model B and on a Linux virtual machine to assess the performance of the proposed approach on different machines. First, we analyzed the method on a CPU-based machine consisting of eight 2.0 GHZ-core processors and 16 GB RAM memory, but no more than 600 MB was required on test time. Then, we migrated to the Raspberry, a single-board microcomputer with a 1.2 GHZ Quad Core CPU and 1 GB RAM memory. Higher frame rates could be achieved with graphical processing units, but it would demand the acquisition of more advanced hardware.

Table 2. APCER and BPCER results (%) on SIW protocols.

Protocol	Method	APCER	BPCER	AVERAGE
1	Deep models [14]	3.58	3.58	3.58
	PLS approach	0.00 ± 0.00	0.67 ± 0.00	0.33 ± 0.00
	SVM approach	0.00 ± 0.00	0.33 ± 0.00	0.16 ± 0.00
2	Deep models [14]	0.57 ± 0.69	0.57 ± 0.69	0.57 ± 0.69
	PLS approach	3.93 ± 4.14	0.66 ± 1.10	2.24 ± 3.30
	SVM approach	8.09 ± 1.02	1.00 ± 0.44	2.88 ± 3.21
3	Deep models [14]	8.31 ± 3.81	8.31 ± 3.80	8.31 ± 3.81
	PLS approach	6.99 ± 1.68	1.16 ± 0.33	1.55 ± 0.05
	SVM approach	6.03 ± 1.22	0.67 ± 0.24	3.36 ± 0.73

Results Analysis. The algorithm proposed in Sect. 3 is evaluated according to the protocols available in the literature and following the datasets instructions. For databases containing only training and test sets, like SIW dataset, we reserve ten percent of all samples available for training to establish an automatic adaptive threshold t. Differently, OULU-NPU and REPLAY-ATTACK contain a development set destined to parameter calibrations.

Table 3. APCER and BPCER results (%) on OULU-NPU protocols.

Protocol	Method	APCER	BPCER	AVERAGE
1	Deep models [14]	1.60	1.60	1.60
	Gradiant [2]	1.30	12.50	6.90
	PLS approach	5.50 ± 2.11	9.79 ± 3.37	7.64 ± 2.74
2	Deep models [14]	2.70	2.70	2.70
	Gradiant [2]	6.90	2.50	4.70
	PLS approach	2.13 ± 1.07	3.61 ± 1.21	2.87 ± 1.14
3	Deep models [14]	2.70 ± 1.30	3.10 ± 1.70	2.90 ± 1.50
	Gradiant [2]	2.60 ± 3.90	5.00 ± 5.30	3.80 ± 2.40
	PLS approach	3.12 ± 2.58	8.51 ± 6.20	5.81 ± 4.39
4	Deep models [14]	9.31 ± 5.60	10.4 ± 6.00	9.50 ± 6.00
	Gradiant [2]	5.00 ± 4.50	15.0 ± 7.10	10.0 ± 5.01
	PLS approach	17.8 ± 9.83	9.37 ± 4.31	13.5 ± 7.07

We evaluate the method's behavior by increasing the number of PLS classification models. According to the results showed in Table 1, as the number of classifiers increases, the method becomes more discriminative. Therefore, in the remaining experiments, we set the number of classification models to 200. Tables 2 and 3 show the results obtained on the SIW and OULU-NPU datasets, respectively. The proposed approach achieves state-of-the-art results on SIW Protocols 1 and 3 and competitive results on Protocol 2. Moreover, the method attains precise results on three out of four OULU-NPU Protocols.

The cross-database analysis provides an insight into countermeasure methods' generalization power. In this sort of scenario, an algorithm is trained and tuned in one of the datasets and tested on the others. Table 4 presents the cross-testing HTER [1] performance for both PLS and SVM methods on the traditional benchmarks. The PLS-based method also achieves a HTER of 34.44 ± 3.91 when trained on SIW and tested on OULU-NPU, and 17.55 ± 1.47 vice versa. Results show that datasets tend to hold some bias regardless of their protocols due to the intrinsic and specific information enclosed in each dataset, culminating in a significant accuracy reduction when compared to same-database evaluations.

Computational Cost Evaluation. In contrast to most recent spoofing detection works in the literature, where deep neural networks benefit from "unlimited computational resources" and high-bandwidth video transmissions, our method is devised towards resource-limited single-board computers in order to reduce network communication. GLCM, HOG and LBP descriptors appear to carry relevant forensic signature information of image and video-based spoofing detection since results show that the combination of spatial and frequency-based descriptors contributes to achieving both competitive and state-of-the art results.

Table 4. Cross-dataset evaluation (%) presenting HTER metric on CASIA-FASD, MSU-MFSD and REPLAY-ATTACK datasets.

Training Set	CASIA-FASD		MSU-MFSD		REPLAY-ATTACK	
Test Set	MSU-MFSD	REPLAY-ATTACK	CASIA-FASD	REPLAY-ATTACK	CASIA-FASD	MSU-MFSD
Color LBP [3]	36.6	47.0	49.6	42.0	39.6	35.2
Color texture [4]	20.4	30.3	46.0	33.9	37.7	34.1
Spectral [19]	-	34.4	-	-	50.0	-
Deep models [14]	-	27.6	-	-	28.4	-
PLS approach	19.2 ± 1.6	30.1 ± 0.7	28.2 ± 0.7	37.1 ± 3.2	35.6 ± 0.4	34.5 ± 2.3
SVM approach	17.3 ± 1.1	42.6 ± 2.5	34.8 ± 0.8	42.6 ± 1.7	38.3 ± 2.0	35.4 ± 1.9

Many researchers have neglected to deliver biometric applications that are able to run on low-power devices [9,13,14,25]. As we take IoT devices into account, the proposed algorithm presents low computational cost, being able to process up to 4.31 ± 0.031 frames per second (FPS) when considering the Raspberry Pi environment. As a comparison, it runs at 32.55 ± 0.96 FPS in the CPU-based computer. Both when the number of classifiers k is set to 100. Such frame rate, 4.31 FPS, make it feasible for tech developers to implement and run biometric IoT technologies in real environments.

When we consider the above frame rate specification and the average amount[1] paid for the following devices: a Raspberry Pi 3 Model B ($35.00), identical to the microcomputer evaluated; an Intel i5 2.8 GHZ processor with 16 GB RAM ($400.00), similar to the virtual machine tested; and an Intel i7 3.2 GHZ CPU with 16 GB RAM and a GeForce GTX 1080Ti ($1600.00), assuming an equivalent frame rate of 32.55, since most quality CCTV cameras record videos between 15 and 30 FPS. Then, the price paid per FPS on the aforementioned machines would be around $8.12, $12.28 and $49.15, respectively. Therefore, running the designed approach on a single-board computer, such as Raspberry Pi, provides better performance per cost than executing in more robust machines.

5   Conclusions

This work[2] proposed a fast and low-memory spoofing detection algorithm and demonstrates how it performs in an experimental setup to emulate real-world scenarios. The proposed algorithm is fast and works well on single-board computers with high-resolution videos and is able to achieve state-of-the-art performance on widely explored databases.

We conduct an objective investigation on how far spatial and frequency-based descriptors can get when combined with multiple classification models. If fact, we work out two approaches (embeddings comprised of either Partial Least Squares

---

[1] Prices taken from official Raspberry Pi resellers and BestBuy Retail Store.

[2] Proposed method available at https://github.com/rafaelvareto/Spoofing-CIARP19.

or Support Vector Machines) to infer that the association of long-established feature descriptors accomplish great performance in same-database settings. An investigation carried out on different datasets show that the accuracy tends to degrade significantly.

Despite the great progress in several biometric research areas, existing anti-spoofing approaches have shown lack of generalization in cross-dataset conditions, which best represents real-world scenarios. As future directions, we plan to add extra feature descriptors, include other relevant spoofing datasets and learn spatial-temporal representations.

Acknowledgments. The authors would like to thank the Brazilian National Research Council – CNPq (Grants #311053/2016-5 and #438629/2018-3), the Minas Gerais Research Foundation – FAPEMIG (Grants APQ-00567-14 and PPM-00540-17), the Coordination for the Improvement of Higher Education Personnel – CAPES (DeepEyes Project), Maxtrack Industrial LTDA and Empresa Brasileira de Pesquisa e Inovacao Industrial – EMBRAPII.

References

1. Information technology - biometric presentation attack detection - part 1: Framework. international organization for standardization. Technical report, ISO/IEC JTC 1/SC 37 Biometrics (2016)
2. Boulkenafet, Z., et al.: A competition on generalized software-based face presentation attack detection in mobile scenarios. In: IJCB, pp. 688–696. IEEE (2017)
3. Boulkenafet, Z., Komulainen, J., Hadid, A.: Face anti-spoofing based on color texture analysis. In: ICIP, pp. 2636–2640. IEEE (2015)
4. Boulkenafet, Z., Komulainen, J., Hadid, A.: Face spoofing detection using colour texture analysis. TIFS **11**(8), 1818–1830 (2016)
5. Boulkenafet, Z., Komulainen, J., Li, L., Feng, X., Hadid, A.: OULU-NPU: a mobile face presentation attack database with real-world variations. In: FG, IEEE (2017)
6. Breiman, L.: Bagging predictors. Mach. Learn. **24**(2), 123–140 (1996)
7. Chingovska, I., Anjos, A., Marcel, S.: On the effectiveness of local binary patterns in face anti-spoofing. In: BIOSIG. No. EPFL-CONF-192369 (2012)
8. Dalal, N., Triggs, B.: Histograms of oriented gradients for human detection. In: CVPR, vol. 1, pp. 886–893. IEEE (2005)
9. Feng, L.: Integration of image quality and motion cues for face anti-spoofing: a neural network approach. JVCIR **38**, 451–460 (2016)
10. Garcia, D.C., de Queiroz, R.L.: Face-spoofing 2D-detection based on moiré-pattern analysis. TIFS **10**(4), 778–786 (2015)
11. Haralick, R.M., Shanmugam, K., et al.: Textural features for image classification. TSMC **6**, 610–621 (1973)
12. Kumar, S., Singh, S., Kumar, J.: A comparative study on face spoofing attacks. In: ICCCA, pp. 1104–1108. IEEE (2017)
13. Li, L., Feng, X., Boulkenafet, Z., Xia, Z., Li, M., Hadid, A.: An original face anti-spoofing approach using partial convolutional neural network. In: IPTA, pp. 1–6. IEEE (2016)
14. Liu, Y., Jourabloo, A., Liu, X.: Learning deep models for face anti-spoofing: Binary or auxiliary supervision. In: CVPR, pp. 389–398 (2018)

15. Määttä, J., Hadid, A., Pietikäinen, M.: Face spoofing detection from single images using micro-texture analysis. In: IJCB, pp. 1–7. IEEE (2011)
16. Menotti, D., et al.: Deep representations for iris, face, and fingerprint spoofing detection. TIFS **10**(4), 864–879 (2015)
17. Ojala, T., Pietikainen, M., Maenpaa, T.: Multiresolution gray-scale and rotation invariant texture classification with local binary patterns. TPAMI **24**(7), 971–987 (2002)
18. Pereira, T., Anjos, A., Martino, J.M., Marcel, S.: Can face anti-spoofing counter-measures work in a real world scenario? In: ICB, pp. 1–8. IEEE (2013)
19. Pinto, A., Pedrini, H., Schwartz, W.R., Rocha, A.: Face spoofing detection through visual codebooks of spectral temporal cubes. TIP **24**(12), 4726–4740 (2015)
20. Pinto, A., Schwartz, W.R., Pedrini, H., de Rezende Rocha, A.: Using visual rhythms for detecting video-based facial spoof attacks. TIFS **10**(5), 1025–1038 (2015)
21. Plataniotis, K.N., Venetsanopoulos, A.N.: Color Image Processing and Applications. Springer, Heidelberg (2013). https://doi.org/10.1007/978-3-662-04186-4
22. Rosipal, R., Krämer, N.: Overview and recent advances in partial least squares. In: Saunders, C., Grobelnik, M., Gunn, S., Shawe-Taylor, J. (eds.) SLSFS 2005. LNCS, vol. 3940, pp. 34–51. Springer, Heidelberg (2006). https://doi.org/10.1007/11752790_2
23. Siddiqui, T.A., et al.: Face anti-spoofing with multifeature videolet aggregation. In: ICPR, pp. 1035–1040. IEEE (2016)
24. Steinwart, I., Christmann, A.: Support Vector Machines. Springer, Heidelberg (2008). https://doi.org/10.1007/978-0-387-77242-4
25. Lucena, O., Junior, A., Moia, V., Souza, R., Valle, E., Lotufo, R.: Transfer learning using convolutional neural networks for face anti-spoofing. In: Karray, F., Campilho, A., Cheriet, F. (eds.) ICIAR 2017. LNCS, vol. 10317, pp. 27–34. Springer, Cham (2017). https://doi.org/10.1007/978-3-319-59876-5_4
26. Varcto, R., Silva, S., Costa, F., Schwartz, W.R.: Towards open-set face recognition using hashing functions. In: IJCB, pp. 634–641. IEEE (2017)
27. Wen, D., Han, H., Jain, A.K.: Face spoof detection with image distortion analysis. TIFS **10**(4), 746–761 (2015)
28. Xiong, Q., Liang, Y.C., Li, K.H., Gong, Y.: An energy-ratio-based approach for detecting pilot spoofing attack in multiple-antenna systems. TIFS **10**(5), 932–940 (2015)
29. Zhang, Z., Yan, J., Liu, S., Lei, Z., Yi, D., Li, S.Z.: A face antispoofing database with diverse attacks. In: ICB, pp. 26–31. IEEE (2012)

Video Iris Recognition Based on Iris Image Quality Evaluation and Semantic Classification

Eduardo Garea-Llano[(⊠)], Annette Morales-González,
and Dailé Osorio-Roig

Advanced Technologies Application Center, 7ma A #21406,
12200 Playa, Havana, Cuba
{egarea,amorales,dosorio}@cenatav.co.cu

Abstract. The use of video in biometric applications has reached a great height in the last five years. The iris as one of the most accurate biometric modalities has not been exempt due to the evolution of the capture sensors. In this sense, the use of on line video cameras and the sensors coupled to mobile devices has increased and has led to a boom in applications that use these biometrics as a secure way of authenticating people, some examples are secure banking transactions, access controls and forensic applications, among others. In this work, an approach for video iris recognition is presented. Our proposal is based on a scheme that combines the direct detection of the iris in the video frame with the image quality evaluation and segmentation simultaneously with the video capture process. A measure of image quality is proposed taking into account the parameters defined in ISO /IEC 19794-6 2005. This measure is combined with methods of automatic object detection and semantic image classification by a Fully Convolutional Network. The experiments developed in two benchmark datasets and in an own dataset demonstrate the effectiveness of this proposal.

Keywords: Iris recognition · Video · Semantic classification · Quality measure

1 Introduction

In biometric recognition, the capture of iris images has been characterized by the use of sensors that work in the near infrared spectrum (NIR). This is mainly because in this wavelength it is possible to capture a better iris textural pattern and its internal structures regardless of the iris color. The use of NIR sensors has largely avoided the problem of low contrast between the iris and the pupil in those people who have dark colored irises. However, NIR sensors has restricted the more extensive use of this biometric that has proven to be very accurate. Generally, sensors used in video protection or access control applications work in the visible spectrum (VS). On the other hand, in recent years there has been a substantial increase in the use of this type of sensors in mobile devices for biometric applications integrated to them. The iris biometric has not been exempt from this phenomenon and there are already applications of this type [1], so it is important to address this topic due to the problems that affect this modality and the wide spectrum of applications that can be developed. Video as a modality for obtaining iris images in applications in real time is a subject that acquires

© Springer Nature Switzerland AG 2019
I. Nyström et al. (Eds.): CIARP 2019, LNCS 11896, pp. 198–208, 2019.
https://doi.org/10.1007/978-3-030-33904-3_18

great relevance in the current context of the use of VS sensors in mobile devices and video surveillance cameras [2, 3] because from the video it is possible to obtain more information about the iris region.

In this work, an approach for video iris recognition is proposed; it is based on a scheme for quality evaluation of the iris image while processing the video capture. For this purpose, a measure of iris image quality is proposed, it takes into account the elements defined in the ISO/IEC 19794-6: 2005 standard [4]. As part of this scheme we also propose a segmentation method that uses the direct iris detection from video frames and combines this process with the semantic classification of the iris image to generate the segmentation mask. This combination ensures that iris images are extracted avoiding the elements that negatively influence the identification process such as closed eyes and out-of-angle look. The work is structured as fallows. Section 2 discusses the related works, Sect. 3 presents the proposed approach, in Sect. 4 the experimental results are presented and discussed, and finally the conclusions of the work are set.

2 Related Works

The quality evaluation of iris images is one of the recently identified topics in the field of iris biometry [5, 6]. In general, quality metrics are used to decide whether the image should be discarded or processed by the iris recognition system.

The quality of iris images is determined by many factors depending on the environmental and camera conditions and on the person to be identified [5]. Some of the quality measures reported in literature [6] focus on the evaluation of iris images after the segmentation process, which results in feeding the system with a significant number of low quality images. Another problem of these approaches is that the evaluation of the iris image quality is reduced to the estimation of a single or a couple of factors [3], such as out-of-focus blur, motion blur, and occlusion. Other authors [6, 7] use more than three factors to evaluate the quality of the iris image: such as the degree of defocusing, blurring, occlusion, specular reflection, lighting and out of angle. They consider that the degradation of some of the estimated parameters below the threshold brings to zero (veto power) the measure that integrates all the evaluated criteria. This may be counterproductive in some systems where the capture conditions are not optimal, preventing the passage to the system of images that can have identification value.

In [8] a quality measure was proposed, it estimates the quality of the iris image within the whole eye image captured in video by a VS camera. This measure considers properties established in the ISO /IEC 19794-6: 2005 [4] and combines them with methods of image sharpness estimation. However, the estimation is carried out on the whole eye image and not on the specific region of the iris. Additionally, the degree of diversity of the iris texture is not considered.

We believe that a quality measure that considers the parameters established in the standard [4] combined with image sharpness estimation, the degree of diversity of the iris texture and evaluates detected iris image before the segmentation, can help reducing errors in the next steps of the system with a consequent increase in recognition rates.

Recently, from the great success reached by the image classification methods based on deep learning, several methods of iris segmentation have been proposed. Fully Convolutional Networks (FCN) have achieved the best results [9, 10]. The main limitation that these methods have is that they are based on the classification of the image into two classes (iris and non-iris) without taking into account other elements in the iris image such as eyelashes and specular reflections that negatively influence the subsequent identification process. In [11] the authors presented a semantic iris segmentation method based on the hierarchical classifier Markov Random Field. The experimental results did not significantly outperform the best state-of-the-art methods, but the authors showed that the idea of semantic segmentation was a promising route for iris recognition in the VS.

In [12] a multi-class approach for iris segmentation was presented. This method is based on the information of different semantic classes in an eye image (sclera, iris, pupil, eyelids and eyelashes, skin, hair, glasses, eyebrows, specular reflections and background regions) by means of a FCN. The experimental results of this work showed that, for iris segmentation, the use of the information of different semantic classes in the eye image is better than the iris and non-iris segmentation. However, the application of this method for a video-based system is not feasible since image classification is performed on the previously detected eye image, which leads to a high calculation time to classify the whole eye image.

3 Proposed Approach

Figure 1 shows the general scheme of the proposed approach. Our proposal is based on direct iris image detection, its combination with a new quality metric and semantic iris classification in 5 classes (iris, pupil, sclera, eyelashes and specular reflections).

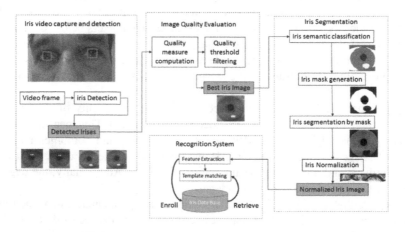

Fig. 1. General scheme of the proposed approach

The application of this approach in a video-based iris recognition system can ensure that the iris images detected are free of elements that negatively influence the identification (illumination, sharpness, blur, gaze, occlusion and low pixel density of the image). On the other hand, the classification applied only to the iris image and not to the whole eye image will reduce the computational time used to classify the iris image, which will allow its use in video-based systems.

3.1 Iris Video Capture and Detection

In [13] the use of white LED light is proposed as a way to attenuate the effect of low contrast between dark colored irises when eye images are captured by mobile devices (cell phones) in the VS. The authors demonstrated that their proposal increases recognition rates to levels similar to those that can be reached by a system that works with NIR images. For implementation of our proposal a similar device was designed, but replacing the cell phone with a webcam for video capture.

For the direct detection of iris images in video a detector was trained through the classic algorithm of Viola and Jones [14]. The image representation called integral image, allows a very fast computation of the features used by the detectors. The learning algorithm based on Adaboost, allows to select a small number of features from the initial set, and to obtain a cascade of simple classifiers to discriminate them. A classifier was trained to detect the region of the iris within the eye image. This region includes the bounding box that encompasses the iris and the pupil (see Fig. 1).

The training set consisted of 3300 positive samples taken from the MobBio [15] (800), UTIRIS [16] (800) and 1700 from frames of a video iris dataset of our own (see Sect. 4.2). The training set was manually prepared by selecting the rectangular regions enclosing the iris region scaled to a size of 24 × 24 pixels. The learning algorithm for iris detection is based on the Gentle stump-based Adaboost. The detector was tested on a test set consisting of 20 videos from our database (see Sect. 4.2), achieving an effectiveness in the detection of 100% of both irises of each individual (at least one couple of iris is detected) and 0.01% of false detections.

3.2 Image Quality Evaluation

One of the most important parameters indicated in the standard [4], the focal distance indicates the optimal distance between the subject and sensor for a normed pixel density. Pixel density is defined in the standard as the sum quantity of the pixels that are on the diagonal length of the iris image. Normed pixel density in the standard should be at least 200 pixels and be composed of at least two pixel lines per millimeter (2 lppmm). For medical literature, it can be assumed that the iris approximately represents 60% of the detected bounding box of iris region under well illuminated environment. Then the pixel density of an iris image (Ird) can be calculated using the classical Pythagorean Theorem by Eq. 1, where w and h are the width and height in pixels of the detected image.

$$Ird = 0.6\left(\sqrt{w^2 + h^2}\right) \tag{1}$$

3.3   Quality Measure for Iris Images

One of the elements that negatively influences the quality of iris images is the degree of image sharpness because depending on it the image may be blurred or out of focus and the iris texture loses the details of the structures that make possible the identification. An evaluation process of NIR eye images using the Kang and Park filter is proposed in [3]. The filter is used to filter the high frequencies and then to estimate the total power based on Parseval's theorem, which establishes that the power is conserved in the spatial and frequency domains. The filter is composed of a convolutional 5×5 kernel. It consists of three functions of 5×5 and amplitude -1, one of 3×3 and amplitude $+5$ and four of 1×1 and amplitude -5. The kernel is able to estimate the high frequencies in the iris texture in the NIR better than other operators of the state of the art at the same time it has a low computational time due to the reduced kernel size. Then we think that it may have a similar behavior for iris images captured in the VS.

The entropy of an iris image has proven to be a good indicator of the amount of information [17]. The entropy only depends on the amount of gray levels and the frequency of each gray level.

Taking into account these elements and considering that the pixel density of the iris image is another of the fundamental elements for the quality of the image, we propose its combination with the Kang & Park filter and the entropy estimation to obtain a measurement of the quality of the iris image ($Qiris$). The proposed $Qiris$ is obtained by Eq. 2.

$$Qiris = \frac{Ird * kpk * ent}{tird * tkpk * tent} \tag{2}$$

Where: kpk is the average value of the image pixels obtained as result of the convolution of the input iris image with the Kang and Park kernel. $tird$ is the threshold established by the standard [4] for the minimum Ird to obtain a quality image. $tkpk$ is the estimated threshold of kpk to obtain a quality image, in [3] the authors, from their experimental results, recommend a threshold = 15. ent is the entropy of the eye image. $tent$ is the estimated threshold of ent with which it will be possible to obtain a quality image. Entropy of an iris image is calculated by Eq. 3, where the pi value is the occurrence probability of a given pixel value inside the iris region and n is the number of image pixels.

$$ent = -\sum_{i=1}^{n} p_i log_2 p_i \tag{3}$$

Experimentally, we have verified that the iris images with a quality according to the international standard have an entropy higher than 4. For this experiment, we took a set of 300 images standardized to a size of 260×260 from the MobBio [15] (150 images) and UBIRIS [16] (150 images) databases and performed the evaluation of their quality by the parameters of pixel density and response of the Kark and Park filter. The experimental results established that those with values greater than the corresponding thresholds have an entropy with a value equal to or greater than 4, so we assume this value as the value of *tent*. *Qiris* can reach values depending on the thresholds selected for *Ird, kpk*, and *ent*. Thus considering the threshold *tird* = 200 established by the standard, *tkpk* = 15 experimentally obtained in [3] and *tent* = 4 experimentally obtained by us, the minimum value of *Qiris* to obtain a quality eye image would be 1, higher values would denote images of higher quality and values less than 1 images with a quality below the standard. One question to explore in this case would be to determine under what minimum values of *Qiris* it is possible to obtain acceptable recognition accuracies for a given configuration of a system.

3.4   Iris Semantic Classification for Segmentation

The main motivation for using the semantic classification of iris image is that the introduction of semantic information can avoid or diminish the false positive classifications of pixels that do not belong to the iris. The objective of the iris semantic classification is to assign to each image region a label of a class describing an anatomical eye structure (iris, pupil or sclera). From a computational point of view this is expressed by assigning the value of the predefined class to each image pixel.

From the promising results obtained in [12], we decided to use a training scheme, where trained FCN was adapted to the case of the semantic classification of the iris through a process of inductive transfer learning [18]. We selected the FCN model. fcn8 s-at-once that was fine-tuned of the pre-trained model VGG-16 [19]. The fine-tuning process allows us to select the most promising hypothesis space to adjust an objective knowledge. The implementation of the fcn8 s-at once architecture was performed in Caffe [20]. Intermediate upsampling layers were initialized by a bilinear interpolation. The output number of the model was set according to the 5 used classes. The difference is that the classification made on the image in our proposal is directly performed on the iris image (Fig. 2b) unlike the proposal of [12] that is performed on the whole eye image (Fig. 2a). This proposal significantly reduces the calculation time with respect to the proposal of [12] due to the smaller size image to be classified and the smaller number of classes to consider.

Due to the need for normalize the texture of the iris from the *Rubbersheet* model, we combine the pixel-based algorithm (FCN) with an edge-based approach that allows the detection of the coordinates of center and radius of the iris and pupil regions.

4   Experimental Results and Discussion

In order to validate the proposal, our experimental design was aimed at verifying the influence of the proposed approach in the iris segmentation and verification tasks by evaluating it using two benchmark iris image databases and an iris video dataset of our own. We compared the performance of our proposal with results obtained using the FCN [12].

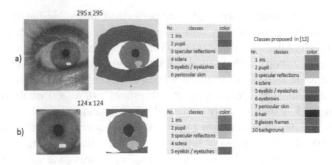

Fig. 2. Segmentation examples. (a) by [12] (b) by proposed method

Three basic functionalities compose the implemented pipeline for experiments: *Iris image acquisition and Image segmentation:* These two modules are based on the approach described in the previous sections. *Feature extraction and comparison:* For the purpose of experiments in this work, we experimented the combination of two feature extraction methods in order to verify the robustness of the proposed approach with respect to the use of different features for recognition. Scale-Invariant Feature Transform (SIFT) [21], and Uniform Local Binary Patterns (LBP) [22]. For the comparison we used dissimilarity scores corresponding to each of these methods. The first estimates the dissimilarity between two sets of key points SIFT [21]. The second one use the Chi-square distance metric [22] which is a non-parametric test to measure the similarity between images of a class. In both cases the minimum distance found between two images gives the measure of maximum similarity between them.

4.1 Iris Datasets

MobBio [15] is a multi-biometric dataset including face, iris, and voice of 105 volunteers. The iris subset contains 16 VS images of each individual at a resolution of 300 × 200. UTIRIS dataset [16] is an iris biometric dataset in VS constructed with 1540 images from 79 individuals.

Our (DatIris) database consists of 82 videos of 41 people taken in two sessions of 10 s each at a distance of 0.55 m. The camera used was a Logitech C920 HD Pro Webcam at resolution of 1920 × 1080. The videos were taken in indoor conditions with ambient lighting and presence of specular reflections to achieve an environment

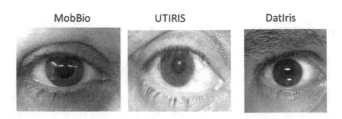

Fig. 3. Samples of eye images from MobBio, UTIRIS and DatIris datasets.

closer to the poorly controlled conditions of a biometric application. The database contains videos of people of clear skin of Caucasian origin, dark skin of African origin and mestizo skin, it is composed by 26 men and 15 women, in a range of ages from 10 to 65 years (see an image example in Fig. 3).

4.2 Experimental Results

The accuracy of the proposed segmentation stage was measured by comparing the results obtained by the FCN approach [12] with our proposal on MobBio database. This database provides manual annotations for every image of both the limbic and pupillary contours, 200 images with $Qiris > 1$ were evaluated and chosen in order to guarantee the quality of the experiment. As evaluation metric we consider the E^1 error proposed by the NICE.I protocol (http://nice1.di.ubi.pt/). This metric estimates the proportion of correspondent disagreeing pixels. The average time (in seconds) that the segmentation method takes to obtain a segmented iris image was calculated using a PC with an Intel Core i5-3470 processor at 3.2 GHz and 8 GB of RAM. Table 1 lists the E^1 obtained by FCN [12] and our approach. The results show a similar performance in terms of E^1 for both methods, however in the case of our proposal a substantial reduction of the computation time for the segmentation process is observed.

Table 1. Experimental results of segmentation stage for MobBio dataset

FCN [12]		Our method	
E^1	t (sec)	E^1	t (sec)
2.5	4	**2.4**	**0.3**

The influence of our proposal on the accuracy of the verification task was estimated by the error of false rejection (FRR) at false acceptance rate (FAR) $\leq 0.001\%$. Comparisons were made in the form of all against all to obtain the distributions of genuine and impostors. For DatIris dataset, the 41 videos of the session 1 were processed taking two images of each iris and comparing them against a database composed of two images of each iris taken from frames of the videos of session 2.

Table 2 shows the comparison of the error of FRR at FAR $\leq 0.001\%$ obtained using the FCN approach [12] and our proposal, on the three experimented datasets, taking two different intervals of $Qiris$ to reject or accept the iris images to be processed. The results show that as the value of the $Qiris$ increases, the system supports high quality images and rejects low quality images. This increase in quality, results in a decrease in the FRR, with a significant result in UTIRIS where an FRR = 0.04 is achieved with the 79.5% of the database and $Qiris > 1$ using LBP features. However, in the MobBio, increase in the $Qiris$ threshold results in a significant decrease in the number of images to be compared. The results obtained on the DatIris dataset show a

Table 2. Comparison of the FRR obtained by the implemented system using the FCN approach [12] and our proposal.

Database	Qiris	% of proc. images*	FCN [12]		Our method	
			SIFT	LBP	SIFT	LBP
MobBio	<1.0	73.7	0.28	0.26	0.27	0.25
	≥1.0	26.3	0.24	0.23	0.25	0.22
UTIRIS	<1.0	20.5	0.06	0.05	0.07	0.06
	≥1.0	79.5	0.07	0.04	0.06	0.04
DatIris	<1.0	10	0.03	0.02	0.04	0.03
	≥1.0	90	0.02	0.01	0.02	0.01

* is the percentage of processed images with quality values specified in column *Qiris*

high performance of the verification process obtaining an FRR of 0.01, which corroborates the relevance of the proposed quality index in a real application.

It is also observed that the levels of FRR remain similar for both methods which shows that the restriction of the semantic classification to the iris region besides reducing the computation time does not affect the accuracy of the recognition process.

Table 3 shows the comparison of the proposed in [8] quality measure (*Qindex*) with our proposal (*Qiris*), at values ≥ 1 using LBP features and the proposed segmentation method. It is appreciated that by including the entropy of the iris image, the system considers a greater percentage of images to be processed by the system while maintaining similar levels of accuracy in the recognition.

Table 3. Comparison of the proposed quality measure in [8] (*Qindex*) with our proposal (*Qiris*)

Database	% of proc. images**		FRR using Qindex [8]	FRR using Qiris
	Qindex [8]	Qiris		
MobBio	19.5	26.3	0.23	0.22
UTIRIS	71.8	79.5	0.05	0.04
DatIris	88	90	0.02	0.01

** is the percentage of processed images with quality value ≥ 1

5   Conclusions

In this paper, we propose a video iris recognition approach based on image quality evaluation and semantic classification for biometric iris recognition in the VS. It combines automatic detection methods, a new image quality measure and semantic classification. We analyzed the relevance of the image evaluation stage as a fundamental step to filter the information generated from the iris video capture. The experimental results showed that the inclusion of the proposed quality measure before iris segmentation limits the passage of low quality images to the system, which results in an increase of recognition rates. The application of the semantic classification on the iris image directly in the process of video capturing allows the reduction of the computational time in the segmentation process maintaining the levels of accuracy in the recognition achieved by the FCN based state of the art method. In future work we will evaluate the use of deep neural networks for the process of iris detection in video.

References

1. Raja, K.B., Raghavendra, R., Vemuri, V.K., Busch, C.: Smartphone based visible iris recognition using deep sparse filtering. Pattern Recog. Lett. **57**, 33–42 (2015)
2. Hollingsworth, K., Peters, T., Bowyer, K.: Iris recognition using signal-level fusion of frames from video. IEEE Trans. Inf. Forensics Secur. **4**, 837–848 (2009)
3. Garea-Llano, E., García-Vázquez, M., Colores-Vargas, J.M., Zamudio-Fuentes, L.M., Ramírez-Acosta, A.A.: Optimized robust multi-sensor scheme for simultaneous video and image iris recognition. Pattern Recog. Lett. **101**, 44–45 (2018)
4. ISO/IEC 19794-6:2005. Part 6: Iris image data, ISO (2005)
5. Schmid, N., Zuo, J., Nicolo, F., Wechsler, H.: Iris quality metrics for adaptive authentication. In: Bowyer, K.W., Burge, M.J. (eds.) Handbook of Iris Recognition, 2nd edn, pp. 101–118. London, Springer-Verlag (2016)
6. Daugman, J., Downing, C.: Iris image quality metrics with veto power and nonlinear importance tailoring. In: Rathgeb, C., Busch, C. (eds.) Iris and Periocular Biometric Recognition. IET Publication, pp. 83–100 (2016)
7. Zuo, J., Schmid, N.A.: An automatic algorithm for evaluating the precision of iris segmentation. In: BTAS 2008, Washington (2008)
8. Garea-Llano, E., Osorio-Roig, D., Hernandez, O.: Image Quality Evaluation for Video Iris Recognition in the Visible Spectrum. Biosensors and Bioelectronics Open Access (ISSN:2577-2260)
9. Liu, N., Li, H., Zhang, M., Liu, J., Sun, Z., Tan, T.: Accurate iris segmentation in non-cooperative environments using fully convolutional networks. In: Proceedings of ICB 2016. IEEE, pp. 1–8 (2016)
10. Jalilian, E., Uhl, A.: Iris segmentation using fully convolutional encoder–decoder networks. In: Bhanu, B., Kumar, A. (eds.) Deep Learning for Biometrics. ACVPR, pp. 133–155. Springer, Cham (2017). https://doi.org/10.1007/978-3-319-61657-5_6
11. Osorio-Roig, D., Morales-González, A., Garea-Llano, E.: Semantic segmentation of color eye images for improving iris segmentation. In: Mendoza, M., Velastín, S. (eds.) CIARP 2017. LNCS, vol. 10657, pp. 466–474. Springer, Cham (2018). https://doi.org/10.1007/978-3-319-75193-1_56
12. Osorio-Roig, D., Rathgeb, C., Gomez-Barrero, M., Morales-González Quevedo, A., Garea-Llano, E., Busch, C.: Visible wavelength iris segmentation: a multi-class approach using fully convolutional neuronal networks. In: BIOSIG 2018. IEEE (2018)
13. Raja, K.B., Raghavendra, R., Busch, C.: Iris imaging in visible spectrum using white LED. In: Proceedings of BTAS 2015. IEEE (2015)
14. Viola, P., Jones, M.: Rapid Object Detection Using a Boosted Cascade of Simple Features. Mitsubishi Electric Research Laboratories Inc., Cambridge (2004)
15. Monteiro, C., Oliveira, H.P., Rebelo, A., Sequeira, A.F.: Mobbio 2013: 1st biometric recognition with portable devices competition. https://paginas.fe.up.pt/~mobbio2013/

16. Hosseini, M.S., Araabi, B.N., Soltanian-Zadeh, H.: Pigment melanin: pattern for iris recognition. IEEE Trans. Instrum. Measur. **59**, 792–804 (2010)
17. Fathy, W.S.A., Ali, H.S.: Entropy with local binary patterns for efficient iris liveness detection. Wireless Pers. Commun. **102**(3), 2331–2344 (2018)
18. Torrey, L., Shavlik, J.: Transfer learning. In: Handbook of Research on Machine Learning Applications (2009)
19. Simonyan, K., Zisserman, A.: Very deep convolutional networks for large-scale image recognition. arXiv preprint arXiv:1409.1556 (2014)
20. Jia, Y., et al.: Caffe: Convolutional architecture for fast feature embedding. In: Proceedings of 22nd ACMMM 2014, pp. 675–678 (2014)
21. Lowe, D.G.: Distinctive image features from scale-invariant key points. Int. J. Comput. Vis. **60**, 91–110 (2004)
22. Liao, S., Zhu, X., Lei, Z., Zhang, L., Li, S.Z.: Learning multi-scale block local binary patterns for face recognition. In: Lee, S.-W., Li, S.Z. (eds.) ICB 2007. LNCS, vol. 4642, pp. 828–837. Springer, Heidelberg (2007). https://doi.org/10.1007/978-3-540-74549-5_87

YOLO-FD: YOLO for Face Detection

Luan P. e Silva$^{(\boxtimes)}$, Júlio C. Batista , Olga R. P. Bellon ,
and Luciano Silva

IMAGO-UFPR Research Group, Departmento de Informática,
Universidade Federal do Paraná, Curitiba, Paraná, Brazil
{luan.porfirio,julio.batista,olga,luciano}@ufpr.br

Abstract. Face detection is a fundamental step for any face analysis approach. However, it remains as an unsolved problem in computer
vision, specially, when it comes to the variability and distractions of in-
the-wild environments. Moreover, a face detector must be accurate and
fast to be used in surveillance/biometrics scenarios. In order to overcome these limitations, this paper proposes a customized version of the
state-of-the-art object detector, YOLOv3, for face detection. The modifications aim at building a real-time, accurate model capable of detecting
faces as small as 16 pixels in 34 FPS. Furthermore, this model was evaluated on three of the most difficult benchmarks for face detection, Wider
Faces, UCCS and UFDD, showing a good score balance across them.
Also, the comparison with the state-of-the-art shown that it was possible to achieve the second best FPS and the fifth best score on Wider
Faces. Finally, the model will be available in https://github.com/luanps/
yolofd.

1 Introduction

Face detection plays an important role in face analysis tasks such as biometrics, surveillance and facial expressions analysis. However, because of the high
variability among people and scenarios, it is far from being a solved problem in
computer vision. For instance, when working on in-the-wild environments, poor
illumination, head pose variation, face scale and low resolution are expected [17].

In the past few years, many computer vision tasks achieved remarkable results
by using Deep Convolutional Networks (DCNs). These tasks include image classification [14], face recognition [10], face alignment [21], facial expression analysis [1], and object detection [12,13]. All of these were possible because of huge
amounts of labelled data and computing power available to optimize these models. Therefore, these models can learn robust representations to handle the variations of the unconstrained environments.

The face detection approach proposed by Viola and Jones [15] was a major
breakthrough back in the day. However, it relies on hand-crafted features which
imposes some limitations to this approach. Hence, the alternative is to use DCNs
which have proven to be able to overcome these limitations. Since face detection can be defined as a special case of generic object detection, the following
paragraphs provide a review of current approaches for this task using DCNs.

© Springer Nature Switzerland AG 2019
I. Nyström et al. (Eds.): CIARP 2019, LNCS 11896, pp. 209–218, 2019.
https://doi.org/10.1007/978-3-030-33904-3_19

Faster R-CNN [13] made important progress on object detection in recent years. This model was built on top of VGG16 for representation learning, but used Region Proposal Networks (RPNs) to generate candidate regions for detection. Moreover, since RPNs are differentiable, it was possible to optimize the model end-to-end. The overall optimization strategy was based on multi-task loss by generating class probabilities for objects and regression coordinates for bounding boxes. However, this models runs in 5 FPS, making it unfeasible for real-time applications. Because of its success for object detection, this model was adapted for face detection in [6] and [20].

Although Faster R-CNN is a robust model, it struggles with performance. It happens because of candidate regions generated for a selective search approach to detect the bounding boxes. To overcome this limitation, models like SSD [7] and YOLO [11] proposed a tilling approach. The idea is to divide an image into $N \times N$ tiles and, for each tile, regress the bounding box coordinates, an objectness score and a class probability. Moreover, it is possible to achieve scale-invariance by selecting the appropriate number of tiles. For instance, by making the tiles larger, it is possible to detect larger objects in the image and vice-versa. This reduction in computations for candidate regions makes SSD achieve 58 FPS and YOLO 78 FPS.

Since SSD and YOLO are state-of-the-art approaches, as for Faster R-CNN, it makes sense to apply these approaches for the specific case of face detection. However, some approaches extend their contributions in order to achieve high accuracy and speed for this task. Zhang et al. [18] proposed a single anchor-based model built on top of SSD, achieving state-of-the-art accuracy and speed. Other examples include HR [4] which proposed an image pyramid approach to find tiny faces; SSH [9] makes use of context by enlarging filters around candidate proposals. Also, Zhu [19] also makes use of context, but by using a new anchor design. Yet, neither one of [4,9,19] achieves a real-time FPS.

Given the current approaches for face/object detection, this paper proposes:

- a real-time, scale-invariant model capable of detecting faces as small as 16 pixels;
- a simple thresholding strategy to improve the detection of smaller faces;
- tuning a generic object detector to achieve results comparable to the face detection state-of-the-art.

The remaining of this paper is organized as follows: Sect. 2 introduces the proposed model and the optimization procedure; Sect. 3 presents the results and comparisons; Finally, Sect. 4 concludes this work.

2   Proposed Model

The proposed model, named YOLO-FD, is based on YOLOv3 [12] and an overview of it is shown in Fig. 1. This model expects an image of size 832×832 as input and computes a representation by applying a series of convolutions grouped into residual blocks [3]. Afterwards, this representation is used in four layers to

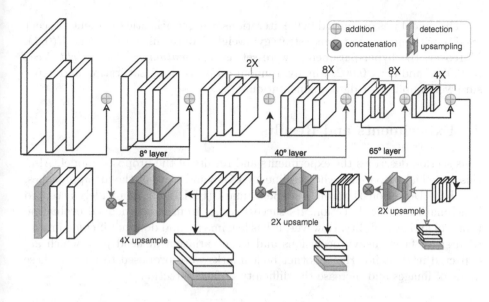

Fig. 1. YOLO-FD. Note that this model is deeper than YOLOv3 and uses four detection layers.

detect faces in different scales. The first layer detects larger faces because it uses 13×13 tiles. The following layers detect smaller faces by upsampling the representation to 26, 52 and 208 tiles.

Since this model is based on YOLOv3, the output for each detection layer is a volume of size $N \times N \times [B * (5+C)]$. This volume contains $B = 3$ bounding box proposals and $C = 1$ class probability for each one of the $N \times N \times$ tiles. For each one of the B proposals, it generates (x, y, w, h) coordinates and an objectness score (p). Through this volume, the proposal with highest p is selected from the B candidates. It is important to note that only candidates with $p > \tau$ are considered, otherwise no bounding box is selected for the given tile. Considering that anchor-based approaches struggle with small regions [18], a simple thresholding strategy was adopted to handle this limitation. For the first two detection layers, $\tau := 0.7$ because bigger faces are easier to detect. Since smaller faces are harder to detect, $\tau := 0.5$ for the last two detection layers. Consequently, the last layers will discard less false negative samples while optimizing.

The architecture of this model is similar to YOLOv3. For feature extraction, a set of six residual blocks with skip connections were stacked. Each one of these blocks is composed by convolutions of size 3×3 and 1×1. Since the input is an image of size 832×832 and each block halves the size of the input tensor, the output of the last layer is a tensor of size $13 \times 13 \times 18$. It is important to note the fusion strategy before the last three detection layers. After computing the detection result, the tensor is upsampled and concatenated with a previous feature map. This allows the model to combine fine-grained information from earlier layers to help it achieve scale-invariance.

The model was optimized in 95k iterations using Stochastic Gradient Descent (SGD) without a fine-tuning strategy; weights were initialized from $\mathcal{U}(0,1)$. Moreover, the hyper-parameters were set as: *momentum* $:= 0.9$, *weight decay* $:= 0.0005$ and $\alpha := 0.002$. Finally, random crops, scaling, saturation and exposure were applied for data augmentation.

3   Experiments and Results

This section describes the experiments and results of the proposed model. Also, this model is compared to the baseline YOLOv3 and state-of-the-art approaches. The Face Detection Data Set and Benchmark (FDDB) [5] is a widely accepted benchmark for face detection. Its success comes from a well defined evaluation protocol and availability. However, it has few images and does not include images with small faces, heavy occlusions and large variation in head pose which are expected in-the-wild. Hence, other benchmarks were proposed to include these kinds of images and increase the difficulty of face detection.

3.1   Wider Faces

The Wider Faces [16] benchmark consists of 32,302 images with 393,703 labeled faces. The faces vary in a large range of scales and the dataset contains three levels of difficulty: easy, medium and hard. The evaluation protocol consists of three random sets denoted: *training* (50%), *validation* (10%) and *test* (40%).

Figure 2 shows the evaluation metrics on the *validation set* while optimizing the model using *training set*.

From Fig. 2 it is possible to note the trade-off among *Precision*, *Recall* and *Intersection Over Union (IOU)*. Although there is a drop in *Precision* and *IOU*, the peak in *Recall* balances the F_1-score. This means that the model accepts some false positives, *i.e.* detecting a face when there is not one, instead of only detecting the ones it is highly confident. It becomes important for smaller faces, since their representation doesn't have the same amount of information as a larger face. The remaining of this section describes the databases and results of the proposed model.

Table 1 compares the proposed model to the state-of-the-art using Wider Faces. Note that the proposed model does not surpass these approaches, except for HR [4] in the Hard subset. However, at inference time this model is approximately 28x faster than Zhu *et al.* [19]. Moreover, computational time is quite similar to SFD [18] despite the difference in input size (640 × 640 instead of 832 × 832) and GPU architecture used for testing. Finally, Fig. 3 shows precision-recall curves of the proposed model.

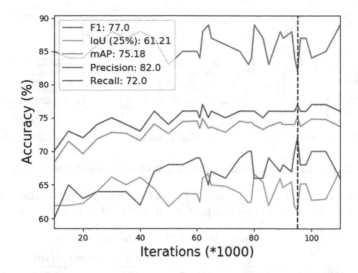

Fig. 2. Evaluation metrics computed on Wider Faces validation subset while optimizing the model. The dashed line marks the iteration number 95k.

Table 1. Comparison of FPS and average precision with the state-of-the-art on wider faces validation subset.

Method	GPU	FPS	Wider faces subsets		
			Easy	Medium	Hard
Zhu *et al.* [19]	TITAN X	1.2	0.949	0.933	0.861
SFD [18]	TITAN X	36	0.942	0.930	0.859
SSH [9]	Quadro P6000	5.5	0.931	0.921	0.845
HR [4]	Not reported	1.5	0.925	0.910	0.806
YOLO-FD τ	GTX 1080 TI	34	0.908	0.894	0.809

Considering this model is an adaptation of YOLOv3, Table 2 compares the proposed model against its baseline, a deep network built on top of Darknet-53 backbone. For a fair comparison, YOLOv3 model with highest accuracy was selected, which was on 150,000 iteration. Because of YOLOv3's multi-scale strategy, it is possible to perform face detection by varying the size of the input image.

As it is possible to note from Table 2, images of size 832 × 832 achieved the best AP in all the subsets. Moreover, this table shows that the proposed model was able to increase the frame rate by 12 FPS over the baseline. Also, it shows that 95 thousand iterations achieved a better score than 150 thousand. This suggest some overfitting of the model as the optimization continues and that it is possible to reduce the optimization time by almost 1/3. Finally, it compares the proposed model using the threshold strategy (YOLO-FD τ) which increased the result for the hard subset (mainly composed by small faces).

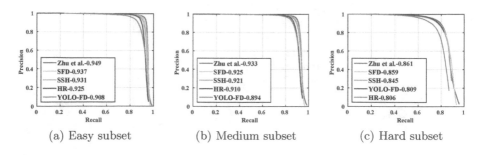

(a) Easy subset   (b) Medium subset   (c) Hard subset

Fig. 3. Precision-Recall curves for each Wider Faces validation subset.

Table 2. Comparison of the input size, FPS and AP against the baseline. The best results are marked in blue and second best results are marked in red.

Method	Input	Iter.	FPS	Wider faces subsets		
				Easy	Medium	Hard
YOLOv3	416	150k	55	0.880	0.848	0.717
YOLOv3	608	150k	37	0.902	0.873	0.754
YOLOv3	832	150k	22	0.907	0.882	0.787
YOLO-FD	832	150k	34	0.918	0.893	0.790
YOLOv3	416	95k	55	0.871	0.838	0.651
YOLOv3	608	95k	37	0.896	0.870	0.749
YOLOv3	832	95k	22	0.896	0.878	0.783
YOLO-FD	832	95k	34	0.896	0.861	0.759
YOLO-FD τ	832	95k	34	0.908	0.894	0.809

3.2 UCCS

The UnConstrained College Students dataset (UCCS) [2] was built using high-resolution images from a surveillance camera at a distance of 100 meters. It contains a total of 48,139 images with 70,611 labeled faces. Since images were acquired by a surveillance camera, it shows different weather conditions and occlusions. The evaluation protocol consists of three sets: *training*, *validation* and *test*. However, the ground truth for the *test set* is not available, hence the *validation set* was used for evaluation. Moreover, its evaluation protocol uses a modified version of the IOU measure because the bounding boxes provided as ground truth are larger than the face regions, so the model was retrained with UCCS *training set* to get concise bounding box predictions.

Figure 4 shows the proportion between detection rate and false accepts. A false accept occurs when the model detects a face and there is not one, in other words, it is the recall. From the 13,582 faces in the *validation set*, this model detected 2,518 false positives and 13,097 true positives. As a comparison, the *baseline* detected 39,053 false positives and 12,785 true positives. These results show that the performance of the baseline was degraded by the huge amount of

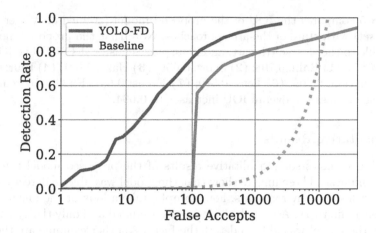

Fig. 4. Free Response Operating Characteristics (FROC) plot for UCCS dataset. False accepts are in logarithmic scale.

false positives. The effect of these wrong detections was eased in the proposed model which decreased the number of false positives by a factor of 15.

3.3   UFDD

The Unconstrained Face Detection Dataset (UFDD) [8] is a newer benchmark. It contains 6,424 images with 10,895 labeled faces in six different conditions: rain, snow, haze, blur, illumination and lens distortion. Moreover, an extra class, distractor, is provided in order to evaluate the models in images without a face.

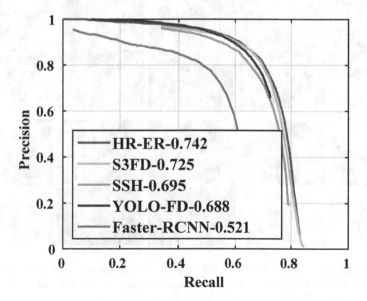

Fig. 5. Intersection over union plot for UFDD dataset.

Figure 5 compares the IOU of the proposed model (trained on Wider Faces *training set*) with state-of-the-art approaches. Note that the proposed model is comparable to the state-of-the-art, and it surpasses Faster R-CNN. The IOU for each subset is: **(1)** Rain: 0.708; **(2)** Snow: 0.656; **(3)** Haze: 0.631; **(4)** Blur: 0.770; **(5)** Illumination: 0.709; **(6)** Lens: 0.510; **(7)** Focus: 0.631. Finally, by removing the distractor class the overall IOU increases to 0.694.

3.4   Qualitative Results

At last, Fig. 6 shows some qualitative results of the proposed model using the three benchmarks. These images show that the model was able to detect faces in different scales and environments. For example, take a look at the image in the fourth image, first row. Although the faces are occluded and only the eyes region is visible, the model was able to detect the faces. Another example are the tiny faces in a foggy weather in the second column of the fifth row. Also, six faces

Fig. 6. Qualitative results of YOLO-FD. The first and second rows show detections on Wider Faces; the third and fourth rows show images from UCCS; the last two rows show images from UFDD.

were detected, and a blurry one was missed, in the fifth column of the second row. All in all, these qualitative results show the generalization capabilities of the model to handle face detection in-the-wild.

4 Conclusions and Future Works

This paper presented an adaptation of YOLOv3 for face detection. The model was improved to achieve scale-invariance, specially for tiny faces, and real-time inference. Although the model was only optimized using Wider Faces, it was evaluated using three benchmarks. However, there is still need for an extensive evaluation on these benchmarks in order to gather insights on appropriate improvements of generic object detectors for face detection. Finally, some future works include: using a different model as baseline, evaluate the effect of increasing the number of detection layers and the number of anchor boxes.

Acknowledgments. This study was financed in part by the Coordenação de Aperfeiçoamento de Pessoal de Nível Superior - Brasil (CAPES) - Finance Code 001 The authors gratefully acknowledge the contribution of reviewers' comments, etc. (if desired). Put sponsor acknowledgments in the unnumbered footnote on the first page.

References

1. Batista, J.C., Albiero, V., Bellon, O.R., Silva, L.: AUMPNet: simultaneous action units detection and intensity estimation on multipose facial images using a single convolutional neural network. In: 2017 12th IEEE International Conference on Automatic Face & Gesture Recognition, FG 2017, pp. 866–871. IEEE (2017)
2. Günther, M., et al.: Unconstrained face detection and open-set face recognition challenge. In: 2017 IEEE International Joint Conference on Biometrics (IJCB), pp. 697–706. IEEE (2017)
3. He, K., Zhang, X., Ren, S., Sun, J.: Deep residual learning for image recognition. In: Proceedings of the IEEE Conference on Computer Vision and Pattern Recognition, pp. 770–778 (2016)
4. Hu, P., Ramanan, D.: Finding tiny faces. In: 2017 IEEE Conference on Computer Vision and Pattern Recognition (CVPR), pp. 1522–1530. IEEE (2017)
5. Jain, V., Learned-Miller, E.: FDDB: a benchmark for face detection in unconstrained settings. Technical report, Technical Report UM-CS-2010-009, University of Massachusetts, Amherst (2010)
6. Jiang, H., Learned-Miller, E.: Face detection with the faster R-CNN. In: 2017 12th IEEE International Conference on Automatic Face & Gesture Recognition, FG 2017, pp. 650–657. IEEE (2017)
7. Liu, W., et al.: SSD: single shot multibox detector. In: Leibe, B., Matas, J., Sebe, N., Welling, M. (eds.) ECCV 2016. LNCS, vol. 9905, pp. 21–37. Springer, Cham (2016). https://doi.org/10.1007/978-3-319-46448-0_2
8. Nada, H., Sindagi, V., Zhang, H., Patel, V.M.: Pushing the limits of unconstrained face detection: a challenge dataset and baseline results. arXiv preprint arXiv:1804.10275 (2018)

9. Najibi, M., Samangouei, P., Chellappa, R., Davis, L.S.: SSH: single stage headless face detector. In: ICCV, pp. 4885–4894 (2017)

10. Parkhi, O.M., Vedaldi, A., Zisserman, A., et al.: Deep face recognition. BMVC. **1**(3), 6 (2015)

11. Redmon, J., Divvala, S., Girshick, R., Farhadi, A.: You only look once: Unified, real-time object detection. In: Proceedings of the IEEE Conference on Computer Vision and Pattern Recognition, pp. 779–788 (2016)

12. Redmon, J., Farhadi, A.: Yolov3: an incremental improvement. arXiv preprint arXiv:1804.02767 (2018)

13. Ren, S., He, K., Girshick, R., Sun, J.: Faster R-CNN: towards real-time object detection with region proposal networks. In: Advances in neural information processing systems, pp. 91–99 (2015)

14. Russakovsky, O., et al.: Imagenet large scale visual recognition challenge. Int. J. Comput. Vis. **115**(3), 211–252 (2015)

15. Viola, P., Jones, M.J.: Robust real-time face detection. Int. J. Comput. Vis. **57**(2), 137–154 (2004)

16. Yang, S., Luo, P., Loy, C.C., Tang, X.: Wider face: a face detection benchmark. In: IEEE Conference on Computer Vision and Pattern Recognition (CVPR) (2016)

17. Zafeiriou, S., Zhang, C., Zhang, Z.: A survey on face detection in the wild: past, present and future. Comput. Vis. Image Underst. **138**, 1–24 (2015)

18. Zhang, S., Zhu, X., Lei, Z., Shi, H., Wang, X., Li, S.Z.: S³3FD: Single shot scale-invariant face detector. In: 2017 IEEE International Conference on Computer Vision (ICCV), pp. 192–201. IEEE (2017)

19. Zhu, C., Tao, R., Luu, K., Savvides, M.: Seeing small faces from robust anchor's perspective. In: Proceedings of the IEEE Conference on Computer Vision and Pattern Recognition, pp. 5127–5136 (2018)

20. Zhu, C., Zheng, Y., Luu, K., Savvides, M.: CMS-RCNN: contextual multi-scale region-based CNN for unconstrained face detection. In: Bhanu, B., Kumar, A. (eds.) Deep Learning for Biometrics. ACVPR, pp. 57–79. Springer, Cham (2017). https://doi.org/10.1007/978-3-319-61657-5_3

21. Zhu, X., Lei, Z., Liu, X., Shi, H., Li, S.Z.: Face alignment across large poses: a 3D solution. In: Proceedings of the IEEE Conference on Computer Vision and Pattern Recognition, pp. 146–155 (2016)

On Using Document Scanners for Minutiae-Based Palmprint Recognition

Manuel Aguado-Martínez[1]([⊠]), José Hernández-Palancar[1],
Katy Castillo-Rosado[1], Christof Kauba[2], Simon Kirchgasser[2],
and Andreas Uhl[2]

[1] Advanced Technologies Application Center,
7ma A #21406 e/214 y 216, Siboney, Playa, 12200 Havana, Cuba
{maguado,jpalancar,krosado}@cenatav.co.cu
[2] Department of Computer Sciences, University of Salzburg, Salzburg, Austria
{ckauba,skirch,uhl}@cs.sbg.ac.at
http://www.cenatav.co.cu/

Abstract. The development of forensic palmprint biometric systems has not been as popular as civilian systems until today, because of the high costs of the required capturing devices and the lack of publicly available high-resolution palmprint databases. The feasibility of using low-cost technologies like document scanners to acquire a high-resolution palmprint database is explored in this work. A database was established using a biometric industry standard scanner and an HP document scanner. Experimental results show the potential of using similar inexpensive technologies to develop high-resolution palmprint systems for forensic applications. Advantages and disadvantages of both technologies are highlighted too.

Keywords: High resolution palmprint matching · Palmprint recognition · Minutiae based recognition · Document scanner

1 Introduction

Palmprint applications are employed for a wide range of scenarios. For civilian applications, palmprint biometrics has been thoroughly explored. Used as a stand-alone biometric or in combination with others, palmprints has proven to be very reliable [3,4,6,23,24] due to the large area of the palm which contains rich and useful features, hence being highly distinctive. Furthermore, the urgent need for palmprint applications in forensic scenarios has been clearly stated by several studies regarding the number of latent palmprints found in crime scenes as summarize in [11].

Nevertheless, the development of forensic palmprint biometric systems has not been as popular as civilian systems until today. Civilian applications usually employ low-resolution images and extract features from the principal lines or texture information. For forensics, more detailed features like minutiae need

© Springer Nature Switzerland AG 2019
I. Nyström et al. (Eds.): CIARP 2019, LNCS 11896, pp. 219–229, 2019.
https://doi.org/10.1007/978-3-030-33904-3_20

to be extracted which require high-resolution images (500 ppi) [17]. Therefore, bigger and more expensive devices are needed to establish a palmprint database for forensic applications. The high cost of these scanners (3000–4000 USD) is one of the main issues refraining the development of palmprint forensic systems. Moreover, the lack of publicly available high-resolution palmprint databases has prevented the research in this area.

This study is the first step towards solving the aforementioned problems. The feasibility of using commercial document scanners for minutiae-based palmprint recognition is explored. Compared to forensics industry standard scanners, document scanners are cheaper and widely distributed [21]. If this type of scanners proves to be feasible for forensic applications, it will open a wide range of possibilities of using similar technologies and therefore, reducing the development cost of forensic palmprint systems. The main contributions of this work are:

1. A feasibility study of using document scanners for minutiae-based palmprint recognition as the first step to apply similar technologies in forensic applications.
2. A new database of high-resolution palmprint images acquired with two different scanners. This database will be made publicly available to the research community.
3. Evaluating the cross-device recognition performance between a forensic palmprint capturing device and the HP document scanner.

Following, a summary of related works is given. Section 2 will introduce the database structure, acquisition protocol, and the challenges faced. Experimental results are discussed in Sect. 3, and finals conclusions are presented in Sect. 4.

1.1 Related Work

In 2006 a set of algorithms to deal with fingerprints acquired using a mobile phone camera was proposed [15]. In order to get rid of fingerprint sensors on mobile phones and thus, to reduce the cost of these devices. The proposed preprocessing algorithms improved the performance of a fingerprint system based on a regular mobile phone camera. In [9] a similar work was presented. Instead of using a mobile phone camera to acquire the fingerprint images, a digital camera Canon PowerShot Pro1, with 8 megapixels of effective resolution and a 7x optical zoom Canon "L" series lens was used. Although the proposal looked promising, no verification results using minutiae were provided. In [2] the use of mobile phone cameras was revisited. Using a Nokia N95 and an HTC Desire the authors achieved an EER of 4.5%. More recently, a webcam-based fingerprint system was proposed in [13]. Unfortunately, results were presented in a way which is very hard to validate. All of the above mentioned works focused only on fingerprints and not on palmprints.

In [14] the use of a digital camera for palmprint recognition was first introduced. Palm principal lines were used in conjunction with hand geometry features to match the palmprints. Inspired by this work, a custom palmprint scanner was presented in [21]. The custom scanner is a modified HP Scanjet 3500c document scanner. This work was revisited in [22] to investigate hand aging effects on palmprint matching. In both works, features regarding the palm shape and texture were used. Additionally, since the scanner was big enough to capture the whole bottom side of the hand, minutiae were extracted from the fingertips. Results were highly encouraging, and the advantages of using similar scanners were nicely highlighted. However, no minutiae were extracted from palms, so in order to test the applicability of this type of technologies on palmprint-based forensics more studies are needed. These previous results motivate the present work.

2 Database Description and Acquisition

To test the feasibility of a document scanner for minutiae-based palmprint recognition and to compare its performance with an industry standard scanner, a database using both types of scanners was established. Scanners used in this study were a Green Bit MC517 palmprint scanner [7] (Fig. 1a) and a HP Scanjet G4010 document scanner [10] (Fig. 1c).

Sixty-four subjects were willing to participate in this study. Both hands from each subject were captured using both scanners. 5–6 imprints were captured from each hand for each scanner. Between each capture the surface of the scanner was cleaned. A total of 657 images were captured with the Greenbit MC517 scanner and 656 images were obtained with the HP Scanjet G4010 scanner. Figures 1b and 1d show examples of collected images.

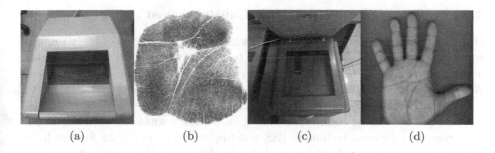

(a)   (b)   (c)   (d)

Fig. 1. Greenbit scanner MC517 (a, b) and a document scanner HP Scanjet G4010 (c, d)

Several problems arose during the acquisition process. While acquiring images with the Greenbit MC517 scanner users needed to apply a considerably high pressure on the surface of the scanner to be able to capture the whole palm. Sometimes the scan process had to be repeated and users needed to help themselves by putting one hand over the other to apply more pressure. Examples of images captured with too little pressure on the scanner device can be seen in Figs. 2a and 2b. On the other hand, for the document scanner the pressure that has been applied to the palm was too high due to the weight of the lid that caused deformations of the palm imprints. This caused e.g. ridges which were not clearly visible on the scanned images and high skin deformation in others (See Figs. 2c and 2d). Leaving the lid open could cause too much over-illumination. All these problems, affect the recognition performance as it can be seen later in the experimental evaluation (Sect. 3). Nevertheless, problems with the HP Scanjet G4010 scanner can be easily overcome by modifying the device as in [21].

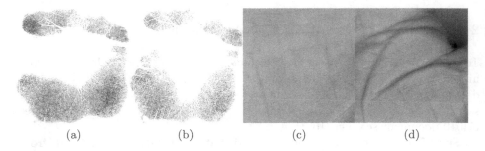

(a)   (b)   (c)   (d)

Fig. 2. Missing information due to non-excessive pressure on the Greenbit scanner (a, b) and high deformations caused by excessive pressure on the HP Scanjet G4010 scanner (c, d)

2.1 Pre-processing of HP Scanjet G4010 Scanned Palms

Before extracting features from the palmprint images, the region of the palm has to be isolated to avoid extracting features from the fingertips. In order to do this a color segmentation algorithm similar to the one proposed in [9] was applied. After the color segmentation algorithm, a set of morphological operations is used to remove the fingers from the final mask.

Two pre-processing techniques were used for the segmented palms. First, a histogram equalization technique [20] was applied to compensate for the light entering the scanner due to the semi-closed lid. Second, a local contrast enhancement using a window size of 17 pixels and an enhancement factor of 0.005 is applied[1]. The idea is to morph the HP Scanjet G4010 scanned palms to become more similar images to the ones obtained with the Green Bit scanner. The impact of the pre-processing techniques is evaluated in the experimental section. Both pre-processing techniques are abbreviated as EQ.1 and EQ.2, respectively. Figures 3b and 3c show examples of the pre-processing techniques applied on the Fig. 3a image.

---

[1] http://homepages.inf.ed.ac.uk/rbf/HIPR2/adpthrsh.htm.

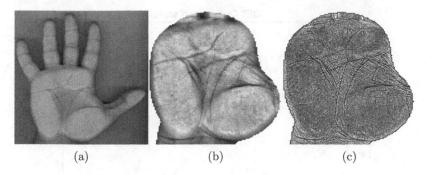

(a) (b) (c)

Fig. 3. Pre-processing techniques: Original image (a), EQ.1 (b), and EQ.2 (c)

3 Experiments

Verification experiments were carried out on each dataset to assess their discriminatory potential and the feasibility of each scanner for biometric recognition. Furthermore, cross-device matching experiments were conducted to evaluate a system where both scanners are employed. Since latent impressions resemble more the ones obtained with the GreenBit scanner, cross-device matching experiments assess the feasibility of using document scanners for forensic applications. Experiments were conducted following the FVC protocol [16].

Several systems were employed to conduct the experiments. Verifinger 4.2 [19] and VeriPalm [18] were used for feature extraction. Verifinger 4.2 (VF) and VeriPalm (VP) minutiae templates were matched using CPIM [8] and Minutiae Cylinder-Code (MCC) [1] algorithms. VeriPalm was used in conjunction with templates extracted by this matcher.

DET curves for each dataset are depicted in Figs. 4 and 5. For convenience, the GreenBit scanner dataset and the document scanner dataset had been named PalmA and PalmB datasets, respectively. The impact of the pre-processing techniques on the PalmB dataset can be seen in Fig. 5. All systems performed better on the pre-processed sets achieving comparable results to the ones obtained in the PalmA dataset. In both datasets, as expected, VeriPalm minutiae templates are more discriminative than Verifinger 4.2 minutiae templates. None of the pre-processing techniques has been proven to be more effective than the other. The best results for Verifinger 4.2 minutiae templates were achieved with the EQ.1 pre-processing but for VeriPalm templates the best results were achieved using the EQ.2 technique. This, of course, is related to the intrinsic characteristics of both minutiae extractors.

Cross-device matching results are shown in Fig. 6. These results were obtained by crossing the PalmA dataset with the pre-processed PalmB datasets. The results for the Verifinger 4.2 minutiae templates and in general for the CPIM algorithm were inferior compared to the single dataset results. Both MCC and the VeriPalm matcher performed well on the VeriPalm minutiae templates. Regarding the pre-processing techniques once more the results showed, that they

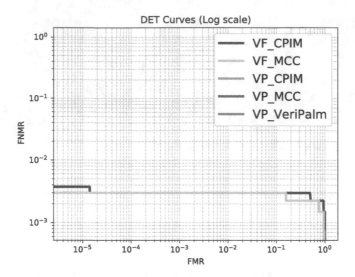

Fig. 4. DET curves on the PalmA dataset (curves for systems with an EER = 0% are not visible).

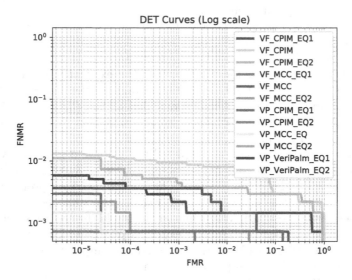

Fig. 5. DET curves on the PalmB dataset (curves for systems with an EER = 0% are not visible).

are system dependent. MCC performed better on the EQ.2 pre-processed images while the rest of the systems performed better with EQ.1 technique. Further details and performance numbers are given in Table 1.

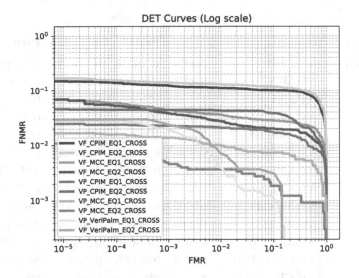

Fig. 6. DET curves for crossmatch.

3.1   Discussion

Experimental results show that the use of document scanners for minutiae-based palmprint recognition is feasible. Intra-datasets results are similar to each other and generally better in the case of the PalmB dataset. Considering the low cost of the HP Scanjet G4010 scanner compared to the Green Bit MC517, this is of great importance for law enforcement agencies and for the development of biometric palm recognition systems in general. However, there are some important issues that need to be taken into account before drawing any conclusions.

Acquiring palmprint images with the HP Scanjet G4010 scanner is considerably slower than with the Green Bit scanner. This can be a problem for civilian applications. Nevertheless, the final goal of this study is to apply this type of scanner in the context of forensic applications, and since palmprints are acquired offline, the acquisition speed is not a major issue. Additionally, the pre-processing stage in the PalmB dataset increases the total time for feature extraction. As mentioned before, the histogram equalization stage is needed to compensate for the over-illumination caused by semi-closed cover. This problem can be overcome by adapting the scanner as in [21].

Cross-device matching results, are applicable for real-world scenarios but worse compared to the intra-dataset results. This can be explained by the different origins of the images, but also by distortions introduced in the images in the acquisition process. As mentioned in Sect. 2, to acquire the PalmA dataset users needed to put a strong pressure on the scanner surface which resulted in high non-linear distortion between minutiae. Additionally, for the PalmB dataset, the weight of the lid by closing over the palm increases the palm pressure over the scanner surface causing non-linear distortion between minutiae too. Furthermore, palm principal lines and creases are more notorious in the PalmB

Table 1. Summarized results.

Dataset	Algorithm	AUC	EER	ZeroFMR	FMR1000
PalmA	VF_CPIM	0.9974	0.2971%	0.3745%	0.2996%
	VF_MCC	0.9978	0.2952%	0.2996%	0.2996%
	VP_CPIM	**1**	**0%**	**0%**	**0%**
	VP_MCC	**1**	**0%**	**0%**	**0%**
	VP_VeriPalm	**1**	**0%**	**0%**	**0%**
PalmB	VF_CPIM_EQ1	0.9988	0.1449%	0.5878%	0.2204%
	VF_CPIM	0.9973	0.7969%	1.3226%	0.9552%
	VF_CPIM_EQ2	0.9979	0.3699%	1.1186%	0.4474%
	VF_MCC_EQ1	0.9999	0.068%	0.0735%	0.0735%
	VF_MCC	0.9998	0.2969%	0.3674%	0.3674%
	VF_MCC_EQ2	0.9999	0.0745%	0.2237%	0.0746%
	VP_CPIM_EQ1	0.9998	0.0732%	0.2983%	0.0746%
	VP_CPIM_EQ2	**1**	**0%**	**0%**	**0%**
	VP_MCC_EQ1	0.9999	0.0037%	0.1491%	**0%**
	VP_MCC_EQ2	**1**	**0%**	**0%**	**0%**
	VP_VeriPalm_EQ1	**1**	**0%**	**0%**	**0%**
	VP_VeriPalm_EQ2	**1**	**0%**	**0%**	**0%**
Crossmatch	VF_CPIM_EQ1	0.9279	9.8607%	15.00%	12.17%
	VF_CPIM_EQ2	0.9167	11.5924%	16.835%	13.9884%
	VF_MCC_EQ1	0.9798	3.136%	6.996%	4.403%
	VF_MCC_EQ2	0.9861	2.2951%	6.9177%	3.7955%
	VP_CPIM_EQ1	0.9890	2.001%	2.4793%	2.2651%
	VP_CPIM_EQ2	0.9799	4.1566%	4.5914%	4.4077%
	VP_MCC_EQ1	0.9954	0.9927%	1.6835%	1.3774%
	VP_MCC_EQ2	0.9988	**0.367%**	**1.1019%**	**0.4591%**
	VP_VeriPalm_EQ1	**0.9997**	0.6463%	3.0303%	1.8365%
	VP_VeriPalm_EQ2	0.9993	0.8414%	3.3058%	2.2957%

dataset. These features are known to cause problems in combination with minutiae extraction algorithms which are based on the gradient like the one used by Verifinger 4.2 [5]. Hence, the development of minutiae extraction algorithms which are more robust to creases as in [12] should improve the quality of the extracted features and therefore the cross-matching results. Table 2 summarizes the characteristics of the two capturing devices including the above mentioned issues.

Table 2. Characteristics of both scanners.

	GreenBit MC517	HP Scanjet G4010
Cost	Very expensive (3000–4000 USD)	Cheap (less than 500 USD)
Speed	Fast	Very slow
Pre-processing	No need for pre-processing; Algorithms can accurately extract minutiae from them	Segmentation and histogram equalization needed
Pressure	Users need to apply high pressure on the scanner	No need to apply pressure
Design	Specifically designed for palmprints	Designed for documents and images

4 Conclusions

The feasibility of using document scanners for minutiae-based palmprint recognition was explored. Document scanners are cheaper and more widely distributed compared to forensic palmprint scanners. The use of similar scanners has been evaluated but to the best of our knowledge this is the first time that minutiae are used for matching. Minutiae matching is mandatory for forensic applications. Results were encouraging and highlighted the issues that still need to be addressed for forensic applications. These issues are going to be the focus of our further work. Particularly, the performance in the case of cross-device matching needs to be studied in-depth. The last is crucial for the use of similar technologies in forensic applications. Moreover, in the context of this study, a new database was established, which will be made publicly available.

Acknowledgements. This project was partially funded by the European Union's Horizon 2020 research and innovation program under grant agreement No. 690907 (IDENTITY).

References

1. Cappelli, R., Ferrara, M., Maio, D.: A fast and accurate palmprint recognition system based on minutiae. IEEE Trans. Syst. Man Cybern. Part B (Cybern.) **42**(3), 956–962 (2012)
2. Derawi, M.O., Yang, B., Busch, C.: Fingerprint recognition with embedded cameras on mobile phones. In: Prasad, R., Farkas, K., Schmidt, A.U., Lioy, A., Russello, G., Luccio, F.L. (eds.) MobiSec 2011. LNICST, vol. 94, pp. 136–147. Springer, Heidelberg (2012). https://doi.org/10.1007/978-3-642-30244-2_12
3. Fei, L., Xu, Y., Teng, S., Zhang, W., Tang, W., Fang, X.: Local orientation binary pattern with use for palmprint recognition. In: Zhou, J., et al. (eds.) CCBR 2017. LNCS, vol. 10568, pp. 213–220. Springer, Cham (2017). https://doi.org/10.1007/978-3-319-69923-3_23

4. Fei, L., Zhang, B., Zhang, W., Teng, S.: Local apparent and latent direction extraction for palmprint recognition. Inf. Sci. **473**, 59–72 (2019)
5. Funada, J.I., et al.: Feature extraction method for palmprint considering elimination of creases. In: Fourteenth International Conference on Pattern Recognition, vol. 2, pp. 1849–1854. IEEE (1998)
6. Genovese, A., Piuri, V., Plataniotis, K.N., Scotti, F.: PalmNet: Gabor-PCAconvolutional networks for touchless palmprint recognition. IEEE Trans. Inf. Forensics Secur. **14**(12), 3160–3174 (2019)
7. Green-Bit: Green-bit mc517 scanner. http://www.greenbit-china.cn/index.php?m=content&c=index&a=show&catid=36&id=17
8. Hernandez-Palancar, J., Munoz-Briseno, A., Gago-Alonso, A.: Using a triangular matching approach for latent fingerprint and palmprint identification. Int. J. Pattern Recogn. Artif. Intell. **28**(07), 1460004 (2014)
9. Hiew, B., Teoh, A.B., Ngo, D.C.: Preprocessing of fingerprint images captured with a digital camera. In: 2006 9th International Conference on Control, Automation, Robotics and Vision, pp. 1–6. IEEE (2006)
10. HP-Inc.: Hp scanjet g4010 series scanner. https://support.hp.com/us-en/document/c00817232
11. Jain, A., Demirkus, M.: On latent palmprint matching. Technical Report 48824, Michigan State University (2008)
12. Jain, A.K., Feng, J.: Latent palmprint matching. IEEE Trans. Pattern Anal. Mach. Intell. **31**(6), 1032–1047 (2009)
13. Khan, S., Waqas, A., Khan, M.A., Ahmad, A.W.: A camera-based fingerprint registration and verification method. Int. J. Comput. Sci. Netw. Secur. **18**(11), 26–31 (2018)
14. Kumar, A., Wong, D.C.M., Shen, H.C., Jain, A.K.: Personal verification using palmprint and hand geometry biometric. In: Kittler, J., Nixon, M.S. (eds.) AVBPA 2003. LNCS, vol. 2688, pp. 668–678. Springer, Heidelberg (2003). https://doi.org/10.1007/3-540-44887-X_78
15. Lee, C., Lee, S., Kim, J., Kim, S.-J.: Preprocessing of a fingerprint image captured with a mobile camera. In: Zhang, D., Jain, A.K. (eds.) ICB 2006. LNCS, vol. 3832, pp. 348–355. Springer, Heidelberg (2005). https://doi.org/10.1007/11608288_47
16. Maio, D., Maltoni, D., Cappelli, R., Wayman, J.L., Jain, A.K.: FVC 2002: Second fingerprint verification competition. In: Object Recognition Supported by User Interaction for Service Robots, vol. 3, pp. 811–814. IEEE (2002)
17. Maltoni, D., Maio, D., Jain, A.K., Prabhakar, S.: Handbook of Fingerprint Recognition. Springer, Heidelberg (2009). https://doi.org/10.1007/978-1-84882-254-2
18. Neurotechnology-Inc: Megamatcher (SDK). https://www.neurotechnology.com/cgi-bin/biometric-components.cgi?ref=mm&component=palm-mat
19. Neurotechnology-Inc: Verifinger 4.2 (2004). http://www.neurotechnologija.com/down-load.html
20. Reza, A.M.: Realization of the contrast limited adaptive histogram equalization (CLAHE) for real-time image enhancement. J. VLSI Sig. Process. Syst. Sig. Image Video Technol. **38**(1), 35–44 (2004)
21. Uhl, A., Wild, P.: Personal recognition using single-sensor multimodal hand biometrics. In: Elmoataz, A., Lezoray, O., Nouboud, F., Mammass, D. (eds.) ICISP 2008. LNCS, vol. 5099, pp. 396–404. Springer, Heidelberg (2008). https://doi.org/10.1007/978-3-540-69905-7_45
22. Uhl, A., Wild, P.: Experimental evidence of ageing in hand biometrics. In: 2013 International Conference of the BIOSIG Special Interest Group (BIOSIG), pp. 1–6. IEEE (2013)

23. Zhong, D., Du, X., Zhong, K.: Decade progress of palmprint recognition: a brief survey. Neurocomputing **328**, 16–28 (2019)
24. Zhou, K., Zhou, X., Yu, L., Shen, L., Yu, S.: Double biologically inspired transform network for robust palmprint recognition. Neurocomputing **337**, 24–45 (2019)

Stacking Fingerprint Matching Algorithms for Latent Fingerprint Identification

Danilo Valdes-Ramirez[1,2]([✉]) [iD], Miguel Angel Medina-Pérez[1] [iD],
and Raúl Monroy[1] [iD]

[1] Tecnologico de Monterrey, School of Science and Engineering,
Carretera al Lago de Guadalupe Km. 3.5, 52926 Atizapán, Estado de México, Mexico
{A01746752,migue,raulm}@tec.mx
[2] Department of Computer Sciences, Universidad de Ciego de Ávila,
65100 Ciego de Ávila, Cuba
dvramirez@gmail.com
https://tec.mx

Abstract. Automatic latent fingerprint identification is still challenging for biometric researchers. One infrequently explored approach for improving the identification rate involves stacking latent fingerprint identification algorithms with a supervised classification algorithm, instead of using a weighted sum or a product of likelihood ratio. A stacking approach fuses the result provided by different base algorithms to achieve higher performance than each individual algorithm. Latent fingerprints present different qualities, causing deviations between the identification rates of various algorithms. Thus, we propose stacking latent fingerprint identification algorithms using a supervised classifier. We use two different minutia descriptors with a global matching algorithm independent of the local matching of the minutia descriptor. Our stacking method improves the identification rate of each base algorithm by 2% when comparing the fingerprints in the database NIST SD27. Furthermore, our proposal achieves a 73.26% rank-1 identification rate when comparing 258 samples in the database NIST SD27 against 29,258 references, and 68.99% against 100,000 references.

Keywords: Latent fingerprint identification · Match score fusion · Minutia descriptor stacking

1 Introduction

Automatic latent fingerprint identification (LFI) remains a challenge for fingerprint matching research [1]. LFI requires high identification rates (IRs) and protection against presentation attacks [7] due to the sensitivity of its applications. For example, a jury often supports its decision in a trial based on the

Supported by the CONACyT under grant PN720 and scholarship grant 638948.

I. Nyström et al. (Eds.): CIARP 2019, LNCS 11896, pp. 230–240, 2019.
https://doi.org/10.1007/978-3-030-33904-3_21

latent fingerprints found at a crime scene, and police apprehend suspects by matching their identities to latent fingerprints. However, the IRs reported in the literature are insufficient for meeting those demands.

The results in Table 1 indicate that there is plenty of room for improvement. Especially as the background database grows, matching algorithms for LFI achieve low IRs. The development of more accurate LFI algorithms, which is the goal of this work, is highly desirable for police departments and justice managers.

Table 1. Identification rates (IRs) output by some recent LFI algorithms. Note: the IRs decrease for larger background databases.

Algorithm	Rank-1 IR	Background database size
Improving automated latent fingerprint identification using extended minutia types [12]	92.7%	151
Latent fingerprint recognition: role of texture template [2]	78.2%	10,000
Automated latent fingerprint recognition [1]	64.7%	100,000

Different fusion schemes have been proposed for LFI [1,9,11,17,19], and they have been coarsely classified as data level, feature level, match score level, and rank level schemes [19]. More details about score fusion methods appear in the paper published by Sankaran *et al.* [19]. Match score level fusion allows every LFI base algorithm fused to address the intrinsic complexities of latent fingerprints, *e.g.*, background noise, nonlinear distortion, and lack of features. Therefore, dealing with those issues is unnecessary at the fusion step. Consequently, match score level fusion is the most commonly used fusion scheme [19]. Authors have proposed match score fusion algorithms for LFI using weighted sums and product rules [1,9,11,17,19]. However, these approaches usually depend on the weights, which must be empirically determined. Furthermore, the fusion score formula remains the same, although the background database grows. Match score fusion based on supervised learning algorithms can determine more complex functions than a weighted sum or a product rule. Moreover, supervised learning algorithms can be trained systematically with new datasets to obtain updated functions without human intervention.

Supervised learning algorithms have been successfully used for computing the match score for fingerprint verification [3,5,10,14]. Nevertheless, they are seldom used for LFI [18], possibly because of the differences between fingerprint verification and LFI; fingerprint verification outputs whether the fingerprint in question matches a template fingerprint, which is similar to the output of a supervised classifier, while LFI returns a list of references fingerprints from a background database with the highest match score to the sample latent fingerprint. Nevertheless, some supervised classifiers compute the probability of belonging to each class (as a value between 0 and 1). Thus, we may take advantage of these supervised classifiers by assuming that the probability of belonging to the matching class is the match score.

In this study, we explore various match score fusion algorithms for LFI. We stack two LFI algorithms by employing two different minutia descriptors and the same global matching algorithm, deformable minutiae clustering (DMC), proposed by Medina-Pérez et al. [16]. Both minutia descriptors match only the minutiae of the fingerprints. Matching minutiae is the basis of several LFI algorithms and the methodology developed by latent examiners [14]. Hence, improving the IR for LFI algorithms matching only minutiae provides a more accurate basis for complex LFI algorithms.

In the next two sections, we describe our method of stacking LFI algorithms and design an experimental study to validate it. Our proposal improves the rank-1 IR of the LFI base algorithms by approximately 2% when comparing 258 samples in the NIST SD27 database [6] against various background databases. Additionally, our proposal achieves a 73.26% rank-1 IR when comparing 258 samples in the NIST SD27 database against 29,258 references (258 of NIST SD27, 2,000 of NIST SD4, and 27,000 of NIST SD14), and 68.99% against 100,000 references (258 of NIST SD27, 2,000 of NIST SD4, 27,000 of NIST SD14, and 70,742 nonpublic references).

The remainder of this paper is organized as follows. We describe our method of stacking LFI algorithms in Sect. 2. Then, we discuss our experimental results in Sect. 3. Finally, we state some conclusions and future work in Sect. 4.

2   Stacking Latent Fingerprint Identification Algorithms

LFI algorithms often perform two-step fingerprint matching (local and global) as a hybrid approach [14]. Global matching algorithms compare two fingerprints using features that describe the whole fingerprint. Global matching algorithms could be dependent [1,20] or independent on the local matching algorithm [4,16]. Conversely, local matching algorithms compute match scores between fragments of fingerprints, e.g., two minutia descriptors [14]. Minutia descriptors are fingerprint feature representations that incorporate information into the minutia standard representation (ISO/IEC 19794-2:2005) of other fingerprint features in the neighborhood. For example, minutia descriptors can incorporate the orientation of ridges, the ridge count, or the relationships with neighboring minutiae.

Our proposal performs a third step fusing the match scores and the matching minutia count between the sample (latent fingerprint) and the references (impressions) computed with different LFI base algorithms. The match score output by each LFI base algorithm is normalized as a continuous value between 0 and 1, where 1 means a perfect match and 0 means a perfect nonmatch. In addition to the match score output by each LFI base algorithm, we have considered the number of matching minutiae (we will provide further details about the matching minutia count in Sect. 3). Next, the fusion algorithm, which can be a supervised classifier, learns a function from these attributes using a training database.

To identify a new sample, we set up a tuple of attributes with the LFI base algorithms and compute the match scores and the matching minutia count between the sample and each reference. Each LFI base algorithm matches the fingerprint pair with DMC [16] as global matching algorithm but with different local matching algorithms, *i.e.*, Minutia Cylinder Code (MCC) [4] and mtriplet [15] in our proposal. The LFI base algorithms output the match scores and the matching minutiae count. Next, the fusion algorithm computes the fused match score (see Fig. 1 and Algorithm 1).

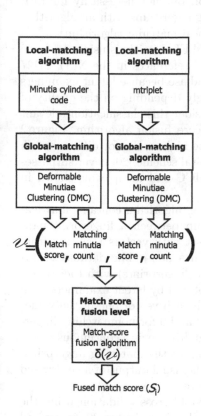

Fig. 1. Diagram of our stacking for computing the match score between a fingerprint pair (sample and reference). Our LFI algorithm returns a list of 20 references with the highest match scores, as described in Algorithm 1.

function identifySample(l, T)
Data:

- l is the sample (latent fingerprint)
- $T = \{T_1, T_2, ..., T_m\}$, is a set of references (impressions)
- $A = \{A_1, A_2\}$ are the LFI base algorithms (DMC as global matching algorithm and MCC and mtriplet as local matching algorithms, respectively), $A_j(l, T_i)$ computes the match score and the matching minutia count between l and T_i
- S is the list of fused match scores between l and each T_i
- $v \in \mathbb{R}^4$ is the tuple of matching features (match score and matching minutia count from each LFI base algorithm A_j)
- δ is the match score fusion algorithm

begin
  $S \longleftarrow \phi$
  foreach $T_i \in T$ **do**
    $v \longleftarrow \phi$
    foreach $A_j \in A$ **do**
      $v \longleftarrow v \cup A_j(l, T_i)$
    end
    $S_i \longleftarrow \delta(v)$
  end
  Return 20 sorted references (T_i) with the highest match scores (S_i)
end

Algorithm 1. Stacking LFI algorithms.

2.1 Selection of LFI Base Algorithms, Attributes, and Fusion Algorithms

A recent study [22] published a performance evaluation of various minutia descriptors with their respective local matching algorithms for LFI. This evaluation showed that MCC [4], mtriplet [15], and a spiral scheme combining the orientations of ridges and minutiae [20] achieved higher IRs than some other descriptors, all evaluated with the NIST SD27 database. These three minutia descriptors were proposed for fingerprint verification. Thus, their global matching algorithm could perform well for verification but not necessarily for LFI. Accordingly, we substitute their global matching algorithms with an algorithm proposed for LFI, which is independent of the local matching algorithm.

We evaluated the performance of these three LFI algorithms using the same global matching algorithm with the samples and references in the NIST SD27 database. We selected this latent fingerprint database because latent examiners have classified the samples into good, bad, or ugly depending on the quality of the observations of ridges and minutiae. We consider this classification relevant for understanding the behavior of our match score fusion algorithm. Figure 2 shows that MCC [4] and mtriplet [15] achieve the highest IRs for different rank values and fingerprint qualities. The IR of the spiral scheme [20] never surpasses those of the previous two. Thus, we selected MCC and mtriplet as shown in Fig. 2.

Our feature tuple used to determine the fusion algorithm contains four features that characterize the matching relationship between a fingerprint pair. With these features, we set up a training dataset with vectors resulting from the comparisons of each sample and reference in a non-public database. This non-public database contains 284 samples (latent fingerprints) with their references (impressions) captured and manually annotated by latent examiners from successful investigations. Next, we associate with each vector one of the values of the class attribute: 1 for matching fingerprints and 0 for nonmatching fingerprints. Since an LFI algorithm must output a match score as a continuous value, we use the probability of an instance of belonging to the matching fingerprints output by the supervised classifier, rather than the usual output of a supervised classifier, which is either 0 or 1, as the match score.

With the training dataset, we performed 10-fold cross-validation using the Auto-WEKA tool [21] in WEKA 3.8.2. Auto-WEKA evaluates 30 supervised classification algorithms with different parameter configurations to optimize a user-defined metric. We used a server with 48 cores, 1 TB of RAM, and a 1-TB hard drive for 48 hours to determine the supervised classifier that maximizes the true positive rate. As a result, Auto-WEKA returned a Bayesian network "weka.classifiers.bayes.BayesNet" optimizing the true positive rate (sensitivity) and "weka.classifiers.functions.MultilayerPerceptron" (MLP) optimizing the area under the receiver operator characteristics (ROC) curve. MLP has one hidden layer with two units, and each node of the Bayesian network has one parent. Additionally, we explored three other as fusion algorithms: a linear regression, a weighted sum, and a product of a likelihood ratio.

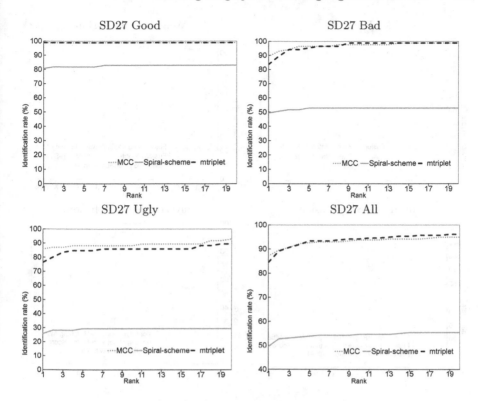

Fig. 2. CMC curves output by LFI base algorithms for good, bad, ugly, and all samples on the NIST SD27 database. We can observe that MCC and mtriplet achieve the highest IR for different rank values and fingerprint qualities of the sample. These results support the idea of an improvement of the IR by stacking these two matching algorithms.

Similar to the way that we built our training dataset, we set up other datasets for testing our algorithm for LFI using the samples in the NIST SD27 database and different background databases. We tested our fusion algorithms for LFI by comparing the cumulative match characteristic (CMC) curves of all our approaches and the LFI base algorithms using these testing datasets. Unlike a machine learning problem, our test consists of computing the CMC curve (ISO/IEC 19795-1). The CMC curve plots the IRs by rank number, *i.e.*, for rank i, the ratio of genuine matching pairs that have received a match score that is greater than the *i-th* match score.

3   Results and Discussion

We experimented with three sets of features: using only the match score, using only the matching minutia count, and using both. The CMC curves in Fig. 3 show that, by using both, we achieved the highest IR comparing the fingerprint in the NIST SD27 database and fusing the scores with MLP and weighted sum.

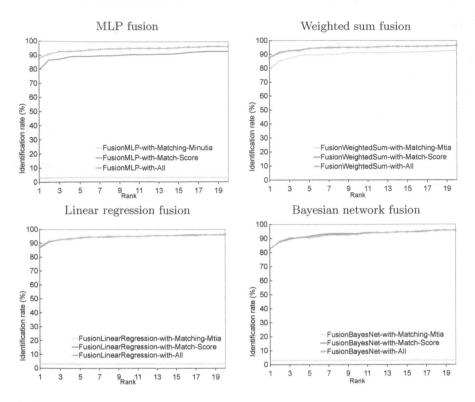

Fig. 3. CMC curve output by the explored fusion algorithms using three sets of features. Note: using both the match score and matching minutia count improves the IR but only for the MLP and the weighted sum fusion algorithms.

However, the matching minutia count does not improve the IR of fusing with linear regression and Bayesian network. Furthermore, MLP outputs the highest rank-1 IR (87.98%), but it is only slightly different from the weighted sum and the linear regression (87.59%) and fusing all the match scores and the matching minutia count.

Since the IR decreases when the background database increases, we evaluated the fusion algorithms using three testing datasets with comparisons between 258 samples (NIST SD27) and 258 (NIST SD27), 2,258 (NIST SD27+NIST SD4), and 29,258 (NIST SD27+NIST SD4+NIST SD14) references, respectively. We found that the fusion algorithms with an MLP, a linear regression, or a weighted sum surpass the IRs of the base algorithms except the rank-1 IR using the background database with 2,258 references (see Fig. 4). We also found that the MLP increases its difference against the remainder fusion models regarding the IR for larger background databases. Although all models were trained with the same dataset, the MLP has learned a function more capable of computing higher match score to the tuples of matching fingerprints than a linear regression or a weighted sum. As a result, we claim that supervised learning algorithms can

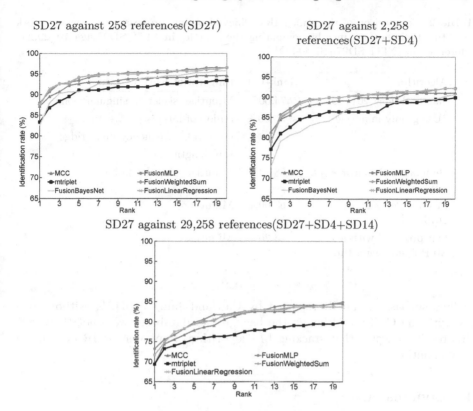

Fig. 4. CMC rank-20 curves output by four evaluated fusion algorithms (MLP, linear regression, Bayesian network, and weighted sum) and the base algorithms MCC and mtriplet. The MLP outputs better CMC curves than the others for larger background databases. The product rule outputs IRs under 70% for all rank values and background databases, and the Bayesian network outputs lower IRs than the base algorithms comparing the samples in the database NIST SD27 against 29,258 references; hence, we do not plot them.

improve the IR for LFI as fusion algorithms because they can learn nonlinear functions unlike linear regression or weighted sum.

We compared our algorithm with previous work that used similar background databases (see Table 2). Our stacking algorithm that fuses two LFI base algorithms with an MLP outperforms most of other algorithms in terms of rank-1 IR, except for the algorithm of Jain and Feng [9], which achieved a rank-1 IR that is similar to but higher than ours. However, our proposal uses only minutiae, whereas Jain and Feng [9] fused several fingerprint features, such as ridge maps, qualities maps, and frequency maps.

Additionally, we tested our proposal using an MLP as fusion algorithm and comparing 258 samples (NIST SD27) against 100,000 references (NIST SD27+SD4+SD14+nonpublic references). We achieved a rank-1 IR equal to

Table 2. Comparison of the rank-1 IRs achieved by our proposal and previous work reporting their performance by comparing the samples in NIST SD27 against 29,257 references on (NIST SD27+SD4+SD14).

Algorithm	Rank-1 IR	Fingerprint features used
Jain and Feng 2011 [9]	74.00%	Minutiae, skeleton, singular point,
*Using only minutiae	≈ 35.00%	region of interest, ridge quality map, ridge frequency map, ridge wavelength map
Hernández-Palancar *et al.* 2014 [8]	58.13%	Minutiae, singular point
Medina-Pérez *et al.* 2016 [16]	68.60%	Minutiae
Our proposal with an MLP fusion algorithm	73.26%	Minutiae

68.99% surpassing the IR reported by Cao and Jain [1] (64.7% without fusing with the COTS). Although the background databases were not the same, this result indicates that stacking LFI algorithms improves the IR even using only minutiae.

4   Conclusions

We developed and tested stacked latent fingerprint identification (LFI) algorithms using supervised learning to compute a fused match score. We found that using the matching minutia count in addition to the match score improves the identification rate (IR). Stacking two LFI algorithms using a multilayer perceptron obtained rank-1 IRs that outperform those of the base algorithms when comparing the samples in NIST SD27 against three different background databases with 258 (NIST SD27), 2,258 (+NIST SD4), and 29,258 (+NIST SD14) references. The MLP fusion algorithm increased its difference regarding the IR against the others for larger background databases. As a result, we claim that stacking LFI with supervised learning algorithms can improve the IR because they can learn more complex functions.

Further feature engineering is to be done to gain insight into the set of features that best characterize the matching relationship between a sample and a reference. Additionally, we will explore other supervised learning algorithms not included in Auto-WEKA tool, and for class imbalanced problems [13].

References

1. Cao, K., Jain, A.K.: Automated latent fingerprint recognition. IEEE Trans. Pattern Anal. Mach. Intell. **41**(4), 788–800 (2019)

2. Cao, K., Jain, A.K.: Latent fingerprint recognition: role of texture template. In: 9th International Conference on Biometrics Theory, Applications and Systems (BTAS), pp. 1–9. IEEE (2019)
3. Cao, K., Yang, X., Tian, J., Zhang, Y., Li, P., Tao, X.: Fingerprint matching based on neighboring information and penalized logistic regression. In: Tistarelli, M., Nixon, M.S. (eds.) ICB 2009. LNCS, vol. 5558, pp. 617–626. Springer, Heidelberg (2009). https://doi.org/10.1007/978-3-642-01793-3_63
4. Cappelli, R., Ferrara, M., Maltoni, D.: Minutia cylinder-code: a new representation and matching technique for fingerprint recognition. IEEE Trans. Pattern Anal. Mach. Intell. **32**(12), 2128–2141 (2010)
5. Feng, J.: Combining minutiae descriptors for fingerprint matching. Pattern Recogn. **41**(1), 342–352 (2008)
6. Garris, M.D.: NIST special database 27: fingerprint minutiae from latent and matching tenprint images. US Department of Commerce, National Institute of Standards and Technology (2000)
7. González-Soler, L.J., Chang, L., Hernández-Palancar, J., Pérez-Suárez, A., Gomez-Barrero, M.: Fingerprint presentation attack detection method based on a bag-of-words approach. In: Mendoza, M., Velastín, S. (eds.) CIARP 2017. LNCS, vol. 10657, pp. 263–271. Springer, Cham (2018). https://doi.org/10.1007/978-3-319-75193-1_32
8. Hernández-Palancar, J., Munoz-Briseno, A., Gago-Alonso, A.: Using a triangular matching approach for latent fingerprint and palmprint identification. Int. J. Pattern Recogn. Artif. Intell. **28**(07), 1460004 (2014)
9. Jain, A.K., Feng, J.: Latent fingerprint matching. IEEE Trans. Pattern Anal. Mach. Intell. **33**(1), 88–100 (2011)
10. Jea, T.Y., Govindaraju, V.: A minutia-based partial fingerprint recognition system. Pattern Recogn. **38**(10), 1672–1684 (2005)
11. Jeyanthi, S., Uma Maheswari, N., Venkatesh, R.: Neural network based automatic fingerprint recognition system for overlapped latent images. J. Intell. Fuzzy Syst. **28**(6), 2889–2899 (2015)
12. Krish, R.P., Fierrez, J., Ramos, D., Alonso-Fernandez, F., Bigun, J.: Improving automated latent fingerprint identification using extended minutia types. Inf. Fus. **50**, 9–19 (2019)
13. Loyola-González, O., Martínez-Trinidad, J.F., Ariel Carrasco-Ochoa, J., García-Borroto, M.: Effect of class imbalance on quality measures for contrast patterns: an experimental study. Inf. Sci. **374**, 179–192 (2016)
14. Maltoni, D., Maio, D., Jain, A.K., Prabhakar, S.: Handbook of fingerprint recognition. Springer, Heidelberg (2009). https://doi.org/10.1007/978-1-84882-254-2
15. Medina-Pérez, M.A., García-Borroto, M., Gutierrez-Rodríguez, A.E., Altamirano-Robles, L.: Improving fingerprint verification using minutiae triplets. Sensors **12**(3), 3418–3437 (2012)
16. Medina-Pérez, M.A., Moreno, A.M., Ballester, M.A.F., García-Borroto, M., Loyola-González, O., Altamirano-Robles, L.: Latent fingerprint identification using deformable minutiae clustering. Neurocomputing **175**, 851–865 (2016)
17. Paulino, A.A., Feng, J., Jain, A.K.: Latent fingerprint matching using descriptor-based hough transform. IEEE Trans. Inf. Forensics Secur. **8**(1), 31–45 (2013)
18. Sankaran, A., Dhamecha, T.I., Vatsa, M., Singh, R.: On matching latent to latent fingerprints. In: International Joint Conference on Biometrics (IJCB), pp. 1–6 (2011)

19. Sankaran, A., Vatsa, M., Singh, R.: Hierarchical fusion for matching simultaneous latent fingerprint. In: IEEE Fifth International Conference on Biometrics: Theory, Applications and Systems (BTAS), pp. 377–382. IEEE (2012)
20. Shi, Z., Govindaraju, V.: Robust fingerprint matching using spiral partitioning scheme. In: Tistarelli, M., Nixon, M.S. (eds.) ICB 2009. LNCS, vol. 5558, pp. 647–655. Springer, Heidelberg (2009). https://doi.org/10.1007/978-3-642-01793-3_66
21. Thornton, C., Hutter, F., Hoos, H.H., Leyton-Brown, K.: Auto-WEKA: Combined selection and hyperparameter optimization of classification algorithms. In: Proceedings of the 19th ACM SIGKDD International Conference on Knowledge Discovery and Data Mining, pp. 847–855. ACM (2013)
22. Valdes-Ramirez, D., et al.: A review of fingerprint feature representations and their applications for latent fingerprint identification: trends and evaluation. IEEE Access 7, 48484–48499 (2019)

Detecting Steading Conversational Groups on an Still Image: A Single Relational Fuzzy Approach

Elvis Ferrera-Cedeño[1](✉), Niusvel Acosta-Mendoza[2], and Andrés Gago-Alonso[2]

[1] DATYS, Ave. 26 de Julio e/ 9 and 11 ,Vigía Sur, Santa Clara, Villa Clara, Cuba
elchago8787@gmail.com
[2] Academia de Ciencias de Cuba,
Cuba No. 460 e/ Amargura and Teniente Rey, Habana Vieja, Havana, Cuba

Abstract. The small group detection has become one of the most important step in crow scene analysis, which has several application in surveillance-video (i.e. for detecting, preventing and predicting dangerous situations). A Steading Conversational Group (a.k.a. F-Formation) is a kind of small group, where their stationary people interact through social signals (i.e. non-verbal expressions). The proposed state-of-the-art methods have reported encouraging results; however, they are based on complex theories. Moreover, these methods have had difficulties for rehearsing and high computational complexity. In this paper, we propose a new method for detecting F-Formation in an image. We introduce a new representation and clustering method, basing our solution on the fuzzy relation theory. The performance of our proposal is evaluated and compared against other reported methods over a synthetic and two real-world databases. The experimental results show the effectiveness of our proposal.

Keywords: F-Formation detection · Small groups ·
Surveillance-video · Fuzzy relations

1 Introduction

In Pattern Recognition and Computer Vision, several works are focused to automate the scene analysis [1,2]. They are based on some social, biological and psychological theories. These works have shown that scenes with crowds are composed by small groups of people and the behavior of the groups are given by the interactions between people [3].

The small groups detection is an important step in scene analysis, allowing to obtain high levels of semantic interpretation. It has several application in surveillance-video, such as: anomaly detection and video classification [4–7].

Steading Conversational Group (a.k.a. F-Formation) is a kind of small groups, which has achieved great interest in the scientific community. An F-Formation is composed by stationary people, which interact through social signals (i.e. non-verbal expressions) [1]. Beside, the people form patterns of space

© Springer Nature Switzerland AG 2019
I. Nyström et al. (Eds.): CIARP 2019, LNCS 11896, pp. 241–250, 2019.
https://doi.org/10.1007/978-3-030-33904-3_22

and orientation between them, when they are interacting. Moreover, they have equal and exclusive access to a space inside of the F-Formation [8].

An F-Formation is composed by three spaces: *O-space*, *P-space* and *R-space* (see Fig. 1(a)). The O-space is an empty space, which is surrounded by oriented people toward it. This is the most important space, because most of the algorithms reported in the literature are based on it. The P-space involves the O-space, while the R-space is the complement of the P-space.

An F-Formation can take different geometrical forms: L-form, Face-to-Face, Side-by-Side and Circular form (see Figs. 1(b), (c), (d) and (e), respectively). When the number of people in an F-Formation is longer than two, the F-Formation has commonly the circular form.

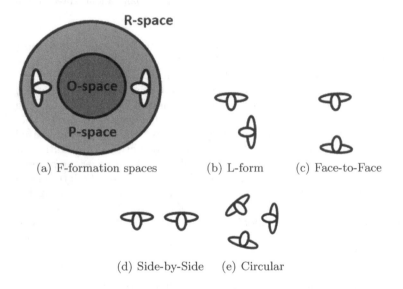

(a) F-formation spaces   (b) L-form   (c) Face-to-Face

(d) Side-by-Side   (e) Circular

Fig. 1. Examples of F-formation spaces (a) and F-Formation geometrical forms (b–e).

Several approaches were proposed to detect F-Formations [9–12]. They have used some features, such as: the people positions on the ground floor and their head orientations.

The first approach is based on the Hough transformation [9,10], where an accumulator space for finding many local maximum by a vote strategy is created. Each local maximum represents an O-space center, where the people are assigned.

Other approaches are based on graph theory [13], where people and their relations are represented by vertices and edges, respectively. In this way, the F-Formation detection is reduced to the maximal clique detection (i.e. dominant set) problem [11,14].

The methods proposed in [12,15] are based on the game-theory [16], where the F-Formation detection is reduced to a clustering problem over an evolutionary environment.

The aforementioned approaches have high computational complexity, are based on complex theories and have had difficulties for rehearsing. In the literature, little efforts have been realized for reducing the mentioned difficulties, but, to the best of our knowledge, only in [17], the authors treat to solver it. However, this method requires large number of parameters and the group detection is not automatic. Furthermore, it is designed for detecting F-Formations in sequence of images. Based on [17], we propose a solution for reducing number of parameter and automatically detecting F-Formations.

The main contributions of this paper are: (1) a new method for detecting F-Formations in an image, (2) a new image representation, where a membership function for computing social people relations is introduced, and (3) an automatic clustering for associating people with their O-space.

The basic outline of this paper is the following. In Sect. 2, some basic concepts are provided. Section 3 contains the description of the proposed method. The experimental results are discussed in Sect. 4. Finally, conclusions and some ideas about future directions are exposed in Sect. 5.

2   Basic Concepts

In this section, we show a set of concepts, which are required for understanding our proposal.

Definition 1 (Fuzzy relation). *Let X and Y be two sets, a fuzzy relation from X to Y is a membership function $\rho : X \times Y \to [0,1]$. If $X = Y$, then ρ is named fuzzy relation on X.*

By Definition 1, the similarity relation can be defined as follows.

Definition 2 (Similarity relation (by Zadeh [18])). *The fuzzy relation ρ on X is a similarity relation if for all $x, y \in X$ the followed properties are fulfilled:*

- *Reflexivity: $\rho(x,x) = 1$*
- *Symmetry: $\rho(x,y) = \rho(y,x)$*
- *Transitivity: $\rho(x,y) \geq \max_{z \in X} \{\min \{\rho(x,z), \rho(z,y)\}\}$.*

Sometimes, a fuzzy relation is represented by a matrix, which is known as fuzzy matrix (see Definition 3).

Definition 3 (Fuzzy matrix). *A matrix M is an $m \times n$ fuzzy matrix if each cell of M has a value in the interval $[0,1]$.*

On a fuzzy matrix M, which represents a similarity relation, we define an F-Formation as follows:

Definition 4 (F-Formation). *An F-Formation is a set of connected cell indexes of M, where their corresponding cell values are greater than a given $\alpha, \in [0,1]$.*

3   The Proposed Method

Given a database of images with people positions on the floor and orientations (i.e. the head or body orientations), our proposal carry out two steps: (1) to build a representation, where people relations are modeled through fuzzy relations, which later are codified in a fuzzy matrix (see Sect. 3.1), and (2) to cluster fuzzy relations for detecting F-Formations (see Sect. 3.2).

3.1   Representation

Let (x_k, y_k) and σ_k, be the position and the orientation of a person p_k, and let $v_k = [x_k + r \cdot \cos \sigma_k, y_k + r \cdot \sin \sigma_k]$ be their vote point [9], where r is the vote length. The visual field interaction between people p_i and p_j, $i, j \in [1, k]$ is computed by their frustum interception (i.e. the vote point interception).

In [17], the authors proposed an idea based on vote points, where each person frustum[1] is represented by a vote point. However, this idea fall on assumption of perfect alienation [9], where some F-Formations detections could be missed (see Fig. 2(a)). For this reason, we propose an alternative of the idea proposed in [17], where we represent each person frustum by three vote points. In this way, we avoid the assumption of perfect alienation (see Fig. 2(b)).

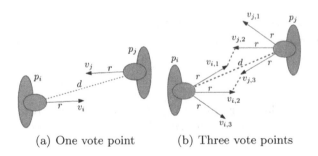

(a) One vote point   (b) Three vote points

Fig. 2. Frustum model.

According to [17], a valid frustum interception between p_i and p_j, $i, j \in [1, k]$, is fulfilled by the following two rules: (1) both vote points v_i and v_j must be on the same side of the segment d, and (2) the distance between p_i and p_j (i.e. length of d) must be longer than the distance between v_i and v_j. Notice that, the previous rules are accomplished only for $r = d/2$ value.

In our proposal (see Algorithm 1), for each people p_i and p_j we compute their votes points $v_{i,l}$ and $v_{j,l}$, $l \in [1, 3]$ by the Eqs. 1, 2 and 3. For searching a valid frustum interception, we use the Eqs. 6 and 7 after build a matrix X where their elements are values of the distances between the vote points.

---

[1] A frustum is a biological area where interactions between people often occur [12].

$$v_{k,1} = [x_k + r \cdot \cos(\sigma_k + \gamma), y_k + r \cdot \sin(\sigma_k + \gamma)] \tag{1}$$

$$v_{k,2} = [x_k + r \cdot \cos\sigma_k, y_k + r \cdot \sin\sigma_k] \tag{2}$$

$$v_{k,3} = [x_k + r \cdot \cos(\sigma_k - \gamma), y_k + r \cdot \sin(\sigma_k - \gamma)] \tag{3}$$

$$\Gamma_{k,l} = (v_{k,l_x} - x_i) * (y_j - y_i) - (v_{k,l_y} - y_i) * (x_j - x_i) \tag{4}$$

$$dv_{i,j} = \sqrt{(v_{i,l_x} - v_{j,l_x})^2 + (v_{i,l_y} - v_{j,l_y})^2} \tag{5}$$

$$\rho_{i,j} = \begin{cases} 1 & \text{if } \Gamma_{i,l} \geq 0 \text{ and } \Gamma_{j,l} \geq 0 \text{ or } \Gamma_{i,l} \leq 0 \text{ and } \Gamma_{j,l} \leq 0, dv_{i,j} < d \\ -1 & \text{if other case} \end{cases} \tag{6}$$

$$\kappa_{i,j} = \rho_{i,j} dv_{i,j} \tag{7}$$

$$X = \begin{pmatrix} \kappa(v_{i,1}, v_{j,1}) & \kappa(v_{i,1}, v_{j,2}) & \kappa(v_{i,1}, v_{j,3}) \\ \kappa(v_{i,2}, v_{j,1}) & \kappa(v_{2,1}, v_{j,2}) & \kappa(v_{2,1}, v_{j,3}) \\ \kappa(v_{i,3}, v_{j,1}) & \kappa(v_{i,3}, v_{j,2}) & \kappa(v_{i,3}, v_{j,3}) \end{pmatrix}$$

$$\mu_{i,j} = (1 - \frac{u}{d}) \exp(\frac{-d}{h}) \tag{8}$$

Algorithm 1. RFRepresentation

---

input : P of size n, with $P(n) = [(x_n, y_n), \sigma_n]$
input : $\gamma \in [0, 60]$
input : $h \in [0, 360]$
output: Fuzzy matrix M
for $i \leftarrow 1$ to n do
$\quad$ $M(i,i) \leftarrow 1$
$\quad$ for $j \leftarrow i+1$ to $n-1$ do
$\quad\quad$ Compute $d_{i,j}$, take (x_i, y_i) and (x_j, y_j) (see Equation 5)
$\quad\quad$ $r \leftarrow d_{i,j}/2$
$\quad\quad$ Compute each $v_{i,l}, v_{j,l}, l \in [1,3]$ (see Equations from 1 to 3)
$\quad\quad$ Compute $\kappa_{i,j}$ for votes points $v_{i,l}$ and $v_{j,l}$ (see Equation 7)
$\quad\quad$ Compute X matrix
$\quad\quad$ Compute $\mu_{i,j}$ (see Equation 8)
$\quad\quad$ $M(i,j) \leftarrow M(j,i) \leftarrow \mu_{i,j}$
$\quad$ end
end

For computing the social relation between two people, we propose the membership function $\mu_{i,j}$ (see Eq. 8), the h values are taken from the Hall theory [19]. The Hall theory characterizes people social interactions by physical distances. The value of u is the minimum value taken from the positive elements of X.

When all element of X are negative values, then $\mu_{i,j} = 0$. However, when $i = j$ (i.e. $d = 0$), $\mu_{i,j} = 1$. Notice that, $\mu_{i,j}$ is a fuzzy relation on a people set, and our representation is a fuzzy matrix M with $\mu_{i,j}$ values.

3.2 Clustering

We propose the Algorithm 2 for clustering, which uses ClusteringRF algorithm [17] for transforming the input fuzzy matrix M in a similarity relation matrix M' (i.e. a fuzzy relation fulfills the reflexivity, symmetry and transitivity properties) [18] and generating a partition $C_\alpha = \{c_1...c_k\}$ by an α-cut.

For determining the number of clusters (i.e. F-Formations number within an image), we use a naive average of scores and select a C_α for the maximal w_α value. Notice that, $|C_\alpha|$ is the cluster number, $|c_k|$ is the number of elements in the cluster k and $M'(i,j)$ is a value of the fuzzy matrix M'.

Algorithm 2. ClusteringARF

---

input : a fuzzy matrix M
output: a set F with F-Formations
$w_\alpha \leftarrow 0$
α-cut $\leftarrow 0$
while α-cut ≤ 1 **do**
  $\{C_\alpha, M'\} \leftarrow$ ClusteringRF(M, α-cut)
  $s \leftarrow 0$
  for $k \leftarrow 1$ **to** $|C_\alpha|$ **do**
    $s \leftarrow s + \frac{1}{|c_k|-1} \sum_{i=1, i\neq j}^{|c_k|} M'(i,j)$, where $i, j \in c_k$
  end
  if $w_\alpha < s/|C_\alpha|$ **then**
    $F \leftarrow C_\alpha$
  end
  α-cut $\leftarrow \alpha$-cut $+ 0.01$
end

---

4   Experimental Results

In this section, we present the experimental evaluation of our proposed method; comparing its results against the best results reported in the literature over two real-world databases (Coffee Break [9] and GDet [9]) and one synthetic database (Synth [9]).

4.1   Databases

Coffee Break database [9], was obtained from a real-world environment in out-door scenario, from a single camera with a resolution of 1440×1080 px. It is composed by social events of people which are interacting and enjoying a cup of coffee. This database has 120 annotated images by psychologists using several questionnaires, where head orientations were estimated considering four directions: front, back, left and right.

GDet database [9], was obtained from an indoor scenario of vending machines area with several occlusions. It has 403 images, which were acquired by two low resolution cameras with 352×328 px, located on opposite angles of the room. Ground truth generations were carried out by psychologists, where the head orientations were estimated considering four directions (front, back, left and right) and people position were computed by a particle filter tracking algorithm [20].

Synth database [9] was generated by a trained expert, contains 100 situations provided by using 10 different based situations and slightly varying the position and head orientations of the people. It is important to highlight that, there are not noise in this database.

4.2   Experiments

For evaluating our proposal, we use the validation protocol proposed in [12], where a group is correctly detected if at least $\lceil T \cdot |G| \rceil$ of its members are found and not more than $\lceil (1-T) \cdot |G| \rceil$ are not members. The value $|G|$ is the cardinality of the labeled group and $T = 2/3$. For each image, the precision p, sensitivity s and the parameter $F1$ are computed for each group formation.

$$p = \frac{tp}{tp + fp}, s = \frac{tp}{tp + fn}, F1 = 2 \cdot \frac{p \cdot s}{p + s}$$

Our experiments were carried out with C++ on Eclipse, using opencv and armadillo libraries, over a personal computer Intel(R) Core(TM) 2 Duo CPU with 1.83 GHz and 2 GB RAM, with the Ubuntu 18.04.2 distribution.

Table 1 shows the obtained results by the related works reported in the literature, as well as, the results achieved by our proposal. In the first column of this table, the name of the methods are shown. In the other three columns, the precision, sensitivity and $F1$ values achieved over Coffee Break, GDet and Synth, highlighting the best results of each columns.

We varied h values between intimate and social space of Hall theory [19] (i.e [0, 360]). For generating each people frustum, the orientation σ_k are token of the people head and the vote points $v_{k,g}, g \in \{1, 3\}$ are computed with angles $\sigma_k \pm \gamma$, where γ values are between 0 and 60 for an effective visual field. Our best results are achieved with $h = 70$ and $\gamma = 30$ in Coffee Break, $h = 10$ and $\gamma = 60$ in GDet and $h = 116$ and $\gamma = 30$ in Synth database.

We obtained different results because theses databases represent different environments, where the crowd level changed in the scene (i.e. the people number, occlusion and distance of interaction between them). For this reason, in

Table 1. F1 measure results achieved by several methods over three databases

Method	Coffee Break			GDet			Synth		
	p	s	F1	p	s	F1	p	s	F1
HFF[9]	0.82	0.83	0.82	0.67	0.57	0.62	0.73	0.83	0.78
MULT[10]	0.82	0.77	0.80	0.71	0.73	0.72	0.86	0.94	0.90
GTCG[15]	0.83	**0.89**	0.86	0.76	0.76	0.76	1	1	1
R-GTCG[12]	0.86	0.88	**0.87**	0.76	0.76	0.76	1	1	1
CRF[17]	0.70	0.53	0.61	0.71	0.70	0.70	0.96	0.77	0.85
Proposed	**0.87**	0.83	0.85	**0.83**	**0.82**	**0.82**	1	1	1

practice, the parameter h and γ must be carefully selected in a pre-processing step. We recommend to decrease value of h, and to increase value of γ, when the crowd level increase.

For showing only an example, in Fig. 3, we show a result of our proposal in Coffee Break database, where circles with the same color over head people represent the same detected Steading Conversational Groups. Notice that, are 4 small groups (green, red, yellow and blue groups), with cardinality between 2 and 3.

Fig. 3. A visual result of our method in Coffee Break database. (Color figure online)

5   Conclusions and Future Works

In this paper, we proposed a new method for detecting Steading Conversational Group (F-Formation) in a still image. We based our proposal on fuzzy relations theory for building a new representation with three vote points. Moreover, we proposed a clustering on fuzzy relations, for obtaining the best number of F-Formation. We evaluate our proposal over two real-world databases (Coffee Break and GDet) and a synthetic one (Synth).

The results archived by our proposal outperform the best ones reported in the literature over the GDet database, keeping similar results over Coffee Break, while in Synth we obtained the best results. Based on our experiments, we can conclude that our proposal is an effective and simple solution. In the future, we will explore several internal index validation clustering for improving as possible the number of F-Formation.

References

1. Vinciarelli, A., Pantic, M., Bourlard, H.: Social signal processing: Survey of an emerging domain. Image Vis. Comput. **27**(12), 1743–1759 (2009)
2. Li, T., Chang, H., Wang, M., Ni, B., Hong, R., Yan, S.: Crowded scene analysis: a survey. IEEE Trans. Circ. Syst. Video Technol. **25**(3), 367–386 (2015)
3. Moussaïd, M., Perozo, N., Garnier, S., Helbing, D., Theraulaz, G.: The walking behaviour of pedestrian social groups and its impact on crowd dynamics. PLoS ONE **5**(4), e10047 (2010)
4. Musse, S.R., Thalmann, D.: A model of human crowd behavior : group inter-relationship and collision detection analysis. In: Thalmann, D., van de Panne, M. (eds.) Computer Animation and Simulation 1997. Eurographics. Springer, Vienna (1997). https://doi.org/10.1007/978-3-7091-6874-5_3
5. Shao, J., Loy, C., Wang, X.: Scene-independent group profiling in crowd. In: Proceedings of the IEEE Conference on Computer Vision and Pattern Recognition, pp. 2219–2226 (2014)
6. Cosar, S., Donatiello, G., Bogorny, V., Gárate, C., Alvares, L.O., Brémond, F.: Toward abnormal trajectory and event detection in video surveillance. IEEE Trans. Circuits Syst. Video Techn. **27**(3), 683–695 (2017)
7. Liu, C., Wang, G., Ning, W., Lin, X., Li, L., Liu, Z.: Anomaly detection in surveillance video using motion direction statistics. In: Proceedings of the International Conference on Image Processing, ICIP 2010, 26–29 September 2010, Hong Kong, China, pp. 717–720 (2010)
8. Kendon, A.: Spacing and orientation in co-present interaction. In: Esposito, A., Campbell, N., Vogel, C., Hussain, A., Nijholt, A. (eds.) Development of Multimodal Interfaces: Active Listening and Synchrony. LNCS, vol. 5967, pp. 1–15. Springer, Heidelberg (2010). https://doi.org/10.1007/978-3-642-12397-9_1
9. Cristani, M., et al.: Social interaction discovery by statistical analysis of F-formations. In: BMVC, vol. 2, p. 4(2011)
10. Setti, F., Lanz, O., Ferrario, R., Murino, V., Cristani, M.: Multi-scale F-formation discovery for group detection. In: 2013 IEEE International Conference on Image Processing, pp. 3547–3551. IEEE (2013)
11. Zhang, L., Hung, H.: Beyond F-formations: Determining social involvement in free standing conversing groups from static images. In: Proceedings of the IEEE Conference on Computer Vision and Pattern Recognition, pp. 1086–1095 (2016)
12. Vascon, S., Mequanint, E.Z., Cristani, M., Hung, H., Pelillo, M., Murino, V.: Detecting conversational groups in images and sequences: a robust game-theoretic approach. Comput. Vis. Image Underst. **143**, 11–24 (2016)

13. West, D.B., et al.: Introduction to Graph Theory, vol. 2. Prentice Hall, Upper Saddle River (2001)
14. Hung, H., Kröse, B.: Detecting F-formations as dominant sets. In: Proceedings of the 13th International Conference on Multimodal Interfaces, pp. 231–238. ACM (2011)
15. Vascon, S., Mequanint, E.Z., Cristani, M., Hung, H., Pelillo, M., Murino, V.: A game-theoretic probabilistic approach for detecting conversational groups. In: Cremers, D., Reid, I., Saito, H., Yang, M.-H. (eds.) ACCV 2014. LNCS, vol. 9007, pp. 658–675. Springer, Cham (2015). https://doi.org/10.1007/978-3-319-16814-2_43
16. Sigmund, K.: Introduction to evolutionary game theory. In: Sigmund, K., (ed.) Evolutionary Game Dynamics, vol. 69, pp. 1–26 (2011)
17. Ferrera, E., Acosta, N., Alonso, A., García, E.: Detecting free standing conversational group in video using fuzzy relations. INFORMATICA 30(1), 21–32 (2019)
18. Zadeh, L.A.: Similarity relations and fuzzy orderings. Inf. Sci. 3(2), 177–200 (1971)
19. Hall, E.T.: The Hidden Dimension. New York (1966)
20. Lanz, O.: Approximate Bayesian multibody tracking. IEEE Trans. Pattern Anal. Mach. Intell. 28(9), 1436–1449 (2006)

Multi-task Learning for Low-Resolution License Plate Recognition

Gabriel Resende Gonçalves[1]($\boxtimes$), Matheus Alves Diniz[1], Rayson Laroca[2],
David Menotti[2], and William Robson Schwartz[1]

[1] Smart Sense Laboratory, Universidade Federal de Minas Gerais,
Belo Horizonte, Minas Gerais, Brazil
{gabrielrg,matheusad,william}@dcc.ufmg.br
[2] Department of Informatics, Universidade Federal do Paraná,
Curitiba, Paraná, Brazil
{rblsantos,menotti}@inf.ufpr.br

Abstract. License plate recognition is an important task applied to a myriad of important scenarios. Even though there are several methods for performing license plate recognition, our approach is designed to work not only on high resolution license plates but also when the license plate characters are not recognizable by humans. Early approaches divided the task into several subtasks that are executed in sequence. However, since each task has its own accuracy, the errors of each are propagated to the next step. This is critical in the last two steps of the pipeline known as segmentation and recognition of the characters. Thus, we employ a technique to perform these two steps at once. The approach is based on a multi-task network where each task represents the recognition of an entire license plate character. We do not address the license plate detection problem in this paper. We also propose the use of a so called generative model for data augmentation of low-resolution images simulating images as if they were acquired farther away from where they actually are. We are able to achieve very promising results with improvements of more than 30% points of accuracy on images with multiple resolutions and a character recognition accuracy on low-resolution images higher than 87%.

Keywords: Multi-task learning · Low-resolution · Deep learning · CNN

1 Introduction

Nowadays, recognize vehicles using their license plate is a well-known task and it is performed by many companiengs from different segments. This task is called *Automatic License Plate Recognition (ALPR)* and it was widely explored by researchers in the last two decades. Hence, there are multiple works proposing techniques to execute this task and improve its results [3,15].

© Springer Nature Switzerland AG 2019
I. Nyström et al. (Eds.): CIARP 2019, LNCS 11896, pp. 251–261, 2019.
https://doi.org/10.1007/978-3-030-33904-3_23

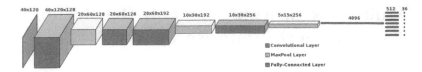

Fig. 1. Architecture of the proposed multi-task CNN to recognize license plate images.

Early approaches divide the license plate recognition into multiple subtasks and execute them in sequence [2]. These subtasks normally are (i) license plate detection; (ii) character segmentation; and (iii) optical character recognition (OCR). According to Gonçalves et al. [3], this has an important drawback since all error are propagated to the next step through the entire ALPR pipeline. Therefore, these approaches might have a large error rate at the end even if each subtask has good performance when evaluated separately. Moreover, cascading multiple approaches can be very time-consuming which is not desirable for real scenarios systems. Therefore, current approaches commonly try to perform ALPR without explicit executing all aforementioned subtasks [1,3,10].

Despite researches focus on proposing an end-to-end ALPR system [3,10,11, 16], there are many papers in the literature proposing techniques to solve only a few tasks of all steps described earlier. In this paper, we address two steps of the license plate recognition that we consider the most important and sensitive in terms of accuracy among all five: segmentation and recognition of the license plate characters. The former is usually approached by the researchers using algorithms that are not learning-based such pixel counting [19], mathematical morphology [13] and template matching [17]. The latter is a step known as optical character recognition (OCR) and is a wide-explored topic on the computer vision research community. Thus, many techniques provide accurate recognition rates on multiples challenging datasets [15,22].

In this work, we propose to perform license plate recognition in low-resolution images, focusing on segmenting and recognizing all characters holistically, avoiding the need to explicit segment the license plates (which is very hard on low-resolution images). Our experiments are carried out using a dataset with Brazilian license plates. Nonetheless, the approach can be further fine-tuned to work with any license plate standards. Our main concerns are to avoid the need to have high-resolution images and to avoid performing multiple tasks in sequence. We focus on a system to handle low-resolution images to reduce the cost of ALPR systems and improve its feasibility on companies or government departments that do not have a large budget to invest on high quality cameras. In addition, this work might be employed to forensics sciences, which have to handle poor quality images captured from a crime scene.

Since many works have provided very promising results in computer vision problems using deep learning techniques [7,8,10] and multi-task learning [12,14, 23,24], we decided to employ a deep convolutional network with multiple tasks, in which each task of our network represents the recognition of one character. This idea was already proved to work before [1,20]. In this case, the character is not explicitly located using our approach as the network only outputs the predicted characters. In addition, since deep learning networks require a large amount of data to learn, we employ two data augmentation techniques to increase the number of training samples, where we are able to train our network using only $2,520$ original license plate samples (later increased to $1,200,000$ samples with the data augmentation). Finally, we also design a network to generate new low-resolution samples to train our proposed multi-task network. There are two main contributions in this work: (i) a multi-task CNN model to segment and recognize low-resolution license plates characters; (ii) a deep generative network trained to create synthetic low-resolution images.

We evaluate our CNN model on two sets containing high-resolution and low-resolution images. Both sets were sampled from the *SSIG-ALPR* dataset [3]. The multi-task network is able to recognize 83.3% and 40.3% of the high and low-resolution sets, respectively. These results represent an improvement of 34.9 and 39.4 percentage points when compared to the best baselines. Moreover, our generative network brought an improvement of 4.9 p.p. in the experiments with the low-resolution images.

2 Proposed Approach

In this section, we detail our techniques. First, we present the architecture of the network used to simultaneously perform segmentation and recognition on low-resolution license plate images. Then, we describe the data augmentation techniques specially designed for low-resolution images.

Our approach consists on a deep convolutional network that receives a detected license plate and outputs the characters predicted considering the multiple tasks. Multi-task networks hypothesize that it is possible to improve the robustness of the network by learning a joint representation that is useful to describe more than one task on the same image [23]. In our case, our hypothesis is that the segmentation of a license plate character is dependent from the segmentation of the remaining characters since most characters are located sequentially in the same line. Moreover, the OCR also depends on how accurate the character is segmented. Hence, we train our deep network to segment and recognize all license plate characters at once to achieve better results than using two separated techniques (i.e., a for segmentation and a for OCR). This removes the need for cascading two approaches and bypasses the aforementioned error propagation. Even though this multi-task technique was already employed in previous approaches in literature [1,3], they do not focus on recognizing low-resolution license plate images.

It is common sense that deep learning techniques demand a great amount of training samples to learn discriminative and representative features to achieve promising results. Thereby, in addition to the proposed CNN, we also design a technique to permute characters aiming at performing data augmentation. Furthermore, since our focus is to increase the robustness of the network to handle low-resolution images, we need to train it with such images. However,

Table 1. Hyperparameters of the CNN trainable layer (upper table) and non-trainable (lower table) layers.

Layer	# Filters/Neurons	Filter size
Conv1	128	$\{5, 5\}$
Conv2	128	$\{3, 3\}$
Conv3	192	$\{3, 3\}$
Conv4	256	$\{3, 3\}$
Shared FC	512	–
Non-shared FCs	512	–
Output layers	36	–
Layer	Size	Stride
Input layer	$\{120, 40\}$	–
MaxPool layers (1, 2 and 3)	$\{2, 2\}$	$\{2, 2\}$

instead of only down-sampling the image with standard pixel-based techniques, we create a generative network that creates synthetic license plate images as if they were acquired farther away from where they actually were.

2.1 Multi-task CNN

As stated earlier, we propose the use of a convolutional neural network with multiple tasks to simultaneously perform segmentation and recognition of license plate characters. Our architecture is designed to implicitly perform the segmentation whereas the recognition are explicitly returned, one character per task. The network is designed to perform the character segmentation in the shared convolutional layers and recognize them on each task afterwards. The architecture is composed by five weighted shared layers: four convolutional layers and one fully-connected layer. There are also three shared max-pooling layers that do not contain any weights to train. Moreover, each task contains additional two fully-connected layers. We also placed dropout layers in between fully-connected layers to prevent overfitting. The input layer receives images of size 120 × 40 pixels and each output layers contains 36 neurons representing each one of the possible characters. This architecture is illustrated in Fig. 1.

The hyperparameters of the network are described in Table 1. Note that the non-shared layers are replicated for the each task, therefore, we have seven tasks with two layers each.

2.2 Data Augmentation

To improve the network robustness, we employed two techniques to increase the number of samples to train the network. The first technique is proposed by Gonçalves et al. [3] and is responsible to generate new artificial images with

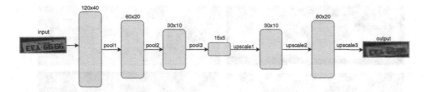

Fig. 2. Architecture of the DGM. The first and last layers contain a single filter. All other layers contains 64 filters.

license plate that were not initially on the dataset. The strategy consists on modifying the license plate images by changing (or permuting) the order of its characters. The authors claim that it helps the network to create an association of the character position with the correspondent task. This strategy was also successfully used in Laroca et al. [9].

Deep Generative Model. The second technique is a Deep Generative Model used to create new low-resolution license plates and is illustrated in Fig. 2. This model (hereinafter called DGM) is trained using the original images that we already had. Instead of using Generative Adversarial Networks [6], we created a network very similar to a variational autoencoder [21]. The goal is to generate low-resolution images from high-resolution ones. We could simply downscale the images in the dataset, but we believe that this does not emulate the actual behavior of low-resolution license plate that contains noises arising usually from long distance captures or by low-resolution cameras. Our model contains six convolution layers where the first three are followed by a max pooling layer and the last three are followed by a upsampling layer, similar to a generic convolutional autoencoder. The network is trained with pairs of images of the same vehicle license plate: one high-resolution image captured close to the camera used as input and a low-resolution image captured far from the camera used as output.

We performed the model training as follows. We recorded a video with multiple vehicles and track them throughout the video. Then, we manually chose a frame where the license plate is recognizable by a human and a frame where the license plate is not recognizable by a human. Afterwards, we had multiple pairs of high-resolution and low-resolution images from the same license plate, which have been use to train the network.

3   Experimental Results

In this section, we present the experiments carried out to evaluate the approach to segment and recognize license plate characters simultaneously. First, in Sect. 3.1, we present the datasets and baselines and then we describe and discuss the results achieved (Sect. 3.2).

Fig. 3. Training sample on top-left; validation sample on top-right; low-resolution and high-resolutions testing samples on bottom-left and bottom-right, respectively.

3.1   Evaluation Protocol

We created an evaluation protocol to measure the effectiveness of our approach. The protocol establishes three datasets, one for each task: training, testing and validation; and also a comparison scheme with four baselines. The images from the first and second dataset were sampled from the *SSIG-ALPR* dataset proposed in Gonçalves et al. [3].

The *first dataset* is used for training. It was originally composed from 2,520 images. However, we applied two data augmentation algorithms described in the previous section to increase the number of samples. We generate 800,000 images using the permutation technique and created other 400,000 using the DGM. Therefore, we presented 1,200,000 samples of license plate images to learn the model on the proposed network architecture. The license plate images have average size of 94 × 35 pixels. The images in the *second dataset* were used to test our proposed approach and are divided into two partitions: high-resolution and low-resolution. The high-resolution one contains 2,360 license plate images having average size of 104 × 39 pixels and the low-resolution one contains 820 images with average size of 49 × 17 pixels. The images of the first and second datasets were acquired in Brazil on two different places of the UFMG campus. The *third dataset*, called *SSIG-SegPlate*, was proposed by Gonçalves et al. [5] and contains 2,000 images of multiple resolutions and was only used for validation purposes. Figure 3 shows some examples from the three described datasets.

We compare our approach with four other techniques used as baselines. The first baseline contains two deep convolutional networks in cascade, in which the first was trained to segment the license plate characters and the second one to recognize them. We train these networks using our train set. We decided to utilize this baseline to demonstrate our hypothesis that training a single CNN to perform both tasks might avoid (and reduce) the error propagation previously mentioned. The second baseline is the approach proposed by Silva and Jung [18] that contains an end-to-end vehicle identification pipeline with two deep network; currently, this approach achieves state-of-the-art results in the Brazilian license plates recognition. Since we are assuming the license plate is already detected, we utilized only their recognition network. The third baseline is a hand-crafted

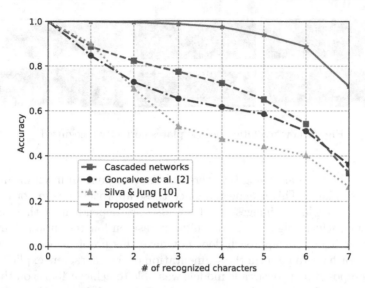

Fig. 4. Accuracy of all baselines as a function of the number of characters in the license plate that were correctly predicted.

Table 2. Recognition rates achieved by the proposed approach compared to baselines.

Approach	High-resolution	Low-resolution
Cascade networks	43.3%	0.9%
Gonçalves et al. [4]	48.3%	0.1%
Silva and Jung [18]	35.4%	0.4%
Proposed network	**83.2%**	**35.4%**

approach proposed by Gonçalves et al. [4] which employs a HOG-SVM classifier. Finally, our fourth baseline is the free version of the system called OpenALPR[1].

3.2   Results and Discussion

We perform two experiments to validate the proposed approach. The first evaluates our technique with the baselines in two different test sets, the partition with high-resolution images and the partition with low-resolution images. In this case, the samples generated by DGM were not considered. Then, the second experiment evaluates the influence of adding the examples of our generative network (DGM). We trained our multi-task network with these new samples and use it to predict the images from these two partitions of the test set once again.

The results of the first experiments are showed in Table 2. According to the results, the proposed multi-task network outperformed all baselines. If we consider only samples with low-resolution, the approaches proposed by Gonçalves

---

[1] Available at http://www.openalpr.com.

Fig. 5. Low-resolution license plates correctly recognized.

et al. [4] and by Silva and Jung [18] were able to recognize an insignificant number of license plates. This is expected since these approaches were not designed to handle low-resolution images and they also are dependent of the character segmentation, which might be a very difficult task on low-resolution characters. On the other hand, our approach does not have this problem because our network does not have to perform the segmentation of the characters explicitly. The baseline composed by two cascade models was able to achieve 43.3% on the high-resolution set and 0.9% on the low-resolution set, the latter result is very low, as the ones achieved by the other baselines. Our approach, on the other hand, was able to outperform the best baselines by 34.9 and 34.5 percentage points in high- and low-resolution images, demonstrating our hypothesis regarding the increasing of error when two steps are performed (i.e., character segmentation followed by character recognition) instead of a single step for both, as executed by our approach.

We also compared our approach with the commercial system OpenALPR. However, we were not able to use only the segmentation and the recognition steps of this technique. Therefore, we only evaluate the images in which the license plate was correctly detected by the OpenALPR system. The system was able to recognize 86.3% of high-resolution license plates with a full pipeline. Analyzing only the samples where the license plates were detected by OpenALPR, our method achieved 87.1% of accuracy on the high-resolution set, outperforming, therefore, the OpenALPR. Regarding the low-resolution license plates, the OpenALPR was able to detect only six of 820 license plates and was not capable of recognizing any of them. On the other hand, our approach was able to recognize correctly three of the six low-resolution license plates detected by the OpenALPR system, giving an edge for our approach on this set.

We also perform an evaluation of the number of characters that each method was able to recognize, combining the samples with high and low resolutions of the second dataset. According to the results showed in Fig. 4, our approach is able to recognize the entire license plate in 70.4% of the cases and predicts all but one character in 88.7% of the times. These results show an improvement of the best baselines in 34.2 (i.e., 70.4–36.2%) and 34.5 (i.e., 88.7–54.2%) percentage points in the case of zero and one misclassified characters, respectively.

The last experiment evaluates the influence of the DGM (Sect. 2.2). We trained our network using the 800,000 achieved by the character permutation and with the addition of 400,000 images from the DGM, resulting in a total of

Table 3. Accuracies achieved with and without the addition of the samples from DGM.

Approach	Accuracy	
	High-resolution	Low-resolution
Without samples from DGM	83.2%	35.4%
With samples from DGM	83.2%	40.3%

1,200,000 samples. According to the results showed in Table 3, the examples generated by the DGM were able to improve the robustness of the network on the low-resolution set in 4.9 percentage points, an improvement of 39.4 percentage points when compared to the best baseline). One also can note that the accuracy with the high-resolution was maintained, which mean that the samples generated by the DGM only improved the network generalization without compromising the results on the high-resolution test set. Figure 5 shows some low-resolution license plates correctly recognized by our approach.

4   Conclusions and Future Directions

In this paper, we introduced an approach to recognize license plates in low-resolution images. We employed a multi-task CNN to perform the segmentation and recognition of the license plate images simultaneously. We also designed a data augmentation techniques to increase the number of samples available to train our network. The DGM model was designed to simulate a license plate as if it was recorded from a long distance.

Our results demonstrated that our approach was able to recognize 83.2% and 40.3% of high- and low-resolution Brazilian license plates, respectively, being the best results on both sets among the baselines. Note that 40.3% of accuracy for license plates with seven characters stands for an accuracy of approximately 87.83% of character recognition accuracy (i.e., $0.8783^7 \approx 40.3\%$), which is a promising result for characters that are in low-resolution. The experiments also demonstrated that the use of the DGM improves our network, increasing the accuracy in the low-resolution images by 4.9 percentage points.

As future works, we intend to evaluate our network with other license plate standards and also investigate how to access the confidence of the recognition for each character in the license plate.

Acknowledgments. The authors would like to thank the Brazilian National Research Council – CNPq (Grants #311053/2016-5, #428333/2016-8, #313423/2017-2 and #438629/2018-3), the Minas Gerais Research Foundation – FAPEMIG (Grants APQ-00567-14 and PPM-00540-17), the Coordination for the Improvement of Higher Education Personnel – CAPES (DeepEyes Project), Maxtrack Industrial LTDA and Empresa Brasileira de Pesquisa e Inovacao Industrial – EMBRAPII.

References

1. Dong, M., He, D., Luo, C., Liu, D., Zeng, W.: A CNN-based approach for automatic license plate recognition in the wild. In: BMVC (2017)
2. Du, S., Ibrahim, M., Shehata, M., Badawy, W.: Automatic license plate recognition (ALPR): A state-of-the-art review. TCSVT (2013)
3. Gonçalves, G., Diniz, M.A., Laroca, R., Menotti, D., Schwartz, W.R.: Real-time automatic license plate recognition through deep multi-task networks. In: SIB-GRAPI. IEEE (2018)
4. Gonçalves, G.R., Menotti, D., Schwartz, W.R.: License plate recognition based on temporal redundancy. In: ITSC (2016)
5. Gonçalves, G.R., da Silva, S.P.G., Menotti, D., Schwartz, W.R.: Benchmark for license plate character segmentation. JEI **25**, 053034 (2016)
6. Goodfellow, I., et al.: Generative adversarial nets. In: NIPS (2014)
7. Hu, C., Bai, X., Qi, L., Chen, P., Xue, G., Mei, L.: Vehicle color recognition with spatial pyramid deep learning. T-ITS **16**, 2925–2934 (2015)
8. Krizhevsky, A., Sutskever, I., Hinton, G.: Imagenet classification with deep convolutional neural networks. In: NIPS (2012)
9. Laroca, R., Barroso, V., Diniz, M.A., Gonçalves, G.R., Schwartz, W.R., Menotti, D.: Convolutional neural networks for automatic meter reading. JEI **28**, 013023 (2019)
10. Laroca, R., et al.: A robust real-time automatic license plate recognition based on the YOLO detector. In: IJCNN (2018)
11. Li, H., Wang, P., Shen, C.: Towards end-to-end car license plates detection and recognition with deep neural networks. arXiv preprint arXiv:1709.08828 (2017)
12. Moeskops, P., et al.: Deep learning for multi-task medical image segmentation in multiple modalities. In: Ourselin, S., Joskowicz, L., Sabuncu, M.R., Unal, G., Wells, W. (eds.) MICCAI 2016. LNCS, vol. 9901, pp. 478–486. Springer, Cham (2016). https://doi.org/10.1007/978-3-319-46723-8_55
13. Nomura, S., Yamanaka, K., Shiose, T., Kawakami, H., Katai, O.: Morphological preprocessing method to thresholding degraded word images. PRL **30**, 729–744 (2009)
14. Ranjan, R., Patel, V.M., Chellappa, R.: Hyperface: a deep multi-task learning framework for face detection, landmark localization, pose estimation, and gender recognition. PAMI **41**, 121–135 (2017)
15. Rao, Y.: Automatic vehicle recognition in multiple cameras for videosurveillance. Vis. Comput. **31**, 271–280 (2015)
16. Rizvi, S.T.H., Patti, D., Björklund, T., Cabodi, G., Francini, G.: Deep classifiers-based license plate detection, localization and recognition on gpu-powered mobile platform. Fut. Internet **9**, 66 (2017)
17. Shuang-Tong, T., Wen-Ju, L.: Number and letter character recognition of vehicle license plate based on edge Hausdorff distance. In: PDCAT (2005)
18. Silva, S.M., Jung, C.R.: Real-time Brazilian license plate detection and recognition using deep convolutional neural networks. In: SIBGRAPI. IEEE (2017)
19. Soumya, K.R., Babu, A., Therattil, L.: License plate detection and character recognition using contour analysis. IJATCA (2014)
20. Špaňhel, J., Sochor, J., Juránek, R., Herout, A., Maršík, L., Zemčík, P.: Holistic recognition of low quality license plates by CNN using track annotated data. In: AVSS (2017)

21. Walker, J., Doersch, C., Gupta, A., Hebert, M.: An uncertain future: forecasting from static images using variational autoencoders. In: Leibe, B., Matas, J., Sebe, N., Welling, M. (eds.) ECCV 2016. LNCS, vol. 9911, pp. 835–851. Springer, Cham (2016). https://doi.org/10.1007/978-3-319-46478-7_51
22. Zhang, X., Zhao, J., LeCun, Y.: Character-level convolutional networks for text classification. In: NIPS (2015)
23. Zhang, Y., Yang, Q.: A survey on multi-task learning. arXiv preprint arXiv:1707.08114 (2017)
24. Zhang, Z., Luo, P., Loy, C.C., Tang, X.: Facial landmark detection by deep multi-task learning. In: Fleet, D., Pajdla, T., Schiele, B., Tuytelaars, T. (eds.) ECCV 2014. LNCS, vol. 8694, pp. 94–108. Springer, Cham (2014). https://doi.org/10.1007/978-3-319-10599-4_7

Colour Feature Extraction
and Polynomial Algorithm
for Classification of Lymphoma Images

Alessandro S. Martins[1(✉)], Leandro A. Neves[2], Paulo R. Faria[4],
Thaína A. A. Tosta[3], Daniel O. T. Bruno[3], Leonardo C. Longo[2],
and Marcelo Zanchetta do Nascimento[5]

[1] Federal Institute of Triângulo Mineiro, Ituiutaba, Brazil
alessandro@iftm.edu.br
[2] Department of Computer Science and Statistics (UNESP),
São Paulo State University, São José do Rio Preto, Brazil
[3] Center of Mathematics, Computing and Cognition,
Federal University of ABC, Santo André, Brazil
[4] Department of Histology and Morphology, Institute of Biomedical Science,
Federal University of Uberlândia, Uberlândia, Brazil
[5] Faculty of Computer Science, Federal University of Uberlândia,
Uberlândia, Minas Gerais, Brazil

Abstract. Lymphomas are neoplasms that originate in the lymphatic
system and represent one of the most common types of cancer found in
the World population. The feature analysis may contribute toward results
of higher relevance in the classification of the lesions. Feature extraction
methods are employed to obtain data that can indicate lymphoma incidence. In this work, we investigated the multiscale and multidimensional
fractal geometry with colour channels and colour models for classification
of lymphoma tissue images. The fractal features were extracted from the
RGB and LAB models and colour channels. The fractal features were
concatenated to form the feature vector. Finally, we employed the Hermite polynomial classifier in order to evaluate the performance of the
proposed approach. The colour channels obtained of histological images
achieved higher accuracy values, the obtained rates were between 94%
and 97%. These results are relevant, especially when we consider the
difficulties of clinical practice in distinguishing the lesion in lymphoma
images.

Keywords: Colour fractal · Hermite polynomial · Classification ·
Lymphoma

1 Introduction

Lymphoma is divided into Hodgkin lymphoma (HL) and non-Hodgkin lymphoma (NHL) [7]. The NHL is divided into several sub-types including man-
tle cell lymphoma (MCL), follicular lymphoma (FL) and chronic lymphocytic

© Springer Nature Switzerland AG 2019
I. Nyström et al. (Eds.): CIARP 2019, LNCS 11896, pp. 262–271, 2019.
https://doi.org/10.1007/978-3-030-33904-3_24

leukemia (CLL), which corresponds to 85% of the lymphomas [6]. The diversity of clinical presentations and the histopathological features have made diagnosis a complex task for specialists. Factors related to subjectivity and workload of specialists does not contribute to a more precise diagnosis. The digitisation of the sample allows the application of computational methods that aid specialists in reaching a more desirable result [20].

Feature extraction methods are employed to obtain data that can indicate the NHL incidence. This stage allows the extraction of highly specific features from the image which can improve the classification performance. However, the features are of a challenging nature for specialists and computer vision methods, especially when attempting to identify and correctly classify each type of NHL [13]. In fact, there is not a universal texture descriptor that always gives the best quantification for all different kinds of images. In Literature, some researches have indicated that fractal techniques can be more effective for texture analysis for histological image quantification, mainly due to the presence of stochastic properties and self-similarities [15]. Moreover, these methods allow the multiscale analysis of images and any other multidimensional objects, by observing the variation of a measure as a function of the scale analysis [4]. However, there are few experiments carried out on colour domain [11] and the images of tissue samples stained with hematoxylin-eosin (H&E) vary significantly in colour, which can impact the classification results. Then, new studies aimed at improving the understanding of features for classification of lymphoma remain as challenges in this area.

Computer-aided diagnosis (CAD) has been developed and can produce consistent results for the diagnosis of diseases. In Literature, there are different systems for the analysis and classification of lymphoma histological images. The researches in [8] developed a method or classification that consists of dividing the image into 25 blocks for the extraction of a set of visual features from each block. In the study of Nascimento et al. [10], an approach was proposed for NHL lesions' classification that employed extraction of sub-bands from the histological image through stationary wavelet transform, application of analysis of variance to clean noise and select the most relevant information, and classification with the support vector machine algorithm. The kernel types Linear, Radial Basis Function and Polynomial were evaluated to 210 images of lymphoma from the National Institute on Aging. In [18] an approach for quantifying and extracting features from lymphoma images database was proposed. The study in [17] presented an approach that used a high-dimensional multimodal descriptor that combines multiple texture features for classification that can be generally applicable to a wide variety of classification problems. The authors in [15] proposed a method to quantify and classify tissue samples of non-Hodgkin lymphomas based on the percolation theory. The proposed works do not explore approaches such as those investigated in this paper, considering the association of fractal techniques with Hermite polynomial (HP) algorithm, as well as exploring the influence that variations of colour models can have on the NHL classification.

In this context, an approach that investigates the multiscale and multidimensional fractal dimension with different colour model and classification is a relevant and unexplored combination. In this paper we present and detail such a method, which extracts features from lymphoma tissue images, and also supports the development of CAD systems. The fractal features were extracted from the channels of the RGB and LAB colour models. The colour fractal features were concatenated to form the feature vector and, finally, we employed the HP classifier in order to evaluate the performance of the proposed approach. The HP can be applied to estimate non-parametric density, with some advantages over kernel density estimation in higher dimensions [19]. This advantage can be applied to evaluate a set of features and to understand the more acceptable dimensions to distinguish such data. This was the main motivation to use the Hermite function as the kernel for an HP classifier. Another innovative contribution of our method is that it defines the best association after applying the HP.

2   Methodology

In Fig. 1 we show the main steps of the proposed approach in this study. The algorithms were developed using MATLAB® language and the experiments were performed on a notebook Intel Core i5, 12 GB RAM and 1 TB.

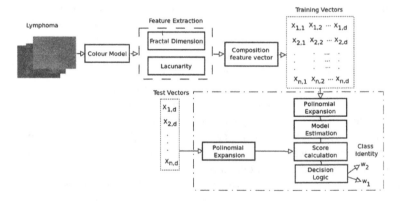

Fig. 1. A summarised version of the proposed method for evaluating lymphoma images considering fractal descriptors and the HP classifier.

2.1   Database

In this study, we employed the public database from National Cancer Institute and National Institute on Aging, in the United States. There is a total of 30 histological slides of lymph nodes stained with H&E, containing 10 cases from the CLL, FL and MCL. The database is available for download at [16].

The microscopic images were digitally obtained, using a light microscope (Zeiss Axioscope) with 20× objective and a colour digital camera (AXio Cam MR5). Regions from each slide were digitally photographed and stored in the tiff format, RGB colour model, resolution of 1388 × 1040, with quantisation of 24 bits. This database have 374 images, containing 113, 139 and 122 regions of CLL, FL and MCL, respectively.

2.2 Fractal Techniques

In this work, we employed the fractal dimension (FD) and lacunarity (Lac) for feature extraction from the colour image and from each channel image. In the Literature, there are different approaches able to calculate the fractal dimension of an image, the main FD technique is the probabilistic method from Ivanovici, Richard and Decean [4]. The probabilistic method was also applied to calculate the Lac [4]. The sizes of the boxes chosen for the obtainment of the features were $L = 3$ to $L = 45$.

Fractal Dimension. A colour image can be represented for a hypersurface in a colour space. This approach was defined as representing a colour image as a hypersurface. For instance, given an image represented with the RGB model, a 5-D Euclidean hyperspace was obtained considering the coordinates and the colours of each pixel for composing a 5-D vector (x, y, r, g, b).

This approach can be summarised representing a colour image as a hypersurface. For instance, considering an image represented with the RGB model, a 5-D Euclidean hyperspace can be obtained considering the coordinates each pixel.

Thus, given a box of size, L positioned over the image with the central pixel (F_c), the colour channel with more relevance was selected by comparing F_c with the remaining pixels F from the box under analysis. The distance of the pixels $F = f(x, y, r, g, b)$ to the central pixel $F_c = f_c(x_c, y_c, r_c, g_c, b_c)$ were obtained by applying the Minkowski distance (Eq. 1), where the values of x and y indicate the location of the pixel on the image and the values of r, g and b correspond to the channel intensities. If the maximum value obtained was less or equal than L, the pixel is considered to belong to the box. The sum and storage on a probability matrix $P(m, L)$ is performed, which represents the probability of m points belonging to a box on its side L:

$$|F - F_c| = max|f(i) - f_c(i_c)| \leq L, \forall i = \overline{1, 5}. \tag{1}$$

After applying the multiscale and multidimensional method, the total number of boxes $N_{FD_p}(L)$ for covering an image was defined (Eq. 2) and calculated for different observation scales L. The probabilistic fractal dimension FD_p was given by the angular coefficient of the linear regression defined by $\log L \times \log N_{FD_p}(L)$. The size of the boxes was $L = 3$ to $L = 45$. These values were sufficient for indicating multiscale observations in the histological image.

Another colour model used in this work was the CIE-LAB. In this case, the representation used to define the RGB hyper-space also was applied to evaluate the CIE-LAB model. Thus, the colour components were represented by (L,a,b), with values of $f(x,y) = (x, y, L, a, b)$. It is important to note that LAB colour space aims to conform to human vision and can describe all the colours visible to the human eye [5].

$$N_{FD}(L) = \sum_{m=1}^{N} \frac{P(m, L)}{m}. \tag{2}$$

Lacunarity. The pixel distribution and organisation were quantified by applying the methods described by [21]. The $Lac(L)$ was obtained from the colour image. The applied approach for multidimensional Lac was the method proposed in [4] by calculating the first order moment (Eq. 3) and second order moment (Eq. 4). These moments were defined based on the probability matrix $P(m, L)$. The calculation of $Lac(L)$ is presented in Eq. 5. The Lac curves were analysed by applying the metrics area under curve (ARC), skewness (SKW), area ratio (AR), maximum point (MP) and the scale of the maximum point (SMP). The ARC was obtained through numerical integration using the trapezoidal method. The SKW consisted of an asymmetry indication compared to the average value. The negative values of skewness indicated that the sample was concentrated to the left of the average value. The positive values of SKW indicated that the sample was concentrated to the right of the average value. Considering a case of the perfectly symmetrical sample, the obtained SKW is 0. The AR considered the ratio between the right side and the left side areas under the function curve. The MP provided the value of the maximum point of each function and SMP the scale of the maximum point.

$$\lambda(L) = \sum_{m=1}^{N} mP(m, L), \tag{3}$$

$$\lambda^2(L) = \sum_{m=1}^{N} m^2 P(m, L), \tag{4}$$

$$Lac(L) = \frac{\lambda^2(L) - (\lambda(L))^2}{(\lambda(L))^2}. \tag{5}$$

2.3   Composition of the Feature Vectors

The features were defined by associating the described approaches: 1 value of FD_p and 5 values of Lac $(ARC, SKW, AR, MP$ and $SMP)$, represented by $Lac(1)$ until Lac (5). In this stage, the feature vectors for each colour model (RGB and LAB) were composed of 6 values (5 features obtained with the lacunarity metric and 1 feature obtained with the fractal dimension). In the case of the analysis of separate colour channels, the characteristic vector was constructed with 18 features from the channels of which 6 features were obtained from each of the channels.

2.4   Hermite Polynomial Classifier

The purpose of the HP classifier is to identify and interpret the information about the lesions based on features extracted from images. The polynomial can be defined by a universal function which can be employed in several studies. One of the main characteristics of this technique is the ability to reduce the redundancy of data in a satisfactory way in several applications. Examples of applications that employ the polynomial are an interpolation, data compression, pattern recognition for feature extraction and classification [12,14].

The orthogonal polynomial of Hermite employs in the choice of the interval $(-\infty, \infty)$, and the weight function $w = e^{-x^2/2}$. Then the HP can be calculated by the recurrence ratio in any order $n > 0$ [9]:

$$H_{n+1}(x) = xH_n(x) - nH_{n-1}(x), \tag{6}$$

where $H_0(x) = 1$ and $H_1(x) = x$. The HP classifier is defined by Eq. (7)

$$g(\mathbf{x}) = \mathbf{a}^T H_n(\mathbf{x}), \tag{7}$$

where $\mathbf{a}$ is a coefficient vector of the polynomial basis function, $H_n(\mathbf{x})$ is the HP basis function and n represents the order or degree of the polynomial function.

The vector of the HP function coefficients $\mathbf{a}$ can be calculated by the Eq. (8):

$$\mathbf{a} = (\mathbf{M}^T\mathbf{M})^{-1}\mathbf{M}^T\mathbf{b} = \mathbf{M}^\dagger\mathbf{b}, \tag{8}$$

where the matrix $\mathbf{M}^\dagger$ of dimension $L \times N$ is known as the pseudo-inverse of $\mathbf{M}$ [3].

If there are only two classes under analysis, the HP classifier will identify them by the following decision rule: the test vector belongs to class ω_1, if $g(\mathbf{x}) > 0$, otherwise, the test vector belongs to class ω_2, that is, $g(\mathbf{x}) < 0$.

The algorithm is divided into two stages, namely training and test, which are detailed in [12]. In this study, an empirical investigation was employed to define the degree of order for the polynomial bases. Empirically, values from $1th$ to $4th$ order with an interval of 1 to 4 features were evaluated. The most significant results were obtained with the $4th$ order polynomial base with 4 features to separate the classes with the data that has a nonlinear behaviours.

The HP classifier was employed with two groups separately (CLL-FL, CLL-MCL and FL-MCL). We employed the 10-fold cross-validation method, where 90% of data were used for training and 10% were used for testing. The classifications were trained on k-1 sets and tested on the one holdout set. The method was evaluated by considering the metrics: area under the ROC curve (AUC), sensitivity (SE), specificity (SP) and accuracy (AC) [1].

3   Experimental Results

For the feature analysis, we employed the Mann-Whitney U-test to measure the significance of the features for discriminating between the information from the CLL, FL and MCL images. In this step, the notations RGB and LAB were

named for the RGB and LAB colour models, respectively. The notations S-RGB
and S-LAB were employed, respectively, for the separate colour channels of the
models RGB and LAB. Figure 2 presents the empirical cumulative distribution
function (CDF) for the p-values. In Fig. 2(a), we note that more than 65% of the
features have p-values close to zero for the features obtained with the CLL-FL
(RGB), FL-MCL (RGB), CLL-FL (S-RGB) and FL-MCL (S-RGB) data groups.
Only in the CLL-MCL (RGB) and CLL-MCL(S-RGB) groups counting p-values
greater than 0.05. In the LAB (Fig. 2(b)), part of the investigated groups has
statistical relevance counting p-values smaller than 0.05, which can be used for
discrimination of the information concerning each class of lesion. Only the CLL-
MCL group (LAB) had a p-values greater than 0.05.

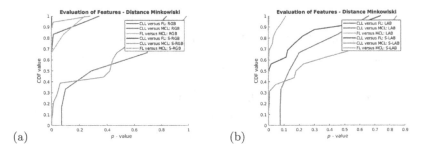

Fig. 2. Empirical cumulative distribution function of the p-values of the feature groups:
(a) RGB and S-RGB; (b) LAB and S-LAB.

The HP classifier was applied to distinguish the groups and the obtained
results are shown in Tables 1 and 2, considering the different combination
between the set of features: the discriminative performances were evaluated by
using the SE, SP and AC metrics. The combinations that provided the best
AC values are highlighted in bold. On Table 1, one observes that the results are
improved using the S-RGB channel. It is evident that the polynomial algorithm
resulted in the best values of AC for the FL-CLL group ($AC = 97.1 \pm 0.02$). In

Table 1. Results obtained (%) with classification algorithms for the lymphoma images
of the dataset with the 10-fold cross-validation method with RGB and S-RGB.

Type	Groups	SE	SP	AC
RGB	FL-CLL	87.6 ± 0.07	87.0 ± 0.04	87.0 ± 0.04
	FL-MCL	83.7 ± 0.10	80.8 ± 0.11	82.3 ± 0.06
	CLL-MCL	81.9 ± 0.09	75.0 ± 0.11	78.3 ± 0.06
S-RGB	FL-CLL	97.2 ± 0.04	97.0 ± 0.05	**97.1 ± 0.02**
	FL-MCL	93.4 ± 0.06	97.5 ± 0.05	**95.3 ± 0.03**
	CLL-MCL	96.4 ± 0.06	93.3 ± 0.06	**94.8 ± 0.02**

Table 2. Performance of classifier (%) with the LAB and S-LAB obtained from the histological images.

Type	Groups	SE	SP	AC
LAB	FL-CLL	83.8 ± 0.15	89.7 ± 0.05	87.0 ± 0.06
	FL-MCL	83.8 ± 0.08	79.1 ± 0.09	81.6 ± 0.06
	CLL-MCL	79.3 ± 0.08	86.6 ± 0.08	83.1 ± 0.04
S-LAB	FL-CLL	98.1 ± 0.03	97.1 ± 0.04	$\mathbf{97.6 \pm 0.02}$
	FL-MCL	95.4 ± 0.06	96.6 ± 0.04	$\mathbf{96.0 \pm 0.04}$
	CLL-MCL	95.4 ± 0.06	94.1 ± 0.05	$\mathbf{94.8 \pm 0.04}$

LAB (see Table 2), the AC metric showed observable relevant results concerning the FL-CLL group ($AC = 97.6 \pm 0.02$) for a histological analysis system.

Figure 3(a) and (b) show a comparison between AUC obtained using the colour channels. The AUC values presented in these Figures show that the fractal features obtained of the S-LAB or S-RGB channels associating with HP classifier are relevant.

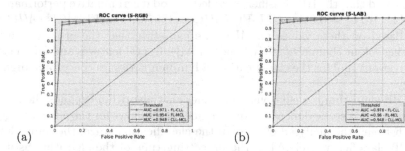

(a)   (b)

Fig. 3. ROC curves with its corresponding AUC values considering the proposed classifier and colour models: (a) S-RGB; (b) S-LAB.

Lymphoma cancer classification from histological images is the subject of several papers available in the literature (see Table 3). In this context, the proposed method presented significant results that contributed with a new solution for the area of analysis of histological images of lymphoma. The results are important for the area and contribute as a new strategy in relation to the other methods shown in this analysis. Part of the methods in Table 3 lead to an almost ideal system.

Table 3. Performance of the proposed method and correlated works by considering AC.

Reference	Features	Classifier	AC (%)
Meng et al. [8]	Colour, Histogram, Moment, Gabor and LBP	C-RSPM	92.70
Nascimento et al. [10]	Stationary wavelet	SVM	100.0
Song et al. [18]	Fisher, LBP and HOG	SVM	96.80
Codella et al. [2]	Histogram, edge histogram, LBP, curvelets, colour, wavelet	SVM	95.50
Roberto et al. [15]	Percolation theory	Logistic	96.40
Proposed method	Fractal features	HP	97.60

4   Conclusions

This study presented a computer-aided biopsy analysis support system of lymphoma for the classification. We presented several experiments that helped to choose adequate combinations in order to develop our technique. The obtained results showed that the HP classifier provided a good discriminative performance: CLL-MCL ($AUC = 0.948$); CLL-FL ($AUC = 0.976$); and FL-MCL ($AUC = 0.960$). Based on these values for AUC, we can state that the proposed technique using colour channels (S-RGB and S-LAB), fractal features and HP classifier is a relevant tool for the classification of lymphoma tissues concerning digital histology.

The HP algorithm is a supervised classifier that algorithm expands the input data in a superior space dimension, in a manner that allows for the adequate separation between the analysed classes. In the classification stage, the limitations of the HP classifier are related to the processing time of the algorithm as it is of exponential order. Future studies can be addressed to investigate the feature sets that allow thorough discrimination for the improvement of the effectiveness and efficiency of the classification stage.

Acknowledgement. The authors gratefully acknowledge the financial support of National Council for Scientific and Technological Development CNPq (Grants 427114/2016-0, 304848/2018-2, 430965/2018-4 and 313365/2018-0), the State of Minas Gerais Research Foundation - FAPEMIG (Grant APQ-00578-18).

References

1. Bradley, A.P.: The use of the area under the ROC curve in the evaluation of machine learning algorithms. Pattern Recogn. **30**(7), 1145–1159 (1997)
2. Codella, N., Moradi, M., Matasar, M., Sveda-Mahmood, T., Smith, J.R.: Lymphoma diagnosis in histopathology using a multi-stage visual learning approach.

In: Medical Imaging 2016: Digital Pathology, vol. 9791, p. 97910H. International Society for Optics and Photonics (2016)

3. Duda, R.O., Hart, P.E., Stork, D.G.: Pattern Classification. Wiley, Hoboken (2012)

4. Ivanovici, M., Richard, N., Decean, H.: Fractal dimension and lacunarity of psoriatic lesions-a colour approach. Medicine **6**(4), 7 (2009)

5. Kuru, K.: Optimization and enhancement of H&E stained microscopical images by applying bilinear interpolation method on lab color mode. Theor. Biol. Med. Model. **11**(1), 9 (2014)

6. Lowry, L., Linch, D.: Non-Hodgkin's lymphoma (2013)

7. Mauriño, B.B., Siqueira, S.A.C.: Classificação dos Linfomas (2011)

8. Meng, T., Lin, L., Shyu, M.L., Chen, S.C.: Histology image classification using supervised classification and multimodal fusion. In: 2010 IEEE International Symposium on Multimedia (ISM), pp. 145–152. IEEE (2010)

9. Moghaddam, V.H., Hamidzadeh, J.: New hermite orthogonal polynomial kernel and combined kernels in support vector machine classifier. Pattern Recogn. **60**, 921–935 (2016)

10. do Nascimento, M.Z., Neves, L., Duarte, S.C., Duarte, Y.A.S., Batista, V.R.: Classification of histological images based on the stationary wavelet transform. In: Journal of Physics: Conference Series. vol. 574, p. 012133. IOP Publishing (2015)

11. Nayak, S.R., Mishra, J., Khandual, A., Palai, G.: Fractal dimension of RGB color images. Optik **162**, 196–205 (2018)

12. Neves, L.A., et al.: Multi-scale lacunarity as an alternative to quantify and diagnose the behavior of prostate cancer. Expert Syst. Appl. **41**(11), 5017–5029 (2014)

13. Orlov, N.V., et al.: Automatic classification of lymphoma images with transform-based global features. IEEE Trans. Inf. Technol. Biomed. **14**(4), 1003–1013 (2010)

14. Padierna, L.C., Carpio, M., Rojas-Domínguez, A., Puga, H., Fraire, H.: A novel formulation of orthogonal polynomial kernel functions for SVM classifiers: the Gegenbauer family. Pattern Recogn. **84**, 211–225 (2018)

15. Roberto, G.F., et al.: Features based on the percolation theory for quantification of non-hodgkin lymphomas. Comput. Biol. Med. **91**(Suppl. C), 135–147 (2017)

16. Shamir, L., Orlov, N., Eckley, D.M., Macura, T.J., Goldberg, I.G.: Iicbu 2008: a proposed benchmark suite for biological image analysis. Med. Biol. Eng. Comput. **46**(9), 943–947 (2008)

17. Song, Y., Li, Q., Huang, H., Feng, D., Chen, M., Cai, W.: Low dimensional representation of fisher vectors for microscopy image classification. IEEE Trans. Med. Imaging **36**(8), 1636–1649 (2017)

18. Song, Y., Cai, W., Huang, H., Feng, D., Wang, Y., Chen, M.: Bioimage classification with subcategory discriminant transform of high dimensional visual descriptors. BMC Bioinform. **17**(1), 465 (2016)

19. Sorensen, J.: An assessment of hermite function based approximations of mutual information applied to independent component analysis. Entropy **10**(4), 745–756 (2008)

20. Tosta, T.A.A., de Faria, P.R., Batista, V.R., Neves, L.A., Nascimento, M.Z.: Using wavelet sub-band and fuzzy 2-partition entropy to segment chronic lymphocytic leukemia images. Appl. Soft Comput. **64**, 49–58 (2018)

21. Voss, R.F.: Random fractals: characterization and measurement. In: Pynn, R., Skjeltorp, A. (eds.) Scaling phenomena in disordered systems, pp. 1–11. Springer, Boston (1991). https://doi.org/10.1007/978-1-4757-1402-9_1

Deep Generic Features for Tattoo Identification

Miguel Nicolás-Díaz, Annette Morales-González,
and Heydi Méndez-Vázquez[✉]

Advanced Technologies Application Center (CENATAV), 7A #21406 b/214 and 216,
Siboney, Playa, 12200 Havana, Cuba
{mnicolas,amorales,hmendez}@cenatav.co.cu

Abstract. Recently, interest has grown in using tattoos as a biometric feature for person identification. Previous works used handcrafted features for the tattoo identification task, such as SIFT. However, deep learning methods have shown better results than this kind of methods in many computer vision tasks. Taking into account that there are little research on tattoo identification using deep learning, we asses several publicly available CNNs models, pre-trained on large generic image databases, for the task of tattoo identification. We believe that, since tattoos mostly depict objects of the real world, their semantic and visual features might be related to those learned from a generic image database with real objects. Our experiments show that these models can outperform previous approaches without even fine-tuning them for tattoo identification. This allows developing tattoo identification applications with minimum implementation cost. Besides, due to the difficult access to public tattoo databases, we created two tattoo datasets and put one of them in public domain.

Keywords: Tattoo identification · Deep learning · Pre-trained models · Tattoo datasets

1 Introduction

In recent years, several forensic techniques have been developed in order to identify victims and criminals in forensic scenarios [7]. These systems are mainly based on biometric traits such as the face, fingerprints and iris. However, there are many situations in which primary biometric traits like these are not available, and therefore it is necessary to resort to other types of information [11]. The so called "soft biometric traits" are physiological or behavioral characteristics that provide some identifying information about an individual [2], but lack distinctiveness and permanence to sufficiently differentiate any two individuals [11]. Eye color, gender, ethnicity, skin color, height, weight, hair color, scars, birthmarks and tattoos are examples of soft biometric traits. Several techniques have been proposed to identify or verify the identity of a person, automatically, based on soft biometric traits [2,8]. In particular, person identification and retrieval

© Springer Nature Switzerland AG 2019
I. Nyström et al. (Eds.): CIARP 2019, LNCS 11896, pp. 272–282, 2019.
https://doi.org/10.1007/978-3-030-33904-3_25

systems based on tattoos have gained much interest in recent years [4,16]. This is due to several reasons, including the fact that the tendency of people to have tattoos has increased [11]. Furthermore, tattoos provide additional information about the person: affiliation to groups or gangs, religious beliefs, years in prison, etc. These have been used to assist law enforcement authorities in investigations that lead to the identification of offenders and victims of natural disasters and accidents [11].

This paper focuses on the identification of individuals based on tattoos. Most of existing works in this topic [9,11] propose to use handcrafted features to describe the images and then evaluate the similarity between two images by comparing their features. However, in recent years deep learning methods have shown better results than this kind of features in similar computer vision tasks [3]. That is the reason why some works [3,4] have proposed the use of deep neural networks for tattoo identification.

Due to the great variability that usually exists between tattoos of different individuals, in [3], it is explored the idea of fine-tuning deep networks trained on large generic databases such as ImageNet [17] for the task of tattoo image classification considering two classes: Tattoo and Not-Tattoo. This database contains many varied images of a large number of object types, something similar to what happens with tattoos. Therefore, it has a certain logic to think that a neural network trained to discern the images that belong to the same class and those that do not, can also do it for tattoo images. This way, it is not necessary to train a network from scratch or count on the large volume of data that this training requires. Nevertheless, for the task of tattoo matching, the authors of [3] used a Siamese network, which has the disadvantage that it is necessary to perform the inference, every time a comparison is made between two images. This incurs in a high computational cost when used in an identification scenario since it is necessary to compare the image to be identified with all the images of a database. In contrast to that approach, if a network is used to extract features from the database images, they can be stored and then used for future comparisons. In this way, the network is executed only once for each requested identification. In this paper we adopt this strategy as a way to perform tattoo identification more efficiently.

On the other hand, recent papers [1] have shown that top layers of a large convolutional neural network (CNN) provide high-level descriptors of the visual content of the image. They proved that these descriptors can be used for different tasks other than that for which the network was trained. This allows using them effectively without the need of having large databases to train the network, which is the case for tattoos, where there are few public databases. Based on the previous ideas, the main contribution of our work is a study to assess the use of some deep neural networks trained on generic databases such as ImageNet, for tattoo identification. We extract the features from intermediate layers of these networks and used them as descriptors of the tattoo images. The difference with previous works [3,4] is that we show that it is possible to achieve competitive results without training or even fine-tuning the networks.

The remainder of this paper is structured as follows. We briefly review related literature in Sect. 2. In Sect. 3, the details of the proposed deep networks for tattoo identification are provided. We show some experimental results in Sect. 4. Finally, we conclude this work in Sect. 5.

2   Related Works

The early practice of tattoo image retrieval relied on keywords or metadata based matching [4]. However, a keyword-based tattoo image retrieval has several limitations in practice: (i) Labels are insufficient for describing all visual information of a tattoo; (ii) multiple keywords may be needed to adequately describe a tattoo image; (iii) human annotation is subjective and different subjects can give dramatically different labels to the same tattoo image [4].

Due to these problems, interest has grown in developing content-based image retrieval techniques (CBIR) to improve the efficiency and accuracy of the tattoo search [9,14]. CBIR aims to extract features such as edges, color and texture, that can reflect the content of an image, and use them to identify images with high visual similarity [4]. The scale-invariant feature transform (SIFT) [13] and its variants have been the most used among this kind of methods for tattoo identification [9,14].

Recently, most researches are focusing on deep learning methods due to its success in many computer vision tasks [3]. In particular, AlexNet [10], which won the ImageNet challenge of 2012, has been successfully used for tattoo vs. non-tattoo classification in [3,22]. Other works [21] focus on tattoo localization using Faster R-CNN. However, little research exists on tattoo image identification. To the best of our knowledge, only [3] and [4] studied the identification of tattoos using deep learning. In [3] a Siamese network pre-trained on ImageNet and fine-tuned for tattoo identification is used. As mentioned in the Introduction, the use of Siamese networks for matching is less efficient because it is required to perform the network inference for every comparison. This is not suitable for an identification scenario where it is necessary to match a query image with thousands of images in an operational database.

In [4] a network based on a Faster R-CNN was used to learn tattoo detection and a compact representation of the tattoo in the same network. The features that return this network are binarized in order to make search efficient. This work obtains comparable results with other state-of-the-art methods and generalizes well to other retrieval tasks. However, this network was trained with hundred of thousands of images, which are not available for everyone neither the computational resources to do it. Moreover, a pre-trained model of this network is not publicly available.

3   Generic Neural Networks for Tattoo Identification

The proposal of this paper is to use the features from intermediate layers of neural networks, trained on large generic databases, to describe the content of a tattoo image. These features should be matched with the features of a tattoo gallery in order to identify the tattoo.

In order to evaluate the proposal, we selected some neural networks that have obtained good results in the ImageNet classification challenge. It is worth noting that none of these networks was fine-tuned with images of tattoos as was done in [3], instead we used their publicly available models trained on ImageNet. We believe that, since tattoos mostly depict objects of the real world, their semantic and visual features might be related to those learned from a generic image database with real objects.

- MobileNetV1 [5]: network designed to be used in mobile and embedded devices thanks to its low complexity. It is trained on ImageNet, where it obtained 70.6% classification accuracy.
- MobileNetV2 [18]: improved version of MobileNetV1. It obtains 74.7% of accuracy, improving the previous version at a similar processing cost.
- Inception21k [6]: it is trained on ImageNet, but with 21841 classes, unlike the 1000 that are generally used. It obtains 68.3% accuracy.
- Resnet50_CVGJ [19]: variant with batch normalization [6] of the ResNet50 network trained on ImageNet.
- VGG_CVGJ [19]: variant with batch normalization [6] of the VGG19 [20] network trained on ImageNet.

We also evaluate some networks that have been proposed for image retrieval. These are designed to return a feature vector as a descriptor of the visual content of the image instead of a classification output.

- DeepBit [12]: network trained in an unsupervised manner to build high-level compressed features. During the training, restrictions were used so that the features met three requirements: invariance to the rotation of the image, high entropy of the features and high standard deviation.
- SSDH [23]: network that constructs hash functions as a latent layer in a deep network. It was designed so that classification and retrieval were unified in a single learning model.

Both networks were designed so that their features were binarized allowing a more efficient matching using the Hamming distance. Both define a function to do that, however, we do not binarize the features because we obtained better results with the real values.

Table 1 shows the layer used for each network, as well as the vector dimension of output features. For DeepBit and SSDH we used the last layer, while for the other networks we report the layers that achieve better results.

All networks have a fixed input of 224×224 pixels, except SSDH that has an input of 227×227. The euclidean distance was used to match the features, except for DeepBit, for which was used the cosine distance that gaves much better results.

Table 1. Deep networks characteristics.

Network	Layer used	Output size
MobileNetV1	pool6	1024
MobileNetV2	pool6	1280
Inception21k	global_pool	1024
Resnet50_CVGJ	global_pool	2048
VGG_CVGJ	fc7	4096
DeepBit	-	32
SSDH	-	32

4   Experimental Evaluation

In this section we aim at evaluating and comparing the selected networks in the task of tattoo identification. Different tattoo databases have been used in the literature [15, 16], but most of them are not public or their access is difficult. In particular, most of the works have used the Tatt-C database [16] for experimenting, training and comparing with other methods, but it is not public any more due to legal issues. The authors of [4] created a large dataset named WebTattoo combining other tattoo datasets that they have access. However, this dataset is not public yet; should be released soon. Therefore, in order to validate our proposal, we have created our own datasets.

The conducted experiments were performed in a PC with a CPU Intel Core i7-4470 with 8 GB of RAM.

4.1   Proposed Databases and Evaluation Protocol

Two databases of tattoo images were created in order to evaluate the proposal. The first one (BIVTatt)[1] was collected by the authors of this paper and contains 210 images belonging to 159 individuals (some individuals have only one image). The second (PinTatt) is composed of 454 images downloaded from Pinterest belonging to 160 individuals. Images from BIVTatt have higher resolution than PinTatt images and their content is sharper. All the images are cropped around the tattoo so there is not much background information. Figure 1 shows some sample images from both databases. For each image, 20 new images were generated applying transformations with two different intensities of illumination, two diffusions, four affine transformations, four aspect ratio transformations, four different rotations and four color changes, as described in [11]. This way, the databases were increased obtaining a total of 4410 images for BIVTatt and 9534 for PinTatt. Figure 2 shows examples of transformations of an image from the BIVTatt dataset.

---

[1] The BIVTatt dataset is available at https://github.com/mnicolas94/BIVTatt-Dataset.

Fig. 1. Sample images of both datasets: (a) BIVTatt, (b) PinTatt.

Fig. 2. Examples of image transformations in the BIVTatt dataset: (a) original; (b) affine transformation; (c) aspect ratio; (d) blurring; (e) color change; (f) illumination; (g) rotation. (Color figure online)

For the aim of identification experiments on each database, the probe set was conformed by the transformed images, while the gallery, was composed of the original images. For every test image, the original image that originate it (by some transformation) was excluded from the comparison. Thus, we simulate a real scenario of forensic identification were different images from the same tattoo can be available, but not exactly the same image. The final BIVTatt probe set consists of 1540 images and the gallery of 209 images. In the case of PinTatt it has 9080 and 453, respectively.

To evaluate the performance of the compared methods, we employed the cumulative match characteristic (CMC) curve. Each point of the CMC curve is the fraction of images of the probe set that were correctly matched with any of its pairs in the gallery, in a given range.

4.2   Experimental Results

Besides the experiments with the networks analyzed in Sect. 3, all the tests were also carried out using SIFT in order to compare the proposal of this work with a general approach used in previous works. The matching method used for SIFT was a knn-based matcher implemented in the Fast Library for Approximate Nearest Neighbors (FLANN) of the OpenCV library with k = 2. The Fig. 3 and the Fig. 4 show the CMC curves for BIVTatt and PinTatt databases respectively. The Table 2 and the Table 3 show the identification recognition rates at different rank values for BIVTatt and PinTatt databases respectively.

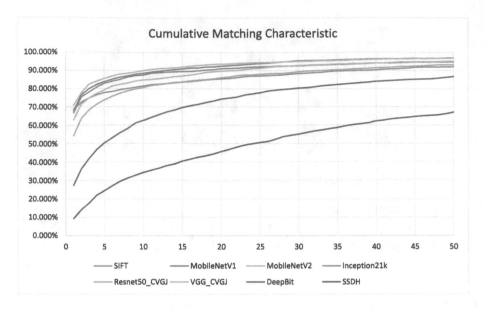

Fig. 3. Curve CMC on the BIVTatt dataset.

As can be seen, the best general performance was achieved by MobileNetV2 on both datasets. The two image retrieval networks, DeepBit and SSDH, obtained poor results. The image classification networks obtain good results in

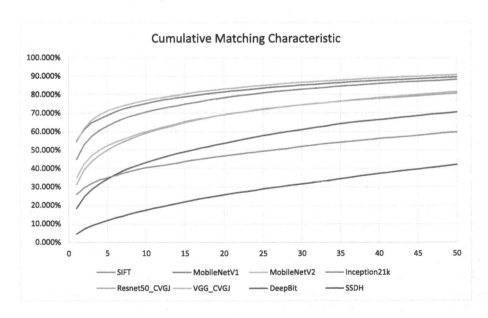

Fig. 4. Curve CMC on the PinTatt dataset.

Table 2. BIVTatt dataset identification results.

Algorithm/network	Rank1 (%)	Rank5 (%)	Rank10 (%)	Rank20 (%)	Rank50 (%)
SIFT	68.571	77.727	81.429	85.519	92.078
MobileNetV1	67.857	83.766	88.442	92.532	96.883
MobileNetV2	**70.909**	**85.519**	**89.935**	**93.636**	**96.818**
Inception21k	66.688	82.792	87.468	91.169	94.740
Resnet50_CVGJ	54.221	73.701	80.584	86.169	93.182
VGG_CVGJ	62.792	79.156	84.805	89.805	95.195
DeepBit	27.338	50.195	62.468	74.351	86.688
SSDH	9.221	24.351	34.221	45.584	67.078

Table 3. PinTatt dataset identification results.

Algorithm/network	Rank1 (%)	Rank5 (%)	Rank10 (%)	Rank20 (%)	Rank50 (%)
SIFT	25.936	34.967	40.485	46.806	59.515
MobileNetV1	**55.033**	68.689	75.154	81.432	89.405
MobileNetV2	54.350	**71.244**	**76.806**	**83.095**	**90.694**
Inception21k	44.901	62.952	70.540	78.304	88.062
Resnet50_CVGJ	31.278	50.000	59.152	69.053	80.617
VGG_CVGJ	34.945	52.390	59.835	69.097	81.465
DeepBit	18.469	34.317	43.249	53.590	70.363
SSDH	4.615	11.861	17.478	25.760	42.026

general. All of them outperformed SIFT, except for Rank-1 on BIVTatt dataset. In the case of Resnet50_CVGJ, it only exceeds SIFT starting from Rank-20 on this dataset.

It is also necessary to evaluate the efficiency of these networks to know if it is feasible to use them in real applications and for which scenarios. We measure the time each method takes to extract the features and the time required by the matching method to match an image against all images in the gallery. The sum of both times is the identification time of a query tattoo image. Figure 5 shows the time for the feature extraction of an image and the time to match it with all images in the gallery, averaged over 100 images in the BIVTatt dataset. Similar results were observed for the same experiment on the PinTatt dataset.

As can be seen in Fig. 5, SIFT is the fastest method for feature extraction but it is too slow on the matching step. In an identification scenario the matching efficiency is critical because the matching algorithm must be executed for each image in the gallery. We think that this fact makes SIFT a questionable alternative for this kind of scenario. On the other hand, MobileNetV2 has the best accuracy/efficiency relation which makes it a good option for tattoo

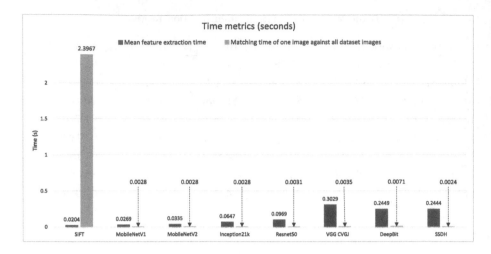

Fig. 5. Mean feature extraction and matching times in BIVTatt dataset.

identification. These experiments show that CNNs trained for image classification on generic databases such as ImageNet, can be extended to the tattoo domain, achieving good results as well.

5   Conclusions

In this article, we studied the use of intermediate features of deep neural networks, trained on generic databases, as descriptors of tattoo images. Unlike previous works where they used transfer learning to adjust the network to the context of tattoo identification, we used the original pre-trained network models. The results of the identification tests showed that by using this approach we can obtain better results, both in efficiency and accuracy, than the handcrafted solutions previously adopted, such as SIFT. In addition, the implementation cost is minimal since there are many publicly available CNNs similar to those used in this work. In future research, we will consider the use of tattoo detection and segmentation methods to extract the background of the image.

References

1. Babenko, A., Slesarev, A., Chigorin, A., Lempitsky, V.: Neural codes for image retrieval. In: Fleet, D., Pajdla, T., Schiele, B., Tuytelaars, T. (eds.) ECCV 2014. LNCS, vol. 8689, pp. 584–599. Springer, Cham (2014). https://doi.org/10.1007/978-3-319-10590-1_38
2. Dantcheva, A., Elia, P., Ross, A.: What else does your biometric data reveal? A survey on soft biometrics. IEEE Trans. Inf. Forensics Secur. **11**(3), 441–467 (2016)

3. Di, X., Patel, V.M.: Deep learning for tattoo recognition. In: Bhanu, B., Kumar, A. (eds.) Deep Learning for Biometrics. ACVPR, pp. 241–256. Springer, Cham (2017). https://doi.org/10.1007/978-3-319-61657-5_10

4. Han, H., Li, J., Jain, A.K., Chen, X.: Tattoo image search at scale: joint detection and compact representation learning. IEEE Trans. Pattern Anal. Mach. Intell. **41**, 2333–2348 (2019)

5. Howard, A.G., et al.: Mobilenets: efficient convolutional neural networks for mobile vision applications. arXiv preprint arXiv:1704.04861 (2017)

6. Ioffe, S., Szegedy, C.: Batch normalization: accelerating deep network training by reducing internal covariate shift. arXiv preprint arXiv:1502.03167 (2015)

7. Jain, A., Nandakumar, K., Ross, A.: 50 years of biometric research: accomplishments, challenges, and opportunities. Pattern Recogn. Lett. **79**, 80–105 (2016)

8. Jain, A.K., Park, U.: Facial marks: soft biometric for face recognition. In: 2009 16th IEEE International Conference on Image Processing (ICIP), pp. 37–40. IEEE (2009)

9. Kim, J., Parra, A., Yue, J., Li, H., Delp, E.J.: Robust local and global shape context for tattoo image matching. In: 2015 IEEE International Conference on Image Processing (ICIP), pp. 2194–2198. IEEE (2015)

10. Krizhevsky, A., Sutskever, I., Hinton, G.: Imagenet classification with deep convolutional neural networks. In: Advances in Neural Information Processing Systems, pp. 1097–1105 (2012)

11. Lee, J.E., Jain, A.K., Jin, R.: Scars, marks and tattoos (SMT): soft biometric for suspect and victim identification. In: Biometrics Symposium, BSYM 2008, pp. 1–8. IEEE (2008)

12. Lin, K., Lu, J., Chen, C.S., Zhou, J.: Learning compact binary descriptors with unsupervised deep neural networks. In: Proceedings of the IEEE Conference on Computer Vision and Pattern Recognition, pp. 1183–1192 (2016)

13. Lowe, D.G.: Distinctive image features from scale-invariant keypoints. Int. J. Comput. Vision **60**(2), 91–110 (2004)

14. Manger, D.: Large-scale tattoo image retrieval. In: 2012 Ninth Conference on Computer and Robot Vision (CRV), pp. 454–459. IEEE (2012)

15. Martin, M., Dawson, J., Bourlai, T.: Large scale data collection of tattoo-based biometric data from social-media websites. In: 2017 European Intelligence and Security Informatics Conference (EISIC), pp. 135–138. IEEE (2017)

16. Ngan, M., Grother, P.: Tattoo recognition technology-challenge (tatt-c): an open tattoo database for developing tattoo recognition research. In: 2015 IEEE International Conference on Identity, Security and Behavior Analysis (ISBA), pp. 1–6. IEEE (2015)

17. Russakovsky, O., et al.: Imagenet large scale visual recognition challenge. Int. J. Comput. Vision **115**(3), 211–252 (2015)

18. Sandler, M., Howard, A., Zhu, M., Zhmoginov, A., Chen, L.C.: Mobilenetv 2: inverted residuals and linear bottlenecks. In: Proceedings of the IEEE Conference on Computer Vision and Pattern Recognition, pp. 4510–4520 (2018)

19. Simon, M., Rodner, E., Denzler, J.: Imagenet pre-trained models with batch normalization. arXiv preprint arXiv:1612.01452 (2016)

20. Simonyan, K., Zisserman, A.: Very deep convolutional networks for large-scale image recognition. arXiv preprint arXiv:1409.1556 (2014)

21. Sun, Z.H., Baumes, J., Tunison, P., Turek, M., Hoogs, A.: Tattoo detection and localization using region-based deep learning. In: 2016 23rd International Conference on Pattern Recognition (ICPR), pp. 3055–3060. IEEE (2016)

22. Xu, Q., Ghosh, S., Xu, X., Huang, Y., Kong, A.W.K.: Tattoo detection based on CNN and remarks on the NIST database. In: 2016 International Conference on Biometrics (ICB), pp. 1–7. IEEE (2016)
23. Yang, H.F., Lin, K., Chen, C.S.: Supervised learning of semantics-preserving hash via deep convolutional neural networks. IEEE Trans. Pattern Anal. Mach. Intell. **40**(2), 437–451 (2017)

Analysis of the Impact of Ear Alignment on Unconstrained Ear Recognition

Elaine Grenot-Castellano[✉], Yoanna Martínez-Díaz,
and Francisco José Silva-Mata

Advanced Technologies Application Center, 7th A Avenue #21406 % 214 and 216,
Siboney, Playa, 12200 Havana, Cuba
{egrenot,ymartinez,fjsilva}@cenatav.co.cu

Abstract. The use of the ear in biometric recognition has been widely covered in controlled environments. However, the advantages of the ear as a biometric characteristic impose the need to know how it behaves in unconstrained scenarios, where it is common the presence of occlusions, pose variations, illumination changes and different resolutions. According to this challenge and considering the experience in other biometric recognition processes, the alignment has shown to be a key step. In this work, we carry out an exhaustive and detailed study of the impact of the alignment on the performance of several state-of-the-art ear descriptors, when the images are captured in uncontrolled conditions. Our analysis is based on identification experiments against different types of variations in ears image of the challenging UERC dataset. The obtained results corroborate the hypothesis of the alignment also improves the efficacy of the ear recognition process and show how this improvement behaves for various factors such as head rotation, occlusions, flipping and resolution.

Keywords: Ear alignment · Unconstrained ear recognition · Covariates

1 Introduction

The face, the iris, and the fingerprint are examples of the most popular biometric objects used for person recognition. In recent times, the ear has become important as an identifying part among people. The rich structure of an ear combined with its stability over time is a promising source of data to identify subject since its collection can be done in a noninvasive way, has a high degree of permanence, distinctiveness and universality [5]. However, some factors such as partial o full occlusions, pose variations and the presence of ear accessories can be affect sensitively the ear recognition performance.

A typical fully automatic ear recognition system follows a traditional pipeline of detection, alignment, feature extraction and classification. Many approaches have been proposed attempting to improve ear recognition capabilities for reliable deployment in surveillance and commercial applications [1,11]. Most of these

© Springer Nature Switzerland AG 2019
I. Nyström et al. (Eds.): CIARP 2019, LNCS 11896, pp. 283–293, 2019.
https://doi.org/10.1007/978-3-030-33904-3_26

works rely on develop feature descriptors that can be resilient to variability found in unconstrained conditions. Depending on the type of feature extraction technique used, ear recognition approaches can be grouped into hand-crafted [5] and deep-learning descriptors [2,8].

As in other modalities such as the face and iris, image alignment plays a crucial role in a recognition system, since most approaches are very sensitive to the pose and scale variations. Even the best performing state-of-the-art descriptors require that images are aligned as good as possible in order to achieve better results. In the case of ear, several methods [12,14,15] have been develop for aligning images but it is still not completely clear how these methods are able to improve the recognition performance in the presence of factors found in unconstrained settings such as head rotation, occlusions or image resolution.

The main contribution of this work is a comprehensive experimental evaluation of several state-of-the-art ear recognition techniques on the challenging UERC dataset with the aim of studying the effect of alignment on uncontrolled conditions. Specifically, we perform a comparative assessment of recent hand-crafted and deep-learning descriptors using both aligned and no aligned images and investigate their robustness in front to unseen data characteristics such rotations, occlusions and image resolution. As result, we present an extensive experimental analysis in terms of recognition rates which contributes to a better understanding of the behavior of the alignment on the evaluated methods, showing its importance on unconstrained ear recognition.

The remainder of this paper is organized as follows. Section 2 describes the existing works related to the ear alignment topic. In Sect. 3 we present the ear alignment method and recognition techniques considered in this work. The experimental setup and the results obtained are provided in Sect. 4. Finally, conclusion and future work are given in Sect. 5.

2   Alignment Methods

Different from other biometric features such as the iris (radial symmetry and approximately circular shape) or the face (it is possible to determine an axis of approximate symmetry that divides the face into two similar parts), the ear lacks symmetrical properties. Therefore, the attempts to align the ear images have depended to a large extent on defining certain axes or parts of its that serve to be taken as reference for the alignment [12,14,15].

Some authors have been used the helix (outer edge of the ear) as reference to align ear images [14,15]. The main difficulty of this approach is that it must use precise methods of edge detection, in order to determine the reference axes or landmark. The elliptical shape of the ear has also exploited by using a cascaded pose regression [12]. This method fits the ear outer rim with an abstract elliptical model and then, transforms it to its normal position given by the main axes of the ellipse. In [13] the Random sample consensus (RANSAC) [7] is used over SIFT descriptors to estimate the transformation in the plane of each image to an average image, which is then applied together with ear mask. Various statistical deformable models with different features descriptors were evaluated in [17] for ear landmark localization on images taken from uncontrolled environments. As result, their best combination was achieved by using a holistic Active Appearance Model based on SIFT features.

Recently, deep convolutional neuronal networks (CNNs) have also been used to detect landmarks in different areas of the ear in order to carry out their alignment [8,16]. A cascading convolution neural network was proposed in [16] to detect six landmark points. These points were defined in accordance with the morphological and geometric characteristics of the ear; three of them were located in its internal region and three in the external contour of the ear. In [8] the authors introduced a two-stage landmark detector based on Convolutional Neural Networks to locate a set of 55 landmarks which are then employed to translate, rotate and scale the input ear image.

Although several methods have been proposed for ear alignment, few works have investigated its role in unconstrained ear recognition. In [13] the authors evaluate the influence of RANSAC method but only in a subset of images of AWE dataset, according to the severity of pose variations. Hansley et al. [8] analyze the benefits of their alignment method by checking the difference in the recognition performance with and without alignment; but only for hand-crafted descriptors. In [17] aligned versus non-aligned ears were compared in ear verification and close identification experiments. All these works demonstrate that in general, the alignment consistently improves the ear recognition performance. However, they do not provided detailed information about its importance in front different covariates present in uncontrolled conditions.

In the present work we evaluate the impact of alignment on ear recognition under novel aspects that were not considered before such as occlusions and image resolution. In addition, we perform an extensive experimental analysis for both hand-crafted and deep-learning descriptors on the challenging UERC database.

3   Baselines

In order to align the ear images we select the method proposed in [8], since unlike others methods it directly attacks one of the more difficult problem: the pose variations. Their solution relies on a two-stage landmark detector based on CNNs to locate a set of 55 landmarks. The first network is used to create an easier landmark detector by reducing scale and translation variations. The coordinates obtained by this network are used to refine the center and orientation of an ear image and then, the rectified image is used as input of the second network to fine-tune small variations. After landmark detection, the ears are normalized by applying PCA on the retrieved landmarks. Finally, in order to diminish the effects of poses variations, different sampling rates are used in such a way the width and the height of the normalized ear are approximately the same. In addition, before the automatic aligning process, the authors use a simple side classifier to detect explicitly whether they are processing left or right ears and then flip the images to a common reference, so that all ears would have the same orientation.

3.1 Ear Recognition Techniques

With the aim of evaluating the benefits of the previous alignment method, we considered seven hand-crafted and two deep-learning descriptors based on their good performance reported for ear recognition [3,5].

Specifically, we used Local Binary Patterns (LBP), Gabor wavelets, Binarized Statistical Image Features (BSIF), Local Phase Quantization Features (LPQ), Rotation Invariant LPQ (RILPQ), Patterns of Oriented Edge Magnitudes (POEM) and Histograms of Oriented Gradients (HOG) as hand-crafted descriptors [5]. In the case of deep-learning descriptors, we selected the MobileNet [10] and the ResNet-18 [9] networks, which cover some of the most popular architectures for ear recognition.

4   Experimental Setup

In this section, we asses the performance of hand-crafted and deep-learning techniques with and without alignment. First, we describe the recognition dataset used and give some implementations details. Then, we present the recognition results obtained, taking into account different covariates.

4.1 Ear Recognition Dataset and Protocols

For the experimental evaluation, we use the UERC 2019 dataset [6], that consists of 11 000 ear images collected from the web of 3 690 subjects, making it the largest publicly available dataset of unconstrained ear images.

The main part of this dataset was taken from the Extended Annotated Web Ears (AWEx) dataset [5] and comprised 3 300 ear images of 330 subjects. Images from this part are annotated with different covariates hence, it was used as the basis for our analysis. The rest of the data was taken from the UERC 2017 dataset [4], which presents characteristics and variability similar to the AWEx images, but with greater variations in the size of the images. Sample images of the UERC dataset are illustrated in Fig. 1.

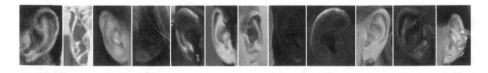

Fig. 1. Examples of images from the UERC 2019 dataset.

In order to develop and test models, the public UERC 2019 dataset was partitioned into disjoint training and testing sets. The training set consists of 2 304 images of 166 subjects from the AWEx dataset, whereas the testing set contains the remaining AWEx data and the rest of the images from UERC 2017.

4.2   Implementations Details

In the case of the alignment method [8], we use the demo and the deep models provided by the authors (http://github.com/maups/ear-recognition), where the two-stage landmark detector and the side classifier are available.

For hand-crafted descriptors it was used the implementations provided in the AWE toolbox [5] with their default values. The MobileNet and ResNet-18 models was used with initial parameters learned on the ImageNet dataset and fine-tune certain layers training using aligned and non-aligned ear images, separately. For this, the UERC training set was used and data augmentation was performed with a 50% chance. For both CNNs we set the learning rate to 0.01, the momentum to 0.75 and the weight decay to 0.005. After training, the last fully-connected layers from the networks were used as feature extractors. Once the representations are computed, the cosine similarity is used for the comparison of test images.

4.3   Experimental Results

Several identification experiments were carried out, taking into account different factors that affect the performance of ear recognition process in unconstrained scenarios, such as sensitivity to same-side vs. opposite-side matching, occlusions, rotations (in terms of yaw, pitch and roll angles) and different resolutions.

Table 1. Recognition rates (%) at rank-1 and rank-5 for all evaluated descriptors using aligned and non-aligned images of AWEx and UERC datasets.

	Rank-1 AWEx		Rank-5 AWEx		Rank-1 UERC		Rank-5 UERC	
	No align	Align	No align	Align	No align	Align	No align	Align
LBP	13.17	28.39	26.17	46.39	8.07	15.39	15.76	**25.50**
GABOR	14.28	22.61	28.44	40.94	5.47	9.26	11.33	17.60
BSIF	15.33	28.72	31.06	47.22	8.55	14.93	16.37	24.33
LPQ	14.94	27.28	28.33	45.00	8.75	14.57	17.01	24.59
RILPQ	15.89	27.61	29.11	44.33	7.26	13.33	13.56	21.61
POEM	18.28	31.39	32.72	52.78	8.36	14.70	15.74	25.30
HOG	19.39	**40.50**	36.00	**59.67**	7.93	**17.27**	15.04	**26.31**
ResNet-18	15.72	20.17	36.56	42.33	6.42	7.91	15.32	17.28
MobileNet	**21.44**	25.89	**44.84**	48.06	**8.99**	10.43	**19.58**	20.53

Ear Identification. Table 1 shows the recognition rates at rank-1 and rank-5 for each descriptor using aligned and non-aligned images from the testing sets of AWEx dataset (involving 180 subjects) and UERC 2019 dataset. As it can be seen, in general, all the results are improved when the images are aligned. However, these improvements were noticeably lower for the case of deep descriptors. We think that this is because these deep models are able to learn in a better way the variations present in the training images. In addition, if these variations are severe it can affect the performance of the alignment method which can introduce some noise information in the learning stage. Consequently, we can said that this kind of descriptors include a partially solution for non-aligned ears. In contrast, the alignment process shows significant improvements for all hand-crafted descriptors, even obtaining better results that deep descriptors.

The best rates at rank-1 for AWEx and UERC datasets using non-aligned images are achieved by the deep descriptor MobileNet with a 21.44% and 8.99%, respectively; while when the images are aligned the best rates are obtained by the hand-crafted HOG descriptor, increasing to 40.50% and 17.27% for AWEx and UERC, respectively. LBP and HOG descriptors are the most benefited when images are aligned, improving their results at least a factor of 1.5x in all cases.

The low recognition rates obtained for the UERC dataset corroborate that all the evaluated descriptors do not present good scalability when the number of subject increases and the ear images have poorer resolution and lower-quality, even using aligned images. The CMC curves shown in Fig. 2 complement the results, being able to observe clearly the increase of the area under the curve when the alignment is used for all the descriptors in AWEx and UERC datasets.

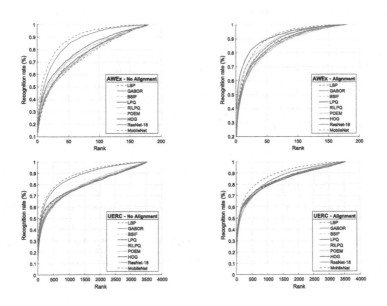

Fig. 2. CMC curves from the AWEx and UERC datasets.

Same-Side vs. Opposite-Side Matching. This experiment leads to know the impact of alignment when we match ear images from the same side (e.g., right-to-right), and from opposite sides of the head. Table 2 presents the recognition rates at rank-1 when the original images, the side classifier (Flip) and the alignment method (Align) are used. The results evidence that determining which side of the head the ear images came from (Flip), improves the recognition especially when the ear images are from opposite sides, but the difference in performance is less than when alignment is applied. When the images are from the same side, the best results are obtained by the HOG descriptor, achieving the highest recognition rates by using aligned images. In case of ear images from opposite sides, deep-learning descriptors outperform hand-crafted descriptors when the images are not aligned, especially, when original images are used. This is because the deep models are trained to learn features that are mostly not affected by the corresponding ear side, although if the ear is misaligned by some error of the alignment method, this stage may work against it. However, using aligned images again HOG descriptor reaches the highest scores.

On the other hand, we can see that sometimes, when we match ears images from the same side, the results of using flipped ears are a little worse than those using the original version. This is due to errors of the classifier used to determine automatically the side of a given ear.

Table 2. Recognition rates (%) at rank-1 for same-side vs. opposite-side matching using original, flipped and aligned images from the AWEx dataset.

	Right-Right			Right-Left			Left-Left			Left-Right		
	Orig.	Flip	Align	Orig.	Flip	Align	Orig.	Flip	Align	Orig.	Flip	Align
LBP	11.42	11.19	21.28	0.89	8.17	17.02	14.88	14.33	22.16	1.10	11.36	19.40
GABOR	13.21	12.77	17.81	1.23	7.84	13.33	15.66	15.10	19.96	1.21	9.37	13.23
BSIF	13.10	12.99	22.84	1.23	11.42	18.93	17.86	17.53	21.61	0.99	12.89	17.86
LPQ	14.56	14.22	22.28	0.78	10.86	16.69	15.66	15.22	21.17	0.88	12.35	18.08
RILPQ	15.23	14.78	21.61	0.89	9.97	14.11	16.65	15.33	22.82	1.43	11.69	16.20
POEM	17.36	16.79	25.87	0.56	11.98	19.26	19.74	18.52	26.79	0.55	12.79	20.62
HOG	**18.25**	**17.58**	**30.35**	0.78	11.86	**23.29**	20.84	**21.06**	**35.94**	1.43	12.89	**23.59**
ResNet-18	12.54	12.65	16.46	9.52	11.31	13.10	15.44	17.64	16.54	9.04	10.58	13.89
MobileNet	16.46	17.36	20.60	**12.43**	**13.77**	17.58	20.07	18.85	21.50	**14.11**	**14.77**	17.75

Occlusion. Figure 3 shows the recognition rates at rank-1 for different occlusion levels. We can see that while the performance of evaluated descriptors for minor and mild occlusions are considerably improved by using the alignment method, this improvement is not the same when severe occlusions are present. The hand-crafted descriptors from aligned images are seem to ensure better performance across all levels. However, for deep-learning techniques using aligned images causes performance degradation in front severe occlusions. One possible hypothesis of this is that, in case of severe occlusions, as the problem consists of a great absence of information in the images to match, the alignment process

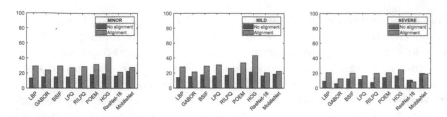

Fig. 3. Recognition rates (%) at rank-1 for different levels of occlusion using aligned and non-aligned images of AWEx dataset.

is not able to solve it. In contrast, the deep learning process is more capable of lead with this phenomenon thanks to the training process, where ear images with several occlusions were included.

Ear Rotation. Figure 4 illustrates the recognition rates at rank-1 under different rotation variations (roll, yaw, pitch), where for most of the descriptors a remarkable improvement is obtained through alignment step in almost all the cases and angles, being the HOG the best overall descriptor.

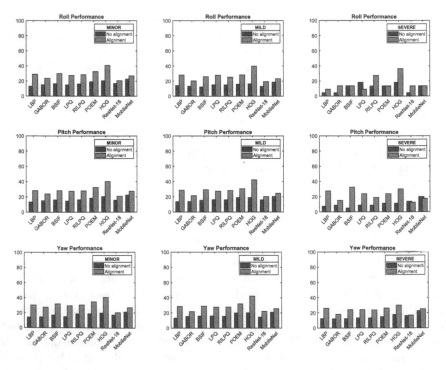

Fig. 4. Recognition rates at rank-1 across the pitch, roll and yaw rotation angles.

It can be seen that, severe roll and pitch angles impact more negatively on the recognition performance. For example, in case of roll rotations, the alignment shows it worse behavior due to this is one of the greatest challenges for any alignment method. However, in the case of pitch angles, the recognition rates of hand-crafted descriptors increase considerably by using aligned images. Similar to previous experiments, the improvements for hand-crafted descriptors are higher than for deep descriptors, especially for extreme variations.

Image Resolution. In this experiment we asses the impact of alignment in front different resolutions on UERC test dataset. In Table 3 presents the results in terms of recognition rates at rank-1 for all the descriptors. As can be seen, aligning images helps to increase the recognition performance in all cases for all tested resolutions. Smallest images with less than 1k pixels inevitably lead to performance degradation, hence, the improvements are smaller. However, as the resolution increases, they become more significant, especially for the hand-crafted descriptors. This fact is due to resolution significantly contributes to represent by the descriptors the details of the ear images.

Results achieved with resolution images between 5K and 10K pixels arc similar to the results with images having more than 10K pixels, which suggests that images of at least 5K pixels are needed to obtain an adequate recognition performance. In these cases, HOG descriptor obtains the best results by using aligned images, while MobileNet is the best one when alignment is not applied.

Table 3. Recognition rates at rank-1 for different resolutions on UERC database.

	<1K (#4573)		1K-5K (#3883)		5K-10K (#412)		>10K (#632)	
	No Align	Align	No Align	Align	No Align	Align	No Align	Align
LBP	6.27	**11.27**	8.12	15.58	11.38	28.33	14.58	28.05
GABOR	3.16	5.92	5.27	8.58	13.08	22.28	13.79	21.87
BSIF	6.77	10.94	**8.29**	14.48	15.01	28.33	15.05	29.64
LPQ	**7.19**	11.06	8.09	13.82	13.32	27.12	17.27	28.68
RILPQ	4.76	9.02	6.50	12.93	14.77	26.88	19.33	29.48
POEM	5.59	9.73	7.62	13.69	18.64	31.23	19.97	35.34
HOG	4.28	10.29	7.72	**16.27**	17.92	**42.37**	21.87	**42.95**
ResNet-18	4.23	4.82	5.67	7.66	17.43	17.92	14.58	19.18
MobileNet	5.71	6.71	7.82	9.38	**22.28**	24.21	**23.45**	26.30

5   Conclusion

In this work an exhaustive analysis was carried out to evaluate the impact of ear alignment on the recognition performance of several state-of-the-art techniques on unconstrained conditions, taking into account different covariates. For this, we conduct several identification experiments for hand-crafted and deep-learning descriptors on the challenge UERC dataset by using both aligned and non-aligned images. As result, we evidence that the alignment is an important step in the ear recognition process to achieve better results. Specifically, we found that for the hand-crafted descriptors, the alignment has a greater impact than for the deep models trained with aligned images. It can be said that when the images are not aligned, the deep-learning descriptors achieve a more discriminative description of ears with severe covariates, although in most cases an improvement is obtained. We argue that deep models are more capable of learning extreme variations, especially when these affect the performance of the alignment method. Among the tested hand-crafted descriptors, the HOG was the most benefited with the use of alignment, obtaining the highest recognition rates in almost all cases. However, it was evidenced that to achieve high recognition rates, not only alignment is sufficient, robust descriptors are also necessary.

With this work we provide a better understanding about the impact of ear alignment step on unconstrained ear recognition and identify the most challenges covariates which affect it. Note that, in cases where deviations are minimal or medium and hand-crafted descriptors are used, we can expect great improvements in recognition by alignment, contrary to severe variations, where there is still work to be done especially for those cases where the images contain high roll rotations, severe occlusions and low resolutions.

References

1. Abaza, A., Harrison, M.A.F.: Ear recognition: a complete system. In: Biometric and Surveillance Technology for Human and Activity Identification X, vol. 8712, p. 87120N (2013)
2. Dodge, S., Mounsef, J., Karam, L.: Unconstrained ear recognition using deep neural networks. IET Biom. **7**(3), 207–214 (2018)
3. Emeršič, Ž., Križaj, J., Štruc, V., Peer, P.: Deep ear recognition pipeline. In: Hassaballah, M., Hosny, K.M. (eds.) Recent Advances in Computer Vision. SCI, vol. 804, pp. 333–362. Springer, Cham (2019). https://doi.org/10.1007/978-3-030-03000-1_14
4. Emeršič, Ž., et al.: The unconstrained ear recognition challenge. In: IEEE IJCB, pp. 715–724 (2017)
5. Emeršič, Ž., Štruc, V., Peer, P.: Ear recognition: more than a survey. Neurocomputing **255**, 26–39 (2017)
6. Emeršič, Ž., et al.: The unconstrained ear recognition challenge 2019-arxiv version with appendix. arXiv preprint arXiv:1903.04143 (2019)
7. Fischler, M.A., Bolles, R.C.: Random sample consensus: a paradigm for model fitting with applications to image analysis and automated cartography. Commun. ACM **24**(6), 381–395 (1981)
8. Hansley, E.E., Segundo, M.P., Sarkar, S.: Employing fusion of learned and hand-crafted features for unconstrained ear recognition. IET Biom. **7**(3), 215–223 (2018)
9. He, K., Zhang, X., Ren, S., Sun, J.: Deep residual learning for image recognition. In: IEEE Conference on Computer Vision and Pattern Recognition, pp. 770–778 (2016)

10. Howard, A.G., et al.: Mobilenets: efficient convolutional neural networks for mobile vision applications. arXiv preprint arXiv:1704.04861 (2017)
11. Oravec, M., et al.: Mobile ear recognition application, pp. 1–4 (2016)
12. Pflug, A., Busch, C.: Segmentation and normalization of human ears using cascaded pose regression. In: Bernsmed, K., Fischer-Hübner, S. (eds.) NordSec 2014. LNCS, vol. 8788, pp. 261–272. Springer, Cham (2014). https://doi.org/10.1007/978-3-319-11599-3_16
13. Ribič, M., Emeršič, Ž., Štruc, V., et al.: Influence of alignment on ear recognition: case study on awe dataset. In: International Electrotechnical and Computer Science Conference (2016)
14. Shu-zhong, W.: An improved normalization method for ear feature extraction. IJSIP **6**(5), 49–56 (2013)
15. Yazdanpanah, A.P., Faez, K.: Normalizing human ear in proportion to size and rotation. In: Huang, D.-S., Jo, K.-H., Lee, H.-H., Kang, H.-J., Bevilacqua, V. (eds.) ICIC 2009. LNCS, vol. 5754, pp. 37–45. Springer, Heidelberg (2009). https://doi.org/10.1007/978-3-642-04070-2_5
16. Yuan, L., Zhao, H., Zhang, Y., Wu, Z.: Ear alignment based on convolutional neural network. In: Zhou, J., et al. (eds.) CCBR 2018. LNCS, vol. 10996, pp. 562–571. Springer, Cham (2018). https://doi.org/10.1007/978-3-319-97909-0_60
17. Zhou, Y., Zaferiou, S.: Deformable models of ears in-the-wild for alignment and recognition. In: 12th IEEE International Conference on Automatic Face & Gesture Recognition, pp. 626–633 (2017)

Breast Lesion Discrimination Using Saliency Features from MRI Sequences and MKL-Based Classification

Henry Jhoán Areiza-Laverde[1], Carlos Andrés Duarte-Salazar[1],
Liliana Hernández[2], Andrés Eduardo Castro-Ospina[1(✉)], and Gloria M. Díaz[1]

[1] Instituto Tecnológico Metropolitano, Medellín, Colombia
{henryareiza135582,carlosduarte128074}@correo.itm.edu.co,
{andrescastro,gloriadiaz}@itm.edu.co
[2] Instituto de Alta Tecnología Médica, Medellín, Colombia
maria.hernandez@iatm.com

Abstract. Breast MRI interpretation requires that radiologists examine several images, depending on the acquisition protocol that is managed in the health institution, a very subjective and time-consuming process that reports large variability, which affects the final diagnosis and prognosis of the patient. In this paper, we present a computational method for classifying lesions detected in breast MRI studies, which aims to reduce physician subjectivity. The proposed approach take advantage of the ability of the Multiple Kernel Learning (MKL) strategy for optimally fusing the features extracted from the different image sequences that compose a breast MRI Study, which describe the grey level distribution of the original image and its saliency map, computed using the Graph-based Visual Saliency (GBVS) algorithm. Breast lesions were classified as positive and negative findings with an accuracy of 85.5% and 84.8% when nine and five sequences were used, respectively.

Keywords: Breast MRI · GBVS · Machine learning · MKL · Visual saliency

1 Introduction

Magnetic resonance imaging (MRI) has proven to be the most sensitive method for breast cancer detection, it has allowed the location of the smaller tumors even in dense breast tissues, which are usually not completely visible in other image modalities such as mammography and ultrasound [6]. However, breast MRI presents a high cost due to the high time it takes to acquire and to interpret the different sequences that make up this study [14], since a typical MRI study consists of around 2000–2500 images. Additionally, a varied variability has

Supported by Colciencias, Instituto Tecnológico Metropolitano, and Instituto de Alta Tecnología Médica. Project RC740-2017.

I. Nyström et al. (Eds.): CIARP 2019, LNCS 11896, pp. 294–305, 2019.
https://doi.org/10.1007/978-3-030-33904-3_27

been reported, especially, in novice radiologists, when differentiating between positive (probably cancer) and negative or benign findings. This variability can seriously affect the patient outcome because a false positive entails an unnecessary biopsy but a false negative prevents early detection of the tumor. Thus, computer-assisted interpretation can potentially reduce the radiologist workload by automating some of the diagnostic tasks, such as lesion detection or classification.

Multimodality involves the simultaneous using of multiple information sources to solve one specific problem [10], for this reason, the lesion detection in breast MRI studies can be taken as a multimodal task, understanding that each MRI sequence is used as one independent information source when the radiologists determine a diagnosis. Recently, a novel method called Multiple Kernel Learning (MKL) [12] has increased the interest in incorporating information from multiples sources over machine learning techniques, by the association of a similarity measure (Kernel) to each information source before its inclusion in the learning task; thus, this strategy allows to take the maximum advantage of the information provided by each source and generates models and results that are easy to interpret [4]. Although MKL has been used over multiple machine learning techniques, it has shown to be special powered when it is implemented over the Support Vector Machines (SVM) for classification tasks [12,20].

In this paper, MKL is used for developing a classifier able to distinguish between positive and negative findings in regions of interest (ROI) from a MRI study with several image sequences. As it is expected, the overall process is composed of two main stages: feature extraction and classification model. In the feature extraction stage, we decided to explore the visual salience analysis since has recently aroused interest in the research area related to automatic identification of ROI in MRI [5,11]. In this analysis, a typical salient region is defined as rare in an image and with high discriminative information, which could be associated to diagnostic findings in medical images [18]. In [16] was tested three popular computational models of salience (Itti-Koch, GBVS and Spectral Residual) to detect abnormalities in chest radiography images and in color retina images, in which, the GBVS [13] presented the best performance for radiography images, likewise, it has proven to be one of the salience models of better prediction performance in eye fixation. For this reason, the GBVS model was herein used to obtain a salient image per sequence, from which, first-order statistical measures were computed, obtaining as many spaces of features as sequences there are. Once the feature extraction process is completed, a MKL model combines the information from the different sequences described by those feature spaces.

2   Materials and Methods

In this section, the techniques implemented for feature generation and classification are presented, detailed specifications of the methods used for generating descriptive features from the MRI sequences are provided for establishing a well

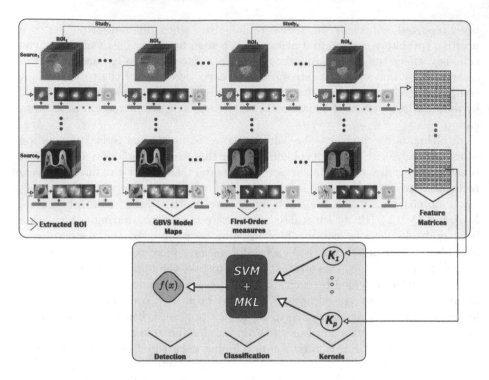

Fig. 1. General scheme of the proposed breast lesion classification method.

understanding of their implementation in the classification process, on which one optimization technique was used with the purpose of obtaining confident and accurate results.

The proposed method for the classification of breast lesions, depicted in Fig. 1, is composed of two main phases. The first one consists of the manipulation of breast MRI to extract the ROI corresponding to the findings selected by the specialists (Radiologists), with the aim of characterizing them by a perceptual relevance model (visual attention) and first-order statistical measures. In the second one, the model of MKL is defined with an SVM, whose aim is identifying between positive (probably malignant) and negative (probably benignant) regions from the feature spaces generated by the first phase.

2.1   Dataset

The dataset of breast MRI studies used in this work was retrospectively obtained from Instituto de Alta Tecnología Médica (IATM) of Medellín, Colombia. This dataset is composed of 189 regions containing 152 breast lesions, 2 undefined regions, and 35 normal tissues, which where extracted from 92 fully anonymized studies. Each study is made up by 16 image sequences *(Axial T1, Axial T2, Stir Coronal, ADC maps, Axial Diffusion B800, 6 Axial Dynamic Contrast Enhanced images, and 5 subtracted images).* The ROI were marked in the most visible sequence by two experimented radiologists, without considering any clinical data. This marking was used to triangulate the position of the same region in the others sequences with the help of Horos software (https://horosproject.org/) for enclosing these in a rectangular area. Correspondingly, the regions were marked in all the sequences of the 92 studies. In the course of this process, the information generated by the radiologists of each ROI was stored, i.e. the BI-RADS category, the type of finding, the sequence in which the finding was initially pointed out, among others. Overall, there were 3024 annotated regions, 560 regions of normal tissue, 36 undefined regions, and 2432 lesions that were classified as probably malignant (1424) or probably benign (1008).

2.2   Extraction of Regions

The ROI extraction process was based on the generation of a CSV (comma-separated values) file by each sequence with the help of Horos, in which spatial information of each marked ROI was stored, such as the coordinates of its four corners, the number of image slice on the volume, the area of the region in mm, among others. The spatial coordinates in terms of pixels and the slice number were used to locate the ROI in the other unmarked sequences. Due differences between width and height of most regions, the rectangular area was regularized to a completely square region by taking the highest value between the height and the width as the new length of the square. A range was established according to the normal distribution of the width values of the sizes of all the regions, as $\mu \pm 2\sigma$, where μ is the mean and σ is the standard deviation. This range was defined to include a little more than 80% of the total regions, to which we add the value of the standard deviation on each side of the already square region, to extend these areas and include a part of the tissue surrounding the finding. Finally, when we have the size of all regions defined, the crop is performed on the image corresponding to the slice of the volume.

2.3   Saliency Detection

The GBVS model was used to determine the saliency level of each pixel in the ROI. The purpose of this task is to emulate an important aspect of the interpretation process, in which, the hypo or hyper intensities of the lesions, with respect to the surrounding tissue, are considered as relevant to differentiate between positive or negative findings [22]. Thus, the computation of the GBVS salient map allows objectively to measure the findings made by radiologists when is considering their perceptual process.

GBVS Model. This model initially works with extracted feature maps at multiple spatial scales such as intensity (I), color (C), and orientation (O) or movement (M). Then, a Gaussian pyramid transform of scale space is derived from each feature, and a completely connected graph is generated over all the grid locations of each feature map, where the weights of two nodes are assigned proportionally to the similarity of the values of the features and their spatial distance. Further, the dissimilarity or difference between two positions (i, j) and (p, q) in the feature map, with their respective value $M(i, j)$ and $M(p, q)$, is defined as:

$$d((i, j) \vee (p, q)) = \left| log \frac{M(i, j)}{M(p, q)} \right| \tag{1}$$

The directed edge from node (i, j) to the node (p, q) is then assigned a weight proportional to its dissimilarity and its distance in the lattice M [7].

$$w((i, j), (p, q)) = d((i, j) \vee (p, q)) \ F(i - p, j - q) \tag{2}$$

$$with \ F(a, b) = \exp\left(\frac{a^2 + b^2}{2\sigma^2}\right) \tag{3}$$

The resulting graphs are treated as Markov chains to normalize the weights of the outbound edges of each node in one and defining an equivalence relation betwee benignantn nodes and states, as well as between the edge weights and the transition probabilities. Its equilibrium distribution is adopted as the activation and saliency maps, considering that in this distributions to the nodes that are very different from the surrounding nodes will be assigned large values. Finally, the activation maps are normalized to emphasize the conspicuous details and then combined into a single salient general map [7].

After an experimental process, we identified that the features maps of I, O and M favored the adjustment of levels of saliency with the regions marked by the radiologists. It is important to consider that on MRI images is not possible to generate the color feature map, therefore, we use the M map, since it allows to predict the fixation of the human eyes, as expressed in [15]. An example of the feature maps that make up the resulting saliency map (SaM) is shown in Fig. 2.

2.4   Feature Extraction

The characterization process was carried out with the first-order statistical measures, because these are basic features that allow us to describe in a simple way the visual information of regions that have greater visibility both on the saliency maps generated by the GBVS model and on the original extracted regions. The first-order statistical measures are defined as:

$$m_1 = E[I^1] = \sum_{I=0}^{N_g - 1} I^1 P(I) \tag{4}$$

$$\mu_k = E\left[(I - E[I])^k\right] = \sum_{I=0}^{N_g - 1} (I - m_1)^k P(I), \ with \ k = 2, 3, 4 \tag{5}$$

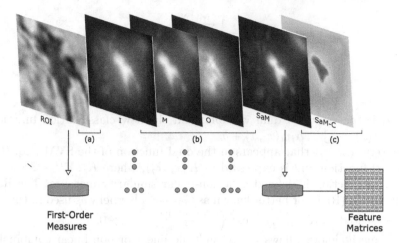

Fig. 2. Feature extraction from the saliency Map. (a) Extracted Region, (b) Feature maps grouping used to generate the saliency Map. (c) SaM-C is a color representation of the SaM.

where, $P(I)$ is the histogram, N_g is the number of possible gray levels, m_1 is the mean and the central moments given by μ_k. With μ_2 representing the variance, which is the most common central moment, and indicates the variability of the data with respect to the mean, μ_3 is the asymmetry or (skewness) that allows establishes the degree of symmetry of the histogram with respect to the average, and the fourth moment μ_4 is the Kurtosis, which indicates the degree of concentration that the data have around the mean in the histogram [1].

2.5   Classification Method

Support Vector Machines are categorized as wide margin classifiers proposed for binary classification problems [9], for this reason are ideal to be applied in the breast cancer diagnosis when only want to differentiate between positive and negative findings. Given a dataset with N training samples $\{(x_i, y_i)\}_{i=1}^{N}$ where x_i is a D-dimensional input vector and $y \in \{-1, +1\}$ is the labels vector of dimension N, the SVM finds the discriminative line with maximal margin M that better separates the samples in the feature space.

The classification function of the SVM is given by $f(x) = \langle w, x \rangle + b$, where w is the weights vector representing the coefficients of each sample x_i, b is the hyperplane separation bias term and the $\langle \cdot, \cdot \rangle$ operator represents the dot product between two vectors. The primal optimization problem of the SVM is given by $w^* = \min_{w} \ \frac{1}{2}\|w\|_2^2 + C \sum_{i=1}^{N} \xi_i$ and must fulfill that $y_i(\langle w, x \rangle + b) \geq 1 - \xi_i$, where C is the regularization parameter and ξ is the slack variables vector that create the *soft margin representation* of the SVM. This quadratic optimization problem with restrictions is solved through Lagrange Multipliers obtaining the expression in Eq. (6), namely *dual function* of the SVM.

$$\max_{0 \leq \alpha \leq C} \sum_{i=1}^{N} \alpha_i - \frac{1}{2} \sum_{i=1}^{N} \sum_{j=1}^{N} \alpha_i \alpha_j y_i y_j \langle x_i, x_j \rangle$$

$$s.t. \quad \sum_{i=1}^{N} \alpha_i y_i = 0$$

(6)

where, α is the vector of dual variables and the new classification function is given by $f(x) = \sum_{i=1}^{N} \alpha_i y_i \langle x_i, x \rangle + b$.

The term $\langle x_i, x_j \rangle$ that appears in the dual function of the SVM (Eq. (6)) is the Kernel function and is expressed as $K(x_i, x_j)$, where $K : \mathbb{R}^D \times \mathbb{R}^D \longrightarrow \mathbb{R}$. The Kernel function is defined as a non-linear similarity measure. The Radial Basis Function (RBF) or better known as Gaussian Kernel was used in this work and is presented as $K(x_i, x_j) = \exp\left(\frac{-\|x_i - x_j\|_2^2}{\sigma^2}\right)$, $\sigma > 0$.

MKL methodology allows to use multiple linear or non linear combinations of Kernels instead of one single Kernel [12]. MKL proposes the use of one combination function composed by P independent feature groups that can be provided by different sources, even varying the acquisition and composition nature or the number of features in each representation. In this work, each MRI sequence can be associated whit an individual Kernel for explode its potential, thus, the combination function is defined as in Eq. (7).

$$f_\eta = K_\eta(x_i, x_j) = \sum_{m=1}^{P} \eta_m K_m(x_i^m, x_j^m)$$

(7)

where η_m represents the weight assigned to each Kernel function K_m.

It is possible to apply a penalization to each Kernel weight, The most common penalization methods used in this type of problems are the ℓ_1-norm and ℓ_2-norm [12]. An efficient optimization strategy was proposed for update the kernel weights when arbitrary ℓ_p-norms with $p \geq 1$ are applied [17,23], this optimization method solves an SVM in each iteration and update the kernel weights using the Eq. (8).

$$\eta_m = \frac{\|w_m\|_2^{\frac{2}{p+1}}}{\left(\sum_{h=1}^{P} \|w_h\|_2^{\frac{2p}{p+1}}\right)^{\frac{1}{p}}}$$

(8)

where $\|w_m\|_2^2 = \eta_m^2 \sum_{i=1}^{N} \sum_{j=1}^{N} \alpha_i \alpha_j y_i y_j K_m\left(x_i^m, x_j^m\right)$ from the dual function of the SVM.

2.6   Parameters Optimization

Optimization algorithms are fundamental in machine learning tasks because they always seek that the results obtained by tasks such as classification, can be improved respect to objective evaluation measures automatically by the selection of the parameters that determine the performance of the algorithms [21].

In this work, were optimized the parameters that determine the behavior of the Kernels associated with the information sources (MRI sequences), i.e. the different parameters σ that determine the bandwidth of the Gaussian Kernels and the regularization parameter C of the SVM. For doing so, we use the Particle Swarm Optimization (PSO) algorithm, which is a metaheuristic algorithm that uses cooperative and stochastic methods to find the optimum working point of the function to be optimized, in this case, the performance measure of the SVM. In addition, the PSO algorithm is generally able to find a global optimum, being less susceptible than other algorithms to fall in a local optima [8].

3   Experiments and Results

After applying the feature extraction process described in Sect. 2.4, two different dataset configurations were used to evaluate the performance of the proposed method, the first configuration was determined by using 9 MRI sequences corresponding to T1, T2, ADC, Diffusion and the subtractions from 1 to 5, this dataset configuration intends to take advantage of all the available information sources. DCE images were not used due that substracted images correspond to the DCE post-processed. In the second dataset only five MRI sequences corresponding to T1, T2, ADC, Diffusion and the second subtraction, were used. This dataset configuration was determinated to reduce the number of information sources used to solve the classification problem, in a similar way to abbreviated protocols proposed in the clinical practice [22]. Additionally, for each dataset configuration, three tests were performed, one of them using the first-order measures from the original regions, another one using the first-order measures from the saliency maps and the last one using the concatenated measures from original regions and saliency maps. During all the tests, a K-Folds cross-validation technique was applied over the classification algorithm with $K = 10$, aiming to avoid biased or overfitted results.

The classification task was developed by SVM, one Gaussian Kernel was assigned to each information source and all the corresponding Kernels were incorporated to the SVM using the MKL method, both with the ℓ_1-norm as with the ℓ_2-norm to take advantage of each independent information source in a grouped way. The PSO algorithm was used to auto-tuning the Kernel parameters (σ) and the regularization parameter of the SVM (C), 40 particles were used and the algorithm was carried out until the convergence in each probe, around of 50 iterations were taken by the algorithm to converge at each test.

The results obtained when the method was applied over the dataset configuration conformed by 9 information sources are presented in Table 1, five performance measures were computed aiming to evaluate in detail the method performance. In this case the penalization by ℓ_1-norm presents the better performance when only the features from saliency maps are used and the penalization by ℓ_2-norm when the saliency maps features are concatenated with the original region features. Table 2 shows the results obtained over the dataset configuration that implemented five information sources, in this case a clear outperforming of the method is exhibited when the concatenated features from original regions and saliency maps are used to solve the classification problem and penalizing the Kernel wieghts with ℓ_2-norm. Additionally, this second configuration shows less variability than the first one, although both result groups show a similar overall performance respect to maximum performance results.

Table 1. Results over the 9-sources dataset configuration

	ℓ_1-Norm			ℓ_2-Norm		
	Region	Saliency	Region + Saliency	Region	Saliency	Region + Saliency
Accuracy	83.49 ± 8.5	$\mathbf{85.41 \pm 12.6}$	84.82 ± 9.5	82.90 ± 10.9	84.16 ± 11.0	84.82 ± 7.1
F1-Score	86.21 ± 6.9	87.19 ± 12.8	87.06 ± 9.3	85.66 ± 9.3	86.02 ± 11.7	$\mathbf{87.35 \pm 6.4}$
Geo-Mean	82.24 ± 9.8	$\mathbf{84.01 \pm 12.4}$	82.32 ± 10.3	81.72 ± 11.9	82.35 ± 11.4	82.62 ± 7.9
Sensitivity	87.61 ± 6.3	89.56 ± 15.8	90.83 ± 14.9	86.64 ± 10.2	88.44 ± 16.2	$\mathbf{90.97 \pm 11.5}$
Specificity	77.62 ± 14.0	$\mathbf{79.52 \pm 14.1}$	76.43 ± 16.9	77.62 ± 15.6	77.86 ± 13.6	76.19 ± 13.8

Table 2. Results over the 5-sources dataset configuration

	ℓ_1-Norm			ℓ_2-Norm		
	Region	Saliency	Region + Saliency	Region	Saliency	Region + Saliency
Accuracy	80.31 ± 11.6	81.57 ± 9.3	83.73 ± 9.4	80.39 ± 11.3	82.90 ± 8.4	$\mathbf{85.57 \pm 8.1}$
F1-Score	83.10 ± 10.7	84.01 ± 9.2	86.22 ± 7.9	83.85 ± 9.0	85.56 ± 8.2	$\mathbf{87.52 \pm 7.5}$
Geo-Mean	79.15 ± 11.8	80.29 ± 9.1	82.01 ± 11.3	78.36 ± 13.2	80.91 ± 9.3	$\mathbf{84.41 \pm 8.7}$
Sensitivity	84.42 ± 14.1	85.36 ± 13.5	87.61 ± 11.4	85.39 ± 9.1	88.58 ± 11.5	$\mathbf{88.61 \pm 12.0}$
Specificity	75.00 ± 14.1	76.67 ± 14.0	78.57 ± 19.7	73.33 ± 21.3	75.00 ± 14.7	$\mathbf{81.67 \pm 14.6}$

It is important to stand out how both the ℓ_1-norm and ℓ_2-norm presented good performances in the tests, showing a slight outperforming with the ℓ_2-norm specially in the Table 2, this behavior may be due to the fact that the ℓ_1-norm aims to generate sparsity between the kernel weights ($\eta_m = 0$), which can be interpreted as the elimination of some kernels, while the ℓ_2-norm aims to assign a value to the kernel weights according to their relevance but without strictly eliminate the kernel, taking full advantage of each kernel even when they have few relevance to the classification task.

Considering some other works that boarded similar problems to the approach of this paper, obtained results show a equivalent performance respect state of the art. Most of the works that implement or propose methods based on visual attention as the saliency analysis are focused on the detection of the ROI into the whole image, which is more similar to the problem of segmentation than classification [19]. On the other hand, the specific classification problem based on the use of saliency information is boarded in a few papers; in [2] an algorithm was applied to a benign-versus-malignant image classification task in MRI images of brain; they randomly divide the dataset of images into a training set, containing 70 benign and 70 premalignant images, and a testing set, containing 30 benign and 30 premalignant images, after that they train an SVM model to classify images from the testing set obtaining a median over 20 runs classification accuracy rate of 80%. Other related work is presented in [3], they propose a hybrid mass detection algorithm that combines unsupervised candidate detection with deep learning-based classification, the detection stage identifies image-salient regions and a convolutional neural network (CNN) is used to classify the detected candidates into true-positive and false-positive masses, the dataset used was composed by breast cancer MRI studies from 171 patients, with 1957 annotated slices of malignant and benign masses and the highest classification accuracy obtained was 0.86 ± 0.02.

4 Conclusions and Future Work

A computational method for classifying lesions detected in breast MRI studies was presented, which takes advantage of MKL and regularization strategy to fuse multimodal sources of features as MRI sequences. An optimization strategy was also performed for SVM and Kernel parameter tuning. First-order measurements from both original and salient images were evaluated as feature descriptors of the image sequences, achieving accuracies over 80% with a 10-Fold cross-validation, in all cases. However, results showed that the use of saliency-based features improves the classification performance respect to use only original based features. Additionally, two different dataset configurations were considered, one containing all the nine sequences that make up the Breast MRI study and the other one using only a subset of them (five sequences). The results have shown that the use of a subset of sequences does not affect significantly the overall performance, which could be useful in the definition of less expensive breast MRI studies. It is important to note that regions were processed without requiring prior manual or automatic segmentation stage. Even more, the use of the saliency distribution is an attempt by capturing the relationship between the lesion and the contextual tissue, which is also relevant in the diagnosis process.

As future work, the use of MKL will be evaluated for fusing more specific and high-level features, such as quantitative image measurements, texture-based features, among others. Additionally, the performance of the method for predicting the final malignancy of a lesion will be also evaluated in a prospective study that will include biopsy-proven lesions.

References

1. Aggarwal, N., Agrawal, R.K.: First and second order statistics features for classification of magnetic resonance brain images. J. Signal Inf. Process. **03**(02), 146–153 (2012). https://doi.org/10.4236/jsip.2012.32019

2. Alpert, S., Kisilev, P.: Unsupervised detection of abnormalities in medical images using salient features. In: Medical Imaging 2014: Image Processing, vol. 9034, p. 903416. International Society for Optics and Photonics (2014)
3. Amit, G., et al.: Hybrid mass detection in breast MRI combining unsupervised saliency analysis and deep learning. In: Descoteaux, M., Maier-Hein, L., Franz, A., Jannin, P., Collins, D.L., Duchesne, S. (eds.) MICCAI 2017. LNCS, vol. 10435, pp. 594–602. Springer, Cham (2017). https://doi.org/10.1007/978-3-319-66179-7_68
4. Areiza-Laverde, H.J., Díaz, G.M., Castro-Ospina, A.E.: Feature group selection using MKL penalized with ℓ_1-norm and SVM as base learner. In: Figueroa-García, J.C., López-Santana, E.R., Rodriguez-Molano, J.I. (eds.) WEA 2018. CCIS, vol. 915, pp. 136–147. Springer, Cham (2018). https://doi.org/10.1007/978-3-030-00350-0_12
5. Banerjee, S., Mitra, S., Shankar, B.U., Hayashi, Y.: A novel GBM saliency detection model using multi-channel MRI. PLoS ONE **11**(1), 1–16 (2016). https://doi.org/10.1371/journal.pone.0146388
6. Bickelhaupt, S., et al.: Fast and noninvasive characterization of suspicious lesions detected at breast cancer X-Ray screening: capability of diffusion-weighted MR imaging with MIPs. Radiology **278**(3), 689–697 (2016). https://doi.org/10.1148/radiol.2015150425
7. Borji, A., Itti, L.: State-of-the-art in visual attention modeling. IEEE Trans. Pattern Anal. Mach. Intell. (TPAMI) **35**(1), 185–207 (2013). https://doi.org/10.1109/TPAMI.2012.89
8. Clerc, M.: Standard particle swarm optimisation (2012)
9. Cristianini, N., Shawe-Taylor, J.: An Introduction to Support Vector Machines and Other Kernel-Based Learning Methods. Cambridge University Press, Cambridge (2000)
10. Culache, O., Obadǎ, D.R.: Multimodality as a premise for inducing online flow on a brand website: a social semiotic approach. Procedia - Soc. Behav. Sci. **149**, 261–268 (2014). https://doi.org/10.1016/j.sbspro.2014.08.227
11. Erihov, M., Alpert, S., Kisilev, P., Hashoul, S.: A cross saliency approach to asymmetry-based tumor detection. In: MICCAI 2015: 18th International Conference on Medical Image Computing and Computer Assisted Intervention, vol. 9351, pp. 636–643 (2015). https://doi.org/10.1007/978-3-319-24574-4
12. Gönen, M., Alpaydın, E.: Multiple kernel learning algorithms. J. Mach. Learn. Res. **12**(8), 2211–2268 (2011)
13. Harel, J., Koch, C., Perona, P.: Graph-based visual saliency. In: Advances in Neural Information Processing Systems, pp. 545–552 (2007)
14. Heller, S.L., Moy, L.: Breast MRI screening: benefits and limitations. Current Breast Cancer Rep. **8**(4), 248–257 (2016). https://doi.org/10.1007/s12609-016-0230-7
15. Itti, L., Dhavale, N., Pighin, F.: Realistic avatar eye and head animation using a neurobiological model of visual attention. In: Applications and Science of Neural Networks, Fuzzy Systems, and Evolutionary Computation VI, vol. 5200, pp. 64–79. International Society for Optics and Photonics (2003)
16. Jampani, V., Sivaswamy, J., Vaidya, V.: Assessment of computational visual attention models on medical images. In: Proceedings of the Eighth Indian Conference on Computer Vision, Graphics and Image Processing - ICVGIP 2012, pp. 1–8 (2012). https://doi.org/10.1145/2425333.2425413
17. Kloft, M., Brefeld, U., Sonnenburg, S., Zien, A.: Non-sparse regularization and efficient training with multiple kernels. arXiv preprint arXiv:1003.0079, vol. 186, pp. 189–190 (2010)

18. Mitra, S., Banerjee, S., Hayashi, Y.: Volumetric brain tumour detection from MRI using visual saliency. PLoS ONE **12**(11), 1–14 (2017). https://doi.org/10.1371/journal.pone.0187209
19. Mitra, S., Banerjee, S., Hayashi, Y.: Volumetric brain tumour detection from MRI using visual saliency. PloS ONE **12**(11), e0187209 (2017)
20. Narváez, F., Díaz, G., Poveda, C., Romero, E.: An automatic BI-RADS description of mammographic masses by fusing multiresolution features. Expert Syst. Appl. **74**, 82–95 (2017)
21. Papadimitriou, C.H., Steiglitz, K.: Combinatorial Optimization: Algorithms and Complexity. Courier Corporation, North Chelmsford (1998)
22. Strahle, D.A., Pathak, D.R., Sierra, A., Saha, S., Strahle, C., Devisetty, K.: Systematic development of an abbreviated protocol for screening breast magnetic resonance imaging. Breast Cancer Res. Treat. **162**(2), 283–295 (2017). https://doi.org/10.1007/s10549-017-4112-0
23. Xu, Z., Jin, R., Yang, H., King, I., Lyu, M.R.: Simple and efficient multiple kernel learning by group lasso. In: Proceedings of the 27th International Conference on Machine Learning (ICML 2010), pp. 1175–1182. Citeseer (2010)

Role of EMG Rectification for Corticomuscular and Intermuscular Coherence Estimation of Spinocerebellar Ataxia Type 2 (SCA2)

Y. Ruiz-Gonzalez[1]($\boxtimes$), L. Velázquez-Pérez[2], R. Rodríguez-Labrada[3], R. Torres-Vega[3], and U. Ziemann[4]

[1] Informatics Research Centre, Universidad Central "Marta Abreu" de Las Villas, Santa Clara, Villa Clara, Cuba
yuselyr@uclv.edu.cu
[2] Cuban Academy of Sciences, Havana, Cuba
[3] Department Clinical Neurophysiology, Centre for the Research and Rehabilitation of Hereditary Ataxias, Holguin, Cuba
[4] Department Neurology and Stroke, and Hertie Institute for Clinical Brain Research, University Tübingen, Tübingen, Germany

Abstract. Corticomuscular and intermuscular coherence are established methods to study connectivity between activity of neurons in sensorimotor cortex measured with electroencephalography (EEG) and muscle measured with electromyography (EMG), or between muscles, in a variety of neurological conditions. However, there is a debate on the importance of EMG signal rectification before coherence estimation. This paper studies the effects of EMG rectification in corticomuscular and intermuscular coherence estimation from SCA2 patients and prodromal SCA2 gene mutation carriers in comparison to healthy controls. EEG and EMG were recorded from 20 SCA2 patients, 16 prodromal SCA2 gene mutation carriers and 26 healthy control subjects during a motor task in upper or lower limbs. Coherence estimations were carried out using the non-rectified raw EMG signal vs. the rectified EMG signal. The results showed that EMG rectification impairs the level of significance of the differences in corticomuscular and intermuscular coherence between SCA2 patients and prodromal SCA2 gene mutation carriers vs. healthy controls in the beta-band, and also results in overall lower coherence values.

Keywords: Corticomuscular coherence · Intermuscular coherence · EMG · Rectification · Raw EMG

1 Introduction

Spinocerebellar ataxia type 2 (SCA2) is caused by an abnormal expansion of the CAG trinucleotide in the ataxin-2 gen [1, 2]. This disease includes a wide range of clinical manifestation like cerebellar syndrome, saccadic slowing, cognitive decline, sensory neuropathy and corticospinal tract damage [3–9]. Signs of corticospinal tract dysfunction had been found on clinical (ex. hyperreflexia and spasticity), anatomic (ex. degeneration of the motor cortex) and electrophysiology studies of SCA2 patients and

© Springer Nature Switzerland AG 2019
I. Nyström et al. (Eds.): CIARP 2019, LNCS 11896, pp. 306–315, 2019.
https://doi.org/10.1007/978-3-030-33904-3_28

prodromal SCA2 gene mutation carriers [7, 10–12]. Electrophysiological studies had demonstrated the increase of resting motor thresholds (RMT) and central motor conduction time (CMCT) by using transcranial magnetic stimulation (TMS) [13, 14]. Corticomuscular coherence (CMC) and intermuscular coherence (IMC) estimation are well-established methods to study connectivity between activity of neurons in sensorimotor cortex (EEG or MEG) and muscle (EMG) or between muscles [15–18]. These measures have been reported in a variety of neurological condition such as stroke [17], Parkinson disease [19], sleep behavior disorder [20] and recently in SCA2 [10, 21] as a marker of corticospinal tract dysfunction.

However, there has been a discussion about the processing steps to estimate the CMC and IMC [22–25]. The discussion mainly focused on the appropriateness of EMG signal rectification prior to coherence estimation. The main concern is about CMC estimation being a frequency domain method used to demonstrate a linear coupling between rhythmic activity from sensorimotor cortex and the activity of voluntarily contracting muscles [26, 27], but rectification is a non-linear operator that changes the frequency components of the signal to which it is applied [24]. Theoretical, simulation and experimental studies had been conducted to study the influence of the rectification on the EMG signal spectrum to understand its effect on neural connectivity [28]. Some authors claimed that the rectification is a necessary step [29–32], while others indicated that rectification alters characterization of oscillatory input to muscle, consequently affecting identification of corticomuscular coherence [33, 34].

In [10, 21], estimation of CMC and IMC was performed to study neurodegeneration of the corticospinal tract in SCA2 patients and prodromal SCA2 gene mutation carriers. Findings revealed a significant CMC and IMC reduction in the beta-band in SCA2 patients and prodromal SCA2 gene mutation carriers in comparison to healthy controls. In the present paper we focus on CMC and IMC estimation, using the same data as in [10, 21] to study the effects of EMG rectification. Also, the previous CMC and IMC analyses are extended to the theta, alpha and gamma frequency bands.

2 Materials and Methods

2.1 Data Description

The study was approved by the institutional ethics committee and conducted according to the declaration of Helsinki. The experimental design and inclusion criteria have been reported elsewhere [10, 21]. Written informed consent was obtained from all subjects prior to participation. Twenty SCA2 patients (mean age (±SD) 45.2 ± 9.8 years, age range 32–64 years) and sixteen prodromal SCA2 gene mutation carriers (mean age (±SD) 41.8 ± 11.8 years; range 26–72 years) were admitted to the Center for Research and Rehabilitation of Hereditary Ataxias in Holguín to participate in this study. Twenty-six healthy non-paid volunteers from Holguin province (mean age (±SD) 43.8 ± 10.4 years, age range 20–66 years) served as age-matched control group.

All subjects performed a motor task as follows: Subjects were seated comfortably in a chair with the examined limbs placed in their natural resting positions for 1 min. Then, repeated simultaneous flexion movements of fingers and wrist, for the upper limb, or dorsal extension movements of the foot, for the lower limb, were executed over a period of 10 min. Onset and offset of each contraction were signaled by a computer-generated tone. The contraction cycles consisted of 3 s contraction and 2 s rest. The force level was kept at 30% of the maximum voluntary force, as determined by using a hand digital dynamometer (Smedley Hand dynamometer, China) for the upper limb. For the lower limb task, the force level was not measured, but previously reported quantitative EMG analysis showed that there was no difference in task-related EMG activity between SCA2 patients, prodromal SCA2 gene mutation carriers and healthy controls [10, 21].

2.2   Signals Recording

EMG and EEG recordings were performed using a Medicid 5 amplifier with Ag/AgCl surface electrodes, electrodes impedances were <5 kΩ throughout the experiment. Bipolar EMG signals were band-pass filtered (2 Hz–100 Hz), amplified (gain, 500–5000) and digitized at 1 kHz. EEG channels were referenced to the left ear lobe, amplified (gain, 1000), band-pass filtered (2 Hz–100 Hz) and digitized at 1 kHz.

EMG was recorded from the first dorsal interosseous (IOD), flexor digitorum superficialis (FDS) and extensor digitorum communis (EDC) muscles of the right upper limb, and from the extensor digitorum brevis (EDB), tibialis anterior (TA) and medial gastrocnemius (MG) muscles of the right lower limb. The EEG was recorded from the sensorimotor cortex of the left hemisphere. For the upper limb, electrode FC3 (placed 30 mm lateral and 20 mm anterior relative to the vertex) and CP3 (30 mm lateral and 20 mm posterior relative to the vertex) according to the International 10/20 system were used to recorded the EEG signal. For the lower limb, EEG electrodes were placed over vertex (Cz) and 20 mm anterior to vertex (Cz′). The electrode position criteria have been reported elsewhere [10, 21].

2.3   Coherence Estimation

The coherence of two signals was defined by Eq. 1

$$Coh_{xy} = \frac{|S_{xy}(f)|}{\sqrt{S_{xx}(f) * S_{yy(f)}}} \tag{1}$$

where Sxy is the cross spectral density between the x(t) and y(t) signals and Sxx and Syy are their respective auto-spectral density functions. The coherence value is bounded between 0 and 1 and provides a measure of the correlation of the time series of the two signals in the frequency domain [26]. The corticomuscular coherence was calculated between the EEG data from the FC3 and CP3 electrodes and the EMG data from the IOD, FDS and EDC electrodes for the upper limb, and between the Cz and Cz' EEG data and the EDB, TA and MG EMG data for the lower limb. Intermuscular coherence was calculated within-limb between the respective three EMG signals from the upper and lower limb. Two coherence estimations were carried out, using the raw EMG signal vs. the rectified EMG.

Data analysis and spectral estimation were performed using the FieldTrip Toolbox [35]. Signals were segmented as 2.6 s trials, defined as the central part of the 3 s contraction periods, in order to avoid unstable EMG during the onset and offset of the contractions [15]. Auto-spectra and cross-spectra were calculated using the multi-taper method based on Hanning tapers with a spectral smoothing of 3 Hz through multi-tapering [36, 37]. Spectral analyses were performed between 5 and 55 Hz, bounded by the EEG interest band and the effects of the line frequency in the coherence at 60 Hz. The spectra were averaged across all trials in a given participant. Statistical significance of differences in CMC and IMC between SCA2 patients vs. healthy controls and prodromal SCA2 gene mutation vs. healthy controls were tested using a non-parametric cluster-based permutation test [38].

3   Results and Discussion

3.1   Corticomuscular Coherence

Previous research [10, 21] reported a CMC increase in the beta-band in healthy controls which is not present in SCA2 patients or prodromal SCA2 gene mutation carriers. This finding has been considered as an early neurodegenerative affection of the corticospinal tract [10, 15, 17]. However, those differences disappear if the rectified EMG is used to estimate the spectral densities and coherence.

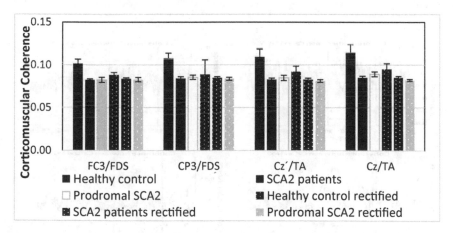

Fig. 1. Mean ± SD CMC in beta-band, differences between SCA2 patients and prodromal SCA2 carriers were significant for the non-rectified raw EMG(p < 0.05), but not for the rectified EMG.

Figure 1 shows the mean CMC values in the beta-band (15–30 Hz for the EEG-EMG channels FC3-FDS, CP3-FDS, Cz'-TA and Cz-TA [10, 21] using the non-rectified raw EMG vs. rectified EMG data. In the case of the rectified EMG the differences were not significant (p > 0.05). The significant differences of CMC to upper and lower limb in

SCA2 patients and prodromal SCA2 gene mutation carriers compared to healthy controls were vanish if the rectified EMG is used (see also Figs. 2 and 3).

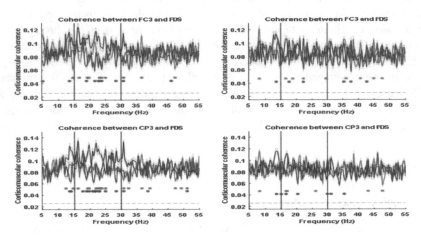

Fig. 2. Upper limb CMC group averages (solid lines) and standard error of the mean (shades areas), left: non-rectified raw EMG, right: rectified EMG, red: SCA2 patients, blue: prodromal SCA2 carriers, black: healthy controls. Pink dashed lines represent the 95% confidence limits of significant CMC. * denote significant differences at single frequencies binned at 0.30 Hz, red: SCA2 patients vs. healthy controls, blue: prodromal SCA2 carriers vs. healthy controls (Color figure online).

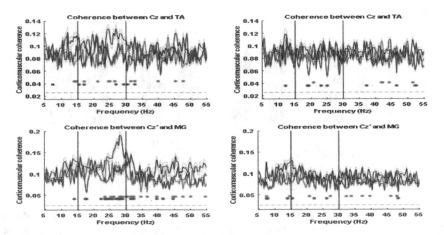

Fig. 3. Lower limb CMC group averages (solid lines) and standard error of the mean (shades areas), left: non-rectified raw EMG, right: rectified EMG, red: SCA2 patients, blue: prodromal SCA2 carriers, black: healthy controls. Pink dashed lines represent the 95% confidence limits of significant CMC. * denote significant differences at single frequencies binned at 0.30 Hz, red: SCA2 patients vs. healthy controls, blue: prodromal SCA2 carriers vs. healthy controls. (Color figure online)

In other frequency ranges, i.e., the theta (5–7 Hz) and alpha bands (8–12 Hz), and the low gamma band (30–55 Hz), significant differences were not previously reported between SCA2 patients or prodromal SCA2 gene carriers and healthy controls [10, 21]. EMG rectification had no impact on these nil findings.

3.2 Intermuscular Coherence

IMC analysis showed lower values with the rectified EMG data than the non-rectified raw EMG data. Figure 4 shows the mean IMC in the beta-band. Significant differences between SCA2 patients and healthy controls were found for the raw EMG data for the upper limb in EMG channels IOD-FDS, FDS-EDC, and for the lower limb in EMG channels EDB-TA, EDB-GNM. Between prodromal SCA2 gene mutation carriers and healthy controls, there was a significant difference only in the upper limb EMG channels IOD-FDS. Significant differences were found for the same EMG channel combinations for the rectified EMG data, but the p-values were less significant than for the non-rectified raw EMG data (see Table 1). There was an exception for IOD-FDS IMC where the difference was not significant for the rectified data.

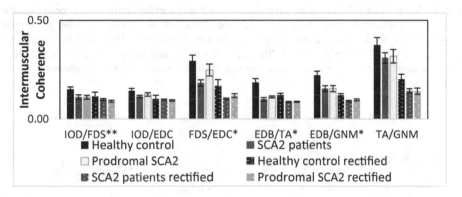

Fig. 4. Mean ± SD IMC in beta-band * significant differences between SCA2 patients vs. healthy controls, ** significant differences between SCA2 patients and prodromal SCA2 carriers vs. healthy controls.

Table 1. p-values of the IMC analyses, comparison of SCA2 patients vs. healthy controls

p-value	IOD-FDS	FDS-EDC	EDB-TA	EDB-GNM
Raw EMG	0.00356	0.0120	0.00223	0.01285
Rectified EMG	0.31966	0.0220	0.01285	0.03107

Figures 5 and 6 show the mean IMC of the three groups for the EMG channel combinations of the upper and lower limb as indicated above, plotted over the analyzed full frequency range. Qualitatively, rectification of the EMG did not produce any important change in the theta, alpha or gamma band coherence, but resulted in less significant differences compared to the analysis of the non-rectified EMG. In general the results do not support the use of EMG rectification as processing step to identify IMC differences between the studied groups.

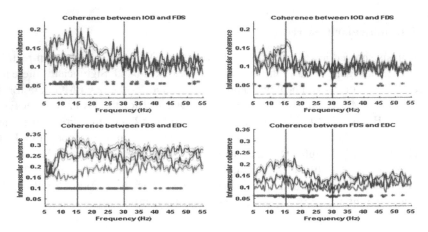

Fig. 5. Upper limb IMC group averages (solid lines) and standard error of the mean (shades areas), left: non-rectified raw EMG, right: rectified EMG, red: SCA2 patients, blue: prodromal SCA2 carriers, black: healthy controls. Pink dashed lines represent the 95% confidence limits of significant CMC. * denote significant differences at single frequencies binned at 0.30 Hz, red: SCA2 patients vs. healthy controls, blue: prodromal SCA2 carriers vs. healthy controls. (Color figure online)

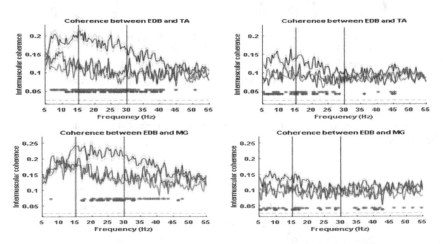

Fig. 6. Lower limb CMC group averages (solid lines) and standard error of the mean (shades areas), left: non-rectified raw EMG, right: rectified EMG, red: SCA2 patients, blue: prodromal SCA2 carriers, black: healthy controls. Pink dashed lines represent the 95% confidence limits of significant CMC. * denote significant differences at single frequencies binned at 0.30 Hz, red: SCA2 patients vs. healthy controls, blue: prodromal SCA2 carriers vs. healthy controls (Color figure online).

4 Discussion

The CMC results suggest that the disappearance of the statistically significant differences in the rectified EMG in the beta-band can largely be attributed to a CMC decrease in the healthy controls which is not present in the SCA2 patients and prodromal SCA2 gene mutations carriers (Fig. 1). The SCA2 patients and prodromal SCA2 gene mutations carriers already had a reduced value of CMC coherence that was not affected by the rectification process, perhaps due to the corticospinal tract damage caused by the disease [7, 14, 15, 39]. However in healthy controls, with an intact corticospinal tract, EMG rectification could have obscured genuine coherence because rectification may have eliminated frequencies that exist in raw signals or introduce power peaks at false frequencies [32, 34].

EMG rectification is a not linear procedure often used to augments the neural information of the EMG. This procedure can introduce power peaks at frequencies not present in the non-rectified raw EMG making possible overestimation of CMC coherence or impose significant CMC at inappropriate frequency bands, but the other way around is not possible [25, 30, 34]. When the rectified EMG is used, it becomes difficult to discern whether a change in CMC values are relates to a genuine physiological change or simply reflects the variable effect of rectification.

In [28, 30] the authors demonstrated that common oscillatory inputs are presents on rectified EMG and non-rectified raw EMG. Also, common oscillatory inputs may be stronger in rectified EMG only at low level of amplitude cancellation [23] and amplitude cancellation changes with contraction level, fatigue, noise level, or across subjects and muscles. Nevertheless non-rectified raw EMG spectrum is less influenced by amplitude cancellation, so the coherence estimation [28, 30] consequently, analysis with the non-rectified raw EMG will be able to detect higher IMC value irrespective of the amplitude cancellation effects.

5 Conclusions

In general, for our data, the rectification process resulted in lower a CMC and IMC value which impairs the significant differences in CMC and IMC between SCA2 patients and prodromal SCA2 gene mutation carriers vs. healthy controls. Based on former arguments and our data reported here, we recommend that future CMC and IMC studies will be carry out on non-rectified raw EMG data.

References

1. Giunti, P., et al.: The role of the SCA2 trinucleotide repeat expansion in 89 autosomal dominant cerebellar ataxia families. Frequency, clinical and genetic correlates. Brain J. Neurol. 121(Pt 3), 459–467 (1998)
2. Velázquez-Pérez, L., Rodríguez-Labrada, R., García-Rodríguez, J.C., Almaguer-Mederos, L. E., Cruz-Mariño, T., Laffita-Mesa, J.M.: A comprehensive review of spinocerebellar ataxia type 2 in cuba. Cerebellum 10(2), 184–198 (2011)

3. Velázquez-Pérez, L., et al.: Saccade velocity is reduced in presymptomatic spinocerebellar ataxia type 2. Clin. Neurophysiol. **120**(3), 632–635 (2009)
4. Velázquez Pérez, L., Rodríguez Labrada, R.: Manifestaciones tempranas de la Ataxia Espinocerebelosa tipo 2. Ediciones Holguin, Cuba (2012)
5. Rodríguez-Labrada, R., et al.: Saccadic latency is prolonged in spinocerebellar ataxia type 2 and correlates with the frontal-executive dysfunctions. J. Neurol. Sci. **306**(1–2), 103–107 (2011)
6. Rodríguez-Labrada, R., et al.: Subtle rapid eye movement sleep abnormalities in presymptomatic spinocerebellar ataxia type 2 gene carriers. Mov. Disord. **26**(2), 347–350 (2011)
7. Velázquez-Pérez, L., et al.: Abnormal corticospinal tract function and motor cortex excitability in non-ataxic SCA2 mutation carriers: a TMS study. Clin. Neurophysiol. **127**(8), 2713–2719 (2016)
8. Linnemann, C., et al.: Peripheral neuropathy in spinocerebellar ataxia type 1, 2, 3, and 6. Cerebellum **15**(2), 165–173 (2016)
9. Velázquez-Pérez, L., et al.: Sleep disorders in spinocerebellar ataxia type 2 patients. Neurodegener. Dis. **8**(6), 447–454 (2011)
10. Velázquez-Pérez, L., et al.: Early corticospinal tract damage in prodromal SCA2 revealed by EEG-EMG and EMG-EMG coherence. Clin. Neurophysiol. **128**(12), 2493–2502 (2017)
11. Velázquez-Pérez, L., et al.: Comprehensive study of early features in spinocerebellar ataxia 2: delineating the prodromal stage of the disease. Cerebellum **13**(5), 568–579 (2014)
12. Velázquez-Pérez, L., Rodríguez-Labrada, R., García-Rodríguez, J.C., Almaguer-Mederos, L. E., Cruz-Mariño, T., Laffita-Mesa, J.M.: A comprehensive review of spinocerebellar ataxia type 2 in Cuba. Cerebellum Lond. Engl. **10**(2), 184–198 (2011)
13. Velázquez-Pérez, L., et al.: Central motor conduction time as prodromal biomarker in spinocerebellar ataxia type 2. Mov. Disord. Off. J. Mov. Disord. Soc. **31**(4), 603–604 (2016)
14. Velázquez-Pérez, L., et al.: Progression of corticospinal tract dysfunction in pre-ataxic spinocerebellar ataxia type 2: a two-years follow-up TMS study. Clin. Neurophysiol. **129**(5), 895–900 (2018)
15. Fisher, K.M., Zaaimi, B., Williams, T.L., Baker, S.N., Baker, M.R.: Beta-band intermuscular coherence: a novel biomarker of upper motor neuron dysfunction in motor neuron disease. Brain **135**(9), 2849–2864 (2012)
16. Bowyer, S.M.: Coherence a measure of the brain networks: past and present. Neuropsychiatr. Electrophysiol. **2**(1), 1 (2016)
17. Mima, T., Toma, K., Koshy, B., Hallett, M.: Coherence between cortical and muscular activities after subcortical stroke. Stroke **32**(11), 2597–2601 (2001)
18. Baker, M.R., Baker, S.N.: The effect of diazepam on motor cortical oscillations and corticomuscular coherence studied in man. J. Physiol. **546**(3), 931–942 (2003)
19. Caviness, J.N., Shill, H.A., Sabbagh, M.N., Evidente, V.G.H., Hernandez, J.L., Adler, C.H.: Corticomuscular coherence is increased in the small postural tremor of parkinson's disease: postural tremor in PD. Mov. Disord. **21**(4), 492–499 (2006)
20. Jung, K.-Y., et al.: Increased corticomuscular coherence in idiopathic REM sleep behavior disorder. Front. Neurol. **3**, 60 (2012)
21. Velázquez-Pérez, L., et al.: Corticomuscular coherence: a novel tool to assess the pyramidal tract dysfunction in spinocerebellar ataxia type 2. Cerebellum **16**(2), 602–606 (2017)
22. Farina, D., Merletti, R., Enoka, R.M.: The extraction of neural strategies from the surface EMG. J. Appl. Physiol. **96**(4), 1486–1495 (2004)
23. Farina, D., Merletti, R., Enoka, R.M.: The extraction of neural strategies from the surface EMG: an update. J. Appl. Physiol. **117**(11), 1215–1230 (2014)

24. Myers, L.J., et al.: Rectification and non-linear pre-processing of EMG signals for cortico-muscular analysis. J. Neurosci. Methods 124(2), 157–165 (2003)
25. McClelland, V.M., Cvetkovic, Z., Mills, K.R.: Inconsistent effects of EMG rectification on coherence analysis. J. Physiol. 592(1), 249–250 (2014)
26. Rosenberg, J.R., Amjad, A.M., Breeze, P., Brillinger, D.R., Halliday, D.M.: The fourier approach to the identification of functional coupling between neuronal spike trains. Prog. Biophys. Mol. Biol. 53(1), 1–31 (1989)
27. Random Data: Analysis and Measurement Procedures, 4th edn. Wiley.com. https://www.wiley.com/en-us/Random+Data%3A+Analysis+and+Measurement+Procedures%2C+4th+Edition-p-9780470248775. Accessed 20 November 2018
28. Negro, F., Keenan, K., Farina, D.: Power spectrum of the rectified EMG: when and why is rectification beneficial for identifying neural connectivity? J. Neural Eng. 12(3), 036008 (2015)
29. Mima, T., Hallett, M.: Electroencephalographic analysis of cortico-muscular coherence: reference effect, volume conduction and generator mechanism. Clin. Neurophysiol. Off. J. Int. Fed. Clin. Neurophysiol. 110(11), 1892–1899 (1999)
30. Farina, D., Negro, F., Jiang, N.: Identification of common synaptic inputs to motor neurons from the rectified electromyogram. J. Physiol. 591(10), 2403–2418 (2013)
31. Halliday, D.M., Farmer, S.F.: On the need for rectification of surface EMG. J. Neurophysiol. 103(6), 3547 (2010)
32. Yao, B., Salenius, S., Yue, G.H., Brown, R.W., Liu, J.Z.: Effects of surface EMG rectification on power and coherence analyses: an EEG and MEG study. J. Neurosci. Methods 159(2), 215–223 (2007)
33. Neto, O.P., Christou, E.A.: Rectification of the EMG signal impairs the identification of oscillatory input to the muscle. J. Neurophysiol. 103(2), 1093–1103 (2010)
34. McClelland, V.M., Cvetkovic, Z., Mills, K.R.: Rectification of the EMG is an unnecessary and inappropriate step in the calculation of corticomuscular coherence. J. Neurosci. Methods 205(1), 190–201 (2012)
35. Oostenveld, R., Fries, P., Maris, E., Schoffelen, J.-M.: FieldTrip: open source software for advanced analysis of MEG, EEG, and invasive electrophysiological data. Comput. Intell. Neurosci. 2011, 1–9 (2011)
36. Thomson, D.J.: Spectrum Estimation and Harmonic Analysis. IEEE Proc. 70, 1055–1096 (1982)
37. Press, W.H., Teukolsky, S.A., Vetterling, W.T., Flannery, B.P.: Numerical Recipes: The Art of Scientific Computing, 3rd edn. Cambridge University Press, Cambridge (2007)
38. Maris, E., Oostenveld, R.: Nonparametric statistical testing of EEG- and MEG-data. J. Neurosci. Methods 164(1), 177–190 (2007)
39. Farmer, S.F., Swash, M., Ingram, D.A., Stephens, J.A.: Changes in motor unit synchronization following central nervous lesions in man. J. Physiol. 463(1), 83–105 (1993)

Face Recognition on Mobile Devices Based on Frames Selection

Nelson Méndez-Llanes[1], Katy Castillo-Rosado[1], Heydi Méndez-Vázquez[1(✉)], Souad Khellat-Kihel[2], and Massimo Tistarelli[2]

[1] Advanced Technologies Application Center,
7A #21406 b/ 214 and 216, P.C., 12200 Playa, Havana, Cuba
{nllanes,krosado,hmendez}@cenatav.co.cu
[2] Computer Vision Laboratory,
University of Sassari, Viale Italia 39, 07100 Sassari, Italy
{skhellatkihel,tista}@uniss.it

Abstract. The easy image capture process on a phone, the ability to move the camera, the non-intrusive characteristic that allows the authentication without interaction of the user, make the face a suitable biometric trait to be used on mobile devices. During the continuous use of the mobile, it is possible to analyze the expression, to determine the gender and ethnic, or to recognize the user. In this paper we propose a robust algorithm for face recognition on mobile devices. First, the best frames of a face sequence are selected based on the face pose, blurness, eyes and mouth expression. Then, a unique feature vector for the best selected frames is obtained using a deep learning model. Finally, a SoftMax function is used for authenticate the user. The experimental evaluation conducted on the UMD-AA dataset shows the robustness of the proposal, that outperforms state-of-the-art methods.

Keywords: Face recognition · Face image quality · Mobile authentication

1 Introduction

Mobile devices are used daily to create and transmit private information. Traditionally, numbers and passwords have been used for mobile devices security, but they are not so secure since users tend to forget and reuse them. Recently, biometric-based authentication has shown to be a convenient and secure option [2]. Among different biometric technologies, face recognition has become a promising security option for mobile devices [10]. Faces are easy to capture and to store in mobile devices; besides, the achieved levels of accuracy of face recognition systems in the past few years, make them more secure. One of the main studied problems in mobile authentication, known as Active Authentication (AA), is to guarantee that the user originally authenticated, is the one that maintain the control of the device.

© Springer Nature Switzerland AG 2019
I. Nyström et al. (Eds.): CIARP 2019, LNCS 11896, pp. 316–325, 2019.
https://doi.org/10.1007/978-3-030-33904-3_29

Different active authentication approaches have been proposed for verifying the user identity based on the facial information [14]. In [5], nine state-of-the-art face recognition methods were evaluated for active authentication, and a challenging evaluation protocols representing real scenarios was introduced. Continuous authentication based on facial attributes was proposed in [15] for fast processing. A bunch of binary attribute classifiers was trained and the authentication was done by comparing the estimated attributes with the enrolled attributes of the original user. Recently, a sparse representation-based method for multiple users authentication, was proposed in [12]. A parameter selection scheme was introduced for extreme value distributions, to make them feasible for an automated mechanism.

There are just a few methods based on deep learning for mobile authentication [4,14]. Although this kind of methods achieve very high recognition rates, they require high computational resources. Besides, existing methods usually use the complete set of frames available in a captured face sequence, that can also be time consuming. On the other hand, it is well known that the degradation of the face quality overcomes in a poor performance of the recognition algorithms. Due to the non-controlled environment of a phone camera, the obtained videos for training the system and the ones that are recognized later, can be very different. Hence, training a method for all possible qualities that can appear, requires too much information [16]. Under this situation, algorithms able of adapting their behavior depending on the quality of the biometric sample are needed [17,18].

In this paper is introduced a method that decides which biometric information is good enough for obtaining a reliable face authentication result. In order to choose which part of a video sequence contains good enough biometric information, a frame selection method that uses face features quality, is introduced. The proposed method takes into account blur, pose and expression variations, which are among the most affecting factors in mobile authentication.

The rest of this paper is organized as follows. Section 2 reviews some of the existing works to determine face images quality. Section 3 introduces the proposed method. In Sect. 4 the experimental evaluation is presented. Finally, conclusions and future works are given.

2   Related Work

The sample quality has an important impact in the accuracy of biometric recognition systems [9]. Particularly for faces, a number of approaches have been proposed to determine the quality of face images and selected a reduced number of frames from a given sequence [1,3,13,16].

Most of the recent proposals that achieve very good results are based on deep learning methods. Yang et. al. [16] proposed to use deep neural networks for classifying the images based on their quality and then, separated face detectors and recognizer algorithms are used depending on the images quality. Two types of image quality problems are considered: JPEG compression, and low-resolution. One drawback of this approach is that these types of quality problems are not always proportional to the quality of the face features that can be extracted from the images. Besides, for each type of quality problem, a large number of samples are needed for training a specialized face detector and a recognizer method. Another approach that uses a deep convolutional neural network (CNN) for predicting the face image quality is [1]. First, features are extracted by using a CNN, and with these features, a prediction model of the face quality is learned by using support vector regression. The main disadvantage of using deep learning in this early step is the high computational cost that implies analyze all the frames in a sequence, which can be not feasible for mobile applications.

Zohra and Gavrilova [17,18] proposed a system where the illumination distortion is normalized by using quality-based normalization approaches. Only quality problems related to illumination are overcome in their approach. The method proposed in [3] considers both image and face characteristics, but it was specifically designed for FPGA architectures.

It is evident from the above analysis, that methods particularly designed for mobile authentication are still needed. They should be able to analyze the most common quality problems on these scenarios in an efficient way.

3   Proposed System

The proposed method for face recognition on mobile devices is composed by three main steps that will be explain on details in this section. The first step is to determine the frames of the sequence which contain the most valuable face information for recognition. The selected frames are then represented by a unique feature vector that is obtained using a CNN model for face feature extraction. In this paper, three different models are evaluated. Finally, the classification is made through a SoftMax function.

3.1   Face Image Quality Assessment

For active authentication in mobile devices, a video sequence can be captured and then the face classification can be done with those frames with the most relevant information. For this aim, a quality value is estimated for each frame of the video. This quality value is calculated by measuring four parameters that describe the most common problems present in face mobile authentication. The proposed quality measures are based on the use of facial landmarks, obtained through a fast and accurate method [7], implemented on Dlib library [8].

Pose Evaluation. The problem of the face pose is that subjects with different identities in different poses are grouped better than the same subject in different poses. Enrolled face images are usually in frontal position, hence, it is desirable that the face pose of the images received for the authentication ranges between $\pm45°$.

The estimation of the pose is determined by calculating the displacement of a set of landmark points with respect to that points on a face with neutral pose

Fig. 1. A graphic example of the points used for estimating the face pose.

in a 3D model. The selected points belong to the edges of the eyes, the edges of the lips, the tip of the nose and the tip of the chin. For a better understanding, a graphic example is illustrated in Fig. 1. The average displacement is computed and normalized between $[0, 1]$, where 1 stands for no displacement and 0 means a large displacement from the frontal position.

Facial Expression Evaluation. Different facial expressions modify the face shape and appearance. The eyes and the mouth, are two of the face regions that have greater changes with different expressions. In the case of the eyes being completely closed (both or one eye), can be seen as a particular case of occlusion. On the other hand, if the mouth is not naturally closed or if it is very open, the facial appearance changes drastically.

For defining a neutral expression based on the eyes and the mouth, the landmark points of the corners of these regions are used. With these points a triangle with side labels A, B and C is formed, where the aperture angle α is that which is opposite to the side A. Then α is calculated as follows:

$$\alpha = arccos(\frac{B^2 + C^2 - A^2}{2 * A * B})\tag{1}$$

The Eyes. A natural eyes expression is defined by determining the angle that is formed with the commissure of the edges of the eyes: the greater the angle, the better the expression. By using a set of 500 face images, the average values of the maximum and minimum aperture is determined, in order to define the classification intervals. The maximum and minimum angles from this set, are used to linearly normalize a given angle between $[0, 1]$. For a better understanding, a visual example is shown in Fig. 2. Both corners of the left and right eyes are taken into account to determine the state of the eyes.

Fig. 2. A graphic example of the eyes expression state. Left: eyes with neutral expression; Right: closed eyes.

The Mouth. The mouth is also analyzed using the angles of the corners. In this case, the smaller the angle, the better the expression. The landmark points detected at the corners of the lips are used as illustrated in Fig. 3. Besides, it is also taken into account if both angles have similar angle values, to determine a neutral expression. The mouth expression can change in different ways, reason why this heuristic is based on determining if the mouse remains closed to ensure a neutral expression.

Fig. 3. A graphic example of the mouth expression state. Left: mouth with neutral expression; Right: opened mouth.

Face Blurness. For detecting how blur the face image is, the variance of the Laplacian is calculated [11]. By using the landmark points it is ensured that only the face region is considered and that background information is not taken into account. The Laplacian kernel, Equation (2), is commonly used for detecting edges. Hence, its variance gives an idea of how normal is the response of the edges in an image, allowing to determine how blurred the face image is.

$$Lap(m,n) = \begin{bmatrix} 0 & -1 & 0 \\ -1 & 4 & -1 \\ 0 & -1 & 0 \end{bmatrix} \tag{2}$$

Final Quality Assessment. Let $p \in [0,1]$, $e \in [0,1]$, $m \in [0,1]$ and $b \in [0,1]$ be the values of the four quality measures defined: pose, eyes, mouth and blur respectively. The global quality value can be estimated by using the following linear equation:

$$q = p * k_p + e * k_e + m * k_m + b * k_b, \tag{3}$$

where $q \in [0,1]$: 1 represents the highest quality and 0 the lowest, and $k_i, i \in \{p,e,m,b\}$ are the weights for each feature with:

$$\sum_{i \in \{p,e,m,b\}} k_i = 1. \tag{4}$$

The weights can vary depending on the scenario and the most affecting factors. A visual example can be seen in Fig. 4, where the general steps of the face image quality assessment method are presented.

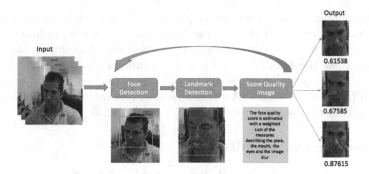

Fig. 4. General steps for selecting the best N frames in a video sequence.

3.2 Feature Extraction

Once the best frames are selected, a face descriptor is needed. In this paper we explore the use of three different CNN models selected from the literature. One of these models is the widely used face deep model from Dlib library [8]. This model is based on a version of the ResNet50 model which has 29 convolutional layers [6]. The output of this network is a 128 dimensional vector, that represents the subject facial features and appearance on every frame. Since the obtained vectors represent a unique face image, it is possible to combine them by applying an average pooling operation across the 128 dimensions. Other model used is the original ResNet50 trained on MS-Celeb-1M dataset [6] and then fine-tuned on VGGFace2 dataset. The last one, is a model specifically designed for high-accuracy real-time face verification on mobile devices, the MobileFaceNet [4]. This network was trained with the refined MS-Celeb-1M dataset using the ArcFace loss.

3.3 Classification

The classification stage is carry out using the SoftMax function. The loss function of SoftMax is based on the cross-entropy loss:

$$L_i = -log\left(\frac{e^{f_{y_i}}}{\sum_j e^{f_j}}\right) \tag{5}$$

where f_j is the j-th element of the feature vector representing subject f, while L_i is the full loss of the dataset over the training examples.

Softmax function gives probabilities for each class and it is commonly used as the final layer at the end of a neural network. Supposing that we have a classification problem with 10 different classes, thus the dimension of the output layer is 10. The ideal goal is to find 1.0 as score for a single output node, and a probability of zero for the rest of the output nodes. The best architecture for such requirement is Max-layer output, which will provide a probability of 1.0

for the maximum output of previous layer and the rest of the output nodes will be considered as zero. But such output layer will not be differentiable, hence it will be difficult to train. Alternatively, if the SoftMax function is used, it will almost work like the Max-layer and it will be differentiable by gradient descent. Exponential function will increase the probability of maximum value of the input compared to the other values. Another special characteristic of SoftMax layer is that the summation of all outputs is always equal to 1.0.

4   Experimental Results

The UMD-AA dataset [5], a very challenging testbed for performing experiments on active authentication for mobile devices, has been used to perform the experimental evaluation. The videos are recorded in different illumination conditions within a laboratory room. The first subset of videos was captured with artificial lighting (Session 1). The second subset was captured without any illumination (Session 2). The last subset was captured under natural sunlight (Session 3). The database is composed by videos from 150 subjects. For each subject 5 videos are available in each session. One out of the five videos, containing different changes in the face position and rotation, is used for enrollment. The remaining four videos are used for testing. The test videos were captured from mobile devices while the user was performing a specific activity, such as looking at a window popup, scrolling test, taking a picture or working on a document.

We use Protocol 1, which the most difficult one. Under this protocol, the training data is composed by the enrollment videos from one session while the test videos belonging to the other two sessions are used for testing. Hence, there are six available scenarios for this protocol: enrollment from Session 1 with testing from Sessions 2 and 3; enrollment from Session 2 with testing from Sessions 1 and 3; and enrollment from Session 3 with testing from Sessions 1 and 2. In our experiments, the landmark points are obtained directly from the images/frames without any preprocessing. Considering that blur is the less affecting factor on this database, we use $k_b = 0.1$ and 0.3 as weights for the other three parameters.

The first experiment conducted, focuses on the selection of the best frames. We evaluate selecting the best three and the best ten frames and the results are compared with respect to use all frames of the sequence and ten random frames. The Rank-1 recognition rates for the three models on each evaluated scenario are shown in Table 1. As can be seen from the table, the results for all cases selecting a given number of frames, are much better that using all frames or randomly selected frames. On the other hand, selecting ten frames is in general better than selecting only three. However, except for Dlib model, the results are very close, so can be an option using only three frames for devices with limited resources. It should be noticed that using all the available frames in a sequence is not feasible in terms of computing time for mobile authentication. The average processing time of an image, in the feature extraction stage is around 352.63 ms for Dlib network, 85.56ms for ResNet50 and 20.65 ms for MobileFaceNet. The videos of the database have an average of 180 frames. Is not possible to process this number of frames on mobile devices in real time using a traditional deep-learning

Table 1. Rank-1 Recognition rates (%) for different frames selection strategies.

Session		Dlib				MobileFaceNet				ResNet50			
Enroll	Test	All frames	10 rand frames	10 best frames	3 best frames	All frames	10 rand frames	10 best frames	3 best frames	All frames	10 rand frames	10 best frames	3 best frames
1	2	13.07	15.34	**93.18**	55.81	75.00	82.56	**92.44**	89.53	97.09	98.26	98.26	**100.00**
1	3	13.07	14.20	**84.09**	76.74	90.12	89.53	**98.26**	**98.26**	97.67	98.26	99.42	**100.00**
2	1	9.66	10.23	**88.64**	48.84	83.72	87.21	**93.02**	91.86	98.26	98.84	100.00	100.00
2	3	15.34	17.61	**81.82**	51.16	83.14	85.47	**94.19**	**94.19**	96.51	98.26	100.00	99.42
3	1	14.20	14.77	**90.91**	81.40	84.88	87.79	97.09	**98.26**	93.02	97.67	100.00	100.00
3	2	10.23	14.20	**88.64**	79.07	79.65	88.37	**90.70**	86.05	93.60	97.09	100.00	100.00

based approach (Dlib and ResNet50). Even for the MobileFaceNet neural network which is designed for mobile environments, processing all the frames could be very time consuming.

The ROC curves for the three models when selecting 10 frames, using each session for enrollment, are shown in Fig. 5. By analyzing the balance between accuracy and efficiency of the three evaluated models, we believe that the most adequate for mobile authentication is the MobileFaceNet network.

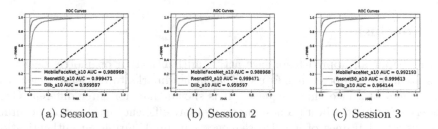

(a) Session 1   (b) Session 2   (c) Session 3

Fig. 5. ROC curves for the proposed method (selecting 10-frames) when each session is used as enrollment.

In Table 2 we compare the results obtained by the MobileFaceNet model in Protocol 1, with the best performing methods in [5]: Fisherfaces (FF), Sparse Representation-based Classification (SRC), and Mean-Sequence SRC (MSSRC). One can see clearly that our proposal outperforms the other methods by a large margin.

On the other hand, by analyzing the Area Under the Curve (AUC) for this model on Fig. 5, it can be seen that it is much more higher than those exhibit in [15]. In Table 3 the EER values for the proposed strategy, using the Mobile-FaceNet model, are compared with those based on attributes presented in [15], and it is corroborated the superiority of the proposal.

Table 2. Rank-1 Recognition rates (%) of state-of-the-art methods on Protocol 1 of UMD-AA dataset.

Session		FF- all frames	SRC- all frames	MSSRC- all frames	MobileFaceNet-Proposal
Enroll	Test				
1	2	54.48	52.79	47.21	**92.44**
1	3	45.27	51.18	46.15	**98.26**
2	1	25.52	44.18	43.06	**93.02**
2	3	56.80	58.58	60.36	**94.19**
3	1	24.77	17.64	17.64	**97.09**
3	2	56.01	51.95	45.85	**90.70**

Table 3. EER for different methods in UMD-AA dataset.

Session	LBP	Attributes [15]	Fusion [15]	MobileFaceNet-Proposal
1	0.13	0.14	0.10	**0.044**
2	0.31	0.18	0.20	**0.044**
3	0.19	0.16	0.14	**0.036**

5   Conclusion

In this paper an approach for face active authentication on mobile devices is presented. The proposal makes use of facial landmarks to efficiently select the best frames of a face video sequence. It is shown that, for three different CNN models, selecting the best frames is not only more efficient but also more accurate than using all frames of a video sequence captured during an authentication session.

Acknowledgment. This research work has been partially supported by a grant from the European Commission (H2020 MSCA RISE 690907 IDENTITY) and by a grant of the Italian Ministry of Research (PRIN 2015).

References

1. Best-Rowden, L., Jain, A.K.: Learning face image quality from human assessments. IEEE Trans. Inf. Forensics Secur. **13**(12), 3064–3077 (2018)
2. Blanco-Gonzalo, R., Lunerti, C., Sanchez-Reillo, R., Guest, R.M.: Biometrics: Accessibility challenge or opportunity? PLOS ONE **13**(3), 1–20 (2018)
3. Chang, L., Rodés, I., Méndez, H., del Toro, E.: Best-shot selection for video face recognition using FPGA. In: Ruiz-Shulcloper, J., Kropatsch, W.G. (eds.) CIARP 2008. LNCS, vol. 5197, pp. 543–550. Springer, Heidelberg (2008). https://doi.org/10.1007/978-3-540-85920-8_66
4. Chen, S., Liu, Y., Gao, X., Han, Z.: MobileFaceNets: efficient CNNs for accurate real-time face verification on mobile devices. In: Zhou, J., Wang, Y., Sun, Z., Jia,

Z., Feng, J., Shan, S., Ubul, K., Guo, Z. (eds.) CCBR 2018. LNCS, vol. 10996, pp. 428–438. Springer, Cham (2018). https://doi.org/10.1007/978-3-319-97909-0_46

5. Fathy, M.E., Patel, V.M., Chellappa, R.: Face-based active authentication on mobile devices. In: 2015 IEEE International Conference on Acoustics, Speech and Signal Processing (ICASSP), pp. 1687–1691. IEEE (2015)

6. He, K., Zhang, X., Ren, S., Sun, J.: Deep residual learning for image recognition. In: Proceedings of the IEEE Conference on Computer Vision and Pattern Recognition, pp. 770–778 (2016)

7. Kazemi, V., Sullivan, J.: One millisecond face alignment with an ensemble of regression trees. In: Proceedings of the IEEE Conference on Computer Vision and Pattern Recognition, pp. 1867–1874 (2014)

8. King, D.E.: Dlib-ml: a machine learning toolkit. J. Mach. Learn. Res. **10**, 1755–1758 (2009)

9. Liu, X., Pedersen, M., Charrier, C., Bours, P.: Performance evaluation of no-reference image quality metrics for face biometric images. J. Electron. Imaging **27**(2), 023001 (2018)

10. Neal, T.J., Woodard, D.L.: Surveying biometric authentication for mobile device security. J. Pattern Recogn. Res. **1**, 74–110 (2016)

11. Pech-Pacheco, J.L., Cristóbal, G., Chamorro-Martínez, J., Fernández-Valdivia, J.: Diatom autofocusing in brightfield microscopy: a comparative study. In: 15th International Conference on Pattern Recognition, ICPR 2000, Barcelona, Spain, pp. 3318–3321 (2000)

12. Perera, P., Patel, V.M.: Face-based multiple user active authentication on mobile devices. IEEE Trans. Inf. Forensics Secur. **14**, 1240–1250 (2018)

13. Qi, X., Liu, C., Schuckers, S.: Cnn based key frame extraction for face in video recognition. In: 2018 IEEE 4th International Conference on Identity, Security, and Behavior Analysis (ISBA), pp. 1–8. IEEE (2018)

14. Rattani, A., Derakhshani, R.: A survey of mobile face biometrics. Comput. Electr. Eng. **72**, 39–52 (2018)

15. Samangouei, P., Patel, V.M., Chellappa, R.: Facial attributes for active authentication on mobile devices. Image Vis. Comput. **58**, 181–192 (2017)

16. Yang, F., Zhang, Q., Wang, M., Qiu, G.: Quality classified image analysis with application to face detection and recognition. In: 24th International Conference on Pattern Recognition, ICPR 2018, Beijing, China, pp. 2863–2868 (2018)

17. Zohra, F.T., Gavrilova, M.L.: Adaptive face recognition based on image quality. In: 2017 International Conference on Cyberworlds, CW 2017, Chester, UK, September 20–22, pp. 218–221 (2017)

18. Zohra, F.T., Gavrilova, M.L.: Image quality-based illumination-invariant face recognition. Trans. Computat. Sci. **32**, 75–89 (2018)

Brain Tumour Segmentation from Multispectral MR Image Data Using Ensemble Learning Methods

Ágnes Győrfi[1,2], Levente Kovács[2], and László Szilágyi[1,2(✉)]

[1] Computational Intelligence Research Group,
Sapientia - Hungarian University of Transylvania, Tîrgu Mureş, Romania
{gyorfiagnes,lalo}@ms.sapientia.ro
[2] University Research, Innovation and Service Center (EKIK),
Óbuda University, Budapest, Hungary
{kovacs.levente,szilagyi.laszlo}@nik.uni-obuda.hu

Abstract. The number of medical imaging devices is quickly and steadily rising, generating an increasing amount of image records day by day. The number of qualified human experts able to handle this data cannot follow this trend, so there is a strong need to develop reliable automatic segmentation and decision support algorithms. The Brain Tumor Segmentation Challenge (BraTS), first organized seven years ago, provoked a strong intensification of the development of brain tumor detection and segmentation algorithms. Beside many others, several ensemble learning solutions have been proposed lately to the above mentioned problem. This study presents an evaluation framework developed to evaluate the accuracy and efficiency of these algorithms deployed in brain tumor segmentation, based on the BraTS 2016 train data set. All evaluated algorithms proved suitable to provide acceptable accuracy in segmentation, but random forest was found the best, both in terms of precision and efficiency.

Keywords: Magnetic resonance imaging · Image segmentation · Tumor detection · Brain tumor · Ensemble learning

1 Introduction

Multi-spectral MRI is the usual imaging modality used to detect, localize and grade brain tumors [1]. Huge effort has been invested lately in the development of automatic MRI data processing techniques [2,3]. A wide range of algorithms were developed that cover the whole arsenal of decision making algorithms. Most

This project was supported by the Sapientia Foundation – Institute for Scientific Research. The work of L. Kovács was supported by the European Research Council (ERC) under the European Union's Horizon 2020 research and innovation programme (grant agreement No. 679681). The work of L. Szilágyi was supported by the Hungarian Academy of Sciences through the János Bolyai Fellowship program.

© Springer Nature Switzerland AG 2019
I. Nyström et al. (Eds.): CIARP 2019, LNCS 11896, pp. 326–335, 2019.
https://doi.org/10.1007/978-3-030-33904-3_30

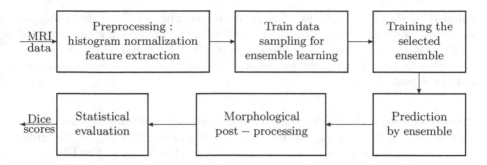

Fig. 1. Block diagram of the evaluation framework.

solutions rely on supervised and semi-supervised machine learning techniques supported by advanced image segmentation methods like: random forest ensembles [4–7], discrete and real AdaBoost [8], extremely random trees [9], support vector machines [10], convolutional neural network [11,12], deep neural networks [13–15], Gaussian mixture models [16,17], fuzzy c-means clustering in semi-supervised context [18,19], tumor growth model [20], cellular automata combined with level sets [21], active contour models combined with texture features [22], and graph cut based segmentation [23]. Earlier brain tumor segmentation solutions were remarkably summarized by Gordillo et al. in [24].

In this study we built an evaluation framework to evaluate ensemble learning algorithms in segmenting brain tumors from volumetric MRI data. We compare the accuracy and efficiency achieved by various decision making techniques, employed within the same scenario to work with the very same pre-processed data originating from the BraTS 2016 database. The rest of the paper is structured as follows: Sect. 2 presents provides the technical details of the framework and the algorithms included in the evaluation. Section 3 analyses and discusses the obtained results. Section 4 concludes the investigation.

2   Materials and Methods

2.1   Framework

Data. This study is based on the whole set of 220 high-grade (HG) tumor records of the BraTS 2016 train dataset [2]. Each record contains four data channels (T1, T2, T1C, FLAIR). All channels are registered to the T1 channel. Volumes consist of $155 \times 240 \times 240$ isovolumetric voxels. Each voxels reflects one cubic millimeter of brain tissues. An average volume contains approximately 1.5 million brain voxels. The human expert made annotations provided by BraTS is used as ground truth within this study.

Processing Steps. The main steps of this application are presented in Fig. 1. Data records need a preprocessing to provide uniform histograms and to generate further features for the classification. Data originating from train records are sampled for the training of ensembles. Trained ensembles are evaluated using the whole test volumes. Post-processing is applied to the prediction result provided by the ensembles, to regularize the shape of the tumor and improve the segmentation quality. Finally, the precision of the segmentation is evaluated using statistical tools.

Pre-processing. There are three main pre-processing problems to handle when working with MRI data: (1) the intensity non-uniformity [25–27]; (2) the great variety of MR image histograms; (3) generating further features. The HG tumor volumes of the BraTS dataset contains no relevant inhomogeneity [2], so its compensation can be omitted. Uniform histograms are provided for each data channel of each MR record, using a context dependent linear transform, which assigns the 25 and 75 percentile to intensity levels 600 and 800, respectively, and forces all transformed intensities to be situated in the predefined range of 200 to 1200. Details of this transform are presented in our previous paper [28]. Beside the 4 observed data channels, 100 further features are generated using morphological, gradient, Gabor wavelet based techniques [28, 29].

Decision Making. The 220 HG tumor records were randomly divided into two equal groups, which served as train and test data during the two-round cross validation. Thus we obtain segmentation accuracy benchmark for each MRI record using ensembles trained with data from the complementary group. Ensemble units were trained using the feature vectors of 10,000 randomly selected voxels from the train records that contained 93% negatives and 7% positives, as described in our previous study [28]. All ensembles were trained to separate two classes: normal tissues and whole tumor lesions.

Post-processing. Our post-processing step relabels each pixel based on the rate of predicted positives situated within a $11 \times 11 \times 11$ cubic neighborhood. The threshold was set empirically at 35%.

Evaluation Criteria. The accuracy indicators involved in this study are based on the amount of true positives (TP), true negatives (TN), false positives (FP), and false negatives (FN). The main accuracy indicators derived from these numbers, namely the Dice score (DS), sensitivity (true positive rate, TPR), specificity (true negative rate, TNR), and accuracy (ACC), are presented in Table 1. These indicators are established for each individual HG tumor record, and then average and median values are computed to characterize the overall accuracy. The evaluation criterion of algorithm efficiency is the average runtime of the whole processing of individual MR records.

Table 1. Criteria to evaluate segmentation quality

Indicator		Values	
Name	Formula	Possible	Ideal
Dice score	$DS = \frac{2 \times TP}{2 \times TP + FP + FN}$	$0 \leq DS \leq 1$	1
Sensitivity	$TPR = \frac{TP}{TP + FN}$	$0 \leq TPR \leq 1$	1
Specificity	$TNR = \frac{TN}{TN + FP}$	$0 \leq TNR \leq 1$	1
Accuracy	$ACC = \frac{TP + TN}{TP + FP + TN + FN}$	$0 \leq ACC \leq 1$	1

2.2 Algorithms

Ensemble learning methods achieve high accuracy in classification from the majority voting of several weak classifiers. In this study we investigate the following algorithms:

- Random forest (RF) classifier, as implemented in OpenCV ver. 3.4.0. RF is an ensemble of binary decision trees. The main parameters are the number of trees and the maximum tree depth. Train data sets of 10,000 items were best learned using maximum depth set to seven.
- Ensemble of real Adaboost classifiers, as implemented in OpenCV ver. 3.4.0.
- Ensemble of perceptron networks (ANN), as implemented in OpenCV ver. 3.4.0., using four layers of sizes 104, 15, 7, and 1, respectively.
- Ensemble of binary decision trees (BDT), using an own implementation [28]. BDTs can be trained to perfectly separate negative from positive samples unless there exist coincident feature vectors with different ground truth. The maximum depth of BDTs was 20.6 ± 3.4 (AVG $\pm$ SD), but decisions were made at average depth of 7.71 ± 2.89.

3 Results and Discussion

The above listed machine ensemble learning techniques were tested using the 220 high-grade tumor records of the BraTS 2016 database. Four ensembles sizes ranging from 5 to 255 were evaluated. Quality indicators shown in Table 1 were extracted for each algorithm and each MRI record separately, together with the average and median value for each indicator for overall accuracy evaluation. Comparisons in group involving all algorithms, and one-against-one tests were carried out, using individual data records and the whole HG data set as well.

Overall average and median values of the four main quality indicators are exhibited in Table 2, for all evaluated ensemble learning algorithms and various ensemble sizes. Median values were found greater than the average, for all indicators and scenarios, because there are a few records of reduced or damaged quality that are likely to be segmented considerably worse than all others. Highest values highlighted in each column of the table indicate that the random forest achieved slightly better results than any other evaluated technique. The accuracy of segmentation rises together with the ensemble size up to 125 units, above which it seems to stabilize or fall slightly. Highest achieved average Dices scores approached 81%, while median values surpass 86%. The accuracy of all evaluated ensemble learning techniques is around 98%, meaning that approximately one pixel out of 50 is misclassified.

Table 2. Various statistical accuracy indicator values achieved by tested techniques and ensemble sizes, expressed in percentage (%). Best performance is highlighted in all columns. AVG stands for average, MED stands for median.

Classifiers in ensemble	Ensemble size	Dice score		Sensitivity		Specificity		Accuracy	
		AVG	MED	AVG	MED	AVG	MED	AVG	MED
ANN	5	79.11	84.82	83.00	**90.62**	98.40	98.88	97.50	97.98
	25	80.02	85.73	83.15	90.48	98.40	98.86	97.52	97.97
	125	80.09	85.62	**83.33**	90.58	98.40	98.85	97.52	97.96
	255	80.05	85.49	83.31	90.56	98.39	98.56	97.52	97.97
Adaboost	5	79.69	85.37	82.24	90.00	98.48	98.92	97.53	97.97
	25	80.00	85.59	82.02	89.58	98.55	99.00	97.58	98.01
	125	79.98	85.54	81.89	89.34	**98.56**	**99.02**	97.59	98.03
	255	80.04	85.77	81.96	89.49	98.55	99.00	97.59	98.03
Random forest	5	80.29	85.59	82.95	89.83	98.45	98.93	97.56	97.99
	25	80.59	85.93	82.73	89.74	98.53	98.98	97.62	98.06
	125	80.71	86.22	82.77	89.94	98.55	99.00	**97.64**	98.08
	255	**80.74**	**86.27**	82.77	89.88	98.55	99.01	**97.64**	**98.09**
Binary decision trees	5	79.22	85.52	82.27	89.93	98.39	98.91	97.46	97.86
	25	79.80	85.36	82.12	90.05	98.50	98.94	97.54	98.04
	125	80.05	85.68	82.03	89.37	**98.56**	99.00	97.59	98.02
	255	80.03	85.73	81.95	89.46	**98.56**	98.99	97.59	98.03

Figure 2 exhibits the Dice score and Sensitivity in the left panel, respectively the Specificity and Accuracy in the right panel, indicator values obtained by the random forest using ensemble of 125, which was identified as the most accurately performing algorithm. Approximately 10% of the records lead to mediocre result. In these cases the classification methods failed to capture the main specific characteristics of the data, probably because the recorded images were of low quality.

Table 3 shows for each test scenario (algorithm and ensemble size) the number of successfully segmented records, where the Dice score exceeded predefined threshold values ranging from 50% to 92%. Highest values highlighted for each threshold value indicate again that random forest achieved the best segmentation quality.

Figure 3 presents the outcome of one-against-one comparison of the tested algorithms, each using ensembles of 125 units. Dice scores shown here were obtained on each individual HG tumor records. Each cross (×) in the graph shows the Dice score achieved by the two ensemble learning techniques on the

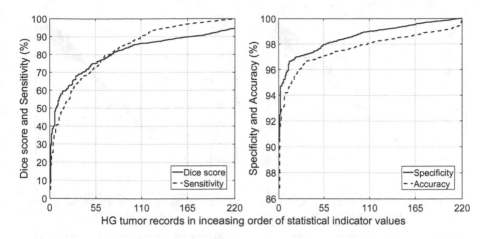

Fig. 2. Main quality indicator values obtained for individual HG tumor volumes, using the random forest method in ensemble of 125, sorted in increasing order.

Table 3. Comparison of the tested ensemble learning techniques using the DS obtained for the 220 individual HG tumor records. Bests scores were identified and highlighted in each row of the table.

Classifier	BDT				Random forest				Adaboost				ANN			
Ensemble size	5	25	125	255	5	25	125	255	5	25	125	255	5	25	125	255
DS > 50%	208	208	209	209	**213**	212	212	212	208	208	208	209	210	210	210	210
DS > 60%	194	198	200	201	197	200	201	201	198	200	200	201	195	201	**202**	201
DS > 70%	173	174	175	175	177	180	**181**	**181**	173	175	175	175	173	175	174	174
DS > 75%	160	163	164	163	164	**168**	166	166	161	163	164	164	159	164	164	164
DS > 80%	138	140	143	143	142	144	**146**	**146**	141	143	143	143	135	143	145	141
DS > 85%	114	111	117	116	117	119	**120**	**120**	112	116	116	116	109	114	114	114
DS > 88%	70	76	75	76	75	**82**	81	**82**	76	76	76	78	67	74	74	74
DS > 90%	38	47	52	51	47	49	51	50	46	51	51	**53**	38	44	44	42
DS > 92%	21	26	25	25	23	24	**27**	26	25	26	26	25	18	23	24	23

very same data. Most crosses are situated in the proximity of the diagonal, indicating that both algorithms obtained pretty much the same accuracy. There are also crosses apart from the diagonal, representing scenarios where one of the methods led to significantly better segmentation quality.

Table 4 exhibits the same results as Fig. 3, but here the one-against-one outcome of tests is organized in a tournament format. The tournament was won by the random forest algorithm, followed by BDT, Adaboost, and ANN. Figure 4 compares the efficiency of the four evaluated algorithms. Total runtimes exhibited here include the duration of histogram normalization and feature generation, segmentation and post-processing of an average sized never seen MR data volume. All tests were performed on a notebook computer, using a single core of a quad-core i7 processor that runs at 3.4 GHz. AdaBoost and ANN proved to be significantly less efficient than RF and BDT.

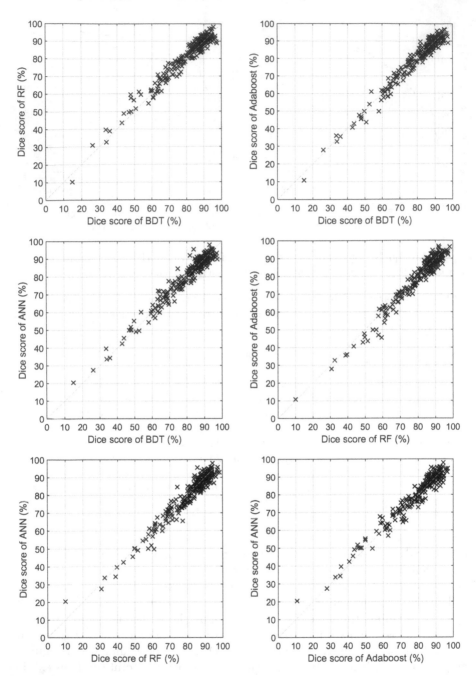

Fig. 3. Dice scores obtained for individual volumes by the four algorithms using ensembles of size 125, plotted one algorithms vs. another, in all possible six combinations.

Table 4. Dice score tournament using the 54 LG volumes: algorithms against each other, each using ensembles of size 125. Here ANN proved to be the weakest.

Algorithm	ANN	Adaboost	RF	BDT	Won:Lost
ANN	N/A	89:131	57:163	86:134	0:3 (232:428)
Adaboost	131:89	N/A	71:149	102:118	1:2 (304:356)
RF	163:57	149:71	N/A	147:73	3:0 (459:201)
BDT	134:86	118:102	73:147	N/A	2:1 (325:335)

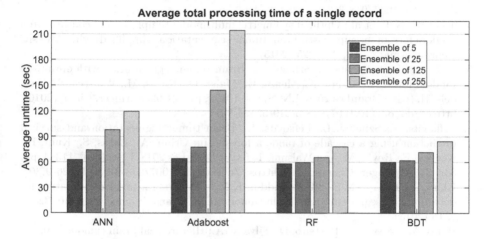

Fig. 4. Runtime benchmarks of the four classification algorithms: the average value of the total processing time in a single record testing problem.

4 Conclusions

This study attempted to compare the accuracy and efficiency of various ensemble learning algorithms involved in a brain tumor segmentation based on multispectral magnetic resonance image data. The performed investigation indicates that publicly available implementations of ensemble learning methods are all capable to detect and segment the tumor with an acceptable accuracy. The small differences in terms of accuracy, and larger ones in terms of efficiency together revealed that random forest is the best decision making algorithm from the investigated ones. Further works will aim at involving more data sets and more machine learning algorithms into the comparative study.

References

1. Mohan, G., Subashini, M.M.: MRI based medical image analysis: survey on brain tumor grade classification. Biomed. Signal Process. Control **39**, 139–161 (2018)

2. Menze, B.H., Jakab, A., Bauer, S., Kalpathy-Cramer, J., Farahani, K., Kirby, J., et al.: The multimodal brain tumor image segmentation benchmark (BRATS). IEEE Trans. Med. Imaging **34**, 1993–2024 (2015)

3. Bakas, S., Reyes, M., Jakab, A., Bauer, S., Rempfler, M., Crimi, A., et al.: Identifying the best machine learning algorithms for brain tumor segmentation, progression assessment, and overall survival prediction in the BRATS challenge. arXiv: 1181.02629v3, 23 April 2019

4. Phophalia, A., Maji, P.: Multimodal brain tumor segmentation using ensemble of forest method. In: Crimi, A., Bakas, S., Kuijf, H., Menze, B., Reyes, M. (eds.) BrainLes 2017. LNCS, vol. 10670, pp. 159–168. Springer, Cham (2018). https://doi.org/10.1007/978-3-319-75238-9_14

5. Tustison, N.J., et al.: Optimal symmetric multimodal templates and concatenated random forests for supervised brain tumor segmentation (simplified) with ANTsR. Neuroinformatics **13**, 209–225 (2015)

6. Lefkovits, L., Lefkovits, S., Szilágyi, L.: Brain tumor segmentation with optimized random forest. In: Crimi, A., Menze, B., Maier, O., Reyes, M., Winzeck, S., Handels, H. (eds.) BrainLes 2016. LNCS, vol. 10154, pp. 88–99. Springer, Cham (2016). https://doi.org/10.1007/978-3-319-55524-9_9

7. Lefkovits, S., Szilágyi, L., Lefkovits, L.: Brain tumor segmentation and survival prediction using a cascade of random forests. In: Crimi, A., Bakas, S., Kuijf, H., Keyvan, F., Reyes, M., van Walsum, T. (eds.) BrainLes 2018. LNCS, vol. 11384, pp. 334–345. Springer, Cham (2019). https://doi.org/10.1007/978-3-030-11726-9_30

8. Islam, A., Reza, S.M.S., Iftekharuddin, K.M.: Multifractal texture estimation for detection and segmentation of brain tumors. IEEE Trans. Biomed. Eng. **60**, 3204–3215 (2013)

9. Pinto, A., Pereira, S., Rasteiro, D., Silva, C.A.: Hierarchical brain tumour segmentation using extremely randomized trees. Pattern Recogn. **82**, 105–117 (2018)

10. Zhang, N., Ruan, S., Lebonvallet, S., Liao, Q., Zhou, Y.: Kernel feature selection to fuse multi-spectral MRI images for brain tumor segmentation. Comput. Vis. Image Underst. **115**, 256–269 (2011)

11. Pereira, S., Pinto, A., Alves, V., Silva, C.A.: Brain tumor segmentation using convolutional neural networks in MRI images. IEEE Trans. Med. Imaging **35**, 1240–1251 (2016)

12. Shin, H.C., et al.: Deep nonvolutional neural networks for computer-aided detection: CNN architectures, dataset characteristics and transfer learning. IEEE Trans. Med. Imaging **35**, 1285–1298 (2016)

13. Kim, G.: Brain tumor segmentation using deep fully convolutional neural networks. In: Crimi, A., Bakas, S., Kuijf, H., Menze, B., Reyes, M. (eds.) BrainLes 2017. LNCS, vol. 10670, pp. 344–357. Springer, Cham (2018). https://doi.org/10.1007/978-3-319-75238-9_30

14. Li, Y., Shen, L.: Deep learning based multimodal brain tumor diagnosis. In: Crimi, A., Bakas, S., Kuijf, H., Menze, B., Reyes, M. (eds.) BrainLes 2017. LNCS, vol. 10670, pp. 149–158. Springer, Cham (2018). https://doi.org/10.1007/978-3-319-75238-9_13

15. Zhao, X.M., Wu, Y.H., Song, G.D., Li, Z.Y., Zhang, Y.Z., Fan, Y.: A deep learning model integrating FCNNs and CRFs for brain tumor segmentation. Med. Image Anal. **43**, 98–111 (2018)

16. Juan-Albarracín, J., et al.: Automated glioblastoma segmentation based on a multiparametric structured unsupervised classification. PLoS One **10**(5), e0125143 (2015)

17. Menze, B.H., van Leemput, K., Lashkari, D., Riklin-Raviv, T., Geremia, E., Alberts, E., et al.: A generative probabilistic model and discriminative extensions for brain lesion segmentation - with application to tumor and stroke. IEEE Trans. Med. Imaging **35**, 933–946 (2016)

18. Szilágyi, L., Szilágyi, S.M., Benyó, B., Benyó, Z.: Intensity inhomogeneity compensation and segmentation of MR brain images using hybrid c-means clustering models. Biomed. Signal Process. Control **6**, 3–12 (2011)

19. Szilágyi, L., Lefkovits, L., Benyó, B.: Automatic brain tumor segmentation in multispectral MRI volumes using a fuzzy c-means cascade algorithm. In: Proceedings of the 12th International Conference on Fuzzy Systems and Knowledge Discovery, pp. 285–291. IEEE (2015)

20. Lê, M., et al.: Personalized radiotherapy planning based on a computational tumor growth model. IEEE Trans. Med. Imaging **36**, 815–825 (2017)

21. Hamamci, A., Kucuk, N., Karamam, K., Engin, K., Unal, G.: Tumor-Cut: segmentation of brain tumors on contranst enhanced MR images for radiosurgery applicarions. IEEE Trans. Med. Imaging **31**, 790–804 (2012)

22. Sahdeva, J., Kumar, V., Gupta, I., Khandelwal, N., Ahuja, C.K.: A novel content-based active countour model for brain tumor segmentation. Magn. Reson. Imaging **30**, 694–715 (2012)

23. Njeh, I., et al.: 3D multimodal MRI brain glioma tumor and edema segmentation: a graph cut distribution matching approach. Comput. Med. Imaging Graph. **40**, 108–119 (2015)

24. Gordillo, N., Montseny, E., Sobrevilla, P.: State of the art survey on MRI brain tumor segmentation. Magn. Reson. Imaging **31**, 1426–1438 (2013)

25. Vovk, U., Pernuš, F., Likar, B.: A review of methods for correction of intensity inhomogeneity in MRI. IEEE Trans. Med. Imaging **26**, 405–421 (2007)

26. Szilágyi, L., Szilágyi, S.M., Benyó, B.: Efficient inhomogeneity compensation using fuzzy c-means clustering models. Comput. Methods Programs Biomed. **108**, 80–89 (2012)

27. Tustison, N.J., et al.: N4ITK: improved N3 bias correction. IEEE Trans. Med. Imaging **29**, 1310–1320 (2010)

28. Szilágyi, L., Iclănzan, D., Kapás, Z., Szabó, Z., Győrfi, Á., Lefkovits, L.: Low and high grade glioma segmentation in multispectral brain MRI data. Acta Univ. Sapientia Informatica **10**(1), 110–132 (2018)

29. Győrfi, Á., Kovács, L., Szilágyi, L.: A feature ranking and selection algorithm for brain tumor segmentation in multi-spectral magnetic resonance image data. In: 41st Annual International Conferences of the IEEE EMBS. IEEE (2019, accepted paper)

Computer Vision Approaches to Detect Bean Defects

Peterson A. Belan, Robson A. G. de Macedo,
and Sidnei A. de Araújo$^{(\boxtimes)}$

Informatics and Knowledge Management Graduate Program,
Universidade Nove de Julho – UNINOVE, Rua Vergueiro, 235/249 – Liberdade,
São Paulo, SP, Brazil
{belan, saraujo}@uni9.pro.br,
robson.gomes10@etec.sp.gov.br

Abstract. In this work are proposed computer vision approaches to detect three of the main defects found in beans: broken, bored by insect (*Acanthoscelides obtectus*) and moldy. In addition, we describe a fast and robust segmentation step that is combined with the proposed approaches to compose a computer vision system (CVS) applicable to the Brazilian beans quality inspection process, to determine the type of the product. The proposed approaches constitute an important practical contribution since, although there are some papers in the literature addressing visual inspection of beans, none of them deals with defects. In the conducted experiments a low-cost equipment, composed by a table made in structural aluminum, a conveyor belt and an image acquisition chamber, was used to simulate the characteristics of an industrial environment. The CVS evaluation was performed in two modes: offline and online. In the offline mode, a database composed by 120 images of bean samples containing grains of different classes and with different defects was employed, while in the online mode the grains contained in a batch were spilled continuously in the conveyor belt of the equipment for the proposed CVS to perform the tasks of segmentation and detection of defects. In the experiments the CVS was able to process an image of 1280×720 pixels in approximately 2 s, with average hit rates of 99.61% (offline) and 97.78% (online) in segmentation, and 90.00% (offline) and 85.00% (online) in detecting defects.

Keywords: Computer vision · Beans · Defects · Inspection · Visual quality

1 Introduction

The visual properties of many agricultural products, including beans, are important factors in determining their market prices and assisting the consumer choice. Thus, visual inspection processes are essential to ensure the quality of these products [1–4].

Brazil is the world's largest producer and consumer of beans, followed by India, China and Mexico. Beans, together with rice, form the basis of the Brazilian people's diet. The visual inspection of Brazilian bean quality is done manually following operating procedures established by the Ministry of Agriculture, Livestock and Supply, which instruct how to frame the beans in a group, according to the botanical species

© Springer Nature Switzerland AG 2019
I. Nyström et al. (Eds.): CIARP 2019, LNCS 11896, pp. 336–345, 2019.
https://doi.org/10.1007/978-3-030-33904-3_31

(Group I comprises the specie *Phaseolus vulgaris* and Group II the specie *Vigna unguiculata*); class, according to the skin color of the grains and type, according to the quantities of defects [5].

Since manual quality inspection processes are usually subject to problems such as high operational costs and difficulty in standardizing results [6, 7], the automation of such processes represents an important alternative to reduce costs and standardize results, generating a competitive differential for companies, especially in the context of agroindustry 4.0, in which is preconized the use of intelligent technologies to optimize decision making and improve production processes, making it possible to obtain products with greater quality, respecting the environment and sustainability [8].

In the literature of the last two decades we can find several works proposing computer vision systems (CVS) for quality inspection of grains. Specifically for beans there are the works [6, 9–16]. Surprisingly none of them addresses defects detection, which is an indispensable task for a CVS developed to classify and typify the beans. Obviously, there are several works in the literature addressing the detection of defects in fruits and vegetables. However, the defects considered in these works are different from those that affect beans.

It is in this context that this work is inserted, with the proposition of computer vision-based approaches to detect three of the main defects found in beans: broken, bored by insect (*Acanthoscelides obtectus*) and moldy. These approaches, besides consisting in a novelty in terms of application, are feasible for embedded systems since they require low computational cost. In addition, we describe a fast and robust segmentation step that is combined with proposed approaches to compose a CVS for quality inspection of Brazilian beans, to determine the type of the product.

2 Related Works

In the literature of last two decades we can find some works addressing automated approaches for visual quality inspection of beans. Kiliç et al. [6], for example, proposed a CVS for beans classification based on color and grain size, using color histograms and a Multilayer Perceptron artificial neural network (MLP-ANN). In Venora et al. [9, 10] computer vision approaches were proposed for the identification of varieties of beans grown in Italy, based on the attributes of grain size, shape, color and texture. Laurent et al. [11] employed image processing techniques to evaluate changes in bean color during storage, in order to correlate such changes with a phenomenon called "hard-to-cook grains".

The works [6, 9–11] show the necessity and importance of computational systems for bean inspection. However, the segmentation approaches proposed in these works fail in images containing glued grains (touching grains). For this reason, in the experiments conducted, the authors manually positioned the grains, spaced from each other, to acquire the image or acquired an image for each grain, in order to facilitate the segmentation step. This is a severe limitation that hinders the practical application of the systems proposed in such works.

The works proposed in the literature in the last five years have been concerned to solve the problem of touching grains. The work of Araújo, Pessota and Kim [12], for example, presents a CVS to classify the most consumed Brazilian beans that includes a robust routine of segmentation capable of solving the problem of touching grains. It employs the k-NN algorithm for removal the background of the image and the normalized cross-correlation granulometry technique, using elliptical kernels, to segment the grains, even though they are touching each other. Despite the high hit rates obtained in segmentation and classification tasks (99.88% and 99.99%), the computational time spent to segment an image (16 s for an image of 800 × 600 pixels), prevents the practical application of the system proposed in [12].

Since then, the works found in the literature addressing the visual inspection of beans such as [13–16], focused on the development of low computational cost approaches that keep the high hit rates obtained in [12].

Although many advances have been achieved in terms of segmentation and classification of beans, there are no works in the literature addressing the detection of defects, which is an important part of the visual quality inspection of the product.

3 Proposed Approaches

Consider as input a color image denoted by a function $I : \mathcal{D} \rightarrow \mathbb{K}$, which maps a rectangular grid $\mathcal{D} \subseteq \mathbb{Z}^2$ into a set of color intensities $\mathbb{K} = \{0, 1, \ldots, 2^{nBits} - 1\}^{nBands}$, where $nBits$ is the number of bits to represent a pixel intensity and $nBands$ the number of bands of employed color system. Consider also that each pixel of I is represented by $p = (p_x, p_y)$ whose values p_x and p_y denote the horizontal and vertical coordinates of p. First, I is segmented and, after that, the approaches for detecting defects are applied.

3.1 Segmentation

First, the input RGB image (I_{RGB}) is converted to the CIELab color system generating I_{Lab}. Each pixel p from I_{Lab} is then mapped for two class of gray in I_{gray}, in which typical bean colors are represented by the darker tone (black) and the typical background colors are represented by the lighter tone (white). For this, differently from the proposal of Araújo, Pessota and Kim [12] in which a training is done for each image using the k-NN algorithm, we employ a look-up table (LUT) to compute this mapping more efficiently. Once the LUT is created it is possible to describe the mapping process to remove the image background, as defined by Eq. 1.

$$\forall p \in D, I_{gray}(p) = LUT(I_{Lab}(p)) \tag{1}$$

For the refinement of the mapping process the morphological opening operator is applied to remove small connected components, considered as noise. The next step consists in the application of the watershed transform (WT), using distance transform (Euclidean distance), to segment the grains in I_{gray}. As WT is very susceptible to the problem of super-segmentation [17], we use some information such as average grain size and distance between the centers of two grains to join two or more fragments erroneously divided by WT.

Finally, the cross-correlation granulometry (CCG) technique described in [12] is applied in regions of I_{gray} where WT was not able to segment correctly the grains. It is identified from connected components considered too large. However, instead of the 162 templates originally used in [12], we employed a set of 72 elliptical templates ($\mathcal{T}p$) defined from 4 scales and 18 rotation angles at each scale. The CCG operation can be mathematically formulated as:

$$\forall \, p \in \mathcal{D}, \left[CCG(CC(I_{gray}), \mathcal{T}p)\right](p) = \arg \max_{\mathcal{T}} \{\left[NCC(CC(I_{gray}, \mathcal{T}))\right](p) : \mathcal{T} \in \mathcal{T}p\} \quad (2)$$

where $CC(I_{gray})$ is a connected component extracted from I_{gray}, NCC is the normalized cross-correlation operation and $\mathcal{T}$ is a template belonging to the set of templates $\mathcal{T}p$. The working of segmentation process is shown in Fig. 1.

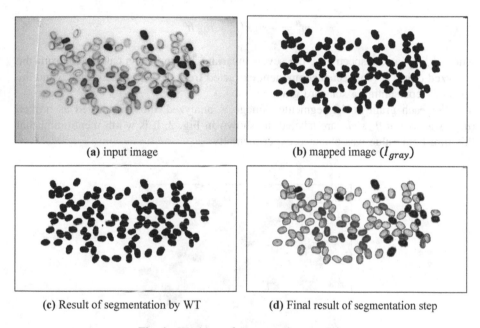

(a) input image   (b) mapped image (I_{gray})

(c) Result of segmentation by WT   (d) Final result of segmentation step

Fig. 1. Working of segmentation process.

It should be noted that the fact of applying CCG only in some regions of the image leads to a large reduction in segmentation processing time.

3.2   Detection of Bean Defects

Broken Beans. As known, the shape of a whole bean grain is very close of an ellipse. Thus, the detection of a broken bean can be done using some algorithm that extracts the signature of the image of a grain and compares it with the signature extracted from the image of an ellipse or from a whole grain, considered as standard.

The verification of the broken grain signature, described by Eqs. 3 to 5, is conducted using the seven invariant moments of Hu [18]. Basically, we compute the difference between signatures (*diff_sign*) of a standard grain and the grain under analysis. If this difference is greater than a threshold *t_diff_sign* (in our experiments *t_diff_sign* = 0.4) then the grain is considered as broken. The choice of this scheme was motivated by the speed and precision of the algorithm and the facility of its implementation. In addition, it was considered that the moments of Hu are invariant to scale and rotation, essential for the purpose explored in this work.

$$diff\_sign(A, B) = \sum_{i=1}^{7} \left| \frac{1}{m_i^A} - \frac{1}{m_i^B} \right| \tag{3}$$

$$m_i^A = -sign\left(h_i^A\right).\log h_i^A \tag{4}$$

$$m_i^B = -sign\left(h_i^B\right).\log h_i^B \tag{5}$$

where h_i^A, h_i^B are, respectively, the seven invariant Hu moments calculated from the analyzed object A (connected component extracted from the segmented image) and the object B representing a standard grain.

After each grain in the segmented image is analyzed, those identified as broken (*diff_sign* > *t_diff_sign*) are labeled, as shown in Fig. 2. It is worth mentioning that this threshold value was obtained experimentally.

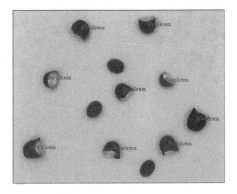

Fig. 2. Example of result of the detection of broken grains (Color figure online)

It is valid to mention that the signature could be extracted using other robust techniques such as Curvature Scale Space (CSS). However, in the tests performed in this work CSS spent approximately 0.73 s to analyze an image of 1280 × 720 pixels while the algorithm employed spent 0.01 s to perform the same task.

Beans Bored by Insect (*Acanthoscelides obtectus*). The Hough transform for circle detection (HTC) was used to detect the holes caused by the insect. The employed algorithm is implemented in the OpenCV image processing library[1]. Firstly, a median blur filter is applied to I_{gray} with a kernel of 3×3 pixels aiming to increase the successful of the HTC, that is applied in the sequence. To avoid problems of noise in the detection of the holes, maximum and minimum radii are stipulated and adopted as parameters of the HTC. These values are easy to obtain since the holes have homogeneous sizes, as can be seen in Fig. 3a.

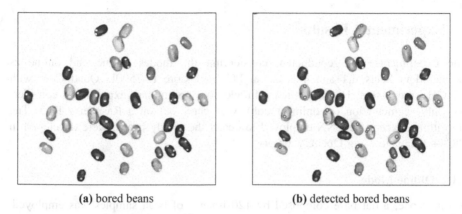

(a) bored beans   (b) detected bored beans

Fig. 3. Detection of bored beans. (Color figure online)

The holes caused by Acanthoscelides obtectus detected by HTC algorithm are marked on the output image with a red circle, as showed in Fig. 3b.

Moldy Beans. For the detection of moldy grains, a convolutional neural network (CNN) was employed. It was trained using a dataset composed by 5920 subimages (central regions of the grains with 30×30 pixels, as showed in Fig. 4) extracted from healthy and defective grains. This dataset was divided into two parts, being 4000 subimages for training and the remaining 1920 for validation of the training.

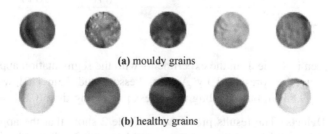

(a) mouldy grains

(b) healthy grains

Fig. 4. Examples of images used for CNN training/validation

---

[1] https://opencv.org/.

The CNN employed in this task was implemented in Phyton, using the Keras library[2], and its parameterization was adjusted in order to receive images of 30×30 pixels in the input layer. It comprises two convolutional layers formed by 16 maps of 3×3 intercalated with two pooling layers with dimension of 2×2, and two fully connected layers with 64 and 1 neurons.

Finally, the following parameters were used for training of described CNN: 8000 steps per epoch and a total of 20 epochs, totaling 160000 iterations.

4   Experimental Results

The experiments were conducted, considering the modes offline and online as explained in Sects. 4.1 and 4.2, on an PC Intel Core i7-7500U Quad Core with 2.7 GHZ processors, 16 GB Memory. In addition, preliminary experiments considering only segmentation (in online mode) were carry out on a Raspberry Pi 3. The algorithms, except the CNN employed to detect the moldy beans, were developed in C/C++ language using OpenCV library.

4.1   Offline Mode

In this mode, a database composed by 120 images of bean samples was employed. From these images, 100 were used to evaluate the segmentation step, since if it fails, the detection of defects will be affected. The remaining 20 images were used to evaluate the approaches for detection of defects.

Segmentation. Table 1 presents the hit rates obtained in the evaluation of the segmentation step. Each of 100 images contains 100 grains of three different classes as showed in Figs. 1a and 3a.

Table 1. Results of segmentation in offline mode

Number of grains in the 100 images	Number of grains detected correctly (True Positives - TP)	False Negative (FN)	False Positive (FP)	Hit rate (%)
10,000	9,961	14	25	99.61

As can be seen in Table 1, in the experiments with the segmentation approach, a hit rate very close to that presented by Araújo, Pessota and Kim [12] was obtained (99.61% versus 99.88%), but spending much less processing time (0.8 s versus 16 s).

Detection of Defects. The results presented in Table 2 show that the approaches for detection of broken and bored beans were not able to detect all the defective grains contained in the images. In the case of the bored beans, some grains have dark brown streaks mixed with its predominant skin color (see Fig. 2a), making it difficult to detect

---

[2] https://keras.io/.

Table 2. Results obtained by the approaches to detect defective grains in offline mode

Number of grains containing each defect	Broken beans		Bored beans		Moldy beans	
	Grains detected	Hit rate (%)	Grains detected	Hit rate (%)	Grains detected	Hit rate (%)
50	40	80.00	45	90.00	50	100.00

the hole caused by *Acanthoscelides obtectus*. In some cases, even for the human eye is a difficult task. On the other hand, the moldy detection approach did not make detection errors, demonstrating the effectiveness of the built CNN.

The computational time consumed by defects detection task was approximately 1.2 s, being 83% of this time (1 s) was consumed by CNN in detecting of moldy beans.

4.2 Online Mode

In the online mode the grains contained in a batch of 1000 grains were spilled continuously in the conveyor belt of the equipment for the proposed CVS to perform the segmentation and detection of defects. In addition, some preliminary experiments considering the segmentation task were conducted with a Raspberry Pi 3, in order to investigate the feasibility of using the proposed approaches in an embedded platform.

Segmentation. Table 3 presents the hit rates and standard deviations (σ) obtained in 3 experiments with 3 replicates in each of them. As can be seen, these results were lower than those obtained with the database of images (offline). This occurs mainly due to loss of image quality since the grains are in movement and also due to the existence of small ranges in the samples of grains between two consecutive acquisitions which are not analyzed. Although the equipment has been adjusted to acquire the images avoiding this last problem, there are still cases of grain that are not processed.

As one can see in Table 3, the results obtained in the experiments with Raspberry were inferior to the results obtained with the use of a PC. This was due to the decrease in the number of calls of the CCG algorithm, to enable the execution of the segmentation task on the Raspberry platform with a suitable time (4.8 s, in average). Finally, the low standard deviations in all experiments indicate good stability of the segmentation approach.

Table 3. Results of segmentation in online mode

Experiment	Number of grains detected			Average	σ	Hit rate
1 (PC)	977	998	979	984.70	9.46	98.47%
2 (PC)	985	969	999	984.33	12.26	98.43%
3 (Raspberry)	961	965	967	964.33	2.49	96.43%

Detection of Defects. Table 4 presents the hit rates obtained by approaches to detect defects in online mode. In these experiments the results are also lower than the results obtained in offline mode. In the case of broken grains, some of them were glued to whole grains or even to other broken grains, generating segmentation errors and, consequently, failure in the detection of this defect. Regarding to defect of bored beans (holes), again the fact that the beans are in movement leads to the small loss in the image quality that is enough to cause errors in the detection of the holes.

Table 4. Results obtained by the approaches to detect defective grains in online mode

Number of grains containing each defect	Broken beans		Bored beans		Moldy beans	
	Grains detected	Hit rate (%)	Grains detected	Hit rate (%)	Grains detected	Hit rate (%)
50	41	82.00	44	88.00	–	–

Unfortunately, as the approach to detect moldy beans was developed on a computing platform different than that one used for the development of other approaches, it was not incorporated into CVS and therefore was not evaluated in online mode experiments.

5 Conclusions

Regarding the segmentation, the approach described in this work presented an excellent result in all aspects, since it obtained an average hit rate of 98.24% (99.61% in offline mode and 97.78% in online mode, including the experiments on a Raspberry), requiring only 0.80 s to segment an image on a PC and 4.8 s on a Raspberry. The results obtained in the detection of defects showed that the approaches developed are good alternatives, since they presented average hit rates of 81% for broken beans, 89% for bored beans and 100% for moldy beans (considering only offline experiments for this last defect). Although the approaches for defect detection have shown reasonable performance and low computational cost, about 0.2 s for detection of broken and bored beans and about 1.0 s for the detection of moldy beans, there is still room for optimization of the processing time of the latter approach. These low processing times, without using multithread or other advanced hardware and software resources, make these approaches feasible to be executed by an embedded system, since they can be optimized. This optimization is a challenge that will be addressed in our futures works.

Acknowledgements. The authors would like to thank FAPESP – São Paulo Research Foundation (Processes 2017/05188-9 and 2019/18389-8) by financial support.

References

1. Aggarwal, A.K., Mohan, R.: Aspect ratio analysis using image processing for rice grain quality. Int. J. Food Eng. **6**(5), 1–14 (2010)
2. Fernández, L., Castillero, C., Aguilera, J.M.: An application of image analysis to dehydration of apple discs. J. Food Eng. **67**(1–2), 185–193 (2005)
3. Stegmayer, G., Milone, D.H., Garran, S., Burdyn, L.: Automatic recognition of quarantine citrus diseases. Expert Syst. Appl. **40**(9), 3512–3517 (2013)
4. Zareiforoush, H., Minaei, S., Alizadeh, M.R., Banakar, A., Samani, B.H.: Design, development and performance evaluation of an automatic control system for rice whitening machine based on computer vision and fuzzy logic. Comput. Electron. Agric. **124**, 14–22 (2016)
5. Knabben, C.C., Costa, J.S.: Manual of classification of the bean: normative instruction n. 12, 28 March 2008. Embrapa Rice and Bean-Folder/Leaflet/Booklet – INFOTECA-E (2012)
6. Kiliç, K., Boyaci, I.H., Köksel, H., Küsmenoglu, I.: A classification system for beans using computer vision system and artificial neural networks. J. Food Eng. **78**, 897–904 (2007)
7. Patil, N.K., Yadahalli, R.M., Pujari, J.: Comparison between HSV and YCbCr color model color-texture based classification of the food grains. Int. J. Comput. Appl. **34**(4), 51–57 (2011)
8. Parronchi, P.: The development pioneers and the new agriculture 4.0: economic development from the countryside? In: XXIII Encontro Nacional de Economia Política, pp. 1–18 (2017)
9. Venora, G., Grillo, O., Ravalli, C., Cremonini, R.: Tuscany beans landraces, on-line identification from seeds inspection by image analysis and linear discriminant analysis. Agrochimica **51**, 254–268 (2007)
10. Venora, G., Grillo, O., Ravalli, C., Cremonini, R.: Identification of Italian landraces of bean (Phaseolus vulgaris L.) using an image analysis system. Sci. Hortic. **121**, 410–418 (2009)
11. Laurent, B., Ousman, B., Dzudie, T., Carl, M.F M., Emmanuel, T.: J. Eng. Technol. Res. **2**(9), 177–188 (2010)
12. Araújo, S.A., Pessota, J.H., Kim, H.Y.: Beans quality inspection using correlation-based granulometry. Eng. Appl. Artif. Intell. **40**, 84–94 (2015)
13. Araújo, S.A., Alves, W.A.L., Belan, P.A., Anselmo, K.P.: A computer vision system for automatic classification of most consumed Brazilian beans. In: Bebis, G., et al. (eds.) ISVC 2015. LNCS, vol. 9475, pp. 45–53. Springer, Cham (2015). https://doi.org/10.1007/978-3-319-27863-6_5
14. Belan, P.A., Araújo, S.A., Alves, W.A.L.: An intelligent vision-based system applied to visual quality inspection of beans. In: Campilho, A., Karray, F. (eds.) ICIAR 2016. LNCS, vol. 9730, pp. 801–809. Springer, Cham (2016). https://doi.org/10.1007/978-3-319-41501-7_89
15. Belan, P.A., Pereira, M.M.A., Araújo, S.A., Alves, W.A.L.: Abordagem Computacional para Classificação Automática de Grãos de Feijão em Tempo Real. In: SeTII 2016, São Paulo, Brazil, pp. 1–4 (2016)
16. Belan, P.A., de Macedo, R.A.G., Pereira, M.M.A., Alves, W.A.L., de Araújo, S.A.: A fast and robust approach for touching grains segmentation. In: Campilho, A., Karray, F., ter Haar Romeny, B. (eds.) ICIAR 2018. LNCS, vol. 10882, pp. 482–489. Springer, Cham (2018). https://doi.org/10.1007/978-3-319-93000-8_54
17. Soille, P.: Morphological Image Analysis: Principles and Applications. Springer, Heidelberg (2003)
18. Hu, M.K.: Visual pattern recognition by moment invariants. IRE Trans. Inf. Theory **8**(2), 179–187 (1962)

Automated Identification of Breast Cancer Using Digitized Mammogram Images

Bryan Chachalo[1], Erik Solis[1], Eddy Andrade[1], Santiago Pozo[1],
Robinson Guachi[3,4], Saravana Prakash Thirumuruganandham[1],
and Lorena Guachi-Guachi[1,2(✉)]

[1] Yachay Tech University, Hacienda San José, 100119 Urcuquí, Ecuador
lguachi@yachaytech.edu.ec
[2] SDAS Research Group, Yachay Tech University, Urcuquí, Ecuador
[3] Department of Mechatronics, Universidad Internacional del Ecuador,
Av. Simon Bolivar, 170411 Quito, Ecuador
[4] Department of Mechanical and Aerospace Engineering, Sapienza University of
Rome, Via Eudossiana 18, 00184 Rome, RM, Italy
http://www.sdas-group.com/

Abstract. In this paper, we present a new algorithm based on Computer-Aided Diagnosis (CAD) to detect breast cancer using digitized mammogram images. Here, we use image processing to make the pre-processing step of the images before we enter them into the classification step, in which we use machine learning for the classification of tissues in two conditions: normal and abnormal, either three conditions: normal, benign or malignant. In our CAD implementation, for pre-processing we are using transformations as binarization, thresholding, smoothing and the main operation, Gabor wavelet to suppress labels and unnecessary information, to obtain the best identifying characteristics. For feature selection and dimensionality reduction, we are using techniques as Principal Component Analysis (PCA), t-Distributed Stochastic Neighbor Embedding (TSNE) and models of analysis of variance. Finally, for classification we discussed k-Nearest Neighbors (k-NN).

Keywords: Breast cancer · k-NN · Image processing

1 Introduction

Breast cancer is the top cancer in women worldwide and is increasing particularly in developing countries where the majority of cases are diagnosed in late stages. It is estimated that worldwide over 508,000 women died in 2011 due to breast cancer [1]. Early diagnosis improves cancer outcomes by providing care at the earliest possible stage and is therefore an important public health strategy in all settings. There is a wide variety of tools and technologies to screen, detect, and diagnose breast cancer. In this sense, mammography is essentially the only

© Springer Nature Switzerland AG 2019
I. Nyström et al. (Eds.): CIARP 2019, LNCS 11896, pp. 346–356, 2019.
https://doi.org/10.1007/978-3-030-33904-3_32

widely used imaging modality for breast cancer screening that has become as an efficient non-invasive tools, so, many optimized CAD systems were created through the use of mammogram images.

Current CAD systems in clinical use serve as a second reader to assist radiologists in the mammographic interpretation process. It works mainly in some common steps such as pre-processing tasks, feature extraction and classification. Pre-processing is performed to enhance the mammogram visual quality and perceptibility of the anomalies present in the breasts. Features extraction obtains a set of discriminative and informative data such as texture, abnormality type, mass ratio, color, shape, spatial relations, among others. Classification allows to predict the class/category of given data features from pixel images, which recently is performed with artificial intelligence techniques [2,3] characterized by a high effectiveness and computational cost. One problem faced by radiologists and the CAD systems is that acquired mammogram images often have low quality having slight dissimilarity between normal, benign, and malignant cancer tissues that leads to inaccurate results. The digitized images took by the mammograph need to be improved, because image features can be distinguished and they can reflect the subtle variance in the order of many degrees. Thereby, pre-processing tasks for mammogram image enhancement become critical before feature extraction.

In this sense some works have been introduced for enhancement and classification of mammogram images. Histogram Equalization (HE) is applied as one of the popular methods for contrast enhancement which modify the gray level histogram of an image to a uniform distribution [4]. But in many cases it produces over enhancement in output image and loss of local information which yields an over enhancement, thus results loss of local information. In order to overcome this limitation methods, models such as LCM-CLAHE [5] are proposed. This method conducts an optimal contrast without losing any local information of the mammogram image. LCM-CLAHE consists of two stages of processing to increase the potentiality of contrast enhancement, and also to preserve the local details in the image. In the mathematical details of algorithms, it is inevitable to discuss about the interpolation techniques such as cubic, nearest-neighbor, and linear [6,7]. Interpolation is the process of estimating the intermediate values of a contiguous event from discrete samples; it is used extensively in digital image processing to magnify or to reduce images and to correct spatial distortions. Because of the amount of data associated with digital images, an efficient algorithm is essential. To note that, cubic convolution interpolation (CCI) was used in response to this requirement. In CCI, fiction is more accurate than the nearest-neighbor algorithm or linear interpolation method, thus it can be performed much more efficiently. Here we recall that, Bottom-hat [8] is a technique used for extraction of various features of a digital image. A Bottom-hat filter enhances black spots in a white background by performing a dilation followed by erosion. The effect is to fill holes and join nearby objects. Two-dimensional spatial linear filters are constrained by general uncertainty relations that limits their attainable information in resolution for orientation, spatial frequency, and 2D spatial position [9]. The theoretical lower limit for the joint entropy, or uncertainty of these variables is achieved by a family of optimal 2D filter, whose spatial weighting functions are generated by exponentiated bivariate second-order polynomials with complex coefficients.

The presented approach of this paper is designed and implemented using the database of the Mammographic Image Analysis Society (MIAS) [10], because it has a better resolution obtained than the images that are enhanced in DDSM [11] approach. Additionally, we applied Gabor wavelet to distinguish clearly the normal and abnormal tissues in digital mammograms. The purpose of Gabor wavelet is to use elliptic generalization of the one-dimensional elementary functions [12]. Also to note that, in this work we revealed the Gabor filter bank with different orientations and scales extracts the texture pattern such as edges, lines, spots and flat areas in the images, which helps to discriminate normal and malignant tissues. Finally, k-NN classifier is discussed for classification.

2   Materials and Method

2.1   Dataset

Automatic learning is one of the most used artificial intelligence techniques in intelligent systems, such as computational vision, language processing and classification. The correct selection of the database that will be used in an automatic learning process is a fundamental stage for the design and implementation of the system. There are many databases of mammogram images. The Mammographic Image Analysis Society is an organization of UK research groups which contains MIAS database of digital mammograms [13], with a resolution of 1024. It is a collection of 322 digitised films that includes normal, benign, and malignant cases and radiologist's "ground truth" with the detection of any abnormalities that may be present. Total of 330 samples were obtained from MIAS dataset for training (207 normal, 69 malignant, 54 benign), while 492 were used for testing (158 normal, 231 malignant, 103 benign) purposes.

In order to provide an effective breast cancer diagnosis classification, it is essential the use of mamogram dataset that contains close number of images for each lesion/class. Thus, we have used data augmentation technique to increase the number of images for each case (207 normal, 207 malignant, 207 begnin) by flipping images horizontally and vertically.

2.2   Method

The proposed method is introduced to classify breast cancer in normal, begnin, and malignant conditions. After the data acquisition condition, rest of the pre-processing tasks are sequentially performed in order to enhance the mammogram images. Further, feature extraction is carried out to extract significant information to distinguish breast cancer of 3 different conditions. Hence, in order to reduce the amount of discriminative features, dimensionality reduction techniques are applied. Finally, classification stage is performed through supervised features learning. The work flow of the proposed method is depicted in Fig. 1. In the following sub-sections, the main computational stages are described.

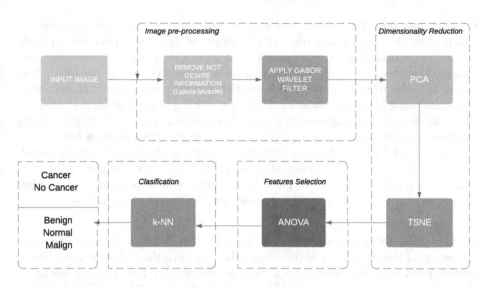

Fig. 1. Description of algorithm.

Algorithm 1 Remove Labels

1: **function** REMOVELBL($Img, ImgBin, color$)
2:    $height = height.Img$
3:    $width = width.Img$
4:    **for** $i = 0$ to $height$ **do**
5:      **for** $i = 0$ to $width$ **do**
6:       **if** $color = 0$ **then**
7:        **if** $ImgB[i, j] = 0$ **then**
8:         $Img[i, j] = 0$
9:        **else if** $ImgB[i, j] = 255$ **then**
10:         $Img[i, j] = 0$
11:        **end if**
12:       **end if**
13:      **end for**
14:    **end for**
15: **end function**

(a)

Algorithm 2 Binarize Through a ROI

1: **function** THRESHROI(Img, Thr)
2:    $height = height.Img$
3:    $width = width.Img$
4:    $ImgRes = ZeroMatrix(height, width)$
5:    $Left = width.Img/2$
6:    $Rigth = width.Img/2$
7:    **for** $i = 0$ to $height$ **do**
8:      **for** $i = 0$ to $width$ **do**
9:      **if** $j < Left$ or $j > Rigth$ **then**
10:       **if** $Img[i, j] > Thr$ **then**
11:        $ImgRes[i, j] = 255$
12:      **else**
13:       $ImgRes[i, j] = 0$
14:      **end if**
15:      **end if**
16:      $ymidI- = 1$
17:      $ymidR+ = 1$
18:      **end for**
19:    **end for**
20: **end function**

(b)

Algorithm 3 Remove Edges

1: **function** RMVEDGES(Img)
2:    $Ret = \textbf{Threshold}(Img, 240, 255, BINARY)$
3:    $edgesTh = \textbf{Threshold}(Img, 240, 255, BINARY)$
4:    $Kernel = \textbf{OnesMatrix}((20, 20), uint8)$
5:    $dilEdges = \textbf{Dilate}(edgesTh, kernel, iter = 1)$
6:    $NoEdges = \textbf{RemoveLbl}(Img, dilEdges, 255)$
7: **end function**

(c)

Algorithm 4 Generate Filter Gabor Wavelet

1: **function** BUILDFILTER(Img)
2:    $height = height.Img$
3:    $width = width.Img$
4:    $ImgRes = ZeroMatrix(height, width)$
5:    $Filter(Img)$
6:    **for** $Kernel = 0$ to $Filter$ **do**
7:      $fimg = \textbf{Filter2D}(Img, CV_8UC3, Kernel)$
8:      $\textbf{Maximum}(accum, fimg, accum)$
9:    **end for**
10:
11: **end function**

(d)

Algorithm 5 Pre Processing Images

1: **function** PREPROCESSING(Img)
2:    $Img = \textbf{Resize}(Img, (1360, 796))$
3:    $Smooth = \textbf{GaussianBlur}(Img, (5, 5), 0)$
4:    $Ret = \textbf{Threshold}(Img, 65, 255, BINARY)$
5:    $SmoothBin = \textbf{Threshold}(Img, 65, 255, BINARY)$
6:    $kernel = \textbf{OnesMatrix}((55, 55), uint8)$
7:    $erosion = \textbf{Erode}(SmoothBin, kernel, iter = 1)$
8:    $dilation = \textbf{Dilate}(SmoothBin, kernel, iter = 1)$
9:    $NoLbl = \textbf{RemoveLbl}(Img, dilation, 0)$
10:    $smoothNoLbl = \textbf{GaussianBlur}(NoLbl, (39, 39), 0)$
11:    $smoothRoi = \textbf{ThreshROI}(smoothNoLbl, 150)$
12:    $NoLbl2 = \textbf{RemoveLbl}(NoLbl, smoothRoi, 255)$
13: **end function**

(e)

Fig. 2. The pseudo-code of the proposed method. (a) Remove labels; (b) ROI binarization; (c) Remove edges; (d) Buid Gabor filter; (e) Pre-processing function.

Pre-processing. The aim of this step is to remove undesirable information which could degrade the accuracy of the proposed approach. Information such as margins, labels, and some regions of tissues were removed from mammogram images. The pre-processing steps starting from resizing the images to 1360×796 pixels. In order to remove labels and not the desired features, first images were smoothed using Gaussian Blurring defining width and height of kernel to 5×5. Then, images are binarized using global thresholding $Th1 = 65$. Additionally, morphological techniques are applied, first erosion and then dilation both using a kernel size of 55×55. Another smoothing is applied to images using a kernel size 39×39 and a second binarization is applied through a region of interest using a threshold of $Th2 = 150$. These parameters value were fixed for the entire dataset. Figure 2a, b and c explain the our algorithmic implementation.

Feature Enhancement and Extraction. It consists of the improvement and extraction of characteristics of the images. In this work the Gabor wavelet filters were applied over the entire image in order to extract discriminant features. Next, all image pixels were casted in to a 32 bit floating point. Gabor function parameters are specified on Table 1.

Dimensionality Reduction (DR). The main objective of DR is to find a low-dimensional representation of the data set that retains as much information as possible. Generally, feature extraction produces enormous data, and it is difficult to analyze. In this approach, Principal Components Analysis (PCA) and t-Distributed Stochastic Neighbor Embedding (TSNE) were used to reduce the dimensionality of mammograms. First, PCA was used saving 95% ratio of variance. Subsequently, TSNE was applied to entire data, holding 2 components. Figure 3a depicts the implementation [14].

Feature Selection (FS). It is a process in which automatically or manually select those features that contribute most of the prediction variable or the output. This work is interested in FS techniques for enhancing classifier performance, computational time and cost effective. It is noteworthy to mention that the obtained features from FS techniques were subjected to feature selection process by using Analysis of Variance (ANOVA).

Classification. This phase provides a procedure for the assignment of a class label (mammogram condition type) to the input pattern (mammogram). This work uses k-nearest neighbors (k-NN), one of the most popular machine learning methods. It is a method of supervised learning, based on instances and that does not require a learning phase. The k-NN algorithm is based on instances and allows the classification of new elements by calculating their distance to all the other elements $dist(X_1, X_2)$. The proper functioning of the algorithm depends on the choice of the distance function used and the value of the parameter k, which represents the number of nearby neighbors to the query x_q. The neighbors

Algorithm 6 Dimensionality Reduction
1: function DIMREDUC(Img)
2:    $pca = PCA(n_components = 0.95)$
3:    $ImgReduced = PCA.FitTransform(Img)$
4:    $tsne = TSNE(n_jobs = 150, n_components = 3, random_state = 42)$
5:    $TSNEImg = TSNE.FitTransform(ImgReduced)$
6: end function

(a)

Algorithm 7 Image Classification
1: function IMGCLA(TSNEImg)
2:    $Fvalue = SelectKBest(Fclassif, k = 3)$
3:    $Xkbest = Fvalue.FitTransform(TSNEImg)$
4:    $knnReg = neighbors.KNeighborsClassifier()$
5:    $knnSearch = RandomizedSearchCV(knnReg)$
6:    $knnSearch = knnSearch.Fit(XTrainImgs, YTrainImgs)$
7:    $Prediction = CrossValPredict(knnSearch, XTestImgs, YTestImgs)$
8: end function

(b)

Algorithm 8 Main Function
1: function MAIN
2:    $Img = ReadImage$
3:    $Img = Preprocessing(Img)$
4:    $filters = BuildFilters()$
5:    $resImg = Process(Img, filters)$
6:    $PPImg = RmvEdges(resImg)$
7:    $dimReducImg = DimReduc(PPImg)$
8:    $clasfImg = ImgCla(dimReducImg)$
9: end function

(c)

Fig. 3. The pseudo-code of the proposed method. (a) Dimensionality reduction; (b) Image classification; (c) Main computational calls.

are weighed by the distance that separates them from the new elements that are classified. This work uses the minkowski distance in order to reach a greater precision with a minimum effect due to the variation of the parameter k. K-NN parameters setting are specified on Table 1. K-NN is effective in noisy training data and suitable for cases of a large number of training samples, however, the computation time is increased as much as we need to compute the distance of each instance to all training samples. The main code sequences of K-NN are reported in Fig. 4.

Algorithm 8 k-NN algorithm.

Choose the best values for the parameters using RandomizedSearchCV, prediction with the best_estimator, and/or score.
Input: X: matrix with features to train, y:vector of labels, hyper parameters (n_neighbors, weights, metrics)
1: **function** RSEARCHCV $(n\_neighbors, weights, metrics)$
2:    Create a dictionary with the n_neighbors, weights, metrics values to try.
3:    Use Randomized search on hyper parameters
4:    **return** best_estimator   ▷ best hyper parameters for knn classifier
5: **procedure** KNN_CLASSIFIER$(X\_train, y\_train, x\_test, y\_test)$
6:    Use RSearchCV to get the best_estimator
7:    Fit the model with X_train,y_train
8:    Predict labels for x_test
9:    **return** predicted values   ▷ Score can also be obtained by using y_test

Fig. 4. The pseudo-code of the k-NN algorithm.

3   Results and Discussion

In order to process our immense data, available computational resources of simple workstation's were not enough to realize the task. So, to enhance the speed of our calculations, we carried-out all the calculations using our 1640 cores HPC facility called Quinde I, and to accomplish our task 150 cores of Quinde I were used. The algorithm was implemented in python 3, that uses libraries such as sci-kit learn to execute the portion of machine learning algorithms; OpenCV for an image pre-processing task. Similarly, feature enhancement and other tools such as panda, matplotlib, numpy were used for data handling task. Here to note that, Quinde I functions under Linux Red Hat Enterprise Linux Server release 7.2 (Maipo) little endian. The remote computer is also capable of running under the Linux environment.

Table 1. Parameter values.

	Parameter	Tested values	Best value
Gabor kernel	ksize	1; 2; 5; 10; 20; 30; 33; 35; 40; 45	10
	sigma	1; 2; 3; 4; 5; 6; 8; 10; 20	4
	lambda	1; 2; 5; 10; 15; 20	10
	gamma	0.1; 0.2; 0.3; 0.4; 0.5; 0.6; 0.7; 1.0; 1.5	0.5
	psi	0; 0.5; 1; 1.5	0
	ktype	CV_32F	CV_32F
K-NN algorithm	k (number of neighbors)	3; 5; 8; 11; 15	8
	weights	uniform; distance	uniform
	algorithm	auto; ball_tree; kd_tree; brute	auto
	metric	euclidean; manhattan; minkowski	minkowski

Recall, precision, specificity and overall accuracy defined by Eqs. 1, 2, 3 and 4, respectively, have been evaluated to measure the effectiveness of the proposed method. Recall or true positive rate (TP) is defined as the proportion of positive cases that were correctly identified through:

$$Recall = \frac{d}{c + d} \tag{1}$$

Precision is the proportion of predicted positive cases that were identified as correct, and this can be measured as:

$$Precision = \frac{d}{b + d} \tag{2}$$

Specificity also called true negative rate, measures the proportion of negatives that are correctly identified:

$$Specificity = \frac{a}{a + b} \tag{3}$$

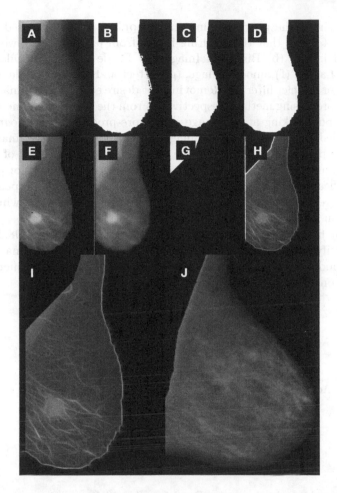

Fig. 5. Steps for pre-processing images: (a) Original image; (b) Binarized Image; (c) Erode image; (d) Dilate image; (e) Remove Labels; (f) Smooth Image; (g) Detect and remove muscle; (h) Image applied Gabor wavelet filter; (i) Removing not desire edges; (j) Image applied original pre-processing method [15].

Accuracy measures the proportion of true and negatives that are correctly identified:

$$Accuracy = \frac{c + d}{a + b + c + d} \tag{4}$$

Where **a** is the number of correct predictions that indicates the instance is negative, **b** is the number of incorrect predictions that infers an instance is positive, **c** is the number of incorrect of predictions that confirms the negative instance, and **d** is the number of correct predictions that verifies the positive instance.

Figure 5 shows the pre-processing stages applied in the proposed model. It is important to note that the resulting image from pre-processing were labeled (a) Original image, (b) Binarized Image, (c) Erode image, (d) Dilate image, (e) Remove Labels, (f) Smooth Image, (g) Detect and remove muscle, (h) Image applied Gabor wavelet filter, (i) Removing not desire edges and (j) Image applied original pre-processing method, respectively. From the above mentioned sequence of images, upon looking at the image of the pre-processed mammogram, after having applied the Gabor filter, i.e image Fig. 5I differs from the final result of the reference paper [15], and this can be noted in the image Fig. 5J of [15]. This is solely because a combination of different parameters were used for the Gabor filter. Specifically, parameters those were implemented in our proposed model allows us a greater deal in realizing morphological characteristics, which intend to increase an accuracy in the CAD system.

From our findings depicted in Table 2, we estimate a value of 98.22% accuracy in classifying malignant condition. While the overall performance values of Table 3, indirectly supports that the choice of a correct classifier depends enormously on the accurate evaluation of its performance.

Table 2. Precision and recall values achieved.

		Predicted			Precision
		Normal	Benign	Malignant	
Actual	Normal	46	8	2	82.14%
	Benign	24	14	1	35.89%
	Malignant	0	0	168	98.22%
	Recall	65.71%	64.63%	97.07%	

Examining the entire Accuracy, Precision, Specificity percentage is very informative presented in Table 3. The reported averages include the mean accuracy which is a harsh metric for the 10 forecasts used in cross-validation. Here to mention that, in 3 and 2 Conditions, averaging is set to total effectiveness, true positives, and true negative, respectively. The proposed approach has been compared with respect to [15] to measure the effectiveness of the introduced pre-processing tasks by using the same classifier. In addition, the proposed method demonstrated to be more robust than those have included deep learning techniques [16,17]. However, Multilayer Artificial Neural Network (ANN) + Mean Square error [18] achieved the highest overall accuracy 99.2%, but provides a classification for only two conditions. Our method allows to classify in 2 and 3 Conditions and outperforms to identify true positive values.

Table 3. Overall results.

Method	3 Conditions			2 Conditions		
	Accuracy	Precision	Specificity	Accuracy	Precision	Specificity
k-NN (Proposed)	89.39%	80.00%	80.00%	98.48%	98.48%	97.00%
k-NN + Pre-processing (Proposed)	86.99%	77.00%	77.00%	99.18%	99.00%	98.00%
k-NN + Pre-processing [15]	–	–	–	98.69%	99.13%	98.26%
Deep Convolutional Neural Network [16]	–	–	–	93.35%	–	–
Multilayer ANN + Mean Square error [18]	–	–	–	99.2%	–	99.15%
Decision tree classifiers with EFDs [19]	–	–	–	97.22%	–	–
Gaussian with texture feature [19]	–	–	–	97.44%	–	–
AlexNet-Polynomial-Based-SVM [17]	–	–	–	85.00%	–	86.00%
GoogLeNet-CNN-Softmax [17]	–	–	–	77.00%	–	71.00%
Hybrid CNN and RBF-Based SVM [17]	–	–	–	92.00%	–	86.00%

4   Conclusions

The proposed approach aims to automate the classification and segmentation process in mammogram analysis. The types of data that need to be classified normal, benign and malignant conditions. It is important to note that the image resulting from the pre-processing, the image of the pre-processed mammogram and after having applied the Gabor filter, image F of Fig. 5, differs from the final result of the reference paper [15], image I of Fig. 5J. This is because a different parameter combination has been used for the Gabor filter. The parameters used in the proposed model allow a greater realization of morphological characteristics, which is intended to increase the accuracy of the computer-assisted diagnostic system.

Hence our model could be improved with techniques of automatic parameterization. Focusing the cases of normal and benign, both are with lower precision and this could improve our model 1. upon implementing more data to these cases 2. engineering new features that will lead the improvement of classifier 3. changing the parametrization of preprocessing stage.

As future work, the effects of machine learning classifiers could be explored in this context to determine their overall accuracy to correctly classify mammogram image conditions.

Source of Algorithm

https://github.com/jer8856/Automated-identification-of-breast-cancer-using-digitized-mammogram-images.git.

Acknowledgements. This work used the supercomputer of the National Supercomputing Service Yachay EP of Ecuador (Quinde I).
Conflicts of Interest. The authors declare that there are no conflicts of interest regarding the publication of this paper.

References

1. World Health Organization Homepage. https://www.who.int/cancer/detection/breastcancer/en/index1.html. Accessed 28 May 2019
2. Guachi, L.: Automatic colorectal segmentation with convolutional neural network. Comput.-Aided Des. Appl. **16**(5), 836–845 (2019)
3. Lorena, G., Robinson, G., Stefania, P., Pasquale, C., Fabiano, B., Franco, M.: Automatic microstructural classification with convolutional neural network. In: Botto-Tobar, M., Barba-Maggi, L., González-Huerta, J., Villacrés-Cevallos, P., S. Gómez, O., Uvidia-Fassler, M.I. (eds.) TICEC 2018. AISC, vol. 884, pp. 170–181. Springer, Cham (2019). https://doi.org/10.1007/978-3-030-02828-2_13
4. Makandar, A.: Pre-processing of mammography image for early detection of breast cancer. Int. J. Comput. Appl. **144**(3), 0975–8887 (2016)
5. Muneeswaran, V., Pallikonda Rajasekaran, M.: Local contrast regularized contrast limited adaptive histogram equalization using tree seed algorithm—an aid for mammogram images enhancement. In: Satapathy, S.C., Bhateja, V., Das, S. (eds.) Smart Intelligent Computing and Applications. SIST, vol. 104, pp. 693–701. Springer, Singapore (2019). https://doi.org/10.1007/978-981-13-1921-1_67
6. Robert, K.: Cubic convolution interpolation for digital image processing. IEEE Trans. Acoust. Speech Signal Process. **29**(6), 1153–1160 (1981)
7. Parker, J.: Comparison of interpolating methods for image resampling. IEEE Trans. Med. Imaging **2**(1), 31–39 (1983)
8. Das, S.: Medical image enhancement techniques by bottom hat and median filtering. Int. J. Electron. Commun. Comput. Eng. **5**(4), 347–351 (2014)
9. Meng, H.: Iris recognition algorithms based on Gabor wavelet transforms. In: 2006 International Conference on Mechatronics and Automation, pp. 1785–1789. IEEE, China (2006)
10. Suckling, J.: The mammographic image analysis society digital mammogram database. Digital Mammo, pp. 375–386 (1994)
11. DDSM: Digital Database for Screening Mammography. http://www.eng.usf.edu/cvprg/Mammography/Database.html. Accessed 14 Jun 2019
12. Daugman, J.: Uncertainty relation for resolution in space, spatial frequency, and orientation optimized by two-dimensional visual cortical filters. JOSA A **2**(7), 1160–1169 (1985)
13. The mini-MIAS database of mammograms. http://peipa.essex.ac.uk/info/mias.html. Accessed 14 Jun 2019
14. Yi, Y.: An improved locality sensitive discriminant analysis approach for feature extraction. Multimed. Tools Appl. **74**(1), 85–104 (2015)
15. Raghavendra, U.: Application of Gabor wavelet and locality sensitive discriminant analysis for automated identification of breast cancer using digitized mammogram images. Appl. Soft Comput. **46**, 151–161 (2016)
16. Arfan, M.: Deep learning based computer aided diagnosis system for breast mammograms. Int. J. Adv. Comput. Sci. Appl. **7**, 286–290 (2017)
17. Alkhaleefah, M.: A hybrid CNN and RBF-based SVM approach for breast cancer classification in mammograms (2018)
18. Tariq, N.: Breast cancer classification using global discriminate features in mammographic images. Int. J. Adv. Comput. Sci. Appl. (IJACSA) **10**(2), 381–387 (2019)
19. Hussain, L., Aziz, W.: Automated breast cancer detection using machine learning techniques by extracting different feature extracting strategies. In: 2018 17th IEEE International Conference on Trust, Security and Privacy in Computing and Communications, pp. 327–331. IEEE, New York (2018)

Sparse-Based Feature Selection for Discriminating Between Crops and Weeds Using Field Images

Daniel Guillermo García-Murillo[1]($\boxtimes$), Andrés M. Álvarez[1],
David Cárdenas-Peña[2], William Hincapie-Restrepo[3],
and German Castellanos-Dominguez[1]

[1] Signal Processing and Recognition Group,
Universidad Nacional de Colombia, Manizales, Colombia
`dggarciam@unal.edu.co`
[2] Automatic Research Group, Universidad Tecnológica de Pereira,
Pereira, Colombia
[3] Physicochemistry of Terrestrial Fluids Group,
Universidad de Caldas, Manizales, Colombia

Abstract. Control of weed growing in yields is a critical task for reducing crop losses. Recently, image-based systems attempt to discriminate between crops and weeds from a set of features. Although some features have a physiological meaning, most of them are redundant or noisy. Therefore, selecting relevant features must result in interpretable and accurate results while reducing the computational complexity of the system. In this work, we introduce a sparse-based feature selection approach using the Lasso operator that eliminates noisy features aiming to improve the classification of crops. We evaluate our proposal on the Crop/Weed Field Image Dataset, for which we tune the parameters by maximizing the accuracy and minimizing feature dimension. Achieved performance results evidence that our proposed approach improves discrimination in comparison with other feature selection approaches, with the benefit of providing interpretability in weed/crop discrimination tasks.

Keywords: Lasso Feature Selection · Weed/crop discrimination · HOG

1 Introduction

The global population grows a rate of 1.09% per year, increasing the demands for food and requiring higher agricultural yields [13]. However, one of the multiple challenges that has a marked correlation with crop yield loss is the management of weeds [15], i.e., the unwanted plants grow randomly in all the field, competing with crops for resources (like water, nutrients, and sunlight) [5].

To deal with the weed spreading, several control strategies have been proposed for several crops and scale fields [19]. In small scale fields, manual weed

© Springer Nature Switzerland AG 2019
I. Nyström et al. (Eds.): CIARP 2019, LNCS 11896, pp. 357–364, 2019.
https://doi.org/10.1007/978-3-030-33904-3_33

control is still being used even that it is a tedious, inefficient, and high labor-cost task. In turn, smart agriculture, including remote sensing and mechanical weed management methods for the tillage of the soil, arises as a better control app-roach as they are more productive and labor-saving [8]. Nevertheless, mechanical management will cause crop damage if the discrimination of intra-row weeds is not accurate enough [6].

Generally, traditional approaches devoted to weed discrimination using remote sensing include the following stages [3]: (i) feature estimation, (ii) fea-ture extraction and selection, and (iii) classification. For the first stage, the Histogram of Oriented Gradients (HOG) is one of the principal scale-invariant feature descriptors used that allocate shape information of regions. However, the selection of relevant features is determinant to obtain interpretable and accurate results reducing computational complexity by removing redundant information concerning the task studied [12]. Among the apporaches to feature selection are attribute measures using decision trees [7], iterative backward elimination by a feature ranking obtained from the Random Forest algorithm [16], decision rule using the chi-square statistic metrics [11], measures based on Random Forest [2], measures of highest normalized information gain [18], dimensionality reduction approach by calculating the Correlation-based Feature Selection,Principal Com-ponent Analysis, Kernel Principal Component Analysis, Linear Discriminant Analysis, and Stepwise Linear Discriminant Analysis [9,14,20], among others. However, for weed discrimination, feature selection methods still have challenges eliminating redundant information without affecting performance [10,19].

In this study, we introduce a sparse-based feature selection approach based on Lasso operator that eliminates redundant information on the Crop/Weed Field Image Dataset (CWFID), aiming of estimating the optimum regularization value that maximizes the accuracy and minimizes feature dimension. As a result, we improve performance and interpretability in weed/crop discrimination tasks.

2   Materials and Methods

2.1   Lasso Sparse Regression Model

The LASSO sparse regression is involved in feature extraction as follows [17]:

$$\mathbf{u} = \arg_{\mathbf{u}} min \frac{1}{2} \left\| \mathbf{G}\mathbf{u} - \mathbf{y} \right\|_2^2 + \lambda \left\| \mathbf{u} \right\|_1 \tag{1}$$

where $\|.\|_1$ denotes the l_1-norm, $\mathbf{y} \in \mathbb{R}^N$ is a vector containing class labels $\{1, 2\}$. To optimize $\mathbf{u}$ in Eq. (1), the coordinate descent algorithm is adopted as [4]:

$$\widetilde{u}_j \leftarrow S \left(\sum_{i=1}^{N} g_{i,j} \left(y_i - \widetilde{y}_i^{(j)} \right), \lambda \right) \tag{2}$$

where $\widehat{y}_i^{(j)} = \sum_{d \neq j} (g_{i,d} \widetilde{u}_d)$ is the fitted value excluding the contribution from $g_{i,j}$, and $S(a, \lambda)$ is a shrinkage-thresholding operator defined as below:

$$\text{sign}(a)(|a| - \lambda)_+ = \begin{cases} a - \lambda \ if \ a > \lambda \\ 0 \quad if \ |a| \leqslant \lambda \\ a + \lambda \ if \ a < -\lambda \end{cases} \tag{3}$$

So, we calculate the optimized sparse vector $\widetilde{u}$, satisfying Eq. (1) by repeating the update until convergence. The column vectors in $\mathbf{G}$ are those zero-entries in $\widetilde{u}$, which are excluded from an optimized feature set $\widetilde{\mathbf{G}}$ under assumption that it has lower dimensionality than $\mathbf{G}$. The real-valued λ determines the sparsity degree of $\widetilde{u}$, and hence it rules the selection of HOG feature sets.

2.2 Feature Selection Algorithm

Since any projection matrix $\mathbf{A} \in \mathbb{R}^{P \times M}$ encode a linear combination of features in each row, we can estimate the relevance of each feature $\varrho \in \mathbb{R}^P$ as follows:

$$\varrho_p = \sum_{m=1}^{M} |a_m| \, ; \forall p \in P \tag{4}$$

Where $a_m \in \mathbb{R}^P$ is the m-th row in $\mathbf{A}$ and largest values of ϱ_p must point out to better input attributes since they exhibit higher overall dependencies. As a result, the calculated relevance vector ϱ can be employed to rank the original features. Furthermore, aiming to estimate a representation space encoding discriminant input patterns, we compute the matrix $\mathbf{X}_S \in \mathbb{R}^{N \times P_s}$, where $P_s \leqslant P$ holding the features that satisfy the condition $\bar{\varrho}_p \leq \zeta$, where $\bar{\varrho}_p$ is the normalized feature relevance vector index, and it is calculated as follows:

$$\bar{\varrho}_p = \frac{\varrho_p}{\sum \varrho_p} \tag{5}$$

3 Experimental Set-Up

Dataset and Preprocessing: We evaluate different approaches on the Crop/Weed Field Image Dataset (CWFID)[1], holding 162 crop and 332 weed plants. Items had been labeled from 60 top-down field images of organic carrots, in the leaf development growth stage, with the appearance of intra-row and close-to-crop weeds. The dataset was acquired at a commercial organic carrot farm just before manual weed control. Also, the dataset included a soil mask for each image.

In the preprocessing stage, we extract all crop and weed plants as individual images, as seen in Fig. 1. Moreover, since our goal is the crop/weed discrimination task, we use the soil mask of each image to ensure only vegetation information. Furthermore, we set the size of all individual images to the one with

---

[1] Available at http://github.com/cwfid.

Fig. 1. Exemplary of items in the CWFID Dataset: weeds are remarked by red rectangles and crops by green rectangles. (Color figure online)

the hugest size. Using the HOG descriptor at different window parameter values $\omega = \{32, 44, 56, 68, 80\}$, we extract relevant morphological information at different scales for each crop and weed plants.

Evaluation Scheme and Performance Assessment: We propose a sparse-based dimensional reduction approach using the well-known Lasso operator, achieving the selection of significant features on a nested cross-validation framework, for which a grid of λ and ω parameters are fixed to find the ones that select fewer features with better performance. Afterward, for the optimum λ and ω values, we calculate the average relevant feature occurrence per trial, yielding the average of the most significant features that feed the classifier based on an incremental learning approach.

For the sake of comparison, we contrast our method with no feature selection (NFS), Principal Component Analysis Selector version (PCA_sel) and centered kernel alignment selector version (CKA_sel), where, we obtain selector versions as shown in Sect. 2.2, also, we find the best ζ using a grid of values between 0 and 1. Moreover, we use a kernelized support vector machine (KVM), for which the ϵ value is tuned via an exhaustive search, and the Gaussian kernel bandwidth is selected as in [1]. Then, we estimate the average accuracy (a_{cc}) and F1-score ($F1_s$) by a nested ten-fold cross-validation scheme as a measure of classifier performance. Where, $F1_s$ is an unbalance binary classification accuracy measure (see Eq. (6)), that considers both the precision and the recall.

$$F1_s = 2\frac{precision \cdot recall}{precision + recall} \tag{6}$$

4   Results and Discussion

Figure 2 depicts the KSVM classification accuracy using different values of window size (ω), showing that the proposed method reach the maximum accuracy score (76.47 ± 5.52) using $\omega = 56$.

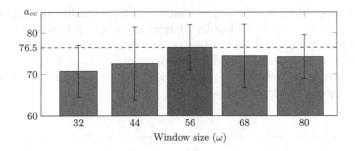

Fig. 2. KSVM accuracy for different ω values

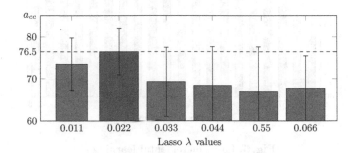

Fig. 3. KSVM accuracy for different λ values

As seen in Fig. 3, large values of λ may exclude several features, while small values do not eliminate redundancy effectively. For reaching a trade-off between both limiting cases, the optimal regularization parameter value is determined by a nested ten-fold cross-validation framework on training data, resulting in $\lambda = 0.022$. Consequently, the proposed dimension reduction approach enhances the crop/weed discrimination task by performing the minimization of the LASSO regression model, for which the regularization parameter λ rules the feature selection effectiveness.

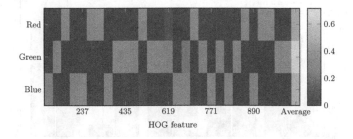

Fig. 4. Estimated relative feature relevance (Color figure online)

Another aspect to consider is the enhanced interpretability that allows identifying the set of relevant HOG features, having a meaningful understanding. For better representation, Fig. 4 depicts the sorted relative feature relevance together with the average per spectral band for the optimum λ value. Note that features extracted from the green band are more relevant due to plants reflect with higher intensity the green spectrum.

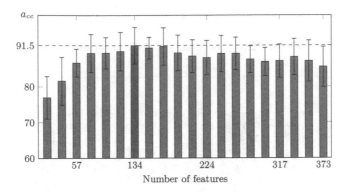

Fig. 5. Lasso incremental learning

Besides improved interpretability, the contribution for the relevant can be assessed as seen in Fig. 5 that depicts performed accuracy by using the incremental learning approach by feeding stepwise the classifier with the feature set ranked in decreasing order or relevance. Adjusting $\lambda = 0.022$, therefore, the first ranked 134 features perform the higher accuracy value 91.52 with relatively high confidence (standard deviation of 5.11). Afterwards, by adding more features, the accuracy tends to decrease since either redundant or noisy information is added.

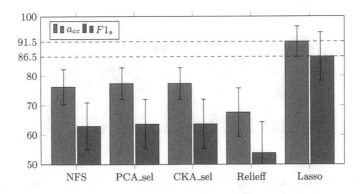

Fig. 6. Accuracy performance assessed by the SVM classifier

Lastly, Fig. 6 depicts the KSVM classification accuracy and F1-score derived by NFS, PCA_sel, CKA_sel, and Relieff, showing that the proposed method outperforms other approaches, reaching 91.52 ± 5.11 and 86.45 ± 8.12 in average accuracy and F1-score, respectively.

5 Concluding Remark

We develop a sparse-based feature selection approach based on Lasso operator that eliminates redundant information using field images. By optimizing the regularization value, we handle the trade-off between two limiting conditions: accuracy maximization versus dimensional feature reduction. As a result, our proposal selects effectively relevant features that avoid redundant information, improving both accuracy and F1-score. Besides, we improve the performance and interpretability in weed/crop discrimination tasks.

For future work, we plan to use deep learning, the one that uses convolutional neural network strategies, aiming to obtain more relevant features from the images, and thus improving weed/crop discrimination. Also, multiple learning kernel approaches can be explored to merge the information, resulting in a more relevant kernel that enhances weed/crop classification.

Acknowledgment. "This work was developed under the research project *CARACTERIZACIÓN DE CULTIVOS AGRÍCOLAS MEDIANTE ESTRATEGIAS DE TELEDETECCIÓN Y TÉCNICAS DE PROCESAMIENTO DE IMÁGENES (Hermes-36719)* funded by Universidad Nacional de Colombia".

References

1. Álvarez-Meza, A.M., Cárdenas-Peña, D., Castellanos-Dominguez, G.: Unsupervised kernel function building using maximization of information potential variability. In: Bayro-Corrochano, E., Hancock, E. (eds.) CIARP 2014. LNCS, vol. 8827, pp. 335–342. Springer, Cham (2014). https://doi.org/10.1007/978-3-319-12568-8_41
2. Duro, D.C., Franklin, S.E., et al.: Multi-scale object-based image analysis and feature selection of multi-sensor earth observation imagery using random forests. Int. J. Remote Sens. **33**(14), 4502–4526 (2012)
3. Fernández-Quintanilla, C., Peña, J., et al.: Is the current state of the art of weed monitoring suitable for site-specific weed management in arable crops? Weed Res. **58**(4), 259–272 (2018)
4. Friedman, J., Hastie, T., et al.: Regularization paths for generalized linear models via coordinate descent. J. Stat. Softw. **33**(1), 1 (2010)
5. Hamuda, E., Mc Ginley, B., et al.: Automatic crop detection under field conditions using the HSV colour space and morphological operations. Comput. Electron. Agric. **133**, 97–107 (2017)
6. Hamuda, E., Mc Ginley, B., et al.: Improved image processing-based crop detection using Kalman filtering and the Hungarian algorithm. Comput. Electron. Agric. **148**, 37–44 (2018)

7. Han, J., Pei, J., et al.: Data Mining: Concepts and Techniques. Elsevier, Amsterdam (2011)
8. Huang, H., Lan, Y., et al.: A semantic labeling approach for accurate weed mapping of high resolution UAV imagery. Sensors **18**(7), 2113 (2018)
9. Ma, L., Cheng, L., et al.: Training set size, scale, and features in geographic object-based image analysis of very high resolution unmanned aerial vehicle imagery. ISPRS J. Photogramm. Remote Sens. **102**, 14–27 (2015)
10. Ma, L., Fu, T., et al.: Evaluation of feature selection methods for object-based land cover mapping of unmanned aerial vehicle imagery using random forest and support vector machine classifiers. ISPRS Int. J. Geo-Inf. **6**(2), 51 (2017)
11. Peña-Barragán, J.M., Ngugi, M.K., et al.: Object-based crop identification using multiple vegetation indices, textural features and crop phenology. Remote Sens. Environ. **115**(6), 1301–1316 (2011)
12. Perez-Sanz, F., Navarro, P.J., et al.: Plant phenomics: an overview of image acquisition technologies and image data analysis algorithms. GigaScience **6**(11), gix092 (2017)
13. Sankaran, S., Khot, L.R., et al.: Low-altitude, high-resolution aerial imaging systems for row and field crop phenotyping: a review. Eur. J. Agron. **70**, 112–123 (2015)
14. Siddiqi, M.H., Lee, S.-W., et al.: Weed image classification using wavelet transform, stepwise linear discriminant analysis, and support vector machines for an automatic spray control system. J. Inf. Sci. Eng. **30**(4), 1227–1244 (2014)
15. Singh, A., Ganapathysubramanian, B., et al.: Machine learning for high-throughput stress phenotyping in plants. Trends Plant Sci. **21**(2), 110–124 (2016)
16. Stumpf, A., Kerle, N.: Object-oriented mapping of landslides using random forests. Remote Sens. Environ. **115**(10), 2564–2577 (2011)
17. Tibshirani, R.: Regression shrinkage and selection via the lasso. J. R. Stat. Soc.: Ser. B (Methodol.) **58**(1), 267–288 (1996)
18. Vieira, M.A., Formaggio, A.R., et al.: Object based image analysis and data mining applied to a remotely sensed landsat time-series to map sugarcane over large areas. Remote Sens. Environ. **123**, 553–562 (2012)
19. Wang, A., Zhang, W., et al.: A review on weed detection using ground-based machine vision and image processing techniques. Comput. Electron. Agric. **158**, 226–240 (2019)
20. Weis, M., Sökefeld, M.: Detection and identification of weeds. In: Oerke, E.C., Gerhards, R., Menz, G., Sikora, R. (eds.) Precision Crop Protection - The Challenge and Use of Heterogeneity, pp. 119–134. Springer, Dordrecht (2010). https://doi.org/10.1007/978-90-481-9277-9_8

Automated Nuclei Segmentation in Dysplastic Histopathological Oral Tissues Using Deep Neural Networks

Adriano Barbosa Silva[1] , Alessandro S. Martins[2]([✉]), Leandro A. Neves[3],
Paulo R. Faria[4], Thaína A. A. Tosta[5], and Marcelo Zanchetta do Nascimento[1]

[1] Faculty of Computer Science, Federal University of Uberlândia,
Uberlândia, Minas Gerais, Brazil
[2] Federal Institute of Triângulo Mineiro, Ituiutaba, Brazil
alessandro@iftm.edu.br
[3] Department of Computer Science and Statistics,
São Paulo State University (UNESP), São José do Rio Preto, Brazil
[4] Department of Histology and Morphology, Institute of Biomedical Science,
Federal University of Uberlândia, Uberlândia, Brazil
[5] Center of Mathematics, Computing and Cognition, Federal University of ABC,
Santo André, Brazil

Abstract. Dysplasia is a common pre-cancerous abnormality that can
be categorized as mild, moderate and severe. With the advance of digital
systems applied in microscopes for histological analysis, specialists can
obtain data that allows investigation using computational algorithms.
These systems are known as computer-aided diagnosis, which provide
quantitative analysis in a large number of data and features. This work
proposes a method for nuclei segmentation for histopathological images
of oral dysplasias based on an artificial neural network model and post-
processing stage. This method employed nuclei masks for the training,
where objects and bounding boxes were evaluated. In the post-processing
step, false positive areas were removed by applying morphological operations, such as dilation and erosion. This approach was applied in a
dataset with 296 regions of mice tongue images. The metrics accuracy,
sensitivity, specificity, the Dice coefficient and correspondence ratio were
employed for evaluation and comparison with other methods present in
the literature. The results show that the method was able to segment the
images with accuracy average value of $89.52 \pm 0.04\%$ and Dice coefficient
of $84.03 \pm 0.06\%$. These values are important to indicate that the proposed method can be applied as a tool for nuclei analysis in oral cavity
images with relevant precision values for the specialist.

Keywords: Convolutional neural network · Dysplasia · Nuclei
segmentation · CAD

I. Nyström et al. (Eds.): CIARP 2019, LNCS 11896, pp. 365–374, 2019.
https://doi.org/10.1007/978-3-030-33904-3_34

1 Introduction

Dysplasia is characterized by alterations on the cell information such as size, shape and brightness intensity. In developing countries, this anomaly is a common type of pre-cancerous lesions that can be classified as mild, moderate and severe [5]. Cancer is the second most common cause of death and severe class dysplasias have a 3% to 36% chance of progressing to this type of malignant lesion [12]. Usually, the diagnosis for these lesions is performed by analysing the size of the lesion and the intensity of morphological alterations in tissue nuclei. However, different size lesions may have similar levels of nuclear alterations [12]. Thus, pathologists may assign different levels of dysplasia to lesions with similar level of alterations [17]. Moreover, there are several malignant lesions that are identified only by the intensity of nuclei alterations [12].

With the advance of digital systems applied in microscopes for histological analysis, specialists can obtain data that allows investigation using computational algorithms. The development of digital tools for analysis can assist pathologists in decision making as well as reduce the error rate caused by subjectivity [3]. These systems are known as computer aided diagnosis (CAD), which provide quantitative analysis in a large number of data and features [7]. This system shows a set of steps ranging from the signal-to-noise ratio improvement, segmentation, feature extraction and data classification. Segmentation is an important step that allows the identification of objects that will be analyzed by feature descriptors and classified in subsequent steps [7]. The main ways of applying this stage are by discontinuity or similarity of pixel brightness intensity. Discontinuity is based on the detection of abrupt variations of the pixel intensity to delimit region borders. The goal of the similarity segmentation is to divide an image into regions with similar features based on established criteria [7].

In literature, there are several studies that investigate the segmentation of cellular structures of malignant lesions based on computer vision algorithms that employed thresholding techniques, region merging and semantic information [11,15]. With the goal of identifying nuclei in epithelial breast tissues, the authors in [10] proposed a method that employs color deconvolution and Otsu thresholding for nuclei segmentation. In [1], the K-means method was used to segment and identify cells in lymphoma images. In [15], the authors proposed a method that segment cell nuclei using neural networks for semantic segmentation where each pixel is assigned to a class region present in the image. In the context of histological images, segmentation of epithelial nuclei is a complex task due to irregular characteristics shown by nuclei, such as dye variation. These nuclei may have color and aspects similar to other structures present in the tissue [11]. This shows that it is an area where there are major challenges related to nuclei segmentation. In the case of dysplasia, this process can be more complex due to the growth of the connective tissue that can invade the epithelial tissue and difficult the nuclei segmentation [12]. The described proposals have not yet considered a method focused to the segmentation of epithelial nuclei as proposed in this work.

This work proposes a segmentation method based on region-based convolutional neural networks (R-CNN) aiming to identify cell nuclei present in oral histological tissues. For this, in the first step individual nuclei masks were generated to train the network. In the training step, bounding boxes were defined for candidate objects and combined with the masks. In the segmentation step, the R-CNN classifies each image pixel in nuclei or background based on these pixels neighborhood. As post-processing step, morphological operations of dilation and erosion were combined with hole-fill operations and small object exclusion to eliminate false negatives and positives. A dataset of mice images was used for the evaluation of the method. The results were evaluated in relation to the gold standard and a comparison was made employing methods used in literature.

This paper is organized as follows. In the Sect. 2 we detail the dataset, the methods used to train the network, the segmentation, post-processing and the method for quantitative analysis. The obtained results and the comparison with other segmentation methods present in literature are shown in Sect. 3. The Sect. 4 presents discussions and our work conclusions.

2 Methodology

2.1 Image Dataset

The dataset was built from tongue slides of 30 mice previously submitted to a carcinogen during two experiments performed between 2009 and 2010, duly approved by the Ethics Committee on the Use of Animals under protocol number 038/09 at the Federal University of Uberlândia, Brazil.

The histological images were obtained using the LeicaDM500 light microscope with a magnification of 400x and saved in TIFF format using the RGB color model and resolution of 2048×1536 pixels. In total, 43 images were obtained and with the aid of a pathologist were classified into healthy tissues, mild dysplasia, moderate dysplasia and severe dysplasia. After digitalization, these images were cropped in regions of interest (ROI) with a size of 450×250 pixels, totalling 296 ROI with 74 ROI for each class. In Fig. 1 some cases of dysplasia and a case of a healthy region of histological images of the buccal cavity are shown. The lesions were manually marked by a specialist and automatically analyzed by the proposed method (gold standard).

2.2 Segmentation via Mask R-CNN

The Mask R-CNN model is based on the Faster R-CNN model, which has two stages: the first is called the region proposal network (RPN), which proposes to define bounding boxes for candidate objects [14]; in the second stage, bounding box regression is used to refine the area of the boxes [8].

Then, in the training step is applied the loss function defined by:

$$L = L_{cls} + L_{box} + L_{mask} \tag{1}$$

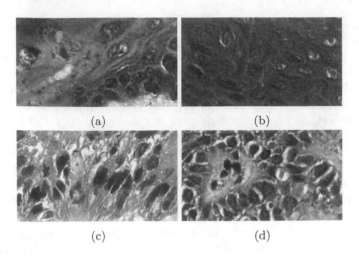

(a)   (b)

(c)   (d)

Fig. 1. Histological images of oral epithelial tissues: (a) healthy tissue, (b) mild oral dysplasia, (c) moderate dysplasia and (d) severe dysplasia.

In this context, we have $L_{cls} = -\log p_u$, where p is the distribution probability of each ROI over the $K+1$ possible classes and u the gold standard of the ROI class [6]. The bounding boxes loss is defined by:

$$box(t^u, v) = \sum_{i \in (x,y,w,h)} smooth_{L_1}(t_i^u - v_i), \tag{2}$$

where:

$$smooth_{L_1}(x) = \begin{cases} 0.5x^2 & if |x| < 1 \\ |x| - 0.5 & otherwise \end{cases}, \tag{3}$$

where t^u is the regression of the boxes, v is the gold standard of the box regression, x and y are the coordinates of the upper left corner of each ROI, and h and w are height and width information for the region [6].

The output of this step is the masks with size Km^2, containing K binary masks of resolution $m \times n$, one for each of the K classes. In this study, the L_{mask} adopted was the average binary cross-entropy loss as described in the work of [8]. The main reason for the choice of masks was the advantage of allowing the spatial representation of the objects. Thus, this information was extracted by pixel-by-pixel correspondence through convolution operations.

In this work, Mask R-CNN was applied to the Resnet-50 convolutional neural network model [9]. This network has 50 layers arranged in the following structure: input layer; 16 blocks of convolutional layers organized into 4 groups called B1, B2, B3 and B4; and an output layer. The network structure can be seen in Fig. 2. Each block has 3 layers, with convolutions of sizes 1×1, 3×3 and 1×1, respectively. The 1×1 convolutions are responsible for reducing and restoring dimensions leaving the 3×3 convolution with smaller input and output dimension

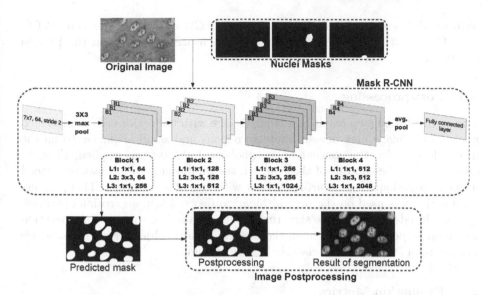

Fig. 2. Proposed method workflow to illustrate the steps used for segmentation of histological images.

size. Between the first layer and blocks B1, a max pooling filter with a 3×3 dimension and stride of 2 is applied, reducing the size of the input by half. To halve input size between each block, in the first layer of groups B2, B3 and B4, the convolution is performed using a moving window with a stride of 2. In the final step of the network, an average pooling filter and a fully connected layer were used for object classification. The network uses the rectified linear unit (ReLU) activation function.

Given the set of 160 ROI obtained from the 4 classes of histological images of the oral cavity, 40 ROI were employed for the network training. In this stage, 32 ROI were used to build the weights of the net and 8 ROI to evaluate each epoch. With the help of a pathologist, 1,220 individual nucleus masks were obtained from these ROI. Since the Mask R-CNN maps the masks over the ROI to extract nuclei features, our set consists of 1,220 nucleus masks, being 1,027 for the training set and 193 for the test set. According to the authors in [14], since the RPN has a magnitude order with few parameters, it has less risk of overfitting on small dataset. Also, by using the Resnet-50 model instead of the Resnet-101 there is an overall overfitting reduction, as explained by the authors in [9]. The network was pre-trained on the ImageNet dataset and fine-tuned using our dataset. For the training, it was defined as a batch size of 9 masks and learning rate of 0.001. It was also used the SGD optimizer with a momentum of 0.9. The fine-tuning of the network was performed using 40 epochs with 142 iterations, which is the number of batches passed to the network in each epoch. These parameters were empirically defined for the analyzed dataset. After this stage, the 120 ROI were employed in the network evaluation stage. The CNN was trained on a computer with an eight-core AMD FX-8320 processor, 8 GB of RAM and Nvidia GTX-1060 GPU with 6 GB of VRAM, using the TensorFlow library in the Python language.

2.3 Post-processing

The segmentation step result in a binary image with the regions of the identified nuclei. This image may have incomplete regions and small artifacts. To fill the incomplete nuclei, the concept of morphological closing was applied. First, in order to close the contour of these nuclei, the dilation operation was performed using a cross-shaped structuring element with size of 3×3 pixels. Then, a hole-fill function was used on the present binary objects. Then, an erosion operation was employed with the same structuring element used in the dilation operation to eliminate noise still present in the images. Finally, objects with an area size smaller than 30 pixels were classified as background.

2.4 Evaluation Metrics

The evaluation of a segmentation method can be performed by calculating the overlapping regions of the segmented image and the regions of a reference image demarcated by a specialist (gold-standard) [4]. In this stage, 30 histological ROI of each class were randomly chosen and first segmented by specialists considered the gold standard. Then, the following metrics were considered: accuracy (A_{CC}), sensitivity (S_E), specificity (S_P), correspondencerRate (C_R) and Dice coefficient (D_C) [2, 13, 16].

The values of S_E and S_P were used to determine the proportion of pixels correctly marked as objects and as background, respectively. The metric A_{CC} was used to measure the amount of true positive and true negative calculated in relation to all positives and negatives. The C_R metric evaluated the correspondence between the result obtained and the gold standard. Then, finally, the D_C measure was used to evaluate the similarity between the gold standard and the result.

3 Experimental Results

The results of the proposed segmentation method for images of mild, moderate and severe dysplasia are shown in Fig. 3. In an evaluation of the images of the mild class (see Figs. 3a, d and g), it is noted that the method was able to detect and segment the regions with the presence of nuclei. In these figures, it is possible to observe, by the arrows marked in red color, that there are objects detected as false-positive in relation to the marking made by the specialist (Fig. 3d). It is also possible to note regions with the presence of false negative, that is, some nuclei were eliminated by the segmentation process (see the arrows marked in green color). However, the method was able to segment obscured and difficult-to-identify nuclei, obtaining close similarity to the gold standard as seen in the

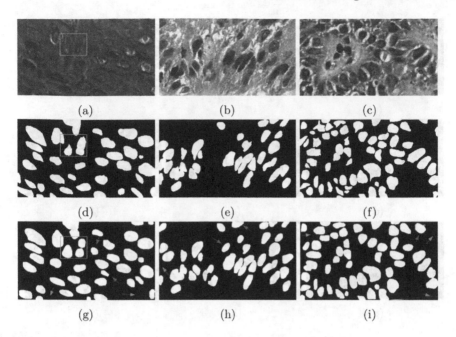

Fig. 3. Histological images of the oral cavity with the presence of dysplasia: (a) mild, (b) moderate and (c) severe. The regions marked by the pathologist: (d) marking of mild dysplasia, (e) marking of moderate dysplasia and (f) marking of severe dysplasia. Segmentation with the proposed method: (g) result for mild dysplasia, (h) result for moderate dysplasia and (i) result for severe dysplasia. (Color figure online)

region marked in blue in Fig. 3g. In a similar way, the images of the moderate dysplasia and severe dysplasia classes (see Figs. 3h and i) also showed some regions with the presence of the false positive and false negative.

Aiming to investigate the method, algorithms based on semantic segmentation by SegNet [15], EM-GMM and K-means [11] were also applied on the image dataset. Some of these methods are extensively used for comparison of histological component segmentation methods [11]. The EM-GMM and K-means algorithms were performed using a cluster number $k = 3$. The SegNet was trained using the same 10 ROI used for the proposed method and using 10 groundtruth masks, each one containing all nuclei from the ROI.

Figures 4a, b, c, d, e, f, g, h and i show the results of the EM-GMM, K-means and SegNet methods, respectively. In the results it is noticed that there are also flaws in relation to false positive regions (red color arrows) and false negative regions (green color arrows). In addition, there are results in which there was a great degradation with respect to the nuclei structures (see Figs. 4a, h and i), where part of these structures was eliminated.

The performance of the proposed method for tissue segmentation is shown in Table 1. In images with severe dysplasia, the S_E and A_{CC} were smaller than the results of the other classes of lesions. This shows that the method has greater

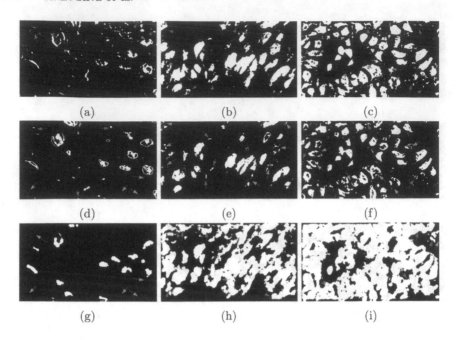

Fig. 4. Results with proposed methods: EM-GMM algorithm: (a) mild dysplasia, (b) moderate dysplasia, and (c) severe dysplasia. The images after application of K-means: (d) mild dysplasia, (e) moderate dysplasia and (f) severe dysplasia. Segmentation with Segnet: (g) mild dysplasia, (h) moderate dysplasia and (i) severe dysplasia. (Color figure online)

difficulty in identifying the nuclei for this class. This may occur because some nuclei of this class have a high-intensity morphological alteration, which makes it difficult for the algorithm to identify it, classifying it as a background region [12].

Table 1. Evaluation of the proposed method in tissues with different levels of dysplasia.

	S_E (%)	S_P (%)	A_{CC} (%)
Healthy	89.00 ± 0.06	91.22 ± 0.04	90.34 ± 0.04
Mild	89.12 ± 0.05	90.47 ± 0.04	90.12 ± 0.04
Moderate	85.99 ± 0.05	90.76 ± 0.04	89.33 ± 0.04
Severe	84.84 ± 0.07	89.30 ± 0.04	88.28 ± 0.04

The results of the algorithms present in literature are presented in Table 2. The proposed method obtained relevant results in relation to the methods present in literature ($A_{CC} = 89.52 \pm 0.04\%$, $C_R = 0.76 \pm 0.10$ and $D_C = 0.84 \pm 0.06$). The K-means methods obtained a value of $A_{CC} = 77.32 \pm 0.05\%$ and this represents a difference of 12% in relation to the proposed method. The SegNet algorithm obtained the values of $C_R = 0.40 \pm 0.24$ and $D_C = 0.60 \pm 0.16$, being 37% and 23% lower than the results of the proposed method. This behavior with lower results can also be noticed in the results presented by the EM-GMM method ($A_{CC} = 72.91 \pm 0.15$ and $C_R = 0.35 \pm 0.30$).

Table 2. Comparison among the method and techniques presented in literature.

	$A_{CC}(\%)$	C_R	D_C
EM-GMM	72.91 ± 0.15	0.35 ± 0.30	0.58 ± 0.15
K-means	77.32 ± 0.05	0.35 ± 0.11	0.52 ± 0.11
SegNet	73.12 ± 0.12	0.40 ± 0.24	0.60 ± 0.16
Proposed method	$\mathbf{89.52 \pm 0.04}$	$\mathbf{0.76 \pm 0.10}$	$\mathbf{0.84 \pm 0.06}$

4   Conclusions

In this study, the proposal was put forward an approach for automatic segmentation of nuclei in oral epithelial tissue images. In literature, methods for segmentation of oral dysplastic images are not yet explored and this solution makes a contribution to specialists in the field. This work presented an algorithm of segmentation in images of oral dysplasias based on a deep learning approach. The EM-GMM, K-means and SegNet methods were applied on images of the dataset for comparison purposes. Through qualitative and quantitative analyzes, the combination of the algorithms used in this approach was able to reach more effective results than the compared techniques. In future studies pre-processing algorithms such as color normalization will be investigated aiming to improve the images in the processing initial stage.

Acknowledgement. The authors gratefully acknowledge the financial support of National Council for Scientific and Technological Development CNPq (Grants 427114/2016-0, 304848/2018-2, 430965/2018-4 and 313365/2018-0), the State of Minas Gerais Research Foundation - FAPEMIG (Grant APQ-00578-18). This study was financed in part by the Coordenação de Aperfeiçoamento de Pessoal de Nível Superior - Brasil (CAPES) - Finance Code 001.

References

1. Amin, M.M., Kermani, S., Talebi, A., Oghli, M.G.: Recognition of acute lymphoblastic leukemia cells in microscopic images using k-means clustering and support vector machine classifier. J. Med. Signals Sens. **5**(1), 49 (2015)
2. Baratloo, A., Hosseini, M., Negida, A., El Ashal, G.: Part 1: simple definition and calculation of accuracy, sensitivity and specificity. Emergency **3**(2), 48–49 (2015)
3. Belsare, A., Mushrif, M., Pangarkar, M., Meshram, N.: Breast histopathology image segmentation using spatio-colour-texture based graph partition method. J. Microsc. **262**(3), 260–273 (2016)

4. Estrada, F.J., Jepson, A.D.: Benchmarking image segmentation algorithms. Int. J. Comput. Vis. **85**(2), 167–181 (2009)
5. Fonseca-Silva, T., Diniz, M.G., Sousa, S.F., Gomez, R.S., Gomes, C.C.: Association between histopathological features of dysplasia in oral leukoplakia and loss of heterozygosity. Histopathology **68**(3), 456–460 (2016)
6. Girshick, R.: Fast R-CNN. In: Proceedings of the IEEE International Conference on Computer Vision, pp. 1440–1448 (2015)
7. Gonzalez, R.C., Woods, R.: Digital Image Processing (2018)
8. He, K., Gkioxari, G., Dollár, P., Girshick, R.: Mask R-CNN. In: Proceedings of the IEEE International Conference on Computer Vision, pp. 2961–2969 (2017)
9. He, K., Zhang, X., Ren, S., Sun, J.: Deep residual learning for image recognition. In: Proceedings of the IEEE Conference on Computer Vision and Pattern Recognition, pp. 770–778 (2016)
10. Husham, A., Hazim Alkawaz, M., Saba, T., Rehman, A., Saleh Alghamdi, J.: Automated nuclei segmentation of malignant using level sets. Microsc. Res. Tech. **79**(10), 993–997 (2016)
11. Irshad, H., Veillard, A., Roux, L., Racoceanu, D.: Methods for nuclei detection, segmentation, and classification in digital histopathology: a review–current status and future potential. IEEE Rev. Biomed. Eng. **7**, 97–114 (2014)
12. Kumar, V., Aster, J.C., Abbas, A.: Robbins and Cotran Patologia-Bases Patológicas das Doenças. Elsevier, Brasil (2010)
13. Ma, Z., Wu, X., Song, Q., Luo, Y., Wang, Y., Zhou, J.: Automated nasopharyngeal carcinoma segmentation in magnetic resonance images by combination of convolutional neural networks and graph cut. Exp. Ther. Med. **16**(3), 2511–2521 (2018)
14. Ren, S., He, K., Girshick, R., Sun, J.: Faster R-CNN: towards real-time object detection with region proposal networks. In: Advances in Neural Information Processing Systems, pp. 91–99 (2015)
15. Tokime, R.B., Elassady, H., Akhloufi, M.A.: Identifying the cells' nuclei using deep learning. In: 2018 IEEE Life Sciences Conference (LSC), pp. 61–64. IEEE (2018)
16. Tran, P.V.: A fully convolutional neural network for cardiac segmentation in short-axis MRI. arXiv preprint arXiv:1604.00494 (2016)
17. Warnakulasuriya, S., Reibel, J., Bouquot, J., Dabelsteen, E.: Oral epithelial dysplasia classification systems: predictive value, utility, weaknesses and scope for improvement. J. Oral Pathol. Med. **37**(3), 127–133 (2008)

A Study on Histogram Normalization for Brain Tumour Segmentation from Multispectral MR Image Data

Ágnes Győrfi[1,2], Zoltán Karetka-Mezei[1], David Iclănzan[1], Levente Kovács[2], and László Szilágyi[1,2(✉)]

[1] Computational Intelligence Research Group,
Sapientia - Hungarian University of Transylvania, Tîrgu Mureş, Romania
{gyorfiagnes,iclanzan,lalo}@ms.sapientia.ro, karetkaz@yahoo.com
[2] University Research, Innovation and Service Center (EKIK),
Óbuda University, Budapest, Hungary
{kovacs.levente,szilagyi.laszlo}@nik.uni-obuda.hu

Abstract. Absolute values in magnetic resonance image data do not say anything about the investigated tissues. All these numerical values are relative, they depend on the imaging device and they may vary from session to session. Consequently, there is a need for histogram normalization before any other processing is performed on MRI data. The Brain Tumor Segmentation (BraTS) challenge organized yearly since 2012 contributed to the intensification of the focus on tumor segmentation techniques based on multi-spectral MRI data. A large subset of methods developed within the bounds of this challenge declared that they rely on a classical histogram normalization method proposed by Nyúl et al. in 2000, which supposed that the corrected histogram of a certain organ composed of normal tissues only should be similar in all patients. However, this classical method did not count with possible lesions that can vary a lot in size, position, and shape. This paper proposes to perform a comparison of three sets of histogram normalization methods deployed in a brain tumor segmentation framework, and formulates recommendations regarding this preprocessing step.

Keywords: Magnetic resonance imaging · Brain tumor detection · Tumor segmentation · Histogram normalization

1 Introduction

The ever growing number of medical imaging devices cannot be followed by the number of human experts who are able to reliably evaluate the image records.

This project was supported by the Sapientia Foundation – Institute for Scientific Research. The work of L. Kovács was supported by the European Research Council (ERC) under the European Union's Horizon 2020 research and innovation programme (grant agreement No 679681). The work of L. Szilágyi was supported by the Hungarian Academy of Sciences through the János Bolyai Fellowship program.

© Springer Nature Switzerland AG 2019
I. Nyström et al. (Eds.): CIARP 2019, LNCS 11896, pp. 375–384, 2019.
https://doi.org/10.1007/978-3-030-33904-3_35

This is causing an intensifying need for automated algorithms that can filter out the surely negative cases and recommend the suspected positives to be investigated by the human experts. The most important requirement for such automated algorithms is to minimize the number of false negatives, which means that they have to be sensitive to any sort of lesions in the tissues.

Magnetic resonance imaging (MRI) is a frequently used technique in the brain tumor detection and segmentation problem, because of its high contrast and relatively good resolution. However, MRI has a serious drawback: the numerical values in its records do not directly reflect the imaged tissues. In order to correctly interpret the observed images, it is necessary to adapt them to the context, which is usually performed via histogram normalization. Without this step, comparing two intensity values from two different MRI records would be like comparing the water amount in two bottles by checking only the depth of the water in them and ignoring the shape of the bottles.

Several solutions have been proposed to normalize or standardize the histograms of MRI records [1–5]. However, none of them were designed to tackle with focal lesions (tumors, gliomas) that might be present. Some brain tumors grow to 20–30% of the brain volume, which strongly distorts the histogram of any data channel of the MRI histograms. Luckily, normal and tumor tissues look differently in some data channels, and thus we are able to identify the presence of tumors. Normalizing the histograms in batch mode, as it is done by the most popular technique proposed by Nyúl et al. [1] (referred to as method A1 in the following), and expecting them to look similar whether they contain tumor or not, is prone to damage the segmentation quality. A1 produces a two-step transformation of intensities using some predefined intensity percentiles as landmark points. Several recent studies report using A1, without giving details of its parametrization [6–16]. Few studies indicate the number of landmark points involved: Soltaninejad et al. [17] mentioned using 12 landmarks, while Pinto et al. [18] seem to be using the S1 setting of the method A1, see details in Sect. 3.1. Tustison et al. [19] remarked that a simple linear transformation based method can provide slightly better accuracy than A1, without giving details of their method. Such simple linear transforms were applied in [20–22], without comparing their effect to other histogram normalization methods.

This paper intends to investigate how suitable the above mentioned most popular histogram normalization method is at preprocessing MRI data in a brain tumor segmentation problem. In this order, three sets of algorithms are compared:

1. Method A1, with several settings schemes that affects the number and position of landmark points;
2. Method A2, which in fact is method A1 with landmark points defined by the fuzzy c-means clustering algorithm [23];
3. Method A3, a simple linear transform, with a single parameter, that generalizes the method employed in [20–22].

The rest of the paper is structured as follows: Sect. 2 presents the necessary details of background works, Sect. 3 gives details of the compared algorithms,

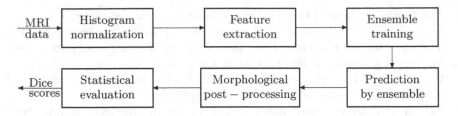

Fig. 1. Block diagram of the evaluation framework.

Sect. 4 provides a detailed analysis of the obtained results, while Sect. 5 concludes the study.

2   Background

2.1   Data

This study relies on the 54 low-grade glioma (LGG) volumes of the BraTS 2016 train dataset [24,25]. All MRI records contain four data channels (T1, T2, T1C, FLAIR), each with 155 slices of 240×240 isovolumetric pixels representing one cubic millimeter of brain tissue. Records contain approximately 1.5 million brain pixels. The ground truth provided by human experts is available for each record, which stands at the basis of training and testing machine learning solution deployed in the segmentation problem.

2.2   Framework

In order to evaluate various histogram normalization techniques, a framework was built that can deploy ensemble learning methods in a tumor segmentation problem based on MRI data. The block diagram of the framework is shown in Fig. 1.

In this study we worked with ensembles of binary decision trees (BDT). Each BDT was trained to separate negative and positive pixels based on the feature vectors of 10000 pixels that were randomly selected from the train data set. During the training process, BDTs were allowed to place nodes at any depth that was necessary. Most BDTs grew to a maximum depth between 20 and 25. On the other hand, when decision were made by the trained BDTs, the average depth of the leaf making the decision was around depth 8.

The tested histogram normalization methods (see Sect. 3) were used as the first step of the data processing, as indicated in Fig. 1. The output of each was fed to the further steps, and statistical evaluation results collected for comparison.

2.3   The A1 Method

The histogram normalization method proposed by Nyúl et al. [1] works as follows:

1. The previously defined target intensity interval is denoted by $[\alpha, \beta]$.
2. A previously defined set of MRI records $\mathcal{R}$ is involved in the process, the number of records is denoted by r. The histogram of each record is extracted.
3. The set of landmark points is defined, for example $\Lambda = \{p_{\text{low}} = 1\%, p_{L1} = 10\%, p_{L2} = 20\%, \dots, p_{L9} = 90\%, p_{\text{high}} = 99\%\}$. Let us denote the number of inner landmark points by λ (in the previous example $\lambda = 9$).
4. For all MRI records with index i, $i = 1 \dots r$, the intensity values corresponding to the landmark points defined in Λ are identified and denoted by $y_{\text{low}}^{(i)}, y_{L1}^{(i)}, y_{L2}^{(i)}, \dots, y_{L\lambda}^{(i)}, y_{\text{high}}^{(i)}$, respectively.
5. A first transformation step is performed: a linear transformation is designed such a way that maps $y_{\text{low}}^{(i)}$ to $\overline{y}_{\text{low}}^{(i)} = \alpha$, $y_{\text{high}}^{(i)}$ to $\overline{y}_{\text{high}}^{(i)} = \beta$, and applies this linear transform to all intensity values situated between $y_{\text{low}}^{(i)}$ and $y_{\text{high}}^{(i)}$ in the original histogram. The two tails of the histogram is cut, meaning that intensity values below $y_{\text{low}}^{(i)}$ are transformed to α, and intensity values above $y_{\text{high}}^{(i)}$ are transformed to β. For any $j = 1 \dots \lambda$, $y_{Lj}^{(i)}$ is transformed to $\overline{y}_{Lj}^{(i)}$.
6. Target intensity values for each inner landmark point with index j ($j = 1 \dots \lambda$) is computed next. These values are the same for all MRI records:

$$\widetilde{y}_{Lj} = \frac{1}{r} \sum_{i=1}^{r} \overline{y}_{Lj}^{(i)}. \tag{1}$$

7. The target intensity values for the two extremes are: $\widetilde{y}_{\text{low}} = \alpha$ and $\widetilde{y}_{\text{high}} = \beta$.
8. A final transformation is applied to the first transformed intensities such a way, that $\overline{y}_{\text{low}}^{(i)}$ is mapped onto $\widetilde{y}_{\text{low}}$, $\overline{y}_{\text{high}}^{(i)}$ is mapped onto $\widetilde{y}_{\text{high}}$, and any $\overline{y}_{Lj}^{(i)}$ is mapped onto $\widetilde{y}_{Lj}$ for any $j = 1 \dots \lambda$. Further on, for any $j = 0 \dots \lambda$, any intensity value $\overline{y}^{(i)} \in [\overline{y}_{Lj}^{(i)}, \overline{y}_{L,j+1}^{(i)}]$ (where $\overline{y}_{L0}^{(i)}$ is an alias for $\overline{y}_{\text{low}}^{(i)}$, and $\overline{y}_{L,\lambda+1}^{(i)}$ is an alias for $\overline{y}_{\text{high}}^{(i)}$) is piecewise linearly transformed to a value $\widetilde{y}$ situated in the interval $[\widetilde{y}_{Lj}, \widetilde{y}_{L,j+1}]$:

$$\widetilde{y} = \widetilde{y}_{Lj} + (\widetilde{y}_{L,j+1} - \widetilde{y}_{Lj}) \times \frac{\overline{y}^{(i)} - \overline{y}_{Lj}^{(i)}}{\overline{y}_{L,j+1}^{(i)} - \overline{y}_{Lj}^{(i)}}. \tag{2}$$

The algorithm is applied to each data channel separately.

3   Methods

Three approaches are compared in this study, each involving several parameter settings. The goal is to establish, which algorithm produces the best final segmentation accuracy and what settings are needed for that. The three approaches are presented in the following subsections.

3.1   Method A1 with Parameter Setting Schemes

The first approach denoted by A1 applies the algorithm presented in Sect. 2.3. Seven different parameter setting schemes were defined, they are denoted by S1...S7, and listed in Table 1. Each setting was involved in testing with values of p_{low} varying between 1% and 5% in steps of 0.5%. The value of p_{high} varied together with p_{low} such a way that the equality $p_{\text{high}} + p_{\text{low}} = 100\%$ was held.

Table 1. Various settings for the Approach A1

Setting	Landmark count (λ)	Landmarks used
S1	11	$p_{\text{low}}, 10\%, 20\%, \ldots 80\%, 90\%, p_{\text{high}}$
S2	9	$p_{\text{low}}, 10\%, 25\%, 40\%, 50\%, 60\%, 75\%, 90\%, p_{\text{high}}$
S3	7	$p_{\text{low}}, 10\%, 25\%, 50\%, 75\%, 90\%, p_{\text{high}}$
S4	5	$p_{\text{low}}, 10\%, 50\%, 90\%, p_{\text{high}}$
S5	5	$p_{\text{low}}, 25\%, 50\%, 75\%, p_{\text{high}}$
S6	4	$p_{\text{low}}, 25\%, 75\%, p_{\text{high}}$
S7	3	$p_{\text{low}}, 50\%, p_{\text{high}}$

3.2   Method A2: Landmarks Established by Fuzzy c-Means

The second approach denoted by A2 employs a very similar mechanism as A1, but the landmark points are established by the use of the fuzzy c-means algorithm. The steps of the algorithm are presented in the following:

1. The previously defined target intensity interval is denoted by $[\alpha, \beta]$. The number of inner landmark points is set as $\lambda \geq 2$. In this study we evaluated cases with $2 \leq \lambda \leq 7$.
2. A previously defined set of MRI records $\mathcal{R}$ is involved in the process, the number of records is denoted by r. The histogram of each record is extracted.
3. The set of landmark points is $\Lambda = \{p_{\text{low}}, p_{L1}, p_{L2}, \ldots, p_{L,\lambda}, p_{\text{high}}\}$, but only p_{low} and p_{high} have predefined fixed values.
4. A first transformation step is performed: a linear transformation is designed such a way that maps $y_{\text{low}}^{(i)}$ to $\overline{y}_{\text{low}}^{(i)} = \alpha$, $y_{\text{high}}^{(i)}$ to $\overline{y}_{\text{high}}^{(i)} = \beta$, and applies this linear transform to all intensity values situated between $y_{\text{low}}^{(i)}$ and $y_{\text{high}}^{(i)}$ in the original histogram. The two tails of the histogram is cut, meaning that intensity values below $y_{\text{low}}^{(i)}$ are transformed to α, and intensity values above $y_{\text{high}}^{(i)}$ are transformed to β. For any $j = 1 \ldots \lambda$, $y_{Lj}^{(i)}$ is transformed to $\overline{y}_{Lj}^{(i)}$.
5. For all MRI records with index i, $i = 1 \ldots r$, the transformed intensity values undergo histogram-based quick fuzzy c-means clustering with $c = \lambda$ clusters. The obtained cluster prototypes sorted in increasing order $v_1, v_2, \ldots v_\lambda$ are then assigned as dynamically established landmark points: $\overline{y}_{Lj}^{(i)} = v_j \, \forall j = 1 \ldots \lambda$.

6. Target intensity values $\widetilde{y}_{Lj}$ for each inner landmark point with index j ($j = 1 \ldots \lambda$) is computed next, using Eq. (1). These values are the same for all MRI records.
7. The target intensity values for the two extremes are: $\widetilde{y}_{low} = \alpha$ and $\widetilde{y}_{high} = \beta$.
8. The final transformation is applied the same way as in the original A1 algorithm, presented in Sect. 2.3.

The algorithm is applied to each data channel separately.

3.3   Method A3: Linear Transform with One Parameter

The third approach denoted by A3 is a generalization of the technique proposed in our previous paper [21]. This method uses a single linear transformation, whose coefficients depend on the histogram of the original MRI volume. In contrast with the previous two approaches, the normalization of any MRI record does not depend on other MRI records.

1. The previously defined target intensity interval is denoted by $[\alpha, \beta]$. The algorithm uses a parameter q which controls the compactness of the final histogram.
2. The histogram of the current MRI record is extracted. The 25-percentile and 75-percentile intensity values are identified, and denoted by y_{25} and y_{75}.
3. The target intensities for the 25-percentile and 75-percentile intensity values are established using the formulas

$$\widetilde{y}_{25} = \frac{1}{2}\left[(\beta + \alpha) - \frac{\beta - \alpha}{q}\right] \quad \text{and} \quad \widetilde{y}_{75} = \frac{1}{2}\left[(\beta + \alpha) + \frac{\beta - \alpha}{q}\right]. \quad (3)$$

4. The coefficients of the linear transform $y \rightarrow ay + b$ are extracted such a way, that y_{25} and y_{75} are transformed to $\widetilde{y}_{25}$ and $\widetilde{y}_{75}$, respectively, using the formulas

$$a = \frac{\beta - \alpha}{q(y_{75} - y_{25})} \quad \text{and} \quad b = \widetilde{y}_{25} - \frac{(\beta - \alpha)y_{25}}{q(y_{75} - y_{25})}. \quad (4)$$

5. Any intensity y from the input MRI volume becomes

$$\widetilde{y} = \begin{cases} \alpha & \text{if } ay + b < \alpha \\ ay + b & \text{if } \alpha \le ay + b \le \beta \; . \\ \beta & \text{if } ay + b > \beta \end{cases} \quad (5)$$

The algorithm is applied to each data channel separately. In our previous works [20–22], this approach was used with parameter setting $q = 5$.

4   Results and Discussion

Each of the three algorithms were tested with various settings using the same evaluation framework, having the target intensity interval bounded by $\alpha = 200$ and $\beta = 1200$, corresponding to an approximately 10-bit resolution. The 54 LGG

Table 2. Overall Dice scores obtained using Approach A1 using various settings

Approach A1 Setting	p_{low} set to						
	1%	1.5%	2%	2.5%	3%	3.5%	4%
S1	80.456%	80.410%	80.577%	80.472%	80.096%	79.824%	79.229%
S2	80.420%	80.489%	80.553%	80.519%	80.251%	79.916%	79.333%
S3	80.409%	80.537%	80.543%	80.612%	80.239%	79.865%	79.183%
S4	79.937%	80.133%	80.213%	80.257%	79.726%	79.317%	78.803%
S5	81.254%	81.459%	81.545%	81.518%	80.939%	80.619%	80.194%
S6	81.051%	81.363%	81.543%	81.404%	81.014%	80.581%	80.370%
S7	80.723%	80.972%	81.474%	81.364%	80.959%	80.631%	80.129%

Table 3. Overall Dice scores obtained using Approach A2 using various settings

Approach A2 Setting	p_{low} set to						
	1%	1.5%	2%	2.5%	3%	3.5%	4%
$\lambda = 2$	81.862%	82.122%	82.058%	81.991%	81.789%	81.561%	81.312%
$\lambda = 3$	81.829%	81.902%	81.958%	82.137%	81.857%	81.612%	81.105%
$\lambda = 4$	80.819%	81.494%	81.707%	81.789%	81.800%	81.548%	81.223%
$\lambda = 5$	79.289%	80.928%	81.150%	81.322%	81.077%	81.137%	81.085%
$\lambda = 6$	80.421%	80.426%	80.527%	80.732%	81.029%	81.012%	80.511%
$\lambda = 7$	80.965%	80.991%	80.668%	80.459%	80.722%	80.665%	80.722%

volumes of the BraTS 2016 data set underwent a ten-fold cross validation using the BDT ensemble based classifier algorithm described in Sect. 2. Each ensemble consisted of 125 BDTs, each trained with 10000 randomly selected feature vectors from the train data, out of which 92% were negatives and 8% positives. From 104 generated features (for each of the 4 observed channels: minimum, maximum and average extracted from $3 \times 3 \times 3$ neighborhood; average and median extracted from planar neighborhoods of size ranging from 3×3 to 11×11; four directional gradients and eight directional Gabor wavelet values) the 13 most relevant features (minimum, maximum and average of T2 and FLAIR, maximum and average of T1C, and minimum of T1 from $3 \times 3 \times 3$ neighborhood; average of T1C, T2 and FLAIR from 11×11 neighborhood; average of FLAIR from 3×3 neighborhood) were included into the feature vector, details are presented in our previous paper [22]. The outcome of the classification produced by the ensemble underwent a post-processing that relabeled each pixel according to the neighbors of the pixel. Those pixels were declared final positives, which had at least one third of its neighbors declared positive by the ensemble. The main evaluation criterion is the Dice score (DS), which is defined as $DS = 2TP/(2TP + FP + FN)$, where TP, FP, and FN represent the number of true positives, false positives, and false negatives, respectively. Average Dice scores for each MRI record were established after the ten-fold cross-validation. Finally, the overall Dice score was

Table 4. Overall Dice scores obtained using Approach A3 using various settings

$q = 2.5$	$q = 3$	$q = 3.5$	$q = 4$	$q = 4.5$	$q = 5$	$q = 5.5$	$q = 6$	$q = 7$
80.844%	81.517%	82.024%	82.350%	82.395%	82.249%	82.233%	82.103%	82.070%

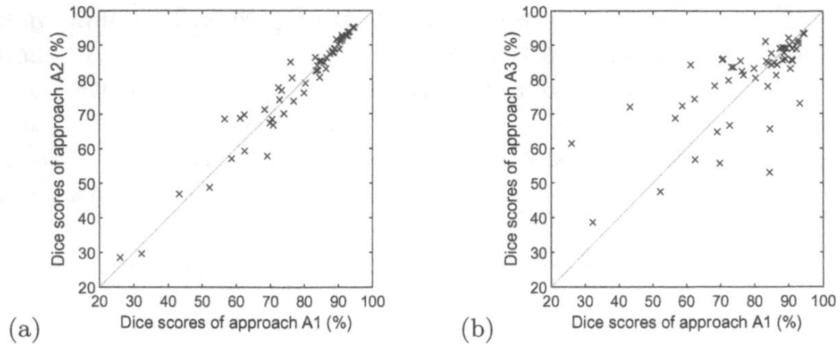

(a)   (b)

Fig. 2. Dice scores obtained on individual LGG records, the best performance of the three approaches plotted one against another: (a) A2 vs. A1; (b) A3 vs. A1.

computed for each approach and setting, based on all pixels from all volumes. Results are exhibited in Tables 2, 3 and 4.

The best achieved overall Dice score (ODS) is 82.395%. Most of the evaluated approaches and settings led to ODS values over 80%. The classical A1 approach hardly achieved 81.5%, with its best setting that used the landmark set $\{p_{\text{low}} = 2\%, 25\%, 50\%, 75\%, p_{\text{high}} = 98\%\}$, where the middle landmark point is virtually optional. A larger number of landmarks, as used for example in [17,18], in our studies led to ODS around 80.5%, which is well below optimal.

The A2 approach achieved best ODS values around 82%, when using 2 or 3 inner landmarks, p_{low} ranging between 1.5% and 2.5%. The accuracy is finer than in case of approach A1, while the best scenario is quite similar.

The A3 approach has a wide interval of its parameter, where the algorithm scores ODS values above 82%. The best accuracy was achieved at $q = 4.5$, which means that in each data channel of the MRI records, all original intensity values are subject to linear transformation into the target interval $[200, 1200]$ such a way, that the 25-percentile is mapped to 578, the 75-percentile to 822, and the tails of the transformed histogram is cut at 200 and 1200. The normalization of any histograms occurs independently, it does not depend on the histograms of other records or other data channels.

Figure 2 exhibits the comparison of the three approaches, when applied to individual MRI volumes. Each approach is represented with its overall best setting. This figure also shows that A3 and A2 can perform slightly better than A1, but the slight superiority comes in average only, because the segmentation accuracy of individual MRI records can be either better of worse, with virtually same probability. Tests have confirmed the observation of Tustison et al. [19], who remarked that a well designed simple linear transformation performs better than previous algorithms like A1, in such tumor segmentation problems. Further tests involving more data, more algorithms, and further quality indicators could provide stronger evidence of this superiority. Our results do not mean that A3 leads to better accuracy than the frequently used A1 in all segmentation problems. But when the goal is tumor detection, it is recommendable to apply histogram normalization via approach A3.

5   Conclusions

This study investigated the effect of various histogram normalization methods upon the final accuracy in an MRI data based brain tumor segmentation problem. Two approaches were proposed and compared to the most frequently used and most cited such algorithm. Tests have revealed a slight superiority of both proposed algorithms, compared to the previous one.

References

1. Nyúl, L.G., Udupa, J.K., Zhang, X.: New variants of a method of MRI scale standardization. IEEE Trans. Med. Imaging **19**(2), 143–150 (2000)
2. Weisenfeld, N.L., Wartfeld, S.K.: Normalization of joint image-intensity statistics in MRI using the Kullback-Leibler divergence. In: IEEE International Symposium on Biomedical Imaging (ISBI), pp. 101–104. IEEE (2004)
3. Jäger, F., Deuerling-Zheng, Y., Frericks, B., Wacker, F., Hornegger, J.: A new method for MRI intensity standardization with application to lesion detection in the brain. In: Kobbelt, L., et al. (eds.) Vision Model, Visualization, pp. 269–276. AKA GmbH, Köln, Germany (2006)
4. Leung, K.K., et al.: Robust atrophy rate measurement in Alzheimer's disease using multi-site serial MRI: tissue-specific intensity normalization and parameter selection. Neuroimage **50**, 516–523 (2010)
5. Shinohara, R.T., Crainiceanu, C.M., Caffo, B.S., Gaitán, M.I., Reich, D.S.: Population-wide principal component-based quantification of blood-brain-barrier dynamics in multiple sclerosis. Neuroimage **57**(4), 1430–1446 (2011)
6. Pereira, S., Pinto, A., Alves, V., Silva, C.A.: Deep convolutional neural networks for the segmentation of gliomas in multi-sequence MRI. In: Crimi, A., Menze, B., Maier, O., Reyes, M., Handels, H. (eds.) BrainLes 2015. LNCS, vol. 9556, pp. 131–143. Springer, Cham (2016). https://doi.org/10.1007/978-3-319-30858-6_12
7. Meier, R., et al.: Clinical evaluation of a fully-automatic segmentation method for longitudinal brain tumor volumetry. Sci. Rep. **6**, 23376 (2016)
8. Ellwaa, A., et al.: Brain tumor segmantation using random forest trained on iteratively selected patients. In: Crimi, A., et al. (eds.) BrainLes 2016. LNCS, vol. 10154, pp. 129–137. Springer, Cham (2017). https://doi.org/10.1007/978-3-319-55524-9_13
9. Bakas, S., et al.: Advancing the cancer genome atlas glioma MRI collections with expert segmentation labels and radiomic features. Sci. Data **4**, 170117 (2017)
10. Rezaei, M., et al.: A conditional adversarial network for semantic segmentation of brain tumor. In: Crimi, A., Bakas, S., Kuijf, H., Menze, B., Reyes, M. (eds.) BrainLes 2017. LNCS, vol. 10670, pp. 241–252. Springer, Cham (2018). https://doi.org/10.1007/978-3-319-75238-9_21

11. Fidon, L., et al.: Scalable multimodal convolutional networks for brain tumour segmentation. In: Descoteaux, M., Maier-Hein, L., Franz, A., Jannin, P., Collins, D.L., Duchesne, S. (eds.) MICCAI 2017. LNCS, vol. 10435, pp. 285–293. Springer, Cham (2017). https://doi.org/10.1007/978-3-319-66179-7_33

12. Chen, X., Nguyen, B.P., Chui, C.K., Ong, S.H.: An automated framework for multi-label brain tumor segmentation based on kernel sparse representation. Acta Polytech. Hung. **14**(1), 25–43 (2017)

13. Soltaninejad, M., Zhang, L., Lambrou, T., Yang, G., Allinson, N., Ye, X.: MRI brain tumor segmentation and patient survival prediction using random forests and fully convolutional networks. In: Crimi, A., Bakas, S., Kuijf, H., Menze, B., Reyes, M. (eds.) BrainLes 2017. LNCS, vol. 10670, pp. 204–215. Springer, Cham (2018). https://doi.org/10.1007/978-3-319-75238-9_18

14. Chen, W., Liu, B.Q., Peng, S.T., Sun, J.W., Qiao, X.: Computer-aided grading of gliomas combining automatic segmentation and radiomics. Int. J. Biomed. Imaging **2018**, 2512037 (2018)

15. Wu, Y.P., Liu, B., Lin, Y.S., Yang, C., Wang, M.Y.: Grading glioma by radiomics with feature selection based on mutual information. J. Ambient Intell. Hum. Comput. **9**(5), 1671–1682 (2018)

16. Chang, J., et al.: A mix-pooling CNN architecture with FCRF for brain tumor segmentation. J. Vis. Commun. Image Represent. **58**, 316–322 (2019)

17. Soltaninejad, M., et al.: Supervised learning based multimodal MRI brain tumour segmentation using texture features from supervoxels. Comput. Methods Programs Biomed. **157**, 69–84 (2018)

18. Pinto, A., Pereira, S., Rasteiro, D., Silva, C.A.: Hierarchical brain tumour segmentation using extremely randomized trees. Pattern Recogn. **82**, 105–117 (2018)

19. Tustison, N.J., et al.: Optimal symmetric multimodal templates and concatenated random forests for supervised brain tumor segmentation (simplified) with ANTsR. Neuroinformatics **13**, 209–225 (2015)

20. Lefkovits, L., Lefkovits, S., Szilágyi, L.: Brain tumor segmentation with optimized random forest. In: Crimi, A., et al. (eds.) BrainLes 2016. LNCS, vol. 10154, pp. 88–99. Springer, Cham (2017). https://doi.org/10.1007/978-3-319-55524-9_9

21. Szilágyi, L., Iclănzan, D., Kapás, Z., Szabó, Z., Győrfi, Á., Lefkovits, L.: Low and high grade glioma segmentation in multispectral brain MRI data. Acta Univ. Sapientia Informatica **10**(1), 110–132 (2018)

22. Győrfi, Á., Kovács, L., Szilágyi, L.: A feature ranking and selection algorithm for brain tumor segmentation in multi-spectral magnetic resonance image data. In: 41st Annual Conference of the IEEE EMBS. IEEE (2019, accepted paper)

23. Bezdek, J.C.: Pattern Recognition with Fuzzy Objective Function Algorithms. Plenum, New York (1981)

24. Menze, B.H., Jakab, A., Bauer, S., Kalpathy-Cramer, J., Farahani, K., Kirby, J., et al.: The multimodal brain tumor image segmentation benchmark (BRATS). IEEE Trans. Med. Imaging **34**, 1993–2024 (2015)

25. Bakas, S., Reyes, M., Jakab, A., Bauer, S., Rempfler, M., Crimi, A., et al.: Identifying the best machine learning algorithms for brain tumor segmentation, progression assessment, and overall survival prediction in the BRATS challenge. arXiv: 1181.02629v3, 23 April 2019

Machine Learning and Neural Networks

Probabilistic Forecasting Using Monte Carlo Dropout Neural Networks

Cristián Serpell[1]([✉]), Ignacio Araya[1], Carlos Valle[2], and Héctor Allende[1]

[1] Universidad Técnica Federico Santa María, Valparaíso, Chile
cserpell@alumnos.inf.utfsm.cl, {iaraya,hallende}@inf.utfsm.cl
[2] Universidad de Playa Ancha, Valparaíso, Chile
carlos.valle@upla.cl

Abstract. Using artificial neural networks for forecasting tasks is a popular approach that has proven to be very accurate. When used to estimate prediction intervals, a normal distribution is usually assumed as the data noise uncertainty term, as in MVE networks, while model parameters uncertainty is often ignored. Because of this, prediction intervals estimated by them are narrow in uncertain regions where train data is scarce. To tackle this problem we apply Monte Carlo dropout, which is a model uncertainty representation technique, to the network parameters of a Long Short-Term Memory MVE network, allowing us to construct better prediction intervals in probabilistic forecasting tasks. We compare our proposal with the pure MVE method in four wind speed and one consumer load real forecasting datasets, showing that our method improves results in terms of the Winkler loss in both one step ahead and multi-step ahead probabilistic forecasting.

Keywords: Probabilistic forecasting · Prediction interval · Monte Carlo dropout · Mean Variance Estimation · Wind speed forecasting · Consumer load forecasting

1 Introduction

Time series forecasting is a main research topic in the machine learning community, as several human activities, such as engineering, economics, social sciences, medicine and others, depend on estimating values of an uncertain future. Usually, point predictions corresponding to expected future values are computed, without conveying information about how accurate such estimates are. In contrast, probabilistic forecasting aims at giving a measure of such uncertainty, allowing decision makers to know how confident they can be on the forecasted values. Commonly, such uncertainty is expressed as a prediction interval, and it is specially important in activities that have big intrinsic uncertainty, such as wind speed and electric consumer load forecasting, as mentioned in [8,9].

Forecasting methods based on artificial neural networks have shown to be very accurate for point forecasts and are currently in use for all kind of tasks.

© Springer Nature Switzerland AG 2019
I. Nyström et al. (Eds.): CIARP 2019, LNCS 11896, pp. 387–397, 2019.
https://doi.org/10.1007/978-3-030-33904-3_36

One way to extend them to build prediction intervals is to assume an output distribution around the point estimate, usually a constant variance normal, which works nicely when the series is homoscedastic, i.e. has a constant variance in time. However, when the variance is non constant, the network can be extended so that it also outputs an estimate of the future variance; an approach that is called Mean Variance Estimation (MVE) and seems to work well in practice as shown in [8,9]. Although data noise is arguably the most important source of uncertainty in forecasting, and is what MVE actually estimates, it is not the only one, as there is also the uncertainty coming from the model parameters themselves. This uncertainty comes from the fact that many different models could fit our finite length time-series, hence that the MVE prediction intervals might not result in proper uncertainty estimates in regions when train data is scarce or not present. Failing to consider all sources of uncertainty can lead to narrow prediction intervals, making systems that depend on these forecasts to be less reliable. An example of this can be found in wind power generation forecasting, where prediction intervals are used to decide which of the other power generators of the power system have to work to satisfy the overall power demand in case wind power is not enough. Consequently, narrow prediction intervals result in higher energy costs and a less reliable electrical grid [11].

Among the many methods that can be used to capture model uncertainty in neural networks, Monte Carlo (MC) dropout holds great appeal given its low computational cost and simplicity. In MC dropout, a network is trained using the standard dropout technique and, at test time, dropout is still used so that, through randomly masking hidden units, different outcomes for test data can be obtained, which are then used to construct prediction intervals. However, this technique alone is not suitable for probabilistic forecasting since it does not capture uncertainty due to data noise: the model is trained to estimate the expected value of the predictions, thus its uncertainty estimates exclusively concern such expectations. To illustrate the difference between these methods, we trained an MVE network and an MC dropout network on a synthetic one dimensional regression task dataset, and show the results in Fig. 1. It can be seen that MVE properly captures the heteroscedastic nature of the train data, but have low uncertainty estimates in other regions, whereas in MC dropout the opposite takes place. Therefore, to capture both sources of uncertainty in forecasting, we propose enhancing MVE by applying the MC dropout technique to its network parameters. Following MVE, our proposal is trained to predict the mean and variance parameters of assumed normally distributed future values, while MC dropout allows us to obtain many such distributions from which to sample point forecasts and construct prediction intervals. Figure 1 shows how such intervals are thick in regions with no train data while properly covering regions where there are. We experimentally support our proposal by embedding it in a Long Short-Term Memory (LSTM) and comparing it with a traditional MVE network for four wind speed and one consumer electrical load forecasting tasks.

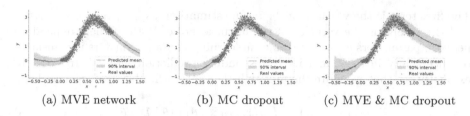

(a) MVE network	(b) MC dropout	(c) MVE & MC dropout

Fig. 1. Neural networks producing prediction intervals for a 1D regression, trained on data in red. (a) An MVE network captures changing variance in data, but has narrow intervals in other regions. (b) MC dropout captures variance of network parameters, but not of train data, giving narrow intervals within it. (c) Our proposal captures both sources of variance, thus producing better prediction intervals. (Color figure online)

The paper is structured as follows: Sect. 2 introduces how prediction intervals are estimated and does a review of related work, Sect. 3 details our proposal, Sect. 4 describe our experiments to validate the proposal, and Sect. 5 makes final remarks and describe future research directions.

2   Prediction Interval Estimation

Given an univariate time-series $\{x_t\}$ of an underlying stochastic process, our goal is to construct good prediction intervals $I_t = [L_t, U_t]$ for any time t, with a given confidence level $c = 1 - 2\alpha$, meaning that $p(x_t < L_t) \leq \alpha$, $p(x_t > U_t) \leq \alpha$ and $p(L_t \leq x_t \leq U_t) \geq c$. Certainly, if we knew the underlying process distribution $p(x_t)$, our task would be over, but since our interest is to find such intervals for future values not yet observed, we use previously observed values and aim at modeling $p(x_{t+1} \mid x_1, \ldots, x_t)$. As the number of past values may be too large, we make the Markov assumption that this distribution may be modelled using just last M values. So, we look forward modeling $p(x_{t+1} \mid \mathbf{x}_t)$, where $\mathbf{x}_t = (x_{t-M+1}, \ldots, x_t)$. To do so, we model the relation between x_{t+1} and $\mathbf{x}_t$ as $x_{t+1} = f(\mathbf{x}_t) + n(\mathbf{x}_t)$, with $n(\mathbf{x}_t)$ an additive noise with normal distribution of mean 0 and a standard deviation depending on $\mathbf{x}_t$. Furthermore, we can write $n(\mathbf{x}_t) = \sigma(\mathbf{x}_t)\epsilon$, with ϵ a standard normal random variable, and $\sigma(\mathbf{x}_t)$ a scale factor function. Thus, x_{t+1} will have a conditional normal distribution $\mathcal{N}\left(f(\mathbf{x}_t), \sigma(\mathbf{x}_t)^2\right)$. To define functions f and σ, we use a neural network. Calling all network parameters, such as weights and biases, as θ, we can rewrite the relation of x_{t+1} and $\mathbf{x}_t$ as

$$x_{t+1} = f(\mathbf{x}_t, \theta) + \sigma(\mathbf{x}_t, \theta)\,\epsilon. \tag{1}$$

Due to lack of train data in the whole domain, learning those parameters θ will undoubtedly come with errors. Indeed, the variance of x_{t+1} is given by:

$$\mathrm{Var}\left[x_{t+1}\right] = \mathrm{Var}\left[f(\mathbf{x}_t, \theta)\right] + \mathbb{E}\left[\sigma(\mathbf{x}_t, \theta)^2\right]. \tag{2}$$

The first term is the variance of the mean estimation, as $\mathbb{E}\left[x_{t+1}\right] = \mathbb{E}\left[f\left(\mathbf{x}_t, \theta\right)\right]$, and the second is the variance of the intrinsic noise. Both of them depend on network parameters θ, so uncertainty in learning those parameters is translated to uncertainty in x_{t+1}, and its corresponding prediction interval. In order to explicitly see this, we write the probabilistic relationship between x_{t+1} and $\mathbf{x}_t$:

$$p\left(x_{t+1} \mid \mathbf{x}_t\right) = \int_{\theta} p\left(x_{t+1} \mid \mathbf{x}_t, \theta\right) p\left(\theta \mid \mathcal{D}\right) d\theta, \tag{3}$$

where $p\left(\theta \mid \mathcal{D}\right)$ if the posterior distribution of parameters θ after training with data $\mathcal{D}$, and $x_{t+1} \mid \mathbf{x}_t, \theta \sim \mathcal{N}\left(f\left(\mathbf{x}_t, \theta\right), \sigma\left(\mathbf{x}_t, \theta\right)^2\right)$. Neural networks based on Mean Variance Estimation (MVE) fix one value for its parameters, θ^*, learned during training using the maximum likelihood criterion. Therefore, they fix $\theta = \theta^*$ in (1), reducing (3) to just $p\left(x_{t+1} \mid \mathbf{x}_t\right) = p\left(x_{t+1} \mid \mathbf{x}_t, \theta^*\right)$, and the variance of (2) only to the second term: $\mathrm{Var}\left[x_{t+1}\right] = \sigma\left(\mathbf{x}_t, \theta^*\right)^2$. Thus, the first term of the variance is completely ignored, and the corresponding prediction intervals miss an important source of uncertainty.

On the other hand, [5] has proposed the use of dropout to handle the aforementioned uncertainty of network parameters θ. Dropout is a regularization technique that modifies the network each time it is evaluated on an input batch, turning off some of its connections. Each active pattern can be seen as a different network, and several evaluations with different connections as an ensemble of networks with shared weights. When applying many of those networks during prediction, we are sampling posterior $p\left(\theta \mid \mathcal{D}\right)$ of (3) using another distribution implicitly defined by the selection of which connections are disabled. As we approximate the integral with a Monte Carlo like approach, this procedure is called Monte Carlo (MC) dropout. Hence, to create prediction intervals using this idea, we sample network parameters K times, having a set of different parameters $\{\theta_1, \ldots, \theta_K\}$, and apply model in (1) for each of them. Then, sort obtained values and finally mark the corresponding α percentage value as lower bound L_{t+1} and the corresponding $1 - \alpha$ as upper bound U_{t+1}. The authors of [13] do something similar, though assuming a constant intrinsic noise variance $\sigma\left(\mathbf{x}_t, \theta\right) = \sigma\left(\theta\right)$. This works well for homoscedastic time series, i.e. that have constant variance in time, but not for heteroscedastic ones.

2.1   Related Work

Estimation of prediction intervals using neural networks has been tackled by several authors. A first group of them does not explore how model parameters uncertainty affects their solution. Among them, we highlight direct estimation of lower and upper bound, similar to quantile regression by [12], and MVE models, introduced in [10]. To include parameters uncertainty, most other works are based on ensemble methods like bootstrap, as mentioned by [8,9], that are heavy to train, as they require training of many models.

There is a branch of research that explore neural network parameters uncertainty. In general, they make assumptions regarding the distribution of those parameters, and use variational inference, as explained by [6]. One example is [2], that assumes Gaussian parameters distributions. Also, as explained by [3], many regularization techniques usual in the deep learning community can be interpreted as assumptions of the parameters distributions. In particular, MC dropout has been adapted to this framework and applied to many kinds of networks, including recurrent ones by [4].

3   Proposal

To build prediction intervals that correctly include both sources of uncertainty, we propose a neural network that combines the MVE method and MC dropout, allowing to create prediction intervals that are wider when presented with previously unseen data, compared to those of MVE alone, and that can handle heteroscedastic time series, a feature not present when working only with dropout.

As mentioned above, the posterior distribution $p(\theta \mid \mathcal{D})$ from (3) is unknown, hence we approximate it using another distribution $q(\theta)$ that we can parameterize. The parameters of such distribution $q(\theta)$ are trained maximizing a lower bound of the train data log-likelihood, called variational bound and given by

$$\log p(\mathcal{D}) \geq \mathbb{E}_{\theta \sim q(\theta)} \left[\log p(\mathcal{D} \mid \theta) \right] - D_{\mathrm{KL}}(q(\theta) \mid\mid p(\theta)). \qquad (4)$$

Thus, we have a bound given by two terms: an approximate likelihood using distribution $q(\theta)$ instead of $p(\theta \mid \mathcal{D})$, and a divergence between an a priori assumption $p(\theta)$ and the posterior approximation. As explained by [5], when using dropout, we are assuming a distribution $q(\theta)$ implicitly given by the selection of dropped out connections, and also that the second term is equivalent to weight decay regularization. Recalling from (3), we assumed a normal distribution for x_{t+1}, given $\mathbf{x}_t$ and θ, so the log-likelihood term for each pair $(\mathbf{x}_t, x_{t+1})$ is

$$\log p(x_{t+1} \mid \mathbf{x}_t, \theta) = -\frac{1}{2} \left[\log 2\pi + \log \sigma(\mathbf{x}_t, \theta)^2 + \frac{(x_{t+1} - f(\mathbf{x}_t, \theta))^2}{\sigma(\mathbf{x}_t, \theta)^2} \right].$$

Considering this, we maximize the variational bound minimizing its negative version using a stochastic gradient optimization algorithm that, for each minibatch of size B, samples a value θ from $q(\theta)$, say θ_k, applies it over the minibatch, and then estimates the negative of (4) as

$$\sum_{b=1}^{B} \left[\log \sigma\left(\mathbf{x}_t^{(b)}, \theta_k\right)^2 + \frac{\left(x_{t+1}^{(b)} - f\left(\mathbf{x}_t^{(b)}, \theta_k\right)\right)^2}{\sigma\left(\mathbf{x}_t^{(b)}, \theta_k\right)^2} \right] + \text{Regularization}(\theta_k).$$

Note that for the original MVE method, without MC dropout, this same estimate is used by the optimization algorithm, but instead of using a sampled θ_k provided by dropout, a single current value for the parameters θ is used and optimized.

To implement this strategy, we used a Long-Short Term Memory (LSTM) like the one in Fig. 2, as it is a recurrent neural network that is known to capture

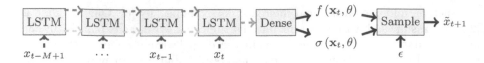

Fig. 2. Our network takes an input $\mathbf{x}_t$, produces $f(\mathbf{x}_t, \theta)$ and $\sigma(\mathbf{x}_t, \theta)$, parameters of a normal distribution, and then gets a sample of x_{t+1} as $\tilde{x}_{t+1} = f(\mathbf{x}_t, \theta) + \sigma(\mathbf{x}_t, \theta)\,\epsilon$. When using MC dropout, the network is modified, deactivating some of the dashed connections and those within LSTM layer, therefore using a sampled θ_k instead of a fixed θ. The same dropout mask is applied in all steps of the LSTM recurrence, thus having the same color in the figure, and a different mask is applied in the other dashed lines. Black solid lines are not affected by dropout. Thus, for different dropout masks, different $\tilde{x}_{t+1}$ are produced, and the prediction interval is created with all those samples.

long term dependencies and has proven to work well for forecasting tasks. As shown, we start the sequence processing M steps before the forecast time step, since adding all time steps in the past makes it slower and it does not help the LSTM to learn to keep more useful information during training.

3.1   Multi Step Forecasting

Given a trained model from our proposal, we compute prediction intervals for a horizon of length ℓ time steps using the following procedure: we fix one dropout mask randomly, thus getting one value θ_k, and then sample a single value of x_{t+1}, from the Gaussian distribution given by the network, $\tilde{x}_{t+1} = f(\mathbf{x}_t, \theta) + \sigma(\mathbf{x}_t, \theta)\,\epsilon$, re-injecting this value as part of a new input $\mathbf{x}_{t+1} = (x_{t-M+2}, \ldots, \tilde{x}_{t+1})$, and sampling again, until a trajectory of sampled values of length ℓ is obtained, as described in Fig. 3. This process is repeated K times with different dropout masks each, generating K trajectories of sampled

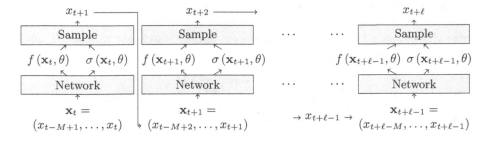

Fig. 3. To simulate a trajectory of ℓ steps, we get one sample of θ using dropout, and then predict one step at a time, reinjecting the output as input for next time step, getting a trajectory of future values $(x_{t+1}, \ldots, x_{t+\ell})$. After K simulations, we build prediction intervals for each time step using gathered samples from all trajectories.

values, each with a different θ_k. Thus, for each future time step, we obtained K sampled values, which are then used to build the prediction interval corresponding to that time step, using the same strategy explained in the one step ahead case. Therefore, the prediction interval at each time-step captures the uncertainty that accumulates through all its previous time-steps, which is given by both the parameters uncertainty and data noise. For the MVE model, the same procedure is applied, but removing the dropout step, computing K trajectories as well, thus comparisons with our proposal are fair.

4   Experiments

To asses our proposal, we considered UCI's individual household electric power consumption data set, and four wind speed series that correspond to measurements taken every 10 min by prospecting towers located in Northern Chile[1]. We considered hourly averages, thus our time series correspond to hourly time steps. Missing data, though rare, was filled repeating last observed value at the same time of the day. For example, if 10 am measurement of a day is missing, we look for the value at 10 am of previous day. In order to validate that our proposal works for different regimes of a series, we divided each dataset in five blocks, considering a small overlap between them, as shown in Fig. 4. For each block, we considered last 1000 measurements as test set. Of the remaining block data, 10% was considered as a validation set, used to select best hyperparameters. We standardized data for each block, considering only the training set to fix those standardization parameters.

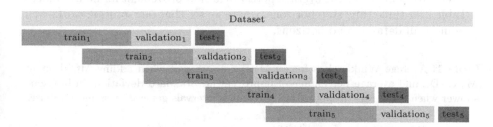

Fig. 4. Each time series is divided into five blocks of train, validation and test sets.

Our assessment of produced prediction intervals quality considers:

- *Coverage*: We measure prediction interval coverage probability (PICP), the actual percentage of measurements in between lower and upper bound of intervals for all time steps, to know how close they are to desired confidence.
- *Sharpness*: We use prediction interval normalized average width (PINAW), the relative size of the interval, compared with the range of values in the series, as we prefer narrower intervals having at least desired confidence.

---

[1] Data can be downloaded from http://walker.dgf.uchile.cl/Mediciones/.

– *Resolution*: As the series can be heteroscedastic, the width of intervals have to change through time accordingly.

To summarize these features in one metric, we used Winkler loss, commonly used in probabilistic forecasting tasks, specially consumer load, as mentioned in [7]. For a $[L, U]$ interval, designed to have a confidence $c = 1 - 2\alpha$, and an observation of x_{t+1}, it is given by

$$\text{Winkler}\,(L, U, x_{t+1}, \alpha) = U - L + \begin{cases} 2\frac{(L-x_{t+1})}{\alpha} & \text{if } x_{t+1} < L \\ 0 & \text{if } L \leq x_{t+1} \leq U \\ 2\frac{(x_{t+1}-U)}{\alpha} & \text{if } x_{t+1} > U \end{cases}.$$

It considers coverage, as it gives a linearly increasing weight to values laying out of the interval, and also sharpness, as it has a base penalization depending on its width. As we create intervals and measure the loss for multiple time steps into the future, resolution is being taken into account during the whole process.

4.1   Results

For each block, we did a random search of hyperparameters, as recommended by [1], and considered those that had better average Winkler loss in the validation set, considering 10 repetitions with different random seeds. Then, we gathered average test metrics using those parameters, considering 10 repetitions as well.

As Winkler loss depends on the range of values of a block, we computed the percentage of improvement to make them comparable among blocks and datasets. We summarize the average percentage of improvement for all five blocks in Table 1 for three time step horizons, where we see that our solution works well for almost all datasets and horizons.

Table 1. Average Winkler loss for an MVE network and when adding MC dropout (MVE+D), and average percentage of improvement. Standard deviation, in brackets, is lower when using MC dropout, thus prediction intervals generation is more robust.

Set	1 step			9 steps			18 steps		
	MVE	MVE+D	%	MVE	MVE+D	%	MVE	MVE+D	%
B08	8.1 (0.4)	**7.1** (0.1)	11.0	13.0 (1.0)	**11.3** (0.4)	10.7	13.6 (1.1)	**12.5** (0.7)	7.5
D05a	8.4 (0.2)	**7.8** (0.1)	6.8	14.0 (0.9)	**13.5** (0.7)	4.3	**14.9** (1.0)	15.1 (1.0)	0.0
D08	8.3 (0.2)	**7.7** (0.1)	6.9	14.8 (1.1)	**13.3** (0.5)	9.9	15.9 (1.6)	**14.7** (0.6)	6.7
E01	10.3 (0.2)	**9.2** (0.2)	10.2	17.2 (1.7)	**14.9** (1.0)	13.5	18.7 (1.8)	**15.5** (1.0)	16.3
UCI	187 (10)	**182** (6)	1.2	238 (15)	**236** (17)	1.9	243 (18)	**240** (13)	0.3

In Table 2 we show detailed test metrics for one wind dataset, for one step forecasting. In this case, we see that our model builds better prediction intervals, and also that the standard deviation of metrics is much lower. This means that our model works better independently of the random seed used to train it. This

Table 2. Side by side comparison of MVE and MC dropout test metrics, for B08 wind speed dataset. As expected, MC dropout makes wider prediction intervals (PINAW), the coverage (PICP) is increased, and standard deviation of all metrics is reduced.

Metric	Model	Block 1	Block 2	Block 3	Block 4	Block 5
RMSE	MVE	2.41 (0.05)	1.66 (0.07)	2.32 (0.09)	2.06 (0.13)	**1.65** (0.04)
	MVE+D	**2.35** (0.02)	**1.66** (0.05)	**2.28** (0.07)	**1.95** (0.00)	1.66 (0.02)
PICP	MVE	83.6 (1.5)	**90.1** (1.2)	79.9 (2.3)	86.7 (3.0)	**90.0** (1.2)
	MVE+D	**89.8** (0.7)	94.0 (0.6)	**85.6** (0.6)	**94.4** (0.6)	91.7 (0.9)
PINAW	MVE	0.29 (0.01)	0.36 (0.02)	0.26 (0.01)	0.31 (0.02)	0.30 (0.01)
	MVE+D	0.34 (0.01)	0.40 (0.01)	0.30 (0.00)	0.41 (0.00)	0.32 (0.00)
Winkler	MVE	9.3 (0.4)	6.8 (0.2)	10.2 (0.5)	7.7 (0.6)	6.5 (0.1)
	MVE+D	**8.0** (0.1)	**6.2** (0.1)	**8.9** (0.0)	**6.6** (0.1)	**6.0** (0.1)

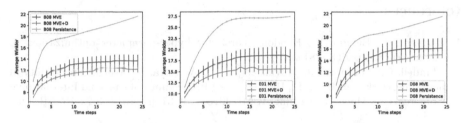

Fig. 5. Average Winkler loss, considering different prediction horizons. We can see that MVE has higher loss and higher standard deviation for most time horizons.

is in line with our consideration of model parameters uncertainty. At the other hand, MVE may have a very different behavior in each run, for the same set of hyperparameters. Also, we note that point predictions are slightly improved, as RMSE decreases, thus our model can be used for point predictions as well.

In Fig. 5 we explore what happens when the prediction horizon growths to many time steps. We also included a strochastic persistence model that performs simulations similarly, repeating last value with added noise of variance adjusted using residuals. As expected, we see that not only Winkler loss is smaller for our model, but also standard deviations. This means that the model is more robust compared to MVE to changes in the random seed used.

5   Conclusion and Future Work

In this work, we showed that using a combination of an MVE model with Monte Carlo dropout improves prediction interval generation for five data sets, supporting the conclusion that considering model uncertainty is important for prediction intervals estimation. Also, we showed that our proposal works for both one step and multi-step forecasting tasks.

Although we explained our proposal for univariate time series, the analysis for multivariate ones is similar, and extending the network for them is straightforward, adding outputs of mean and variance estimations for each dimension. Also, the use of MC dropout for multi-step forecasting can be explored for other kinds of networks, such as feedforward, convolutionals, extreme learning machines, etc. Further research directions include using other techniques than dropout to handle model parameters uncertainty, and, on the other hand, change the normal distribution assumption. This last point is specially important for time series that are asymmetric in nature, requiring different variance estimates under and over the mean prediction point.

Acknowledgments. This work was supported in part by Conicyt doctoral scholarship 21170109, Fondecyt Grant 1170123, and Basal Project FB0821.

References

1. Bergstra, J., Bengio, Y.: Random search for hyper-parameter optimization. J. Mach. Learn. Res. **13**, 281–305 (2012)
2. Blundell, C., Cornebise, J., Kavukcuoglu, K., Wierstra, D.: Weight uncertainty in neural networks. In: Bach, F., Blei, D. (eds.) Proceedings of the 32nd International Conference on Machine Learning (ICML 2015), vol. 37, pp. 1613–1622. PMLR, Lille (2015)
3. Gal, Y., Ghahramani, Z.: On modern deep learning and variational inference. In: Advances in Approximate Bayesian Inference Workshop, Neural Information Processing Systems (NIPS), vol. 2, pp. 1–9 (2015)
4. Gal, Y., Ghahramani, Z.: A theoretically grounded application of dropout in recurrent neural networks. In: Lee, D.D., Sugiyama, M., Luxburg, U.V., Guyon, I., Garnett, R. (eds.) Advances in Neural Information Processing Systems 29 (NIPS 2016), pp. 1019–1027. Curran Associates Inc., Barcelona (2016)
5. Gal, Y., Ghahramani, Z.: Dropout as a Bayesian approximation: representing model uncertainty in deep learning. In: Balcan, M.F., Weinberger, K.Q. (eds.) Proceedings of the 33rd International Conference on Machine Learning (ICML 2016), vol. 48, pp. 1050–1059. PMLR, New York (2016)
6. Graves, A.: Practical variational inference for neural networks. In: Shawe-Taylor, J., Zemel, R.S., Bartlett, P.L., Pereira, F., Weinberger, K.Q. (eds.) Advances in Neural Information Processing Systems 24 (NIPS 2011), pp. 2348–2356. Curran Associates Inc., Granada (2011)
7. Hong, T., Fan, S.: Probabilistic electric load forecasting: a tutorial review. Int. J. Forecast. **32**(3), 914–938 (2016)
8. Kabir, H.M., Khosravi, A., Hosen, M.A., Nahavandi, S.: Neural network-based uncertainty quantification: a survey of methodologies and applications. IEEE Access **6**, 36218–36234 (2018)
9. Khosravi, A., Nahavandi, S., Creighton, D., Atiya, A.F.: Comprehensive review of neural network-based prediction intervals and new advances. IEEE Trans. Neural Netw. **22**(9), 1341–1356 (2011)
10. Nix, D.A., Weigend, A.S.: Estimating the mean and variance of the target probability distribution. In: Proceedings of 1994 IEEE International Conference on Neural Networks (ICNN 1994), vol. 1, pp. 55–60. IEEE, Orlando (1994)

11. Quan, H., Srinivasan, D., Khosravi, A.: Incorporating wind power forecast uncertainties into stochastic unit commitment using neural network-based prediction intervals. IEEE Trans. Neural Netw. Learn. Syst. **26**(9), 2123–2135 (2015)
12. Taylor, J.W.: A quantile regression approach to estimating the distribution of multiperiod returns. J. Deriv. **7**(1), 64–78 (1999)
13. Zhu, L., Laptev, N.: Deep and confident prediction for time series at uber. In: 2017 IEEE International Conference on Data Mining Workshops (ICDMW), New Orleans, LA, pp. 103–110 (2017)

Hermite Convolutional Networks

Leonardo Ledesma[1,2], Jimena Olveres[2],
and Boris Escalante-Ramírez[2(✉)]

[1] Posgrado en Ciencia e Ingeniería de la Computación,
Universidad Nacional Autónoma de México, Mexico City, Mexico
[2] Facultad de Ingeniería, Universidad Nacional Autónoma de México,
Mexico City, Mexico
`boris@unam.mx`

Abstract. Convolutional Neuronal Networks (CNNs) have become a fundamental methodology in Computer Vision, specifically in image classification and object detection tasks. Artificial Intelligence has focused much of its efforts in the different research areas of CNN. Recent research has demonstrated that providing CNNs with a priori knowledge helps them improve their performance while reduce the number of parameters and computing time. On the other hand, the Hermite transform is a useful mathematical tool that extracts relevant image features useful for classification task. This paper presents a novel approach to combine CNNs with the Hermite transform, namely, Hermite Convolutional Networks (HCN). Furthermore, the proposed HCNs keep the advantages of CNN while leading to a more compact deep learning model without losing a high feature representation capacity.

Keywords: Hermite transform · Convolutional neural networks · Kernel modulation · Orientation · Scale · Feature map

1 Introduction

One of the challenges of image classification tasks is robustness against geometric transformations, such as rotation, deformation and scale. A large number of models and techniques of CNNs has been developed to deal with these issues, for example models based on attention mechanism or actively rotating filters [1], even hand-crafted filters with no learning processes involved. Moreover, DCNNs (Deep CNNs) have emerged as an alternative of feature extraction [10], however, these techniques derive in expensive training and complex model parameters, without solving the problem of geometric transformations.

Recently, a priori learning approaches have emerged as a way to help CNNs improve their performance by using mathematical transformations based on HVS (Human Visual System), such is the case of the Gabor CNN (GCNs) [2].

In [2] Gabor Orientation Filters (GoF) were proposed as the result of a new process called Modulation. This paper is inspired on the same idea, but instead of Gabor transform, we introduce the Hermite transform, which is also a human perception relevant image representation model [3, 4].

© Springer Nature Switzerland AG 2019
I. Nyström et al. (Eds.): CIARP 2019, LNCS 11896, pp. 398–407, 2019.
https://doi.org/10.1007/978-3-030-33904-3_37

1.1 CNN Using Mathematical Transforms

Some important models of CNN's based on mathematical transforms have been described in the literature:

(1) *Scattering Networks* [5, 6]: This kind of architecture uses a wavelet scattering based on the expression of receptive fields in CNNs. The wavelets are another way to introduce mathematical transforms in an CNN.

(2) *Oriented Response Networks (ORN):* In [1] ORN resembles a hierarchical orientation encoder using Actively Rotation filters in order to discriminate structures.

(3) *Gabor filters:* Two approaches; the first one as a pre-processing model in order to improve the accuracy of CNNs [7], or by applying the Gabor filters in the 1st convolutional layer instead of using random filter kernels.

The second approach; modulating the learned convolution filters using the Gabor filters [2]. This idea inspires this paper.

1.2 Motivation

In [2], it was demonstrated that after the training process in an Alex Net architecture [8], the learned filters resemble very much Gabor filters oriented towards different directions, which are known to be good detectors of image structures, such as edges.

This idea is of great importance, since it suggests that the convolutional stages of a CNN aim at learning primitive image features, that we have known for years, to be relevant for image analysis and characterization, i.e., edges, lines, corners. These features can be obtained by designing appropriate filters for this task. Nevertheless, the learned filters behave like original Gabor filters after several epochs of training.

Recent research use Gabor filters in the convolution layers, for example; in [9], Gabor filters are used in the first or second convolution layers as a feature extractor such as a pre-processing stage in a CNN [10].

From the point of view of image processing this transform has been widely used in image segmentation, classification and texture detection.

In this paper, we introduce the Hermite transform as a model for building CNNs. The HT has the following characteristics:

- It allows multiscale analysis
- It is useful to perform directional analysis
- Its filter functions are based on Gaussian derivatives, which in turn are good detectors of image primitives, such as edges, lines, corners, etc.
- The HT allows perfect reconstruction and is shift invariant.

We present this work as a novel attempt to improve the feature extractor and performance of a CNN, taking advantage of Hermite transform and its benefits.

1.3 Contributions

We propose to use Hermite filters in a CNN basic architecture in such a way that the number of parameters is reduced and on the other hand, robustness of orientation and scale changes are enhanced.

Based on the idea given by [2], we will use the HT to generate a modulation process and a modified convolutional process to improve the performance of a CNN.

Therefore, the contributions of this papers are:

- The first attempt to incorporate Hermite filters into a CNN architecture in the state of the art.
- Prove the advantages of using a mathematical transform based on HVS in comparison to traditional methods.
- Develop modulation and modified convolutional processes that can be easily extended to more complex architectures like ResNet or Wide Residue Network (WRN) in the near future.

2 Related Work

2.1 Hermite Transform

The Hermite transform is defined by the analysis functions called D_n [3, 4] that comprise a set of filters obtained from the product of Gaussian window g and the Hermite polynomials H_n,

$$D_n(x) = \frac{(-1)^n}{\sqrt{2^n n!}} \frac{1}{\sigma \sqrt{\pi}} H_n\left(\frac{x}{\sigma}\right) e^{\left(-\frac{x^2}{\sigma^2}\right)} \tag{1}$$

where n is the polynomial order.

It is worth mentioning that the analysis functions are spatially separable and rotationally symmetric. A straight forward extension provides the 2D expression for the analysis functions:

$$D_{n-m,m}(x,y) = H_{n-m,m}(-x,-y) \cdot g^2(-x,-y) \tag{2}$$

where $(n - m)$ and m are the polynomial order in x and y axes for $n = 0, \ldots, \infty$ and $m = 0, \ldots, n$.

The associated orthogonal polynomials are defined:

$$H_{n-m,m}(x,y) = \frac{1}{\sqrt{2^n m!(n-m)!}} H_{n-m}\left(\frac{x}{\sigma}\right) H_m\left(\frac{y}{\sigma}\right) \tag{3}$$

In the case of a discrete implementation of the Hermite filters, we use parameter N equal to the finite length of the filter that in turn is related to the scale, i.e., the standard deviation of the Gaussian window as follows: $N = 2\sigma^2$ (Fig. 1).

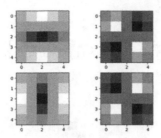

Fig. 1. The image illustrates of the Hermite Filters of length $N = 5$ specifically $D_{2,0}, D_{1,1}$, $D_{0,2}$ y $-D_{1,1}$ displayed as follows: $\begin{bmatrix} D_{0,2} & D_{1,1} \\ D_{2,0} & -D_{1,1} \end{bmatrix}$

3   The Proposed Method

Hermite Convolutional Networks (HCN) are created with the purpose of being used in the following tasks in computer vision:

- Image Classification
- Object Detection
- Texture Discrimination
- Segmentation

During the training stage learned convolution filters are adjusted followed by modulation process with a Hermite filter bank.

Similar to GCNs, Hermite orientation filters (HoFs) are the result of this modulation process and are used to generate feature maps in each layer.

The three main process are:

3.1   Hermite Orientation Filters (HoFs) and Modulation Process

HoFs are obtained by element product between Hermite Filter bank (see Fig. 2) and the learned filters in each convolutional layer. The HoFs have the size:

$$C_{out} \times C_{in} \times U \times N \times N$$

where C_{out}, C_{in} are the channel of output and input feature map respectively, and U is the number of orientation filters of the Hermite bank selected, in this paper $U = 4$.

Let the modulation process be:

$$C_{i,n-m,m}^n = C_{i,o} \cdot D_{n-m,m} \tag{4}$$

The $C_{i,n-m,m}^n$ is a modulated filter of the learned filter $C_{i,o}$ It is important to mention that the scale parameter N in one convolutional network is the same for all orientation filters in the Hermite bank but changes between layers.

During the modulation process Hermite filters adjust the kernels of the learned filters in order to extend the properties of the transformation to the back-propagation.

HoFs are defined by:

$$C_i^n = \left\{ C_{i,0,2}^n, C_{i,1,1}^n, \ldots, C_{i,2,0}^n \right\} \tag{5}$$

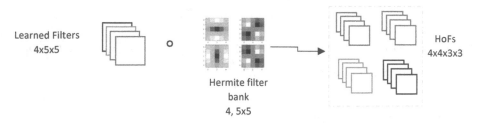

Fig. 2. The modulation process of HoFs

3.2   HCN Convolution Process

The HoFs are used in this process to produce an output feature map $\widehat{F}$, according to equation:

$$\widehat{F}_{i,n,n-m} = \sum_{j=1}^{U} F^{(j)} \otimes C_{i,n,n-m}^{(j)} \tag{6}$$

Where F is an input feature map and $C_{i,nn-m}^j$ is set of HoFs (Fig. 3).

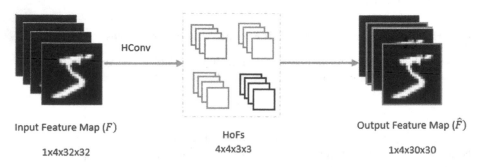

Fig. 3. The convolution process of HoFs and Feature Map

3.3   Back-Propagation Process

The HoFs only affect in forward calculation in HCNs. However, the back-propagation process updates the learned filters in each layer and the HoFs are recalculated in each epoch.

$$C_{i,o}^{t+1} = C_{i,o}^t - \eta \frac{\partial L}{\partial C_{i,o}^t} \qquad (7)$$

Where L is the loss function, in this case, cross-entropy.

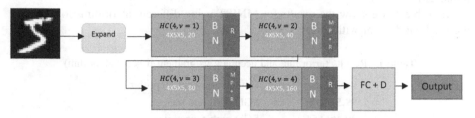

Fig. 4. Basic architecture of the Hermite Convolutional Network (HCN) where BN: Batch Normalization, MP: Max Pooling, FC: Fully Connect, D: Dropout, R: ReLu Activation, HC: Hermite convolutional layer.

4 Experiments and Implementation

The general implementation is based on [2] but is extended to support Hermite transform, so the module of Gabor Filter Bank is substituted by Hermite Filter Bank. The basic architecture HCN is shown in Fig. 4.

The proposed HCN architecture was evaluated with the datasets: MNIST [11, 12], Fashion-MNIST [13] y E-MINIST [14]. We evaluated these experiments in a GPU NVIDIA GeForce 1080 with an isolated environment in order to work with other versions of Python and PyTorch.

Algorithm 1 Hermite Convolutional Networks (HCN)

---

1. Set hyper-parameters and network structure.
2. Design the Hermite filters bank given a filter order n keeping the rule: $[n - m, m], n \leq N - 1$, and $m \leq n$
3. Start training
4. **Repeat**
5.   Select a mini batch set of training images.
6.   Produce HoFs using Eq. 4
7.   Apply Forward convolution using Eq. 6
8.   Calculate the cross-entropy and run back-propagation process using Eq. 7
9.   Update the learned filters
10. **Until** the maximum epoch

---

5 Results

5.1 MNIST

MNIST dataset was divided in 10,000 samples for validation and 50,000 samples for training. We evaluated different optimizers: SGD, ADAM and ADEDELTA. We used a batch size of 128 and 50 epochs for training and testing.

In the case of ADEDELTA optimization the initial learning rate was 0.01, the learning weight decay was set as 0.00003, these hyper-parameters were the same for ADAM and SGD.

Based on [2], we designed filters in four orientation and sizes of 5×5 and 7×7, we did not use 3×3, because it is too small to represent all the properties of Hermite transform.

In Table 1, we show the results for MNIST using different algorithms to calculate the gradient descent with filters 5×5.

Table 1. Results (error rate and accuracy vs gradient descent algorithm)

%Accuracy/%error	Gradient descent algorithm
99.51/0.49	SGD with one scale
99.46/0.54	SGD with four scales
99.42/0.58	ADAM with four scales
99.34/0.66	Adedelta with four scales

Table 2. Results (error rate and accuracy vs time and different filter sizes with SGD algorithm)

%Accuracy/%error	Time	Filter size per layer
99.51/0.49	9 min 30 s	SGD (5, 5, 5, 5)
99.39/0.61	9 min 10 s	SGD (3, 3, 3, 3)
99.31/0.58	25 min 9 s	SGD (7, 7, 7, 7)
99.27/0.66	12 min 10 s	SGD (3, 3, 5, 5)

Table 3. Results comparison on MNIST

% error	# network stage kernel	Architecture	Year
0.21	5 CNN/6-layer: 784-50-100-500-1000-10-10	Regularization of Neural Networks using DropConnect [15]	2013
0.23	35 CNNs, 1-20-P-40-P-150-10	Multi-column Deep Neural Networks for Image Classification [16]	2012
0.23	6-layer 784-50-100-500-1000-10-10	APAC: Augmented PAttern Classification with Neural Network [17]	2015
0.57	10-20-40-80	ORF4 (ORAlign) [1]	2017
0.48[a]	20-40-80-160	GCN4 (7×7) [2]	2018
0.49	10-20-40-80	HCN4 (5×5)	2019

[a]In this case the authors reported an error of 0.42 but we were not able reproduce it.

We also tested different filter sizes per layer in order to find the best performance, Table 2 shows the results for this experiment. Regarding computing time, as shown, the best result is obtained when filter size is 5×5.

The architectures reported in the state of the art have more millions of parameters and consequently need more computing time. In contrast, HCN and GCN reduce considerably the computing time and the complexity of parameters. Moreover, the HCN and GCN architectures are more compact in comparison with the others.

5.2   Fashion-MNIST and E-MNIST

Both datasets have the same number of samples as MNIST; we separated the sets into two groups of validation and training as in the first experiment; the hyper-parameters were the same (Fig. 5).

It was necessary to normalize the datasets before training to avoid gradient fading problems and divergence of the CNN. In Table 3, the state-of-the-art results are shown using the same this simple image database with their corresponding architectures and parameters (Tables 4 and 5).

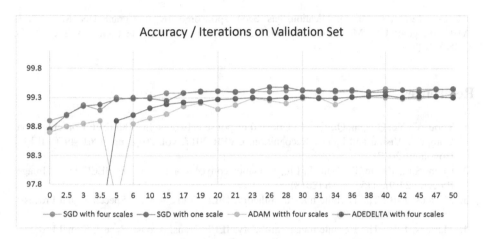

Fig. 5. In this graphic the x-axis is the number of iterations (x 1K) and y-axis is the accuracy for the test set, it's shown the performance of HCN with several algorithm gradient descent, being SGD with one scale the best performance.

Table 4. Results (error rate and accuracy rate vs time and different filter sizes with SGD algorithm on fashion MNIST)

%Accuracy/%error	Time	Filter size per layer
97.09/2.91	10 m 12 s	SGD (5, 5, 5, 5)
95.01/4.99	10 m 2 s	SGD (3, 3, 3, 3)
90.13/9.87	30 min 2 s	SGD (7, 7, 7, 7)
97.19/2.81	**14 min 14 s**	**SGD (3, 3, 5, 5)**

Table 5. Results (error rate and accuracy rate vs time and different filter sizes with SGD algorithm on E-MNIST)

%Accuracy/%error	Time	Filter size per layer
99.67/0.33	**9 m 50 s**	**SGD (5, 5, 5, 5)**
99.60/0.40	8 m 55 s	SGD (3, 3, 3, 3)
99.41/0.59	25 min 29 s	SGD (7, 7, 7, 7)
99.46/0.54	10 min 14 s	SGD (3, 3, 5, 5)

6 Conclusions

This paper presents a novel deep model incorporating Hermite filters to CNN, aiming at improving the performance with a high feature representation capacity. The proposed Hermite Convolutional Networks is advantageous tool for image classification. Results show that HCN got an accuracy near to state of the art of the MNIST database with lower number of parameters. As future work, more image databases will be tested on HCN, such as ImageNet and CIFAR-10. Furthermore, new hybrid architectures, e.g. similar to ResNet, will be developed in order to tackle more complex problems.

Acknowledgments. This publication has been sponsored by the grant UNAM PAPIIT IA103119, grant UNAM PAPIIT IN116917 and Consejo Nacional de Ciencia y Tecnología (CONACyT).

References

1. Zhou, Y., Ye, Q., Qiu, Q., Jiao, J.: Oriented response networks. In: 30th IEEE Conference on Computer Vision and Pattern Recognition, CVPR 2017, vol. 2017, pp. 4961–4970. IEEE, Honolulu (2017)
2. Chen, S.L., Zhang, B., Han, J., Liu, J.: Gabor convolutional networks. IEEE Trans. Image Process. **27**(9), 4357–4366 (2018)
3. Martens, J.-B.: The Hermite transform-theory. IEEE Trans. Acoust. Speech Signal Process. **38**(9), 1595–1606 (1990)
4. Martens, J.-B.: The Hermite transform-theory. IEEE Trans. Acoust. Speech Signal Process. **38**(9), 1607–1618 (1990)
5. Bruna, J., Mallat, S.: Invariant scattering convolution networks. IEEE Trans. Pattern Anal. Mach. Intell. **45**(8), 1872–1886 (2013)
6. Sifra, L., Mallat, S.: Rotation, scaling and deformation invariant scattering for texture discrimination. In: 26th IEEE Conference on Computer Vision and Pattern Recognition, CVPR 2013, pp. 1233–1240. IEEE, Portland (2013)
7. Calderón, A., Roa, S., Victorino, J.: Handwritten digit recognition using convolutional neural networks and Gabor filters. In: Proceedings of the International Congress on Computational Intelligence CIIC 2003, Medellin (2003)
8. Krizhevsky, A., Sutskever, I., Hinton, G.E.: ImageNet classification with deep convolutional neural networks. In: Proceedings of the Advances in Neural Information Processing Systems, Nevada, vol. 1, pp. 1097–1105 (2012)
9. Sarwar, S.S., Panda, P., Roy, K.: Gabor filter assisted energy efficient fast learning convolutional neural networks. In: Proceedings of the IEEE/ACM International Symposium on Low Power Electronics and Design, Taipei, pp. 1–6 (2017)
10. Yao, H., Chuyi, L., Dan, H., Weiyu, Y.: Gabor feature based convolutional neural networks. In: Proceedings of the International Conference on Information Science and Control Engineering, pp. 386–390. IEEE, Beijing (2016)
11. LeCun, Y.: The MNIST Database of Handwritten Digits (1998). http://yann.lecun.com/exdb/mnist/. Accessed 22 Mar 2019
12. LeCun, Y., Bottou, L., Bengio, Y., Haffner, P.: Gradient-based learning applied to document recognition. Proc. IEEE **86**(11), 2278–2324 (1998)

13. Xiao, H., Rasul, K., Vollgraf, R.: Fashion-MNIST: a novel image dataset for benchmarking machine learning algorithms. https://github.com/zalandoresearch/fashion-mnist. Accessed 18 Mar 2019
14. The E-MNIST Database of Handwritten character. https://www.nist.gov/itl/iad/image-group/emnist-dataset. Accessed 18 Mar 2019
15. Wan, L., Zeiler, M., Zhang, S., LeCun, Y., Fergus, R.: Regularization of neural networks using dropconnect. In: Proceeding ICML 2013 Proceedings of the 30th International Conference on Machine Learning, Atlanta, vol. 28, pp. 1058–1066 (2013)
16. Ciregan, D., Meier, U., Schmidhuber, J.: Multi-column deep neural networks for image classification. In: 2012 IEEE Conference on Computer Vision and Pattern Recognition, pp. 3642–3649. IEEE Computer Society, Washington (2012)
17. Sato, I., Nishimura, H., Yokoi, K.: APAC: Augmented PAttern Classification with Neural Networks. https://arxiv.org/abs/1505.03229. Accessed 24 Mar 2019

Analytical Solution for the Optimal Addition of an Item to a Composite of Scores for Maximum Reliability

Carlos A. Ferrer[1,2]([✉]) [iD], Idileisy Torres-Rodríguez[1] [iD],
Alberto Taboada-Crispi[1] [iD], and Elmar Nöth[2] [iD]

[1] Informatics Research Center, Central University "Marta Abreu" de las Villas,
Santa Clara, Cuba
{cferrer,itrodriguez,ataboada}@uclv.edu.cu
[2] Pattern Recognition Lab, Friedrich Alexander University
Erlangen-Nuremberg, Erlangen, Germany
elmar.noeth@fau.de

Abstract. This paper presents a derivation of the optimal weight to be assigned for an item so that it maximally increases the reliability of the aggregate. This aggregate is the best estimate of the underlying true repeating pattern. The approach differs from previous solutions in being analytical, based on the Signal to Noise Ratio (SNR) instead of the reliability itself, and the ability to visually inform the researcher about the relevance of the weighting strategy and the gains produced in the SNR. Optimal weighting of repetitive phenomena is a bonus not only in the behavioral sciences, but also in many engineering fields. Its uses may include the selection or discarding of raters, judges, repetitions, or epochs, depending on the field.

Keywords: Reliability · Signal-to-Noise Ratio · Composites · Ensemble Averages

1 Introduction: Reliability and Classic Test Theory

Within statistics, the reliability of a measure has been dealt-with mostly within the framework of Classic Test Theory (CTT) [1, 2]. Measures dealt-with in CTT differ from common physical magnitudes measurements in that CTT is usually oriented to simultaneously obtaining several measures (the different items/questions in the test/questionnaire) over several objects (the subjects taking the test) [2]. The framework developed for CTT can, however, be extrapolated to certain engineering problems, outside of the behavioral sciences, as will be described later. A brief exposition of the basics of CTT, required for the rest of the paper, is presented, followed by the objectives of this paper.

In CTT, an individual measure x_i, such that $1 \leq i \leq I$ being I the number of available measures, is considered to be the sum of a true value τ and a random measurement error e_i, when applied to K subjects, such that $1 \leq k \leq K$, and could be expressed as:

I. Nyström et al. (Eds.): CIARP 2019, LNCS 11896, pp. 408–416, 2019.
https://doi.org/10.1007/978-3-030-33904-3_38

$$x_i(k) = \tau(k) + e_i(k) \tag{1}$$

It is assumed that any individual error e_i is uncorrelated to τ, as well as to other errors e_j, for $i \neq j$. The reliability of the individual item is then defined as:

$$\rho_i = \frac{\sigma_\tau^2}{\sigma_{x_i}^2} = \frac{\sigma_\tau^2}{\sigma_\tau^2 + \sigma_{e_i}^2} = \frac{1}{1 - \frac{1}{\sigma_\tau^2/\sigma_{e_i}^2}} \tag{2}$$

The reliability is the fraction of the observed variance accounted for by the true variance, and as such ranges between 0, for completely erroneous measurements, and 1 in the absence of error. It is actually a correlation coefficient, between the true component τ and the measured value x_i. The third expression for ρ_i in Eq. (2) is useful to show the relationship between reliability and the Signal-to-Noise ratio (SNR) given by the quotient between the variances of τ and e_i. As seen from (2), SNR has a monotonic relationship with ρ_i. SNR is preferred in engineering fields to characterize x_i, and several advantages of interpreting test items in terms of SNR were described in [3]. Nevertheless, ρ remains the most relevant feature within the CTT framework.

For any given subject, the set of I individual x_i outcomes need to be aggregated to form a single score. This aggregated score X is called the composite, and can be expressed in the more general way as a weighted sum of the items, with weights w_i:

$$X(n) = \sum_{i=1}^{I} w_i x_i(n) \tag{3}$$

The aggregate score is known to have a reduced noise variance [2, 4], a fact that makes it useful in many fields where noisy repetitions of a supposedly invariant true pattern/component are available (e.g. geodesic sensing [5], electrocardiography [6] or voiced signals [7]). The former equation is equivalent to the typical averaging operation if all weights are equal or set to $1/I$. There is no point in giving different weights to items with equal reliabilities, so if Eq. (1) holds, averaging results the aggregate of choice. However, all x_i can be expected to have different reliabilities, resulting from different proportions between the variances of true and error components. To reflect this, the expression for the individual items given by (1) should be revised [8]. It is common to arbitrarily assign unitary variances to the true and error variances, and account for the differences in individual reliabilities by using different loadings (β_i and ε_i) for the true and error components, respectively. An offset term μ_i can also be included:

$$x_i(n) = \beta_i \tau(n) + \varepsilon_i e_i(n) + \mu_i \tag{4}$$

There are different models [8] to obtain the reliability of X depending on certain restrictions imposed on β_i (the scale on which the true score is measured), ε_i (the magnitude of error in the measurement), and μ_i in Eq. (4). The unrestricted model for the x_i set is called *congeneric* (all items measure the true score on different scales with different errors). If β_i is assumed to be constant for all i, the model is called *essentially tau-equivalent* (all items measure the true score on the same scale, but with different errors). If, additionally, μ_i is assumed to be zero for all i, the model is called *tau-equivalent*. Since reliability is based in variances and correlations, not influenced by mean values, *essentially tau-equivalent* and *tau-equivalent* models yield identical estimates of composite reliability. If additionally to constant β_i and zero μ_i the value of ε_i is also considered constant, the model is called *parallel* (all items measure the true score in the same scale with the same amount of error). The *parallel* model is the most restrictive one, and could de completely described by Eq. (1).

The most commonly used estimate of an un-weighted composite's reliability is Cronbach's α [9], which is based on an *essentially tau-equivalent* model:

$$\alpha = \frac{Ic}{(v + (I - 1)c)} \tag{5}$$

Here v stands for the average variance of all x_i items, and c for the average covariance across all x_i items (excluding same-item covariances). Popularity of α is due to not requiring estimates of individual ρ_i: the assumption made for the *essentially tau-equivalent* model allows obtaining the value in (5) only from the observed x_i scores.

The reliability in the *congeneric* model for an un-weighted composite is:

$$\rho = \frac{\left(\sum\limits_{i=1}^{I} \beta_i\right)^2}{\left(\sum\limits_{i=1}^{I} \beta_i\right)^2 + \sum\limits_{i=1}^{I} (\varepsilon_i^2)} = \frac{\left(\sum\limits_{i=1}^{I} \beta_i\right)^2}{\left(\sum\limits_{i=1}^{I} \beta_i\right)^2 + \sum\limits_{i=1}^{I} \phi_i} \tag{6}$$

The term ε_i^2 has been frequently denoted as ϕ_i being equal to the unique/error variance of the ith item. The use of (6) requires to go beyond the observed scores x_i, and perform estimates of the individual true and error loadings, which is a drawback compared to (5). This can be performed by means of covariance matrix analysis [10, 11], of the kind performed in Common Factor Analysis (CFA), in this case for a single factor, τ [12]. For the case of a weighted composite, it has been shown [13] by algebraic methods that the weights producing maximum reliability of the composite in (3) using the items described in (4) are:

$$w_i = \frac{\beta_i}{\phi_i} \tag{7}$$

In spite of theoretically offering maximum reliability, differential weights as given by (7) are far from being widely adopted [14, 15]. A main concern is the adequacy of a single-factor CFA to estimate the loadings [16] together with reports that unity weights work rather well [17, 18]. On the other hand, the use of unity/constant weights and composite's reliability measures like α are heavily criticized for conducting to several misleading beliefs [19], like "Reliability increases with I (test length)", "Reliability increases with the individual item reliabilities" or "Reliability increases with the correlation between items", all of which seem sound at first sight. Contradictory examples of all these common sense beliefs are given in Appendix A of [19].

In this paper, we attempt to provide the readers with analytical tools to understand the effect, in the composite's reliability, of the chosen weight for a particular item.

2 Analytical Derivations

Our focus is the effect of differential weighting of x_i items in the composite's reliability. We depart from reported results in optimal reliability in two ways:

- First, our approach is analytical: looking for the effect of the weighting of an individual item in the resulting composite, and the optimal value for weight w will then be shown within the continuum of possible weight values (including unity, as the alternative approach). This is different from the widely used algebraic solution providing the optimal values of all w_i as the solution for an equation system, which prevents the user to grasp what is being gained, and if it is relevant.
- Second, we perform our analysis in terms of SNRs instead of reliabilities. There is no additional information in SNR as compared to reliability, since they are monotonically related as shown in Eq. (2). However their known limits (reliability values from $0\dots1$ map into SNR values from $0\dots\infty$) favor visualizing the influence of w_i in terms of SNR. Besides, the analytical derivations of the influence of the value of w are simpler to obtain for SNR than for ρ, since we depart from a simpler quotient.

Let's assume we have $I + 1$ items, from where we extract one item to analyze the effect of performing a weighted addition to the existing composite consisting of the other I items. For all the $I + 1$ items we have estimates of the true and error loadings, presumably obtained by CFA. The items x_i, $1 \le i \le I$, are not necessarily weighted in the existing composite of our assumption. For the now extended composite, with $I + 1$ items including the newly added item weighted by weight w, the SNR^{I+1} is:

$$SNR^{I+1} = \frac{\left(\sum_{i=1}^{I} \beta_i + w\beta_{I+1}\right)^2}{\sum_{i=1}^{I} (\phi_i) + w^2 \phi_{I+1}} = \frac{T^I + 2w\beta_{I+1}\sum_{i=1}^{I} \beta_i + w^2 \beta_{I+1}^2}{E^I + w^2 \phi_{I+1}} \tag{8}$$

For simplicity we have represented by T^I the true variance in the original (I items) composite, and its error variance by E^I. For $w = 0$ (i.e. not adding the $I + 1$ item), the SNR^{I+1} would still be the original $SNR^I = T^I/E^I$. It is straightforward from Eq. (8) that SNR^{I+1} has no singularities, and a double zero located at a negative value of w:

$$w^{SNR=0} = -\frac{\sum_{i=1}^{I} \beta_i}{\beta_{I+1}} = -\frac{\sqrt{T^I}}{\beta_{I+1}} \tag{9}$$

The double zero is also the location of the minimum value of SNR^{I+1}, but we are actually interested in the location of the maximum, corresponding to the optimal value of w. We obtain the derivative of SNR^{I+1} as expressed in (8) with respect to w as:

$$SNR^{I+1'} = 2 \frac{-w^2 \phi_{I+1} \beta_{I+1} \sum\limits_{i=1}^{I} \beta_i + w\left(E^I \beta_{I+1}^2 - \phi_{I+1} T^I\right) + E^I \beta_{I+1} \sum\limits_{i=1}^{I} \beta_i}{\left(E^I + w^2 \phi_{I+1}\right)^2} \quad (10)$$

The roots of Eq. (10) follow those from a quadratic form, and the values are:

$$w_a = -\frac{\sum\limits_{i=1}^{I} \beta_i}{\beta_{k+1}} \qquad w_b = \frac{E^I \beta_{I+1}}{\phi_{I+1} \sum\limits_{i=1}^{I} \beta_i} \quad (11)$$

The root w_a is the already known position for the double-zero, minimum value of SNR^{I+1} given by Eq. (9), while w_b is the one corresponding to the maximum value of SNR^{I+1}, i.e. the optimal value of w. Substituting w_b as given by Eq. (11) in the expression for the SNR^{I+1} given by Eq. (8), yields, after some manipulation:

$$SNR^{I+1}(w_b) = \frac{T^I}{E^I} + \frac{\beta_{I+1}^2}{\phi_{I+1}} = SNR^I + SNR^{item} \quad (12)$$

Equation (12) shows that the maximum increment that an item can produce to a composite's original SNR^I is limited to its own SNR^{item}. A final element of interest could be to obtain the intersection of the SNR^{I+1} curve with the original SNR^I value:

$$SNR^{I+1} - SNR^I = \frac{w\left(w\left(\beta_{I+1}^2 E^I - \phi_{I+1} T^I\right) + 2\beta_{I+1} E^I \sum\limits_{i=1}^{I} \beta_i\right)}{\left(E^I + w^2 \phi_{I+1}\right) E^I} = 0 \quad (13)$$

One of the solutions is $w = 0$ (the no-addition case), while the other occurs at:

$$w^{SNR_{I+1}=SNR_I} = \frac{2\beta_{I+1} E^I \sum\limits_{i=1}^{I} \beta_i}{\left(\phi_{I+1} T^I - \beta_{I+1}^2 E^I\right)} \quad (14)$$

The sign of w depend on the denominator, and will be positive if:

$$\frac{T^I}{E^I} > \frac{\beta_{I+1}^2}{\phi_{I+1}} \quad \text{i.e. :} \quad SNR^I > SNR^{item} \quad (15)$$

According to Eq. (14), the SNR^{I+1} curve will show a different behavior whether the SNR of the item to be added (SNR^{item}) exceeds the SNR of the existing composite

(SNR^I) or not. With the expressions for all the relevant points obtained, we can plot representative examples of both cases, shown in top and bottom panes in Fig. 1.

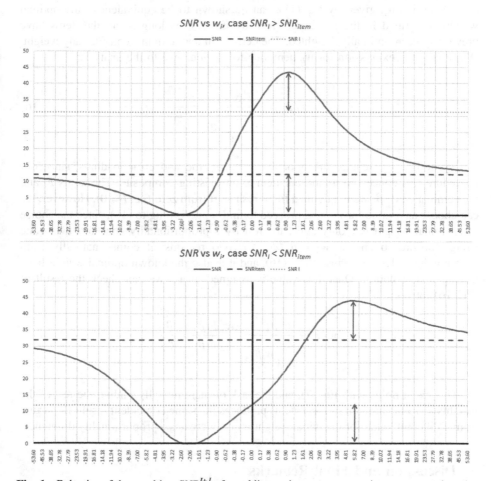

Fig. 1. Behavior of the resulting SNR^{I+1} after adding an item to a composite, top: case when the additional intersect of $SNR^{I+1} = SNR^{item}$ occurs for positive w; bottom: case of the intersect occurring for negative w. Dotted horizontal line corresponds to the SNR^I of the composite, dashed horizontal line corresponds to the SNR^{item}. Arrows placed at the optimum value of w to show that the resulting SNR^{I+1} at that point is the sum of the levels of both horizontal lines.

Both figures clearly depict the behaviors analytically described above. The graphs were obtained interchanging values of 11.5 and 31.5 between the SNR^I and SNR^{item}. The horizontal axis has been horizontally compressed. A third behavior not shown in Fig. 1, for the case $SNR^I = SNR^{item}$ (i.e. with both dotted and dashed lines at the same level) leaves the $w = 0$ as the only intersect.

2.1 Equivalence with the Previous Algebraic Solution

The expression we obtained for the optimal weight of an item to be added to a composite, i.e. w_b given by Eq. (11), can be shown to be equivalent to the optimal weights described in the literature, given by Eq. (7), as long as all the items have previously been optimally weighted. The derivation above didn't consider any weight. We rewrite the expression for w_b here, due to convenience to the analysis:

$$w_b = w_{I+1} = \frac{E^I \beta_{I+1}}{\phi_{I+1} \sum\limits_{i=1}^{I} \beta_i} = \frac{\sqrt{\frac{T_{item}}{T^I}}}{\frac{E_{item}}{E^I}} \tag{16}$$

By design, we require the prior existence of true and error loadings, since our procedure was conceived to add our item to an existing composite. The terms E^I and the summation of β_i are not defined for a fresh start from zero (i.e. $I = 0$). However, we can choose any value for w_1, and proceed to find w_2 according to (16), since the choice of an initial weight can only affect the rest by a proportionality constant. It is quite straightforward to choose $w_1 = \beta_1/\phi_1$, for two reasons: it comes naturally from ignoring the undefined terms in Eq. (16), and it is also the known optimal weight from the algebraic solution. Once this value is assigned to w_1, we can check the result of Eq. (16) for w_2:

$$w_2 = \frac{E_1 \beta_2}{\phi_2 \sum\limits_{i=1}^{1} w_i \beta_i} = \frac{w_1^2 \phi_1 \beta_2}{\phi_2 w_1 \beta_1} = \frac{\beta_2}{\phi_2} \tag{17}$$

This result holds with Eq. (7), and the task of incrementally finding the rest of the weights up to the amount of $I + 1$ items produces the same coincident result. Departing from a different value for the initial weight w_1 only creates a proportionality difference.

3 Discussion and Final Remarks

The analytical approach developed here provides more insights to the researcher than the previously reported batch-oriented, simultaneous solution of the values for all the weights. With the set of equations provided, the researcher can evaluate whether the use of the optimal or unitary weight modifies or not the value of SNR in a significant way (so as to trust the loadings provided by CFA), whether the gain in SNR that an item can provide to the composite will be significant or not (so as to keep it in or out of the composite) among other uses. In particular, the functional dependency obtained for $SNR^{I+1}(w_{I+1})$ and depicted in Fig. 1 allows to understand the causes of the apparently contradictory examples provided in [19]. The corresponding explanations are not provided here for space reasons, but the readers can readily attain them by means of Eq. (10) and the behaviors depicted in Fig. 1.

The differential weighting strategies in composites are not, by far, exclusive from the behavioral sciences. Optimal weighting of repetitive patterns immerse in noise is a goal in many areas in order to recover the true component. Different strategies for recovering this pattern have been evaluated in many engineering fields, like geodesics [5], acoustics [20], evoked potentials [21], glottal pulses [22], electrocardiography [6], or perceptual judgments of pathological symptoms [23].

The analysis of the influence of the differential weighting in the resulting SNR (and consequently in the reliability of the composite/pattern recovered) and its visual comparison with the unitary weighting, made possible by the procedure described here, can be of great help for researchers in any of those areas.

Future research by the authors will address the development of procedures for selecting items/judges to be discarded from the composite.

Acknowledgements. This work was partially supported by an Alexander von Humboldt Foundation Fellowship granted to one of the authors (Ref 3.2-1164728-CUB-GF-E).

References

1. Gulliksen, H.: Theory of Mental Tests. Routledge, New York (2013)
2. Lord, F.M., Novick, M.R.: Statistical Theories of Mental Test Scores. Information Age Publishing, Charlotte (2008)
3. Cronbach, L.J., Gleser, G.C.: The signal/noise ratio in the comparison of reliability coefficients. Educ. Psychol. Meas. **14**(3), 467–480 (1964)
4. Rompelman, O., Ross, H.H.: Coherent averaging technique: a tutorial review. I. Noise reduction and the equivalent filter. J. Biomed. Eng. **8**(1), 24–29 (1988)
5. Kotsakis, C., Tziavos, I.N.: Parametric versus non-parametric methods for optimal weighted averaging of noisy data sets. In: Sansò, F. (ed.) A Window on the Future of Geodesy. IAG SYMPOSIA, vol. 128, pp. 434–439. Springer, Heidelberg (2005). https://doi.org/10.1007/3-540-27432-4_74
6. Hashimoto, K., et al.: A novel signal-averaged electrocardiogram and an ambulatory-based signal-averaged electrocardiogram show strong correlations with conventional signal-averaged electrocardiogram in healthy subjects: a validation study. J. Electrocardiol. **51**(6), 1145–1152 (2018)
7. Ferrer, C., González, E., Hernández-Díaz, M.E.: Correcting the use of ensemble averages in the calculation of harmonics to noise ratios in voice signals (L). J. Acoust. Soc. Am. **118**(2), 605–607 (2005)
8. Graham, J.M.: Congeneric and (essentially) tau-equivalent estimates of score reliability what they are and how to use them. Educ. Psychol. Meas. **66**(6), 930–944 (2006)
9. Cronbach, L.J.: Coefficient alpha and the internal structure of tests. Psychometrika **16**(3), 297–334 (1951)
10. Jöreskog, K.G.: A general method for analysis of covariance structures. Biometrika **57**, 239–251 (1970)
11. Jöreskog, K.G.: Statistical analysis of sets of congeneric tests. Psychometrika **36**(2), 109–133 (1971)
12. de Winter, J.C.F., Dodou, D.: Common factor analysis versus principal component analysis: a comparison of loadings by means of simulations. Commun. Stat. - Simul. Comput. **45**, 299–321 (2014)

13. Knott, M., Bartholomew, D.J.: Constructing measures with maximum reliability. Psychometrika **58**(2), 331–338 (1993)
14. Sočan, G.: Assessment of reliability when test items are not essentially t-equivalent. Dev. Surv. Methodol. **15**, 23–35 (2000)
15. Chang, S.-W.: Choice of weighting scheme in forming the composite. Bull. Educ. Psychol. **40**, 489–510 (2009)
16. Lee, S.-Y.: Handbook of Latent Variable and Related Models. Elsevier, Amsterdam (2007)
17. Streiner, D.L., Goldberg, J.O., Miller, H.R.: MCMI-II item weights: their lack of effectiveness. J. Pers. Assess. **60**(3), 471–476 (1993)
18. Lindell, M.K., Whitney, D.J.: Accounting for common method variance in cross-sectional research designs. J. Appl. Psychol. **86**(1), 114–121 (2001)
19. Li, H., Rosenthal, R., Rubin, D.B.: Reliability of measurement in psychology: from spearman-brown to maximal reliability. Psychol. Methods **1**(1), 98–107 (1996)
20. Telle, A., Vary, P.: A novel approach for impulse response measurements in environments with time-varying noise. In: Proceedings of the 20th International Congress on Acoustics, ICA 2010, Sydney, Australia, 23–27 August 2010, pp. 1–5, July 2010
21. Pander, T., Przybyla, T., Czabanski, R.: An Application of the LP-norm in robust weighted averaging of biomedical signals. J. Med. Inform. Technol. **22**(2), 1–8 (2013)
22. Ferrer, C., González, E., Hernández-Díaz, M.E., Torres, D., Del Toro, A.: Removing the influence of shimmer in the calculation of harmonics-to-noise ratios using ensemble-averages in voice signals. EURASIP J. Adv. Signal Process. **2009**, 784379 (2009)
23. Shrivastav, R., Sapienza, C.M., Nandur, V.: Application of psychometric theory to the measurement of voice quality using rating scales. J. Speech Lang. Hear. Res. **48**(2), 323–335 (2005)

A New Method to Evaluate Subgroup Discovery Algorithms

Lisandra Bravo Ilisástigui⬤, Diana Martín Rodríguez⬤,
and Milton García-Borroto(✉)⬤

Universidad Tecnológica de la Habana José Antonio Echeverría, CUJAE,
Havana, Cuba
{lbravo,dmartin,mgarciab}@ceis.cujae.edu.cu

Abstract. A Subgroup Discovery algorithms is usually considered better than other method if the average of all its mined subgroups is higher, with respect to some predefined quality measures. This process has some drawbacks: it ignores the redundancy in mined patterns and it might hide important differences among algorithms that return subgroup sets with the same averaged value. In this paper, we propose a new method to evaluate and compare subgroup discovery algorithms. This method starts by removing redundancy using a novel procedure based on the examples covered by the patterns and the statistical redundancy between them. Then, a new similarity and quality methods is used to compared the algorithms based on their ability to detect the patterns and the quality of the mined patterns, respectively. The experimental results obtained show some interesting results that would be unnoticed by the traditional approach.

Keywords: Subgroup discovery · Pattern mining · Algorithms evaluation

1 Introduction

Subgroup Discovery (SD) [1,9,16] is a pattern recognition task that identifies subsets descriptions of a dataset that show different behavior with respect to certain interestingness criteria. SD searches local pattern generally in the form of rules, where the body has constrains applied to data and the head represents the best supported class. Different approaches have been presented to discovery subgroup sets [3,8].

Most of these papers evaluate their proposals with an experimental study that only uses the estimation of some quality measures through a 10-fold cross-validation. To summarize the results, the average for all subgroups mined in each fold are computed and then the average of the results obtained by all partition is calculated. Notice that the average is highly sensitive to extreme values, and it might hide important differences between miners that returns the same averaged value. Common SD quality measures are: unusualness or weighted relative accuracy, sensitivity or recall, and confidence or precision [5].

© Springer Nature Switzerland AG 2019
I. Nyström et al. (Eds.): CIARP 2019, LNCS 11896, pp. 417–426, 2019.
https://doi.org/10.1007/978-3-030-33904-3_39

In typical comparison methodology, subgroup's redundancy is ignored, even when it introduces errors on the computation of the average metrics. Subgroups are represented as patterns, and the redundant ones may present similar quality when they cover overlapped sets of instances in a dataset. On the other hand, this methodology does not consider the individual quality of the subgroups mined and the similarity between the sets obtained by the different algorithms.

This paper proposes a new method to evaluate and compare SD algorithms taking into account the redundancy, quality and similarity between the subgroup sets obtained by them. To do so, we first apply a novel algorithm to remove the redundant subgroups, which is based on the examples covered by the patterns and the statistical redundancy between them. Then, the similarity and quality procedures proposed can be applied over the subgroups obtained in the previous step. The quality evaluation procedure allows to consider the quality distribution of the subgroup set. Also, the similarity approach allows the user to select the algorithm that provide more different information of the dataset. Finally, different graphics for each method are presented to improve the comprehensibility of the results.

2   Subgroup Discovery: Redundancy and Comparison

The main objective of SD task is to identify interesting group of individuals, where interestingness is defined as a distributional unusualness with respect to a certain property of interest. In most SD algorithms, a set of the best qualified subgroups are provided, where its quality is defined as the mean values of the measures obtained by all the subgroups mined [6]. So, this typical comparison methodology involves the summarization of the results using average, and then the subgroup set with the highest values is selected as the best ones [1,7,8].

This comparison approach has some drawbacks related with the natures of the average. It is known that the average is affected by outliers and data that not follow a central distribution. For instance, let us consider two subgroup sets A and B, where A has almost all subgroups with high quality except a few of them and most of the subgroups from B present lower quality that the best ones from A. For this reason, the average of the quality measure obtained by B would be higher than the one obtained by A. Then, using the traditional methodology the B set would be better than A, even when A present more subgroups with better quality than B. The distribution of the subgroups among the quality measure domain is also important to the comparison of subgroup sets.

Other major problem that affect not only the average measure calculation, but the comprehensibility of the results is redundancy. Dependencies between the non-target attributes lead to large numbers of variations of a particular subgroup. Since many descriptions can have a similar coverage of the given data, this may lead to many redundant evaluations of the quality function and to a subgroup set that contains multiple descriptions of the same subpopulation. Moreover, the average of measures of a subgroup set with high redundancy level can be affected, as is shown in Fig. 1. So, the redundancy is an important factor to take into account in subgroup sets comparison [2,10].

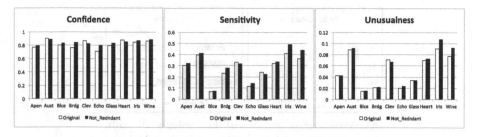

Fig. 1. Average of measure calculate with redundancy and without it for different dataset.

Redundant subgroups are those that cover a subset or a similar set of data records of some other subgroups [10]. Several approaches have been presented to detect and remove redundancy. Li et al. in [12] propose an interesting approach to detect and prune the redundant subgroups using an heuristic search and the error bounds of the *OddsRatio* measure [11]. In [2] a closure system is used to represent a subgroup by its coverage of a dataset. Van Leeuwen et al. [10] propose some selection strategies in order to eliminate redundancy in heuristic search algorithms. In general, these proposals employ one of two search space for redundancy detection: description or coverage space. The first one is more efficient but less precise that the second one.

3 A New Method to Evaluate Subgroups Discovery Algorithms

In this section, we present a new method to evaluate and compare SD algorithms, analyzing redundancy, quality and similarity of the subgroups obtained by different approaches. First, this method removes the redundant subgroups using a novel procedure based on the examples covered by the patterns and the statistical redundancy between them. Then, the quality and similarity procedures can be applied over the subgroups obtained in the first step. All their characteristics are presented in detail in the following.

Evaluating and Removing Redundancy

We propose a new procedure to identify whether two patterns are redundant using the follows properties: the ratio of examples covered by the two patterns and the statistical redundancy between them. We use the covered example ratio presented in [13], which represents the maximum percentage of covered examples by the two patterns regarding the examples covered for each pattern. If this ratio is higher than a threshold value $CovRat_{min}$, then these patterns would appear to provide us with similar information of the search space. However, these patterns could be describing different class distribution that can be interesting for the users. Because of this, the statistical redundancy proposed in [12] is also calculated, which is based on the confidence intervals of *OddsRatio*. If the

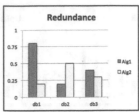
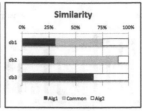

(a) Percentage of redundant subgroups obtained by the SD algorithms on each dataset.

(b) Percentage of subgroups mined by Alg1 and Alg2 and the common ones for each dataset

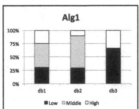
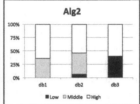

(c) Percentage of subgroups from each quality interval.

Fig. 2. Graphics designed for the evaluated methods. (a) Redundancy Method, (b) Similarity Method, (c) Quality Method

confidence intervals of the *OddsRatio* of the patterns overlap, then they are redundant.

The ratio of examples covered takes values in the range $[0,1]$, where values close to 0 show that the rules cover a few common examples and values close to 1 that the rules cover almost the same examples. Notice that the $CovRat_{min}$ threshold allows the user to determine the overlap degree of the compared subgroups. This is defined as $CovRat(P_1, P_2) = MAX\left[\frac{cov(P_1 \wedge P_2)}{cov(P_1)}, \frac{cov(P_1 \wedge P_2)}{cov(P_2)}\right]$, where $CovRat(P_1 P_2)$ represents the number of common examples covered by both subgroups P_1 and P_2, and $CovRat(P_1)$ and $CovRat(P_2)$ represent the number of examples covered by P_1 and P_2, respectively.

The *OddsRatio* of subgroup P is defined as $OR(P) = \frac{TP*TN}{FP*FN}$, where TP, FP, FN, TN are the terms of the contingency table. The confidence interval of the *OddsRatio* is calculated as $[OR(P)e^{-w}, OR(P)e^{w}]$, where $w = z_{\alpha/2} * \sqrt{\frac{1}{TP} + \frac{1}{FP} + \frac{1}{FN} + \frac{1}{TN}}$. The critical value of the confidence interval for a 95% confidence is $z_{\alpha/2} = 1,96$.

Notice that, algorithms with a large percentage of redundant subgroups are less efficient and the subgroup set mined bring information that can potentially decrease the user ability to understand the results. To analyze the redundancy detected by this method, we propose to use a bar chart graphic that shows the percentage of redundant subgroups obtained by the algorithms considered on each dataset as can be seen in Fig. 2(a).

Similarity Between Mined Subgroup Sets

Similarity of subgroup sets can be defined by the amount of commons or similar subgroups between the sets obtained by the algorithms analyzed, where two patterns are common when they are redundant. To do so, all pattern mined by both algorithms are added to a pool. Then, the similar subgroups are identified using the method presented previously. In this way, we can obtain the set of common patterns and the subgroup set obtained only by each one of the algorithms analyzed. Notice that, the subgroup sets obtained on the same dataset partition are the ones to be considered in this method.

To better analyze this results, we propose to use a stacked bar graphic for each pair of algorithms in all the datasets as can be seen in Fig. 2(b). This figure shows a similarity comparison between *alg1* and *alg2*, each bars represent the total number (100%) of patterns founded by both algorithms by dataset. The gray color represents the proportion of common subgroups found, and the black and white color are the ones obtained only by *alg1* and *alg2* respectively. We can see how the *alg1* and *alg2* obtain very similar results for the db2 dataset since the percentage of commons patterns is large. Moreover, it can be seen how *alg1* extracts more information from the db2 dataset that *alg2* since it can obtain all the common subgroups and it gets more different subgroups than the ones obtained by *alg2*. Notice that, this similarity analysis allows the user to select the algorithm that provide more different information of the dataset.

Comparing the Quality of the Mined Subgroups

The quality of a subgroup is defined by the values of different quality measures proposed in the literature as confidence, sensitivity and unusualness. For all of these measures the highest values are the better ones. Then, we can divide the range of the values obtained in a number of intervals N_{Interv} to identify the quality of a subgroup depending on the interval it belongs to. The N_{Interv} and its limits are determined by the user. In this work, we empirically set $N_{Interv} = 3$ to identify the lowest, middle and highest quality intervals. The range of the values is divided into three equal parts to define the intervals limits. Then, the quality of a subgroup set can be also analyzed considering the percentage of patterns from each of these quality intervals, which allow us to consider the quality distribution of the subgroup set. Finally, the lower and upper bound of the range of values are defined by the minimum and maximum value found from all the patterns obtained by the algorithms analyzed.

To represent the results of this method, we employ a composite graphic like the one in Fig. 2(c). This comparison is pairwise, so the representation employs two bar graphics that have to be interpreted as a whole, since the intervals boundaries for both algorithms are determined by the conjunction of its results. Each subgraphic represents the results obtained by each algorithm, where each bar show the percentage of pattern that belong to each quality interval by dataset. The black, gray and white colors represents the lowest, middle and highest intervals of the domain of metrics respectively. The higher the percentage of patterns in the upper interval, the better is the subgroup set. Figure 2(c) shows how *alg2* has more subgroups with high quality than *alg1*.

4   Experimental Validation

To validate the new evaluation method, we compare 3 well known SD algorithms: SD-map [1], Apriori-SD [8] and NMEEF-SD [3]. We have considered the follows 20 datasets from the UCI Repository of machine learning databases [4]: Appendicitis, Australian,Balance,Brest Cancer,Bridges, Bupa, Cleveland, Diabetes, Echo, German, Glass, Haberman, Heart, Hepatitis, Ionosphere, Iris, Led, Primary Tumor, Vehicle, Wine. The parameters of the analyzed algorithms are presented in Table 1. These parameters were selected using the recommendations of the authors. Apriori-SD and SD-map implementation don't allow continuous variables, so a ID3 [15] discretization was applied. To develop the different experiments, the parameters of our proposal are defined as $CovRat_{min} = 0.75$ and $N_{Interv} = 3$. We consider the average results 10-fold cross-validation. In addition, as NMEEF-SD is stochastic, three runs are performed.

Table 1. Parameters of the algorithms.

Algoritmo	Parámetros
NMEEF-SD	RulesRep = can; nLabels = 3,5; nEval = 10000; popLength = 50; crossProb = 0.6; mutProb = 0.1; diversity = crowding; ReInitCob = yes porcCob = 0.5; Obj1 = comp; Obj2 = unus; Obj3 = null; minCnf = 0.6; StrictDom = yes
SD-map	MinSupp = 0.1; minConf = 0.8; RulesReturn = 10
Apriori-SD	MinSupp = 0.03; minConf = 0.8; Number-of-Rules = 5; Postpruning-type = SELECT-N-RULES-PER-CLASS

In these experiments, first we apply the traditional methodology to evaluate the algorithms analyzed. Then, we use the new evaluation methods to show how they can improve the quality of the comparison, providing more information about it. In this study, we show a pairwise comparison between the algorithms considered. To apply the traditional methodology, we statically compare the average of the quality measures obtained by the algorithms in all datasets. We analyzed these results considering all the subgroups discovered and the ones found when the redundant subgroups are removed. We have used a Wilcoxon's test [14] with a level of significance of 0.05, Table 2 shows the results obtained.

The redundancy analysis also is shown in Fig. 3, where the percentage of redundant subgroups found are described for each dataset. The similarity between the algorithms studied is presented in Fig. 4. To consider the quality distribution of the subgroup sets mined using different quality intervals, we show in Fig. 5 the percentage of patterns for each algorithm than belong to the low, middle and high quality interval.

Table 2. Wilcoxon's test ($\alpha = 0.05$) on the different measures for the SD algorithms, where • indicates the algorithm in the row improves the algorithm of the column and ∘ the algorithm in the column improves the algorithm of the row. Upper diagonal shows the test's results removing redundant subgroups, lower diagonal shows the test's results with all the subgroups obtained.

Alg	Confidence			Sensitivity			Unusualness		
	(1)	**(2)**	**(3)**	**(1)**	**(2)**	**(3)**	**(1)**	**(2)**	**(3)**
SD-map (1)	-		•	-		•	-		
Apriori-SD (2)		-		∘	-	∘		-	∘
NMEEF-SD (3)			-		•	-	•	•	-

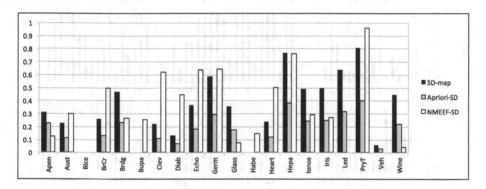

Fig. 3. Percentage of redundant subgroups obtained by each algorithm for all the datasets

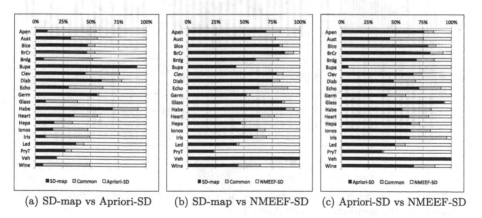

(a) SD-map vs Apriori-SD   (b) SD-map vs NMEEF-SD   (c) Apriori-SD vs NMEEF-SD

Fig. 4. Percentage of common and individual patterns

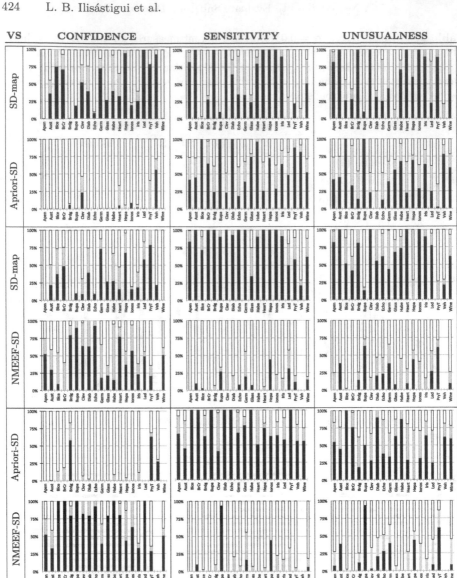

Fig. 5. Pairwise comparison of the average measures distribution of the subgroup sets obtained for all algorithms for each dataset.

We can draw the following conclusions based on an analysis of the results presented of the pairwise comparison between the algorithms considered:

- SD-map vs Apriori-SD: We can see from Fig. 3 how Apriori-SD obtain less redundant subgroups than SD-map. The similarity analysis presented in

Fig. 4(a) shows that Apriori-SD discovers more knowledge than SD-map since it can obtain more than 50% of the total number of subgroups mined by both methods in most of the datasets. The statistical analysis shows that there are not significant difference between the average results of the confidence and unusualness measures. However, Fig. 5 shows how more subgroups mined by Apriori-SD obtain better values for these measures than the ones obtained by SD-map. So, Apriori-SD can be considered better than SD-map because it provides more diverse knowledge with better quality in most of the measures considered.

– SD-map vs NMEEF-SD: The redundancy analysis shows that NMEEF-SD mines more redundant subgroups than SD-map. Table 2 shows how the statistical results between these two algorithms change when the redundant subgroups are removed. Moreover, if we analyze only the test's results once the redundant subgroups are removed, we can see how the difference for the measures unusualness and sensibility are not significant. However, most of the subgroups obtained by NMEEF-SD get higher values for these measures than the ones mined by SD-map, as can be seen in Fig. 5. The confidence values obtained by SD-map are better than the values obtained by NMEEF-SD as the statistical results and Fig. 5 show. The similarity analysis shows how the subgroups obtained are very different, having few subgroups in common. Finally, both algorithms provide different knowledge to the users, where the subgroups mined by NMEEF-SD present better values for the measures unusualness and sensibility, who are more interesting in the SD task.

– Apriori-SD vs NMEEF-SD: The statistical comparison between these algorithms don't change when the redundant subgroups are removed, being NMEEF-SD better than Apriori-SD for unusualness and sensibility measure as is shown in Table 2. Figure 5 also shows that most of the subgroups mined by NMEEF-SD belong to the highest quality interval. These algorithms present low similarity between them. NMEEF-SD can be considered better than Apriori-SD because it provides more diverse knowledge with better quality in most of the measures considered.

5   Conclusions

In this paper, we propose a new method to evaluate and compare SD algorithms considering redundancy, quality and similarity between the subgroup sets obtained by them. First, this method removes the redundant subgroups using a novel procedure based on the examples covered by the patterns and the statistical redundancy between them. Then, despite previous researches, which estimate the quality of the algorithms using the average of the quality measures obtained from 10 fold cross validation, we perform a paired comparison between subgroup set obtained from the same chunk of the dataset to determine the quality distribution of the subgroup sets mined. Moreover, the method proposed can also determines how much a subgroup set is similar to another using a new procedure which gets the common subgroups between the sets obtained by the algorithms analyzed, where two patterns are common when they are redundant. Finally, the experimental validation shows how our proposal and its associated graphics can provide more useful information to the users in order to select the best algorithm for their SD's problems.

References

1. Atzmueller, M., Puppe, F.: SD-map – a fast algorithm for exhaustive subgroup discovery. In: Fürnkranz, J., Scheffer, T., Spiliopoulou, M. (eds.) PKDD 2006. LNCS (LNAI), vol. 4213, pp. 6–17. Springer, Heidelberg (2006). https://doi.org/10.1007/11871637_6
2. Boley, M., Grosskreutz, H.: Non-redundant subgroup discovery using a closure system. In: Buntine, W., Grobelnik, M., Mladenić, D., Shawe-Taylor, J. (eds.) ECML PKDD 2009. LNCS (LNAI), vol. 5781, pp. 179–194. Springer, Heidelberg (2009). https://doi.org/10.1007/978-3-642-04180-8_29
3. Carmona, C.J., González, P., del Jesus, M.J., Herrera, F.: NMEEF-SD: non-dominated multiobjective evolutionary algorithm for extracting fuzzy rules in subgroup discovery. IEEE Trans. Fuzzy Syst. 18(5), 958–970 (2010)
4. Dheeru, D., Karra Taniskidou, E.: UCI machine learning repository (2017). http://archive.ics.uci.edu/ml
5. García-Borroto, M., Loyola-González, O., Martínez-Trinidad, J.F., Carrasco-Ochoa, J.A.: Evaluation of quality measures for contrast patterns by using unseen objects. Expert Syst. Appl. 83, 104–113 (2017)
6. Grosskreutz, H., Rüping, S.: On subgroup discovery in numerical domains. Data Mining Knowl. Discov. 19(2), 210–226 (2009)
7. del Jesus, M., Gonzalez, P., Herrera, F.: Multiobjective genetic algorithm for extracting subgroup discovery fuzzy rules. In: IEEE Symposium on Computational Intelligence in Multicriteria Decision Making (2007)
8. Kavsek, B., Lavrac, N.: APRIORI-SD: adapting association rule learning to subgroup discovery. Appl. Artif. Intell. 20(7), 543–583 (2006)
9. Klösgen, W.: Explora: a multipattern and multistrategy discovery assistant. In: Advances in Knowledge Discovery and Data Mining, pp. 249–271. American Association for Artificial Intelligence (1996)
10. van Leeuwen, M., Knobbe, A.: Diverse subgroup set discovery. Data Mining Knowl. Discov. 25(2), 208–242 (2012)
11. Li, H., Li, J., Wong, L., Feng, M., Tan, Y.P.: Relative risk and odds ratio: a data mining perspective. In: Proceedings of the Twenty-Fourth ACM SIGMOD-SIGACT-SIGART Symposium on Principles of Database Systems, pp. 368–377. ACM (2005)
12. Li, J., Liu, J., Toivonen, H., Satou, K., Sun, Y., Sun, B.: Discovering statistically non-redundant subgroups. Knowl.-Based Syst. 67, 315–327 (2014)
13. Martín, D., Alcalá-Fdez, J., Rosete, A., Herrera, F.: NICGAR: a niching genetic algorithm to mine a diverse set of interesting quantitative association rules. Inf. Sci. 355, 208–228 (2016)
14. Quinlan, J.: Induction of decision trees. Mach. Learn. 1, 81–106 (1986)
15. Wilcoxon, F.: Individual comparisons by ranking methods. Biometrics 1, 80–83 (1945)
16. Wrobel, S.: An algorithm for multi-relational discovery of subgroups. In: Komorowski, J., Zytkow, J. (eds.) PKDD 1997. LNCS, vol. 1263, pp. 78–87. Springer, Heidelberg (1997). https://doi.org/10.1007/3-540-63223-9_108

Mathematical Theory of Pattern Recognition

KAdam: Using the Kalman Filter to Improve Adam algorithm

José David Camacho, Carlos Villaseñor, Alma Y. Alanis, Carlos Lopez-Franco, and Nancy Arana-Daniel[✉]

Universidad de Guadalajara Centro Universitario de Ciencias Exactas e Ingenierías, Blvd. Marcelino García Barragán # 1421, Guadalajara, Jalisco, Mexico
nancyaranad@gmail.com

Abstract. Nowadays, the Adam algorithm has become one of the most popular optimizers to train feed-forward neural networks because it takes the best features of other gradient-based optimizers, such as working well with sparse gradients, in online and non-stationary settings, and also it is very robust to the rescaling of the gradient. The above makes Adam the best choice to solve problems with non-stationary objectives, very noise gradients, and with large data inputs. In this work, we enhanced the Adam algorithm by using the Kalman filter, and the novel proposal is called *KAdam*. Instead of using the computed gradients directly from the cost function, we first apply the Kalman filter on them. As a result, the filtered gradients allow the algorithm to explore new (and potentially better) solutions on the cost function. The results obtained when applying our proposal and other state-of-the-art optimizers to solve classification problems show that KAdam is able to obtain better accuracies than its competitors in the same execution time.

Keywords: Gradient descent optimization · Backpropagation · Kalman filter

1 Introduction

Gradient descent is an iterative algorithm to perform function optimization. Also, it is by far one of the most popular and common methods used in the training of neural networks. The gradient descent (GD) algorithm has three variants: Vanilla gradient descent (a.k.a Batch gradient descent), Stochastic gradient descent (SGD), and Mini-batch gradient descent. Which differs in the amount of training data used to compute the gradients (see [9] for further details). However, the GD algorithm and all its variants may present slow convergence time or heavy oscillations in the cost function [10]. As a result, there have been many proposals to improve the conventional gradient descent algorithms.

Momentum [8] is one of these methods, which accelerate the SGD algorithm in a relevant direction, even though its weakness is to show the behavior of a blind-rolling ball down the hill. Then AdaGrad [2] was designed to solve the

© Springer Nature Switzerland AG 2019
I. Nyström et al. (Eds.): CIARP 2019, LNCS 11896, pp. 429–438, 2019.
https://doi.org/10.1007/978-3-030-33904-3_40

blind-rolling problem by using non-constant learning rates. Unfortunately, it presents a radically diminishing learning rate, in which point the algorithm is no longer able to keep learning. After that, Adadelta [12] and RMSProp [11], both introduced the concept of adaptive learning rates, solving AdaGrad's problem. On the other hand, Adam [4] (described by its authors as a combination of AdaGrad and RMSProp) is one of the most popular optimizers in nowadays neural network frameworks like [1,7]. The Adam algorithm is commonly used because it presents high performance, is straightforward to implement, works well with sparse gradients and in online and non-stationary settings, and also it is very robust to the rescaling of the gradient. The above makes Adam the best choice to solve problems with non-stationary objectives, very noise gradients, and with large data inputs.

The methods mentioned early (see [13] for a detailed introduction) have shown great empirical results. However, we propose to enhance Adam algorithm using the Kalman filter [3] because we can obtain significant variations by using the estimated gradients instead of the computed ones. This change may help to explore and reach better solutions on the cost function, like other works have done by adding Gaussian noise to the gradients [6]. Hence, in this paper, it is introduced the *KAdam* algorithm, an extension of Adam using the Kalman filter.

The structure of this paper is described next. First, Sect. 2 describes the Adam algorithm. Then, Sect. 3 provides a brief introduction to the Kalman filter. After that, Sect. 4 describes the *KAdam* algorithm. Subsequently, Sect. 5 shows different carried out experiments and performance comparisons between the proposed method and other gradient-based optimizers. Finally, Sect. 6 shares conclusions from the authors and their future work.

2   Adam

The first step on the Adaptive Moment Estimation (Adam) algorithm [4], is to save the exponentially decaying averages of past gradients (first moment) and past squared gradients (second moment). This is done by computing the first moment estimate v_t (the mean) and the second moment estimate m_t (the uncentered variance) in the following equations:

$$m_t = \beta_1 m_{t-1} + (1 - \beta_1)g_t \tag{1}$$

$$v_t = \beta_2 v_{t-1} + (1 - \beta_2)g_t^2 \tag{2}$$

where β_1 and β_2 are the decay rates for the first and second moment (which the authors of Adam suggest to be set to 0.9 and 0.999 respectively), $g_t \in \mathbb{R}^n$ is the computed gradient of the cost function and g_t^2 is the squared (element-wise) gradients.

As v_t and m_t are initialized as zero vectors, the authors of Adam observe that they are biased towards zero, especially during the initial time steps and when the decay rates are small. Thus, they counteract these biases by computing bias-corrected first and second moment estimates.

$$\hat{m}_t = \frac{m_t}{1 - \beta_1^t} \tag{3}$$

$$\hat{v}_t = \frac{v_t}{1 - \beta_2^t} \tag{4}$$

Finally, the parameters update rule is given by:

$$\theta_{t+1} = \theta_t - \frac{\eta}{\sqrt{\hat{v}_t} + \epsilon} \hat{m}_t \tag{5}$$

where, η is the learning rate and ϵ is the smooth term (used to ensure algorithmic stability), which the authors of Adam suggested to be set to 0.001 and a value on the order 10×10^{-10} respectively.

3 Kalman Filter

The Kalman filter [3] is a recursive state estimator for linear systems. The algorithm consist in a group of equations that works in a two-steps process: prediction and update. The prediction phase is described by the following equations.

$$\hat{\mathbf{x}}_{k|k-1} = \mathbf{F}_k \hat{\mathbf{x}}_{k-1|k-1} + \mathbf{B}_k \mathbf{u}_k \tag{6}$$

$$\mathbf{P}_{k|k-1} = \mathbf{F}_k \mathbf{P}_{k-1|k-1} \mathbf{F}_k^\mathsf{T} + \mathbf{Q}_k \tag{7}$$

These equations gives a prediction of the state estimate and the covariance error but based only on information from the previous time step. In Eq. (6), the Kalman filter computes an a *priori* state estimate $\hat{\mathbf{x}}_{k|k-1}$ where, $\hat{\mathbf{x}}_{k-1|k-1}$ is the past predicted state, $\mathbf{F}_k$ is the state transition model and $\mathbf{B}_k$ is the control-input model with its respective input vector $\mathbf{u}_k$. In Eq. (7), the predicted a *priori* error covariance $\mathbf{P}_{k|k-1}$ is computed, where, $\mathbf{P}_{k-1|k-1}$ is the previous covariance error, and $\mathbf{Q}_k$ is the covariance of the process noise. On the other hand, the update phase is described by the following equations.

$$\mathbf{K}_k = \mathbf{P}_{k|k-1} \mathbf{H}_k^\mathsf{T} \left(\mathbf{H}_k \mathbf{P}_{k|k-1} \mathbf{H}_k^\mathsf{T} + \mathbf{R}_k \right)^{-1} \tag{8}$$

$$\hat{\mathbf{x}}_{k|k} = \hat{\mathbf{x}}_{k|k-1} + \mathbf{K}_k (\mathbf{z}_k - \mathbf{H}_k \hat{\mathbf{x}}_{k|k-1}) \tag{9}$$

$$\mathbf{P}_{k|k} = (\mathbf{I} - \mathbf{K}_k \mathbf{H}_k) \mathbf{P}_{k|k-1} \tag{10}$$

These equations gives an updated prediction of the state estimate and the covariance error, computed with a correction based on observed information and measurements $\mathbf{z}_k$ from the true state in the current time step. In Eq. (8) the optimal Kalman gain matrix $\mathbf{K}_k$ is computed, where, $\mathbf{H}_k$ is the measuring matrix and $\mathbf{R}_k$ is the covariance of the observation noise. In Eqs. (9) and (10) the Kalman filter computes an updated (a *posteriori*) state estimate $\hat{\mathbf{x}}_{k|k}$ and an updated (a *posteriori*) estimated covariance $\mathbf{P}_{k|k}$, respectively.

4   KAdam

The KAdam algorithm uses a Kalman filter to estimate the gradients of the cost function. Considering the dynamics of the gradients as unknown, the matrices $\mathbf{F}_k$, $\mathbf{H}_k$, $\mathbf{Q}_k$ and $\mathbf{R}_k$ are used as identities and the state vector $\hat{\mathbf{x}}_{k|k}$ initialized as a zero vector, with adequates dimensions according to the gradients vector. Moreover, the gradients g_t of the cost function are used as the measurements $\mathbf{z}_k$ from the true state vector in the Kalman filter. Thus, the estimated gradients $\hat{g}_t$ can be written as the post-fit measurements $\mathbf{H}_k\hat{\mathbf{x}}_{k|k}$ from the filter. The steps to calculate the estimated gradients $\hat{g}_t$ with the Kalman filter are summarized as a function $K(\bullet)$.

$$\hat{g}_t = K(g_t) \tag{11}$$

Hence, the equations to calculate the first and second moment are the following:

$$m_t = \beta_1 m_{t-1} + (1 - \beta_1)\hat{g}_t \tag{12}$$

$$v_t = \beta_2 v_{t-1} + (1 - \beta_2)\hat{g}_t^2 \tag{13}$$

The original equations from Adam to compute the bias-correction of the moments (see Eqs. (3) and (4)) and the update rule (see Eq. (5)) were not modified.

5   Experiments

To empirically evaluate the accuracy and efficiency of the proposal, two experiments (with two types trainings for each one) were carried out using feed-forward neural networks to solve some of the most popular benchmark problems in machine learning. For each experiment, there is a comparison between the proposed algorithm and the following algorithms: GD, Momentum, RMSProp, and Adam. We also include the stochastic and the batch experimentation, where stochastic implies that for every patron in the training set, we adapt the parameters of the model, while in the batch training the full training set is used to calculate one adaptation of the parameters. The comparison criterion is the cost reduction using the mean squared error (MSE) through the training phase and the test phase.

In the experiments, each neural network was configured with the same architecture (experimentally selected) and the same weights initialization. The settings for the hyper-parameters used in the experiments are listed in Table 1 except the learning-rate, which is $\eta = 0.01$ for all the experiments.

5.1   Experiment: Moons

The experiment deals with the classification problem of two interleaving half circles, using a dataset with $12,000$ samples ($10,000$ for training and $2,000$ for

Table 1. Algorithms hyperparameters.

Algorithm	Hyperparameters
GD	$\eta = 0.01$
Momentum	$\gamma = 0.9$
RMSProp	$\gamma = 0.999, \epsilon = 1 \times 10^{-13}$
Adam	$\beta_1 = 0.9, \beta_2 = 0.999, \epsilon = 1 \times 10^{-13}$
KAdam	$\beta_1 = 0.9, \beta_2 = 0.999, \epsilon = 1 \times 10^{-13}, \mathbf{F}_k = \mathbf{H}_k = \mathbf{Q}_k = \mathbf{R}_k = I \in R^{n \times n}$

test) generated by a function[1] from the scikit-learn python package [7]. The architectures for the neural networks were fixed to: (10, 1) layers, with a $\tanh(\bullet)$ function in the hidden layer and a sigmoid function for the output layer.

In the stochastic training, the parameters of the neural network are adapted with each patron. In Fig. 1, we show the error function in the training phase. Notice that GD and Momentum have different behavior than RMSprop, Adam, and KAdam due to the second-moment dynamics. The second moment allows the algorithms to speed-up in an early stage of the training, as is shown in the left image, where KAdam has the fastest descend. On the other hand, these algorithms show a noisy behavior in the long run, where stochastically some low cost can be achieved.

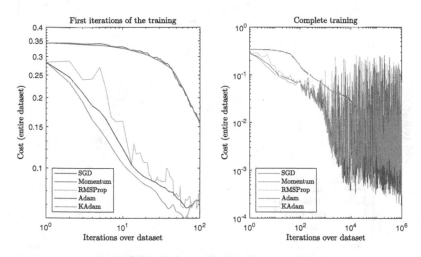

Fig. 1. Cost reduction comparison - Moons experiment (Stochastic training). Left: the first one hundred iterations, Right: Full experiment with 10^6 iterations

In Table 2, we show the results of this experiment, where Adam have the best performance. Notice that in the long run, all the algorithms have close results.

---

[1] https://scikit-learn.org/stable/modules/generated/sklearn.datasets.make_moons. html – using a noise factor of 0.5.

Table 2. Moons experiment - Stochastic Training results.

Algorithm	Training phase minimum cost	Training phase cost variance	Test phase minimum cost
GD	3.17×10^{-4}	3.03×10^{-5}	2.67×10^{-4}
Momentum	3.17×10^{-4}	3.02×10^{-5}	2.67×10^{-4}
RMSProp	2.08×10^{-4}	8.11×10^{-5}	5.64×10^{-4}
Adam	1.86×10^{-4}	3.43×10^{-5}	2.40×10^{-5}
KAdam	2.11×10^{-4}	6.89×10^{-5}	1.50×10^{-3}

In the batch training, the algorithms performed the weights update using all the samples from the data-set for each iteration. Figure 2 shows how RMSprop, Adam, and KAdam do not present the stochastic behavior in the batch training. Moreover, the proposed method showed an improvement compared with Adam and the other presented methods. Table 3 shows the results of the experiment.

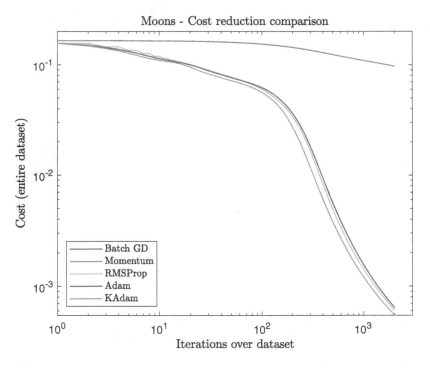

Fig. 2. Cost reduction comparison - Moons experiment (Batch training).

Table 3. Moons experiment - Batch training results.

Algorithm	Training phase minimum cost	Training phase cost variance	Test phase minimum cost
GD	0.0969	2.9694×10^{-4}	0.59002
Momentum	0.0969	2.9491×10^{-4}	0.59047
RMSProp	6.0198×10^{-4}	4.3800×10^{-4}	0.49701
Adam	6.3738×10^{-4}	4.4134×10^{-4}	0.49698
KAdam	5.4726×10^{-4}	3.7443×10^{-4}	0.49704

5.2   Experiment: MNIST

This experiment deals with the MNIST classification problem. Before the training, the entire dataset was embedded into a 2D space (see Fig. 3) using a t-SNE [5] implementation[2].

The architectures for the neural networks were fixed to: (10, 10) layers, with a $\tanh(\bullet)$ function in the hidden layer and a sigmoid function for the output layer.

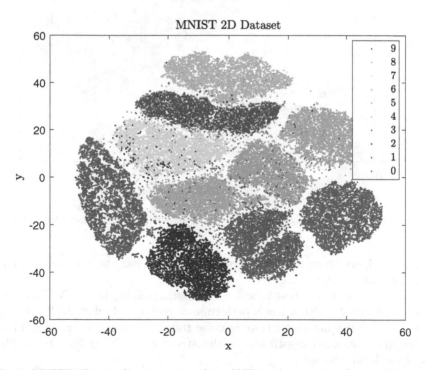

Fig. 3. MNIST 2D visualization using the t-SNE implementation from scikit-learn.

---

[2] https://scikit-learn.org/stable/modules/generated/sklearn.manifold.TSNE.html.

Table 4. MNIST experiment - Stochastic Training results.

Algorithm	Training phase minimum cost	Training phase cost variance	Test phase minimum cost
GD	0.0413	5.6927×10^{-4}	0.0569
Momentum	0.0365	3.0551×10^{-4}	0.0512
RMSProp	0.0345	1.5349×10^{-4}	0.0802
Adam	0.0241	1.5555×10^{-4}	0.0771
KAdam	0.0226	2.2447×10^{-4}	0.0608

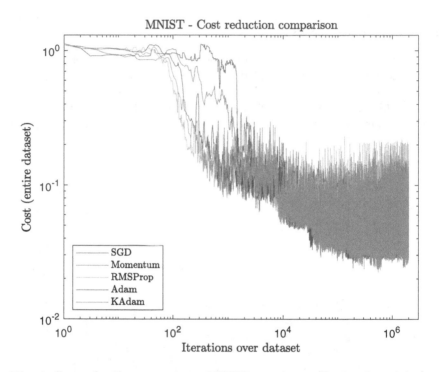

Fig. 4. Cost reduction comparison - MNIST experiment (Stochastic training).

In Table 4, we show the results, where KAdam has the best result in the training phase.

In Fig. 4, we show the cost function stochastic training for the MNIST dataset. The KAdam algorithm has a performance comparable with RMSProp and Adam. The MNIST data-set has more noise than the moons experiment. Therefore the gradient-based algorithms in the stochastic training tend to oscillate around the local minimum.

We also present the batch training experiment for the MNIST data-set. In Fig. 5 and Table 5, We present the results for this experiment where Adam and KAdam have a tight competence and they overcome the other algorithms.

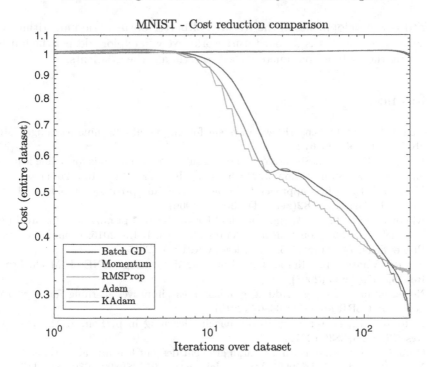

Fig. 5. Cost reduction comparison - MNIST experiment (Batch training).

Table 5. MNIST experiment - Batch training results.

Algorithm	Training phase minimum cost	Training phase cost variance	Test phase minimum cost
GD	0.9863	3.6546×10^{-5}	0.9782
Momentum	0.9945	1.8659×10^{-5}	0.98597
RMSProp	0.3310	0.0242	0.33178
Adam	0.2654	0.0310	0.27059
KAdam	0.2718	0.0285	0.28383

6   Conclusion

In this work, we presented a proposal to improve the performance of Adam optimizer. As we have shown, when the Kalman filter is used, the estimate gradients keep following the original ones but adding relevant enough variations, which allow exploring new and probably better solutions in the cost function.

We present two empirical results with two classical data-sets, the moons, and the MNIST, and with the stochastic and batch training. We have shown that our approach presents an excellent performance in both the training phase and the testing phase. On the other hand, we think this algorithm opens the door to new developments in the research of better optimization algorithms for artificial neural networks. In our future works, we will explore deeper the impact of varying the Kalman parameters used to estimate the gradients.

References

1. Abadi, M., et al.: TensorFlow: a system for large-scale machine learning. CoRR abs/1605.08695 (2016)
2. Duchi, J., Hazan, E., Singer, Y.: Adaptive subgradient methods for online learning and stochastic optimization. J. Mach. Learn. Res. **12**, 2121–2159 (2011)
3. Kalman, R.E.: A new approach to linear filtering and prediction problems. Trans. ASME-J. Basic Eng. **82**(Series D), 35–45 (1960)
4. Kingma, D.P., Ba, J.: Adam: a method for stochastic optimization. In: 3rd International Conference on Learning Representations, ICLR 2015. Conference Track Proceedings, 7–9 May 2015, San Diego, CA, USA (2015)
5. van der Maaten, L., Hinton, G.: Visualizing data using t-SNE. J. Mach. Learn. Res. **9**, 2579–2605 (2008)
6. Neelakantan, A., et al.: Adding gradient noise improves learning for very deep networks. CoRR abs/1511.06807 (2015)
7. Pedregosa, F., et al.: Scikit-learn: machine learning in python. J. Mach. Learn. Res. **12**, 2825–2830 (2011)
8. Qian, N.: On the momentum term in gradient descent learning algorithms. Neural Netw. **12**(1), 145–151 (1999). https://doi.org/10.1016/S0893-6080(98)00116-6
9. Ruder, S.: An overview of gradient descent optimization algorithms. CoRR abs/1609.04747 (2016)
10. Sutton, R.S.: Two problems with backpropagation and other steepest-descent learning procedures for networks. In: Proceedings of the Eighth Annual Conference of the Cognitive Science Society (1986)
11. Tieleman, T., Hinton, G.: Lecture 6.5 - RMSProp. Technical report, COURSERA: Neural Networks for Machine Learning (2012)
12. Zeiler, M.D.: ADADELTA: an adaptive learning rate method. CoRR abs/1212.5701 (2012)
13. Zhang, J.: Gradient descent based optimization algorithms for deep learning models training. CoRR abs/1903.03614 (2019)

Evaluating Restrictions in Pattern Based Classifiers

Andy González-Méndez[iD], Diana Martín-Rodríguez[iD],
and Milton García-Borroto[(✉)][iD]

Universidad Tecnológica de la Habana José Antonio Echeverría, CUJAE,
114 No. 11901, e/ Ciclovía y Rotonda, Marianao, Cuba
{andyg,dmartin,mgarciab}@ceis.cujae.edu.cu

Abstract. Generalizations, also known as patterns, are in the core of
many learning systems. A key component for automatically mine generalizations is to define the predicate to select the most important ones.
This predicate is usually expressed as a conjunction of restrictions. In
this paper, we present an experimental study of some of the most used
restrictions: the minimal support threshold, the jumping pattern and
the minimal pattern. This study that uses 93 databases and two different classifiers reveals some interesting results, including one that should
be very useful for building better classifiers: using minimal patterns could
degrade the accuracy.

Keywords: Pattern based classifiers · Restrictions · Experimental
evaluation

1 Introduction

One capability usually associated to learning is the ability to analyze a large
number of specific observations to extract and retain the important common
features that characterize classes of these observations. In 1982, Mitchel [10]
formalizes this as the problem of generalization: determining the most important
generalizations that allows to differentiate a collection of positive objects from a
collection of negative objects. Generalizations are frequently known as patterns.

To select important generalizations, a predicate formed by a collection of
restrictions should be defined. According to the used restrictions, generalizations
take particular names like *subgroups* [12], *emerging patterns* [4], *contrast sets* [2],
supervised descriptive rule [11], *contrast patterns* [3], and *discriminative patterns*
[9].

The diversity in generalizations pose some challenges to select the appropriate one for a given problem. Firstly, comparative studies evaluate types of
generalizations, obscuring the contribution of each individual restriction. Secondly, some types of generalization are introduced as novel although they are
composed by already known restrictions or have minor changes with respect to
previously defined ones.

I. Nyström et al. (Eds.): CIARP 2019, LNCS 11896, pp. 439–448, 2019.
https://doi.org/10.1007/978-3-030-33904-3_41

In 2018, García-Borroto [7] performs a theoretical study about generalizations from the perspective of their restrictions. For a better understanding, he grouped restrictions based on intuitions, which are insights about how a good generalization should be. Although some interesting conclusions were suggested, no experimental evidence was provided to support them.

In this paper, we present an experimental study to evaluate how some of the most used restrictions impacts in the quality of the generalizations. Different analysis performed with the results allow to extract interesting conclusions, including some novel findings.

The structure of the paper is the following. Firstly, Sect. 2 presents the most common intuitions and some of their corresponding restrictions. Secondly, Sect. 3 shows the results of the experiments and discuss the main findings. Finally, Sect. 4 presents the conclusions and some limitations of the results that might lead to future research lines.

2 Restrictions in Pattern Based Classifiers

The predicate to select the important generalizations is usually composed by a set of restrictions, where each restriction imposes a limitation to the generalization. Restrictions are used for two main purposes: guarantee the quality of the generalization collection (removing noise and redundancy) and speed up the mining process.

To simplify their analysis and comparative evaluation, restrictions with a common objective can be grouped by the intuition they materialize [7].

2.1 Intuition 1. Each Pattern Must Be Frequent Enough to Guarantee It Is Not Due to Chance

This intuition is commonly measured using the pattern support ($p(P) > \mu$). In some extreme cases, even the support in the negative class must be greater than a threshold ($p(P \neg C) > \mu$).[1] Some generalizations do not include these restrictions because setting a good threshold is a hard task that directly impacts the quality of the classification result [7].

Minimal support restrictions are also used to early stop the mining procedures, reducing the computational cost and the number of mined patterns, which improve the understandability of results.

---

[1] The probability of finding an object with a given generalization or pattern P is denoted by $p(P)$, while the probability of not finding an object with a given pattern is denoted as $p(\neg P)$. With respect to a given class C, probabilities of finding an object of a given class or from a different class are denoted as $p(C)$ and $p(\neg C)$, respectively. Joint probabilities are then denoted as $p(PC)$, $p(P \neg C)$, and so on.

2.2   Intuition 2. Each Pattern Must Be Distinctive of Its Representing Class

This intuition lies in the core of the generalization concept, therefore it is materialized by many different restrictions (In [7] the author presents 15). Maybe the first restriction following this intuition was confidence ($p(C|P) \geq \mu$)) that contrast the probability of the joint appearance of patterns and class with the a-priori probability of the pattern. The emerging restriction $\left(\frac{p(P|C)}{p(P|\neg C)} \geq \mu \right)$ is another example, which contrasts the probabilities of the pattern to appear in the positive and negative classes.

The jumping restriction is an extreme case of the emerging restriction, where the pattern is not allowed to appear in the negative class ($p(P\neg C) = 0$). It is widely used for two main reasons: patterns are as much discriminative as they can be and the minimal support threshold does not need to be set. Nevertheless, using the jumping restriction makes patterns more sensitive to noise, decreasing the quality of the collection.

2.3   Intuition 3. More General Patterns Are Preferred, Because They Contains the Irreducible Relations Among Attributes that Determine the Class

Generalizations found in a given problem are usually related by general/specific relationships. For example, the pattern $(Sex = \text{``}F\text{''}) \wedge (Age > 23)$ is a more general pattern than $(Sex = \text{``}F\text{''}) \wedge (Age > 23) \wedge (Hair = \text{``}Red\text{''})$ but more specific than $(Sex = \text{``}F\text{''})$. Since there could be some redundancy in patterns, some types of generalizations contains restrictions to limit it.

Restrictions following Intuition 3 remove more specific patterns if a more general pattern is found. Notice that these restrictions cannot be applied in isolation, because then only one-item patterns would remain. The simplest example of restriction in this intuition is to remove all non-minimal patterns fulfilling the remaining restrictions. More complex examples include contrasting some quality evaluation like support or growth rate; a more specific pattern is retained only if it is significantly better.

3   Experimental Evaluation of Restrictions

In this section, we experimentally evaluate one important restriction per intuition: minimal support threshold, jumping restriction, and minimal pattern.

To evaluate the restrictions we used 93 datasets with different characteristics from the UCI repository [1][2]. Patterns were mined using a random forest based miner available in the platform introduced in [8].

---

[2] Number of instances ranges from 15 to 20 K, number of attributes range from 3 to 280, and number of classes ranges from 2 to 57.

We estimate the quality of a pattern collection using the accuracy of two pattern based classifiers: CAEP [5] and BCEP [6]. CAEP is a simple emerging patterns based classifier that finds the class of query objects using a voting scheme. Each pattern P matching the query object votes equal to $\frac{GR(P)}{GR(P)+1}Sup(P)$, where $GR(P) = \frac{p(P|C)}{p(P|\neg C)}$ is the growth rate and $Sup(P) = p(PC)$ is the support. BCEP is an extension of the Naive Bayes classifier relaxing the independence assumption by estimating probabilities using the mined patterns. Accuracy is estimated using 10-fold cross validation sampling procedure.

3.1   Minimum Support Threshold Restriction

In order to evaluate the influence of this restriction, we analyze the accuracy obtained by both classifiers using pattern sets with different and increasing minimum thresholds: 0, 0.02, 0.06, 0.1, 0.15, 0.18, 0.2, 0.25, and 0.3.

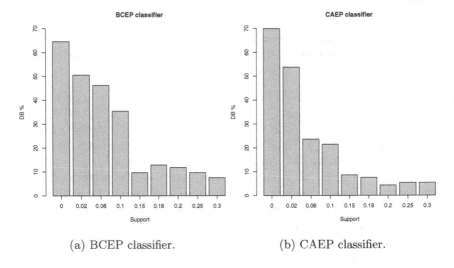

(a) BCEP classifier.   (b) CAEP classifier.

Fig. 1. Count of best accuracy results by support.

Figure 1a and b show, for each support threshold, the percent of datasets where the classifier gets the highest accuracy compared to the remaining support thresholds. If the highest accuracy value is obtained for more than one support, it is counted for each one of them. Moreover, Table 1 shows the number of datasets where each classifier outperforms the other classifier per minimal support threshold. We can present the following conclusions from the analysis of these figures and tables:

– Results obtained by both classifiers are similar. We can see how the classifier accuracy degrades when the minimal support threshold increases. Notice that the highest results are achieved when the minimal support is not used.

Table 1. Number of datasets where the result of each classifier is higher than the results of the other classifier

Support	BCEP	CAEP
0	50	43
0.02	48	45
0.06	46	47
0.10	53	40
0.15	58	35
0.18	56	36
0.20	58	34
0.25	61	30
0.30	61	28

- CAEP is more sensitive to the threshold value than BCEP since its results are degraded faster. It is clear than the performance of BCEP is always better than CAEP, and the difference grows when the minimal support threshold increases.

Table 2. Number of times where the results of a classifier increases/decreases using one threshold with respect to the other threshold

	BCEP		CAEP	
	Increases	Decreases	Increases	Decreases
$acc(0) \implies acc(0.02)$	0	4	1	5
$acc(0.02) \implies acc(0.06)$	5	13	5	23
$acc(0.06) \implies acc(0.1)$	2	15	5	31
$acc(0.1) \implies acc(0.15)$	1	15	5	38
$acc(0.15) \implies acc(0.18)$	3	18	7	31
$acc(0.18) \implies acc(0.2)$	0	11	4	15
$acc(0.2) \implies acc(0.25)$	3	18	3	31
$acc(0.25) \implies acc(0.3)$	2	15	7	22

Table 2 presents a different view of the results, where each row contains a sum up of the differences between using two consecutive thresholds. Differences are also split by those where the accuracy increases or decreases, considering as similar results closer than 0.01. These results show how the increase of the minimal support usually has a negative impact in the classification accuracy. This behavior is similar for both classifiers, but it is more perceptible in the CAEP classifier.

3.2 Jumping Emerging Patterns Restriction

In this section, we evaluate the use of the jumping restriction: a pattern is important if it is supported by a single class. To do so, we introduce different noise levels to the database (0.05, 0.08, 0.12, 0.15, 0.18, and 0.22) since it has been reported that the noise could deteriorate the results when this restriction is used. To introduce noise, a random object subset randomly change their classes.

Table 3. Evaluation of the jumping restriction for CAEP classifier

Noise ratio	Jumping patterns best						All patterns best					
	0.05	0.08	0.12	0.15	0.18	0.22	0.05	0.08	0.12	0.15	0.18	0.22
All	0.45	0.43	0.35	0.35	0.39	0.34	0.43	0.41	0.51	0.55	0.46	0.55
Above avg.	0.31	0.29	0.21	0.21	0.27	0.27	0.56	0.48	0.65	0.67	0.56	0.63
Below avg.	0.60	0.58	0.51	0.51	0.51	0.42	0.29	0.33	0.36	0.42	0.36	0.47

Table 4. Evaluation of the jumping restriction for BCEP classifier

Noise ratio	Jumping patterns best						All patterns best					
	0.05	0.08	0.12	0.15	0.18	0.22	0.05	0.08	0.12	0.15	0.18	0.22
All	0.35	0.31	0.26	0.31	0.30	0.33	0.43	0.41	0.52	0.47	0.49	0.49
Above avg.	0.24	0.16	0.20	0.22	0.24	0.31	0.53	0.53	0.61	0.57	0.63	0.57
Below avg.	0.50	0.50	0.33	0.43	0.38	0.36	0.31	0.26	0.40	0.36	0.33	0.40

Tables 3 and 4 show the results for CAEP and BCEP classifiers, respectively. These tables present the ratio of databases where the patterns mined with the jumping restriction are better than all the patterns and vice-versa. We consider a result better if the accuracy difference is greater than 0.01. Each column present the results obtained by each noise ratio. The first table row presents the results for all the databases, while the second and third rows present the results for those databases where the accuracy is above or below the average accuracy, respectively. The analysis of these tables lead to the following conclusions:

- Considering all databases, increasing of the noise level degrades the results of patterns mined with jumping restriction, as we expected.
- For the databases where the classifier performs better, those having the accuracy level above the average, the differences between using and not using the jumping restriction are more significant: almost twice of the databases performs better without using the jumping restriction. Actually, we do not have an explanation of this behavior so it deserves a deeper study.
- Results for both classifiers are similar.

Table 5. Average difference of accuracies between the original pattern collection and the patterns with different noise levels

Noise levels	0.05	0.08	0.12	0.15	0.18	0.22
CAEP, Jumping	0.06	0.06	0.08	0.12	0.14	0.18
CAEP, All EPs	0.04	0.05	0.07	0.09	0.11	0.14
BCEP, Jumping	0.06	0.05	0.07	0.09	0.12	0.15
BCEP, All EPs	0.02	0.02	0.04	0.05	0.08	0.12

Table 5 shows the average difference between the accuracies obtained by the classifiers when they use the patterns mined from the datasets with different noise levels and the patterns obtained from the original dataset (called original pattern collection). Each row represents a combination of classifier and the use or not of the jumping restriction. We can highlight the following facts:

- The average difference increases as expected with the noise level, showing that the results deteriorate with noise.
- For both classifiers, the difference is always larger when the jumping restriction is used. This confirms the sensibility of using this restriction in noisy databases.
- The accuracy drop is always lower in the BCEP classifier, so it is a superior classifier to deal with noise.

3.3 Minimal Patterns Restriction

In this section, we present an experimental evaluation of the minimal pattern restriction. It consists on keeping only minimal patterns from those fulfilling the other restrictions. Since restrictions in Intuition 3 cannot be used in isolation, we added two additional restrictions: the support difference greater than cero ($p(PC) - p(P\neg C) > 0$), and the minimal support restriction with values 0, 0.02, 0.06, 0.1, 0.15, 0.18, and 0.2. We compare three pattern sets: all the patterns (all), only the minimal patterns (minimal) and all but minimal patterns (non minimal). For all comparisons, we assume a result is better to other if the accuracy difference is greater than 0.01.

Table 6 presents the pairwise comparison of the behavior of classifiers using each evaluated pattern sets. Notice that using all the patterns and non-minimal patterns lead to very similar for thresholds below or equal to 0.1, but for higher values BCEP obtains better results. For minimal patterns, BCEP is clearly superior for all thresholds.

Tables 7 and 8 present a pairwise comparison of the evaluated sets using the BCEP and CAEP classifiers, respectively. Table 7 shows that for the BCEP classifier the results obtained using all and minimal patterns are similar because the differences are always below of the 10% of databases. Nevertheless, as can be seen in Table 8, the results obtained by the CAEP classifier show that using

Table 6. Ratio of databases where each classifier wins with respect to the other classifier

Minimal support	0	0.02	0.06	0.1	0.15	0.18	0.2
CAEP all	0.41	0.40	0.35	0.33	0.30	0.29	0.30
BCEP all	0.41	0.40	0.33	0.31	0.37	0.37	0.37
CAEP minimal	0.30	0.31	0.31	0.27	0.26	0.26	0.27
BCEP minimal	0.43	0.43	0.39	0.33	0.33	0.34	0.33
CAEP non-minimal	0.31	0.32	0.28	0.27	0.26	0.25	0.23
BCEP non-minimal	0.28	0.29	0.24	0.28	0.32	0.31	0.31

Table 7. Ratio of databases where each method wins with respect to the other paired method using the BCEP classifier

Minimal support	0	0.02	0.06	0.1	0.15	0.18	0.2
All Patterns Win	0.06	0.06	0.06	0.08	0.06	0.05	0.04
Minimal Patterns Win	0.10	0.10	0.09	0.08	0.06	0.06	0.06
All Patterns Win	0.41	0.41	0.40	0.49	0.53	0.56	0.56
Non-minimal Win	0.40	0.35	0.33	0.30	0.34	0.34	0.35
Minimal Patterns Win	0.42	0.43	0.43	0.51	0.53	0.57	0.57
Non-minimal Win	0.42	0.38	0.37	0.31	0.34	0.35	0.37

Table 8. Ratio of databases where each method wins with respect to the other paired method using the CAEP classifier

Minimal support	0	0.02	0.06	0.1	0.15	0.18	0.2
All Patterns Win	0.38	0.39	0.30	0.31	0.25	0.25	0.25
Minimal Patterns Win	0.25	0.27	0.24	0.24	0.19	0.16	0.19
All Patterns Win	0.47	0.47	0.47	0.59	0.62	0.61	0.63
Non-minimal Win	0.41	0.39	0.37	0.37	0.37	0.40	0.39
Minimal Patterns Win	0.46	0.45	0.46	0.56	0.58	0.58	0.61
Non-minimal Win	0.42	0.40	0.38	0.40	0.43	0.43	0.43

all patterns seems to be slightly more accurate than using only the minimal patterns.

On the other hand, when the non-minimal patterns are considered, there are more difference between the results analyzed. Notice that the results obtained by the non-minimal set wins in more than 30% of the databases for all support thresholds. This is, as far as we know, the first time that it is experimentally shown that minimal patterns can significantly deteriorate the pattern set quality in many databases.

4 Conclusions and Future Work

In this paper, we present an experimental study to evaluate some of the most used restrictions in pattern based classifiers: the minimal support threshold, the jumping pattern, and the minimal pattern. To obtain more general results, we used two different pattern based classifiers, CAEP and BCEP. The results attained over the 93 real-world datasets have shown how BCEP classifiers are equal or better than CAEP for most the evaluated tasks.

The minimal support restriction degrades the quality of mined patterns in most tested datasets. Then, it should be only used in cases where limiting the amount of patterns or lowering mining time is required. To use the jumping restriction makes the pattern set sensitive to noise in the database. The sensitivity increases in databases where the classifiers perform better.

When using an accurate classifier like BCEP, there is almost no difference between using or not using the minimal restriction. With a simpler classifier like CAEP, results are different, finding better results when the minimal restriction is not used.

One of the found results appears to be specially interesting: there are many databases where removing minimal patterns significantly improves the quality of the pattern collection. As far as we know, this is the first paper where this result is shown. This result could help to build in the future more accurate pattern based classifiers.

This paper has some limitations that should be addressed in future studies. First, only three restrictions are experimentally evaluated from all that are reported. Second, only two classifiers are used, which limits the ability to extract more general conclusions. Finally, novel results found deserve to be explored in more depth.

References

1. Bache, K., Lichman, M.: UCI Machine Learning Repository (2013). http://archive.ics.uci.edu/ml
2. Bay, S.D., Pazzani, M.J.: Detecting change in categorical data: mining contrast sets. In: Proceedings of the Fifth ACM SIGKDD International Conference on Knowledge Discovery and Data Mining, KDD 1999, pp. 302–306. ACM, New York (1999)
3. Dong, G., Bailey, J.: Contrast Data Mining. Concepts, Algorithms, and Applications. Taylor & Francis, Abingdon (2013)
4. Dong, G., Li, J.: Efficient mining of emerging patterns: discovering trends and differences. In: Proceedings of the Fifth ACM SIGKDD International Conference on Knowledge Discovery and Data Mining, KDD 1999, pp. 43–52. ACM, New York (1999)
5. Dong, G., Zhang, X., Wong, L., Li, J.: CAEP: classification by aggregating emerging patterns. In: Arikawa, S., Furukawa, K. (eds.) DS 1999. LNCS (LNAI), vol. 1721, pp. 30–42. Springer, Heidelberg (1999). https://doi.org/10.1007/3-540-46846-3_4

6. Fan, H., Ramamohanarao, K.: A Bayesian approach to use emerging patterns for classification (2003)
7. García-Borroto, M.: A restriction-based approach to generalizations. In: Hernández Heredia, Y., Milián Núñez, V., Ruiz Shulcloper, J. (eds.) IWAIPR 2018. LNCS, vol. 11047, pp. 239–246. Springer, Cham (2018). https://doi.org/10.1007/978-3-030-01132-1_27
8. García-Vico, A., Carmona, C., Martín, D., García-Borroto, M., del Jesus, M.: An overview of emerging pattern mining in supervised descriptive rule discovery: taxonomy, empirical study, trends, and prospects. Wiley Interdiscip. Rev.: Data Mining Knowl. Discov. (2017). https://doi.org/10.1002/widm.1231
9. Liu, X., Wu, J., Gu, F., Wang, J., He, Z.: Discriminative pattern mining and its applications in bioinformatics. Brief. Bioinform. 16(16), 884–900 (2015)
10. Mitchell, T.M.: Generalization as search. Artif. Intell. 18(1982), 203–226 (1982)
11. Novak, P.K., Lavrač, N., Webb, G.I.: Supervised descriptive rule discovery: a unifying survey of contrast set, emerging pattern and subgroup mining. J. Mach. Learn. Res. 10, 377–403 (2009)
12. Wrobel, S.: An algorithm for multi-relational discovery of subgroups. In: Komorowski, J., Zytkow, J. (eds.) PKDD 1997. LNCS, vol. 1263, pp. 78–87. Springer, Heidelberg (1997). https://doi.org/10.1007/3-540-63223-9_108

BRIEF-Based Mid-Level Representations for Time Series Classification

Renato Souza[1], Raquel Almeida[1], Roberto Miranda[1],
Zenilton Kleber G. do Patrocinio Jr.[1], Simon Malinowski[2],
and Silvio Jamil F. Guimarães[1(✉)]

[1] Computer Science Department, Pontifical Catholic University of Minas Gerais,
Belo Horizonte, MG, Brazil
sjamil@pucminas.br
[2] Université de Rennes 1, IRISA, Rennes, France

Abstract. Time series classification has been widely explored over the last years. Amongst the best approaches for that task, many are based on the Bag-of-Words framework, in which time series are transformed into a histogram of word occurrences. These words represent quantized features that are extracted beforehand. In this paper, we aim to evaluate the use of accurate mid-level representation called BossaNova in order to enhance the Bag-of-Words representation and to propose a new binary time series descriptor, called BRIEF-based descriptor. More precisely, this kind of representation enables to reduce the loss induced by feature quantization. Experiments show that this representation in conjunction to BRIEF-based descriptor is statistically equivalent to traditional Bag-of-Words, in terms time series classification accuracy, being about 4 times faster. Furthermore, it is very competitive when compared to the state-of-the-art.

Keywords: Time series · Mid-level representations · BRIEF-based descriptors

1 Introduction

Time series can be seen as series of ordered measurements. They contain temporal information that needs to be taken into account when dealing with such data. Time series classification (TSC) could be defined as follows: given a collection of unlabeled time series, one should assign each time series to one of a predefined set of classes. TSC is a challenge that is receiving more and more attention recently due to its most diverse applications in real life problems involving, for example, data mining, statistics, machine learning and image processing.

An extensive comparison of TSC approaches is performed in [3] and an evaluation of mid-level representations in TSC is given by [1]. Three particular methods stand out from other core classifiers for their accuracies: COTE [4], BOSS [13] and D-VLAD [1]. According to [4], COTE contains classifiers constructed in the

© Springer Nature Switzerland AG 2019
I. Nyström et al. (Eds.): CIARP 2019, LNCS 11896, pp. 449–457, 2019.
https://doi.org/10.1007/978-3-030-33904-3_42

time, frequency, change, and shapelet transformation domains combined in alternative ensemble structures. BOSS is a dictionary-based approach that adopts an extraction of Fourier coefficients from time series windows, and D-VLAD is also a dictionary-based in which SIFT-based descriptors are assembled by the mid-level presentation called VLAD [11]. Many other dictionary-based approaches have been proposed and used recently [1,5,6]. These methods share the same overall steps: (i) extraction of feature vectors from time series; (ii) creation of a codebook (composed of codewords) from extracted feature vectors; and (iii) representation of time series using extracted codewords.

The first and third steps are very important to design an accurate TSC scheme. In this paper, we propose a dictionary-based approach for TSC, that follows these steps. For the feature vector extraction, we propose to adapt a binary descriptor that was designed for image description, which is called BRIEF [9]. This descriptor has the main advantage to be fast to compute, while providing an accurate description. For the representation step, it has been shown in [1], that the classical Bag-of-Word representation (in which codewords are quantized) could be enhanced by using more discriminative methods. In this paper, we make use of BossaNova [2], which is a method that keeps more information than a traditional Bag-of-Word approach.

The main contributions of this paper are hence two-folds: (i) adaptation of a binary descriptor for describing time series, called BRIEF, and (ii) the use of BossaNova in order to enrich the final time series representation.

This paper is organized as follows. Section 2 describes some related works about time series classification. In Sect. 3, we present a methodology for time series classification by using powerful mid-level representation built on BRIEF-based descriptors. Section 4 details the experimental setup and results to validate the method, and finally, some conclusions are drawn in Sect. 5.

2   Related Work

In this section, we give an overview about the related work on TSC. One of the earliest methods for that task is the combination of 1-nearest-neighbor classifier with Dynamic Time Warping. It has been a baseline for TSC for many years thanks to its good performance. Recently, more sophisticated approaches have been designed for TSC.

Shapelets, for instance, were introduced in [14]. They represent existing subsequences able to discriminate classes. Hills et al. proposed the shapelet transform [10], which consists in transforming a time series into a vector whose components represent the distance between time series and different shapelets, extracted beforehand. Classifiers, such as SVM, can then be used with these vectorial representations of time series.

Numerous approaches have been designed based on the Bag-of-Word (BoW) framework. This framework consists in extracting feature vectors from time series, creating a dictionary of words using these extracted features, and then representing each time series as a histogram of words occurrence. The different approaches proposed in the literature differ mainly on the kind of features that are extracted. Local features such as mean, variance and extrema are considered in [6], Fourier coefficients in [13]; while SAX coefficients are used in [12]. Recently, SIFT-based descriptors adapted to time series have been considered as feature vectors in [5].

All the methods based on the BoW framework create a dictionary of words by quantizing the set (or a subset) of extracted features. This quantization step induces some loss when representing time series as a histogram of words occurrence. In [1], in order to improve the accuracy of time series representations, the authors have studied several more discriminative mid-level representations, such as VLAD for instance.

3   Time Series Classification Based on BRIEF Descriptor

In this section, we describe the proposed TSC scheme. We aim at improving classical BoW representation for time series in two ways: we make use of a binary descriptor adapted from BRIEF, that is very fast to compute; and we use BossaNova in order to enrich the final representation of time series. The use of BRIEF is motivated by the study made in [8] in which a comparison of different low-level descriptors to classify pornography videos presented competitive results taking much less time.

Due to the low complexity, binary descriptors is mostly used in real-time applications due to the simplicity of computational procedure not only for the descriptor itself but also for its similarity measure. The basic idea of binary descriptors is to encode some information of a path into a binary sequence, by comparing the intensity of the points present in that path. In the case of BRIEF, there is neither sampling pattern nor orientation compensation.

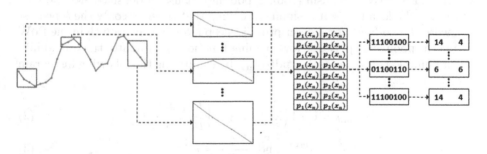

Fig. 1. Firstly, using a time series S, it's created windows according to the size s and the distance between the key points. With the windows and selected n pairs randomly. The binary sequence constructed are concatenate into bit-string. Then, the binary strings are converted to a integer-string where a integer is computed for each k bits.

The proposed approach is composed of the following steps: (i) keypoints selection; (ii) keypoints description; (iii) generation of final mid-level representation of time series; and (iv) classification. These steps are detailed below, and first two are illustrated in Fig. 1. In the following, let $X = x_1, \dots, x_n$ be a time series of length n.

Keypoints Selection: We start by selecting the keypoints that will be described at the next step. Dense selection of keypoints have shown to be more efficient than other methods. We hence select keypoints regularly inside the time series: one keypoint is selected every time interval of τ instants, in which τ is a parameter of the method. At the end of this step, the set $\{x_1, x_{1+\tau}, \dots\}$ of keypoints is selected.

Keypoints Description: Let x_k be the keypoint that we want to describe. For that purpose, the window $W = s_1, \dots, s_w$ of length w is selected around x_k. Then p pairs of numbers $(i_1, i_2) \in [1, w]^2, i_1 \neq i_2$, are then randomly selected. Note that the same pairs are kept to describe each keypoints. For each pair (i_1, i_2), a binary number $b_{(i_1, i_2)}$ is computed as follows:

$$b_{(i_1, i_2)} = \begin{cases} 1 \text{ , if } s_{i_1} < s_{i_2}; \\ 0 \text{ , otherwise.} \end{cases} \tag{1}$$

When all pairs have been processed, a binary vector of length p is generated and represents the description (feature vector) of the keypoint x_k.

Mid-Level Representation: Let $\mathbb{X} = \left\{ \mathbf{x}_j \in \mathbb{R}^d \right\}_{j=1}^N$ be an unordered set of d-dimensional descriptors $\mathbf{x}_j$ extracted from the data. Let also $\mathbb{C} = \{ \mathbf{c}_m \in \mathbb{R}^d \}_{m=1}^M$ be the codebook learned by an unsupervised clustering algorithm, composed by a set of M codewords, also called prototypes or representatives. Consider $\mathbb{Z} \in \mathbb{R}^M$ as the final vector mid-level representation. As formalized in [7], the mapping from $\mathbb{X}$ to $\mathbb{Z}$ can be decomposed into three successive steps: (i) coding; (ii) pooling; and (iii) concatenation. In order to keep more information than BoW during pooling step, we have used BossaNova [2] as mid-level representation which follows BoW formalism (coding/pooling). It uses a density-based pooling strategy and a localized soft-assignment coding that considers only the k-nearest codewords for coding a local descriptor. To keep more information than the BoW during the pooling step, BossaNova pooling function, g, estimates the probability density function of α_m: $g(\alpha_m) = \text{pdf}(\alpha_\mathbf{m})$, by computing the following histogram of distances $z_{m,b}$:

$$z_{m,b} = \text{card}\left(\mathbf{x}_j \mid \alpha_{m,j} \in \left[\frac{b}{B}; \frac{b+1}{B} \right] \right), \tag{2}$$

$$\frac{b}{B} \geq \alpha_m^{min} \text{ and } \frac{b+1}{B} \leq \alpha_m^{max}, \tag{3}$$

in which:

- B denotes the number of bins of each histogram z_m;
- $\alpha_{m,j}$ represents a dissimilarity measure between codewords and feature points x_j; and
- $\alpha_m^{min}, \alpha_m^{max}$ limits the range of distances for the descriptors considered in the histogram computation.

The final BossaNova representation is in the form:

$$\mathbf{z} = [[z_{m,b}], t_m]^{\mathrm{T}}, \quad (m, b) \in \{1, \dots, M\} \times \{1, \dots, B\}, \tag{4}$$

in which t_m scalar value for each codeword as an approximation of the traditional BoW representation.

Classification: The mid-level representation for each time series is then passed to a classifier to learn how to discriminate classes using this description.

4   Experimental Analysis

In this section, we describe our experiments in order to investigate the impact, in terms of classification performances, of more powerful encoding methods applied to dense extracted features for TSC.

4.1   Experimental Setup

Experiments are conducted on the 84 currently available datasets from the UCR repository, the largest on-line database for time series classification. Due to problems on feature extraction, we ignored the 12 largest datasets. All datasets are splitted into training and test sets, whose sizes vary between less than 20 and more than 8,000 time series. For a given dataset, all time series have the same length, ranging from 24 to more than 2,500 points. In order to compute the mid-level representation, we have extracted SIFT-based descriptors as proposed in [5] and BRIEF-based descriptors by using dense sampling. Codebooks have been generated using the following number of clusters $\{16, 32, 64, 128, 256, 512\}$. The sizes of the window and the numbers of pairs of the BRIEF-based descriptor are selected from $\{16, 32, 64, 128, 256, 512, 1024\}$, respectively. The representations are normalized by a L2-norm and signed square root. Moreover, we have used d-BRIEF and d-SIFT for indicating the step used in the keypoint extraction (sampling rate); for example, 1-BRIEF means that all points of the time series are extracted. The best sets of parameters are obtained by a 5-fold cross-validation to be used with a RBF kernel SVM model during classification step.

4.2   Quantitative Analysis

Despite the interest in classifying time series with high accuracy, it is very important to propose methods with low computational time. In that sense, feature extraction for TSC using BRIEF-based descriptor presents very competitive results when compared to SIFT-based descriptor. Depending on the sampling rate, the speed-up is 3.64 and 5.91 times, for 1-BRIEF and 2-BRIEF, respectively, as can be seen in Table 1.

In Fig. 2, we present the accuracy results for compared methods. As we can see, the 1-VLAD (SIFT-based dense sampling and VLAD representation) when compared to COTE presents very competitive results (when COTE and 1-VLAD

Table 1. Average time for describing the datasets using BRIEF-based and SIFT-based descriptors. Here, 2-BRIEF means that a BRIEF descriptor is computed for each 2 keypoints and 1-BRIEF (1-SIFT) for all points of the time series.

Time (in ms)	1-BRIEF	2-BRIEF	1-SIFT
Total	1,012,776.00	623,836.00	3,691,278.00
Average per dataset	14,677.91	9,041.10	53,496.78
Speed-up	3.64	5.91	—

are computed by using the same experiment protocol). Regarding 1-SIFT, the results are competitive to BOSS. When 1-BRIEF is compared to BoTSW [5] and 1-SIFT, the proposed binary descriptor to time series have presented very competitive results in terms of accuracy (as illustrated in Fig. 2) but it is about 4 times faster for keypoint description. Furthermore, they are statistically equivalent when the results are compared by using a t-student test with 95% of confidence level.

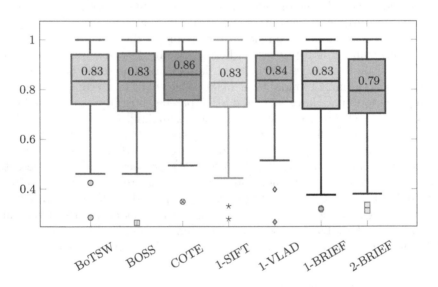

Fig. 2. Accuracy comparison of different methods applied on TSC.

We also have compared the performance of the methods by using a pair-wise distribution, as illustrated in Fig. 3. Taking into account these distribution of points, it is possible to argue that the BRIEF-based and SIFT-based descriptors present similar results.

4.3   Comparison to the State-of-the-art Methods

In order to study the impact of specially designed mid-level representations on TSC. We focus on two different analysis. In the first one, we compare the studied representations, namely d-BRIEF and 1-SIFT to BoTSW and 1-VLAD, which are our baselines. In the second one, we present a comparative analysis between the state-of-the-art, namely BoTSW [5], BOSS [13] and COTE [4], and the new proposed use of mid-level description.

In both cases, we used the average accuracy rate and rank that are summarized in Table 2. As illustrated, 1-VLAD obtained the best results among the mid-representation, and it is very competitive to the state-of-the-art, being better than BoTSW and BOSS. When compared to our baselines BoTSW and 1-VLAD, the 1-BRIEF is statistically equivalent taking into account the paired t-test with 95% of confidence level. 1-BRIEF is statistically better than 2-BRIEF, and equivalent to 1-SIFT. Furthermore, concerning the comparison of 1-BRIEF to BoTSW and 1-SIFT, we have observed: (i) the binary descriptor in conjunction to BossaNova presented competitive performances but it is faster, which

Table 2. Comparison to the state-of-the-art in terms of classification rates and ranking.

Method	State-of-the-art				Mid-level representations		
	BoTSW	BOSS	COTE	1-VLAD	1-SIFT	2-BRIEF	1-BRIEF
Rate	0.819	0.809	0.836	0.827	0.807	0.780	0.813
Ranking	3.78	4.21	3.04	3.27	3.55	4.71	4.01

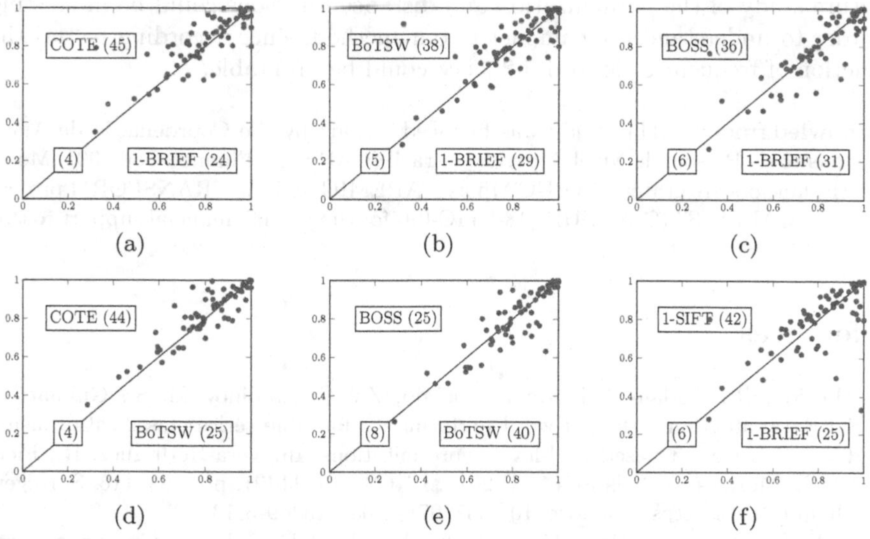

(a)   (b)   (c)

(d)   (e)   (f)

Fig. 3. Pairwise comparison of classification rates between BRIEF-based mid-level representations and the state-of-the-art.

confirms our initial assumptions; and (ii) 1-BRIEF presented good results in terms of classification rates and average rank but it is worse than 1-SIFT and better than BOSS.

5   Conclusions

Time series classification is a challenge task due to its most diverse applications in real life. Among the several kind of approaches, dictionary-based ones have received much attention in the last years. In a general way, these ones are based on the extraction of feature vectors from time series, creation of codebook from the extracted feature vectors and finally representation of time series as traditional Bag-of-Words. In this work, we studied the impact of more discriminative and accurate mid-level representation, called BossaNova, for describing the time series taking into account SIFT-based descriptors [5] and the proposed binary descriptor for time series, called BRIEF-based descriptor.

According to our experiments, 1-BRIEF is statistically equivalent to 1-SIFT and BoTSW but the binary keypoint description is about 4 times faster than the non-binary one. Moreover, we achieve competitive results when compared to the some state-of-the-art methods, mainly with BOSS and 1-VLAD. However, despite the pairwise comparison (Fig. 3) involving both methods, COTE is slightly better than 1-VLAD in terms of average rank but both are statistically equivalent when the comparison is done by the paired t-test with 95% of confidence. Thus, the use of more accurate mid-level representation in conjunction with BRIEF-based descriptor seems to be a very interesting approach to cope with time series classification. From our results and observations, we believe that a future study of the normalization and distance functions could be interesting in order to understand their impact in our method, since according to [11] the reduction of frequent codeword influence could be profitable.

Acknowledgments. This study was financed in part by the Coordenação de Aperfeiçoamento de Pessoal de Nível Superior – Brasil (CAPES) – Finance Code 001. Moreover, the authors are grateful to PUC Minas, FAPEMIG and the TRANSFORM project funded by CAPES/STIC-AMSUD (18-STIC-09) for the partial financial support to this work.

References

1. Almeida, R., Herlanin, H., do Patrocinio, Z.K.G., Malinowski, S., Guimarães, S.J.F.: Evaluation of bag-of-word performance for time series classification using discriminative sift-based mid-level representations. In: Vera-Rodriguez, R., Fierrez, J., Morales, A. (eds.) CIARP 2018. LNCS, vol. 11401, pp. 109–116. Springer, Cham (2019). https://doi.org/10.1007/978-3-030-13469-3_13
2. Avila, S., Thome, N., Cord, M., Valle, E., AraúJo, A.D.A.: Pooling in image representation: the visual codeword point of view. Comput. Vis. Image Underst. **117**(5), 453–465 (2013)

3. Bagnall, A., Lines, J., Bostrom, A., Large, J., Keogh, E.: The great time series classification bake off: a review and experimental evaluation of recent algorithmic advances. Data Mining Knowl. Discov. **31**(3), 606–660 (2017)
4. Bagnall, A., Lines, J., Hills, J., Bostrom, A.: Time-series classification with cote: the collective of transformation-based ensembles. IEEE Trans. Knowl. Data Eng. **27**(9), 2522–2535 (2015)
5. Bailly, A., Malinowski, S., Tavenard, R., Chapel, L., Guyet, T.: Dense bag-of-temporal-SIFT-words for time series classification. In: Douzal-Chouakria, A., Vilar, J.A., Marteau, P.-F. (eds.) AALTD 2015. LNCS (LNAI), vol. 9785, pp. 17–30. Springer, Cham (2016). https://doi.org/10.1007/978-3-319-44412-3_2
6. Baydogan, M.G., Runger, G., Tuv, E.: A bag-of-features framework to classify time series. IEEE PAMI **35**(11), 2796–2802 (2013)
7. Boureau, Y.L., Bach, F., LeCun, Y., Ponce, J.: Learning mid-level features for recognition. In: Proceedings of the CVPR 2010, pp. 2559–2566 (2010)
8. Caetano, C., Avila, S., Guimaraes, S., Araújo, A.D.A.: Pornography detection using Bossanova video descriptor. In: Proceedings of the EUSIPCO 2014, pp. 1681–1685. IEEE, Lisbon (2014)
9. Calonder, M., Lepetit, V., Strecha, C., Fua, P.: BRIEF: binary robust independent elementary features. In: Daniilidis, K., Maragos, P., Paragios, N. (eds.) ECCV 2010. LNCS, vol. 6314, pp. 778–792. Springer, Heidelberg (2010). https://doi.org/10.1007/978-3-642-15561-1_56
10. Hills, J., Lines, J., Baranauskas, E., Mapp, J., Bagnall, A.: Classification of time series by shapelet transformation. Data Mining Knowl. Discov. **28**(4), 851–881 (2014)
11. Jegou, H., Perronnin, F., Douze, M., Sánchez, J., Perez, P., Schmid, C.: Aggregating local image descriptors into compact codes. IEEE PAMI **34**(9), 1704–1716 (2012)
12. Lin, J., Khade, R., Li, Y.: Rotation-invariant similarity in time series using bag-of-patterns representation. J. Intell. Inf. Syst. **39**(2), 287–315 (2012)
13. Schäfer, P.: The BOSS is concerned with time series classification in the presence of noise. Data Mining Knowl. Discov. **29**(6), 1505–1530 (2014)
14. Ye, L., Keogh, E.: Time series shapelets: a new primitive for data mining. In: Proceedings of the 15th ACM SIGKDD International Conference on Knowledge Discovery and Data Mining, pp. 947–956. ACM (2009)

Continuous Hyper-parameter Configuration for Particle Swarm Optimization via Auto-tuning

Jairo Rojas-Delgado[1] , Vladimir Milián Núñez[1(✉)] ,
Rafael Trujillo-Rasúa[1] , and Rafael Bello[2]

[1] Universidad de las Ciencias Informáticas, Havana, Cuba
{jrdelgado,vmilian,trujillo}@uci.cu
[2] Universidad Central "Marta Abreu" de Las Villas, Santa Clara, Cuba
rbellop@uclv.edu.cu

Abstract. Hyper-Parameter configuration is a relatively novel field of paramount importance in machine learning and optimization. Hyper-parameters refers to the parameters that control the behavior of algorithms and are not tuned directly by such algorithms. For hyper-parameters of an optimization algorithm such as Particle Swarm Optimization, hyper-parameter configuration is a nested optimization problem. Usually, practitioners needs to use a second optimization algorithm such as grid search or random search to find proper hyper-parameters. However, this approach forces practitioners to know about two different algorithms. Moreover, hyper-parameter configuration algorithms also have hyper-parameters that need to be considered. In this work we use Particle Swarm Optimization to configure its own hyper-parameters. Results show that hyper-parameters configured by PSO are competitive with hyper-parameters found by other hyper-parameter configuration algorithms.

Keywords: Hyper-parameter · Optimization · Meta-heuristic · Particle Swarm Optimization

1 Introduction

Search is a core concept of Artificial Intelligence and Pattern Recognition. In this field, meta-heuristic algorithms are critical as they do not require any prior on the fitness function, and allows to explore large search spaces. This ability is extensively used for model fitting and several important tasks in Artificial Intelligence. However, meta-heuristics are very dependent on their parameterization and often, require experts to determine which hyper-parameters[1] to modify and how should they be tuned, meaning that for a non-expert it might be hard to find good settings.

---

[1] Hyper-parameters refers to the parameters that control the behavior of the meta-heuristic algorithm and are not tuned directly by such algorithm.

© Springer Nature Switzerland AG 2019
I. Nyström et al. (Eds.): CIARP 2019, LNCS 11896, pp. 458–468, 2019.
https://doi.org/10.1007/978-3-030-33904-3_43

For meta-heuristics, hyper-parameters appear frequently and have a direct impact in the convergence speed and accuracy of the algorithms [7]. Also, hyper-parameters affect the execution time and memory cost of running the algorithm, the quality of the model resulting from the training process, or its ability to generalize to the unseen data. For these reasons, finding the best possible hyper-parameters becomes crucial [13].

Hyper-parameter Optimization (HPO), is a field of paramount importance in machine learning and optimization [5,17]. HPO basically stands for using optimization algorithms in order to perform automatic hyper-parameter config-uration. HPO is a difficult optimization problem. Until recently, HPO in several areas, such as machine learning, was considered a combination of science and art due to the high computational cost of such task [1].

There are two main approaches to HPO: manual or automatic methods. Manually approach assumes that there exists an understanding of how the hyper-parameters affect the algorithm. On the other hand, automatic approach greatly reduce the need for this understanding, but they come at the expense of the costlier computation.

In the latest years there has been an increase in the efforts to address auto-matic HPO with very good results in contrast with manually choosing the hyper-parameters [3,14]. In the literature, among the most popular algorithms are Ran-dom Search (RS) [2], Sequential Model-based Algorithm Configuration (SMAC) [8] and Tree Parzen Estimators (TPE) [3]. HPO involves two nested cycles of optimization when configuring hyper-parameters for an arbitrary meta-heuristic algorithm. Here we refer to the regular optimization algorithm simply as base-algorithm. We call parent-algorithm to the optimization algorithm that deals with choosing the hyper-parameters of the base-algorithm.

When performing HPO, the base-algorithm may need to deal with thousands of dimensions and expensive fitness functions. For example, training a neural net-work is an NP-hard optimization problem [12] with tens of thousands of dimen-sions. Setting the hyper-parameters for such training algorithms is a challenge for practitioners because standard recommendations in the literature are most of the time useless. Trial and error is very tedious, is not reproducible for others and is prone to produce over-fitting. Additionally, insights on the hyper-surface structure usually are not available due to the large amount of dimensions, hence common sense may not be applicable.

When sufficient computational resources are provided, practitioners may choose to use a parent-algorithm (such as RS, SMAC or TPE) to perform HPO. There are several examples of successful applications of such parent-algorithms [3]. However, for the engineering mind this approach faces to major drawbacks:

- the practitioner needs to know the specific details of two (probably different) optimization algorithms and
- parent-algorithms also have hyper-parameters that need to be configured, manually most of the time.

For meta-heuristics, that do not impose restrictions (such as derivatives or convexity) on the fitness function, tuning its own hyper-parameters may be a good approach to perform HPO. In this case, the parent-algorithm hyper-parameters also need to be somehow configured. However, the small number of hyper-parameters, usually less than ten, makes standard recommendations in the literature more suitable. Even if standard recommendations do not produce good results, the practitioner may use its experience about the meta-heuristic and three-dimensional insights to come up with a working set of hyper-parameters for the parent-algorithm.

In this paper, we deal with the problem of automated HPO of Particle Swarm Optimization (PSO) meta-heuristic algorithm by using the same meta-heuristic as parent-algorithm. We tested this approach with several artificial benchmark functions for optimization. Our meta-heuristic HPO of meta-heuristics approach is able to find hyper-parameters that leads to more stable and accurate optimization of the base-algorithm when compared with RS, SMAC and TPE as parent-algorithms. The main limitation of this work is that we only consider continuous hyper-parameters.

In the Sect. 2 we discuss the state of the art on hyper-parameter optimization algorithms. In Sect. 3 we describe PSO algorithm in detail and in Sect. 4 we present the benchmark test functions to be optimized by PSO. Section 5 presents the experimental results when comparing the stability and accuracy of our HPO approach. Finally, some conclusions and recommendations are given. Throughout this article we use x for scalars, w for vectors and X for sets.

2   Automatic Hyper-parameter Optimization

There are at least two kind of methods to perform automatic HPO: model-free and model-based methods. Model-based techniques build a surrogate model of the hyper-parameter space through its careful exploration and exploitation. Alternatively, model-free algorithms do not utilize the knowledge about the solution space extracted during optimization.

2.1   Model-Free Methods

Model-free approaches are characterized by not utilizing the knowledge about the solution space extracted during the optimization. This lack of adaptability makes them simple to implement at the expense of poor results for large hyper-parameter spaces.

RS: is a trivial to implement alternative to Grid Search [2], more convenient to use and faster to converge to an acceptable set of hyper-parameters. There are approaches designed to enhance RS capabilities, for example, the random sampling can be intensified in the neighborhood of the best hyper-parameters. Finally, there are hybrid algorithms to couple RS with other techniques for refining its performance,for example, manual updates provided by an expert [2, 9].

2.2   Model-Based Methods

Model-based methods build a model of the fitness function, and then elaborate hyper-parameter values by performing optimization within this model. Most of such techniques use a Bayesian regression model turning this problem into a trade-off between the exploration and exploitation.

SMAC is a model-based method for optimizing algorithm hyper-parameters [8]. SMAC is effective for HPO of machine learning algorithms, scaling better to high dimensions and discrete input dimensions than other algorithms [11].

TPE method sequentially construct models to approximate the performance of hyper-parameters based on historical measurements, and then subsequently choose new hyper-parameters to test based on this model. This optimization approach is described in detail in [3].

3   Particle Swarm Optimization

PSO is a population based meta-heuristic algorithm [4]. PSO has several hyper-parameters such as: global contribution (α), local contribution (β), maximum speed (v_{max}) and minimum speed (v_{min}). Let q be the number of particles in the population, $w_i \in \mathbb{R}^n$ a particle's position in the search space and $f(w_i)$ the fitness value of the particle.

The speed of a particle in the population is given by Eq. 1 where $w_* \in \mathbb{R}^n$ is the position of the particle with the lowest fitness function value in the population and $w_{i*} \in \mathbb{R}^n$ is the position where w_i achieved its lowest fitness function value. Particle w_* is known as *best-global* particle and w_i^* as *best-so-far* particle of w_i.

$$v_i \leftarrow \gamma v_i + \alpha \omega_1 \odot w_* + \beta \omega_2 \odot w_{i*} \tag{1}$$

The parameter γ is known as speed parameter such as $v_{min} \leq \gamma \leq v_{max}$. The vectors ω_1 and ω_2 are randomly generated by means of an uniform distribution such as $\omega_{1,i} \sim U(0,1)$ and $\omega_{2,i} \sim U(0,1)$. Finally, a particle's position is updated according to Eq. 2:

$$w_i \leftarrow w_i + v_i \tag{2}$$

Due to the diversity of PSO algorithms in the literature, we provide further details on the implementation used in this work as described in Algorithm 1. In step 1 the population is randomly generated. After that, in steps 2 and 3 the best-so-far positions of each particle and the best-global position of the population are set. The algorithm is considered in a convergence state after reaching a given number of fitness function evaluations determined by the value of η in step 5.

For each iteration of PSO algorithm, the best-so-far and best particles are updated in steps 10 and 13. Then, for each particle in the population a new position is calculated in steps 7–8 and the speed parameter is linearly decreased in step 17. Finally, the best particle of the population is returned in step 20.

Algorithm 1. Particle Swarm Optimization Algorithm.

1: Initialize population randomly
2: Set best-so-far particles such as $w_{i*} \leftarrow w_i$
3: Set best-global particle $w_* \leftarrow argmin_{w_i}(f(w_i))$ for $1 \leq i < q$
4: Set speed param value at the maximum speed: $\gamma \leftarrow v_{max}$
5: **while** FITNESS_FUNCTION_EVALUATIONS( ) $< \eta$ **do**
6:     **for** j = 1 : q **do**
7:         Calculate speed of particle j according to Equation 1
8:         Update position of particle j according to Equation 2
9:         **if** $f(w_j) < f(w_{j*})$ **then**
10:            Update $w_{j*} \leftarrow w_j$
11:        **end if**
12:        **if** $f(w_j) < f(w_*)$ **then**
13:            Update $w_* \leftarrow w_j$
14:        **end if**
15:    **end for**
16:    **if** $\gamma > v_{min}$ **then**
17:        Reduce linarly the speed param: $\gamma \leftarrow \gamma \cdot 0.99$
18:    **end if**
19: **end while**
20: **return** Best particle in the population w_*

PSO have at least four continuous hyper-parameters: local contribution, global contribution, maximum speed and minimum speed. In addition, PSO also have several other hyper-parameters of non-continuous nature, e.g. the number of particles in the population, the number of evaluations of the fitness function to archive convergence, the lower and upper bounds of the search space, etc. Here we will only deal with the four continuous hyper-parameters while tuning the others manually. Let $\bar{w}_i \in \mathbb{R}^4$ be a vector that contains the hyper-parameters of PSO algorithm and $\bar{f}(\bar{w}_i)$ the quality of a given set of hyper-parameters, the hyper-parameter configuration problem can be defined as:

$$\bar{w}_* \leftarrow argmin_{\bar{w}_i \in \mathbb{R}^4} \left[\bar{f}(\bar{w}_i) \right] \tag{3}$$

In Eq. 3, $\bar{f}(\bar{w}_i) = f(w_*)$ where w_* is the result of optimizing $f(w_j)$ by means of PSO with hyper-parameters $\bar{w}_i$. Notice here the two nested cycles of optimization: in the outer cycle we try to minimize $\bar{f}(\bar{w}_i)$ while in the inner cycle we try to minimize $f(w)$. While the outer cycle may be solved with GS, TPE or SMAC, here both optimization cycles are solved by PSO.

Several standard recommendations for PSO hyper-parameters are available in the literature. Here we will consider such recommendations for hyper-parameter configuration in order to compare such recommendation with automatic HPO. Some of the PSO recommendations come in the form of a general heuristic, for example, $\alpha + \beta \leq 4$ [16]. Here we will consider the following more specific recommendations:

- From [16], we denote STD.R.1 to the hyper-parameter set: $\alpha = 0.5, \beta = 0.5, v_{min} = 0, v_{max} = 1.2$.
- From [6], we denote STD.R.2 to the hyper-parameter set:, $\alpha = 2, \beta = 2, v_{min} = 0.9, v_{max} = 0.9$.

4 Optimization Benchmark Problems

In this section we describe a set of optimization problems to be solved by base-algorithms. In the literature, several benchmark test functions for optimization have been suggested. We considered five properties regarding to the test functions: continuous/discontinuous, differentiable/non-differentiable, separable/non-separable, scalable/non-scalable and uni-modal/multi-modal [10, 15]. The following list enumerate benchmark test function used in this work and its properties, in all cases consider that $w \in \mathbb{R}^n$.

1. De Jung function:

$$f_1(w) = \sum_{i=0}^{n} w_i^2 \tag{4}$$

has a global minima in $f_1(0, ..., 0) = 0$. De jung is continuous, differentiable, separable, scalable and multi-modal. A commonly used search domain is $0 \le w_i \le 10$.

2. Ackley function:

$$f_2(w) = -20e^{0.02\sqrt{n-1\sum_{i=0}^{n} w_i^2}} - e^{n-1\sum_{i=0}^{n} \cos(2\pi w_i)} + 20 + e \tag{5}$$

has a global minima in $f_2(0, ..., 0) = 0$. Ackley is continuous, differentiable, non-separable, scalable and multi-modal. A commonly used search domain is $-35 \le w_i \le 35$.

3. Grienwangk function:

$$f_3(w) = \sum_{i=0}^{n} w_i^2 / 4000 - \prod_{i=0}^{n} \cos(w_i / \sqrt{i}) + 1 \tag{6}$$

has a global minima in $f_3(0, ..., 0) = 0$. Griendwangk is continuous, differentiable, non-separable, scalable and multi-modal. A commonly used search domain is $-100 \le w_i \le 100$.

4. Rastrigin function:

$$f_4(w) = 10n + \sum_{i=0}^{n} w_i^2 - 10\cos(2\pi w_i) \tag{7}$$

has a global minima in $f_4(0, ..., 0) = 0$. Rastrigin is continuous, differentiable, separable, scalable and multi-modal. A commonly used search domain is $-5.12 \le w_i \le 5.12$.

5. Rosenbrock function:

$$f_5(\boldsymbol{w}) = \sum_{i=0}^{n-1} [100(\boldsymbol{w}_{i+1} - \boldsymbol{w}_i^2)^2 + (\boldsymbol{w}_i - 1)^2] \tag{8}$$

has a global minima in $f_5(1, ..., 1) = 0$. Rosenbrock is continuous, differentiable, non-separable, non-scalable and multi-modal. A commonly used search domain is $-30 \le \boldsymbol{w}_i \le 30$.

6. Schwefel function:

$$f_6(\boldsymbol{w}) = -1/n \sum_{i=0}^{n} \boldsymbol{w}_i \sin \sqrt{|\boldsymbol{w}_i|} \tag{9}$$

has a global minima in $f_6() = -418.983$. Schwefel is continuous, differentiable, separable, scalable and multi-modal. A commonly used search domain is $-500 \le \boldsymbol{w}_i \le 500$.

7. Styblinski-Tang function:

$$f_7(\boldsymbol{w}) = 0.5 * \sum_{i=0}^{n} \boldsymbol{w}_i^4 + 16\boldsymbol{w}_i^2 + 5\boldsymbol{w}_i \tag{10}$$

has a global minima in $f_7(-2.903534, ..., -2.903534) = -78.332$. Styblinski-Tang is continuous, differentiable, non-separable, non-scalable and multi-modal. A commonly used search domain is $-5 \le \boldsymbol{w}_i \le 5$.

8. Step function:

$$f_8(\boldsymbol{w}) = \sum_{i=0}^{n} (\lfloor \boldsymbol{w}_i + 0.5 \rfloor)^2 \tag{11}$$

has a global minima in $f_8(0, ..., 0) = 0$. Step is discontinuous, non-differentiable, separable, scalable and uni-modal. A commonly used search domain is $-100 \le \boldsymbol{w}_i \le 100$.

9. Alpine function:

$$f_9(\boldsymbol{w}) = \sum_{i=0}^{n} |\boldsymbol{w}_i \sin(\boldsymbol{w}_i) + 0.1\boldsymbol{w}_i| \tag{12}$$

has a global minima in $f_9(0, ..., 0) = 0$. Alpine is continuous, non-differentiable, separable, scalable and uni-modal. A commonly used search domain is $10 \le \boldsymbol{w}_i \le 10$.

5 Results and Discussion

This section describes the experimental setup and results of the proposed PSO continuous hyper-parameter optimization approach. We compare quality of PSO hyper-parameters from configured with RS, SMAC, TPE and PSO it self. Moreover, we present a comparison of automatic hyper-parameter optimization with standard recommendation in the literature: STD.R.1 and STD.R.2.

Base PSO was configured with a population size of 40 particles and a search domain based on the literature recommendation (see Sect. 4). For PSO as base-algorithm, we consider convergence when the number of fitness function evaluations is above 4.0E+4. For parent-algorithms (RS, SMAC, TPE and PSO), we consider convergence when number of fitness function evaluations is above 100. For PSO as parent-algorithm the population size is of 10 particles and the search domain is defined in the interval [0, 2]. Hyper-parameters for PSO as parent-algorithm are configured according to STD.R.1.

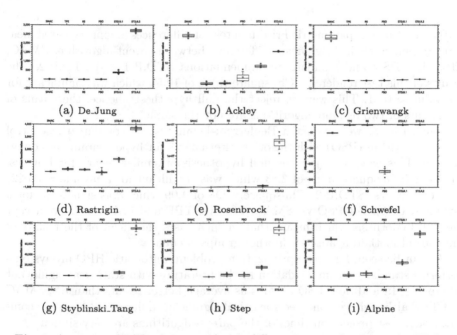

(a) De_Jung   (b) Ackley   (c) Grienwangk

(d) Rastrigin   (e) Rosenbrock   (f) Schwefel

(g) Styblinski_Tang   (h) Step   (i) Alpine

Fig. 1. Accuracy of PSO when considering different parent-algorithms for HPO.

Figure 1 shows the accuracy of PSO when considering different benchmark optimization problems after performing hyper-parameter configuration by means of different parent-algorithms. The bottom and top lines in the box plots represents the first and third quartiles, the line in the middle represents the median of 10 measurements and the whiskers represent standard deviation. The small square represents the average of the measurements and the (x) marks represent the maximum and minimum value.

In Table 1 we a summary of the accuracy of PSO when considering different benchmark optimization problems. The presented accuracy values are the averages of 10 measurements. In addition, we show the global best for each optimization problem.

Table 1. Accuracy of PSO algorithm with hyper-parameters optimized by different parent-algorithms.

Fitness function	BEST	SMAC	TPE	RS	PSO	STD.R.1	STD.R.2
De_jung	0.00E+00	0.00E+00	6.63E−01	2.63E+00	**0.00E+00**	3.96E+02	2.69E+03
Ackley	0.00E+00	2.08E+01	2.56E+00	**1.98E+00**	3.26E+00	8.75E+00	1.45E+01
Grienwangk	0.00E+00	5.40E+01	1.54E−01	**5.56E−02**	7.02E−01	1.12E+00	1.67E+00
Rastrigin	0.00E+00	2.35E+02	5.06E+02	4.10E+02	**1.28E+02**	1.24E+03	3.67E+03
Rosenbrock	0.00E+00	4.28E+04	**8.44E+02**	1.11E+03	1.85E+05	4.43E+05	1.42E+07
Schwefel	−4.18E+02	−5.31E+01	−3.41E+00	−3.35E+00	**−2.42E+02**	−2.04E+00	−1.05E+00
Styblinski_tang	−7.81E+01	−1.08E+01	−3.00E+00	1.57E+01	**−1.45E+01**	5.79E+03	8.92E+04
Step	0.00E+00	4.24E+2	**1.30E+01**	1.16E+02	4.43E+02	5.27E+02	2.61E+03
Alpine	0.00E+00	2.13E+02	2.54E+01	1.82E+01	**5.33E+00**	8.11E+01	2.42E+02

We use the non-parametric Friedman test of differences among repeated measures to find statistical significant differences between parent-algorithms SMAC, TPE, RS, PSO and standard recommendations STD.R.1 and STD.R.2. The results of this test rendered a Chi-square value of 10.54 which was significant for a p-value < 0.01. This way we reject the null hypothesis, hence, the means of the results of two or more algorithms are not the same.

Furthermore, we conduct a Benferroni-Dunn test to compare the control parent-algorithm (PSO) with the other alternatives for hyper-parameter configuration. This way, we reject the null hypothesis of similar means for PSO vs. STD.R.1 (Chi-square value of 2.83 which was significant for a p-value $= 0.022$) and for PSO vs. STD.R.2 (Chi-square value of 4.09 which was significant for a p-value < 0.01). For PSO vs. SMAC, PSO vs. TPE and PSO vs. RS we accept the null hypothesis which means that in such cases the mean of the results of the control method against each other groups is equal.

As can be seen, for each optimization problem, automatic HPO always outperforms standard recommendation in the literature with statistical significant differences. In addition, PSO obtains as average better results than RS, SMAC and TPE although we don't see statistical significant differences. Also, the computational cost (execution time) of the parent-algorithms are very similar.

6   Conclusions and Recommendations

In this paper, we compare different strategies to perform HPO of PSO. We observe that automatic HPO provides better results than standard recommendation in the literature. Moreover, optimizing PSO hyper-parameters with PSO it self proved to be a competitive approach when compared with popular HPO algorithms such as RS, SMAC or TPE. This approach has the advantage that the practitioner do not need to consider several optimization algorithms in order to find proper hyper-parameters. However, the main limitation of this work is that only continuous hyper-parameters can be configured with this approach. Future research should investigate hybrid meta-heuristic approaches that allow to consider not only continuous hyper-parameters.

References

1. Bengio, Y.: Practical recommendations for gradient-based training of deep architectures. In: Montavon, G., Orr, G.B., Müller, K.R. (eds.) Neural Networks: Tricks of the Trade, pp. 437–478. Springer, Heidelberg (2012). https://doi.org/10.1007/978-3-642-35289-8_26
2. Bergstra, J., Bengio, Y.: Random search for hyper-parameter optimization. J. Mach. Learn. Res. **13**(Feb), 281–305 (2012)
3. Bergstra, J., Yamins, D., Cox, D.D.: Making a science of model search: hyper-parameter optimization in hundreds of dimensions for vision architectures. In: Proceedings of the 30th International Conference on International Conference on Machine Learning, ICML 2013, Atlanta, GA, USA, vol. 28, pp. I-115–I-123. JMLR.org (2013). http://dl.acm.org/citation.cfm?id=3042817.3042832
4. Bonyadi, M.R., Michalewicz, Z.: Particle swarm optimization for single objective continuous space problems: a review. Evol. Comput. **25**, 1–54 (2017)
5. Candelieri, A., et al.: Tuning hyperparameters of a SVM-based water demand forecasting system through parallel global optimization. Comput. Oper. Res. **106**, 202–209 (2019)
6. Elegbede, C.: Structural reliability assessment based on particles swarm optimization. Struct. Saf. **27**(2), 171–186 (2005)
7. Goodfellow, I., Bengio, Y., Courville, A.: Deep Learning. MIT Press, Cambridge (2016)
8. Hutter, F., Hoos, H.H., Leyton-Brown, K.: Sequential model-based optimization for general algorithm configuration. In: Coello, C.A.C. (ed.) LION 2011. LNCS, vol. 6683, pp. 507–523. Springer, Heidelberg (2011). https://doi.org/10.1007/978-3-642-25566-3_40
9. Larochelle, H., Erhan, D., Courville, A., Bergstra, J., Bengio, Y.: An empirical evaluation of deep architectures on problems with many factors of variation. In: Proceedings of the 24th International Conference on Machine Learning, pp. 473–480. ACM (2007)
10. Liang, J.J., Suganthan, P.N., Deb, K.: Novel composition test functions for numerical global optimization. In: Proceedings of 2005 IEEE Swarm Intelligence Symposium, SIS 2005, pp. 68–75. IEEE (2005)
11. Lindauer, M., Hutter, F.: Warmstarting of model-based algorithm configuration. In: Thirty-Second AAAI Conference on Artificial Intelligence (2018)
12. Livni, R., Shalev-Shwartz, S., Shamir, O.: On the computational efficiency of training neural networks. In: Advances in Neural Information Processing Systems, pp. 855–863 (2014)
13. Lorenzo, P.R., Nalepa, J., Ramos, L.S., Pastor, J.R.: Hyper-parameter selection in deep neural networks using parallel particle swarm optimization. In: Proceedings of the Genetic and Evolutionary Computation Conference Companion, pp. 1864–1871. ACM (2017)
14. Maclaurin, D., Duvenaud, D., Adams, R.: Gradient-based hyperparameter optimization through reversible learning. In: International Conference on Machine Learning, pp. 2113–2122 (2015)
15. Momin, J., Yang, X.S.: A literature survey of benchmark functions for global optimization problems. J. Math. Model. Numer. Optim. **4**(2), 150–194 (2013)

16. Parsopoulos, K.E., Vrahatis, M.N.: Recent approaches to global optimization problems through particle swarm optimization. Nat. Comput. **1**(2–3), 235–306 (2002)
17. Probst, P., Wright, M.N., Boulesteix, A.L.: Hyperparameters and tuning strategies for random forest. Wiley Interdiscip. Rev. Data Mining Knowl. Discov. **9**(3), e1301 (2019)

Feature Extraction of Long Non-coding RNAs: A Fourier and Numerical Mapping Approach

Robson Parmezan Bonidia[✉], Lucas Dias Hiera Sampaio,
Fabrício Martins Lopes, and Danilo Sipoli Sanches[✉]

Bioinformatics Graduate Program (PPGBIOINFO),
Federal University of Technology - Paraná, UTFPR,
Cornélio Procópio, PR, Brazil
robservidor@gmail.com,{ldsampaio,fabricio,danilosanches}@utfpr.edu.br

Abstract. Due to the high number of genomic sequencing projects, the number of RNA transcripts increased significantly, creating a huge volume of data. Thus, new computational methods are needed for the analysis and information extraction from these data. In particular, when parts of a genome are transcribed into RNA molecules, some specific classes of RNA are produced, such as mRNA and ncRNA with different functions. In this way, long non-coding RNAs have emerged as key regulators of many biological processes. Therefore, machine learning approaches are being used to identify this enigmatic RNA class. Considering this, we present a Fourier transform-based features extraction approach with 5 numerical mapping techniques (Voss, Integer, Real, EIIP and Z-curve), in order to classify lncRNAs from plants. We investigate four classification algorithms like Naive Bayes, Random Forest, Support Vector Machine and AdaBoost. Moreover, the proposed approach was compared with 4 competing methods available in the literature (CPC2, CNCI, PLEK, and RNAplonc). The experimental results demonstrated high efficiency for the classification of lncRNAs, providing competitive performance.

Keywords: Feature extraction · Long non-coding RNAs · Numerical mapping techniques · Fourier · Bioinformatics

1 Introduction

In recent years, the number of RNA transcripts has grown greatly due to thousands of sequencing projects [13,21], creating a huge volume of data for analysis. Hence, advances in complete transcriptome sequencing have offered new challenges for discovering novel functional transcriptional elements [24], for instance,

All authors thank the Federal University of Technology - Paraná (UTFPR), CNPq, Fundação Araucária/SETI and CAPES for supporting this study.

© Springer Nature Switzerland AG 2019
I. Nyström et al. (Eds.): CIARP 2019, LNCS 11896, pp. 469–479, 2019.
https://doi.org/10.1007/978-3-030-33904-3_44

the Long Non-Coding RNAs (lncRNAs). The lncRNAs are a new type of Non-Coding RNA (ncRNA) with a length greater than 200 nucleotides [14]. According to recent studies, play essential roles in several critical biological processes [32], such as transcriptional regulation and immune response. The discovery of lncRNAs as essential gene regulators in many biological contexts have motivated the development of new machine learning approaches in order to extract relevant information from lncRNAs. In this context, some methods were successfully applied: CPC [12], CPAT [26], CNCI [23], PLEK [13], LncRNApred [20], RNAplonc [19], BASiNET [10], and LncFinder [9].

The CPC measures the protein-coding potential of a transcript based on two feature categories. The extent and quality of the Open Reading Frame (ORF), and derivation of BLASTX [2] search. As a prediction method, the authors used the LIBSVM package to train a Support Vector Machine (SVM) model, using the standard radial basis function kernel. CPAT classifies transcripts of coding and non-coding using Logistic Regression (LR). This model uses four sequence features: ORF coverage, ORF size, hexamer usage bias, and Fickett TESTCODE statistic. CNCI was modeled with SVM and uses profiling Adjoining Nucleotide Triplets (ANT - 64*64) and most-like CDS (MLCDS).

In contrast, PLEK (2014) is based on the k-mer scheme ($k = 1 - 5$) to predict lncRNA, also applying the SVM classifier. LncRNApred classified lncRNAs with Random Forest (RF) and features based on ORF, signal to noise ratio, k-mer ($k = 1 - 3$), sequence length, and GC content. RNAplonc considered 16 features (ORF, GC content, K-mer scheme ($k = 1 - 6$), sequence length), besides classifying sequences with the REPtree algorithm. BASiNET classifies sequences based on the feature extraction from complex network measurements. Finally, LncFinder uses five classifiers (LR, SVM, RF, Extreme Learning Machine, and Deep Learning), to apply the algorithm that obtains the highest accuracy. Moreover, the authors use features of ORF, secondary structural, and electron-ion interaction.

Some of these works [9,20] have explored Genomic Signal Processing (GSP) techniques, which according to Abo-Zahhad et al. [1], is defined as the analysis of genomic signals, whose purpose is to obtain and translate biological knowledge into systems-based applications. To use GPS techniques it is necessary to apply a numeric representation for transformation or mapping of genomic data (represented in DNA by the letters A (adenine), T (thymine), G (guanine) and C (cytosine)) [17]. In literature, distinct DNA Numerical Representation (DNR) techniques have been developed [1]. According to Mendizabal-Ruiz et al. [17], these representations can be divided into three categories: single-value mapping (e.g., integer representation [8,17], real number representation [5], real number representation, Electron-Ion Interaction Pseudopotential (EIIP) [18]), multidimensional sequence mapping (e.g., Voss representation [25]), and cumulative sequence mapping (e.g., Z-curve representation [31]).

As previously shown, some works used this approach in the lncRNAs classification, Pian et al. [20] applied Voss representation and Han et al. [9] EIIP representation. Nevertheless, the authors used these approach in conjunction with other features extraction techniques, and without testing other numerical mappings. Furthermore, according to Abo-Zahhad et al. [1], the Voss representation (one of the most applied methods) may be redundant. Therefore, considering that it is not yet clear what the properties of each DNR and how the selection of these distinct techniques can affect the results in a signal processing approach [17], we elaborated a study with 5 numerical mapping techniques (*Voss, Integer, Real, EIIP, Z-curve*), in order to classify lncRNAs.

2   Materials and Methods

This section describes the methodological procedures used to achieve the proposed objectives. Fundamentally, we divided our approach into five stages: (1) Data selection and preprocessing; (2) Feature extraction; (3) Training; (4) Test; (5) Performance analysis.

2.1   Data Selection

Sequences of plant species (*Arabidopsis thaliana*), obtained from CPC2 [11], were adopted in order to validate the proposed method. Following the literature methods, this work also adopts two classes for the datasets: positive class, with lncRNAs, and negative class, with protein-coding genes (mRNAs). The mRNA data were obtained from the RefSeq database with protein sequences annotated by Swiss-Prot [11], and lncRNA data from the Ensembl ($v87$) and Ensembl Plants ($v32$) database. We used only sequences longer than $200nt$ [13], and we also removed sequence redundancy (identity $\geq 90\%$), using CD-HIT-EST tool ($v4.6.1$) [15].

2.2   Feature Extraction

At this stage, a Fourier transform based features extraction approach is performed on input samples (to detect lncRNAs and mRNAs). Thus, we adopt five representations: Voss [25], Integer [8,17], Real [5], Z-curve [31], and EIIP [18]. Fundamentally, we denote a biological sequence $S = (S[0], S[1], \ldots, S[N-1])$ such that $S \in \{A, C, G, T\}^N$.

Fourier Transform: To generate features based in a Fourier approach, we apply the Discrete Fourier Transform (DFT), widely used for digital image processing and digital signal processing, that can reveal hidden periodicities after transformation of time domain data to frequency domain space [27]. According to Yin and Yau [28], the DFT of a signal with length N, $x[n]$ ($n = 0, 1, \ldots, N-1$), at frequency k, can be defined by Eq. (1):

$$X[k] = \sum_{n=0}^{N-1} x[n]\, e^{-j\frac{2\pi}{N}kn}, \qquad k = 0, 1, \ldots, N-1. \tag{1}$$

This method is extensively studied in bioinformatics, mainly for analysis of periodicities and repetitive elements in DNA sequences [3] and protein structures [16].

Voss Representation: This representation can use single or multidimensional vectors. Fundamentally, this approach transforms a sequence $S \in \{A, C, G, T\}^N$ into a matrix $\mathbf{V} \in \{0, 1\}^{4 \times N}$ such that $\mathbf{V} = [\mathbf{v}_1, \mathbf{v}_2, \mathbf{v}_3, \mathbf{v}_4]^T$, where T is the transpose operator and each $\mathbf{v}_i$ array is constructed according to the following relation:

$$v_i[n] = \begin{cases} 1, & S[n] = \alpha[i] \\ 0, & S[n] \neq \alpha[i] \end{cases}, \text{ where } \alpha = (A, C, G, T), \qquad n = 0, 1, \ldots, N - 1. \quad (2)$$

As a result, each row of matrix $\mathbf{V}$ may be seen as an array that marks each base position such that the first row denotes the presence of base A, row two for base C, row three base G and the last row for base T. For example, let $S = (G, A, G, A, G, T, G, A, C, C, A)$ be a sequence that needs to be represented using Voss representation, therefore, $\mathbf{v}_1 = (0, 1, 0, 1, 0, 0, 0, 1, 0, 0, 1)$, which represents the locations of bases A, $\mathbf{v}_2 = (0, 0, 0, 0, 0, 0, 0, 0, 1, 1, 0)$ for bases C, $\mathbf{v}_3 = (1, 0, 1, 0, 1, 0, 1, 0, 0, 0, 0)$ for the G bases, $\mathbf{v}_4 = (0, 0, 0, 0, 0, 1, 0, 0, 0, 0, 0)$ for T bases. Then, using the DFT in the indicator sequences shown above, we obtain (see Eq. 3):

$$V_i[k] = \sum_{n=0}^{N-1} v_i[n] e^{-j\frac{2\pi}{N} kn}, \ \forall \ i \in [1, 4], \qquad k = 0, 1, \ldots, N - 1. \quad (3)$$

The power spectrum of a biological sequence can be obtained by Eq. (4):

$$P_V[k] = \sum_{i=1}^{4} |V_i[k]|^2, \qquad k = 0, 1, \ldots, N - 1. \quad (4)$$

Integer Representation: This representation is one-dimensional [8, 17]. This mapping can be obtained by substituting the four nucleotides (G, A, C, T) of a biological sequence for integers (0, 1, 2, 3), respectively, e.g., let $S = (G, A, G, A, G, T, G, A, C, C, A)$, thus, $d = (3, 2, 3, 2, 3, 0, 3, 2, 1, 1, 2)$, as exposed in Eq. (5). The DFT and power spectrum are exposed in Eq. (6).

$$d[n] = \begin{cases} 3, & S[n] = G \\ 2, & S[n] = A \\ 1, & S[n] = C \\ 0, & S[n] = T \end{cases}, \qquad n = 0, 1, \ldots, N - 1. \quad (5)$$

$$D[k] = \sum_{n=0}^{N-1} d[n] e^{-j\frac{2\pi}{N} kn}, \qquad P_D[k] = |D[k]|^2, \qquad k = 0, 1, \ldots, N - 1. \quad (6)$$

Real Representation: In this representation, Chakravarthy et al. [5] use real mapping based on the complement property of the complex mapping of [3]. This mapping applies positive decimal values for the purines (A, G), and negative decimal values for the pyrimidines (C, T), e.g., let $S = (G, A, G, A, G, T, G, A, C, C, A)$, thus, $r = (-0.5, -1.5, -0.5, -1.5, -0.5, 1.5, -0.5, -1.5, 0.5, 0.5, -1.5)$, as Eqs. (7) and (8).

$$r[n] = \begin{cases} -0.5, & S[n] = G \\ -1.5, & S[n] = A \\ 0.5, & S[n] = C' \\ 1.5, & S[n] = T \end{cases} \quad n = 0, 1, \ldots, N-1. \tag{7}$$

$$R[k] = \sum_{n=0}^{N-1} r[n] e^{-j\frac{2\pi}{N}kn}, \quad P_R[k] = |R[k]|^2, \quad k = 0, 1, \ldots, N-1. \tag{8}$$

Z-Curve Representation: The Z-curve scheme is a three-dimensional curve presented by [31], to encode DNA sequences with more biological semantics. Essentially, we can inspect a given sequence $S[n]$ of length N, taking into account the n-th element of the sequence $(n = 1, 2, \ldots, N)$. Then, we denote the cumulative occurrence numbers A_n, C_n, G_n and T_n for each base A, C, G and T, as the number of times that a base occurred from $S[1]$ up until $S[n]$. Therefore:

$$A_n + C_n + G_n + T_n = n \tag{9}$$

Where the Z-curve consists of a series of nodes $P_1, P_2, \ldots, P_N$, whose coordinates $x[n]$, $y[n]$, and $z[n]$ $(n = 1, 2, \ldots, N)$ are uniquely determined by the Z-transform, shown in Eq. (10):

$$P[n] = \begin{cases} x[n] = (A_n + G_n) - (C_n + T_n) \equiv R_n - Y_n \\ y[n] = (A_n + C_n) - (G_n + T_n) \equiv M_n - K_n, \\ z[n] = (A_n + T_n) - (C_n + G_n) \equiv W_n - S_n \end{cases} \tag{10}$$

$$x[n], y[n], z[n] \in [-n, n], \quad n = 1, 2, \ldots, N.$$

Where R, Y, M, K, W and S denote the bases of purine $(R = A, G)$, pyrimidine $(Y = C, T)$, amino $(M = A, C)$, keto $(K = G, T)$, weak hydrogen bonds $(W = A, T)$ and strong hydrogen bonds $(S = G, C)$, respectively [22, 30]. The coordinates $x[n]$, $y[n]$, and $z[n]$ represent three independent distributions that completely describe a sequence [1]. Therefore, we will have three distributions with definite biological significance: (1) $x[n] = $ purine/pyrimidine, (2) $y[n] = $ amino/keto, (3) $z[n] = $ strong hydrogen bonds/weak hydrogen bonds [31], e.g., let $S = (G, A, G, A, G, T, G, A, C, C, A)$, thus, $x = (1, 2, 3, 4, 5, 4, 5, 6, 5, 4, 5)$; $y = (-1, 0, -1, 0, -1, -2, -3, -2, -1, 0, 1)$; $z = (-1, 0, -1, 0, -1, 0, -1, 0, -1, -2, -1)$. Essentially, the difference between each dimension at the n-th position and the previous $(n-1)$ position can be either 1 or -1 [31]. Finally, the DFT and the power spectrum of the Z-Curve representation may be defined as [30]:

$$X[k] = \sum_{n=1}^{N} x[n]e^{-j\frac{2\pi}{N}kn}, \quad Y[k] = \sum_{n=1}^{N} y[n]e^{-j\frac{2\pi}{N}kn}, \quad Z[k] = \sum_{n=1}^{N} z[n]e^{-j\frac{2\pi}{N}kn}. \quad (11)$$

$$P_C[k] = |X[k]|^2 + |Y[k]|^2 + |Z[k]|^2, \qquad k = 1, 2, \ldots, N. \quad (12)$$

EIIP Representation: Nair and Sreenadhan [18] proposed EIIP values of nucleotides to represent biological sequences and to locate exons. According to the authors, a numerical sequence representing the distribution of free electron energies can be called *"EIIP indicator sequence"*, e.g., let $S =$ (G, A, G, A, G, T, G, A, C, C, A), thus, $b =$ (0.0806, 0.1260, 0.0806, 0.1260, 0.0806, 0.1335, 0.0806, 0.1260, 0.1340, 0.1340, 0.1260), as shown in Eq. (13). The DFT and power spectrum of this representation are presented in Eq. (14).

$$b[n] = \begin{cases} 0.0806, & S[n] = G \\ 0.1260, & S[n] = A \\ 0.1340, & S[n] = C \\ 0.1335, & S[n] = T \end{cases} \quad n = 0, 1, \ldots, N - 1. \quad (13)$$

$$E[k] = \sum_{n=0}^{N-1} b[n]e^{-j\frac{2\pi}{N}kn}, \quad P_E[k] = |E[k]|^2, \quad k = 0, 1, \ldots, N - 1. \quad (14)$$

Features: Finally, we apply the feature extraction in all representations, adopting Signal to Noise Ratio (SNR - [22]), average power spectrum, median, maximum, minimum, sample standard deviation, population standard deviation, percentile (15/25/50/75), amplitude, and variance. The SNR uses the statistical phenomenon known as period-3 behavior or 3-base periodicity [29].

2.3   Normalization, Sampling, Training and Evaluation Metrics

We adopt the min-max normalization method, which fits the data range to 0 and 1 (or -1 to 1, if there are negative values) for each feature, in order to use them on classification step. Moreover, the sampling method was adopted in our dataset, since we are faced with the *imbalanced data problem* (*A. thaliana* (2,540 lncRNA/13,973 mRNA)). Thus, we applied SMOTE [6], an over-sampling approach (to adjust the class distribution), in which "synthetic" examples are created, over-sampling the minority class. Next, we investigate four classification algorithms, like Naive Bayes (NB), Random Forest (RF), SVM and AdaBoost. To induce our models, we used 70% of samples for *training* (with 10-fold cross-validation) and 30% for *testing*. Finally, the representations were evaluated with sensitivity (SE - correctly predicted lncRNAs), specificity (SPC - correctly classified mRNAs), accuracy (ACC), and Cohen's kappa coefficient [7].

3   Results and Discussion

First, we induced our models with the NB, RF, SVM, and AdaBoost classifiers in the training set. Then, to estimate the real accuracy of this set, we used 10-fold cross-validation, as exposed in Table 1. Evaluating each classifier individually, we observed that the best performance was of the Random Forest with Z-curve (0.9605), followed by AdaBoost (EIIP - 0.9521), SVM (EIIP - 0.9476), and NB (Real - 0.9300). After training, the predictive models induced by NB, RF, SVM, and AdaBoost were applied to the test set, in which Fig. 1 summarizes in a polar chart, the SE, SPC, kappa and ACC metrics for each representation.

As can be seen, in Fig. 1, the RF classifier maintained the best performance in the test set using Z-curve (ACC = 0.9553), followed by AdaBoost (ACC = 0.9526) adopting EIIP. In general, the best results are contained in the Real, Z-curve and EIIP representation. However, if we use the AdaBoost classifier as an example, the greatest difference in accuracy between the mappings is approximately 0.0072. Although this, we noted that the mappings have higher peaks of ACC, SPC in NB. We also evaluated the performance of our best predictive

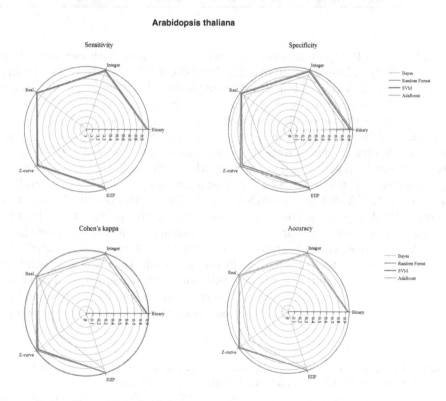

Fig. 1. Polar Chart: This figure compares the sensitivity, specificity, kappa and accuracy metrics for each representation in the test set. In addition, it shows how the classifiers (NB, RF, SVM, and AdaBoost) are behaving in each representation.

Table 1. Real accuracy estimates for the training set using 10-fold cross-validation.

Species	DNR	NB	RF	SVM	AdaBoost
	Binary	0.9237	0.9574	0.9398	0.9439
	Integer	0.9258	0.9575	0.9475	0.9508
A. thaliana	Real	0.9300	0.9595	0.9448	0.9474
	Z-curve	0.7520	0.9605	0.9266	0.9428
	EIIP	0.9260	0.9589	0.9476	0.9521

model (RF with Z-curve) against other four state-of-the-art tools; CPC2 [11] (an updated version of the CPC method), CNCI, PLEK and RNAplonc (specifically for plants), as shown in Table 2.

Table 2. Comparative performance between our approach and state-of-the-art tools.

Metric	CPC2	CNCI	PLEK	RNAplonc	Proposed
ACC	0.9574	0.8997	0.6649	0.9443	0.9553

CPC2 (0.9574) reported a similar performance along with our predictive model (0.9553), followed by RNAplonc (0.9443), CNCI (0.8997), and PLEK (0.6649). Nevertheless, it is important to emphasize that CPC2 and RNAplonc use the ORF descriptor, a highly employed feature for discovering coding sequences and which, according to Baek et al. [4] is an essential guideline for distinguishing lncRNAs from mRNA. Considering this, our approach has an advantage in terms of generalization to distinguish other classes of ncRNA, since this would not be possible only with the ORF. To evaluate this hypothesis, we apply a second experiment with CPC2 using a new dataset with only non-coding sequences (lncRNA and Small ncRNA - A. thaliana - also obtained from [11]) without mRNA sequences. For such, we used the features provided by CPC2 to construct a model with similar procedures to our approach. However, we eliminated the sequence length descriptor provided by CPC2 and also any attribute that would generate this information in our approach, since that any explicit bias to this feature may facilitate the prediction of these sequences. Therefore, we applied new experiments according to the same methodology described in this work (70% training and 30% test) and using the RF classifier, as shown in Table 3.

Table 3. Comparative between CPC2 and our approach to a new dataset with only non-coding sequences (lncRNA and Small ncRNA - A. thaliana).

Method	SE	SPC	ACC	Kappa
CPC2	0.8428	0.7715	0.8071	0.6143
Proposed	0.9607	0.9582	0.9595	0.9189

The tests confirm again the hypothesis that the proposed method is efficient, in which we reached an ACC of 0.9595 against 0.8071 of the features provided by CPC2 (e.g., ORF). That is, our approach is robust in terms of generalization to distinguish lncRNA from mRNA, as well as other classes of ncRNA.

4 Conclusion

In this work, we investigated five numerical mapping techniques (Voss, Integer, Real, EIIP, Z-curve) with the Fourier transform, for feature extraction and classification of lncRNAs[1]. Thereby, sequences of plant species (*A. thaliana*), obtained from [11] were adopted in order to validate the proposed method. As results, we conclude that the RF and AdaBoost classifiers presented the best performance using the Z-curve and EIIP representations, respectively. Furthermore, to validate our study, we also compared with other available methods in the literature (CPC2, CNCI, PLEK, and RNAplonc). The proposed approach presented suitable results, being superior or competitive to other methods, and robust in terms of generalization. Finally, as future works we will analyze these representations more deeply, in order to propose a new numerical mapping with nucleotides triplets and amino acid features, e.g., molar mass, acidity, Van Der Waals volume, to consider more RNA classes and different organisms for the feature extraction analysis.

References

1. Abo-Zahhad, M., Ahmed, S.M., Abd-Elrahman, S.A.: Genomic analysis and classification of exon and intron sequences using dna numerical mapping techniques. Int. J. Inf. Technol. Comput. Sci. **4**(8), 22–36 (2012)
2. Altschul, S.F., et al.: Gapped blast and PSI-BLAST: a new generation of protein database search programs. Nucleic Acids Res. **25**(17), 3389–3402 (1997)
3. Anastassiou, D.: Genomic signal processing. IEEE Sig. Proc. Mag. **18**(4), 8–20 (2001)
4. Baek, J., Lee, B., Kwon, S., Yoon, S.: LncRNAnet: long non-coding RNA identification using deep learning. Bioinformatics **1**, 9 (2018)
5. Chakravarthy, N., Spanias, A., Iasemidis, L.D., Tsakalis, K.: Autoregressive modeling and feature analysis of DNA sequences. EURASIP J. Appl. Sig. Process. **2004**, 13–28 (2004)
6. Chawla, N.V., Bowyer, K.W., Hall, L.O., Kegelmeyer, W.P.: Smote: synthetic minority over-sampling technique. J. Artif. Intell. Res. **16**, 321–357 (2002)
7. Cohen, J.: A coefficient of agreement for nominal scales. Educ. Psychol. Meas. **20**(1), 37–46 (1960)
8. Cristea, P.D.: Conversion of nucleotides sequences into genomic signals. J. Cell. Mol. Med. **6**(2), 279–303 (2002)
9. Han, S., et al.: LncFinder: an integrated platform for long non-coding RNA identification utilizing sequence intrinsic composition, structural information and physicochemical property. Brief. Bioinform. **19**, 1–19 (2018). https://doi.org/10.1093/bib/bby065

---

[1] Data and materials: github.com/Bonidia/FourierFeatureExtraction.

10. Ito, E.A., Katahira, I., da Vicente, F.F.R., Pereira, L.F.P., Lopes, F.M.: BASiNET-biological sequences network: a case study on coding and non-coding RNAs identification. Nucleic Acids Res. **46**, e96 (2018)
11. Kang, Y.J., et al.: CPC2: a fast and accurate coding potential calculator based on sequence intrinsic features. Nucleic Acids Res. **45**(W1), W12–W16 (2017)
12. Kong, L., et al.: CPC: assess the protein-coding potential of transcripts using sequence features and support vector machine. Nucleic Acids Res. **35**(suppl-2), W345–W349 (2007)
13. Li, A., Zhang, J., Zhou, Z.: PLEK: a tool for predicting long non-coding RNAs and messenger RNAs based on an improved k-mer scheme. BMC Bioinform. **15**(1), 311 (2014)
14. Li, A., Zang, Q., Sun, D., Wang, M.: A text feature-based approach for literature mining of lncrna-protein interactions. Neurocomputing **206**, 73–80 (2016)
15. Li, W., Godzik, A.: Cd-hit: a fast program for clustering and comparing large sets of protein or nucleotide sequences. Bioinformatics **22**(13), 1658–1659 (2006)
16. Marsella, L., Sirocco, F., Trovato, A., Seno, F., Tosatto, S.C.: REPETITA: detection and discrimination of the periodicity of protein solenoid repeats by discrete fourier transform. Bioinformatics **25**(12), i289–i295 (2009)
17. Mendizabal-Ruiz, G., Román-Godínez, I., Torres-Ramos, S., Salido-Ruiz, R.A., Morales, J.A.: On DNA numerical representations for genomic similarity computation. PloS One **12**(3), e0173288 (2017)
18. Nair, A.S., Sreenadhan, S.P.: A coding measure scheme employing electron-ion interaction pseudopotential (EIIP). Bioinformation **1**(6), 197 (2006)
19. da Negri, T.C., Alves, W.A.L., Bugatti, P.H., Saito, P.T.M., Domingues, D.S., Paschoal, A.R.: Pattern recognition analysis on long noncoding RNAs: a tool for prediction in plants. Brief. Bioinform. **20**, 682–689 (2018)
20. Pian, C., et al.: LncRNApred: classification of long non-coding rnas and protein-coding transcripts by the ensemble algorithm with a new hybrid feature. PloS One **11**(5), e0154567 (2016)
21. Schneider, H.W., Raiol, T., Brigido, M.M., Walter, M.E.M., Stadler, P.F.: A support vector machine based method to distinguish long non-coding rnas from protein coding transcripts. BMC Genomics **18**(1), 804 (2017)
22. Shao, J., Yan, X., Shao, S.: SNR of DNA sequences mapped by general affine transformations of the indicator sequences. J. Math. Biol. **67**(2), 433–451 (2013)
23. Sun, L., et al.: Utilizing sequence intrinsic composition to classify protein-coding and long non-coding transcripts. Nucleic Acids Res. **41**(17), e166–e166 (2013)
24. Ventola, G.M., Noviello, T.M., D'Aniello, S., Spagnuolo, A., Ceccarelli, M., Cerulo, L.: Identification of long non-coding transcripts with feature selection: a comparative study. BMC Bioinform. **18**(1), 187 (2017)
25. Voss, R.F.: Evolution of long-range fractal correlations and 1/f noise in dna base sequences. Phys. Rev. Lett. **68**(25), 3805 (1992)
26. Wang, L., Park, H.J., Dasari, S., Wang, S., Kocher, J.P., Li, W.: CPAT: coding-potential assessment tool using an alignment-free logistic regression model. Nucleic Acids Res. **41**(6), e74 (2013)
27. Yin, C., Chen, Y., Yau, S.S.T.: A measure of dna sequence similarity by fourier transform with applications on hierarchical clustering. J. Theor. Biol. **359**, 18–28 (2014)
28. Yin, C., Yau, S.S.T.: A fourier characteristic of coding sequences: origins and a non-fourier approximation. J. Comput. Biol. **12**(9), 1153–1165 (2005)
29. Yin, C., Yau, S.S.T.: Prediction of protein coding regions by the 3-base periodicity analysis of a DNA sequence. J. Theor. Biol. **247**(4), 687–694 (2007)

30. Zhang, C.T.: A symmetrical theory of dna sequences and its applications. J. Theor. Biol. **187**(3), 297–306 (1997)
31. Zhang, R., Zhang, C.T.: Z curves, an intutive tool for visualizing and analyzing the dna sequences. J. Biomol. Struct. Dyn **11**(4), 767–782 (1994)
32. Zhang, W., Qu, Q., Zhang, Y., Wang, W.: The linear neighborhood propagation method for predicting long non-coding RNA-protein interactions. Neurocomputing **273**, 526–534 (2018)

Combinatorial Optimization Approach for Arabic Word Recognition Based on Adaptive Simulated Annealing

Zeineb Zouaoui[✉], Imen Ben Cheikh[✉], and Mohamed Jemni[✉]

Latice Laboratory, ENSIT, University of Tunis, Tunis, Tunisia
{zeineb.zouaoui,imen.becheikh}@gmail.com,
mohamed_jemni2000@yahoo.fr

Abstract. The present paper proposes an approach based on **a** combinatorial optimization technique for Arabic word recognition, distinguished by its flexional nature and significant topological variability. We treat a large vocabulary of Arabic decomposable words, which we choose to factorize them by their roots and schemes. We adopt a structure that resembles a molecular cloud. This design rhymes well with the Arabic linguistic philosophy of constructing words **from their roots.** Each sub-vocabulary, corresponding to a sub-cloud, embodies neighboring words, which are derived from one root and follow different schemes and forms of derivation, flexion, and agglutination (proclitic and enclitic). Therefore, we propose to use the metaheuristic simulated annealing (SA) method, as a recognition approach, in this wide cloud. It's an algorithm based on elastic comparisons between their structures and primitives. As an extension of previous works, we opt to implement the SA algorithm by integrating linguistic knowledge. Preliminary experiments were conducted on Arabic word corpus including samples and agglutinated words from APTI database and yielded interesting outcomes.

Keywords: Simulated annealing · Levenshtein distance · Morphological peculiarities · Combinatorial optimization · APTI database

1 Introduction

Arabic word recognition has become a very popular research area in recent years with a large number of possible applications (handwritten postal addresses, banks checks, etc.). The complexity of the recognition process depends on the script's type (printed or handwritten), the approach (holistic, pseudo-global and analytic) [1, 2] and the vocabulary size (reduced, large). Several approaches have been suggested, handling letters and/or pseudo-word levels, and numerous works have experienced statistical, neural, stochastic methods on different kinds of Arabic documents. Some works operated programmatically based on computation ability; others preferred to handle words cognitively while emulating human reading models.

In this work, we want, on one hand, to emphasize the fact that since humans see lexical and syntactic information in written words, it would be promising to underline linguistic knowledge integration in the recognition process. On the other hand, we aim

© Springer Nature Switzerland AG 2019
I. Nyström et al. (Eds.): CIARP 2019, LNCS 11896, pp. 480–489, 2019.
https://doi.org/10.1007/978-3-030-33904-3_45

to exploit combinatorial optimization, which had been proposed as a solution to artificial intelligence problems and test its efficiency in Arabic word recognition since Arabic is a morphologically complex and diverse language. Indeed, due to the fact that Arabic is a highly inflected and derived language, the number of effective forms go past 60 billion [3]. In this context, we intend to propose a novel approach based on a technique of combinatorial optimization that aims to recognize a large vocabulary of Arabic decomposable words (derived from roots), presenting different morphological aspects (derivational, inflectional and agglutinative).

This paper is organized as follows. Section 2 describes the major characteristics of the Arabic script. Section 3 is devoted to the description of the proposed approach as well as the lexicon organization. Then, the forth section reveals preliminary experiments and results to evaluate the approach. Finally, Sect. 5 summarizes the paper and proposes some perspectives.

2 Arabic Script Characteristics

2.1 Topological Specificities

Arabic script is written from right to left. Arabic writing is semi-cursive either in a printed or handwritten form. The Arabic alphabet contains 28 letters, which are primarily consonants. Most of them can change their forms according to their positions in the word; at the beginning, in the middle or at the end of the word. Some letters have four different forms like the letter (غ,ـغـ,ـخ,ـغ) in Fig. 2. Furthermore, six letters have only two shapes, which are «ذ» (Thal), «د» (Dal), «ر» (Ra), «و» (Waw), «ز» (Zai) and «ا» (Alif). They are related to the letters that precede them, rather than to those that follow them. They form words composed of one or more parts called PAW (Peace of Arabic Word) or «Pseudo-Word» [16]. In addition, more than half of the Arabic characters include diacritics (one, two or three diacritical points) in their forms (see Fig. 1). These dots may be disposed either above or below the body of the letter, but they were never perceived simultaneously. The presence or the absence of these dots in their positions help to distinguish between letters that have exactly the same main shape.

Feature	Designation	Description	
A	Ascender	Ascending characteristic over the baseline	ا
D	Descender	Descending characteristic below the baseline	ر
L	Loop	Intersection between 2 closed boundaries	ه
Da	Diacritics above	Diacritical points over the word body	ت
Db	Diacritics below	Diacritical points below the word body	ب
O	Otherwise	No primitives listed above	د
B,M,E,I	(B: beginning, M : Middle, E: End, I: Isolated)	Primitive position in the word	ش ـشـ ـش شـ

Fig. 1. Global features

Fig. 2. Different forms of the same Arabic letter

2.2 Morphological Specificities

The complexity and the richness of the Arabic script is attributed to the derivational structure of its morphological system. Most of the Arabic words are derived from roots by inserting prefixes, infixes, and suffixes. These words can be either conjugated verbs or specific names, such as machine names, lawsuit names, time and place names, preference nouns, and analogue adjectives. They are known as decomposable words [2]. Non-decomposable words, however, correspond for instance to the names of countries, numbers, etc. This diversity is due to Arabic derivational aspect, which consists of the fact that Arabic words are derived from a root according to a given pattern (scheme). This procedure allows the generation of new words by adding prefixes that precede a base (is the set of root consonants), by injecting infixes that appear between the basic letters and by adding suffixes coming at the end. The product of the derivation of a root following a pattern is called radical. This morphological richness appears, also, on the inflectional plan. Indeed, a radical undergoes inflected conjugations that can change the prefixes and suffixes and add additional letters. In this way, the elements of derivational and inflectional conjugations of Arabic comprise the root, scheme, personal pronoun, time, gender, number, function, etc. Finally, Arabic is also characterized by its agglutinative morphology. There are two types of agglutination: proclitic and enclitics [5]. We designate simple proclitic (morphemes of one letter) and compound proclitic (morphemes of several letters). The first are coordinating, conjunctions, prepositions and the latter are obtained by combining first ones. The enclitic is a supplement pronoun that can be single or double and is attached to the word that precedes it. Figure 3 shows the example of the word «سيراجعونه» that derives from the root «رجع», conjugated according to the scheme «فاعل » with the pronoun « هم» (they: plural masculine) and adding the proclitic «س» (designating future tense) and the enclitic «ه» (designating the object complement: him). Furthermore, a single Arabic word can substitute a completely English sentence; as the word «أسيراجعونه» in Fig. 4, can be translated to the sentence (**Will they review it?**).

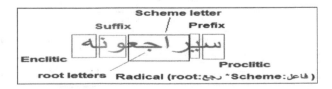

Fig. 3. Morphems of Arabic word

Fig. 4. Sample of agglutinated Arabic word

3 Approach Proposal

3.1 Motivation

Our work tries to benefit from the combinatorial optimization techniques to solve pattern recognition problems known for their strongly exponential complexity. These techniques had been generally proposed for solving artificial intelligence problems in large solution spaces. Thus, we attempt to evaluate how these techniques could be generally effective with word recognition and especially with Arabic one. We focused especially on SA algorithm. It is a powerful probabilistic metaheuristic algorithm originally proposed by Metropolis et al. [6]. This method has proven to be quite versatile and efficient for wide range of combinatorial optimization problems. For example, in the travelling Salesman problem, experiments proved that SA becomes efficacious above 800 cities [7]. In addition, it seems robust with data mining applications particularly when clustering large categorical data sets [8]. Furthermore, it has been proven efficient for partitioning graph problem, pattern detection of seismic applications and job-shop scheduling, etc. [9, 10].

Therefore, SA method looks highly effective in exploring and exploiting the large search space associated with problems having particular properties [11]. Why is the writing recognition problem, characterized by its wide variability and particular peculiarities, not highlighted? (1) So, topologically, we opt to adjust the SA elastic comparison by calculating the Levenshtein (or editing) distance. It is a metric method suggested by Levenshtein [6], to compare two words, two strings, and more generally, two sequences. The edit distance is defined as the minimum number of operations to convert the first word into the second word, by taking into account the presence of only three possible operations:

- replacing a character with another,
- deleting a character,
- inserting a character.

Edit distance algorithm is simple to code in a short calculation time. Hence, several studies have tested edit distance with structural methods of pattern recognition [12, 8 and 13]. Carbonnel in [14] presented an automatic learning method to calculate the edit distance in order to adapt it to online handwriting. Moreover, Gueddah in [13] proposed to integrate a measurement of proximity and similarity between Arabic characters in Levenshtein algorithm to better suggest and schedule automatic correction of spelling errors detected in typed Arabic documents. He achieved interesting outcomes. Inspired by these studies, we thought of adjusting this elastic comparison algorithm to

Arabic word recognition in order to privilege the global over the local and to absorb the writing variability.

(2) Morphologically, we attempt to integrate Arabic linguistic peculiarities in SA neighborhood. It can deeply expand the performance and the accuracy of SA [14]. We have chosen to organize the SA neighborhood around morpho-phonological concepts (root, prefix, suffix, enclitic, proclitic, etc.).

3.2   Arabic Vocabulary as Molecular Cloud

Recall that we have decided to organize our vocabulary of decomposable words as a molecular cloud, whose organization emulates the Arabic linguistic philosophy of word construction around the roots. Each sub-cloud assembles neighboring words derived from the same root according to multiple derivational forms, such as flexion and agglutination (proclitic and enclitic) (see Figs. 5 and 6). This factorization highlighted common morphological entities, like roots, schemes, with various conjugating elements (e.g. present, past, singular, dual, plural, masculine, feminine, etc.) and agglutination, etc. Figure 5 gives an overview of the sub-cloud that represents the root « بعد » (i.e. to go far) linked to its derivatives, which are presented as molecules. On the other hand, Fig. 6 represents the union of several sub-clouds (black balls represent sub-cloud nuclei), which relatively correspond to the roots, in order to constitute the global molecular cloud.

This morpheme-based structuring model ensures the learning phase that characterizes the majority of pattern recognition systems. In a first step, we have prepared an initial cloud of morphologically related words enriched by diverse rules of verb conjugations in different tenses, with various persons, flexion, pronouns, and functions. In the second step, we have integrated some morphological instructions in order to connect the generated words and sub-clouds. Finally, the first obtained cloud represents the initial solution space that is automatically stabilized in accordance with SA iterations itself.

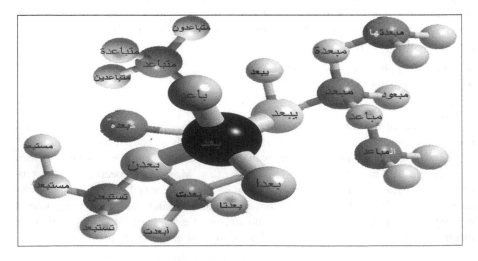

Fig. 5. Part of the sub-cloud of the root « بعد » (i.e. to go far)

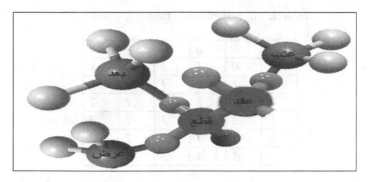

Fig. 6. Molecular cloud: union of various sub-clouds (black balls represent sub-cloud nuclei)

3.3 Recognition

As shown above, we have opted for SA metaheuristic method, whose major advantage is to escape local optima and converge to the global optima (see algorithm in Fig. 7). To adapt the SA algorithm to a particular problem, we must specify the state space, the temperature, the neighbor selection method, the probability transition function and the annealing schedule. These choices have a considerable effect on the results [5]. The chosen parameters were then updated or not based on the decision rule of our SA algorithm.

To handle our open vocabulary, we adopted structures of primitive sequences (e.g. LB AE denote the loop at the beginning followed by an ascender at the end (see Fig. 1)) extracted from the word images in previous works [15]. Therefore, we chose Levenshtein comparative method, between words and their features, that operates as follows:

We look for a recursive solution by introducing ED(i, j) the optimal distance between the first i characters of the first string and the first j characters of the second string. We deduce initial values:

$$ED\,(0,0) = 0$$
$$ED(0, i) = ED(i, 0) = i, \qquad \text{and the recursive relationship:}$$
$$ED\,(i, j) = \text{Minimum}\{ED\,(i-1, j) + 1, \qquad //\text{deletion}$$
$$ED\,(i, j-1) + 1, \qquad //\text{ insertion}$$
$$ED\,(i-1, j-1) + \text{cost}\,\} \qquad //\text{ permutation}$$
$$\text{With } cost = \begin{cases} 0 & \text{if } c(i) = c(j) \\ 1 & \text{else} \end{cases}$$

To reach the ED(i, j) point, you have to insert a letter, delete a letter, replace a letter, or do nothing if the two characters c (i) and c (j) are equal.

Table 1 shows an example of the Levenshtein calculation between the word «يبعد» and the word «يباعد», the edit distance is equal to 2.

Table 1. An example of edit distance calculation

		ي	ب	ع	د
	0	1	2	3	4
ي	1	0	1	2	3
ب	2	1	0	1	2
ا	3	2	1	1	2
ع	4	3	2	2	2
د	5	4	3	3	2

Until then, the calculation of the edit distance is classic (Classic ED). It deals with primitives as normal strings, i.e. it does not differentiate between the root consonant or scheme, and the conjugated letters that appear in prefixes and suffixes. Therefore, this classic ED calculation method can cause confusion, when we choose the optimal solution. Indeed, as mentioned in Table 2 below, the edit distance between the word "يبعد" derived from the scheme "فعل " and all the proposed solutions, which are derived from different schemes, is the same.

To remedy this problem, we propose in this work to integrate a new cost in the editing operations, which enhance the presence or the absence of scheme letters (ShL). This cost differs depending on letter position schemes; whether it is in the ideal location or not. The pseudo-code is as follow:

$$EDsh\,(0,0) = 0$$

and the recursive relationship:

$$EDsh\,(0,i) = EDsh(i,0) = i$$

$$EDsh\,(i,j) = Minimum\,\{\,EDsh\,(i-1,j)+1, \qquad // \text{ deletion}$$
$$EDsh\,(i,j-1)+Cost1, \qquad // \text{ insertion}$$
$$EDsh\,(i-1,j-1)+Cost2\} \qquad // \text{ permutation}$$

With

$$Cost\,1 = \begin{cases} 1.7 & \text{if } (c(i) \text{ is not a ShL}) \\ 0.3 & \text{else} \end{cases}$$

$$Cost\,2 = \begin{cases} 0 & \text{if } (c(i) \text{ is not a ShL \&\& } c(i) = c(j)) \\ 0.2 & \text{if } (c(i) \text{ is a ShL \&\& } c(i) = c(j) \text{ \&\& } i = j) \\ 0.5 & \text{if } (c(i) \text{ is a ShL \&\& } c(i) = c(j) \text{ \&\& } i \neq j) \\ 0.3 & \text{if } (c(i) \text{ is not a ShL \&\& } c(i) \neq c(j)) \\ 1.7 & \text{if } (c(i) \text{ is a ShL \&\& } c(i) = c(j)) \end{cases}$$

$$NewED\,(i,j) = ED\,(i,j) + EDsh(i,j)$$

Table 2. Samples of proposed solutions with the new edit distance calculation

Word to recognize	Solutions	Scheme	Classic ED	New ED
يبعد فعل	يبتعد	افتعل	1	2,9
	مبعد	مفعل	1	2,6
	يبعده	فعل	1	1,3

```
Current-solution = W =select a random word from the cloud
NewED(w)= New Edit distance between the unknown word and
W (current-solution)
Set the initial temperature initial_T
While (temperature >0 and non-convergence)
Neighbor = select the best neighborhood word solution
Calculate delta = NewED (neighbor) - NewED(current-
solution)
If (delta <= 0)
  Current-solution = neighbor
Else
  Select new neighbor with probability e^{-(delta/t)}
End
Decrease the temperature
End
Output the final solution
```

Fig. 7. Simulated annealing algorithm with a new Edit distance

4 Preliminary Experimental Results

The proposed approach has been tested using a database of printed Arabic word images in different fonts and sizes derived from tri-consonantal healthy roots (see Fig. 8), which embody samples extracted from APTI database [16].

Our training corpus is a set of structural primitives that we extract from Arabic word images. The extraction process is expanded in the previous work [15]. We handled 3200 samples derived from 41 tri-consonantal roots by following different schemes. They are also enriched by a variety of agglutinative and flexional features.

Experimental tests showed that the highest recognition rates (99.84%) are achieved at a temperature equal to 100 with stable cooling equal to 0.9. Furthermore, Fig. 9 demonstrates that the higher the cooling is, the more important the recognition rate is, with initial temperature fixed at 100 (for small response time).

Fig. 8. Word samples extracted from APTI Database

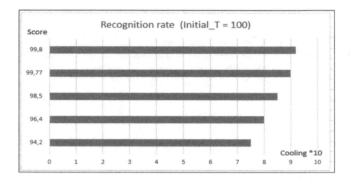

Fig. 9. Evolution of the recognition rate according to cooling

Moreover, the problem of long stagnation in local wells, that we may encounter while using Levenshtein distance, is resolved due to the integration of new costs in the edit distance calculating.

5 Conclusion

To summarize, in this work we proposed a novel approach based on combinatorial optimization and precisely simulated annealing algorithm for the recognition of wide vocabulary of decomposable Arabic words. We chose to integrate linguistic knowledge in both vocabulary structure and the recognition phase through the SA algorithm. In fact, we have organized our vocabulary as a molecular cloud, whose organization matches the Arabic philosophy of constructing words around roots. Each sub-cloud includes neighboring words derived from one root. Preliminary experimentation has been held on a morphological cloud of more than 1000 nodes structured on 41 sub-clouds that correspond to 41 tri-consonantal Arabic roots. We have used simulated annealing on structural primitive vectors extracted from word images including APTI samples. We started by classic elastic comparison. Initial results are promising but with stagnation limit in local wells. To remedy this weakness, we have integrated linguistic knowledge in the edit distance calculation and we have precisely highlighted the scheme letters to help the SA algorithm to achieve the optimal solution, which corresponds to the vocabulary structure.

As perspectives, first, we will try to extend our training corpus in order to generate a very wide molecular morphological cloud (MMC). Second, we propose to integrate a whole model (neural or markovian) in each node, so that a node would be able to neither calculate an ED nor take part in the root research solution within the eventually enormous cloud.

References

1. Touj, S., Ben Amara, N., Amiri, H.: A hybrid approach for off-line Arabic handwriting recognition based on a planar Hidden Markov Modeling. In: ICDAR 2007, Brazil, pp. 964–968 (2007)
2. Avila, J.M.: Optimisation de modèles markoviens pour la reconnaissance de l'écrit. Ph.D., University of Rouen (1996)
3. Cheriet, M., Beldjehem, M.: Visual processing of Arabic handwriting: challenges and new directions. In: SACH 2006, India, pp. 1–21 (2006)
4. Kanoun, S., Alimi, A.M., Lecourtier, Y.: Natural language morphology integration in off-line Arabic optical text recognition. IEEE Trans. Syst. Man Cybern.—Part B: Cybern. **41**(2), 579–590 (2011)
5. Ben Cheikh, I., Allagui, I.: Planar Markovian approach for the recognition of a wide vocabulary of Arabic decomposable words. In: ICDAR, pp. 1031–1035 (2015)
6. Levenshtein, V.: Binary codes capable of correcting deletions, insertions and reversals. SOL Phys. Dokl. **10**, 707–710 (1966)
7. Anigbogu, J.: Reconnaissance de textes imprimés multifontes àl'aide de modèles stochastiques et métriques. Ph.D., University of Nancy 1 (1992)
8. Rani, S., Singh, J.: Enhancing Levenshtein's edit distance algorithm for evaluating document similarity. In: Sharma, R., Mantri, A., Dua, S. (eds.) ICAN 2017. CCIS, vol. 805, pp. 72–80. Springer, Singapore (2018). https://doi.org/10.1007/978-981-13-0755-3_6
9. Huang, K., Hsieh, Y.: Very fast simulated annealing for pattern detection and seismic. In: IEEE International Geoscience and Remote Sensing Symposium, IGARSS 2011, Vancouver, BC, Canada, 24–29 July 2011
10. Goyal, A., Sourav, P.A., Thangavelu, A.: A comparative analysis of simulated annealing based intuitionistic fuzzy k-mode algorithm for clustering categorical data. Int. J. Comput. Inf. Syst. Ind. Manag. Appl. **9**, 232–240 (2017)
11. Baptiste, A.: Les métaheuristiques en optimisation combinatoire. Thesis to obtain probative exam on computing, National Conservatory of Arts and Crafts, Paris (2006)
12. Gueddah, H.: La correction orthographique des textes arabes: Contribution àla résolution d'ordonnancement et de l'insuffisance des lexiques. Ph.D. University of Mohamed V, Rabat (2017)
13. Carbonnel, S., Anquetil, E.: Apprentissage automatique d'une distance d'édition dédié à la reconnaissance d'écriture manuscrite. In: CIFED (2004)
14. Xinchao, Z.: Simulated annealing algorithm with adaptive neighborhood. Appl. Soft Comput. **11**, 1827–1836 (2011)
15. Ben Cheikh, I., Zouaoui, Z.: HMM based classifier for the recognition of roots of a large canonical Arabic vocabulary. In: ICPRAM, pp. 244–252 (2013)
16. https://diuf.unifr.ch/diva/APTI/

Pattern Recognition and Applications

Revisiting Machine Learning from Crowds a Mixture Model for Grouping Annotations

Francisco Mena[(✉)] and Ricardo Ñanculef[(✉)]

Federico Santa María University, Valparaíso, Chile
francisco.mena@alumnos.inf.utfsm.cl, jnancu@inf.utfsm.cl

Abstract. Today, supervised learning is widely used for pattern recognition, computer vision and other tasks. In this setting, data need to be explicitly annotated. Unfortunately, obtaining accurate labels can be difficult, expensive and time-consuming. As a result, many machine learning projects rely on labelling processes that involve *crowds*, i.e. multiple subjective and inexpert annotators. Handling this noise in a principled way is an important challenge for machine learning, called learning from crowds. In this paper, we present a model that learns patterns of label noise by grouping annotations. In contrast to previous art, we do not model specific labeling patterns for each annotator but explain the data using a fixed-size mixture model. This approach allows to handle a sparse distribution of labels among annotators and obtain a model with less parameters that can scale better to large-scale scenarios. Experiments on real and simulated data illustrate the advantages of our approach.

Keywords: Learning from crowds · Mixture model ·
Multiple annotations · Clustering

1 Introduction

In the last years, artificial intelligence has been widely spread into several areas of science and industry. Many of these applications rely on *supervised learning*, i.e., methods capable to realize an input-output mapping from large amounts of data (inputs) annotated with *ground-truth* labels (output). In many real-world tasks however, obtaining accurate labels can be difficult or infeasible. Consider, for instance, the problem of classifying a bio-medical image into a set of clinical conditions of interest. The ground-truth label could only be obtained after performing slow, expensive and invasive experiments in physical labs. Collecting multiple subjective, but possibly inaccurate labels, from annotators of varying levels of expertise, is often more feasible and cheaper [10]. Current crowdsourcing platforms, such as Amazon Mechanical Turk (AMT), are making this procedure more common, especially in computer vision and natural language processing tasks. Unfortunately, as inexpert annotators can be inaccurate, spammer or even malicious, training a traditional supervised model, with annotations

© Springer Nature Switzerland AG 2019
I. Nyström et al. (Eds.): CIARP 2019, LNCS 11896, pp. 493–503, 2019.
https://doi.org/10.1007/978-3-030-33904-3_46

collected in this way, is often ineffective. Similarly, simple aggregation rules such as majority voting, that reduce crowd annotations into a single label, can fail if the expertise of the different annotators vary significantly [12] or if, as usual in crowd-sourcing platforms, the distribution of labels among annotators is sparse. The problem of learning from annotations of varying reliability can be traced back to [2]. Here, Dawid and Skene proposed a method, based on the EM algorithm, that automatically detects the ability of each annotator and estimates a consensus label that can be used to train a standard classifier. Many subsequent methods are extensions of this framework. For instance, Raykar et al. [7] proposed to directly train the ground-truth predictor in the maximization step of the EM algorithm. Kajino et al. [4] proposed to train a separate model for each annotator and then infer a consensus model rather than consensus label. More recently, Albarqouni et al. [1] have proposed the use of deep learning to implement the ground-truth predictor of [7] and Rodrigues et al. [9] introduced more simple training procedures based on back-propagation method.

A common limitation of the current models is that the number of learnable parameters becomes very large as the number of annotators increases, limiting their scalability to massive crowd-sourcing scenarios. In this paper, we present a model for learning from crowds that detects patterns of label noise by grouping/clustering annotations together. In contrast to previous work, we do not model specific labeling patterns for each annotator, but explain the annotations using a fixed-size generative mixture model. As preliminary experiments confirm, this allows to obtain a method that can scale better to large-scale scenarios and can improve the state-of-the-art in sparse annotation scenarios.

The remainder of this paper is organized as follows: Sect. 2 formalizes the problem and introduces the notation used in for the proposed method, that is present in Sect. 3; Sect. 4 provides a discussion of related works; Sect. 5 experimentally compares our approach with baseline methods; finally, Sect. 6 summarizes the conclusions of this work.

2   Problem Statement and Notation

Consider an input pattern $x \in \mathbb{X}$ and a ground-truth label $z \in \mathbb{Z}$ observed with probability distribution $p(x)$ and $p(z|x)$ respectively. The goal of a supervised learning algorithm is to estimate the conditional $p(z|x)$ from a set of examples of the form $S = \{(x^{(1)}, z^{(1)}), \ldots, (x^{(N)}, z^{(N)})\}$, where $(x^{(i)}, z^{(i)}) \sim p(x, z) \ \forall i \in [N]$. More specifically, given a loss function $Q : \mathbb{Z} \times \mathbb{Z} \to \mathbb{R}$ and a hypothesis space $\mathcal{H} \subset \mathbb{Z}^{\mathbb{X}}$, a supervised learning algorithm attempts to minimize $\mathbb{E}_{x,z}(Q(f(x), z))$ in $\mathcal{H}$ when $p(x, z)$ is unknown and only S is given.

In *learning from crowds*, one has the same objective, but the ground-truth labels $z^{(i)}$ corresponding to the input patterns $x^{(i)}$ are not observed. Instead, one is given multiple noisy labels $\mathcal{L}_i = \{y_i^{(1)}, \ldots, y_i^{(T_i)}\}$, $y_i^{(\ell)} \in \mathbb{Z}$ for each training pattern $x^{(i)}$. These labels have been collected from T_i annotators and do not follow the ground-truth distribution $p(z|x)$, but are observed according to an unknown labelling process $p(y^{(\ell)}|x, z)$, which is another objective to study.

In general, the annotations $\mathcal{L}_i$ for a pattern $\boldsymbol{x}^{(i)}$ come from a subset $\mathcal{A}_i$ of the set of all the annotators $\mathcal{A}$ participating on the labelling process. A common assumption is that $\mathcal{A}_i = \mathcal{A}$ (**dense** scenario). A more challenging problem is the scenario in which a variable number of labels is collected by data point and annotator, i.e. $|\mathcal{A}_i| \neq |\mathcal{A}_j| < |\mathcal{A}| = T$ (the **sparse** scenario). Furthermore, we are interested in the so-called **Global** scenario, in which we are given $\mathcal{L}_i$, but we do not know which annotators provided the labels i.e., we know $|\mathcal{A}_i|$ but not $\mathcal{A}_i$. The opposite scenario, referred to as **Individual** , allows to study the properties of each annotator separately.

Focus. For sake of simplicity, we concentrate in the pattern recognition case, that is, we let $\mathbb{Z}$ be a small set of K categories or classes $\{c_1, c_2, \ldots, c_K\}$.

3   Proposed Method

3.1   Model Specification

As in previous works, we represent the ground-truth label as a latent/hidden variable z with (unknown) probability distribution $p(z|\boldsymbol{x})$. To explain an annotation $\boldsymbol{y} \in \mathbb{Z}$, assigned to an input pattern $\boldsymbol{x}$, we propose a generative finite mixture model (GMM) of the form

$$p(\boldsymbol{y} \mid \boldsymbol{x}) = \sum\nolimits_{m=1}^{M} p(\boldsymbol{y} \mid \boldsymbol{x}, \boldsymbol{g} = m) \cdot p(\boldsymbol{g} = m \mid \boldsymbol{x}) = \sum\nolimits_{m=1}^{M} p_m(\boldsymbol{y}|\boldsymbol{x}) \cdot \alpha_m, \quad (1)$$

where $p_m(\boldsymbol{y}|\boldsymbol{x})$ represents one of M possible sub-models, $\boldsymbol{g}$ is a categorical random variable with values in $[M] = \{1, 2, \ldots, M\}$ identifying the group/component that generated the observation $\boldsymbol{y}$, and $\alpha_m = p(\boldsymbol{g} = m)$ is the *a-priori* probability that pattern $\boldsymbol{x}$ is annotated according to $p_m(\boldsymbol{y}|\boldsymbol{x})$. Note that we are assuming that the mixing coefficients α_m are independent of $\boldsymbol{x}$. If we relax this assumption, we obtain a mixture of experts model (MOE) with gating functions $\alpha_m(\boldsymbol{x})$.

The components $p_1(\boldsymbol{y}|\boldsymbol{x}), \ldots, p_M(\boldsymbol{y}|\boldsymbol{x})$ in (1) represent different *annotation patterns* that can occur in the labelling process. They may correspond to clusters/groups of annotators that follow similar rules to annotate data or groups of annotations for which similar mistakes were made. The relationship between an annotation $\boldsymbol{y}$ and the ground-truth z for $\boldsymbol{x}$ is obtained as follows

$$p_m(\boldsymbol{y} = j \mid \boldsymbol{x}) = \sum\nolimits_{k=1}^{K} p(\boldsymbol{y} = j, z = k \mid \boldsymbol{x}, \boldsymbol{g} = m) \qquad (2)$$

$$= \sum\nolimits_{k=1}^{K} p(\boldsymbol{y} = j \mid \boldsymbol{g} = m, z = k) \cdot p(z = k \mid \boldsymbol{x}),$$

where K is the number of classes and the second line was obtained by assuming that $\boldsymbol{y}$ is conditionally independent of $\boldsymbol{x}$ given z, in order to keep a simple model. Indeed, this simplification allow us to parametrize $p_m(\boldsymbol{y}|\boldsymbol{x})$ using only K^2

parameters per sub-model and a single predictive model $f(x; \theta)$ that approximates the ground-truth distribution $p(z|x)$, as Table 1 summarized. Substituting (2) into (1), we obtain the specification of the proposed model for annotation y

$$p(y \mid x) = \sum_{k=1}^{K} \sum_{m=1}^{M} p(y \mid g = m, z = k) \cdot p(z = k \mid x) \cdot p(g = m) \quad (3)$$

3.2 Learning Objective

We start by introducing a data representation that full-fills the requirements of the **Global** and **sparse** scenarios defined in Sect. 2. We define $r^{(i)}$ to be the K-dimensional vector whose components $r_j^{(i)}$ are the frequencies of label c_j among the annotations $\mathcal{L}_i$ of a pattern $x^{(i)}$. If we assume that those annotations are conditionally independent given $x^{(i)}$, we obtain that $r^{(i)}$ follows a Multinomial distribution with sample size T_i and probabilities $p_{ij} = p(y = j|x^{(i)})$ given by the model parametrization (see Table 1) on Eq. (3). The conditional log-likelihood of the data $G = \{(x^{(i)}; r^{(i)})\}_{i=1}^{N}$ is thus given by

$$\ell(\Theta) = \sum_{i}^{N} \log p_{\Theta}(r^{(i)} \mid x^{(i)}) = \sum_{i}^{N} \log \left(\text{const} \cdot \prod_{j=1}^{K} p_{\Theta}(y = j \mid x^{(i)})^{r_j^{(i)}} \right)$$

$$= \text{const} + \sum_{i}^{N} \sum_{j}^{K} r_j^{(i)} \cdot \log p_{\Theta}(y = j \mid x^{(i)})$$

$$= \text{const} + \sum_{i}^{N} \sum_{j}^{K} r_j^{(i)} \cdot \log \left(\sum_{m,k} \beta_{k,j}^{(m)} \cdot f_k(x^{(i)}; \theta) \cdot \alpha_m \right). \quad (4)$$

The parameters of the proposed model, called *Crowd Mixture Model* (CMM), can be learnt to maximize $\ell(\Theta)$. Unfortunately, due to the log-sum, this optimization is not straightforward. We address this issue using the EM algorithm [3].

3.3 Training Procedure

By the Jensen inequality, we can consider any bi-variate distribution $q_{ij}(g, z)$ assigning annotations among groups and ground-truth categories, to obtain the following lower bound of $\ell(\Theta)$

Table 1. Model parametrization. Entry (k, j) of $\beta^{(m)}$ represents the probability that a pattern of class $z = k$ is annotated as $y = j$ by the group/component $g = m$. The model $f(x; \theta)$, used to predict the ground-truth of x, may have many parameters $|\theta|$, but this number is independent of the number of annotators T.

Term	Model	# Parameters
$p(g)$	Mixing coefficients $\alpha_m = p(g = m)$	$M - 1$
$p(y\|g, z)$	Confusion matrix $\beta^{(m)}$ for group m	$MK(K - 1)$
	$p(y = j\|g = m, z = k)$	
$p(z\|x)$	Neural net $f(x; \theta)$	Indep. of T

$$\ell(\Theta) \geq \text{const} + \sum_{i,j} r_j^{(i)} \left[\sum_{m,k} q_{ij}(m,k) \cdot \log \left(\frac{\beta_{k,j}^{(m)} \cdot f_k(\boldsymbol{x}^{(i)};\theta) \cdot \alpha_m}{q_{ij}(m,k)} \right) \right]. \quad (5)$$

The EM algorithm now follows easily. In one step, we improve our estimate of $q_{ij}(\cdot)$ to make the bound tight. Then, we optimize the lower bound in the model parameters Θ. The iteration of these two steps is guaranteed to converge to a local maximum of $\ell(\Theta)$. Exact solutions for our model are provided below.

E-step. For grouping the annotations based on ground-truth, we obtain

$$q_{ij}(m,k) = \tfrac{1}{N_{ij}} \beta_{k,j}^{(m)} f_k(\boldsymbol{x}^{(i)};\theta)\alpha_m, \text{ with } N_{ij} = \sum_{m',k'} \beta_{k',j}^{(m')} f_{k'}(\boldsymbol{x}^{(i)};\theta)\alpha_{m'}.$$

M-step. For the mixing coefficients and confusion matrices, we obtain

$$\alpha_m = \frac{\sum_{ij} r_j^{(i)} \cdot q_{ij}(m,\cdot)}{\sum_{ij} r_j^{(i)}}, \quad \beta_{k,j}^{(m)} = \frac{\sum_i q_{ij}(m,k) \cdot r_j^{(i)}}{\sum_{ij'} q_{ij'}(m,k) \cdot r_{j'}^{(i)}}, \quad (6)$$

where $q_{ij}(m,\cdot) = \sum_k q_{ij}(m,k)$, then confusion matrix is a weighted average of annotations of that group. Now, defining $q_{ij}(\cdot,k) = \sum_m q_{ij}(m,k)$ and $\bar{r}_k^{(i)} = \sum_j q_{ij}(\cdot,k) r_j^{(i)}$, we obtain the following objective to minimize for the neural net:

$$J(\theta) = \sum_{i,k} -\bar{r}_k^{(i)} \cdot \log f_k(\boldsymbol{x}^{(i)};\theta) \propto \sum_i \mathbb{H}\left(\bar{\boldsymbol{p}}^{(i)}, f(\boldsymbol{x}^{(i)};\theta) \right), \quad (7)$$

where $\mathbb{H}(\cdot,\cdot)$ is the *categorical cross entropy loss* between the neural net and a "consensus" distribution on the categories, $\bar{\boldsymbol{p}}_k^{(i)} = \bar{r}_k^{(i)} / \sum_{k'} \bar{r}_{k'}^{(i)}$, which has been computed for $\boldsymbol{x}^{(i)}$ considering the confusion matrices and the mixing coefficients.

3.4   Group Assignment

Our model allows to cluster annotations and annotators, even outside the training data. Given any set of annotations $\mathcal{L} = \{\mathcal{L}_i\}$ for a pattern $\boldsymbol{x}^{(i)}$, we can compute the probability that these annotations were generated by the component p_m in our model as

$$p(\boldsymbol{g} = m | \mathcal{L}, X) = \frac{p(\mathcal{L}|\boldsymbol{g} = m, X)\, p(\boldsymbol{g} = m|X)}{\sum_{m'} p(\mathcal{L}|\boldsymbol{g} = m', X)\, p(\boldsymbol{g} = m'|X)} = \frac{p_m(\boldsymbol{y}_i^{(\ell)}|\boldsymbol{x}^{(i)})\alpha_m}{\sum_{m'} p_{m'}(\boldsymbol{y}_i^{(\ell)}|\boldsymbol{x}^{(i)})\alpha_{m'}}.$$

The probability $p(\boldsymbol{g} = m|\boldsymbol{a})$ that an annotator $\boldsymbol{a}$ belongs to the group m can be estimated with all her annotations. In addition, we can estimate the confusion matrix of an annotator as $\beta_{\boldsymbol{a}} = \sum_m p(\boldsymbol{g} = m|\boldsymbol{a}) \cdot \beta^{(m)}$.

4   Related Work

Existing methods to deal with multiple annotations can be grouped as follows.

Simple Aggregation Methods. These methods use simple summary statistics to reduce the crowd annotations into a single label that can be accepted by standard classifiers. The most used technique of this type is *Majority Voting* (MV), which has two versions in classification problems [8]: *hard-MV*, that selects the most frequent class among the annotations, and *soft-MV*, that defines the output of prediction as the relative frequency of the classes. As shown in [12], the accuracy of MV methods is limited if the annotators have very different levels of accuracy or in cases in which data points do not have many annotations.

Methods Without Predictive Model. These techniques also reduce crowd annotations into a single label and train a predictive model in a separate step. However, they devise specialized techniques to deal with annotators of varying expertise. A pioneer method is the algorithm of Dawid and Skene (DS) [2]. Here, the ability of each annotator is represented using a confusion matrix that can be learnt, together with the ground-truth of the training data, using the EM algorithm. Recently, [13] proposed an initialization method for the EM algorithm that allows to speed-up DS.

Methods with Predictive Model. These methods learn the ground-truth of the training data and the predictive model f approximating $p(z|x)$ jointly, which avoids a second learning stage and allows the model f to learn labelling patterns that depend on x. For instance, Raykar et al. [7] extended DS, using a logistic regression model to implement $p(z|x)$. Almost simultaneously, Yan et al. [11] proposed to use a logistic model to predict the ability of the annotators. A method that avoids the use of the EM algorithm is presented by Kajino et al. [4]. It trains a logistic model for each annotator and then creates a consensus model. Unfortunately the complexity of [4,11] is increased considerably due to the large number of parameters per annotator. Addressing this issue, [8] proposed to change the latent variable of [6], modeling the reliability of each annotator. The main assumption is that an annotator provides completely random labels or annotates data according to a common baseline model. Unfortunately, this assumption represents two extreme possibilities that rarely take place in practice.

Deep Learning. Recent works have proposed the use of neural network models to implement $p(z|x)$. For example, [1] extends [7] using a convolutional net and applies the model to a real cancer detection problem. [5] presents two methods that avoid the effect of *label noise* in neural network training. Unfortunately, a single confusion matrix is considered and it needs to be known before training. It is not evident how to use this method with multiple inaccurate annotators. Rodrigues et al. [9] encode the confusion matrices as additional weights of the neural network, avoiding the use of the EM algorithm. Unfortunately, the size of the so-called "crowd layer" grows linearly in the number of annotators.

Discussion. As pointed out in [14], nowadays there is no method that is superior to the others in all the cases, because different assumptions have to be fulfilled to achieve good results. However, algorithms using a confusion matrix, as our method, to represent the ability of the annotators perform experimentally better than the others [14]. However, while almost all methods focus on modeling each annotator separately (**Individual** scenario), we propose a model with a fixed number of components, into which annotations and annotators can be allocated. As shown in Table 1 this makes the number of parameters independent of T. Besides *computational efficiency*, grouping annotations allows to increase *statistical efficiency*, especially in scenarios where annotators provide a small number of labels and so the estimation of the confusion matrices has to be performed using very few data.

5 Experiments

We evaluate our method on real and simulated scenarios, comparing it against four baselines from the state-of-the-art: *DL-DS* [2], *DL-EM* ([1] and generalized in [9]), and both versions of MV [8]: *hardMV* and *softMV*. We also include the upper bound performance of a model trained with the ground-truth, referred to as *Ideal*. In the vein of latest works, all the methods employ neural networks to implement the ground-truth predictor. All our code is made publicly available[1].

Simulated Scenario. To compare the methods on a controlled scenario, we simulated a crowd-sourcing process with annotators of varying expertise. Following [4,8], we simulated M levels of ability, by training a neural net on the ground-truth and randomly perturbing its weights with different levels of noise. As we use a confusion matrix to represent the ability of annotators, the matrix of each perturbed model was first calculated. Then, we created T annotators by selecting one of the M ability levels according to a probability distribution $p(g)$. To simulate sparse annotations, each data point is labelled by a random subset of the annotators T_i such that, in average, we obtain $\bar{T}_i$ annotators per point and a density of $D_t \approx N \cdot \bar{T}_i / T$ labels per annotator. Each annotator provides a label based on the ground-truth and her ability, i.e, the confusion matrix of the group. This annotation process is applied in two different flavors. In Setup (1), we simulate three uncorrelated isotropic Gaussians (representing classes), with 1000 data points each, centered on $(-0.5; 0)$, $(0.5; 0)$ and $(0; 0.5)$, with $\sigma^2 = 0.4^2$ (homocedasticity). We set $\bar{T}_i = 5$, $M = 3$ (experts, inexperts, spammers), and $p(g) = (0.25; 0.55; 0.20)$. In Setup (2): we use the well-known CIFAR-10 dataset, composed of 60000 real images, classified into 10 categories, and set $\bar{T}_i = 3$, $M = 4$ (experts, inexperts, highly inexpert, spammers), $p(g) = (0.20; 0.45; 0.15; 0.20)$.

Real Data. To evaluate the methods on a real crowd-sourcing scenario, we followed the setup of [9] on the LabelMe dataset. It contains 2688 images of 256×256 resolution, labelled into 8 possible classes by $T = 59$ annotators on Amazon Mechanical Turk. Each image has $\bar{T}_i = 2.6$ annotations in average, which leads to a density of $D_t = 43.2$ labels per annotator.

---

[1] https://github.com/FMena14/MixtureofGroups.

Table 2. Test accuracy of the different methods on a simulated crowd-sourcing scenario for values of T (columns) ranging from $T = 100$ to $T = 10000$. Marker † represents that the method could not be executed due to insufficient memory (16 GB available).

Method	Setup (1)						Setup (2)					
	100	500	1500	3500	6000	10000	100	500	1500	3500	6000	10000
softMV	69.34	66.21	66.87	68.48	67.00	66.49	63.35	65,90	63.59	60.07	63.21	64.20
hardMV	79.57	82.49	80.51	81.57	74.30	79.07	71.09	69,50	68.48	69.09	70.08	66.01
DL-DS	94.66	93.89	92.28	90.00	89.69	85.13	71.33	68,49	68.08	66.86	†	†
DL-EM	93.97	93.99	92.18	88.27	76.47	67.01	81.38	80,42	77.81	69.81	†	†
CMM	90.53	91.07	91.66	90.45	90.26	90.46	78.83	78,36	79.35	77.92	78.45	78.96
Ideal	94.75						83.77					

Table 3. Performance of the different methods in a real crowd-sourcing scenario (LabelMe). Marker ◇ represents no change with respect to the **Individual** setting. Acc. stands for Accuracy. Iters stands for iterations to converge.

Method	Individual setting					Global setting			
	Iters	Train Acc.	Test Acc.	I-JS	G-JS	Iters	Train Acc.	Test Acc.	G-JS
softMV	9.2	83.32	81.69	0.216	0.024		◇	◇	◇
hardMV	11.8	80.34	79.95	0.225	0.035		◇	◇	◇
DL-DS	10.6	84.30	83.57	0.153	0.036	4.1	12.63	14.08	0.473
DL-EM	3.9	85.18	83.07	0.295	0.259	3.0	78.02	75.92	0.467
CMM	7.2	84.58	83.10	0.234	0.054		◇	◇	◇
Ideal	8	97.90	92.09				◇	◇	

Training and Evaluation Details. All the methods are trained until convergence (change in loss or parameters below a threshold) up to a maximum of 50 iterations. To obtain more significant results, we perform 20 runs of each experiment and average the results. The initialization of the EM algorithm is done with *softMV* and, for our model, a K-means clustering is previously done over annotations. Multiple restarts (20) was applied for *DL-EM* and our method. In the M step, the neural nets are executed one epoch using the Adam optimizer. To implement the predictive model $f(z; \theta)$, we choose an architecture that, according to previous works, is known to be appropriate for each dataset. As, for all the datasets, the ground-truth is available, we evaluate the methods measuring the **Accuracy** of the predictive model on the test set. To evaluate the ability of the method to estimate the confusion matrices, on the train set, we compute the *Jensen-Shannon* divergence in two variants. We measure the **I-JS**, the average divergence between the real and the predicted matrices of each annotator, as well as **G-JS**, the divergence between the real and predicted global matrices, that represent the behavior of all the annotators/annotations in the labelling process. On real dataset, the M chosen is the one with the highest log-likelihood.

Table 4. Metrics on confusion matrices found on LabelMe dataset.

Group	α_m	I_{sim}	$\mathbb{H}$
1	0.99	0.91	0.48
2	0.01	0.02	2.08
3	0.00	0.03	2.08

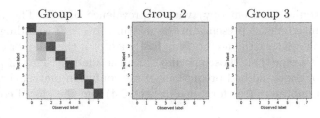

Fig. 1. Confusion matrices found on LabelMe dataset.

Results on the Simulated Data. Table 2 shows the accuracy obtained by the different methods in the simulated scenario, as we vary the number of annotators T. Consistent with previous results [14], we observe that learning-based methods, can significantly improve on simple aggregation techniques such as MV. It can also be seen that, as T grows and thus the number of labels per annotator $(D_t \propto \bar{T}_i/T)$ decreases, the methods DL-EM and DL-DS suffer a sharp fall in performance. In contrast, in both setups, the accuracy of our method is more robust to a change in the density of annotations. We attribute this result to the fact that DL-EM and DL-DS need to estimate a separate sub-model for each annotator (confusion matrix) and thus require that D_t keeps high in order to maintain their accuracy. In contrast, the number of estimated components in our method is independent of T. When the number of annotators is small, our method is competitive, but it is outperformed by more complex models. However, when T is greater than some threshold, in this case 3500, our method achieves the best performance. In some extreme cases, existing learning-based methods cannot be executed due to the large number of parameters in the formulation.

Results on Real Data. We report the results of the LabelMe dataset (using $M = 3$) in Table 3. We experiment with the **Individual** and **Global** settings introduced in Sect. 2. In the first scenario, we know which annotators provided which labels, thus having a quite dense setting. In the second scenario, we do not have that information and thus the annotations are treated independently, leading to a density of $D_t = 1$ (where T grows to 2547 and $\bar{T}_i$ keeps). In the denser case, all the learning-based methods achieve a similar test accuracy ($\sim$83%). In the sparse setting however, the accuracy of DL-EM and DL-DS suffers an important decrease ($\sim$78% and $\sim$13% respectively), while the accuracy of our method and MV is robust to this change. This shows the disadvantage of methods that model each annotator separately compared to methods based on a **Global** representation that can group annotations together.

Groups Analysis. We visualize in Fig. 1 the confusion matrices found by our method in the LabelMe dataset. We also show in Table 4 the entropy of the confusion matrices $\mathbb{H}$, their similarity I_{sim} with respect to the identity matrix (computed as 1 minus the normalized JS divergence, to obtain a number in $[0, 1]$) and the value of the mixing coefficients α_m (prevalence of each group). We conclude that the method found a group of annotators with a quite expert behavior (high I_{sim}, low $\mathbb{H}$) with a presence of 99%, and a group of spammers (quite high entropy) with a prevalence of 1%. The third component has an insignificant presence in the mixture ($\alpha_m = 0$ rounding at two decimals) which shows that the method can easily adapt if the number of real groups in the data is lower than those specified into the model. Figure 2 presents visual examples of good and bad predictions of the confusion matrix corresponding to individual annotators (see formulae in Sect. 3.4) and the global confusion matrix.

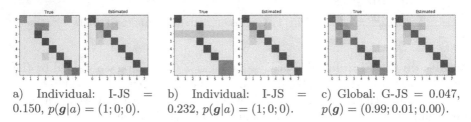

a) Individual: I-JS = 0.150, $p(\boldsymbol{g}|a) = (1; 0; 0)$.

b) Individual: I-JS = 0.232, $p(\boldsymbol{g}|a) = (1; 0; 0)$.

c) Global: G-JS = 0.047, $p(\boldsymbol{g}) = (0.99; 0.01; 0.00)$.

Fig. 2. Examples of confusion matrices (True vs Estimated) on LabelMe dataset.

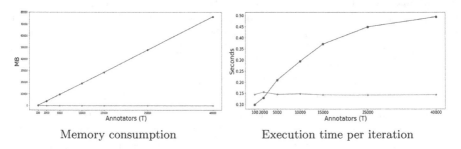

Memory consumption

Execution time per iteration

Fig. 3. Comparison by increasing T on simulated data setup (1). *DL-EM* is presented in blue and our (*CMM*) is presented in green. (Color figure online)

Computational Efficiency. In Fig. 3, we compare the execution time and memory consumption of *CMM* and *DL-EM* in the simulated setup (1) scenario. In contrast to *CMM*, the computational complexity of *DL-EM* increases monotonically with the value of T. This shows that our method can scale better to scenarios with a large number of annotators as expected from its formulation.

6   Conclusions

We presented a model for learning from crowds that, in contrast to existing methods, does not represent annotators separately but has a fixed number of components into which annotations can be grouped together. Our experiments show that this model achieves competitive accuracy in scenarios with several labels per annotator, but can outperform the baselines when the distribution of labels is sparse. The method is more scalable than other approaches in cases with large number of annotators, also adapts naturally when the amount cannot be determined because the individual annotations are not present. In future work, we plant to extend our method in order to avoid the use of the EM algorithm.

References

1. Albarqouni, S., Baur, C., Achilles, F., Belagiannis, V., Demirci, S., Navab, N.: AggNet: deep learning from crowds for mitosis detection in breast cancer histology images. IEEE Trans. Med. Imaging **35**(5), 1313–1321 (2016)
2. Dawid, A.P., Skene, A.M.: Maximum likelihood estimation of observer error-rates using the EM algorithm. JSTOR: Ser. C (Appl. Stat.) **28**(1), 20–28 (1979)
3. Dempster, A.P., Laird, N.M., Rubin, D.B.: Maximum likelihood from incomplete data via the EM algorithm. JSTOR: Ser. B **39**(1), 1–22 (1977)
4. Kajino, H., Tsuboi, Y., Kashima, H.: A convex formulation for learning from crowds. Trans. Jpn. Soc. Artif. Intell. **27**, 133–142 (2012)
5. Patrini, G., Rozza, A., Menon, A.K., Nock, R., Qu, L.: Making deep neural networks robust to label noise: a loss correction approach. In: Proceedings of IEEE Conference Computer Vision Pattern Recognition (CVPR), pp. 2233–2241 (2017)
6. Raykar, V.C., Yu, S.: Eliminating spammers and ranking annotators for crowd-sourced labeling tasks. JMLR **13**(Feb), 491–518 (2012)
7. Raykar, V.C., et al.: Learning from crowds. J. Mach. Learn. Res. **11**, 1297–1322 (2010)
8. Rodrigues, F., Pereira, F., Ribeiro, B.: Learning from multiple annotators: distinguishing good from random labelers. Pattern Recogn. Lett. **34**, 1428–1436 (2013)
9. Rodrigues, F., Pereira, F.C.: Deep learning from crowds. In: The Thirty-Second AAAI Conference on Artificial Intelligence (AAAI) (2018)
10. Snow, R., O'Connor, B., Jurafsky, D., Ng, A.Y.: Cheap and fast–but is it good?: evaluating non-expert annotations for natural language tasks. In: Proceedings of Conference on Empirical Methods in Natural Language Processing, pp. 254–263 (2008)
11. Yan, Y., et al.: Modeling annotator expertise: learning when everybody knows a bit of something. In: Proceedings of the XXX AISTATS, pp. 932–939 (2010)
12. Zhang, J., Wu, X., Sheng, V.S.: Imbalanced multiple noisy labeling. IEEE Trans. Knowl. Data Eng. **27**(2), 489–503 (2015)
13. Zhang, Y., Chen, X., Zhou, D., Jordan, M.I.: Spectral methods meet EM: a provably optimal algorithm for crowdsourcing. JMLR **17**(1), 3537–3580 (2016)
14. Zheng, Y., Li, G., Li, Y., Shan, C., Cheng, R.: Truth inference in crowdsourcing: is the problem solved? Proc. VLDB Endow. **10**(5), 541–552 (2017)

A Model Based on Genetic Algorithm
for Colorectal Cancer Diagnosis

Daniela F. Taino[1]($\boxtimes$), Matheus G. Ribeiro[1], Guilherme Freire Roberto[2],
Geraldo F. D. Zafalon[1], Marcelo Zanchetta do Nascimento[2], Thaína A. Tosta[3],
Alessandro S. Martins[4], and Leandro A. Neves[1]

[1] Department of Computer Science and Statistics,
São Paulo State University (UNESP), R. Cristovão Colombo, 2265,
São José do Rio Preto, São Paulo 15054-000, Brazil
dani_taino@hotmail.com
[2] Faculty of Computation (FACOM), Federal University of Uberlândia (UFU),
Av. João Naves de Ávila, 2121, Uberlândia, Minas Gerais 38400-902, Brazil
[3] Center of Mathematics, Computing and Cognition,
Federal University of ABC (UFABC),
Av. dos Estados, 5001, Santo André, São Paulo 09210-580, Brazil
[4] Federal Institute of Triângulo Mineiro (IFTM),
R. Belarmino Vilela Junqueira S/N, Ituiutaba, Minas Gerais 38305-200, Brazil

Abstract. In this paper we present a method based on genetic algorithm capable of analyzing a significant number of features obtained from
fractal techniques, Haralick texture features and curvelet coefficients, as
well as several selection methods and classifiers for the study and pattern recognition of colorectal cancer. The chromosomal structure was
represented by four genes in order to define an individual. The steps for
evaluation and selection of individuals as well as crossover and mutation
were directed to provide distinctions of colorectal cancer groups with
the highest accuracy rate and the smallest number of features. The tests
were performed with features from histological images H&E, different
values of population and iterations numbers and with the k-fold cross-
validation method. The best result was provided by a population of 500
individuals and 50 iterations applying relief, random forest and 29 features (obtained mainly from the combination of percolation measures
and curvelet subimages). This solution was capable of distinguishing the
groups with an accuracy rate of 90.82% and an AUC equal to 0.967.

Keywords: Genetic algorithm · Colorectal cancer · Feature selection ·
Feature classification

1 Introduction

Colorectal cancer is a malignant tumour that develops on the internal wall of the
intestine (colon) or rectum [2]. The main reasons for studying this disease are
the number os cases and mortality. The International Agency for Research on

© Springer Nature Switzerland AG 2019
I. Nyström et al. (Eds.): CIARP 2019, LNCS 11896, pp. 504–513, 2019.
https://doi.org/10.1007/978-3-030-33904-3_47

Cancer (IARC) presented a study in which colorectal cancer was defined as the third most common cancer in men (746,000 cases) and the second most common one for women (614,000 cases). The number of mortalities was 694,000 and the highest incidence of 52% of deaths occurred in less developed regions of the world [17]. The diagnosis for colorectal cancer can be made through sigmoidoscopy or by colonoscopy. Confirmation occurs by biopsies of the tissues stained with hematoxylin and eosin (H&E) and microscopically analyzed by pathologists.

The main difficulty for a medical diagnosis is the evaluation of the severity of abnormal findings when there are different opinions between inter and intraobservers [7,10]. This fact has motivated the development of systems known as computer-aided diagnosis (CAD) [36] to support specialists in research and decision-making. A common challenge observed in proposals of CAD systems is to indicate the best combination between the selection and classification algorithms to achieve the highest success rates using the smallest features number [10,15]. In this context, solutions obtained from metaheuristics models were relevant for different sorts of medical images. The techniques inspired by analogies found in nature or in evolutionary processes are worth mentioning, such as the methods based on genetic algorithms (GA) for the diagnosis of esophagus cancer [28], lung cancer, brain tumors, prostate cancer and leukemia [21].

A GA is a metaheuristic widely known in the literature and its main advantage in comparison with other evolutionary strategies is to have a structure that makes it possible to represent plausible new organizational forms (individuals) from a successful previous organizational construct (crossover) [5] without losing critical information from the problem [26,34]. Despite of the different strategies considering genetic algorithms for the study and development of CAD, such as diagnosis of cardiac diseases [3] and lung cancer [24], the models available in the literature did not explore the method in order to determine the best combination of features, selection algorithms and classifiers [13,22] in the context of histological images and diagnosis of colorectal cancer. Therefore, in this work we present a method based on a GA capable of analyzing a significative number of features obtained from fractal techniques, Haralick texture features and curvelet coefficients, as well as selection methods and classifiers in order to indicate an acceptable solution for the diagnosis of colorectal cancer. This type of study contributes significantly to the literature focused on the theme, especially with the development and improvement of CAD systems. The main contributions of the proposal are:

1. A method based on genetic algorithm capable of analyzing a significative number of features, selection methods and classifiers for the study and pattern recognition of colorectal cancer;
2. An approach capable of indicating the best features in order to separate benign and malignant colorectal cancer groups;
3. Information about methods and features which support development and enhancement of CAD systems.

2 Methodology

Each individual (genetic's code bearer) was defined as a chromosome structure composed by four genes, represented by integer numbers. The information stored in each gene or genetic code defines a specific combination: (G_{id}) (the individual's identification), selection method (G_{sel}), classification method (G_{clf}) and the number of features considered in the classification process (G_{num}). The initial values attributed to the G_{sel}, G_{clf} and G_{num} genes were random. The structure described is illustrated in Fig. 1. Considering this structure, a population was defined. Each combination is unique and associated with an identifier G_{id} to define an acceptable solution.

$$\boxed{G_{id}}\boxed{G_{sel}}\boxed{G_{clf}}\boxed{G_{num}}$$

Fig. 1. Chromosome structure defined to represent an individual (G_{id})

It is important to emphasize that each gene G_{sel} identifies a method capable of producing a ranking of the most significant features in order to distinguish the datasets under investigation. The explored methods were: T-statistics [11], information Gain [9], relief [23], gain ratio [9] and chi-squared [37]. Features were evaluated by each classifier indicated in the gene (G_{clf}): decision tree [32], J48 [29], random tree [4], random forest [4], multilayer perceptron [12], support vector machine (SVM) [33], K-nearest neighbors (KNN) [14] and KStar (K*) [8]. These techniques were applied on each training set and tested using k-folds cross-validation, with $k = 10$.

The structure of the model requires some parameters as inputs to define the best association, such as: population size (P), maximum number of generations or iterations $(Iter)$, selection threshold (t) (representing who will be selected for reproduction—crossover), genetic mutation probability (m) and maximum number of features $(MaxF)$ defined from the initial set of features. The $MaxF$ parameter allows to limit the number of features that constitutes an individual G_{id}. Considering the input parameters, the proposed method processes the information based on population evaluation, selection of the most fit individuals, reproduction (crossover) and mutation.

2.1 Population Evaluation and Selection

Population evaluation consists in calculating the mean accuracy (the fitness function) produced by each individual, based on the selection and classification techniques indicated in their genes. Therefore, for each individual, a selection method G_{sel} was applied on each of the k training fold and the results were the indexes of the N best features, being N defined by the value drawn for the G_{num} parameter.

The classifier indicated in G_{clf} was trained considering the selected N features. This process was performed for each training file constructed by the $k = 10$

cross-validation technique. The classification was executed in each correspondent test file, composed by chosen features. Therefore, for each individual G_{id}, L accuracy values were obtained, one for each k training test. The average accuracy $MeanAcc(G_{id})$ was calculated by applying Eq. 1:

$$MeanAcc(G_{id}) = \frac{\sum_{i=1}^{L} Acc(G_{id})(i)}{L}. \tag{1}$$

The natural selection behavior proposed by Darwin was considered in the method presented by sorting the accuracy rates $(MeanAcc(G_{id}))$ and selecting individuals with greater values than the selection parameter t $(t = 0.7)$. This model was developed considering the proposal described by Yang and Honavar [35]. Also, it is important to emphasize that chosen individuals were defined as parents in the next generation, by gathering genes (methods and features) capable of providing an acceptable solution: better combinations of features, selection methods and classifiers.

2.2 Crossover and Mutation

Reproduction is responsible for complementing the population on the current generation with individuals produced from those selected in the previous step. This type of approach simulates the sexual reproduction found in several species in nature. The genetic operation of crossover was implemented using the two-point approach, aiming the search for the best solution by replacing both selection and classification methods. Two-point crossover consists in choosing two locus of a chromosome as points of exchange (or pivots) and alternately making the copy of the genes of the parents for the two children generated. In the chromosome structure used in the model proposed, with the exception of the G_{id} gene, the other parts were used to determine the next generations. The mutation operator was applied on children to define the next iteration's population. The mutation operation consisted of a few steps:

- for each new born individual, a random number α is drawn to indicate whether the child should be mutated, considering $\alpha \in \mathbb{R} \mid 0 \leq \alpha \leq 1$;
- if $\alpha > m$, being m mutation probability, which was defined as 0.05% [25], the individual is not mutated. Otherwise, $\alpha \leq m$, the individual will be mutated.

When an individual is submitted to the mutation process, an index β is drawn to indicate which gene must be mutated. The variable β can assume 1, 2 or 3 indexes, which represent the G_{sel}, G_{clf} and G_{num} genes, respectively. Flip mutation was applied on the genes representing lists (G_{sel} and G_{clf}), as well as the creep mutation for the gene that indicates a number (G_{num}). In the flip mutation, a method was replaced by another of the same type. In the creep mutation, a value was subtracted or added to the gene. Considering these mutation processes, the following steps were performed:

- if $\beta = 1, 2$, "take-the-next" strategy was applied on selection methods and classification methods lists.

- if $\beta = 3$, the chosen gene is G_{num}. In this case, a new random number γ is drawn, given by $\gamma = 0$ or $\gamma = 1$. If $\gamma = 1$, G_{num} is incremented by one. Otherwise ($\gamma = 0$), G_{num} is decremented by one. The maximum value for G_{num} is delimited by $MaxF$ (maximum number of features).

The procedures previously described was repeated until the maximum number of generations or if 99% (or more) individuals provide a $MeanAcc$ rate equal to 100%. A summary of the proposed method is shown in Algorithm 1.

Algorithm 1. Proposed Method

1: *generation* ← 0
2: **while** *generation* is less or equal to *Iter* **do**
3:     *Population* ← Initial Population
4:     **for** every individual **do**
5:         Applies G_{sel} and G_{clf} methods on G_{num} features
6:     **end for**
7:     Calculates *MeanAcc*
8:     **if** *MeanAcc* equals 100 OR *generation* equals *MaxIter* **then**
9:         *generation* ← *MaxIter* + 1
10:         Saves Results
11:     **else**
12:         Parents ← Fittest Individuals
13:         Children ← Crossover(Parents)
14:         Children ← Mutation(Children)
15:         *Population* ← *Population* + Children
16:         *generation* ← *generation* + 1
17:     **end if**
18: **end while**

2.3   Colorectal Database and Feature Set

The tests were performed from features extracted from a dataset of histological colorectal cancer images. They were defined by the method described in [30]. The dataset consists of samples derived from 16 H&E colon histology sections from stage T3 or T4 of colorectal adenocarcinoma. Each section belongs to a patient. Areas with different histological architectures were extracted from the sections and the samples were stained with H&E.

For each input image, features were defined by two Fractal Dimension values DF_p [18] and DF_f [27], five lacunarity values (Lac), obtained by area under curve metrics (ARC), skewness (SKW), area ratio (AR), maximum point (MP) and scale of the maximum point SMP, represented by $Lac(1)$ to $Lac(5)$; 14 Haralick texture features (Har) [16], represented by Har(1) up until Har(14), such as angular second moment, correlation and sum of squares; and 15 percolation features (Perc(1) up until Perc(15)) [31], in which ARC, SKW, AR, MP and

SMP metrics were also applied for each percolation function, given by cluster average (C), percolating box ratio (Q) and average coverage ratio of the largest cluster (Γ). Mentioned features were also calculated for the curvelet subimages [6]. The curvelets were calculated through observations made on 4 levels of resolution and a sequence of 8 rotation angles. This approach resulted in 41 curvelet subimages for each colorectal image given and a feature set composed by 1.512 features.

3 Results

The feature set was given as input to our method and analyzed from tests defined with different values for P (numbers of individuals) and $Iter$ (iterations or generations). The purpose was to verify the method's behavior under different situations, as well as identify possible patterns in the context of colorectal images. Results provided by the method are available from Tables 1 and 2, obtained by population values of $P = 50$ and $P = 500$. These values were defined considering works available in the literature [1] [5] and in order to indicate the best combinations in each scenario. The results represent a set of (random) possible solutions involving: selection method, classifier, number of features $(NumF)$ and average accuracy rate $(MeanAcc)$. Area under the ROC curve (AUC) was also collected in each test to complement the performance comparisons of our proposal.

Table 1. Best combination obtained by $P = 50$ for iterations defined as 50, 100 and 500.

Iterations	Selection method	Classifier	NumF	MeanAcc	AUC
50	Relief	J48	146	84.86%	0.852
100	Relief	Random forest	145	87.97%	0.880
500	Gain ratio	J48	506	87.97%	0.896

Table 2. Best combination obtained by $P = 500$ for iterations defined as 50, 100 and 500.

Iterations	Selection method	Classifier	NumF	MeanAcc	AUC
50	Relief	Random forest	29	90.82%	0.967
100	Relief	Random forest	119	90.82%	0.963
500	Relief	Random forest	387	90.82%	0.961

Analyzing the results it is possible to observe considering $P = 50$, the best result was found with 100 iterations. The solution was indicated by relief (selection method) and random forest (classifier). In this case, the accuracy rate was

87.97%. The best case was determined with a significant number of individuals ($P = 500$). The highest accuracy rate with the lowest number of features was indicated with 50 iterations. The solution defined by our method provided an accuracy of 90.82%, computed with 29 features, Relief (selection method) and random forest (classifier). In this case, the value of AUC was 0.967. It is important to mention that despite the same accuracy in the tests performed with 100 and 500 iterations, even with indications of the same selection method and classifier, the difference in the number of features is significant.

Colorectal cancer classification from histological images is the subject of several papers available in the literature, such as those described in [20] and [19]. Therefore, a performance overview obtained with our proposal is presented in Table 3, based on AUC rate (which was measured by all the works used for this verification), total features and classification methods.

Table 3. AUC performance provided by related works developed for the study and classification of colorectal cancer from histological images.

Models	Classifier	Features	AUC
Kather et al. [20]	KNN, SVM and decision tree	74	0.976
Jorgensen et al. [19]	RaF	9	0.960
Proposed method	**Relief and random forest**	29	**0.967**

It is important to observe that direct comparisons cannot be performed to indicate the best approach, since different methodologies and databases were used. Nevertheless, considering the rate provided by our proposal and what was found in the literature, we believe the method is promising and capable of providing an acceptable solution (indication of the highest distinction rate considering the least number of features as possible). Our solution indicated a 0.967 AUC rate, with 29 of 1.512 features, values compatible with important works in the literature directed to the development of CAD systems and colorectal cancer.

One of the advantages of our proposal is identifying the most relevant features and its values (Table 4). It is possible to observe that most features were selected by percolation descriptors and subimages association, totalizing 16 features. Lacunarity attribute was the second most selected type, totalizing 9 features, with measurements obtained (total of 8), mainly of curvelet subimages. Lastly, Haralick's measures contributed with four metrics. On the other hand, multiscale and multidimensional fractal dimension measurements (DF_p and DF_f) were not selected by our strategy to classify colorectal cancer from the H&E images. We believe that this information is important for the CAD system development area. It is possible to observe that area under the curve (ARC), skewness (SKW), area ratio (AR), maximum point (MP) and maximum point scale (SMP) were the most used features for lacunarity and percolation attributes. These metrics were obtained mainly from the combination with curvelet subimages.

Table 4. Discrimination of the selected features obtained in the best result.

DF_p		DF_f		Lac		Har		Perc		Total
Image H& E	Sub image	Image H& E	Sub image	Image H& E	Sub image	Image H& E	Sub image	Image H& E	Sub image	
0	0	0	0	1	8	3	1	0	16	29

4 Conclusion

In this work, a method based on GA capable of finding the best combination of features, selection methods and classifier was proposed in order to provide information for the diagnosis of colorectal cancer from H&E images. This methodology was built from a structured model of evaluation, selection, crossover and mutation. The method presented relevant results. The best solution was determined from 500 individuals and 50 iterations, resulting in 29 features, Relief selection algorithm and random forest classifier. The accuracy rate obtained was 90.82% and the AUC rate was 0.967. Performance was compared to important works available in the literature. The results were relevant, especially when considering the use of comparisons under similar conditions and the number of features considered. As an overview was given from studies developed for colorectal cancer classification, the performance was similar to which is available in the literature, with the differential of discriminating and detailing possible patterns of features indicated for separation of benign and malignant groups of colorectal cancer. In future works we intend to explore different values for the parameters required by our model and types of images. At last, we intend to test our model for pattern recognition in H&E images of lymphomas and breast cancer, with or without normalization of the dyes present in the slides.

Acknowledgments. The authors gratefully acknowledge the financial support of National Council for Scientific and Technological Development CNPq (Grants #427114/2016-0, #304848/2018-2, #430965/2018-4 and #313365/2018-0), the State of Minas Gerais Research Foundation - FAPEMIG (Grant #APQ-00578-18).

References

1. Al-Rajab, M., Lu, J., Xu, Q.: Examining applying high performance genetic data feature selection and classification algorithms for colon cancer diagnosis. Comput. Methods Programs Biomed. **146**, 11–24 (2017)
2. Alteri, R., Kramer, J., Simpson, S.: Colorectal Cancer Facts and Figures 2014–2016, pp. 1–30. American Cancer Society, Atlanta (2014)
3. Anbarasi, M., Anupriya, E., Iyengar, N.: Enhanced prediction of heart disease with feature subset selection using genetic algorithm. Int. J. Eng. Sci. Technol. **2**(10), 5370–5376 (2010)
4. Breiman, L.: Random forests. Mach. Learn. **45**(1), 5–32 (2001)
5. Bruderer, E., Singh, J.V.: Organizational evolution, learning, and selection: a genetic-algorithm-based model. Acad. Manag. J. **39**(5), 1322–1349 (1996)

6. Candès, E.J., Donoho, D.L.: New tight frames of curvelets and optimal representations of objects with piecewise c^2 singularities. Commun. Pure Appl. Math. **57**(2), 219–266 (2004)

7. Chan, H.P., Charles, E., Metz, P., Lam, K., Wu, Y., Macmahon, H.: Improvement in radiologists' detection of clustered microcalcifications on mammograms. Arbor **1001**, 48109–0326 (1990)

8. Cleary, J.G., Trigg, L.E.: K*: an instance-based learner using an entropic distance measure. In: Machine Learning Proceedings, pp. 108–114. Elsevier (1995)

9. Dai, J., Xu, Q.: Attribute selection based on information gain ratio in fuzzy rough set theory with application to tumor classification. Appl. Soft Comput. **13**(1), 211–221 (2013)

10. Doi, K.: Computer-aided diagnosis in medical imaging: historical review, current status and future potential. Comput. Med. Imaging Graph. **31**(4–5), 198–211 (2007)

11. Eltoukhy, M.M., Faye, I., Samir, B.B.: A statistical based feature extraction method for breast cancer diagnosis in digital mammogram using multiresolution representation. Comput. Biol. Med. **42**(1), 123–128 (2012)

12. Gardner, M.W., Dorling, S.: Artificial neural networks (the multilayer perceptron)-a review of applications in the atmospheric sciences. Atmos. Environ. **32**(14–15), 2627–2636 (1998)

13. Gonçalves, E.C., Freitas, A.A., Plastino, A.: A survey of genetic algorithms for multi-label classification. In: 2018 IEEE Congress on Evolutionary Computation (CEC), pp. 1–8. IEEE (2018)

14. Gou, J., Ma, H., Ou, W., Zeng, S., Rao, Y., Yang, H.: A generalized mean distance-based k-nearest neighbor classifier. Expert Syst. Appl. **115**, 356–372 (2019)

15. Gurcan, M.N., et al.: Lung nodule detection on thoracic computed tomography images: preliminary evaluation of a computer-aided diagnosis system. Med. Phys. **29**(11), 2552–2558 (2002)

16. Haralick, R.M.: Statistical and structural approaches to texture. Proc. IEEE **67**(5), 786–804 (1979)

17. IARC: Cancer fact sheets: Colorectal cancer. Technical report, International Agency for Research on Cancer, Lyon, France (2012)

18. Ivanovici, M., Richard, N., Decean, H.: Fractal dimension and lacunarity of psoriatic lesions-a colour approach. Medicine **6**(4), 7 (2009)

19. Jørgensen, A.S., et al.: Using cell nuclei features to detect colon cancer tissue in hematoxylin and eosin stained slides. Cytometry Part A **91**(8), 785–793 (2017)

20. Kather, J.N., et al.: Multi-class texture analysis in colorectal cancer histology. Sci. Rep. **6**, 27988 (2016)

21. Kečo, D., Subasi, A., Kevric, J.: Cloud computing-based parallel genetic algorithm for gene selection in cancer classification. Neural Comput. Appl. **30**(5), 1601–1610 (2018)

22. Khan, A., Qureshi, A.S., Hussain, M., Hamza, M.Y., et al.: A recent survey on the applications of genetic programming in image processing. arXiv preprint arXiv:1901.07387 (2019)

23. Kira, K., Rendell, L.A.: A practical approach to feature selection. In: Machine Learning Proceedings, pp. 249–256. Elsevier (1992)

24. Lu, C., Zhu, Z., Gu, X.: An intelligent system for lung cancer diagnosis using a new genetic algorithm based feature selection method. J. Med. Syst. **38**(9), 97 (2014)

25. Mitchell, M.: An Introduction to Genetic Algorithms. MIT Press, Cambridge (1998)

26. Muni, D.P., Pal, N.R., Das, J.: Genetic programming for simultaneous feature selection and classifier design (2006)
27. Nikolaidis, N., Nikolaidis, I., Tsouros, C.: A variation of the box-counting algorithm applied to colour images. arXiv preprint arXiv:1107.2336 (2011)
28. Paul, D., Su, R., Romain, M., Sébastien, V., Pierre, V., Isabelle, G.: Feature selection for outcome prediction in oesophageal cancer using genetic algorithm and random forest classifier. Comput. Med. Imaging Graph. **60**, 42–49 (2017)
29. Quinlan, J.R.: C4. 5: Programs for Machine Learning. Elsevier, Amsterdam (2014)
30. Ribeiro, M.G., Neves, L.A., do Nascimento, M.Z., Roberto, G.F., Martins, A.S., Tosta, T.A.A.: Classification of colorectal cancer based on the association of multidimensional and multiresolution features. Expert Syst. Appl. **120**, 262–278 (2019). https://doi.org/10.1016/j.eswa.2018.11.034, http://www.sciencedirect.com/science/article/pii/S0957417418307541
31. Roberto, G.F.: Features based on the percolation theory for quantification of non-hodgkin lymphomas. Comput. Biol. Med. **91**, 135–147 (2017)
32. Safavian, S.R., Landgrebe, D.: A survey of decision tree classifier methodology. IEEE Trans. Syst. Man Cybern. **21**(3), 660–674 (1991)
33. Vapnik, V.N.: An overview of statistical learning theory. IEEE Trans. Neural Netw. **10**(5), 988–999 (1999)
34. Whitley, D.: A genetic algorithm tutorial. Stat. Comput. **4**(2), 65–85 (1994)
35. Yang, J., Honavar, V.: Feature subset selection using a genetic algorithm. In: Liu, H., Motoda, H. (eds.) Feature Extraction, Construction and Selection, vol. 453, pp. 117–136. Springer, Heidelberg (1998). https://doi.org/10.1007/978-1-4615-5725-8_8
36. Yu, S., Guan, L.: A CAD system for the automatic detection of clustered micro-calcifications in digitized mammogram films. IEEE Trans. Med. Imaging **19**(2), 115–126 (2000)
37. Zheng, Z., Wu, X., Srihari, R.: Feature selection for text categorization on imbalanced data. ACM SIGKDD Explor. Newslett. **6**(1), 80–89 (2004)

A Spatiotemporal Analysis of Taxis Demand: A Case Study in the Manizales City

Andres Felipe Giraldo-Forero[1,2(✉)], Sebastian Garcia-Lopez[1],
Paula Andrea Rodriguez-Marin[2], Juan Martinez[1],
Yohan Ricardo Céspedes-Villar[1], Oscar Cardona[1], Juan Camilo Acosta[3],
and Luis Carlos Trujillo[3,4]

[1] Centro de Bioinformática y Biología Computacional - BIOS, Manizales, Colombia
[2] Instituto Tecnológico Metropolitano - ITM, Medellín, Colombia
felipegiraldo@itm.edu.co
[3] Centro de Excelencia y Apropiación en Internet de las Cosas,
Bogota, Colombia
[4] Pontificia Universidad Javeriana, Bogotá, Colombia

Abstract. The analysis of urban dynamics has taken on a fundamental role in recent years, even more so considering the accelerated population growth of cities throughout the world. Within this dynamic, one of the most important tasks is urban planning, being able for example to give solution to important problems such as the flow of transport for the improvement of citizen welfare. The following study will present a spatiotemporal analysis of demand flow in taxi service requests in the city of Manizales - Colombia during the year 2016. The study carries out three types of analysis: a spatial analysis that exposes the behavior of requests for taxis throughout the communes of the city, then a temporal analysis is conducted to show the hours of greatest demand and finally the spatiotemporal analysis that gives a general forecast regarding the behavior of taxi requests comprising the second quarter of 2016.

Keywords: Forecasting · Operational dynamics · Smart cities · Time series

1 Introduction

Nowadays, technology has so much impact on society that influences everyday's life decisions. Accordingly, the advent and massification of personal devices that incorporate global geopositioning technology (GPS) have facilitated access to daily required services like public transportation. Hence, in the last years, new research opportunities have emerged intended for acquiring and processing information coming from such devices, and that might help to improve the operation of what has been called smart cities. According to [1], the main research topics using the above-mentioned technologies can be categorized as social dynamics [2],

© Springer Nature Switzerland AG 2019
I. Nyström et al. (Eds.): CIARP 2019, LNCS 11896, pp. 514–524, 2019.
https://doi.org/10.1007/978-3-030-33904-3_48

traffic dynamics [3], and operational dynamics [4,5]. Social dynamics is devoted to the study of populations, its behavior and external factors that influence them. Traffic dynamics studies the population flow and their mobility patterns in order to estimate travel times, design real-time indicators of traffic, among many other tasks. Finally, operational dynamics focus on the taxi drivers operation modes, including collection points, unloading travel times, employed trajectories, among many other variables. As a result, the analysis of these types of dynamics can be directly applied to treat urban problems like traffic management, the design of improved displacement trajectories, ensuring a more fluid transit.

In recent years, operational dynamics has attracted growing attention, producing a diverse number of conducted studies around it. As an example, in Harbin, China, analysis and estimation of hotspots were carried out by means of DBSCAN clustering algorithms by using large-scale taxi GPS data, taking into account features such as the grouping of pick-up and drop-off locations [5]. In Italy, a stochastic model employing a special version of accelerated random walks in combination with exponential decay distributions was introduced to describe the trajectories of 780,000 private vehicles [6]. In Copenhagen, Denmark, multiscale spatiotemporal analysis of mobility patterns was made combining GPS and Wi-Fi scans data [7]. However, it should be noticed that all the mentioned studies use different information highly correlated with the peculiarities of each city. Consequently, although the proposed methodologies may be scalable, the influence of endemic factors such as topography, demography, city distribution, as well as the implicit nature of the data hinders the reproducibility of the analyses and their transfer to other populations.

In this paper, we present a spatiotemporal analysis applied to the forecast of taxi service demands per communes for the city of Manizales (Colombia) by using generalized additive models (GAM). The proposed analysis is based on three components: trend, seasonality, and holidays [8]. The study was conducted on data collected from taxi service requests for the second four-month period of 2016. The rest of this paper is organized as follows: Sect. 2 describes the theoretical aspects of the regressive model employed. Section 3 describes the employed data and its preprocessing, along with the experimental framework. Later, in Sect. 4 we describe and discuss the achieved results. Finally, Sect. 5 presents the conclusions of this work.

2   Methods

2.1   Time Series Forecasting Model

The approach of the forecasting model is based on the use of generalized additive models (GAM) [9], being this a form of regressive model with non-linear smoothing. This model allows the addition of various functions of linear/non-linear nature as components in the final form of the equation. This allows a flexible model analysis by being treated as separate components. The forecasting model is decomposed into three components: trend, seasonality, and holidays [8]. The general model function is given by the following equation:

$$y(t) = g(t) + s(t) + h(t) + \epsilon_t \tag{1}$$

Where $g(t)$ describes the changes in trend of time series, $s(t)$ represents the variations in seasonality along the weeks, while $h(t)$ gives information related to holidays, and the ϵ_t provides error attributed to external factors [8]. The following sections will give a more detailed description of each of the components involved in the model.

Trend Function. In contrast to trend models based on saturated growth, piecewise linear models often allow a simpler and more useful prediction when no abrupt or exponential growth is exhibited. The form that describes such a trend model corresponds to the following equation:

$$g(t) = (k + a(t)^T \delta)t + (m + a(t)^T \gamma) \tag{2}$$

Being k the growth rate, δ the adjustment rate, m the mismatch parameter, and γ is defined as the multiplication of the change points in time s_j and the adjustment rate δ, i.e. $-s_j\delta_j$ to ensure the continuity of the trend function [8]. Finally, the component vector $a(t)$ is defined as:

$$a_j(t) = \begin{cases} 1, & t \geq s_j, \\ 0, & \text{otherwise} \end{cases} \tag{3}$$

It is important to point out that to select the right number of changepoints, we assume a $\delta_j \sim \text{Laplace}(0, \tau)$, being τ the parameter that directly controls the flexibility of the model.

Seasonality Function. When allusion is made to seasonality, it refers to the measurement of periodic behavior in a series of times. In our case, we use seasonality to measure whether there is any periodical behavior of the phenomenon to be evaluated over the course of the days of the week. In order to obtain a smoothed approximation of the effects of seasonality, a Fourier series estimation can be made [10], exhibiting the following equation:

$$X(t) = \sum_{i=1}^{N} \left(a_n cos \left(\frac{2\pi nt}{P} \right) + b_n sin \left(\frac{2\pi nt}{P} \right) \right) \tag{4}$$

Where P is the period used in the time series and will depend on the analysis window, while Fourier coefficients can be expressed as $\beta = [a_1, b_1, ..., a_N; b_N]^T$. However, this form of estimation requires a parameter adjustment. For this purpose, it is necessary to construct a matrix composed by seasonality vectors for each time t from the Fourier series, obtaining:

$$X(t) = \left[cos \left(\frac{2\pi(1)t}{P} \right), \cdots, sin \left(\frac{2\pi(N)t}{P} \right) \right] \tag{5}$$

Where N allows to adjust seasonality patterns to rapid changes. Finally, the estimation can be redrafted as $s(t) = X(t)\beta$, being $\beta \sim \text{Normal}(0, \sigma^2)$.

Holiday Function can be considered as perturbation factors as these can usually affect the dynamics of taxi service requests by depending directly on the behaviour of the population during such dates, e.g. local migration, or leisure activities that would allow a decrease/increase in demand for taxis during the course of the day.

The way to model the holiday function is throughout the generation of a matrix of regressors, by using the following equation:

$$h(t) = Z(t)\kappa \text{ Where } Z(t) = [1(t \in D_1), \cdots, 1(t \in D_L)] \tag{6}$$

Being D_i is the set of past and future dates for each holiday i, and $\kappa \sim$ Normal$(0, \nu^2)$ concurring with the changes in the forecasts.

2.2   Model Fitting

Due to the large number of parameters in all components, i.e. parameters existing in trend, seasonality and holidays functions; L-BFGS is used as optimizer. Limited-memory BFGS (L-BFGS) is an optimization technique derived from the quasi-Newton family of methods, which are known for their great versatility and ease of unrestricted optimization solutions [11–13]. L-BFGS is based on the approximation of the Broyden-Fletcher-Goldfarb-Shanno algorithm (BFGS) having as condition the minimization of completely differentiable functions. This algorithm is based on gradients and has been popular for parameter estimation in many machine learning problems [14,15].

3   Experimental Setup

In this section, we present the experimental framework used for the predictive model generation that corresponds to the demand for taxi services requested by communes in the Manizales city. Initially, a description of the database is made. Then, two types of preprocessing are developed in this study: a spatial preprocessing which explains the form and criteria used to sectorize the city, and a temporal preprocessing which defines the time intervals used for the construction of the time series. Finally, a description of the nature of the model and its generation is carried out.

3.1   Database

The database is conformed by 709531 records of requests for taxi services provided by the company Mobility Solutions S.A.S. through the CityTaxi application. The data collected correspond to the period from April 1, 2016 to July 31, 2016. It should be noted that in accordance with the Colombian law 1581 for protection of personal data, the database does not include sensitive personal information of users and/or drivers. Table 1 shows the fields available in each of the records. It is important to highlight that such records corresponds to hour, date and place where the requests were generated, i.e. places in which the users solicited the service.

Table 1. Example of a record

Longitude	Latitude	Year	Month	Day	Hour	Minute	Second
−75.5089312	5.06279172	2016	05	15	13	50	12

3.2   Spatial Preprocessing

Usually, in order to facilitate data processing, a common practice is to decompose the area of a city by independent sectors. A wide variety of methodologies to carry out these decompositions have been proposed in the literature [4, 16]. However, in several of the approaches, fractionation is done in a uniform way and consequently, this does not always guarantee a correct distribution of the city, in addition to not contemplating information apriori of the city. In the present work we have used the territorial divisions proposed by the governmental entities, which for the case study are the territorial divisions proposed by the Manizales mayor's office, available on the Manizales geoportal website[1]. The information in the geoportal is in turn fed with data supplied by the Agustín Codazzi Geographic Institute, being the entity in charge of producing the official maps and basic cartography of Colombia. Although cities in Colombia present mainly two forms of partitioning, i.e. by commune and by neighbourhood, we have chosen to work with the division by communes, due to the risk of loss of temporal resolution that supposes the use of segmentation by neighborhoods. To make an initial estimation of the behavior regarding the demand in taxis by commune, it is necessary to know the critical mass present in each commune, i.e. the size of the sample, in this case, by sectors. That is why Table 2, presented below, reflects the number of dwellings present per commune.

Table 2. Number of dwellings per commune.

ID	Commune	Strata 1	Strata 2	Strata 3	Strata 4	Strata 5	Strata 6	Non-residential
1	San Jose	1247	1117	903	10	0	0	1963
2	Cumanday	17	125	5196	1108	128	25	6105
3	Estacion	0	27	3322	2214	866	668	3085
4	C. del Norte	2044	11112	2911	139	0	0	1901
5	Cerro de Oro	2	1195	3895	1894	720	535	3198
6	Palogrande	29	19	92	540	3277	6844	9801
7	La Fuente	2014	1924	4813	1510	31	0	1063
8	Universitaria	604	2470	4414	0	0	0	623
9	Tesorito	154	303	4374	614	120	635	966
10	La Macarena	1414	1211	4527	224	0	31	955
11	Atardeceres	295	1199	1558	6829	80	280	3359

---

[1] https://geodata-manizales-sigalcmzl.opendata.arcgis.com/datasets/
2096bb8b949b43fa83888bf6e8ba450f_0.

As can be seen, Table 2 also presents the segmentation of the number of dwellings segmented by socio-economic strata, the above for the purpose of obtaining information about the dynamics of transport according to socio-economic income for being used in the discussion.

3.3 Temporal Preprocessing

Variables such as the demand corresponding to taxi services requests can be considered time-dependent factors. Since the time variable presents a continuous nature and cannot be analyzed directly, a common practice in the literature is the adoption of time intervals. The definition of time intervals or also known as time windows are an essential part of the experimental process. In order to provide an adequate time window, the data were analyzed on the basis of the following criteria: (i) The temporal resolution must capture the population dynamics. (ii) A representative sample, enough large for each of the communes, must be ensured for a posterior spatiotemporal analysis to carried out. After evaluating these criteria, we empirically find that a time window of 1 h guarantees their fulfillment. Each record is assigned exclusively to one of the 2928 temporary windows.

3.4 Model Generation

The models for the forecast of demand of services by commune were generated, by means of GAM as described in Sect. 2.1, using the prophet package [8] available in.[2] Trend change points (changepoints) were automatically adjusted by algorithm. Nevertheless, the amount of changepoints used is determined by the parameter τ, where high values will cause that the trend be adjusted excessively while the choice of small values will increase the generalization capacity of the model. In consequence, a correct choice of τ parameter will positively influence the performance of the model. We adjusted the value of τ through an interval search using exponentially growing sequences $\{10^{-3}, 5 \times 10^{-3}, 10^{-2}, ..., 5 \times 10^{-1}, 10^{0}\}$, in order to minimize mean absolute error (MAE). Additionally only the first 80% of the time series was used to infer changepoints, with the aim of have plenty of runway for projecting the trend forward and to avoid overfitting fluctuations at the end of the time series. The models for each commune were fit using a initial history of 3 months, and a forecast was made on 5 weeks horizon with cutoff per week. We used the classical "rolling origin" forecast evaluation [17], because it is not possible to use cross validation method that is employed in classification problems, given the observations are not exchangeable.

4   Results and Discussion

The results in this section will be shown in the following order: Initially, a detailed explanation about the observations found in the spatial analysis will be given.

---

[2] https://facebook.github.io/prophet/.

For this, we are going to use heat maps taking into account the rate of services per commune. Subsequently, a temporal analysis will be carried out on the behavior of the demand from time series. Finally, a spatiotemporal analysis will be conducted, considering the two previous analyses, focusing more on the prediction of time series.

4.1   Spatial Analysis

The spatial analysis allows to have a wide knowledge about the influence of the population concentrations and their relationship with the demand of the taxi services. That being said, Fig. 1 shows a heat map of the concentrations in the demand for taxis by communes in the city of Manizales.

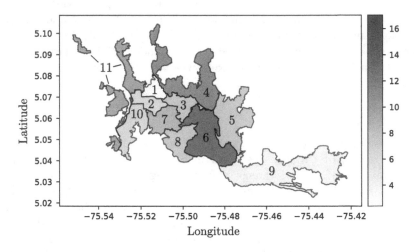

Fig. 1. Heat map of Manizales city with the rate of services per commune.

If we look more closely at the heat map, the colors are related to the demand for taxi services, the greater the intensity of color, the greater the number of requests present in a given commune. It is important to emphasize that the commune with the greatest amount of demand is the commune 6 (Palogrande), concentrating 17.03% of the requests that were generated in the city, positioning itself as the main point of confluence. Presumably due to the fact that the commune 6 presents a greater concentration of banks than others communes, furthermore this holds the greatest quantity of commercial establishments with a total of 9801 entities, as can be evidenced in the Table 2. In contrast, it can be observed that commune 1 (San Jose), shown in the softest color, presents the lowest percentage of requests in the demand for taxis. Another factor regarding the commune 1 is that this commune has the least number of dwellings and these belong to low socioeconomic strata.

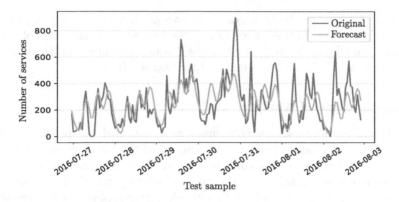

Fig. 2. Forecast and real time series of demand for services. (Color figure online)

4.2   Temporal Analysis

The temporal analysis of the present work considers the behavior over time of the dynamics of taxi requests throughout the city. Figure 2 shows the analysis between the original behavior (blue line) and the forecast values (yellow line) for the periods between July 27, 2016 and August 03, 2016, considering a one-week analysis window. It is easily noticeable that the predicted values have a very high correspondence with the original behaviour of the time series analysed in the above-mentioned time frame, evidencing a high level of accuracy in the prediction. However, it is important to make some additional observations shown in the Fig. 2. The additional observations turn around the highest values exhibited by the demand for taxi services in the time series. One of the explanations for this phenomenon is based on two facts of interest: Both days correspond to weekends and also point out the end of the month. In Colombia, both half and the end of the month are conventional pay days for workers, and if we also note that weekends are usually days of recreation and night-time activities, the peak values in the figure are completely predictable. The second additional observation highlights the non-existence of holidays during the time frame of analysis. In view of the above, a possible cause for the high values in the time series may lie in climatic factors that have propitiate such behaviour in the requests of taxi services.

In addition to the analysis on the dynamics of taxi services along the week, an additional analysis was performed which consisted in the observation of demand of services in range of hours for days contemplated in the analysis. All days in the analysis showed a periodic pattern in their dynamics, revealing low and high demand activities, as exhibit Fig. 3. If we observe in detail, Fig. 3 presents peak values in 3 time ranges: 7:00–8:00, 13:00–14:00, 19:00–20:00; these values

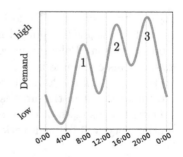

Fig. 3. Demand on an average day.

coincide perfectly with the hours of entry to workday, lunch hours and end of workday respectively. In these time ranges the highest concentrations of vehicular traffic throughout the work week are commonly presented and therefore this phenomenon correlates with the demand for taxi services.

4.3   Spatiotemporal Analysis

After making the forecasts for communes were obtained the predictions with their respective confidence intervals. We computed the following statistics: root mean squared error (RMSE), mean absolute error (MAE), and coverage of the confidence intervals. Additionally, the mean absolute percent error (MAPE) also was computed. However, in order to avoid calculation problems when the signal takes zeros values, these were replaced with one. Each statistic is calculated through a rolling window including 10% of the predictions. Table 3 shows mean and standard deviation calculated between windows, where ($\downarrow$) indicates that the smaller is better, and ($\uparrow$) indicates that the larger is better.

Table 3. Performance measures per commune mean $\pm$ sd.

Commune	RMSE ($\downarrow$)	MAE ($\downarrow$)	MAPE ($\downarrow$)	Coverage ($\uparrow$)
San Jose	4.3747 $\pm$ 0.659	3.258 $\pm$ 0.439	1.122 $\pm$ 0.226	0.859 $\pm$ 0.037
Cumanday	9.0818 $\pm$ 1.302	6.844 $\pm$ 0.716	1.000 $\pm$ 0.286	0.913 $\pm$ 0.034
Estacion	10.7709 $\pm$ 1.225	8.214 $\pm$ 0.824	0.878 $\pm$ 0.207	0.893 $\pm$ 0.028
C. del Norte	18.8648 $\pm$ 3.491	13.681 $\pm$ 2.049	1.060 $\pm$ 0.334	0.935 $\pm$ 0.033
Cerro de Oro	11.5815 $\pm$ 1.229	8.852 $\pm$ 0.939	1.084 $\pm$ 0.312	0.914 $\pm$ 0.032
Palogrande	23.0539 $\pm$ 6.421	16.259 $\pm$ 2.390	1.162 $\pm$ 0.501	0.859 $\pm$ 0.044
La Fuente	14.5140 $\pm$ 2.417	10.368 $\pm$ 1.434	0.957 $\pm$ 0.265	0.915 $\pm$ 0.031
Universitaria	12.7262 $\pm$ 1.643	9.513 $\pm$ 1.274	1.262 $\pm$ 0.320	0.881 $\pm$ 0.036
Tesorito	5.6995 $\pm$ 0.684	4.285 $\pm$ 0.473	0.845 $\pm$ 0.175	0.869 $\pm$ 0.033
La Macarena	9.9410 $\pm$ 2.168	7.190 $\pm$ 1.119	0.962 $\pm$ 0.230	0.905 $\pm$ 0.034
Atardeceres	14.8209 $\pm$ 2.313	11.221 $\pm$ 1.369	1.058 $\pm$ 0.413	0.856 $\pm$ 0.048

As can be observed in Table 3 the models by communes present a very good fit, given the low levels reported in MAPE, which oscillate around 1%. It should also be noted that Palogrande commune has the highest values in RMSE and MAE, since this commune has the highest demand rate of the city. On the other hand, the zone with the lowest error rates is commune 9 (Tesorito). This may be due to the fact that in this area is located not only the most industrialized part of the city, but also two of the most important institutions of higher education, presenting very well defined work schedules.

5   Conclusions

In this paper, we presented a spatiotemporal analysis using taxi pick up points information of the second four-month period of 2016. We studied demand per communes and the predictability of services taxis demand. From the conducted analysis we can conclude that commune with high demand correspond with *Zona rosa*. On the other hand, the temporal analysis revealed three peaks of high demand during the day, these peak coincide perfectly with hours of entry to workday, lunch hours and end of workday. In addition, the good performance of the generalized additive models for prediction horizons of a few weeks could be evidenced.

As future work, climate information will be included in order to explain some high demand peaks that are not explained by the stationarity component of the signal. Moreover, we would like to expand analysis period to some years as well as improve spatial resolution.

Acknowledgements. The authors would like to acknowledge the cooperation of all partners within the *Centro de Excelencia y Apropiación en Internet de las Cosas (CEA-IoT)* project. The authors would also like to thank all the institutions that supported this work: the Colombian Ministry for the Information and Communications Technology (*Ministerio de Tecnologías de la Información y las Comunicaciones - MinTIC*) and the Colombian Administrative Department of Science, Technology and Innovation (*Departamento Administrativo de Ciencia, Tecnología e Innovación - Colciencias*) through the *Fondo Nacional de Financiamiento para la Ciencia, la Tecnología y la Innovación Francisco José de Caldas* (Project ID: FP44842-502-2015). The authors would also like to thank to *Mobility Solutions SAS* for its commitment to the region, and with the great common goal of building intelligent cities through cooperation and strategic alliances linking the productive sectors with digital transformation.

References

1. Castro, P.S., Zhang, D., Chen, C., Li, S., Pan, G.: From taxi GPS traces to social and community dynamics: a survey. ACM Comput. Surv. (CSUR) **46**(2), 17 (2013)
2. Zhang, W., Li, S., Pan, G.: Mining the semantics of origin-destination flows using taxi traces. In: UbiComp, pp. 943–949 (2012)
3. Li, D., et al.: Percolation transition in dynamical traffic network with evolving critical bottlenecks. Proc. Natl. Acad. Sci. **112**(3), 669–672 (2015)
4. Phithakkitnukoon, S., Veloso, M., Bento, C., Biderman, A., Ratti, C.: Taxi-aware map: identifying and predicting vacant taxis in the city. In: de Ruyter, B., et al. (eds.) AmI 2010. LNCS, vol. 6439, pp. 86–95. Springer, Heidelberg (2010). https://doi.org/10.1007/978-3-642-16917-5_9
5. Tang, J., Liu, F., Wang, Y., Wang, H.: Uncovering urban human mobility from large scale taxi GPS data. Phys. A: Stat. Mech. Appl. **438**, 140–153 (2015)
6. Gallotti, R., Bazzani, A., Rambaldi, S., Barthelemy, M.: A stochastic model of randomly accelerated walkers for human mobility. Nat. Commun. **7**, 12600 (2016)
7. Alessandretti, L., Sapiezynski, P., Lehmann, S., Baronchelli, A.: Multi-scale spatio-temporal analysis of human mobility. PloS One **12**(2), e0171686 (2017)
8. Taylor, S.J., Letham, B.: Forecasting at scale. Am. Stat. **72**(1), 37–45 (2018)

A. F. Giraldo-Forero et al.

9. Hastie, T., Tibshirani, R.: Generalized additive models: some applications. J. Am. Stat. Assoc. **82**(398), 371–386 (1987)
10. Harvey, A., Koopman, S.: Structural time series models. Wiley StatsRef: Statistics Reference Online (2014)
11. Dennis Jr., J.E., Schnabel, R.B.: Numerical Methods for Unconstrained Optimization and Nonlinear Equations, vol. 16. SIAM (1996)
12. Fletcher, R.: Practical methods of optimization (1987)
13. Nocedal, J., Wright, S.: Numerical Optimization. Springer, Heidelberg (1999). https://doi.org/10.1007/b98874
14. Malouf, R.: A comparison of algorithms for maximum entropy parameter estimation. In: proceedings of the 6th Conference on Natural Language Learning, vol. 20, pp. 1–7. Association for Computational Linguistics (2002)
15. Andrew, G., Gao, J.: Scalable training of l 1-regularized log-linear models. In: Proceedings of the 24th International Conference on Machine Learning, pp. 33–40. ACM (2007)
16. Zhang, D., Li, N., Zhou, Z.H., Chen, C., Sun, L., Li, S.: iBAT: detecting anomalous taxi trajectories from GPS traces. In: Proceedings of the 13th International Conference on Ubiquitous Computing, pp. 99–108. ACM (2011)
17. Tashman, L.J.: Out-of-sample tests of forecasting accuracy: an analysis and review. Int. J. Forecasting **16**(4), 437–450 (2000)

A Supervised Laplacian Eigenmap Algorithm for Visualization of Multi-label Data: SLE-ML

Mariko Tai[✉] and Mineichi Kudo

Graduate School of Information Science and Technology, Hokkaido University,
Sapporo, Japan
marikotai@ist.hokudai.ac.jp

Abstract. A novel supervised Laplacian eigenmap algorithm is proposed especially aiming at visualization of multi-label data. Supervised Laplacian eigenmap algorithms proposed so far suffer from hardness in the setting of parameters or the lack of the ability of incorporating the label space information into the feature space information. Most of all, they cannot deal with multi-label data. To cope with these difficulties, we consider the neighborhood relationship between two samples both in the feature space and in the label space. As a result, multiple labels are consistently dealt with as the case of single labels. However, the proposed algorithm may produce apparent/fake separability of classes. To mitigate such a bad effect, we recommend to use two values of the parameter at once. The experiments demonstrated the advantages of the proposed method over the compared four algorithms in the visualization quality and understandability, and in the easiness of parameter setting.

Keywords: Supervised Laplacian eigenmap · Multi-Label data ·
Feature and label spaces

1 Introduction

In recent years, various kinds of information, such as location information, search history, and videos, have been converted to numerical/categorical/binary data. Those data are often expressed by vectors of a high dimension. Therefore, it is difficult for us to observe directly the data in order to grasp how data are distributed and what relationship exists among data. To make use of our high-order brain functions and intuition to analyze such data, dimension reduction into a two- or three-dimensional space is effective. Dimension reduction, not limited to two- or three-dimensional, is also useful to avoid the "curse of dimensionality", a common obstacle in regression and classification. Many visualization methods proposed so far are categorized into two of unsupervised methods and supervised methods. They are furthermore divided into two of linear and nonlinear methods. The unsupervised methods do not use class labels as seen in principal

© Springer Nature Switzerland AG 2019
I. Nyström et al. (Eds.): CIARP 2019, LNCS 11896, pp. 525–534, 2019.
https://doi.org/10.1007/978-3-030-33904-3_49

component analysis (PCA) and multidimensional scaling (MDS). PCA is a linear mapping and maximizes the variance of mapped data. MDS is a nonlinear mapping and preserves the distance between data, before and after mapping, as much as possible. On the other hand, the supervised methods use class labels as supervision information. Fisher linear discriminant analysis (FLDA) is a representative example. FLDA is a linear mapping and minimizes the ratio of within-class variance to between-class variance in the mapped data. The visual neural classifier [5] is an example of supervised nonlinear method.

Unsupervised methods are useful for revealing hidden structure, typically manifolds formed from data. On the contrary, supervised methods are effective for revealing the separability of classes. Linear-methods keep the linear structure of data but cannot express the manifold structures with varying curvature. On the other hand, nonlinear-methods can effectively catch the manifold structure but may produce fake structure which can mislead the analysts. Laplacian Eigenmaps (LEs), our main concerns, are, originally unsupervised, nonlinear mappings and preserve the neighbor relationship of data by graph Laplacians over adjacency graphs. In this paper, we propose a novel *supervised* LE, which combines feature and label information into a single neighborhood relation between data.

2   Related Works

In this section, we provide an overview of supervised LEs. So far, CCDR [2], Constraint Score [8], S-LapEig [4] and S-LE [6] have been proposed. In the following, the detail of each algorithm will be introduced. Note that some parameter symbols are changed from the original papers for keeping consistency through this paper. In fact, k is used in common for the number of nearest neighbors, σ^2 for a variance of an exponential, τ^2 for a variance of a second exponential, β for a parameter on label-agreement, and λ for a parameter on the balance between feature space and label space information. In addition, necessary parameters of each algorithm are also shown with the name.

2.1   Laplacian Eigenmaps: LE(k) (Original LE)

Given n data points $\{x_i\}_{i=1}^n$ in a high-dimensional space $\mathbb{R}^M$, the original LE [1] maps them into points $\{z_i\}_{i=1}^n$ in a low-dimensional space $\mathbb{R}^m$ on the basis of a neighbor relation represented by $\{w_{ij}(\geq 0)\}_{i,j=1}^n$ over $\{x_i\}_{i=1}^n$ in such a way to minimize

$$J_{\text{LE}} = \sum_i \sum_j \|z_i - z_j\|^2 w_{ij}. \tag{1}$$

This formulation corresponds to graph Laplacian with the adjacency relation $W = (w_{ij})$. Typically, W is given by

$$w_{ij} = \begin{cases} \exp(-\|x_i - x_j\|^2/\sigma^2) & (x_i \in \text{kNN}(x_j) \vee x_j \in \text{kNN}(x_i)) \\ 0 & (\text{otherwise}) \end{cases},$$

where $x_i \in \text{kNN}(x_j)$ shows that x_i is a member of k nearest neighbors of x_j.

Let Z be a matrix of $n \times m$ and let z_i^T ('T' denotes the transpose) be the ith row. Then J_{LE} becomes $J_{\text{LE}} = 2\text{tr } Z^T L Z$ ('tr' denotes the trace), where $L = D - W$ with $D = \text{diag}(\sum_j w_{1j}, \ldots, \sum_j w_{nj})$. We can find $\{z_i\}_{i=1}^n$ by minimizing $\text{tr } Z^T L Z$, subject to $Z^T D Z = I$. The solution is given by solving the generalized eigenvalue problem, $LZ = DZ\Lambda$, and, avoiding the trivial eigenvector of 1 with $\lambda = 0$, the second to $(m+1)$th smallest (in the corresponding eigenvalue) eigenvectors are used for Z. Note that L is positive semi-definite.

2.2 Classification Constrained Dimensionaly Reduction: CCDR(k, σ^2, λ)

CCDR [2] introduces a hypothetical node for each class, called a *class center*, and requires the points of the same class to gather around the class center in the mapped space. Let $\mu_k \in \mathbb{R}^m$ be the class center of class k in the mapped space and $C = (c_{ki})$ be the class membership matrix, i.e., $c_{ki} = 1$ if $x_i \in \mathbb{R}^M$ has label k and $c_{ki} = 0$ otherwise. CCDR minimizes the cost function

$$J_{\text{CCDR}} = \lambda \sum_{i,j} ||z_i - z_j||^2 w_{ij} + (1 - \lambda) \sum_{k,i} ||\mu_k - z_i||^2 c_{ki}, \tag{2}$$

where

$$w_{ij} = \begin{cases} \exp(-||x_i - x_j||^2/\sigma^2) & (x_i \in \text{kNN}(x_j) \vee x_j \in \text{kNN}(x_i)) \\ 0 & (\text{otherwise}) \end{cases}, c_{ki} = \begin{cases} 1 & (y_i = k) \\ 0 & (y_i \neq k) \end{cases}.$$

Here y_i is the class label of x_i and λ ($0 \leq \lambda \leq 1$) is a balance parameter between feature space information and label space information. In [2], σ^2 is determined as ten times the average of the squared nearest neighbor distances and $\lambda = 1/2$.

2.3 Constraint Score: CS(β)

The Constraint Score [8] is not proposed directly for dimension reduction nor visualization, but for feature selection. However, we can use the criterion for LE. In fact, it is a naïve way to deal with sample pairs of different classes: if the classes are the same, then multiply $+1$ to (1), otherwise -1.

Although two cost functions, division type and subtraction type, are shown in [8], we consider only the subtraction type that minimizes the cost function

$$J_{\text{CS}} = \sum_{i,j} ||z_i - z_j||^2 w_{ij}^M - \beta \sum_{i,j} ||z_i - z_j||^2 w_{ij}^C,$$

where $w_{ij}^M = \mathbb{1}(y_i = y_j)$ and $w_{ij}^C = \mathbb{1}(y_i \neq y_j)$ ($\mathbb{1}(\cdot)$ is the indication function that takes 1 if the argument is true, 0 otherwise).

2.4   S-LapEig(k, σ^2, τ^2)

S-LapEig [4] modifies the distance between data points $\{x_i\}_{i=1}^n$ in the original space such that data of the same class label become closer and data of the different class labels become more distant. The criterion to minimize is the same as the original LE: $J_{\text{S-LapEig}} = J_{\text{LE}}$. However, the weight is determined at two stages as

$$
w_{ij} = \begin{cases} \exp(-d^2(x_i, x_j)/\sigma^2) & (x_i \in \text{kNN}(x_j) \vee x_j \in \text{kNN}(x_i)) \\ 0 & (\text{otherwise}) \end{cases},
$$

where

$$
d^2(x_i, x_j) = \begin{cases} 1 - \exp(-\|x_i - x_j\|^2/\tau^2) & (y_i = y_j) \\ \exp(\|x_i - x_j\|^2/\tau^2) & (y_i \neq y_j) \end{cases}.
$$

Here τ^2 is taken as the square of the average Euclidean distance between all pairs of data points in [4].

2.5   S-LE(σ^2, β)

S-LE [6] computes the adjacency matrix W as follows. Let $AS(x_i) = 1/n \cdot \sum_{j=1}^n s(x_i, x_j)$, where $s(x_i, x_j) = \exp\left(-\|x_i - x_j\|^2/\sigma^2\right)$. If $(s(x_i, x_j) > AS(x_i)) \wedge (y_i = y_j)$, then x_j is judged as the neighbor of x_i and denoted by $x_j \in N_w(x_i)$. On the contrary, if $(s(x_i, x_j) > AS(x_i)) \wedge (y_i \neq y_j)$, then $x_j \in N_b(x_i)$. Under these definitions, S-LE maximizes (not minimizes)

$$
J_{\text{SLE}} = \beta \sum_{i,j} \|z_i - z_j\|^2 w_{ij}^B - (1 - \beta) \sum_{i,j} \|z_i - z_j\|^2 w_{ij}^W,
$$

where

$$
w_{ij}^W = \begin{cases} s(x_i, x_j)(x_i \in N_w(x_j) \vee x_j \in N_w(x_i)) \\ 0 \qquad (\text{otherwise}), \end{cases}, w_{ij}^B = \begin{cases} 1(x_i \in N_b(x_j) \vee x_j \in N_b(x_i)) \\ 0(\text{otherwise}) \end{cases}.
$$

3   Supervised Laplacian Eigenmaps

Almost all supervised LE algorithms that we refer to in Sect. 2 basically separate a pair (x_i, x_j) into a same-class pair or a different-class pair and evaluate them separately. Therefore, we need to pay a special attention to the difference of the number of two kinds of pairs. In addition, some algorithms cannot control the degree to which we mix the label information and the feature information. Most of all, they cannot deal with multi-label datasets where a single data is associated with multiple class labels. Only CCDR can deal with multi-label data, if we want to do that, but it has its own problem as will be discussed later. For the other three algorithms, it is also not easy to extend because they deal with sample pairs differently depending on if they share the same class or not. To cope with these limitations, we propose a novel supervised LE, called the *Supervised Laplacian Eigenmaps for Multi-Label datasets* (shortly, SLE-ML), for visualization mainly. We combine neighbor information in the feature space and that in the class-label space into one with a balance parameter λ $(0 \leq \lambda \leq 1)$.

3.1   Supervised Laplacian Eigenmaps for Multi-label Datasets: SLE-ML(k, λ)

SLE-ML minimizes the same cost function as the original LE using a different weight

$$J_{\text{SLE-ML}} = \lambda \sum_{i,j} ||\boldsymbol{z}_i - \boldsymbol{z}_j||^2 w_{ij}^F + (1 - \lambda) \sum_{i,j} ||\boldsymbol{z}_i - \boldsymbol{z}_j||^2 w_{ij}^L, \qquad (3)$$

where

$$w_{ij}^F = \left(\mathbb{1}(\boldsymbol{x}_i \in kNN(\boldsymbol{x}_j)) + \mathbb{1}(\boldsymbol{x}_j \in kNN(\boldsymbol{x}_i)) \right) / 2, \text{ and } w_{ij}^L = \frac{|\boldsymbol{y}_i \wedge \boldsymbol{y}_j|}{|\boldsymbol{y}_i \vee \boldsymbol{y}_j|}.$$

Here, superscript 'F' stands for "feature space" and 'L' stands for "label space". In addition, w_{ij}^L is the Jaccard similarity coefficient, the ratio of common labels to the union of their labels, and takes a value between 0 and 1. For a single label problem, $w_{ij}^L = 1$ if data points i and j share the same label, and $w_{ij}^L = 0$ otherwise. The original (unsupervised) LE is a special case of SLE-ML with $\lambda = 1.0$. Unlike many of previous supervised LEs that take a trade-off in the feature space between same-class pairs and different-class pairs, SLE-ML take a trade-off of similarity between the feature space and the label space.

3.2   Parameters

All algorithms have their own parameters: CCDR(k, σ^2, λ), CS(β), S-LapEig(k, σ^2, τ^2), SLE(σ^2, β), and SLE-ML(k, λ). It is often critical to choose an appropriate value for each parameter. We first discuss how to determine the values and how sensitive they are to the results. The variance parameters σ^2 and τ^2 are often determined from data. A typical way is to use the average squared Euclidean distance between all pairs of data points. As for the value of k, we need to use the same value in common to all algorithms. In the following experiments, the value of k is set to 1.5 times the average sample size per class in order to relate each sample to other samples of different classes. As for the other parameter, β (as for the label agreement) and λ (as for trade-off between feature and label information), we need to be more careful about the setting. Let us consider β in CS(β) and S-LE(σ^2, β). When the number of classes is large, the cases when two samples have the same label are far less than the counter part. So, we have to set the value of β in accordance with the given dataset. When we consider supervised LEs, the most important thing is how we incorporate the label information into feature information. In contrast to LE that uses the feature information only, if we use the label information only, then all the points of the same label concentrate on a single point in the mapped space, as seen in SLE-ML($k, \lambda = 0.0$). Therefore, we need to be careful about the value of λ more than the other parameters. CCDR(k, σ^2, λ) has the same parameter λ, but it has another problem. The criterion (2) has two terms: the size of the first term is $O(n^2)$ and the size of the second term is $O(Kn)$ where K is the number of classes. Therefore, if $K \ll n$ or its converse (as seen in extreme multi-label problems), the effect of the same value of λ changes. So, it needs to be set carefully. In SLE-ML(k, λ), the two terms in (3) have the same size of $O(n^2)$. Therefore, we do not need to be careful about the number of classes and can consider the value of λ independently of datasets. That is, SLE-ML is problem-independent. The algorithms except for CCDR(k, σ^2, λ) and SLE-ML(k, λ) do not have even a trade-off parameter between feature and label information. This means we cannot control it.

Table 1. Datasets. (A) stands for artificial datasets and (N) for natural datasets. In the artificial datasets, garbage features are added; 7 for Torus and 8 for Clusdat.

Dataset	#samples	#classes	#samples in each class	#features (Intrinsic)
Torus (A)	1000	2	500 500	10 (3)
Clusdat (A)	1600	2	800 800	10 (2)
Digits (N)[†]	1797	10	178 182 $\cdots$ 180	64 (64)
Scene (N)[‡]	1211	6(14)	194 165 $\cdots$ 1	294(294)

†:UCI Machine Learning Repository [3]
‡:Mulan: A Java Library for Multi-Label Learning [7]

4   Experiments

We evaluated the performance of the proposed method on several high-dimensional datasets (Table 1). The dataset digits consist of 1797 images of hand-written digits (0–9). In our experiments, the parameter k for nearest neighbors was set to 1.5 times the average number of samples of each class as described before.

Figure 1 is the visualization result of digits by SLE-ML. To confirm the effect of the parameter λ, we varied the value from 0 to 1 by step 0.2. We see that $\lambda = 1.0$ (the feature space only) derives the same mapping as LE, and $\lambda = 0.0$ (the label space only) derives the class-isolated mapping. For a middle value of λ, we can see the result by a trade-off between feature and label spaces. It should be noted that a smaller value of λ tends to enhance the separability among classes more than the reality. So, we recommend to use two values of λ as $\lambda = 0.5$ and 0.9 at once for analyzing data. We compared four algorithms, CCDR, CS, S-LapEig and S-LE, with SLE-ML. The parameters were chosen so as to produce almost the best results except for SLE-ML. In Fig. 2, the results of

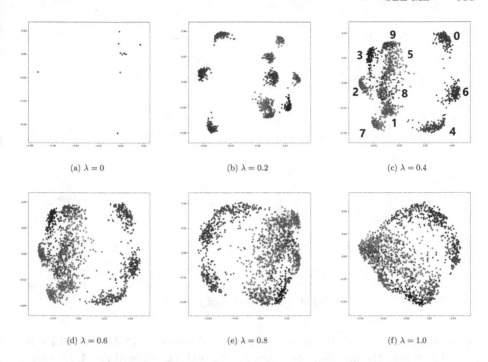

Fig. 1. Effect of the balance parameter λ in the proposed SLE-ML in `digits` dataset (Each color corresponds to a class as seen in the case of $\lambda = 0.4$). (Color figure online)

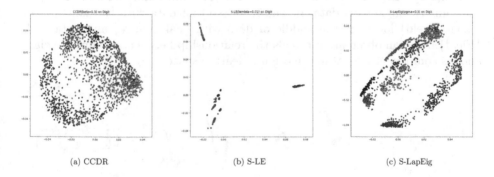

Fig. 2. Visualization of `digits` by three algorithms

CCDR, S-LE and S-LapEig are shown. Since CS did not produce any good result, the result is not shown. We see that a high separability of classes is visualized by CCDR, S-LapEig and SLE-ML($\lambda = 0.4, 0.6$). The other algorithms fail to reveal the separability that actually exists. In the following, therefore, we compared these three only.

To make clear the difference of those algorithms, we visualized two artificial datasets `Torus` and `Clusdat`. Note that these data are contaminated by garbage

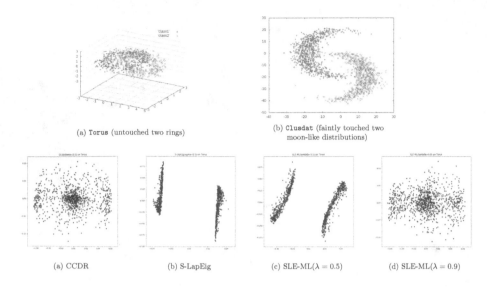

(a) Torus (untouched two rings)

(b) Clusdat (faintly touched two moon-like distributions)

(a) CCDR   (b) S-LapEig   (c) SLE-ML($\lambda = 0.5$)   (d) SLE-ML($\lambda = 0.9$)

Fig. 3. Visualization of Torus

features. The results are shown in Figs. 3 and 4. We see that CCDR and SLE-ML ($\lambda = 0.9$) expose the manifold structure to some extent, while SLE-ML ($\lambda = 0.5$) and S-LapEig succeed to show the separability.

Next we dealt with multi-label datasets. Figure 5 is the visualization result of scene by SLE-ML ($\lambda = 0.5$). We observe that multi-label data are mapped the same as single-label data. In Fig. 5, we see that data with two labels {Fall foliage, Field} locate in the middle of data with {Fall foliage} and data with {Field}. Such an observation reveals the relationship between a composite class and its component classes in the original feature space.

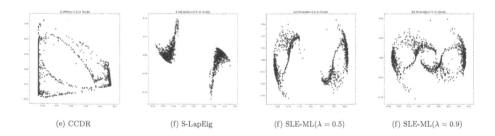

(e) CCDR   (f) S-LapEig   (f) SLE-ML($\lambda = 0.5$)   (f) SLE-ML($\lambda = 0.9$)

Fig. 4. Visualization of Clusdat

5   Discussion

The proposed SLE-ML is advantageous to the compared four algorithms in the sense that the results give more information than the others. This is mainly

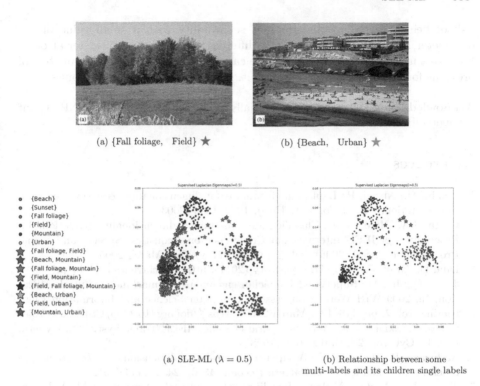

(a) {Fall foliage, Field} ★   (b) {Beach, Urban} ★

(a) SLE-ML ($\lambda = 0.5$)   (b) Relationship between some
                                      multi-labels and its children single labels

Fig. 5. Visualization of **scene** with multiple labels.

because the control parameter λ is intuitive and the multiple results with different values of it help us to analyze data. However, there still remain many more
challenges that the original LEs had and maybe many LEs still have. First of all,
we need to resolve the "out-of-sample" problem. Since the mapping in SLE-ML
is not explicit, we cannot apply this mapping to a newly arrived data. We are
now thinking to simulate the mapping linearly or nonlinearly. If it is succeeded,
we may choose the parameter value under which separability is held high. Next,
we need to cope with "imbalance problem." SLE-ML needs to be modified to
emphasize minority classes. Last, we have to devise some way to visualize a
hundred of thousands of data and data with a large number of multiple labels.

6   Conclusion

In this paper, we have proposed a novel supervised Laplacian eigenmap algorithm
that can handle multi-label data in addition to single-label data. The experiment
demonstrated the advantages of the algorithm over the compared the state-of-
the-art algorithms in the visualization quality and understandability, and in the
easiness of parameter setting.

In the proposed algorithm, we combine the feature information and the label
information into one, and control the balance by a parameter. To mitigate the

risk of being cheated by an apparent separability with a small value of the parameter, we recommend to use two different values of the parameter at once. We also analyzed how appropriately we can give the values in the parameters of previous four algorithms and, as a result, pointed out some careful points.

Acknowledgment. This work was partially supported by JSPS KAKENHI Grant Number 19H04128.

References

1. Belkin, M., Niyogi, P.: Laplacian eigenmaps for dimensionality reduction and data representation. Neural Comput. **15**(6), 1373–1396 (2003)
2. Costa, J.A., Hero, A.O.: Classification constrained dimensionality reduction. In: Proceedings of IEEE International Conference on Acoustics, Speech, and Signal Processing, (ICASSP 2005), vol. 5, pp. v/1077–v/1080, March 2005
3. Dua, D., Graff, C.: UCI ML repository (2017). http://archive.ics.uci.edu/ml
4. Jiang, Q., Jia, M.: Supervised Laplacian eigenmaps for machinery fault classification. In: 2009 WRI World Congress on Computer Science and Information Engineering, vol. 7, pp. 116–120, March 2009. https://doi.org/10.1109/CSIE.2009.765
5. Ornes, C., Sklansky, J.: A visual neural classifier. IEEE Trans. Syst. Man Cybern. Part B (Cybern.) **28**(4), 620–625 (1998)
6. Raducanu, B., Dornaika, F.: A supervised non-linear dimensionality reduction approach for manifold learning. Pattern Recogn. **45**(6), 2432–2444 (2012)
7. Tsoumakas, G., et al.: Mulan: a Java library for multi-label learning. J. Mach. Learn. Res. **12**, 2411–2414 (2011)
8. Zhang, D., Chen, S., Zhou, Z.H.: Constraint score: a new filter method for feature selection with pairwise constraints. Pattern Recogn. **41**(5), 1440–1451 (2008)

Material Fracture Life Prediction Using Linear Regression Techniques Under High Temperature Creep Conditions

Roberto Fernandez Martinez[1]($\boxtimes$) (iD), Pello Jimbert[1] (iD),
Jose Ignacio Barbero[2], Lorena M. Callejo[2], and Igor Somocueto[2]

[1] College of Engineering in Bilbao, University of the Basque Country
UPV/EHU, Bilbao, Spain
`roberto.fernandezm@ehu.es`
[2] Industry and Transport Division, Fundacion TECNALIA Research
& Innovation, Foundry & Steelmaking Area, Parque Tecnologico de Bizkaia,
Geldo Street Building 700, 48160 Derio, Bizkaia, Spain

Abstract. 9–12% Cr martensitic steels are widely used for critical components of new, high-efficiency, ultra-supercritical power plants because of their high creep and oxidation resistances. Due to the time consuming effort of obtaining creep properties for new alloys under high temperature creep conditions, in both short-term and long-term testing, it is often dealt with simplified models to assess and predict the future behavior of some materials. In this work, the total time to produce the material fracture is predicted according to models obtained using several linear techniques, since this property is really relevant in power plants elements. These models are obtained based on 344 creep tests performed on modified P92 steels. A multivariate analysis and a feature selection were applied to analyze the influence of each feature in the problem, to reduce the number of features simplifying the model and to improve the accuracy of the model. Later, a training-testing validation methodology was performed to obtain more useful results based on a better generalization to cover every scenario of the problem. Following this method, linear regression algorithms, simple and generalized, with and without enhanced by gradient boosting techniques, were applied to build several linear models, achieving low errors of approximately 6.75%. And finally, among them the most accurate model was selected, in this case the one based on the generalized linear regression technique.

Keywords: Linear regression · Generalized linear regression · Enhanced linear regression

1 Introduction

Steel mills are highly involved in improving mechanical and creep resistance under conditions of high temperatures and long service times in the materials used to produce steels, among them martensitic steels of high percentage in chromium [1, 2]. Since long service times under creep conditions will produce creep damage in steel used to steam turbine in power plants, this improvement can enhance the efficiency of these plants.

© Springer Nature Switzerland AG 2019
I. Nyström et al. (Eds.): CIARP 2019, LNCS 11896, pp. 535–544, 2019.
https://doi.org/10.1007/978-3-030-33904-3_50

Optimizing of creep damage evolution can improve mechanical properties until fracture, prolonging the service life of equipment operating [3–5].

The purpose of this work is to generate knowledge that allows a better understanding of the different elements influence in the alloy and in the metallurgical mechanisms, allowing to improve martensitic steels materials creep properties. This knowledge will allow easier development of new advanced steels.

Specifically, steels with a content of 9–12% Cr are studied, focusing this study to improve creep resistance [6]. In this approach, several models will be developed to predict the short-long-term creep behavior of new steels based on previous creep behaviors of similar materials.

These models use a previous knowledge of the influence of each composition element of the alloy and the thermal treatment on the fracture time, applied to predict this life expectancy. These models have a direct application in the high-chromium steels studied in this work, but it is also useful for a wide range of steels such as the rest of stainless steels, microalloys, high strength and new generation steels.

Several techniques of linear regression were applied to predict material fracture life, specifically 9–12% Cr martensitic steels based on modified P92 steels. Also, to obtain a clear knowledge of the influence of each feature that define the composition of the material, the thermal pretreatment applied to the material, and the creep test conditions on the final material fracture life. In order to get this purpose, a study of the literature [7–19] focus on creep test of Cr martensitic steels were performed, obtaining a dataset with 344 instances that was used to train and test the models. Then, several techniques like data pre-processing, outlier detection, analysis of variance, analysis of covariance, analysis of correlation, multivariate data visualization, and principal components analysis were applied. This step studies the influence of each attribute with the output, the material fracture life, to finally select the most significant variables to perform the most accurate models to solve the proposed problem. And then several linear regression techniques were used and validated to build and select an accurate linear model that improves the amount of knowledge of the creep behavior of 9–12% Cr martensitic steels.

Linear regression algorithms, simple (simple linear regression) and generalized (generalized linear regression), with and without using enhanced by gradient boosting techniques (enhanced linear regression and enhanced generalized linear regression) were applied in this work.

These kind of methodologies that applies data analysis techniques and machine learning algorithms is gaining interest in industrial engineering fields [20–22] where a mathematical model that represents the problem can be helpful in the decision making. And within of the possible engineering problems, this methodology can be used for the development of new materials (advanced modified P92 steels) improving final properties and reducing time and money during the process.

2   Materials

In this work, short-term and long-term testing were studied. From this study, 344 instances based on these testing were contemplated based on significant information of the composition of the material, the previous thermal treatment performed on the material, and the parameters that control the creep test. More specifically the studied features were:

- Composition of the material: weight percent (wt%) of the most applied elements of the chemical composition of modified P92 type steel according to ASTM A335 (C, Mn, Si, Cr, Ni, Mo, P, S, W, Nb, V, N, Al, B, Co, Cu).
- Heat treatment of this steel before the creep test: temperature (NormTemp) and time (NormTime) for normalizing, and temperature (TempTemp) and time (TempTime) for tempering.
- Control parameters of creep test: Temperature (Temp) and strain (Strain) applied on creep test performance.
- Time to fracture (TimeFractPoint), that is the predicted value based on the rest of features.

Figure 1 shows the relationships between the variables under study.

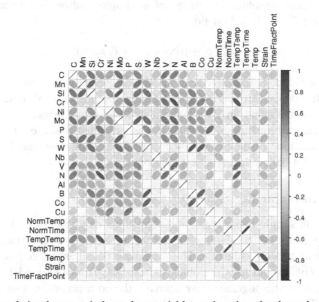

Fig. 1. Correlation between independent variables and against the dependent variable.

3 Methodology

Based on the high number of variables and the complexity of the studied problem (prediction of fracture time of P92 steels or similar materials), the proposed methodology is applied in two steps, first a multivariate analysis and later a linear regression. With the final goal of predicting the material fracture time, but with an easy to understand model that, at the same time, could show information of the influence of the independent variables on the fracture time.

3.1 Multivariate Analysis

A multivariate analysis was performed to simplify the final models, building simpler models and as result, making a better generalization of the problem to improve the results in the validation stage. This analysis was conducted in five steps in order to detect wrong instances and redundant features:

- Data pre-processing.
- Outlier detection [23].
- Analysis of correlation, variance and covariance [24].
- Multivariate data visualization.
- Principal component analysis [25, 26].

3.2 Simple Linear Regression

A probabilistic model of the expected value of an output variable is built based on several independent variables by means of a simple multivariate linear regression (LR) technique [27, 28] based on Eq. 1.

$$\eta = \beta_1 x_1 + \beta_2 x_2 + \ldots + \beta_n x_n + \varepsilon \tag{1}$$

Where η is the dependent variable; x_i are the independent variables; β_i are the weight of each independent variable and indicates the influence of these variables on the dependent variable; and finally ε is the bias based on a Gaussian distribution.

In this case, fracture time of the material is predicted based on 16 features that define the composition of the material, 4 features that define the steel heat treatment, and 2 features that define the creep test control.

3.3 Enhanced Linear Regression

Models based on simple linear regression algorithms can be improved applying gradient boosting [29, 30], combining the results of several simple linear regression models according to a cost function. This technique is call enhanced linear regression (LRB) and calculates and uses the obtained residuals from one model evaluation to train another model, that together make decrease the error according to a squared error loss function. This process is repeated until the training convergence is achieved [31]. The training stage tunes two significant parameters of the algorithm: the number of repetitions (mstop) and the shrinkage factor (shrink), to reduce a possible overtraining.

3.4 Generalized Linear Regression

Another applied technique is the generalized linear regression (GLM), which is based in a flexible generalization of the simple linear regression. In this technique, a linking function allows the optimization of the regression reducing the residual error but assuming different distributions [31–36]. Beforehand, the assumption of different distributions can built more accurate models in a not linear problem, like the case of short and long-term creep tests.

The application of the generalized regression algorithm is based on three components: a prediction based on a simple linear regression (Eq. 1); a linking function that is reversible and transforms the variables according to Eq. 2; and a random component based on the chosen distribution (Gaussian, binomial, gamma, etc.), that based on the independent features, specifies the possible distribution of the dependent feature, η.

$$\eta = g^{-1}(\mu) = g^{-1}(\varepsilon + \beta_1 x_1 + \beta_2 x_2 + \cdots + \beta_n x_n) \tag{2}$$

To apply this technique, it was assumed that the distribution of the original features of the dataset studied in this work were Gaussian, and that subsequently the three link functions shown in Table 1 were applied.

Table 1. Link functions that were applied on this technique

	$\eta = g(\mu)$	$\mu = g^{-1}(\eta)$
Log	$\log_e \mu$	e^{η}
Logit	$\log_e \frac{\mu}{1-\mu}$	$\frac{1}{1+e^{-\eta}}$
Inverse	μ^{-1}	η^{-1}

3.5 Enhanced Generalized Linear Regression

As in the enhanced simple linear regression, models based on generalized linear regression technique can be improved using boosting techniques. In this case, L2 boosting technique [37, 38] was applied to maximize the accuracy minimizing the descending gradient function error. The boosted generalized linear regression algorithm (GLRB) [39–42] obtains the residuals in an iterative way according to a squared error loss function until the training convergence is achieved. Also, the training stage tunes two significant parameters of the algorithm: the number of repetitions (mstop) and the number of variables (based on the Akaike criterion), to reduce a possible overtraining and to simplify the model.

3.6 Validation Method

A validation method to obtain comparable results from the built models was performed. This method started with a normalization between 0 and 1 of the instances that formed the dataset. The benefits of this normalization were two: first that the range of the weight that each feature had in the output variable were equal in such a way that the impact of the features in the prediction could be observed in an easy way; and second that the final accuracy could be improved. Later, a training-testing simple validation method was conducted, where an 80% of the original dataset was randomly selected to build and train the models and the remaining 20% of the original dataset were used to test and validate the models. In the training stage, trying to avoid overtraining, 10-fold cross validation was performed. This cross validation allows having a more trustful error to make a preliminary selection of the obtained models from the parameters tuning. In this way, the most accurate models during training stage were selected to be tested and validated with the testing dataset. And finally based on these last results, the model with lower error in the testing stage was selected.

The work conducted in this study was performed using the statistical software tool R x64 v3.4.1 [43].

3.7 Accuracy Criteria

Models accuracy must be measured in order to evaluate the performance of the model prediction. There are several criteria to measure this accuracy when models predict a numeric feature. Some of the most used criteria are computational validation errors, that calculate the relation between the values predicted by the model and the real values measured during the performance of creep tests. In this case, the following criteria were applied:

- Mean absolute error (MAE) (Eq. 3)

$$MAE = \frac{1}{n} \sum_{k=1}^{d} |m_k - p_k| \tag{3}$$

- Root mean squared error (RMSE) (Eq. 4)

$$RMSE = \sqrt{\frac{1}{n} \sum_{k=1}^{d} (m_k - p_k)^2} \tag{4}$$

- Correlation coefficient (CORR) (Eq. 5)

$$CORR = \frac{\sum_{k=1}^{d} \frac{(p_k - \bar{p}) \cdot (m_k - \bar{m})}{n-1}}{\sqrt{\sum_{k=1}^{d} \frac{(p_k - \bar{p})^2}{n-1} \cdot \sum_{k=1}^{d} \frac{(m_k - \bar{m})^2}{n-1}}} \tag{5}$$

Where d is the number of instances used from the database to validate the model, m are the real values, p are the predicted values, $\bar{m} = \frac{1}{n} \sum_{k=1}^{d} m_k$ and $\bar{p} = \frac{1}{d} \sum_{k=1}^{d} p_k$.

4   Results and Discussion

The dataset of 9–12% Cr martensitic steels based on modified P92 standard specification was formed from a set of creep experiments obtaining from the related literature. The studied features and ranges were selected according to the proposed in the standard specification of the rule. Previously to apply the validation method, a multivariate analysis was performed on the dataset. Then, based on the results of the multivariate analysis the final dataset was set, and like it was commented previously the dataset was split in two new datasets, the training dataset and the testing dataset. Using the training dataset, the models were built applying 10-fold cross validation process combined with a tuning of the most significant parameters of each algorithm. This process allows to optimize the final accuracy that subsequently will be tested and validate using the testing dataset. One example of the results obtained during the training process, using cross validation and parameter tuning is shown in Fig. 2.

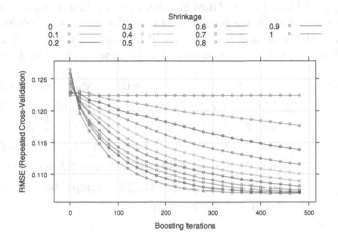

Fig. 2. RMSE obtained during the training stage for the boosted linear regression algorithm. Boosting iterations and shrinkage are tuned.

After the analysis of this parameter tuning, the final results obtained during the training stage are listed in Table 2.

Table 2. Results obtained during training and testing stage

Method	Training		Testing		
	RMSE (%)	CORR (%)	MAE (%)	RMSE (%)	CORR (%)
LR	11.59	28.56	7.35	10.22	16.54
LRB	10.80	28.73	7.21	10.20	15.60
GLM	8.74	56.29	4.08	6.75	59.91
GLRB	10.91	26.84	6.77	9.56	17.20

After the selection of the parameter values to be used in each algorithm during the training stage to improve the accuracy criteria, the models were built. And then were tested with the 20% of the instances that were split previously from the initial dataset and were never used on the training stage. The validation with this testing dataset gives a better idea of the generalization of the problem since these instances were not used to built the models. The results obtained during the testing and validation stages are shown in Table 2.

In this case, the model that gets the most accurate results during the training stage gets the most accurate results during the testing stage. That means that a possible overfitting during the training stage is avoided using 10-fold cross validation, and also that the results obtaining during the validation indicate that the model based on GLM algorithm has an accurate generalization of the problem. That shows that the prediction of the fracture time on modified P92 steels is reliable according to the obtained error.

Then, the model based on the GLM algorithm gets the best performance during all the proposed stages. For this reason, the linear regression model that is defined in Eq. 6 can be considered like an accurate predictor of the material fracture life expectancy, with a RMSE of 6.75% and a correlation of 59.91% based on the testing stage results.

$$
\begin{aligned}
g(\text{TimeFractPoint}) = \ & 27.0059 - 14.9172 \cdot C + 6.0532 \cdot Mn - 13.4758 \cdot Si \\
& + 13.6218 \cdot Cr - 8.7845 \cdot Ni - 15.7346 \cdot Mo + 4.7883 \cdot P \\
& - 2.6898 \cdot S - 11.5633 \cdot W + 1.0100 \cdot Nb + 2.7595 \cdot V \\
& - 7.8088 \cdot N - 9.5136 \cdot Al - 0.8844 \cdot B - 10.0950 \cdot Co \\
& - 0.8578 \cdot Cu - 1.8980 \cdot NormTemp - 6.5165 \cdot NormTime \\
& + 3.1603 \cdot TempTemp + 11.1225 \cdot TempTime - 13.4885 \\
& \cdot Temp - 17.4532 \cdot Strain
\end{aligned} \tag{6}
$$

5   Conclusions

The results show that linear modeling techniques can predict some material properties, such as fracture time in creep conditions, with high accuracy. And not only a good prediction can be obtained, also and based on the join of multivariate data analysis techniques and linear model prediction a hidden knowledge of the process can be obtained. In this case, obtained models predict fracture time of 9–12% Cr martensitic steels with low errors of approximately a RMSE of 6%. Also the most influential variables in the process have been detected, and their weight in the process was determined. Therefore, it can be concluded that these techniques are useful in industrial problems, such as the one presented, and of significant help for developers of new materials in order to improve the final properties of the product.

Acknowledgment. The authors wish to thanks to the Basque Government through the KK-2018/00074 METALCRO.

References

1. Sachadel, U.A., Morris, P.F., Clarke, P.D.: Design of 10% Cr martensitic steels for improved creep resistance in power plant applications. J. Mater. Sci. Technol. **29**(7), 767–774 (2013)
2. Morris, P.F., Sachadel, U.A., Clarke, P.D.: Design of heat treatments for 9–12% Cr steels to optimise creep resistance for power plant applications. In: Proceedings of 9th Liège

Conference on Materials for Advanced Power Engineering, Liège, Belgium, pp 554–564 (2010)

3. Mayer, K.H., Bendick, W., Husemann, R.V., Kern, T., Scarlin, R.B.: International Joint Power Generation Conference, PWR, vol. 33, pp. 831–841. ASME, New York (1998)

4. Gold, M., Jaffee, R.I.: Materials for advanced steam cycles. J. Mater. Energy Syst. 6(2), 130–145 (1984)

5. Viswanathan, R., Bakker, W.: Materials for ultrasupercritical coal power plants - boiler materials: Part 1. J. Mater. Eng. Perform. 10(1), 81–95 (2001)

6. Lanin, A.A., Grin, E.A.: An approach to assessment of the lifetime characteristics of steels under creep conditions using fracture mechanics criteria. Therm. Eng. 65(4), 239–245 (2018)

7. Hald, J.: Prospects for martensitic 12% Cr steels for advanced steam power plants. Trans. Indian Inst. Met. 69(2), 183–188 (2016)

8. Kimura, K., Kushima, H., Sawada, K.: Long-term creep deformation property of modified 9Cr–1Mo steel. Mater. Sci. Eng., A 510, 58–63 (2009)

9. Sklenička, V., Kuchařová, K., Svoboda, M., Kloc, L., Buršík, J., Kroupa, A.: Long-term creep behavior of 9–12%Cr power plant steels. Mater. Charact. 51(1), 35–48 (2003)

10. Fujita, T., Asakura, K., Sawada, T., Takamatsu, T., Otoguro, Y.: Creep rupture strength and microstructure of Low C-10Cr-2Mo heat-resisting steels with V and Nb. Metall. Trans. A 12(6), 1071–1079 (1981)

11. Liu, Y., Tsukamoto, S., Sawada, K., Abe, F.: Role of boundary strengthening on prevention of type IV failure in high cr ferritic heat-resistant steels. Metall. Mater. Trans. A 45(3), 1306–1314 (2014)

12. Mishnev, R., Dudova, N., Kaibyshev, R.: On the origin of the superior long-term creep resistance of a 10% Cr steel. Mater. Sci. Eng., A 713, 161–173 (2018)

13. Abe, F.: Creep behavior, deformation mechanisms, and creep life of Mod.9Cr-1Mo steel. Metall. Mater. Trans. Phys. Metall. Mater. Sci. 46(12), 5610–5625 (2015)

14. Aghajani, A., Somsen, Ch., Eggeler, G.: On the effect of long-term creep on the microstructure of a 12% chromium tempered martensite ferritic steel. Acta Mater. 57(17), 5093–5106 (2009)

15. Tamura, M., Kumagai, T., Miura, N., Kondo, Y., Shinozuka, K., Esaka, H.: Effect of martensitizing temperature on creep strength of modified 9Cr steel. Mater. Trans. 52(4), 691–698 (2011)

16. Sawada, K.: Effect of W on recovery of lath structure during creep of high chromium martensitic steels. Mater. Sci. Eng., A 267(1), 19–25 (1999)

17. Sklenička, V., Kucharova, K., Svobodova, M., Kral, P., Kvapilova, M., Dvorak, J.: The effect of a prior short-term ageing on mechanical and creep properties of P92 steel. Mater. Charact. 136, 388–397 (2018)

18. Haney, E.M., et al.: Macroscopic results of long-term creep on a modified 9Cr–1Mo steel (T91). Mater. Sci. Eng., A 510–511, 99–103 (2009)

19. Fedoseeva, A., Dudova, N., Kaibyshev, R.: Creep strength breakdown and microstructure evolution in a 3% Co modified P92 steel. Mater. Sci. Eng., A 654, 1–12 (2016)

20. Fernandez Martinez, R., Iturrondobeitia, M., Ibarretxe, J., Guraya, T.: Methodology to classify the shape of reinforcement fillers: optimization, evaluation, comparison, and selection of models. J. Mater. Sci. 52(1), 569–580 (2017)

21. Fernandez Martinez, R., Lostado Lorza, R., Santos Delgado, A.A., Piedra Pullaguari, N.O.: Optimizing presetting attributes by softcomputing techniques to improve tapered roller bearings working conditions. Adv. Eng. Softw. 123, 13–24 (2018). https://doi.org/10.1016/j.advengsoft.2018.05.005

22. Fernandez Martinez, R., Jimbert, P., Ibarretxe, J., Iturrondobeitia, M.: Use of support vector machines, neural networks and genetic algorithms to characterize rubber blends by means of

the classification of the carbon black particles used as reinforcing agent. Soft. Comput. (2018). https://doi.org/10.1007/s00500-018-3262-2

23. Hair, J.F., Black, W.C., Babin, B.J., Anderson, R.E.: Multivariate Data Analysis, 7th edn. Pearson, Upper Saddle River (2010)

24. Fernandez Martinez, R., Martinez-de-Pison Ascacibar, F.J., Pernia Espinoza, A.V., Lostado Lorza, R.: Predictive modeling in grape berry weight during maturation process: comparison of data mining, statistical and artificial intelligence techniques. Span. J. Agric. Res. **9**(4), 1156–1167 (2011)

25. Fernandez Martinez, R., Okariz, A., Ibarretxe, J., Iturrondobeitia, M., Guraya, T.: Use of decision tree models based on evolutionary algorithms for the morphological classification of reinforcing nano-particle aggregates. Comput. Mater. Sci. **92**, 102–113 (2014)

26. Jolliffe, I.T.: Principal Component Analysis, 2nd edn. Springer, New York (2002). https://doi.org/10.1007/b98835

27. Wilkinson, G.N., Rogers, C.E.: Symbolic descriptions of factorial models for analysis of variance. Appl. Stat. **22**, 392–399 (1973)

28. Chambers, J.M.: In: Chambers, J.M., Hastie, T.J. (eds.) Statistical Models in S. Wadsworth & Brooks/Cole (1992)

29. Friedman, J.H.: Greedy function approximation: a gradient boosting machine. Technical report, Department of Statistics, Sequoia Hall, Stanford University, Stanford California 94305 (1999)

30. Friedman, J.H.: Stochastic gradient boosting. Technical report, Department of Statistics, Sequoia Hall, Stanford University, Stanford California 94305 (1999)

31. Wang, Z.: HingeBoost: ROC-based boost for classification and variable selection. Int. J. Biostat. **7**(1), 13 (2011)

32. McCullagh, P., Nelder, J.A.: Generalized Linear Models. Chapman and Hall, London (1989)

33. Dobson, A.J.: An Introduction to Generalized Linear Models. Chapman and Hall, London (1990)

34. Hasti, T. J., Pregibon, D.: In: Chambers, J.M., Hastie, T.J (eds.) Statistical Models in S. Wadsworth & Brooks/Cole (1992)

35. Venables, W.N., Ripley, B.D.: Modern Applied Statistics with S. Springer, New York (2002). https://doi.org/10.1007/978-0-387-21706-2

36. Fox, J.: Applied Regression Analysis and Generalized Linear Models, 3rd edn. McMaster University, SAGE Publications, Inc, Los Angeles (2015)

37. Freund, Y., Schapire, R.E.: Experiments with a new boosting algorithm. In: Proceedings of 13th International Conference on Machine Learning, San Francisco, CA, pp. 148–156 (1996)

38. Buehlmann, P.: Boosting for high-dimensional linear models. Ann. Stat. **34**, 559–583 (2006)

39. Buehlmann, P., Yu, B.: Boosting with the L2 loss: regression and classification. J. Am. Stat. Assoc. **98**, 324–339 (2003)

40. Buehlmann, P., Hothorn, T.: Boosting algorithms: regularization, prediction and model fitting. Stat. Sci. **22**(4), 477–505 (2007)

41. Hothorn, T., Buehlmann, P., Kneib, T., Schmid, M., Hofner, B.: Model-based boosting 2.0. J. Mach. Learn. Res. **11**, 2109–2113 (2010)

42. Hofner, B., Mayr, A., Robinzonov, N., Schmid, M.: Model-based boosting in R: a hands-on tutorial using the R package mboost. Comput. Stat. **29**(1–2), 3–35 (2014)

43. R Development Core Team: R: a language and environment for statistical computing. R Foundation for Statistical Computing, Vienna, Austria (2017). http://www.R-project.org/

Knowledge Extraction from Vector Machine Support in the Context of Depression in Children and Adolescents

Thiago Lima[1]([✉]), Renata Santana[1], Maycoln Teodoro[2], and Cristiane Nobre[1]

[1] Pontifical Catholic University of Minas Gerais, Belo Horizonte, Brazil
thiagohenriqueslima@gmail.com, renata.cris.santana@gmail.com,
nobre@pucminas.br
[2] Federal University of Minas Gerais, Belo Horizonte, Brazil
mlmteodoro@hotmail.com

Abstract. Depression is, worldwide, the main cause of diseases and disabilities during the adolescence. This disorder ails over 300 million people, and can interfere with an individual's professional performance and education. Therefore, it is essential to conduct research that contributes in the correct diagnosis and treatment of depression, especially on children and adolescents. The Support Vector Machines (SVM) classifier has shown great performance 3 and generalization capabilities when compared to other classifiers, in the context of depression diagnosis. The objective of this study is to explore the depression disorder on children and adolescents, using this classifier. Since the SVM is a black box method, to better understand the model generated we employed the SHAP approach to help explain the model's output based on feature importance. The final model obtained F-measure results above 87% during training and 82% in its testing. We concluded that the predictive model had satisfactory results and, using the SHAP framework, we explored how the features influenced the results.

Keywords: Depression · Children and Adolescents · Knowledge extraction · Support Vector Machines · SHAP framework

1 Introduction

The depression disorder, according the World Health Organization (WHO)[1], is different from common mood swings and short-lived feelings in response to daily life challenges. It is pointed by WHO as the worldwide main cause of diseases and disabilities in adolescents, potentially evolving into a serious health issue when lasting long with moderate or high intensity. Depression, as a disorder, afflicts about 300 million people, an increase of over 18% between the years of 2005 and 2015. When untreated, this disorder causes great suffering to the individual,

---

[1] Available at http://www.who.int/mediacentre/factsheets/fs369/en/.

© Springer Nature Switzerland AG 2019
I. Nyström et al. (Eds.): CIARP 2019, LNCS 11896, pp. 545–555, 2019.
https://doi.org/10.1007/978-3-030-33904-3_51

and interferes with their professional performance, education, relationships, and possibly leads to suicide, in the worst cases. Every year, about 800 thousand people commit suicide, and this is considered the second greatest cause of death of young people between 15 and 29 years old [1]. Some studies [21] claim that half of people who suffer mental disorders show their first symptoms before completing 15 years of age. The American Psychiatric Association (APA)[2] claims that one in every six people will suffer depressive episodes during their lives, which indicates over a billion possible victims worldwide. Therefore, caring for children and adolescents with mental health issues is important to avoid death and suffering throughout their lives.

A precise diagnosis is fundamental before administering psychological and pharmacological treatments for depression. Thus, there's a need for research on the diagnosis of depression, as well as its treatment. Some researchers conducted studies to aid in depression diagnosis, and among those studies some promising results came from applying machine learning techniques. The literature showed satisfactory results from combining machine learning techniques and pattern recognition to characterise diseases, being particularly efficient for mental health issues such as depression [17].

The Support Vector Machines (SVM) classifier has shown superior performance and generalisation capabilities when compared to other classification techniques, in various different applications, including depression diagnosis [6]. We performed some preliminary experiments training different classification algorithms with the dataset used in our study, and the SVM outperformed all others, namely: C4.5 decision tree [18], CART [4], Multilayer Perceptron Neural Network and Random Forests [5] (results not shown). The SVM is considered a black box method, as it somewhat conceals its internal logic from the user, creating models that are difficult to interpret. Because of that limitation, and because it was especially important in this application to interpret the generated model, we employed the SHAP framework [14] to help explain the output of the SVM classifier, and characterise the diagnosed individuals using a feature importance metric. The classifier was training with a dataset with data from 377 patients between 10 and 16 years of age. The data was obtained through a partnership with the Cognition and Behavioural Psychology Postgraduate Research Program of a University.

2   Background

2.1   Black Box Models

As machine learning algorithms become more complex and precise, they often become less comprehensible and generate models that are harder to interpret. A model is said to be a black box if it keeps its internal structure unknown or hard to interpret, making its classifications hard to explain. The behaviour of a black box model can be described as such: given an input, the model calculates

---

[2] Available at https://www.psychiatry.org/.

the output based on an internal function, without providing an explanation of how it reached its result.

Although they usually have superior generalisation capabilities, when compared to other classifiers, the non-intuitive solutions provided by black box models can become an obstacle to their practical use, especially when it is vital to the project to explain how the classification was made.

Lately, black box classifiers such as the SVM and Artificial Neural Networks have been achieving good results in several applications, but the low interpretability of their models hinders their applicability for cases where the classification process needs to be understood, such as medical applications. In [8], the authors argue that even a limited explanation can positively influence the likelihood of these methods being applied in such cases.

Some rule induction techniques, such as Quinlan's C4.5 decision trees [18], build highly interpretable models, but are likely to lose performance for doing so, being outperformed by more complex classifiers such as the SVM. Therefore, efforts for extracting rules that may help explain black box models have been made, to maintain their superior performance and gain some interpretability. The ultimate goal is to have the performance of black box models associated to a transparent and easily interpretable model that can, for example, be modelled as a decision tree or a rule set [10].

A black box model can be explained either from a global or a local perspective. A global explanation considers the internal functioning of the whole model [11]. The local explanations aim to elucidate the reasoning behind a single prediction. The rule extraction algorithms that are used for this purpose can be classified as "pedagogical" or "decomposition". The pedagogical approach extracts rules directly related to the inputs and outputs of a classifier. The approach utilises the trained model as an oracle to produce a set of examples of inputs and outputs, and then applies pattern search strategies to construct its model (a decision tree, for instance). The decomposition approach is interwoven with the internal structure of the SVM and its hyperplanes, aiming to explain the individual computation of the internal components in the model [10].

2.2 Support Vector Machines

The SVM [7] classifier is based on the Statistical Learning Theory. It constructs hyperplanes with a decision surface, maximising class separation. Several hyperplanes can be constructed to separate the instances, each of them defining a separation margin, where points situated in the limits make the support vectors, and the middle point of the margin is the optimal hyperplane. It is expected that broader hyperplanes with broader margins will be able to classify unseen data better than those with narrower margins.

These classifiers frequently show good generalisation capabilities when compared to others. However, their models are non-intuitive and hard to interpret [9]. To overcome that limitation, several techniques to extract knowledge from SVM were developed, to help interpret them and explain their classifications, such as: SVM+Prototype [16], Barakat [3], Fung [9], SHAP [14], and others [10].

2.3   The SHAP Framework

The SHAP (SHapley Additive exPlanations) is a unified approach to interpret predictions and explain the outputs of any machine learning model. The SHAP connects game theory with local explanation, to attribute an importance measure to each feature, for a given prediction [14], with larger values indicating a greater participation of a feature in a prediction. The SHAP calculates this feature importance measure using Shapley's values, introduced in 1953 in the game theory field [20], but only recently being applied in this context.

The SHAP framework has unified six existing feature importance measures and, according to the author, ensures three desired properties to methods of the same class, with better computational performance and interpretability than other approaches [14]. These properties are: (1) *Local Precision*: the sum of the feature importance's attributions is equal to the model's output, (2) *Missingness*: missing features are not attributed any impact on the model's output, and (3) *Consistency*: altering a model so that a feature gains a bigger impact in it will never reduce the feature importance attribution of that feature.

The calculations of the SHAP values are simple, but computationally expensive. The idea is to re-train the model for all feature subsets $S \subseteq F$, where F is the set of all features. The Shapley values attribute an importance metric to each attribute, representing its impact on the model's predictions. In order to calculate this impact, it compares the predictions of models trained with and without the feature. As the impact of a feature also depends on other features included in the model, the previous comparisons are made for all possible feature subsets, and the Shapley values are a weighted average of all comparisons.

Figure 1 illustrates the SHAP approach to explain the output of a machine learning model. The framework, thus, obtains the model generated by a machine learning method and outputs its feature importance measures.

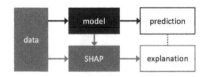

Fig. 1. Diagram of the SHAP framework.

3   Related Work

In [12] the authors analysed data from a city in India serviced by a phone-call based screening system for tuberculosis patients, used to help the patient screening process. The dataset had close to 17 thousand patients and 2.1 million registered phone calls. The authors report that the technique with best predictive performance was a deep learning approach, considered a black box. They employed the SHAP framework to generate visualisations and help explain the model, providing insight to the medical researchers. Their conclusions were that, in a real-time application scenario, the model would be able to support health professionals in making precise interventions on high risk patients.

Yan et al. [23] utilised data from the Acute Myocardial Infarction record, from China, and applied the XGBoost machine learning method to generate a risk prediction model for hospital mortality among patients that had suffered a myocardial heart attack. They employed the SHAP framework to explain the impact of their features in the predictions and, from its results, were able to find new relations between clinical variables and hospital mortality. One example of such relations was the blood glucose level, which showed a nearly linear relationship with hospital mortality in the patients. The authors concluded that the new prediction model had a good discrimination capability and offered individualised explanations of how the clinical variables had influenced the results.

4 Materials and Methods

4.1 Dataset Description

The dataset utilised in this study was obtained from a partnership with the Cognition and Behavioural Psychology Postgraduate Research Program of Federal University of Minas Gerais. The dataset hold information from 377 children and adolescents between 10 and 16 years of age (158 male and 219 female), and has 75 features[3] representing different symptoms of possible depression disorder.

The dataset stores broader demographic data such as the patient's age and gender, and more specific data such as schooling, who they live with, use of medication, Youth Self-Report (YSR) scores, and questions of the Children's Depression Inventory (CDI) [13]. The dataset also stores information about the patient's relationship with their parents, such as hours a week they spend with their parents, whether the patient or the parents has had psychological or psychiatric treatment, and the parents' schooling. Other features deemed important by the mental health study community were included, such as anxiety factors, social problems, lack of attention, aggressiveness, and behaviour issues. Most features in the dataset have ordered categorical values.

4.2 Data Preprocessing

In order to obtain a more robust model, before training the classification models, we did some preprocessing on the dataset. The goal of the preprocessing was to remove features that were unrelated to the problem, merge features when necessary, encode the features and handle missing data and outliers. All the data preprocessing was done using Python on the Jupyter Notebook framework. The preprocessing tasks, executed sequentially, were: *removal of irrelevant features*

---

[3] A complete description of the features of this dataset can be found at https://goo.gl/z2wUKg.

and *treating inconsistencies in the data.* For the second task, two instances had unexpected values for some features (greater than the limit in the presented context, which was daily time spent with the parents). For these cases, we assumed the greatest possible value of the context, 24.

Continuing preprocessing, the following tasks were performed: *nominal to numeric values encoding* and *identification of the class feature.* We observed that, among the patients in the dataset, the "CDI score" feature had values between 0 and 46. The CDI score does not determine the diagnosis of depression, but it does show evidence that can help the precise diagnosing, and is calculated by evaluations made by professionals. However, there was no unanimous threshold to determine a depression diagnosis, as this value can vary on different samples. The recommendation by Kovacs [13] is to utilise a 85 percentile threshold to indicate high symptomatology. Thus, in our dataset, 63 patients had high enough CDI scores to be classified as *HIGH* symptomatology, and the others were classified as *LOW*.

The dataset has 314 individuals of the *LOW* class and 63 of the *HIGH*. In order to avoid the classifier from having a bias towards the majority class, we employed a class balancing strategy of random undersampling, until the number of patients of both class was the same.

To validate the generated classification models, we divided the dataset into a training and a test set. The model is created using the training instances and, then, evaluated in the unseen test instances. The division is shown in Table 1.

Table 1. Number of instances per class.

Class	Before balancing	After balancing	Model building	Model test
HIGH	63	63	53	10
LOW	314	63	53	50
Sum	377	126	106	60

With the balanced dataset, with randomly selected 10 instances of the HIGH class and 50 of the LOW class for testing. The number of instances of each class maintained the original proportions of classes. The model was trained with 53 instances of each class using 10-fold cross-validation.

4.3   Methods

The SVM experiments were conducted using the libSVM implementation on Python's scikit-learn open-source library[4]. The algorithm was selected based on its frequent use in the literature and for meeting the standards required for our study. Three parameters were adjusted when training the classifier: the C

---

[4] Available at https://scikit-learn.org/stable/modules/generated/sklearn.svm.SVC.html.

value ($C = 12$), a smoothing parameter for the hyperplane margins, the *gamma* (*gamma* = 0.001), which is the width of the Gaussian, and the kernel type (*kernel* = *rbf*). These parameters are highly relevant to the performance of the model, as they are directly related with the training times and prediction performance. The SVM parameters were adjusted using the Grid Search algorithm. The Grid Search performs and exhaustive search over specified values of parameters for a classifier, and finds the best combination of values for these parameters, based on a quality criterion.

Given the SVM's complexity and the need to understand the importance of each feature for the predictions, after training and validating the SVM model, we employed the SHAP framework [14] to help interpret the classifier's output.

5   Results and Discussion

In this section, we present the results obtained from the SVM model trained with the preprocessed dataset. Table 2 shows the average values of the evaluation metrics[5].

Table 2. Training set results (in percentage).

Class	Precision	Recall	F-Measure
HIGH	90.0	84.9	87.4
LOW	85.7	90.6	88.1
Average	87.9	87.7	87.7

The training set results showed a greater precision for the *HIGH* class, meaning the model correctly classified 90% of the instances of that class. There is also a noticeably higher recall value for the *LOW* class, meaning the model was correct when predicting that class 90.6% of the times it did.

Table 3 shows the test set results. For these experiments, the test sets had 10 instances of the *HIGH* class and 50 of the *LOW* class, keeping the proportions of the original dataset, and all instances were previously unseen by the model. Analyzing the F-measure metric, the harmonic mean between precision and recall, we see that SVM performance was good, despite observing a reduction in the HIGH class.

In the following part of the study, we identified the most relevant features in the model based on the test set results. Figure 2 shows the SHAP values for the best features and all the test set examples, classified based on their impact on the classifier's output.

The horizontal axis represents the impact (SHAP value) of a feature, with positive values meaning that the values of the feature will increase the likelihood

---

[5] $Precision = \frac{TP}{TP+FP}$; $Recall = \frac{TP}{TP+FN}$; $F-measure = \frac{2 \times Precision \times Recall}{Precision + Recall}$.

Table 3. Test set results (in percentage).

Class	Precision	Recall	F-Measure
HIGH	60.0	90.0	72.0
LOW	97.8	88.0	92.6
Average	78.9	89.0	82.3

of the positive class (HIGH), whereas negative values mean the opposite. The attributes were ordered vertically by their average impact, the highest impact at the top. The points in each distribution for a feature represents a single patient, with high density represented by stacking of points. The colours of the points represent high (red) and low (blue) values of the features. The Figure clearly shows how high or low values of the features impact their SHAP values.

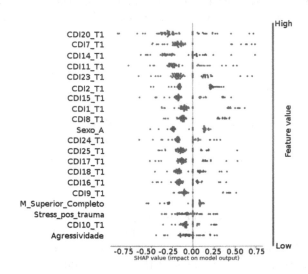

Fig. 2. Impact of the most relevant features for the classification of test set instances. For each feature, the vertical dispersion represents data points with the same SHAP value for that feature. Higher SHAP values mean greater likelihood of a positive class prediction (HIGH). (Color figure online)

To better understand the Fig. 2, let's look at the CDI20_T1 attribute, which is the highest impact attribute for the model. The points represented by the red color indicate higher values for the attribute, while the points represented by the blue color represent lower values. The CDI20_T1 attribute, which characterizes the feeling of loneliness, has the following values: *(0) I don't feel lonely, (1) I feel lonely often* and *(2) I always feel lonely*, therefore, the higher the value of the attribute, the greater the feeling of loneliness. High values of this attribute (red dots) indicate a higher probability of a HIGH class prediction, while low values (blue dots) indicate a higher probability of a LOW class prediction.

An interesting information represented in the Figure is that, among the 20 most relevant features in the model, 16 of them are from the CDI, being CDI20_T1 (loneliness feelings), CDI7_T1 and CDI14_T1 (both low self-esteem) the three features with greater impact in the prediction.

These results corroborate the findings [2,22,24] that self-esteem is an important factor for depression and suggest continuous interventions to increase self-esteem in adolescence, which can greatly reduce the degree of depression. In [15] the authors propose an investigation in young adulthood about the association between social isolation and loneliness and how they relate to depression and conclude that both were associated with depression.

6 Conclusions and Future Work

The objective of this study[6] was to explore the diagnosis of depression disorder in children and adolescents, utilising a SVM classifier. Although there are several studies utilising machine learning classification algorithms to tasks related to depression diagnosis, there are few studies focused on younger individuals. Patients of young age are usually harder to evaluate on a deeper level, reaching complex analysis results that can lead to a more precise diagnosis, hence the need for precise predictions based on high-level data.

Other authors have considered classification metrics above 75% as satisfactory and meaningful [19]. Our model surpasses that threshold, and has a good success rate in discriminating the classes. With the SHAP framework, we were able to analyse the predictions made by the SVM model, and discuss feature importance to an extent.

The model uses the symptomatology feature as its class value. This feature is calculated based on data from the CDI. As future work proposal we suggest the application of classifiers in a dataset that utilises other means of classifying young individuals with depression symptoms, and the discussion of other thresholds to classify the disorder. We also recommend deeper classification analysis, in order to possibly reach more robust models (using other algorithms or parameter combinations). Another possible future direction is to use techniques of rule extraction from SVM models to analyse the predictions made by the classifier, and further understand the diagnosis of depression on children and adolescents.

References

1. Anderson, R.N., Smith, B.L., et al.: Deaths: leading causes for 2002. Natl. Vital Stat. Rep. **53**(17), 1–90 (2005)

---

[6] This study was financed in part by the Coordenação de Aperfeiçoamento de Pessoal de Nível Superior - Brasil (CAPES) - Finance Code 001, by Foundation for Research Support of the State of Minas Gerais/Brazil (FAPEMIG) and the Brazilian National Council for Scientific and Technological Development (CNPq).

2. Babore, A., Trumello, C., Candelori, C., Paciello, M., Cerniglia, L.: Depressive symptoms, self-esteem and perceived parent-child relationship in early adolescence. Front. Psychol. **7**, 982 (2016)
3. Barakat, N., Diederich, J.: Eclectic rule-extraction from support vector machines. Int. J. Comput. Intell. **2**(1), 59–62 (2005)
4. Breiman, L., Friedman, J., Olshen, R., Stone, C.: Classification and Regression Trees. Wadsworth International Group, Monterey (1984)
5. Breiman, L.: Random forests. Mach. Learn. **45**(1), 5–32 (2001)
6. Byun, H., Lee, S.-W.: Applications of support vector machines for pattern recognition: a survey. In: Lee, S.-W., Verri, A. (eds.) SVM 2002. LNCS, vol. 2388, pp. 213–236. Springer, Heidelberg (2002). https://doi.org/10.1007/3-540-45665-1_17
7. Cortes, C., Vapnik, V.: Support-vector networks. Mach. Learn. **20**(3), 273–297 (1995)
8. Davis, R., Buchanan, B., Shortliffe, E.: Production rules as a representation for a knowledge-based consultation program. Artif. Intell. **8**(1), 15–45 (1977)
9. Fung, G., Sandilya, S., Rao, R.B.: Rule extraction from linear support vector machines. In: Proceedings of the Eleventh ACM SIGKDD International Conference, pp. 32–40. ACM, New York (2005)
10. Guidotti, R., Monreale, A., Ruggieri, S., Turini, F., Giannotti, F., Pedreschi, D.: A survey of methods for explaining black box models. ACM Comput. Surv. (CSUR) **51**(5), 93 (2018)
11. Hall, P., Gill, N.: Introduction to Machine Learning Interpretability. O'Reilly Media Incorporated, Sebastopol (2018)
12. Killian, J.A., Wilder, B., Sharma, A., Choudhary, V., Dilkina, B., Tambe, M.: Learning to prescribe interventions for tuberculosis patients using digital adherence data. arXiv preprint arXiv:1902.01506 (2019)
13. Kovacs, M.: Children's Depression Inventory (CDI): technical manual update. Multi-Health Systems (1992)
14. Lundberg, S.M., Lee, S.I.: A unified approach to interpreting model predictions. In: Advances in Neural Information Processing Systems 30. Curran Associates, Inc. (2017)
15. Matthews, T., et al.: Social isolation, loneliness and depression in young adulthood: a behavioural genetic analysis. Soc. Psychiatry Psychiatr. Epidemiol. **51**, 339–348 (2016)
16. Núñez, H., Angulo, C., Català, A.: Rule-based learning systems for support vector machines. Neural Process. Lett. **24**(1), 1–18 (2006)
17. Orrù, G., Pettersson-Yeo, W., Marquand, A.F., Sartori, G., Mechelli, A.: Using support vector machine to identify imaging biomarkers of neurological and psychiatric disease: a critical review. Neurosci. Biobehav. Rev. **36**(4), 1140–1152 (2012)
18. Quinlan, J.R.: C4.5: Programs for Machine Learning. Morgan Kaufmann Publishers Inc., San Francisco (1993)
19. Sacchet, M.D., Prasad, G., Foland-Ross, L.C., Thompson, P.M., Gotlib, I.H.: Support vector machine classification of major depressive disorder using diffusion-weighted neuroimaging and graph theory. Front. Psychiatry **6**, 21 (2015)
20. Shapley, L.S.: A value for n-person games. Contrib. Theory Games **2**(28), 307–317 (1953)
21. Sunmoo, Y., Basirah, T., et al.: Using a data mining approach to discover behavior correlates of chronic disease: a case study of depression. Stud. Health Technol. Inform. **201**, 71 (2014)

22. Ticusan, M.: Low self-esteem, premise of depression appearance at adolescents. Procedia Soc. Behav. Sci. **69**, 1590–1593 (2012). International Conference on Education & Educational Psychology (ICEEPSY 2012)
23. Yang, J., Li, Y., Li, X., Chen, T., Xie, G., Yang, Y.: An explainable machine learning-based risk prediction model for in-hospital mortality for Chinese STEMI patients: findings from China myocardial infarction registry. J. Am. Coll. Cardiol. **73**, 261 (2019)
24. Yoon, M., Cho, S., Yoon, D.: Child maltreatment and depressive symptomatology among adolescents in out-of-home care: the mediating role of self-esteem. Child. Youth Serv. Rev. **101**, 255–260 (2019)

Multi-domain Aspect Extraction Based on Deep and Lifelong Learning

Dionis López[1,2(✉)] [ID] and Leticia Arco[2,3] [ID]

[1] Informatics Department, Faculty of Engineering in Telecommunications,
Informatics and Biomedical, Universidad de Oriente, Avenida Las Américas s/n,
90900 Santiago de Cuba, Cuba
dionis@uo.edu.cu
[2] Department of Computer Science, Faculty of Mathematics,
Physics and Computer Science, Universidad Central "Marta Abreu" de Las Villas,
Carretera a Camajuaní km 5 1/2, 54830 Santa Clara, Cuba
larcogar@vub.be
[3] AI Lab, Computer Science Department, Vrije Universiteit Brussel,
Pleinlaan 9, 1050 Brussels, Belgium

Abstract. Opinions concerning features or aspects of people, entities, products or services are some of the most important textual information. Several methods try to solve the aspect extraction task needed in sentiment analysis by using Deep Learning techniques in specific domains. However, catastrophic forgetting appears when these methods are used to learn aspects of multi-domains. In this paper, we propose a new approach to achieve aspect extraction in multi-domains based on Deep and Lifelong Learning techniques. Our proposal reduces catastrophic forgetting and improves one of the principal state-of-the-art results.

Keywords: Opinion mining · Aspect extraction · Deep learning · Lifelong learning

1 Introduction

Opinion identification and their automatic classification in digital documents have been addressed in several papers since the beginning of 2000 [2,3], which belong to the Sentiment Analysis or Opinion Mining tasks. Aspect Based Sentiment Analysis (ABSA) is one of the Sentiment Analysis subtasks, which allows extracting more information from opinions [2]. An aspect term is associated with the possible characteristics about products, services, events or people. In the ABSA subtask, an effective extraction of aspects is very important to achieve a correct sentiment classification [3].

Some methods perform aspect extraction in a single knowledge domain [7,13]. These methods have had good results for a single domain, but when they have been applied in different domains, their effectiveness decreases. Several of these proposals use Deep Learning techniques [7,13,14] for aspect extraction, but they

© Springer Nature Switzerland AG 2019
I. Nyström et al. (Eds.): CIARP 2019, LNCS 11896, pp. 556–565, 2019.
https://doi.org/10.1007/978-3-030-33904-3_52

only deal with a single domain or at most two domains. This limitation in terms of the number of domains prevents the learning of common features in different datasets and, therefore, reduces the scope of these proposals.

It is important during the learning process not to lose effectiveness in each knowledge domain and to obtain the patterns or features that may be common to several domains (e.g.; the *price* aspect is common for restaurant, hotel and electronic device domains). For this reason, Lifelong Learning strategy is useful in aspect extraction. It takes advantage of the local learning of several domains by identifying the common features or patterns found in the previous learning process without losing the effectiveness when learning a new domain [3].

When using Lifelong Learning with neural networks, it is necessary to avoid catastrophic forgetting. This occurs when networks are sequentially trained in many tasks; because the network weights estimated for a task A can be modified in the learning process of a task B [10]. Some proposals have tried to overcome the catastrophic forgetting; for instance, in image classification [10] and game strategy learning [9]. Nevertheless, there are few proposals devoted to solve the catastrophic forgetting in ABSA subtasks [5].

In this paper, we propose a new model for extracting aspects in multi-domains based on the combination of Convolutional Neural Networks (CNN) [4] and Lifelong Learning [3]. The main contribution of this paper is to reduce the catastrophic forgetting in a Lifelong Learning framework for aspect extraction. The rest of this paper is organized as follow. The principal aspect extraction methods based on Deep Learning techniques are presented in Sect. 2. Section 3 explains the proposed model for aspect extraction based on Deep and Lifelong Learning. Section 4 presents the evaluation of our model with respect to another proposal of the state of the art. Towards the end, we provide concluding remarks and future research directions.

2   Related Work

There are several works which use Deep Learning for solving the ABSA subtask [2,5,7,13]. Some researchers model the ABSA subtask as a sequence of characteristics to be learned by a method. This point of view has motivated the use of the Long Short Term Memory (LSTM) and Gated Recurrent Units (GRU) for aspect extraction [5]. This type of models overcome the vanish or explode gradient problem during the backpropagation process. As advantage these networks capture long-term dependencies of context information, but they can be slower than other techniques as CNN [5]. Usually, LSTM or GRU are combined with a method known as Attention Mechanism (AM) [8]. AM helps improve the training of the neural network by relating the closest words to an aspect in a sentence [5]. However, LSTM, GRU or AM were not selected because there is no clear connection with our Lifelong Learning strategy.

Aspect extraction is also performed by using a CNN [7,13] since it can extract salient n-gram features to create an informative latent semantic representation of sentences [4,5]. Specifically, the approach presented in [13] is one of the best results in ABSA, showing the F-measure value equal to 0.86 in only two separate domains [5]. It is important to note that several linguistic rules are included in this proposal to label aspects in an unsupervised way, which influences the effectiveness of the Deep Learning methods, but their creation is expensive in time and human effort. Our Deep Learning proposal was inspired in [13], due to, its successful results.

There are some proposals for solving the ABSA subtask using Continual or Lifelong Learning for multi-domains [5,14], being the approach presented in [14] one of those that achieve the best results. Besides, the authors created a Lifelong Learning proposal based on Conditional Random Field (CRF). Its main disadvantages are the required engineering of features and the possible error propagation when learning other domains. Despite the mentioned disadvantages of this approach, we have selected it to be compared with our proposal, due to a Lifelong strategy is used, and CRF has reported good results in ABSA [2,14].

We explored the main approaches that deal with catastrophic forgetting to consider their ideas for extracting aspects in multi-domains. In [6] the authors apply the fine tuning for reducing the learning rate to prevent significant changes in the network parameters while training with new data. This solution avoids losing previous knowledge but limits new learning. Another approach applies feature extraction techniques where the network parameters in a previous task are not changed and the output layer is used to extract features in a new task [10]. The Elastic Weight Consolidation (EWC) approach improves other catastrophic forgetting proposals, but it is not capable of learning new categories incrementally [9].

The use of the output layers in a Lifelong model is more simple and allows the lower layers to learn the common word features in a multi-domain training. The following section explains how Deep and Lifelong Learning are amalgamated for extracting aspects and reducing catastrophic forgetting in multi-domains.

3   New Multi-domain Aspect Extraction Model Description

In this section, we introduce a new model for aspect extraction based on Deep and Lifelong Learning through catastrophic forgetting reduction in multi-domains. The Deep Learning architecture used is based on the CNN solution presented in [13], whereas the selected Lifelong Learning technique follows some features of the model proposed in [10].

Our model is composed of four main stages, as shown in Fig. 1. For each domain, an output layer connected to the last layer of the CNN is created to define the domain-dependent characteristics and parameters. The final stage offers a Deep Learning machine ready for use in diverse information systems.

Stage 1: Textual Representation. This stage receives the original textual opinions and returns the Word Embedding [11] vector per word of each sentence, through the following steps:

- Pre-process textual opinions by applying a sentence splitter and a Part-of-Speech (POS) tagger.
- Obtain the word vector model from the pre-trained Word Embeddings.

Stage 2: Basic Knowledge Extraction. This stage is in charge to learn the knowledge to be included in the Knowledge Base (KB) depending on the CNN training process for each current domain. The outputs of this stage are the new parameters obtained in the training process.

Stage 3: Knowledge Base Upgrade. In this stage, catastrophic forgetting is avoided through the analysis of the training process results. The loss or errors obtained in the current domain and the previous ones are evaluated, i.e., an analysis of the results corresponding to the output layer associated with each domain is done. The KB is enriched by the training process and the new aspects extracted from different domains.

Stage 4: Aspect Extractor Creation. This stage makes available the Deep Learning machine for solving the ABSA subtask in multi-domains. The final configuration of the CNN machine is obtained from the common parameters in the KB for multi-domains.

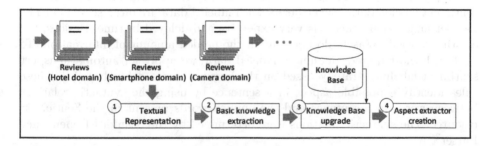

Fig. 1. Multi-domain aspect extraction model based on deep and lifelong learning.

The seven layer CNN architecture used in our approach is very similar to the model presented in [13]. After pre-processing the sentences, we add for each word in a sentence its corresponding POS tag to the Word Embedding vector. We used six basic POS tags (noun, verb, adjective, adverb, preposition, other) encoded as a six-dimensional binary vector. This binary vector is concatenated to the 300-dimensional vector corresponding to each Word Embedding obtained from a pre-trained Word2Vec[1]. Thus, the feature vector associated with each word is 306-dimensional. Consequently, for each sentence to train, the input of the CNN model is a matrix of 82 rows and 306 columns, where the number of rows means the maximum number of words in the training sentences. The number

---

[1] https://code.google.com/archive/p/word2vec/.

of columns represents the Word Embedding vector size and the six-dimensional POS tag binary vector.

The second layer in our CNN architecture is a convolutional layer with 100 feature maps by using a filter equal two. Its output was computed using the hyperbolic tangent. The third and fifth layers are maxpooling layers with pool size equal two. The fourth layer is a convolutional layer and it has 50 feature maps with a filter equal three. Its output was also computed through an hyperbolic tangent function. The sixth layer is a fully connected layer. In our proposal, each domain defines its fully connected seventh layer. The stride for each convolutional layer is one for considering the relation between each word [13].

We used regularization with dropout in the penultimate layer and a restriction of type L_2 for computing the weight vector. Besides, we exploit the combination of convolution and maxpooling layers as well as other neural network parameters as are defined in most of CNN approaches [4,5,7]. Other architectures in our model did not show significant different results.

In the learning process of our CNN model, a windows of five words surround is taken for representing each word in a sentence (e.g.; the analyzed word and two words on the right and left sides). This strategy responds to the possible relationship between the aspect terms and the words most closed to them. Other window sizes did not contribute to increase the accuracy. The error estimation is made by applying the Viterbi algorithm [15] to the output layer of each domain.

Applying Deep Learning models in ABSA requires a large amount of aspect-sentiment labeled data. Nevertheless, such labeled data are often scarce [5]. The dataset annotation process is very expensive and delayed in time. Besides, the existing labeled datasets do not contain the needed amount of information. To deal with the required amount of labeled data, we apply an automatic aspect sentiment labeling strategy based on the linguistic rules defined in [13]. These rules identify a possible aspect in a sentence by using the syntactic relations between words in a sentence and the opinion words present in the SenticNet[2] resource. The syntactic relations were determined by the Stanford Dependency Parser[3].

Finally, we present a new Lifelong Learning method to avoid catastrophic forgetting called Learning without Forgetting with Linguistic Rules (Lwf-CNN-lgR). Our propposal is inspired in the approach proposed in [10], due to it was successful used to avoid catastrophic forgetting during a CNN learning process. Because of the approach proposed in [10] was applied to image multi-classification, we had to adapt it for labelling the words in the sentences. We fit the loss function by a cross-entropy between the Viterbi algorithm output and the correct sequence of tags in a sentence. The main goal of this upgrade is to achieve a more effective ABSA learning. Besides, a set of linguistic rules was added to obtain more aspects from unlabeled datasets and enrich the Knowledge Base.

---

[2] https://sentic.net/.
[3] https://nlp.stanford.edu/software/lex-parser.shtml.

Lwf-CNN-lgR is shown in Algorithm 1, where the subindex s indicates the parameters that are shared by all domains (CNN model), the subindex c is related to specific parameters of the current domains and the subindex p is associated to the specific parameters of the last domains. The constant λ_{prev} is a penalty to the influence on the loss in the previous domains. $\textbf{\textit{Loss}}_{prev}()$ and $\textbf{\textit{Loss}}_{cnt}()$ functions represent the loss to previous and current domains in each training moment. The $\textbf{\textit{R}}()$ function is responsible for adjusting the neural network regularization. Θ_s is a set of parameters shared across all domains (weights in our CNN model), Θ_p is a set of parameters learned specifically from previous domains, and Θ_c is a randomly initialized parameter in the current domain (weights in the output layer for each current domain).

Algorithm 1. Lifelong learning algorithm to reduce catastrophic forgetting

Input:
Θ_s ▷ CNN model with shared parameters
Θ_p ▷ Parameters of the previous domains
X_c, Y_c ▷ Sentences and tagged aspects in the current c domain
Output:
Θ_s ▷ CNN model
Θ_c ▷ The output layer in the last domain
1: $\Theta_c=$ **RandInit** ()   ▷ New parameters randomly initialized
2: **for** $x_i, y_i \leftarrow X_c, Y_c$ **do**
3:      **Define** $Y_p = $ **TrainCNN**$(x_i, \Theta_s, \Theta_p)$ ▷ Outputs of previous domains
4:      **Define** $y_c = $ **TrainCNN**$(x_i, \Theta_s, \Theta_c)$ ▷ New domain output
5:      $\Theta_s, \Theta_p, \Theta_c = $ **argmin**$(\lambda_{prev} * \textbf{\textit{Loss}}_{prev}(y_i, Y_p) + \textbf{\textit{Loss}}_{cnt}(y_i, y_c) + \textbf{R}(\Theta_s, \Theta_p, \Theta_c))$
6: **end for**

In Algorithm 1, the parameters corresponding to the current domain output layer are randomly initialized, as shown in line 1. The parameters of each output layer from previous domains are stored. The CNN model is trained by using each sentence in the dataset. Outputs are obtained for each previous domain and the current one, as shown in lines 3 and 4.

Line 5 shows how the loss values between the current domain and the previous ones are minimized. This process updates the parameters of the neural network by means of the regularization and descendant gradient. The loss combinations between the current and previous domains permit to avoid the catastrophic forgetting. A high value of the λ_{prev} constant determines more influence of previous domains loss values. The two final results are the trained CNN model and the output layer of the last domain. This output layer contains the common knowledge among all output layers. These two results are contained in the KB.

Our proposal requires the use of tools able to provide a grammatical structure of the text. By analyzing the state-of-the-art we noticed that one of the most prominent libraries is the one provided by Stanford Dependency Parser, which also affords a POS tagger and a sentence splitter. Additionally, the model was trained by using the Tensorflow[4] framework. The trained model can be used for analyzing opinions on public services or products as a service to third parties or as part of an information retrieval module in areas such as electronic government or business intelligence.

4   Experimental Results

We select the seven datasets used in [14] to evaluate our proposal. Besides, two more datasets about restaurants [12] and hotel reviews from TripAdvisor[5] are included to explore in-depth the performance of our model in diverse domains, not only about electronic devices. The linguistic rules were applied to the unlabeled datasets used in [1] to increase the training dataset. This dataset provides 1000 reviews tracked from Amazon for each of 50 domains about electronic devices such as keyboards, car stereo, tablets, etc. We design the experiments to compare the performance of our approach to the state-of-the-art models published in [2,14]. We name the proposals to be compared as follow:

- **Lifelong CRF**: The approach presented in [14] which uses a CRF model in a Lifelong learning scheme.
- **CRF**: A linear chain CRF model evaluated in a multitasking learning scheme to consider all domains at the same time [2].
- **Lwf-CNN**: Our proposal using only CNN and Lifelong Learning without linguistic rules, i.e., the unlabeled datasets are not used.
- **Lwf-CNN-lgR**: Our proposal by combining CNN and the Lifelong Learning scheme enriched with linguistic rules.

For the CNN model, we use the pre-trained Google News with Skip–gram model as Word Embeddings [11]. We randomize other model parameters from a uniform distribution $U(-0.05, 0.05)$. The learning rate used was 0.01 (other values did not show better performances) with the Gradient Descent Optimizer. The selected value for the epoch is 100. The batch training was not used because we want to obtain more effective results. The λ_{prev} value is equal to 0.0056 for controlling the influence of the previous learning in the current learning.

The selected evaluation measures are precision (P), recall (R), and F1-score (F1) due to they are widely used in ABSA [2,5] and they were applied in [14]. We conducted both cross-domain and in-domain tests. Our problem setting is cross-domain; nevertheless, in-domain is used for completeness as done in [14].

The cross-domain experiments combine six labeled domain datasets for training and test in the seventh remaining domain (not used in training). The in-domain experiments train and test on the same six domains excluding one of the seven. Figure 2 shows the F1-score results corresponding to the cross-domain Deep and Lifelong Learning evaluation results, whereas Fig. 3 those corresponding to the in-domain experiments. In Fig. 2, each domain in the x-axis means

---

[4] https://www.tensorflow.org/.
[5] http://times.cs.uiuc.edu/~wang296/Data/.

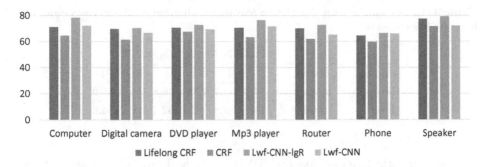

Fig. 2. Cross-domain F1-score results corresponding to the Deep and Lifelong Learning evaluation results.

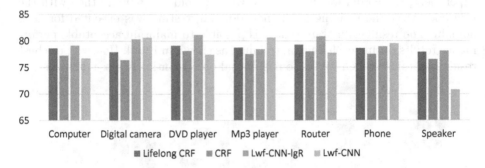

Fig. 3. In-domain F1-score results corresponding to the Deep and Lifelong Learning evaluation results.

that it was not used in training, while it means that the other six domains were used in both training and testing (thus in domain) in Fig. 3.

The best F1 results are obtained by the Lwf-CNN-lgR model, as shown in Fig. 2. The Lwf-CNN-lgR model benefits from the word coincidence cross similar domains, as well as the unlabeled datasets associated with the electronic device domain used at the beginning of the training process, which justifies the achieved results. In the in-domain experiments, the Lwf-CNN-lgR model is mostly the winner, although for some domains it is surpassed by Lwf-CNN, as shown in Fig. 3.

As mention before, we include two new datasets (restaurant and hotel reviews) to analyze in depth how our proposal behaves when the domains are more diverse. Then, the Lwf-CNN and Lwf-CNN-lgR models were evaluated in a cross-domain scheme using in total nine domains, i.e., the seven previous ones and the two additional added. The Precision, Recall, and F1-score values behave below the average results achieved in the previous experimentation, which evidences that our proposal is still sensitive to very diverse domains, as shown in Table 1. Low values are caused by the existence of new aspects in the restaurant and hotel review datasets that did not appear in the datasets associated to electronic device domains.

Table 1. Cross-domain deep and lifelong learning evaluation results in restaurant and hotel review datasets.

Training	Testing	Lwf-CNN-lgR			Lwf-CNN		
		P	R	F1	P	R	F1
-Restaurant	Restaurant	74.8	53.7	62.5	69.8	46.8	56.0
-Hotel	Hotel	78.6	59.8	67.9	62.4	48.5	54.6

Two semantically close domains (DVD player and Mp3 player) and not too close ones (Computer and DVD player) were selected to determine the overcoming catastrophic forgetting during a new domain training. To perform the experiment, the model was first trained with the old domain and then with the new one. After the training of the new domain, testing was executed for each domain. The results illustrate that it is possible to maintain acceptable performance results during each domain training, as shown in Fig. 4. However, the best result is obtained when we test our proposal on semantically close domains.

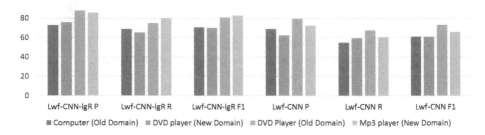

Fig. 4. Experiment results about overcoming catastrophic forgetting.

Our proposal, in general, outperforms the results obtained in [14]. The use of the linguistic rules to increase the training dataset allows to improve the results of Deep Learning methods where there are few data. The deep CNN, which is non-linear in nature, improves the CRF model results. The main advantage of our framework is that it does not need any feature engineering in the Lifelong learning model. Our proposal highlights the importance of combining a trained model in a supervised way with linguistic patterns in the ABSA task.

5   Conclusions

In this work, the multi-domain aspect extraction subtask was performed by combining CNN and Lifelong Learning. Our model reduces catastrophic forgetting in the multi-domain context. The achieved results improve one of the most significance state-of-the-art proposal. The combination of CNN and Lifelong learning techniques in ABSA subtask constitutes a novel proposal in the Sentiment analysis research field. Although the obtained results are promising, the future work will be oriented to test with other algorithms to avoid catastrophic forgetting and evaluate the Lifelong Learning model with other Deep Learning techniques.

References

1. Topic modeling using topics from many domains, lifelong learning and big data. In: Proceedings of the 31st International Conference on International Conference on Machine Learning (ICML 2014), Beijing, China. pp. 703–711 (2014)
2. Bouras, D., Amroune, M., Bendjenna, H.: A review of recent aspect extraction techniques for opinion mining systems. In: Proceedings of the Second International Conference in Natural Language and Speech Processing (ICNLSP), Algiers, Algeria, pp. 1–6 (2018)
3. Chen, Z., Liu, B.: Lifelong machine learning. Synth. Lect. Artif. Intell. Mach. Learn. **12**, 1–207 (2018)
4. Cun, Y.: Generalization and network design strategies. Department of Computer Sciente, University of Toronto, Canada, Technical report (1989)
5. Do, H.H., Prasad, P., Maag, A., Alsadoon, A.: Deep learning for aspect-based sentiment analysis: a comparative review. Expert Syst. Appl. **118**, 272–299 (2019)
6. Girshick, R., Donahue, J., Darrell, T., Malik, J.: Rich feature hierarchies for accurate object detection and semantic segmentation. In: Proceedings of the IEEE Conference on Computer Vision and Pattern Recognition, Columbus, Ohio, USA, pp. 580–587 (2014)
7. Gu, X., Gu, Y., Wu, H.: Cascaded convolutional neural networks for aspect-based opinion summary. Neural Process. Lett. **46**(2), 581–594 (2017)
8. He, R., Lee, W.S., Ng, H.T., Dahlmeier, D.: An unsupervised neural attention model for aspect extraction. In: Proceedings of the 55th Annual Meeting of the Association for Computational Linguistics, Vancouver, Canada, pp. 388–397 (2017)
9. Kirkpatrick, J., Pascanu, R., Rabinowitz, N., Veness, J., Desjardins, G., Rusu, A.A., Milan, K., Quan, J., Ramalho, T., Grabska-Barwinska, A., Hassabis, D., Clopath, C., Kumaran, D., Hadsell, R.: Overcoming catastrophic forgetting in neural networks. Proc. National Acad. Sci. (PNAS) **114**(13), 3521–3526 (2017)
10. Li, Z., Hoiem, D.: Learning without forgetting. IEEE Trans. Pattern Anal. Mach. Intell. **40**(12), 2935–2947 (2018)
11. Mikolov, T., Sutskever, I., Chen, K., Corrado, G., Dean, J.: Distributed representations of words and phrases and their compositionality. Adv. Neural Inf. Process. Syst. **26**, 3111–3119 (2013)
12. Pontiki, M., Galanis, D., Pavlopoulos, J., Papageorgiou, H., Manandhar, S., Androutsopoulos, I.: Semeval-2014 task 4: aspect based sentiment analysis. In: Proceedings of the 8th International Workshop on Semantic Evaluation (SemEval), pp. 27–35 (2014)
13. Poria, S., Cambria, E., Gelbukh, A.: Aspect extraction for opinion mining with a deep convolutional neural network. Knowl.-Based Syst. **108**, 42–49 (2016)
14. Shu, L., Liu, B., Xu, H., Annice, K.: Lifelong-RL: Lifelong relaxation labeling for separating entities and aspects in opinion targets. In: Proceedings of the Conference on Empirical Methods in Natural Language Processing (EMNLP), Austing, Texas, USA, pp. 225–235. Association for Computational Linguistics (2016)
15. Viterbi, A.J.: Error bounds for convolutional codes and an asymptotically optimum decoding algorithm. IEEE Trans. Inf. Theor. **13**(2), 260–269 (1967)

A Framework for Distributed Data Processing

José Kadir Febrer-Hernández[(⊠)] and Vitali Herrera Semenets[(⊠)]

Advanced Technologies Application Center (CENATAV),
7a ♯ 21406, Rpto. Siboney, Playa, C.P. 12200 Havana, Cuba
{jfebrer,vherrera}@cenatav.co.cu

Abstract. Nowadays, the data generated in the telecommunications networks tend to grow exponentially leading to a Big Data challenges, which makes it necessary to discover different ways to safely process this data. The reported strategies aim to provide reliable and flexible services for asynchronous data exchange. The parallel and distributed processing of large volumes of data plays a fundamental role in scenarios that require a response as soon as possible, such as detecting fraud in telecommunications services or carrying out security controls. In this paper, we present a strategy that allows to distribute data and manage several instances of the same application, which are executed in a distributed way. An aspect to be highlighted is that heterogeneity is not required in the computational units, that is, both conventional PCs and blade clusters can participate. Another important advantages of this tool are its flexibility and its adaptability. The data are distributed depending on the workload of the different application instances. Finally, a case study is presented for the distributed processing of the Windows Operating System logs.

Keywords: Distributed systems · Parallel processing · Data stream

1 Introduction

The technological advances have led the telecommunications industry to generate huge amounts of data known as Big Data [11]. For example, the AT & T Company reported that an average of 300 million international calls is performed in one normal day using its services [3]. The improper use of these services to perform malicious activities such as fraud in telecommunication services, network intrusion, among others, makes it necessary to analyze the data in real time [12].

A system that performs this type of processing can prevent or at least reduce any possible damage caused by the execution of malicious activities. Regarding a system which allows processing such data, there is no reference model that has been universally accepted, as it happens with the technologies in its first phases of life. However, there are industrial and academic proposals that try to define what are the components that make up a Big Data system, including its relationships and properties [13].

© Springer Nature Switzerland AG 2019
I. Nyström et al. (Eds.): CIARP 2019, LNCS 11896, pp. 566–574, 2019.
https://doi.org/10.1007/978-3-030-33904-3_53

In the last years, the distributed systems have been one of the most used proposals for the efficient analysis of large volumes of data [7]. In this paper, we present a framework that allows the distributed processing of data streams in real time. Our proposal is designed to be able to support any application for data processing, which wants to be executed in a distributed way. The experimental results show how much an application can improve its efficiency using our proposal.

The rest of the paper is organized as follows. The next section describes the related work. Our proposal is presented in Sect. 3. The study case is exposed in Sect. 4. Finally, the conclusions and future works are given in Sect. 5.

2   Related Work

Generally, the distributed systems have a strategy to provide communication among different system modules facilitating the data exchange [14]. Some of these strategies aim to provide reliable and flexible services for asynchronous data exchange.

These include the Java Message System (JMS) [4], Internet Communication Engine (ICE) [6], Common Object Request Broker Architecture (CORBA) [8], or Data Distribution System (DDS) [9]. JMS relies on the Java platform and allows distributed communication using a common API for the development of Java applications based on messages. ICE, CORBA, and DDS use an Object Request Agent (ORB) which consists of an object exchange interface. However, the most advanced of these is DDS that was developed to standardizing the data distribution in different platforms.

There are other more specific strategies, which are basically oriented to real-time processing. For example, the Open RObot COntrol Software (OROCOS) platform [2] for application control provides tools for data-exchange and event-based services. More recent strategies are iLAND [10] which consists in a data distribution service for real-time applications that incorporates a reconfiguration logic.

On the other hand, DREQUIEMI [10] is based on Java Language and the experiences obtained by iLAND. There is also an architecture presented in [1], which is based on Java Language and designed for real-time analysis.

Several of the above strategies are used to integrate distributed enterprise applications [5]. However, none of them is generic enough to assimilate any application that is required to process data streams in real time or very close to it.

There are some frameworks that have been proposed to deal with Big Data such as: Apache Hadoop, Apache Storm, Apache Spark and Apache Flink, which are briefly described as follows. Hadoop [15] is an open source software for reliable, scalable, distributed computing. It was the first one to enable big data processing and uses batch processing.

Spark [16] combines a distributed computing system through computer clusters. It is an open source framework with a very active community; it is a fast tool; it has a comfortable interactive console for developers; and it also has an

API to work with big data. Despite this, Spark has some limitations such as: no support for real time processing, it simulates the streaming through micro-batch and each iteration is scheduled and executed separately; it requires lots of memory to run and problems arise when working with small files.

Storm [17] is a distributed framework for real-time data processing. Built to be scalable, extensible, efficient, easy to administer and fault-tolerant.

Flink [18] is a distributed framework to process data efficiently and for general use, just like Spark. Unlike Flink, Spark cannot handle a data set larger than the memory it has available. In addition, Spark works in batches of data unlike Flink which can simulate the data stream.

The previous proposals are very efficient and widely used today. However, in order to use these frameworks, many computational resources are needed, or a large amount of time has to be devoted to creating an entire infrastructure for their use, without taking into account the required programming time. On the other hand, the proposal presented in this article tries to give a quick and efficient solution to a big data problem without using large computing resources.

3   Our Proposal

This section presents a framework that allows to distribute data and manage several instances of the same application running in parallel. As part of the strategy presented in this paper, a rule engine is used as the application to be managed.

The rule engine and the framework are described below.

3.1   Rule Engine

A rule engine (RE) usually evaluates rules expressed with the notation "*if X then Y*" ($X \Rightarrow Y$), where X is an interest conditions set and Y is the action to take when X is fulfilled. The Fig. 1 shows the basic model of an RE for the processing of data stream.

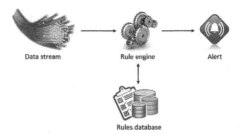

Fig. 1. Basic scheme of a rules engine.

The rule base shown in the Fig. 1 provides persistent storage for a set of rules. When the system begins to process a data stream, immediately the existing rules are evaluated and in the case that the conditions of some rule are fulfilled, the corresponding action "triggers".

3.2 Proposed Framework

Our proposal consists of a framework that from now on we will refer as Task Manager (TM). Its main function is to manage tasks that must be processed in a distributed way by several instances of the same application. The Algorithm 1 shows the general process followed by our Task Manager.

Algorithm 1. Task Manager, pseudo-code

$cf \leftarrow ReadConfigurationFile()$

while *(TM is running)* **do**

    if *(CheckForTask(cf) != false)* **then**

        foreach *(app in cf.applicationsList)* **do**

            if *(CheckStatus(app) != false)* **then**

                app.available $\leftarrow$ *false*

                CopyData(cf)

                RunApplication(app)

                break

            else

                Sleep(cf.sleepTime)

            end

        end

    else

        Sleep(cf.sleepTime)

    end

end

The TM initially reads the configuration file, where several parameters are defined by the analyst. Note that the TM runs as a daemon in the operating system, so it is running continuously until it is stopped by the analyst. The next step is to search in the data directory if there is any dataset to be processed. If it does not exist, the TM waits for a time (defined by the analyst in the configuration file) and re-check the directory. In case there is exist a dataset, the TM conforms the task. When multiple datasets exist, the TM selects the oldest (line 3).

After the task is generated, the TM checks if any of the application instances running are available to process a new task. An application is available when you are not processing a task. The TM instances can run on different computers and the location of each one is defined by the analyst in a configuration file. In case there is no available instance of the task manager, the TM waits for a time defined by the analyst and re-check for a task. If an instance of the application is available, the TM allocates the selected task to that instance and registers it (line 5).

When the TM assigns a task, it begins to check its status. To do this, it checks if the application instance to which you assigned a task has ended. An application ends when it is registered as being assigned a task and it is not running. If the task continues to run, the TM waits for the specified time and rechecks if any of the registered instances are finished (lines 6–8).

In the Fig. 2, we show a scheme where we use a rule engine as the application to be distributed. In this case, a task consist of a dataset to be processed and a set of rules to be evaluated. As can be seen, there are two directories from which the Task Manager is fed. If there is exist a dataset, the TM searches in the rules directory for those rules that share the same identifier that the selected dataset and conforms the task. When a task is generated, the TM checks if any of the application instances running are available to process a new task.

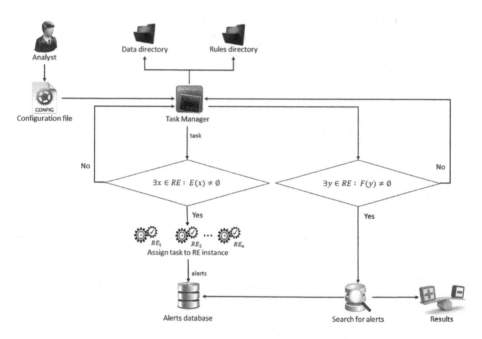

Fig. 2. Task Manager operations diagram for processing the Windows Operating System logs by a rules engine.

In this case, if the application instance ended, the TM searches in the alerts database for those that correspond to the identifier of the ended task. If there are alerts, they are reported. Otherwise, it is reported that no alerts were issued for that task.

4 Study Case

As a distributed processing platform, the application (in this case the rule engine) can be located in a cluster with an unrestricted number of processing nodes as well as on a single computer. The experiment included five personal computers (PCs) equipped with a Quad-Core processor and 4 GB of RAM. In one of them, the TM was installed, and in the remaining four the RE was installed.

A test scenario was created in a laboratory with 20 PCs with the Windows 8 operating system installed. In order to show the improvement in terms of efficiency that the proposal can provide. The experiment consists in analyze the Windows event logs generated by each PC (see Fig. 3). For this, an agent was used to capture each new log generated and create a dataset D to send it to the TM data directory. This agent was installed in each of the 20 PCs used in the experiment. The analyzed logs are associated with Windows application, system, and security events.

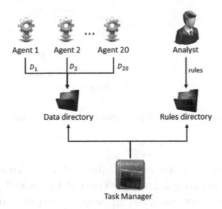

Fig. 3. Test scenario for the analysis of Windows Operating System event logs.

When the agent is executed for first time, it sends an initial dataset with all the event logs stored by the system and then waits for new logs to be sent to the TM data directory. Taking this into account, it was defined that the size of each dataset to be sent does not exceed a threshold defined by the analyst (in this case 20 000). In this way, the first dataset represents a data flow around 20,000 event logs by PC. Running all the agents the same time once generates 20 datasets to be processed. The time to process each dataset is measured since the dataset is copied into the directory until the TM reports the alerts associated to the dataset processed.

Five rules were designed for the experiments, which evaluate regular expressions on the different fields of the event logs. The rules were created intentionally to match with some test logs in order to validate the operation of the system.

In order to evaluate the advantage of the proposed task manager, two experiments were designed. In the first, an instance of the rule engine was executed, without the TM, and in the second, 4 instances of the distributed rule engine were executed on different PC using the TM.

Figure 4 shows a comparison between the time taken by a rule engine instance, without the TM (experiment 1), to process the generated event logs, regarding to the time taken by the TM with 4 instances of the RE (Experiment 2).

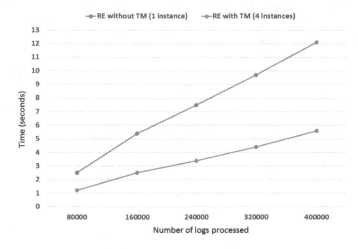

Fig. 4. Comparison of the rule engine processing time with and without Task Manager.

In experiment 1, the time taken to evaluate the five rules created to process the first four sets of generated data (80 000 event log) was 2.5 s. In experiment 2, the generated datasets were analyzed in parallel and the time taken to process the same four datasets was 1.2 s.

The results achieved show that using TM the time taken to process 400 000 event logs was reduced by more than 50% regarding to a single instance of the rule engine.

Following the above idea, it can be estimated that using 8 PCs to increase the distributed processing capacity, one PC per RE, around 800 000 logs could be processed in less than 6 s.

The experiments performed show that our proposal is scalable and has a high efficiency making feasible its application in scenarios where it is necessary to analyze large volumes of data in real time or very close to it. It is important to note, that in some scenarios, the real time is in dependence of the requirements or demand of the user or what is for him a real time analysis.

5   Conclusions

In this paper we have presented a framework called Task Manager for data streams distributed processing.

The study case showed the improvement, in terms of efficiency, of the RE application using our proposal for large-scale filtering of Windows event logs. As can be seen in the experiments performed, the proposed system is scalable, which implies that the computing capacity will be directly related to the number of processing units included. Other aspect to be highlighted is that heterogeneity is not required in the computational units, that is, both conventional PCs and blade clusters can participate. Another important advantages of this tool are its flexibility and its adaptability.

Our next step is to use a dedicated hardware for high performance tasks, which could exponentially increase the performance of the proposal, making it competitive with other distributed systems.

References

1. Basanta-Val, P., García-Valls, M.: A distributed real-time Java-centric architecture for industrial systems. IEEE Trans. Industr. Inf. **10**(1), 27–34 (2014)
2. Bruyninckx, H.: The real-time motion control core of the Orocos project. In: International Conference on Robotics and Automation (ICRA 2003), vol. 2, pp. 2766–2771 (2003)
3. Cortes, C., Pregibon, D.: Signature-based methods for data streams. Data Min. Knowl. Disc. **5**(3), 167–182 (2001)
4. Hapner, M., Burridge, R.: Java message service. Technical report, Sun Microsystems Inc. (2002)
5. He, W., Da Xu, L.: Integration of distributed enterprise applications: a survey. IEEE Trans. Industr. Inf. **10**(1), 35–42 (2014)
6. Henning, M., Spruiell, M.: Distributed programming with ice. Technical report, ZeroC Inc. (2003)
7. Kambatla, K., Kollias, G., Kumar, V., Grama, A.: Trends in big data analytics. J. Parallel Distrib. Comput. **74**(7), 2561–2573 (2014)
8. OMG: The common object request broker (CORBA): architecture and specification. Technical report, Object Management Group (OMG) (1995)
9. OMG: Data distribution service for real-time systems. Version 1 edn. Object Management Group (OMG) (2007)
10. Valls, M.G., Val, P.B.: Comparative analysis of two different middleware approaches for reconfiguration of distributed real-time systems. J. Syst. Architect. **60**(2), 221–233 (2014)
11. Wu, X., Zhu, X., Wu, G.Q., Ding, W.: Data mining with big data. IEEE Trans. Knowl. Data Eng. **26**(1), 97–107 (2014)
12. Habeeb, R.A.A., Nasaruddin, F., Gani, A., Hashem, I.A.T., Ahmed, E., Imran, M.: Real-time big data processing for anomaly detection: a survey. Int. J. Inf. Manag. **45**, 289–307 (2018)
13. Gupta, S., Kar, A.K., Baabdullah, A., Al-Khowaiter, W.A.: Big data with cognitive computing: a review for the future. Int. J. Inf. Manag. **42**, 78–89 (2018)
14. Teixeira, F.A., Pereira, F.M., Wong, H.C., Nogueira, J.M., Oliveira, L.B.: SIoT: Securing Internet of Things through distributed systems analysis. Future Gener. Comput. Syst. **92**, 1172–1186 (2019)
15. Malik, M., et al.: Big vs little core for energy-efficient Hadoop computing. J. Parallel Distrib. Comput. **129**, 110–124 (2019)

16. Meng, X., et al.: MLlib: machine learning in apache spark. J. Mach. Learn. Res. **17**(1), 1235–1241 (2016)
17. Iqbal, M.H., Soomro, T.R.: Big data analysis: apache storm perspective. Int. J. Comput. Trends Technol. **19**(1), 9–14 (2015)
18. Carbone, P., Katsifodimos, A., Ewen, S., Markl, V., Haridi, S., Tzoumas, K.: Apache Flink: stream and batch processing in a single engine. Bull. IEEE Comput. Soc. Tech. Committee Data Eng. **36**(4) (2015)

Articulation Analysis in the Speech of Children with Cleft Lip and Palate

H. A. Carvajal-Castaño[1,2] and Juan Rafael Orozco-Arroyave[1,3(✉)]

[1] Research Group on Applied Telecommunications - GITA, Electronic Engineering
and Telecommunications Department, Faculty of Engineering,
Universidad de Antioquia UdeA, Calle 70 No. 52–21, Medellín, Colombia
rafael.orozco@udea.edu.co
[2] Bioinstrumentation and Clinical Engineering Research Group - GIBIC,
Bioengineering Department, Faculty of Engineering,
Universidad de Antioquia UdeA, Calle 70 No. 52–21, Medellín, Colombia
[3] Pattern Recognition Lab, University of Erlangen-Nüremberg, Erlangen, Germany
http://www.udea.edu.co, http://www.fau.eu

Abstract. Hypernasality is a speech deficit that affects children with cleft lip and palate (CLP). It is characterized by the lack of control of the velum, which causes problems when controlling the amount of air passing from the oral to the nasal cavity while speaking. The automatic evaluation of hypernasality could help in the monitoring of speech-language therapies and in the design of better oriented exercises. Several articulation features have been used for the automatic detection of hypernasal speech. This paper evaluates the suitability of classical articulation features for the automatic classification of hypernasal and healthy speech recordings. Two different databases are considered with recordings collected under different acoustic conditions and with different audio settings. Besides the evaluation of the proposed approach upon each database separately, non-parametric statistical tests are performed to evaluate the possibility of merging features from the two databases with the aim of finding more robust systems that could be used in different acoustic conditions. The results indicate that the proposed approach has a high sensitivity, which indicates that it is suitable to detect hypernasal speech samples. We believe that promising results could be obtained with this approach in future experiments where the degree of hypernasality is evaluated.

Keywords: Cleft lip and palate · Hypernasality · Articulation measures · Classification

1 Introduction

Clef and Lip Palate (CLP) is a craniofacial malformation that occurs in about one in every 700 live births [1]. This malformation is characterized by the incomplete formation of tissues that separate oral and nasal cavities, which generates

© Springer Nature Switzerland AG 2019
I. Nyström et al. (Eds.): CIARP 2019, LNCS 11896, pp. 575–585, 2019.
https://doi.org/10.1007/978-3-030-33904-3_54

several speech disorders such as hypernasality, hyponasality, glottal stops, and others. During speech production it is necessary to control the amount of air that comes out through the nasal cavity. The amount of air is controlled by the velum, which opens and closes the connection between the oral and nasal cavity depending on which sound is intended to be produced by the speaker (e.g., nasal or no-nasal). When there is excess of air coming out through the nasal cavity the speech is perceived as hypernasal, which is the speech pathology suffered by the majority of CLP children [2,3].

The evaluation of CLP patients is subjective and time-consuming because highly depends on the speech and language therapist's expertise. The research community has been interested since several decades in the development of systems that allow the objective evaluation of speech in children with CLP. One of the most suitable approach has been the use of features that reflect articulation deficits. The most common ones include Mel-frequency cepstral coefficients (MFCCs), the first two vocal formants, vowel space area (VSA), and non-linear measures like the Teager Energy Operator (TEO). For instance, in [4] speech recordings of CLP patients are studied with the aim of discriminating the severity of hypernasality. The extracted features were based on the fundamental frequency of voice, energy content, and MFCCs. The authors reported accuracies of up to 80.4% when discriminating four different grades of hypernasality. In [5] the five Spanish vowels are considered to evaluate hypernasality in children with CLP considering acoustic and noise features. The authors reported accuracies of up to 89%. In [6] the authors introduced a method for the classification of hypernasal patients and healthy control subjects using the VSA and 13 MFCCs with their corresponding first and second derivatives. According to the authors, the proposed approach has an accuracy of 86.89%. In [7] features based on acoustic, noise, cepstral analysis, nonlinear dynamics [8] and entropy measures are used for hypernasality detection. Accuracies of around 92% are reported when the five Spanish vowels and the words *coco* and *gato* are considered.

Although articulation-related features have been extensively used in the literature to evaluate hypernasality in the speech of CLP patients, all of the works report results on single databases. Thus, there is no evidence about the robustness of those features when are used in different databases recorded in different acoustic conditions and with different recording settings. This study evaluates the suitability of classical articulation-related features to discriminate between hypernasal and healthy speech signals collected from children with CLP from two different clinics for children in Colombia, using different recording settings and under different acoustic conditions. The features set extracted for study includes 12 MFCCs and their first and second derivatives, the first two vocal formants (F_1 and F_2) and their first and second derivatives, TEO formant centralization ratio (FCR), and VSA. Four statistical functionals are calculated upon each feature vector: mean, standard deviation, kurtosis and skewness, and two different classifiers are evaluated: a support vector machine (SVM) and a Random Forest (RF).

2   Participants

2.1   CLP Manizales

This database was provided by Grupo de Control y Procesamiento Digital de Señales - (GCyPDS) at Universidad Nacional de Colombia, Manizales. This database contains recordings of the five Spanish vowels pronounced by children between 5 and 15 years old. A total of 140 audio registers were collected, 84 labeled as hypernasal by a phoniatry expert and the remaining 56 were labeled as healthy. The signals were recorded in a quiet room but under non-controlled acoustic conditions using a non-professional audio setting with a sampling frequency of 44100 Hz and 16 bit-resolution.

2.2   CLP Clínica Noel

This data set was recorded in the Clínica Noel from Medellín and contains recordings of the five Spanish vowels pronounced by children between 5 and 15 years old. The data contain 95 audio recordings of 53 children with CLP and 42 healthy controls (HC). All of the 53 children included in the CLP group were labeled as hypernasal by a phoniatry expert. The recordings were collected in a quiet room using a professional audio setting with a sampling frequency of 44100 Hz and 16 bit-resolution. Further details of this corpus can be found in [9].

3   Methods

3.1   Feature Extraction

Mel-Frequency Cepstral Coefficients (MFCCs). Since several decades, these features have been widely used in applications for automatic speech recognition and for speaker identification. However, since about a decade the MFCCs started to be used in applications of pathological speech analysis like laryngeal pathologies [10] and dysarthria in Parkinson's disease [11]. The coefficients are based on the Mel scale which considers the perceived frequency of a tone and the actual measured frequency. The scale emulates the frequency-response of the human hearing system.

To estimate MFCCs considers that $s_i(k)$ is the ith frame of the speech signal and $S_i(k)$ is the Discrete Fourier Transform (DFT), the mel-spectrum is [12]:

$$\text{MF}_i[r] = \frac{1}{A_r} \sum_{k=L_r}^{U_r} |V_r[k]S_i[k]|^2, \qquad r = 1, 2, \dots, R \tag{1}$$

where R is the number of filters, $V_r[k]$ is a weigh function for each mel filter, L_r and U_r are the lower and upper range of the filter, and A_r is a normalization factor for the rth filter. The MFCC are finally found as:

$$\text{MFCC}_i[n] = \frac{1}{R} \sum_{r=1}^{R} \log(\text{MF}_i[r]) \cos\left[\frac{2\pi}{R}\left(r + \frac{1}{2}\right)n\right] \tag{2}$$

$\text{MFCC}_i[n]$ is evaluated for $n = 1, 2, \ldots, N_{\text{MFCC}}$ where N_{MFCC} is the number of desired coefficients ($N_{\text{MFCC}} < R$). Besides, the first and second derivatives of MFCCs is compute. First and second derivatives of the MFCCs are also extracted before computing the four statistical functionals, forming a 144-dimensional feature vector.

First and Second Vocal Formants (F_1 and F_1). These two frequencies correspond to the first two peaks that appear, when the envelope of a voice spectrum is calculated (typically by a linear predictive filter). F_1 and F_2 provide information about resonances that occur in the vocal tract during a phonation. Thus they are related to the shape of organs and tissues in the vocal cavity, e.g., the tongue. Typically, the first two formants are used to evaluate the capability of a speaker to keep the tongue in a certain position during the phonation of a given vowel [6]. In this study, we calculate F_1 and F_2 considering the spectral envelope found with a Linear Predictive Coding (LPC) filter. This method assumes that each frame, $s_i(k)$, of the speech signal, can be approximated as a linear combination of the past p samples:

$$\hat{s}_i = \sum_{k=1}^{p} a_k s_{i-k} \tag{3}$$

where $\mathbf{a} = a_1, \ldots, a_p$ is a vector with p coefficients. The main aim is to minimize the mean-square-error such that:

$$\mathbf{a} = \arg \min_{\mathbf{a}} \frac{1}{N} \sum_{n=1}^{N} (\hat{s}_i - s_i)^2 \tag{4}$$

The optimal vector $\mathbf{a}$ that minimizes the MSE comprises the envelope of the speech spectrum and the value of p determines the shape of such an envelope. As in the case of the MFCC, the first and second derivatives of F_1 and F_2 are also calculated before computing the four statistical functional, forming a 24-dimensional feature vector.

Teager Energy Operator (TEO). Consider the signal $x(n)$. Its associated TEO was defined in [13] as:

$$\Psi[x(n)] = x^2(n) - x(n+1)x(n-1) \tag{5}$$

One of the most important characteristics of TEO is the sensibility to composed signals. If we consider a composed signal as $x(n) = s(n) + g(n)$, the TEO is computed as:

$$\Psi[x(n)] = [s(n) + g(n)]^2 - [s(n+1) + g(n+1)][s(n-1) + g(n-1)] \tag{6}$$

if a cross-correlation term ($\Psi_{cross}[s(n), g(n)]$) between the two signals is considered as $\Psi_{cross}[s(n), g(n)] = s(n)g(n) - g(n+1)s(n-1)$, we obtain:

$$\Psi[x(n)] = \Psi[s(n)] + \Psi[g(n)] + \Psi_{cross}[s(n), g(n)] + \Psi_{cross}[g(n), s(n)] \tag{7}$$

Equation 7 shows that the superposition theorem does not apply to TEO. This property is useful in cases where the original signal is composed by several components. A regular speech spectrum has a typical profile where its peaks are characterized by the vocal formants. In the case of a hypernasal speech spectrum, additional peaks (additional formants) appear and also additional anti-formants (additional valleys in the speech spectrum) appear. TEO seems to be a good strategy to model such additional components that result in the spectrum of hypernasal signals [14]. The TEO is extracted from windows of 40 ms-length, then the four statistical functionals are computed to create a 4-dimensional feature vector per speaker.

Formant Centralization Ratio (FCR) is an alternative measure to represent articulatory problems in speakers. It offers the advantage of maximizing the sensitivity to vowel centralization and minimizing the sensitivity to inter-speaker variability. Thus, provides more robust and stable information of the vowel space produced by a speaker. The FCR was proposed in [15] as:

$$\text{FCR} = \frac{F_{2u} + F_{2a} + F_{1i} + F_{1u}}{F_{2i} + F_{1a}} \tag{8}$$

Where F_{1a}, F_{2a}, F_{1i}, F_{2i}, F_{1u} and F_{2u} are the first and second format of the corner vowels /a/, /i/ and /u/ respectively.

Vowel Space Area (VSA) is the most common way of measuring vowel centralization using F_1 and F_2 of corner vowels (/a/, /i/, /u/). It is given by the area of the triangle formed by the vertexes (F_1, F_2) in the vowel space created for the three corner vowels. VSA is computed as [15]:

$$\text{VSA} = \left| \frac{F_{1i}(F_{2a} - F_{2u}) + F_{1a}(F_{2u} - F_{2i}) + F_{1u}(F_{2i} - F_{2a})}{2} \right| \tag{9}$$

Figure 1 shows the resulting vocal triangle for two different children from the two databases. Note that in both cases the triangle of the patients is compressed which is a typical indicator of reduced articulation capability.

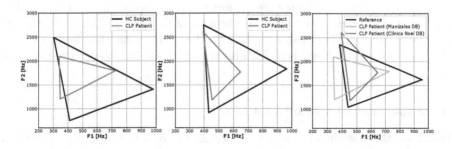

Fig. 1. Vocal triangle computed for two different children from the two databases.

At the end of the feature extraction procedure, each vowel pronounced by each speaker is modeled by a 172-dimensional feature vector. Additionally, FCR and VSA are calculated per speaker. Since the five Spanish vowels are considered together, each speaker is finally represented by a 862-dimensional feature vector.

3.2   Classification

With the aim of comparing the robustness of two different classification approaches, two classifiers are considered: a Support Vector Machine (SVM) with a Gaussian kernel, and a Random Forest (RF). The parameters of the classifiers were optimized following a 5-fold cross-validation strategy, where 4 folds were used for training and the remaining one for test. Within the 4 folds used for training, another 5-fold cross validation was performed. The optimization criterion is based on the accuracy in training. The parameters were optimized using a grid-search over the training folds. For SVM, C and $\gamma \in \{10^{-6}, 10^{-5}, \ldots, 10^{4}\}$ and for RF, the number of trees $N \in \{5, 10, 20, \ldots, 100\}$ and the depth of the decision trees $D \in \{2, 5, 10, 20, \ldots, 100\}$. Optimal parameters are selected according to the mode across the 5-fold cross-validation procedure.

3.3   Statistical Analysis and Merging of Features

Apart from calculating the features and performing the classification between the two classes, we wanted to evaluate the possibility of merging those features extracted from two different datasets that were collected under different acoustic conditions. The analysis to decide which features were able to be merged between datasets was performed according to Kruskal-Wallis statistical tests. The null hypothesis was that the given feature has the same distribution in both databases. Thus, if the p-value < 0.05 the null hypothesis is rejected. The test was applied over all features of both databases and only the features that successfully passed the test (p-value ≥ 0.05) were included in the merging process. At the end of the procedure a total of 508 features passed the test.

4   Experiments and Results

In this study three experiments are performed: (1) classification of CLP vs. HC with the Manizales database, (2) classification of CLP vs. HC with the Clínica Noel database and (3) classification of CLP vs. HC with the fusion of both databases only considering those features that successfully passed the statistical test.

4.1   Experiment 1 - Manizales DB

Results for the classification of CLP patients vs HC subjects are presented in Table 1. Note that the best results are always obtained using the SVM.

Figure 2 shows histograms of the scores obtained during the classification process. Additionally, Receiver Operating Characteristic (ROC) curves obtained with the two classifiers are also included. Note that the scores in an SVM are the distance of each sample to the separating hyperplane. For RF the scores are the probability of a sample to belong to the selected class. From the histograms it is possible to observe that most of the CLP patients are correctly classified, which confirms the high sensitivity obtained in the experiments (96.4%). For the HC subjects the result is not as high but still competitive with a specificity of 89.6%. The values of AUROC allow to perform a more compact analysis of the system's performance considering the classification accuracy of the two classes at the same time (sensitivity and specificity).

Table 1. CLP vs. HC using the Manizales DB

Classifier	ACC [%]	Sen [%]	Spe [%]	AUROC	Parameters
SVM	93.6 ± 6	96.4 ± 5	89.6 ± 9	0.99	$C = 10, \gamma = 10^{-4}$
RF	86.4 ± 5	95.2 ± 6	73.0 ± 16	0.96	$D = 2, N = 10$

ACC: accuracy in the test set, AUROC: Area under ROC curve. Sen: Sensitivity,
Spe: Specificity.
C and γ: complexity parameter and bandwidth of the kernel in the SVM classifier.
N and D: Number of trees and depth of the decision trees in the RF classifier

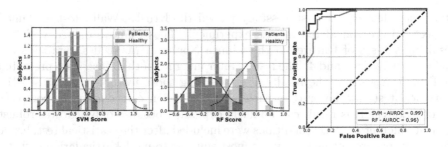

Fig. 2. Histograms of the scores and ROC curves (Manizales DB)

4.2 Experiment 2 - Clínica Noel DB

The results obtained with the two classifiers with data of the Clínica Noel are indicated in Table 2. Similar to Experiment 1, Fig. 3 the histograms of the scores and the ROC curves are presented. Note that as in the experiments with the other dataset, the best results are obtained with the SVM. This result confirms the robustness of these classifiers, which have been extensively used in the literature in problems of pathological speech processing. When this results are compared to those presented in Fig. 2, the histograms have a larger overlapping which reduces the performance of the classifiers.

Table 2. CLP vs. HC using the Clínica Noel DB

Classifier	ACC [%]	Sen [%]	Spe [%]	AUROC	Parameters
SVM	84.3 ± 7	90.8 ± 6	76.6 ± 11	0.90	$C = 10,\ \gamma = 10^{-4}$
RF	76.9 ± 7	88.4 ± 16	62.6 ± 19	0.85	$D = 10,\ N = 70$

ACC: accuracy in the test set, AUROC: Area under ROC curve. Sen: Sensitivity,
Spe: Specificity.
C and γ: complexity parameter and bandwidth of the kernel in the SVM classifier.
N and D: Number of trees and depth of the decision trees in the RF classifier

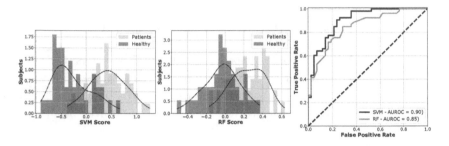

Fig. 3. Histograms of the scores and ROC curves (Clínica Noel DB)

4.3   Experiment 3 - Fusion of both DB

Only those features that successfully passed the Kruskal-Wallis test explained in Sect. 3.3 are considered in this experiment. The results obtained with the resulting sub-set of features in the Clínica Noel and Manizales databases are included in Tables 3 and 4. The results after merging both datasets are presented in Table 5. Note that the results after selecting those features that passed the statistical tests are lower than those obtained when considering the complete set of features. This result can be likely explained because only features robust against different acoustic conditions were included after the statistical test, hence those features are not necessary the most suitable to model articulation deficits. On the other hand, Table 5 indicates that when the two databases are merged the results improve. It seems like the difference in the acoustic conditions of the two databases allow the selected features to complement among them and the result is the improvement in the classification accuracies compared to those obtained when the selected features are used upon each database separately. Figure 4 shows the histograms of the scores obtained in the classification process and the ROC curves. Similarly to the previous experiments, there is a high sensitivity obtained with both classifiers. This result indicates that the proposed approach is suitable to detect CLP patients then it seems to be sensitive to articulation deficits exhibited by children with CLP. We think that it could be used in future studies to evaluate the degree of nasalization. We are currently collecting more data with those labels with the aim of performing these kinds of experiments.

Table 3. CLP vs. HC using the Manizales DB with selected features

Classifier	ACC [%]	Sen [%]	Spe [%]	AUROC	Parameters
SVM	79.2 ± 4	90.4 ± 8	62.8 ± 6	0.86	$C = 10, \gamma = 10^{-4}$
RF	70.1 ± 7	92.8 ± 2	36.0 ± 16	0.80	$D = 90, N = 30$

ACC: accuracy in the test set, AUROC: Area under ROC curve. Sen: Sensitivity.
Spe: Specificity.
C and γ: complexity parameter and bandwidth of the kernel in the SVM classifier.
N and D: Number of trees and depth of the decision trees in the RF classifier

Table 4. CLP vs. HC using the Clinica Noel DB with selected features

Classifier	ACC [%]	Sen [%]	Spe [%]	AUROC	Parameters
SVM	73.5 ± 5	84.8 ± 8	59.2 ± 11	0.78	$C = 10^2, \gamma = 10^{-5}$
RF	68.2 ± 12	79.0 ± 10	54.8 ± 19	0.79	$D = 60, N = 60$

ACC: accuracy in the test set, AUROC: Area under ROC curve. Sen: Sensitivity,
Spe: Specificity.
C and γ: complexity parameter and bandwidth of the kernel in the SVM classifier.
N and D: Number of trees and depth of the decision trees in the RF classifier

Table 5. CLP vs. HC using the merged DB with selected features

Classifier	ACC [%]	Sen [%]	Spe [%]	AUROC	Parameters
SVM	83.4 ± 2	91.2 ± 5	72.4 ± 9	0.90	$C = 10, \gamma = 10^{-3}$
RF	80.0 ± 4	92.8 ± 2	62.2 ± 9	0.89	$D = 2, N = 70$

ACC: accuracy in the test set, AUROC: Area under ROC curve. Sen: Sensitivity,
Spe: Specificity.
C and γ: complexity parameter and bandwidth of the kernel in the SVM classifier.
N and D: Number of trees and depth of the decision trees in the RF classifier

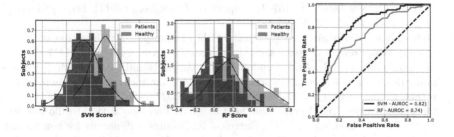

Fig. 4. Histograms of the scores and ROC curves (merged DB)

5   Conclusions

The proposed approach, based on articulation measures is effective for the classification of hypernasality in children. High accuracies were obtained with the SVM classifier (above 90%) in the two databases. When Kruskal-Wallis tests are applied as the selection criterion to include features before merging the two databases, accuracies of around 80.0% are obtained. The scores obtained with the Clínica Noel DB show less separability between CLP patients and HC patients. Conversely, the results obtained with the Manizales DB are higher and the associated histograms show less overlapping between the two classes. This can be likely explained due to the difference in the acoustic conditions of both databases. The sensitivity of the model was consistently high along the three experiments presented in this paper. This result may indicate that the proposed approach is suitable to evaluate the degree of hypernasality. Further research with more data and additional labels are required to confirm this hypothesis. Our team is currently working on the collection of more speech samples such that allow the evaluation of different degrees of hypernasality considering sustained vowel phonations and continuous speech signals.

Acknowledgement. This work was partially funded by CODI at UdeA grant # PRG2018-23541 and SOS18-2-01_ES84180137.

References

1. World Health Organization: Global registry and database on craniofacial anomalies. Report of a WHO registry meeting on craniofacial anomalies (2001)
2. Golabbakhsh, M., et al.: Automatic identification of hypernasality in normal and cleft lip and palate patients with acoustic analysis of speech. J. Acoust. Soc. Am. **141**(2), 929–935 (2017)
3. Kummer, A.W.: Cleft Lip, Palate and Craniofacial Anomalies. CENGAGE Learning, Boston (2014)
4. He, L., Zhang, J., Liu, Q., Yin, H., Lech, M., Huang, Y.: Automatic evaluation of hypernasality based on a cleft palate speech database. J. Med. Syst. **39**, 61 (2015)
5. Rendón, S.M., Orozco Arroyave, J.R., Vargas Bonilla, J.F., Arias Londoño, J.D., Castellanos Domínguez, C.G.: Automatic detection of hypernasality in children. In: Ferrández, J.M., Álvarez Sánchez, J.R., de la Paz, F., Toledo, F.J. (eds.) IWINAC 2011. LNCS, vol. 6687, pp. 167–174. Springer, Heidelberg (2011). https://doi.org/10.1007/978-3-642-21326-7_19
6. Dubey, A.K., Tripathi, A., Prasanna, S.R.M., Dandapat, S.: Detection of hypernasality based on vowel space area. J. Acoust. Soc. Am. **143**(5), 412–417 (2018)
7. Orozco Arroyave, J.R., Arias Londoño, J.D., Vargas Bonilla, J.F., Nöth, E.: Automatic detection of hypernasal speech signals using nonlinear and entropy measurements. In: Proceedings of INTERSPEECH, pp. 2027–2030 (2012)
8. Orozco Arroyave, J.R., Vargas Bonilla, J.F., Arias Londoño, J.D., Murillo Rendón, S., Castellanos Domínguez, C.G., Garcés, J.F.: Nonlinear dynamics for hypernasality detection in Spanish vowels and words. Cogn. Comput. **5**(4), 448–457 (2012)

9. Carvajal Castaño, H.A.: Metodología para la reducción de ruido aditivo de fondo en sistemas basados en procesamiento de voz. Master's thesis, Universidad de Antioquia (2013)
10. Godino Llorente, J.I., Gómez Vilda, P.: Automatic detection of voice impairments by means of short-term cepstral parameters and neural network based detectors. IEEE Trans. Biomed. Eng. **51**(2), 380–384 (2004)
11. Orozco Arroyave, J.R., Hönig, F., Arias Londoño, J.D., Vargas Bonilla, J.F., Nöth, E.: Spectral and cepstral analyses for Parkinson's disease detection in Spanish vowels and words. Expert Syst. **32**(6), 688–697 (2015)
12. Rabiner, L., Schafer, R.W.: Theory and Applications of Digital Speech Processing. Prentice Hall, Upper Saddle River (2011)
13. Kaiser, J.F.: On a simple algorithm to calculate the 'energy' of a signal. In: Proceedings of ICASSP, pp. 381–384 (1990)
14. Cairns, D., Hansen, J., Riski, J.: A noninvasive technique for detecting hypernasal speech using a nonlinear operator. IEEE Trans. Biomed. Eng. **43**(1), 35–45 (1996)
15. Sapir, S., Ramig, L.O., Spielman, J.L., Fox, C.: Formant centralization ratio (FCR): a proposal for a new acoustic measure of dysarthric speech. J. Speech Lang. Hear. Res. **53**(1), 114–125 (2010)

Signals Analysis and Processing

A Wavelet Entropy Based Methodology for Classification Among Healthy, Mild Cognitive Impairment and Alzheimer's Disease People

Jorge Esteban Santos Toural[1]([⊠]), Arquímedes Montoya Pedrón[2], and Enrique Juan Marañón[1]

[1] Universidad de Oriente, Las Américas, Av., Santiago de Cuba, Cuba
{jsantos,enriquem}@uo.edu.cu
[2] Juan Bruno Zayas Alfonso General Hospital, Carretera del Caney esquina 23, Pastorita, Santiago de Cuba, Cuba
arqui@medired.scu.sld.cu

Abstract. Alzheimer's disease (AD) and Mild Cognitive Impairment (MCI) are cognitive diminished conditions that requires neuropsychological, images and complementary test for diagnosis. The electroencephalogram equipment is a less expensive, less invasive and a more portable option that image ones so there is an increased interest in an EEG based methodology for cognitive impairment diagnosis. In this work is presented an eyes closed resting state condition EEG signal diagnosis tool, based on wavelet decomposition and wavelet entropy. The methodology allows discriminating among Healthy, Mild Cognitive Impairment and Alzheimer's disease people. For this purpose theta band-EEG power ratio, beta band-EEG power ratio and entropy values distribution through time in 14 electrodes are used. Wavelet decomposition is performed on five levels using Haar wavelet mother on two seconds windows. After decomposition wavelet power ratio and entropy distribution calculation are performed. The characteristics are used in a Healthy–MCI, Healthy-AD and MCI-AD classification using Support Vector Machine with polynomial kernel providing six inputs to a neural network (two layer, 13 neurons in the hidden layer) in charge of the final classification. Data base is composed of 17 healthy, nine Mild Cognitive Impairment and 15 Alzheimer's disease people registers. A precision of 92.68% to 97.56% is achieved, better or equal to other entropy-based methods with the advantage of separating the three groups and use a bigger database. This methodology reveals as a potential quantitative diagnosis-support tool especially between Healthy people and Mild Cognitive Impairment where some of the conventional test fails.

Keywords: Wavelet entropy · Mild Cognitive Impairment · Alzheimer's disease

© Springer Nature Switzerland AG 2019
I. Nyström et al. (Eds.): CIARP 2019, LNCS 11896, pp. 589–598, 2019.
https://doi.org/10.1007/978-3-030-33904-3_55

1   Introduction

German physician Alois Alzheimer first described the Alzheimer's disease. It is a neurodegenerative disorder caused by senile plaques and neurofibrillary tangles, which provokes loss of neuron connectivity, brain inflammation, oxidative stress and cell death [1, 2]. Symptoms involve loss of recent memory, disorientation, poor judgment, apraxia and aphasia. Is preceded by a condition known as Mild Cognitive Impairment in which cognitive deterioration is not as large as in dementia but greater that the expected for the age. Diagnosis is based on family history, family description of patient behavior, cognitive, blood and images test for discarding other types of dementia or dementia-like syndromes. At this point, based on the Diagnostic and Statistical Manual of Mental Disorders version five (DSM 5) a MCI, AD or other type of dementia can be diagnosed [1–3].

The EEG signals is a register of the electrical activity in the brain through time. Since AD causes loss of connectivity and neurons death, it is expected that the EEG signal get affected. Known alterations are frequency diminution [4–6] (power increase in theta and delta band as well as a power decrease in alpha and beta) and complexity reduction in form of a loss of randomness [4, 6].

Entropy-based classification methodologies has been implemented using Tsallis, Multiscale, Sample, Permutation and Fuzzy entropies [7–13]. The best reported sensibilities and specificities lies between 72.73% to 90.91% [8] and 81.82% [13]. However only the MCI groups with healthy people, or AD with healthy people are studied.

In [14] a specificity of 100.0% and sensitivity of 98.0% is reached separating a healthy group and a MCI+AD group (named AD) with a single electrode (Fp1). The methodology is based on the absolute and relative energy of theta rhythm in eyes open condition (characteristic with the best performance), and wavelet entropy of the alpha rhythm in eyes closed (second best performance characteristic). Despite the results and the fact that Fp1 guarantees the recording of theta rhythm in the frontal region, the alpha rhythm is predominant in the occipital area. This leads to the fact that a displacement of the alpha wave towards frontal regions can be better detected with the use of electrodes in the frontal and occipital regions than with a single one. The number of people with MCI in the cognitive disorder group is not specified as being completely defined as Alzheimer's. The total number of registers is 24. It is also considered that both groups are not completely paired since in the group of AD there are no individuals between 50 and 60 years old (3 in the control group) and few between 60 and 70 years old (1 in AD against four in the control group).

In this work, the main interest is to provide a methodology that could discriminate among the three possible clinical cases: Healthy people, MCI and AD. For this purpose wavelet entropy, theta band-EEG and beta band-EEG power ratio are used as classification characteristics. Its importance lies on the possibility of develop an auxiliary tool for specialists. This tool may help them to eliminate the uncertainty in the diagnosis, especially in MCI where some of the psychological tests provides specificity values of 17.1% to 65.7% and introduces false positives due to the subtle differences between normal ageing and MCI [15–17].

2 Materials and Methods

2.1 Subjects and EEG Recording

The EEG database is composed of 17 registers of healthy people (55 to 75 years old, mean 63.03 and standard deviation of 5.75), nine MCI registers (57 to 78 years old, mean 67.20 and standard deviation of 7.37) and 15 AD registers (57 to 85 years old, mean: 70.00 and standard deviation of 6.51). They were collected in Neurophysiology department at Juan Bruno Sayas General Hospital, Santiago de Cuba. A group of neuropsychologists using as exclusion criteria: psychiatric disorder, epilepsy, severe head trauma, stroke or ischemia performed the classification in Healthy, MCI and AD. The study was approved by the hospital Ethics Committee.

The EEG was acquired with subjects in supine position, resting state, alternating eyes closed and eyes open for eight minutes. It was used a 19 electrode, 10–20 international system referential montage (Fp1, Fp2, F3, F4, C3, C4, P3, P4, O1, O2, F7, F8, T3, T4, T5, T6, Fz, Cz and Pz). The equipment consisted on a MEDICID 5 with 50 Hz/60 Hz programmable notch filter, sampling frequency of 200 Hz, analog to digital converter of 16 bits and band-pass of 0.5 Hz to 30 Hz. Signals were analyzed by a specialist physician to remove sections with artefacts. Finally, a register free of artefacts, two minutes long, resting state eyes closed was obtained.

2.2 Wavelet Entropy

The Wavelet Entropy is a measure of the degree of order of a signal [1]. Like Shannon entropy, is based in a probability distribution. In this case the distribution of the relative power of a frequency interval with respect to the entire signal power. Therefore, an ordered process could be seen as a very few frequency components that comprises the entire signal power (narrow power spectrum). Then a disordered process contains a several number of frequencies components and has a wide power spectrum. It is expected that in case of cognitive disorder (loss of randomness) the power distribution be less widespread that in case of healthy person. Then the greater the cognitive perturbation, the lower entropy value.

In Wavelet Entropy, the power ratio is calculated by means of the wavelet transform. This is a multi-resolution signal analysis with time-frequency information. As bases functions are used the so called "wavelet functions". They are short duration, quickly attenuated and zero mean arbitrary functions. The decomposition is based in comparing the signal with the wavelet function displaced in time and for different scales. This leads to coefficients that inform about the degree of similarity of the wavelet with the signal. The use of scales and time translation leads to the more general concept of wavelet family that is represented in (1).

$$\psi_{j,k}(t) = \frac{1}{\sqrt{j}}\psi\left(\frac{t-k}{j}\right) \tag{1}$$

Where j, k are the scaling and shifting values respectively and ψ is the wavelet mother.

The original signal $s(t)$ can be written as in (2) [18].

$$S(t) = \sum_{j=-DL}^{-1} \sum_{k} C_j(k)\psi_{j,k}(t) \tag{2}$$

Where $C_j(k)$ is the level j and scale k coefficient, $\psi_{j,k}(t)$ is the wavelet family and DL the decomposition level. If (2) is analyzed in the time-frequency domain, the decomposition implementation will be a filter bank of successive high-pass and low-pass filter which provides detail (D) and approximation (A) coefficients respectively (see Fig. 1).

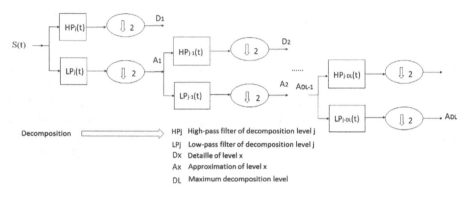

Fig. 1. Wavelet decomposition using filter bank implementation.

The relative power p_j for the decomposition level j, can be calculated by (3) and the Wavelet Entropy by (4) [18].

$$p_j = \frac{E_j}{E_{total}} = \frac{\sum_k |C_j(k)|^2}{\sum_{j<0} \sum_k |C_j(k)|^2} \tag{3}$$

$$H(p) = -\sum_{j<0} p_j \ln(p_j) \tag{4}$$

Wavelet mother selection has no rules. In general is desired orthogonality and similarity to the analyzed signal. In this work, Symlets9 wavelet mother is used based on the results obtained in [19]. The Haar wavelet mother is used based on its similarity with theta and delta rhythms.

2.3 Data Processing

The EEG artifact-free signal is divided in windows of 400 samples (2 s). Then Wavelet Entropy and EEG rhythms-EEG signal power ratio are calculated for Symlets9 and Haar with five to seven decomposition levels. This values guarantee cover all principal EEG rhythms. The relationship between the decomposition level and each EEG rhythm (frequency) can be calculated using (5) where W_s is the sampling frequency, DL is the decomposition level and $|W|$ is the band pass frequencies. Table 1 shows the relationship.

$$\frac{W_s}{2^{DL+1}} \leq |W| \leq \frac{W_s}{2^{DL}} \tag{5}$$

Table 1. Frequency range and its equivalent EEG rhythm for each wavelet decomposition level wavelet

Decomposition level	Frequency range (Hz)	EEG rhythm
1	50.0 to 100.0	
2	25.0 to 50.0	Gamma
3	12.5 to 25.0	Beta
4	6.25 to 12.5	Alpha
5	3.12 to 6.25	Theta
6	1.56 to 3.12	Delta
7	0.78 to 1.56	Delta

Once calculated the aforementioned parameter, a statistical analysis is performed using Student's t-test to determine in which channels and EEG rhythms exists significative differences among the three groups for both wavelets mothers. The objective is to select the wavelet mother who has the greatest number of channels with significative statistical differences and use it for implement a diagnosis methodology.

Entropy is calculated for different time windows. Comparison among groups is mainly made based on its average value [7, 8, 10, 12]. In this work other approach for analyze the time-varying values of entropy is proposed: the maximal difference between entropy histogram peak and its adjacent neighbors (histogram difference hereafter). This is calculated as follow:

1. Calculate the time varying entropy histogram.
2. Normalize the histogram.
3. Select the maximum value of the normalized histogram.
4. Calculate de difference between the maximum value and its right and left neighbors.
5. Select the maximum difference.
6. If the maximum value is the first/last item of the histogram, select the difference with its right/left neighbor.
7. Repeat in all channels of interest.

Once calculated the histogram differences for all channels and both wavelet mothers they are grouped in intervals defined by thresholds with the intention of separate among Healthy, MCI and AD. A proportion test is performed (6) where e_x is the number of people that their histogram difference lies in an interval. T_x is the total number of people in the interval x [20].

$$z = \frac{\frac{e_1}{T_1} - \frac{e_2}{T_2}}{\sqrt{\frac{e_1 + e_2}{T_1 + T_2}\left(1 - \frac{e_1 + e_2}{T_1 + T_2}\right)\left(\frac{1}{T_1} + \frac{1}{T_2}\right)}} \tag{6}$$

2.4   Evaluated Methodology

The intended classification methodology is shown in Fig. 2. Once determined the decision characteristics (channel and parameter), they are submitted to the classification block. This is composed of two stages: a pre-classification stage using SVM and the decision stage implemented with neural network. The pre-classification stage is used to reduce the number of inputs of the neural network. The idea is to group channels according to the calculated parameter and their discrimination capability. The neural network present the final decision: Healthy, MCI or AD patients.

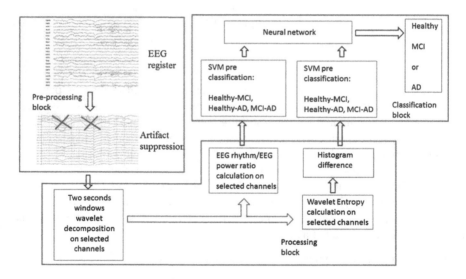

Fig. 2. Evaluated methodology flux diagram

3   Results and Discussion

The statistical analysis of EEG rhythm-EEG signal power ratio consider as null hypothesis that Healthy, MCI and AD groups has equal mean but different variance. The alternative hypothesis is that the mean values are different. The results showed that Haar with five decomposition levels and Symlets9 with seven decomposition levels have the best results. In Fig. 3 is shown that Symlets9 detects changes mainly in alpha rhythm while Haar does in beta and theta. Haar also detects changes in a greater number of channels, 14 vs. 9. For this reason, Haar wavelet mother is selected for implement the classification methodology.

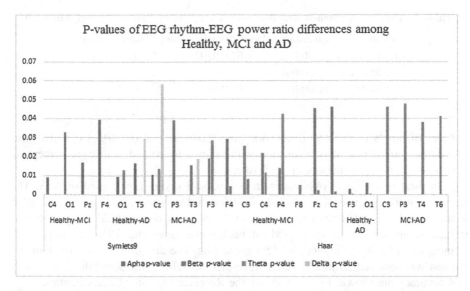

Fig. 3. P-values of EEG rhythm-EEG power ratio differences among Healthy, MCI and AD

The analysis of histogram difference is shown in Fig. 4. The red curve corresponds to AD, blue to MCI and black to Healthy group. In frontal, parietal and central electrodes, the numerical values are different for the three groups so it shows a possible discriminating criterion. The proportion test results are presented in Table 2.

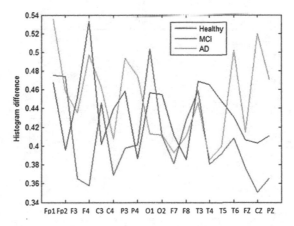

Fig. 4. Histogram difference among the three groups. It is observed the possibility of discrimination on frontal, parietal, temporal, Fz, Cz and Pz electrodes. (Color figure online)

These results allows selecting 27 characteristics to implement the classification methodology. The characteristics are: beta-EEG and theta-EEG power ratio on F3, F4, C3, C4, P4, Fz, Cz, F3, O1; theta-EEG power ratio on F8, C3, P3, T4, T6 and histogram difference on Fp1, Fp2, F3 and Cz. The use of SVM allows reducing the

Table 2. P-values of channels with discrimination capability base on the proportion test

Group	Channel/p-value	Channel/p-value	Channel/p-value
Healthy-MCI	Fp2/0.0367	F3/0.0293	Cz/0.0127
Healthy-AD	Fp2/0.001		
MCI-AD	Fp1/0.002	P4/0.001	

number of inputs of the neural network from 27 to seven. The SVM classifier was tested with polynomial kernel function with order 3 to 10 and multilayer perceptron kernel function with constants of ± 1.0 to ± 2.0.

The best results were obtained with polynomial kernel function of order 8. The neural network classifier is a two-layer feed-forward network with sigmoid activation function and softmax output neurons with 16 neurons on the hidden layer. It was trained using scaled conjugate gradient backpropagation with 70% of samples for training and 15% for validation and 15% for testing. The data division was random and the test was repeated 20 times averaging results. In Table 3 the sensibility, specificity and accuracy are shown. Figure 5 shows the Receiver Operator Characteristics.

Table 3. Classification methodology performance.

Group	Sensibility (%)	Specificity (%)	Accuracy (%)
Healthy	94.11	91.66	92.68
MCI	88.88	100.00%	97.56
AD	93.33	96.15	95.12

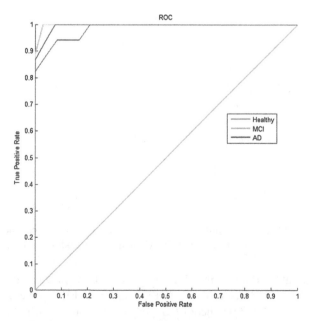

Fig. 5. ROC curves of the classification methodology

4 Conclusions

It is observed that wavelet decomposition and wavelet entropy have the ability to be used in a classification methodology that allows separating among Healthy, MCI and AD patients using beta-EEG, theta-EEG signal power ratio and the entropy values distribution through time. Haar wavelet mother detects changes in beta and theta rhythms and in a greater number of channels than Symlets9. The latter is sensible mainly to alpha rhythm power changes. The proposed classification methodology has better performance that [8, 13] and comparable with [14] with a greater number of cases. It also has the advantage that classifies among the three possible groups present in a clinical environment: Healthy, MCI and AD people. Its importance lies in the possibility of develop a tool oriented to diagnosis support, which helps to reduce the uncertainty in the diagnosis; especially in MCI where some of the psychological tests fails.

The methodology introduces a new analysis approach for entropy values: histogram analysis of entropy values through time instead the mean value. It also uses the beta-EEG power ratio that has not been used in others studies as far as we know. Is interesting the detection of beta power changes in C3, C4, Fz and Cz in MCI with respect to the healthy group. The reported changes in this rhythm occurs in the posterior regions of the brain in presence of AD [4].

As future work is intended to select the characteristics that more contribute in the classification process in order to reduce the calculation requirements of the processing stage.

References

1. Alzheimer's Association: 2017 Alzheimer's Disease Facts and Figures. Alzheimers Dement, New York (2017)
2. Alzheimer' Society: What is Alzheimer's Disease?, July 2014. https://www.alz.co.uk. Accessed Feb 2016
3. Bello, V.M.E., Schultz, R.R.: Prevalence of treatable and reversible dementias. Dement Neuropsychol. 5(1), 44–47 (2011)
4. Al-Qazzaz, N.K., Ali, S.H.B., Ahmad, S.A., Chellappan, K., Islam, M., Escudero, J.: Role of EEG as biomarker in the early detection and classification of dementia. Sci. World J. 2014, 16 (2014)
5. Wang, R., Wang, J., Li, S., Yu, H., Deng, B., Wei, X.: Multiple feature extraction and classification of electroencephalograph signal for Alzheimer's with spectrum and bispectrum. Chaos 25, 013110 (2015)
6. Mittal, S.H., et al.: Abnormal levels of consciousness and their electroencephalogram correlation: a review. EC Neurol. 4(1), 30–35 (2016)
7. Abásolo, D., Hornero, R., Espino, P., Alvarez, D., Poza, J.: Entropy analysis of the EEG background activity in Alzheimer's disease patients. Physiol. Meas. 27, 241–253 (2009)
8. Hornero, R., Abásolo, D., Escudero, J., Gómez, C.: Nonlinear analysis of electroencephalogram and magnetoencephalogram recordings in patients with Alzheimer's disease. Phil. Trans. R. Soc. 367, 317–336 (2009)

9. De Bock, T.J., et al.: Early detection of Alzheimer's disease using nonlinear analysis of EEG via Tsallis entropy. In: Biomedical Sciences and Engineering Conference (2010)
10. Al-nuaimi, A.H., Jammeh, E., Sun, L., Ifeachor, E.: Tsallis entropy as a biomarker for detection of Alzheimer's disease. In: 37th Annual International Conference of the IEEE Engineering in Medicine and Biology Society (2015)
11. Escudero, J., Acar, E., Fernández, A., Bro, R.: Multiscale entropy analysis of resting-state magneto encephalogram with tensor factorizations in Alzheimer's disease. Brain Res. Bull. **119**(Part B), 136–144 (2015)
12. Morabito, F.C., Labate, D., La Foresta, F., Bramanti, A., Morabito, G., Palamara, I.: Multivariate multi-scale permutation entropy for complexity analysis of Alzheimer's disease EEG. Entropy **14**, 1186–1202 (2012)
13. Simons, S., Espino, P., Abásolo, D.: Fuzzy entropy analysis of the electroencephalogram in patients with Alzheimer's disease: is the method superior to sample entropy? Entropy **20** (21), 13 (2018)
14. Ghorbanian, P., Devilbiss, D.M., Hess, T., Bernstein, A., Simon, A.J., Ashrafiuon, H.: Exploration of EEG features of Alzheimer's disease using continuous wavelet transform. Med. Biol. Eng. Comput. **53**, 843–855 (2015)
15. Edmonds, E.C., et al.: Susceptibility of the conventional criteria for mild cognitive impairment. Alzheimer's Dementia **11**, 1–10 (2014)
16. Charernboon, T.: Diagnostic accuracy of the overlapping infinity loops, wire cube, and clock drawing tests for cognitive impairment and dementia. Int. J. Alzheimer's Dis. **2017**, 5 (2017)
17. Allan, C.L., Behrman, S., Ebmeier, K.P., Valkanova, V.: Diagnosing early cognitive decline —when, how and for whom? Maturitas **2017**, 103–108 (2017)
18. Rosso, O.A., et al.: Wavelet entropy: a new tool for short duration brain electrical signals. J. Neurosci. Methods **105**, 65–75 (2001)
19. Al-Qazzaz, N., Hamid Bin Mohd Ali, S., Ahmad, S., Islam, M., Escudero, J.: Selection of mother wavelet functions for multi-channel EEG signal analysis during a working memory task. Sensors **15**, 21 (2015)
20. Castillo, A.J.S.: Apuntes de Estadística para Ingenieros. Creative Commons, Jaén (2012)
21. Jeong, D.H., Kim, Y.D., Song, I.U., Chung, Y.A., Jeong, J.: Wavelet energy and wavelet coherence as EEG biomarkers for the diagnosis of Parkinson's disease-related dementia and Alzheimer's disease. Entropy **18**(8), 17 (2015)

Trainable COPE Features for Sound Event Detection

Nicola Strisciuglio$^{(\boxtimes)}$ and Nicolai Petkov

Bernoulli Institute for Mathematics, Computer Science and Artificial Intelligence,
University of Groningen, Groningen, Netherlands
`n.strisciuglio@rug.nl`

Abstract. Systems for automatic analysis of sounds and detection of events are of great importance as they can be used as substitutes of or complement to video analytic systems. In this paper we describe a flexible system for the detection of audio events based on the use of trainable COPE (Combination of Peaks of Energy) features. The structure of a COPE feature is determined in an automatic configuration process on a single prototype example. Thus, they can be adapted to different kinds of sounds of interest. We configure a set of COPE features in order to account for robustness to variations of the characteristics of sounds within a specific class. The proposed system is flexible as new features (also configured on examples drawn from new classes) can be easily added to the feature set. We performed experiments on the MIVIA road events data set for road surveillance applications and compared the results that we achieved with the ones of other existing methods.

Keywords: Few shot training · Sound event detection · Trainable features

1 Introduction

Automatic detection of sound events of interest gained a great attention in the past years [8], because of its various applicative aspects. In social robotics, for instance, sound signals are used to improve the interaction with the environment and people [9]. Security and surveillance are also of great importance and applications of sound analysis to road and traffic monitoring were explored [10]. In the latter case, sound analysis can complement or substitute video analytic systems. Video analytic systems have to deal with varying illumination conditions (e.g. day and night), changing weather conditions, and the presence of occlusions in the field of view of the camera. In some cases the installation of cameras is not allowed due to privacy issues (e.g. a public toilet) or too expensive (e.g. in the monitoring of long roads). Furthermore, certain abnormal events, such as gun shots, screams or skidding tires are very difficult to detect on video only. In [10] a practical system for surveillance of roads by audio analysis was designed as an alternative to video-based analytic systems.

© Springer Nature Switzerland AG 2019
I. Nyström et al. (Eds.): CIARP 2019, LNCS 11896, pp. 599–609, 2019.
https://doi.org/10.1007/978-3-030-33904-3_56

In this paper, we present a method for audio event detection that is based on trainable feature extractors, called COPE (Combination of Peaks of Energy), recently proposed in [27]. The COPE algorithm is based on the analysis of local maxima in a time-frequency representation of the input audio signal, which have been demonstrated to be robust to additive noise [31]. From an applicative point of view, we focus on the case study of event detection for road surveillance.

A comprehensive survey on audio surveillance was published in [8]. Early approaches to sound event detection were based on the extraction of a set of features to describe important properties of the input audio stream. Temporal and frequency features were combined in hand-engineered sets, e.g. energy, pitch, bandwidth, etc. Feature vectors were then used to train a classifier model, subsequently used in the operating phase for the detection and classification of events of interest in unknown audio streams [6,11,14,23,30]. Subsequently, more complex architectures or data representations were proposed to strengthen the robustness of the system to high variability of the background noise: multi-stage classifiers [19], a classification rejection module [7], representations of the audio signal based on the bag of features approach [12,20]. In [25], the performance of bag-of-features methods were measured on live audio streams. The temporal arrangement of instantaneous features was taken into account in [15], and in [22] with a pyramidal approach. In [11], instead the audio stream was represented as a time-frequency distribution of its energy and object-detection techniques were employed to detect the events of interest. All these methods involve a feature engineering step, in which a proper set of features for the problem at hand is defined. Such process requires knowledge of the specific domain of application and is critical for the design of a robust recognition system. Deep learning and Convolutional Networks based approaches are able to learn effective features from training data [1,24], but require large amount of data. Similarly, AENET [28] is based on the VGG architecture and was designed to optimize the trade-off between required computational resources and performance accuracy of sound event detection. In [13], an audio event detection method based on the MobileNet architecture was proposed that outperformed the detection performance of AENET. In the context of audio event recognition, the DCASE challenge is of great importance and focuses on the benchmark of new methods for polyphonic event detection and localization [18] and specific applications in domestic environments [29]. Co-occurring sound detection is also receiving a growing interest [16].

The COPE features that we propose differ from existing approaches as they are trainable with few data, in that they can be configured to detect any sound pattern of interest by presenting a single example to a configuration algorithm. This allows to determine the important features directly from the training data, also when few examples are available, so reducing the effort required by a feature engineering step. The concept of trainable filters was introduced in previous works with successful application to image processing [2,3] and also integrated into Convolutional Networks [17]. In this work, we deploy the COPE features in a flexible sound recognition system that allows for the extension of the COPE feature set, and compare the results with its counterpart with a fixed set of COPE features. The results that we achieved on the MIVIA road events data set demonstrate the effectiveness of the COPE features for sound event detection.

2 Method

The event detection method that we propose is based on COPE feature extractors. Given a prototype sound of interest, a COPE feature can be configured by an automatic process that determines a model of the constellation of energy peaks in a time-frequency representation (we use the Gammatonegram) of such sound. In the testing phase, it is then able to detect the prototype sound event and also modified versions of it due to noise or distortion. We configure a set of COPE features on few training sound examples for each class of interest, in order to account for intra-class variability.

2.1 Gammatonegram

We process the input audio signal $s(t)$ by a Gammatone filterbank, which outputs a spectrogram-like image $S_{gt}(t, f)$ called gammatonegram. In Fig. 1a, we show the gammatonegram of a car crash sound. The gammatone filterbank is a linear model of the response of the cochlea membrane of the human auditory system, which converts the incoming sound pressure waves into neural stimuli on the auditory nerve [21].

In contrast with the spectrogram, in which the energy distribution over frequency is computed by band-pass filters all with the same bandwidth, in the gammatonegram the bandwidth of the band-pass filters is linear with their central frequency. Thus, at low frequency the band-pass filters have a narrower bandwidth than the ones at high frequency. This accounts for different resolution at low and high frequency, similarly to the way the auditory system processes the sound. We refer the reader to [21,27] for details.

2.2 COPE Features

The COPE algorithm receives as input the energy peaks (i.e. local maxima, in a 8-connected neighborhood) extracted from the Gammatonegram of a sound and responds strongly to constellations of points similar to that included in the COPE model during a configuration process performed on a single prototype sound. In Fig. 1b, we show the constellation of energy peaks corresponding to the gammatonegram of the sound in Fig. 1a.

For the configuration of a COPE feature, one has to choose the size of its support, i.e. the size of the time interval in which to consider the position of the energy peak points. Within the support area we take as reference point the peak with the highest energy (circle dot in Fig. 1b) and describe the remaining points by tuples of three parameters $(\Delta t_i, f_i, e_i)$: Δt_i is the temporal offset of the considered point with respect to the reference point, f_i represents the frequency

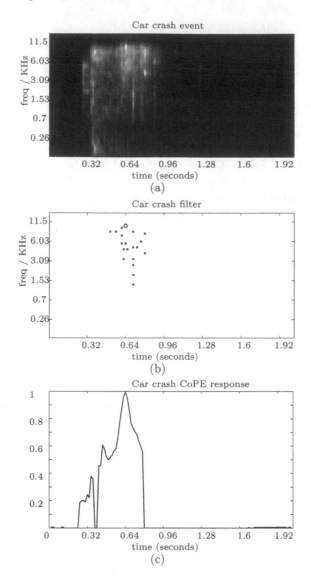

Fig. 1. Gammatonegram representation of a (a) car crash. We detect and use the (b) constellation of energy local maxima points in the Gammatonegram for the configuration a COPE feature. In the testing phase the COPE feature is applied with a sliding window approach and provides a response during time, which is high when the constellation of energy local maxima is similar to the one determined in the configuration.

channel and e_i is the energy contained in its bandwidth. The set of tuples results in the model $S = \{(\Delta t_i, f_i, e_i) \mid i = 1, \ldots, N\}$, where N is the total number of considered points.

In the application phase, given an audio stream of a certain duration, we compute a response for each energy peak included in the model. Such response indicates a score of the presence of each point (i.e. a certain amount of energy in a given frequency band) along the signal to analyse, and is defined as:

$$s_i(t) = \max_{t',f'} \{\psi(t, f_i)G_{\sigma'}(t', f')\} \tag{1}$$

$$t - \Delta t \le t' \le t + \Delta t, f_i - \Delta f \le f' \le f_i + \Delta f$$

where $G_{\sigma'}(\cdot, \cdot)$ is a Gaussian weighting function that accounts for some tolerance in the position of the i-th energy peak within the constellation. The tolerance introduced in the application phase makes the filter robust to deformation of the pattern of interest due to distortion or noise. The function $\psi(t, f)$ represents a measure of similarity between the detected energy peak in the unknown audio stream and the one contained in the model of the COPE feature. In this work we consider $\psi(t, f) = S_{gt}(t, f)$, that takes into account only the position of each point in the constellation and its energy content.

We formally define the response of a COPE feature as the geometric mean of the responses of its tuples (i.e. the responses in Eq. 1):

$$r(t) = \left| \left(\prod_{i=1}^{|S|} s_i(t) \right)^{1/|S|} \right|_{th}, \tag{2}$$

where th is a threshold fraction of the highest possible COPE feature response. We determine the value of th for a class of events of interest by cross-validation experiments. We apply the COPE features with a sliding window approach and achieve a continuous response during time. As an example, in Fig. 1c, we show the response of the COPE feature configured in Fig. 1d and applied on the same samples used for configuration.

2.3   Event Detection and Classification

Given M sounds of interest of classes in the set $\{C_0, C_1, \ldots, C_{N-1}\}$, we configure a set of M COPE features, one for each such samples. We perform the detection of the events of interest in a time window $W_{[t_1, t_2]}$ that forward shifts on the input audio signal by one third of its length, where t_1 and t_2 are the boundaries of the time interval. In this work, we deploy two different classification approaches. The first approach considers the COPE feature that achieves the highest response in the interval $W_{[t_1, t_2]}$:

$$\hat{r}_i = \max_{t \in W_{[t_1, t_2]}} r_i(t), i = 1, \ldots, M \tag{3}$$

and assign to the sound under analysis the class $C_{out} = c(\hat{r}_i)$ that is associated with the COPE feature with highest response over a certain threshold. We determine a threshold value for each class of interest by cross-validation experiments. We refer at this classification approach as *COPE+max*.

The second classification approach involves the use of a multi-class Support Vector Machine (SVM) classifier, trained using the COPE feature vectors constructed by computing the response of the M configured COPE features. We train one SVM for each of the N classes of interest. For the i-th SVM we use the feature vectors computed on the samples of class C_i as positive and the remaining samples as negative. In the classification phase, each SVM provides a score s_i, which we use to determine the class of the sound of interest as:

$$
C_{out} = \begin{cases} C_0, & if \ s_i < 0 \quad \forall i = 0, \ldots, N-1 \\ \arg\max_i s_i, & else. \end{cases}
$$
(4)

where C_0 is the background sound class. We refer at this classification approach as *COPE+SVM*.

3   Experiments

3.1   MIVIA Road Events Data Set

We carried out experiments on the MIVIA road events data set [10], which is publicly available for research purpose at the url http://mivia.unisa.it. The data set is composed of 57 audio clips of about one minute, sampled at 32 KHz and with a resolution of 16 bits per sample. In order to simulate the occurrence of events of interest in various environments, each audio clip contains a number of events that are superimposed to different combinations of typical background sounds of roads (moving cars, traffic jam, crowds, etc.). The data set contains a total of 400 events of interest, namely 200 car crash and 200 tire skidding events, which occur in different situations with different background sounds. The audio clips are divided in four independent folds each containing 50 events of interest per class. Such organization of the data is suitable for cross-validation experiments. In the following we refer to car crash with *CC*, to tire skidding with *TS* and to background noise with *BN*.

3.2   Performance Evaluation

For the evaluation we adopt the experimental protocol proposed in [12]. We perform the detection of events of interest within a time window of 3 s that forward shifts on the audio stream by steps of 1 s. An event of interest is correctly detected if it is detected in at least one of the time windows that overlap with it. When only background noise is present in the considered time window and an event of interest is wrongly detected, we count it as a false positive. In the case a false detection happens in two consecutive time windows, only one false positive event is counted.

We compute the following measures: the true positive rate (TPR) or recognition rate, error rate (ER), miss detection rate (MDR) and false positive rate (FPR). The MDR and FPR provide information about the robustness of the method to various types of background noise.

Table 1. Classification matrices achieved by the (a) COPE+max and (b) COPE+SVM approaches on the MIVIA road events data set.

COPE+max					*COPE+SVM*			
		Guessed class					Guessed class		
		TS	CC	Miss			TS	CC	Miss
True	TS	84%	6%	10%	True	TS	92%	2%	6%
class	CC	7%	89.5%	3.5%	class	CC	0.5%	96%	3.5%
	(a)					(b)			

3.3 Results and Comparison

In Table 1, we report the classification matrices achieved by *COPE+max* and *COPE+SVM* methods on the MIVIA road events data set. The *COPE+max* approach achieved an average recognition rate of 86.75% with standard deviation 3.4 on the 4-folds cross-validation experiments, while the *COPE+SVM* approach achieved an average recognition rate of 94% with standard deviation 4.3. The rate of missed detection is low (*COPE+max*: 6.75%, *COPE+SVM*: 4.75%) and, coupled with a false positive rate of $FPR = 7.01\%$ for *COPE+max* and 3.94% for *COPE+SVM*, indicates that the proposed method based on trainable COPE features is robust to various types of background noise.

In Table 2, we report the results achieved by the proposed method and compare them with the ones achieved by other existing approaches. We consider the feature sets used in [5,10] that were combined with classification architectures based on the bag of features (BoF) classification approach. The performance of three sets of features to describe short-time properties of the audio signal were studied and compared. The recognition rate achieved by COPE features is generally higher than the ones achieved by other approaches. Furthermore, the rate of miss detected events is considerably lower (more than 12%) with respect to existing methods. Most of the classification errors are due to misclassification of the target class of the event of interest. The error rate is, indeed, higher than for other methods. From a practical point of view, it means that the occurrence of an abnormal event is however detected and an alert to the human operator could be raised anyway. The FPR achieved by the proposed approach (7.01%) is higher than the method proposed in [5] based on a combination of temporal and spectral, but it is lower than the ones achieved by the system that uses MFCC and Bark feature descriptors for instantaneous properties of the audio signal.

4 Discussion

The system presented in this work, based on a bank of COPE features, achieved high performance results on the MIVIA road events data set. The higher detection capabilities achieved by COPE features with respect to other approaches

Table 2. Comparison of the results achieved on the MIVIA road events data set with respect to existing methods. TPR refers to true positive rate, MDR to miss detection rate, ER to error rate and FPR to false positive rate.

Results comparison

	TPR	MDR	ER	FPR
COPE + max	86.75%	6.75%	6.5%	7.01%
σ	3.91	4.2	5.23	1.43
COPE + SVM	94%	4.75%	1.25%	3.94%
σ	4.3	4.92	1.25	1.82
BoF [5]	82%	17.75%	0.25%	2.85%
σ	7.79	8.06	1	2.52
BoF (Bark) [10]	80.25%	21.75%	3.25%	10.96%
σ	7.75	8.96	2.5	8.43
BoF (Mfcc) [10]	80.25%	19%	0.75%	7.69%
σ	11.64	11.63	0.96	5.92

are attributable to the fact that they implicitly encode information about the temporal evolution of instantaneous characteristics of the audio stream. This is in contrast with bag of features approaches, where the temporal information is completely lost. The trainable character of the COPE features contributes to the construction of a flexible system for detection of events of interest. Indeed, we configured a number of COPE features on training events of interest (i.e. various car crash and tire skidding events) and used them to detect the occurrence of similar events in unknown audio streams. In principle, one can configure filters that are able to detect other kinds of event and add them to the filterbank in order to extend the detection capabilities of the system. It is worth noting that a COPE feature is configured by only one training sample from which it is able to extract important features, which are later detected in unknown audio streams. This is in contrast with deep learning approaches, which usually require a large number of training samples to learn a robust set of features [4]. In the proposed *COPE+max* system, one can add more COPE feature detectors to an operating detection system without retraining the classifier. The performance of the *COPE+max* are, however, lower than the *COPE+SVM* approach that combines COPE feature vectors with an SVM classifier. In the latter case, the classification system cannot be extended with more feature detectors (e.g. for other classes of sounds of interest) as it need to be retrained. One can, thus, choose a more flexible system in cases where errors can be better handled, while the *COPE+SVM* system is suitable for situations where more precision and reliability is required.

The input features to a COPE feature extractor, namely the energy local maxima in the Gammatonegram of the sound, are reliable and robust to noise characteristic of audio signals [31]. The use of a Gammatone filterbank is not strictly coupled with COPE features, since one can decide to use a time-frequency representation of the sound that is more suitable for the particular application at hand (i.e. a spectrogram, wavelets, etc.). Moreover, the tolerance introduced in the application phase of a COPE feature accounts for detection of the prototype sound used for configuration and also modified versions of it due to noise and sound distortion. This implies that the proposed filters are robust to noise and locally provide good generalization capabilities. They can be employed together with machine learning techniques to further improve the performance and reliability of the system. In this work, indeed, we considered the filters in the filterbank to give equal contribution in the detection task and determined threshold values by analyzing the ROC curves achieved on the training set. However, different filters could contributed in different ways (i.e. with different weight) to the detection task. Thus, the use of machine learning techniques to determine a set of weights for the filters allows to construct a more robust and effective filterbank. As an example, the filter selection procedure proposed in [26] can be employed to reduce the number of concerned filters in the filterbank by selecting only the most relevant ones. Such procedure provides an optimized filterbank with high effectiveness and lower computational resource demand.

The computation of the response of a COPE feature can be paralleled, since the responses of its tuples are independent from each other. Moreover, given the Gammatonegram transformation of the input audio signal, the responses of the filters in the filterbank can also implemented on a parallel architecture.

5   Conclusions

In this paper, we proposed a system for audio event detection that is based on COPE features to extract important features from the input audio signals. COPE features are trainable and rely on the detection of specific constellation of energy peaks in a time-frequency representation of the audio signal. Their structure is determined during an automatic configuration process, given prototype audio events of interest. The flexibility introduced by COPE features results in an adaptive system that can be easily extended to detect various other events of interest. We also demonstrated that COPE feature extractors can be used to design more reliable sound recognition systems (with less scalability) by combining their responses with a multi-class classifier. The performance results that we achieved on the MIVIA road events data set are promising and higher than other existing methods. The good generalization capabilities observed in the experiments make the system suitable for use in real-world applications.

References

1. Aytar, Y., Vondrick, C., Torralba, A.: SoundNet: learning sound representations from unlabeled video. In: NIPS 2016 (2016)
2. Azzopardi, G., Petkov, N.: Trainable COSFIRE filters for keypoint detection and pattern recognition. IEEE Trans. Pattern Anal. Mach. Intell. **35**(2), 490–503 (2013). https://doi.org/10.1109/TPAMI.2012.106

3. Azzopardi, G., Strisciuglio, N., Vento, M., Petkov, N.: Trainable COSFIRE filters for vessel delineation with application to retinal images. Med. Image Anal. **19**(1), 46–57 (2015)

4. Bengio, Y.: Learning deep architectures for AI. Found. Trends® Mach. Learn. **2**(1), 1–127 (2009). https://doi.org/10.1561/2200000006

5. Carletti, V., Foggia, P., Percannella, G., Saggese, A., Strisciuglio, N., Vento, M.: Audio surveillance using a bag of aural words classifier. In: IEEE AVSS, pp. 81–86 (2013)

6. Clavel, C., Ehrette, T., Richard, G.: Events detection for an audio-based surveillance system. In: ICME, pp. 1306–1309 (2005)

7. Conte, D., Foggia, P., Percannella, G., Saggese, A., Vento, M.: An ensemble of rejecting classifiers for anomaly detection of audio events. In: IEEE AVSS, pp. 76–81 (2012)

8. Crocco, M., Cristani, M., Trucco, A., Murino, V.: Audio surveillance: a systematic review. ACM Comput. Surv. **48**(4), 52:1–52:46 (2016)

9. Do, H.M., Sheng, W., Liu, M.: Human-assisted sound event recognition for home-service robots. Robot. Biomim. **3**(1), 7 (2016). https://doi.org/10.1186/s40638-016-0042-2

10. Foggia, P., Petkov, N., Saggese, A., Strisciuglio, N., Vento, M.: Audio surveillance of roads: a system for detecting anomalous sounds. IEEE Trans. Intell. Transp. Syst. **17**(1), 279–288 (2016). https://doi.org/10.1109/TITS.2015.2470216

11. Foggia, P., Saggese, A., Strisciuglio, N., Vento, M.: Cascade classifiers trained on gammatonegrams for reliably detecting audio events. In: IEEE AVSS, pp. 50–55 (2014)

12. Foggia, P., Petkov, N., Saggese, A., Strisciuglio, N., Vento, M.: Reliable detection of audio events in highly noisy environments. Pattern Recogn. Lett. **65**, 22–28 (2015). https://doi.org/10.1016/j.patrec.2015.06.026

13. Foggia, P., Saggese, A., Strisciuglio, N., Vento, M., Vigilante, V.: Detecting sounds of interest in roads with deep networks. In: Ricci, E., Rota Bulò, S., Snoek, C., Lanz, O., Messelodi, S., Sebe, N. (eds.) ICIAP 2019. LNCS, vol. 11752, pp. 583–592. Springer, Cham (2019). https://doi.org/10.1007/978-3-030-30645-8_53

14. Gerosa, L., Valenzise, G., Tagliasacchi, M., Antonacci, F., Sarti, A.: Scream and gunshot detection in noisy environments. In: Proceedings of the EURASIP European Signal Processing Conference, Poznan, Poland (2007)

15. Grzeszick, R., Plinge, A., Fink, G.A.: Temporal acoustic words for online acoustic event detection. In: Gall, J., Gehler, P., Leibe, B. (eds.) GCPR 2015. LNCS, vol. 9358, pp. 142–153. Springer, Cham (2015). https://doi.org/10.1007/978-3-319-24947-6_12

16. Imoto, K., Kyochi, S.: Sound event detection using graph Laplacian regularization based on event co-occurrence. CoRR abs/1902.00816 (2019). http://arxiv.org/abs/1902.00816

17. López-Antequera, M., Leyva Vallina, M., Strisciuglio, N., Petkov, N.: Place and object recognition by CNN-based COSFIRE filters. IEEE Access **7**, 66157–66166 (2019)

18. Mesaros, A., Heittola, T., Virtanen, T.: Metrics for polyphonic sound event detection. Appl. Sci. **6**(6), 162 (2016). https://doi.org/10.3390/app6060162

19. Ntalampiras, S., Potamitis, I., Fakotakis, N.: An adaptive framework for acoustic monitoring of potential hazards. EURASIP J. Audio Speech Music Process. **2009**, 13:1–13:15 (2009). https://doi.org/10.1155/2009/594103

20. Pancoast, S., Akbacak, M.: Bag-of-audio-words approach for multimedia event classification. In: Proceedings of the Interspeech 2012 Conference (2012)

21. Patterson, R.D., Robinson, K., Holdsworth, J., Mckeown, D., Zhang, C., Allerhand, M.: Complex Sounds and auditory images. In: Auditory Physiology and Perception, pp. 429–443 (1992)
22. Plinge, A., Grzeszick, R., Fink, G.A.: A bag-of-features approach to acoustic event detection. In: IEEE ICASSP, pp. 3704–3708 (2014)
23. Rabaoui, A., Davy, M., Rossignol, S., Ellouze, N.: Using one-class SVMs and wavelets for audio surveillance. IEEE Trans. Inf. Forensics Secur. 3(4), 763–775 (2008). https://doi.org/10.1109/TIFS.2008.2008216
24. Ravanelli, M., Bengio, Y.: Speaker recognition from raw waveform with SincNet. In: 2018 IEEE Spoken Language Technology Workshop (SLT), pp. 1021–1028 (2018)
25. Saggese, A., Strisciuglio, N., Vento, M., Petkov, N.: Time-frequency analysis for audio event detection in real scenarios. In: AVSS, pp. 438–443, August 2016. https://doi.org/10.1109/AVSS.2016.7738082
26. Strisciuglio, N., Azzopardi, G., Vento, M., Petkov, N.: Supervised vessel delineation in retinal fundus images with the automatic selection of B-COSFIRE filters. Mach. Vis. Appl. 1–13 (2016). https://doi.org/10.1007/s00138-016-0781-7
27. Strisciuglio, N., Vento, M., Petkov, N.: Learning representations of sound using trainable COPE feature extractors. Pattern Recogn. 92, 25–36 (2019). https://doi.org/10.1016/j.patcog.2019.03.016
28. Takahashi, N., Gygli, M., Gool, L.V.: AENet: learning deep audio features for video analysis. IEEE Trans. Multimed. 20(3), 513–524 (2018). https://doi.org/10.1109/TMM.2017.2751969
29. Turpault, N., Serizel, R., Parag Shah, A., Salamon, J.: Sound event detection in domestic environments with weakly labeled data and soundscape synthesis, June 2019
30. Valenzise, G., Gerosa, L., Tagliasacchi, M., Antonacci, F., Sarti, A.: Scream and gunshot detection and localization for audio-surveillance systems. In: IEEE AVSS, pp. 21–26 (2007)
31. Wang, A.L., Th Floor Block F: An industrial-strength audio search algorithm. In: Proceedings of the 4th International Conference on Music Information Retrieval (2003)

Incremental Training of Neural Network for Motor Tasks Recognition Based on Brain-Computer Interface

Nayid Triana Guzmán[1]([✉]) [iD], Álvaro David Orjuela-Cañón[2] [iD], and Andrés Leonardo Jutinico Alarcon[1] [iD]

[1] Universidad Antonio Nariño, Bogotá, D.C., Colombia
natriana@uan.edu.co
[2] Universidad del Rosario, Bogotá, D.C., Colombia
alvaro.orjuela@urosario.edu.co

Abstract. Brain-computer interfaces (BCI) based on motor imagery tasks (MI) have been established as a promising solution for restoring communication and control of people with motor disabilities. Physically impaired people may perform different motor imagery tasks which could be recorded in a non-invasive way using electroencephalography (EEG). However, the success of the MI-BCI systems depends on the reliable processing of the EEG signals and the adequate selection of the features used to characterize the brain activity signals for effective classification of MI activity and translation into corresponding actions. The multilayer perceptron (MLP) has been the neural network most widely used for classification in BCI technologies. The fact that MLP is a universal approximator makes this classifier sensitive to overtraining, especially with such noisy, non-linear, and non-stationary data as EEG. Traditional training techniques, as well as more recent ones, have mainly focused on the machine-learning aspects of BCI training. As a novel alternative for BCI training, this work proposes an incremental training process. Preliminary results with a non-disabled individual demonstrate that the proposed method has been able to improve the BCI training performance in comparison with the cross-validation technique. Best results showed that the incremental training proposal allowed an increase of the performance by at least 10% in terms of classification compared to a conventional cross-validation technique, which indicates the potential application for classification models of BCI's systems.

Keywords: Brain-computer interface · Electroencephalography · Motor imagery · Multilayer perceptron · Incremental training · Cross-validation

1 Introduction

A Brain-Computer Interface (BCI) is a communication and control system that enables users to operate devices with their brain signals [1]. These systems can be used to command different technologies that focus on the assistance and rehabilitation of people with motor disabilities, such as prostheses, orthoses, robotic exoskeletons, and motor neuroprostheses, among others [2–5].

© Springer Nature Switzerland AG 2019
I. Nyström et al. (Eds.): CIARP 2019, LNCS 11896, pp. 610–619, 2019.
https://doi.org/10.1007/978-3-030-33904-3_57

BCI detects the movement intentions of a user from a variety of different electrophysiological signals, for example, with electroencephalography (EEG) [6]. BCI literature presents a growing number of publications on the acquisition of these signals to extract and to classify information based on motor imagery (MI) for upper limbs [7]. In this sense, it has been demonstrated that movement intentionality can be decoded using EEG signal processing, which has inherent advantages in neural engineering applications [8].

In recent years, research studies have considered machine learning as a useful tool for BCIs [9]. In the conventional approach, the BCI is first trained to decode a subject's mental intentions [10] mainly through the use of machine learning methods such as support vector machines (SVM), artificial neural networks (ANN) such as multilayer perceptron (MLP), or random forest (RF) strategies [11–14].

Different efforts focused on incremental training methods for machine learning models have considerably improved the performance and strategies for modifying the parameters of the models to speeding up the process of learning [15–17]. Thus, these alternative proposals have been applied to BCI environments with specific objectives, such as the use of magneto-encephalography (MEG) information [18], experiments for P300 specific task [19], and adaptive space filtering processes [20]. However, these incremental training strategies have been used in particular case with few applications in motor imagery (MI) tasks and small datasets contexts.

Present work establishes a novel alternative for the training of a classifier model in the context of MI-BCI from EEG signals recorded in a real experiment. The experimental study was devised to obtain EEG signals from a non-disabled user performing MI tasks in a BCI environment. The goal is to implement and evaluate the classification performance of the proposed incremental training process in comparison with the traditional cross-validation technique in order to find the best model from different architectures of ANN. For this, the MLP network was employed, according to its most widely used for BCI technologies [21].

The paper is organized as follows: Sect. 2 describes the experiments developed to acquire EEG signals in an MI-BCI session, the feature extraction of the EEG signals, the pattern recognition with the MLP, the incremental training and the evaluation process; Sect. 3 presents and discusses the results; Sect. 4 presents the conclusions.

2 Methods and Materials

The first Subsect. 2.1 describes the EEG dataset obtained from a volunteer who performed the MI tasks of the upper limbs in the BCI experiment. The second Subsect. 2.2, provides the analysis and preprocessing of the data. In the third Subsect. 2.3, the feature extraction and selection process are described. In Subsect. 2.4, the classification model based on MLP that was used to recognize motor imagery tasks from EEG signals and the incremental training process are evaluated. Lastly, in Subsect. 2.5, the performance evaluation is provided.

2.1   EEG Dataset from MI-BCI Experiment

The MI-BCI experiment to obtain the EEG dataset was carried out on the baseline of [22]. Before beginning the experiment, the participant was duly informed of the aim of the study and freely signed a consent form. The subject (a 30-year-old man) was comfortably seated about 1 m away from a 32" screen size LCD TV displaying the training protocol. Thirty-two active EEG channels over the sensorimotor cortex were wirelessly recorded, employing the 10-20 standard setup (FP1, FP2, AF3, AF4, F7, F3, Fz, F4, F8, FC5, FC1, FC2, FC6, T7, C3, Cz, C4, T8, CP5, CP1, CP2, CP6, P7, P3, Pz, P4, P8, PO7, PO3, PO4, PO8, Oz, GND, reference on right earlobe), with a g.Nautilus system (g.Tec medical engineering GmbH, Schiedlberg, Austria) at 250 Hz. The presentation of the visual stimuli and the recording of the EEG signals were performed from the g.Nautilus system via BCI2000 software (version 2.0). The electrodes impedance measurement was kept below 30 kΩ with the g.NEEDaccess application. Finally, signal processing was developed in MATLAB (ver. R2019a, The MathWorks, Inc., Natick, MA, USA).

The experiment lasted approximately 20 min, and it took place in 1 MI-BCI session that consisted of a motor imagery task performed with either the left hand or the right hand. The subject had to conduct many trials composed of three consecutive stages: a resting stage, a motor imagery stage, and a relaxing stage. Each of these stages had a duration of three seconds; therefore, the length of a trial is nine seconds. Figure 1 illustrates the temporal sequence of the two types of trial. At Stage-1, in the rest period, the screen displays the word "Rest" for three seconds, and the subject must maintain a natural body position without performing or imagining any body movement. Once three seconds have passed, the subject proceeds with Stage-2 of motor imagery, where the screen displays the image of an arrow pointing, either right (Fig. 1a) or left (Fig. 1b), and the subject must imagine closing and opening the right or the left hand, respectively. The subject was requested to avoid blinking or performing any movement during the rest and the motor imagery periods. After the three seconds of motor imagery, the subject can proceed with the relax period at Stage-3, where the word "Relax" appears on the screen, and the subject can move and blink. Immediately after, another trial starts at Stage-1 and so on, until 12 trials are completed.

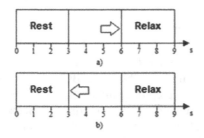

Fig. 1. The temporal sequence of the two types of trial.

The training protocol was executed in runs of 12 trials, and the subject could take a break for a few seconds following each run. Each run lasted about two minutes, including the break. In total, ten runs were recorded, so in turn, 120 trials were registered, which required 20 min. The measurements of the dataset were recorded randomly in the same proportion for the stages of motor imagery (MI), closing the left and the right hand.

2.2 Preprocessing

After the experimental session, to eliminate the noise of blinking or real movement, the relax stage was discarded and not used in the rest of the study. The EEG signals of MI closing the left hand, MI closing the right hand, and "Rest" were separated into 3-classes of BCI, respectively. C3, Cz, and C4 channels were selected for signal processing due to the electrodes' low impedance (≤ 5 kΩ) and their relationship with the most significant sensorimotor areas for discrimination between different motor imagery tasks with the right and the left hand [23]. Then, a Butterworth bandpass filter from 0.5 to 60 Hz was applied to the recorded EEG signals in such channels.

2.3 Feature Extraction and Selection

The act of imagining voluntary movements can be detected in the amplitude of the α (7 Hz–13 Hz), β (14 Hz–35 Hz), or both frequency ranges, from the EEG recordings over the primary motor areas (electrode positions C3, Cz, and C4) [24, 25]. Therefore, for each of these selected electrodes, the EEG signal was bandpass filtered to extract α and β frequency bands obtaining six features (three electrodes $\times$ two frequency bands). Feature extraction was implemented at a rate of 1 Hz (one-second-long window with 50% overlapping), where the number of spectral power values by electrode was 30. These spectral power values $P(v)$ were computed for every window $v \in \{1, 2, 3, 4, 5\}$ of $N - 245$ samples, using the expression:

$$P(v) = \frac{\sum_{n=1}^{N} |B_n(v)|}{\sum_{m=1}^{M} |S(m)|} \tag{1}$$

where $B_n(v)$ is the EEG signal filtered in the α or β frequency band, and $S(m)$ is the EEG signal filtered between 0.5 and 60 Hz of $M = 744$. samples. As a result, the feature vector has three classes corresponding to the periods of rest, right-hand motor imagery, and left-hand motor imagery; each one of the periods has 6 features $\times$ 30 spectral power values.

2.4 Classification and Incremental Training

Once the feature vector was obtained, a multilayer perceptron (MLP) was employed to classify the patterns or changes of spectral power values for each of the three mental tasks performed by the BCI user. The MLP is an artificial neural network (ANN) which has only feedforward connections and is trained in a supervised way. MLP architecture consists of an input, a hidden layer, and an output layer. The number of nodes in the input is given by the number of variables - six features in the present study. The number of neurons in the hidden layer was established through adopting a heuristic approach, from experimenting through one to ten units. Lastly, in the output layer, we used three neurons representing the BCI classes: rest, right-hand motor imagery, and left-hand motor imagery. Typically, one hidden layer and one output layer are enough to solve classification problems [26].

The resilient backpropagation algorithm (Rprop) was applied for the MLP training due to its fast and stable convergence [27]. Hyperbolic tangent functions were used in the hidden neurons as activation functions allowing a wider range of the output from the network. In the output layer, the softmax transfer function was used to calculate the probability to have the events +1 or −1 of each BCI class. In this way, the neuron output value is +1 when the input belongs to one BCI class, and −1 when it does not belong. For each hidden neuron, 100 initializations of the synaptic weights were assigned randomly to avoid the effect of initial conditions in the network training. From all the experimented hidden layers, testing from 1 to 10 the number of neurons in the hidden layer, the best models were selected from the maximum values in terms of performance (classification rate) given by all of them.

Searching to improve the performance in terms of classification of the MLP, an incremental training process was tested. Figure 2 describes the incremental training for Run 1.

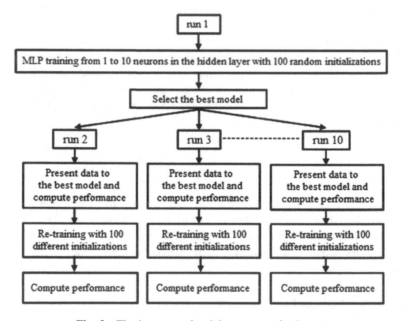

Fig. 2. The incremental training process for Run 1.

The first step consists of the MLP training from 1 to 10 neurons in the hidden layer with 100 random initializations of the synaptic weights, a training set of 20 samples, and measuring the generalization with ten samples to obtain the best model of a first-run independently. The second step was developed after the training with the first run. For this, after the first run training, the best model was stored and employed the data in the next run to be developed an analysis. First, the new data was presented to the model, and the performance was computed. Next, a re-training with 100 different initializations was implemented to see if the classification rate increased or did not. The second step was repeated with all the following nine runs. At the end of the process, the incremental training begins again at the first step with the data of Run 2. Once the sequence finishes, the process restarts, but with the data of Run 3 and so on, up until Run 10.

2.5   Evaluation

The MLP classification was evaluated from the 20 samples assigned to the training set - 70% for training, and 30% for validation and early stopping - because the MLP must generalize its performance to new inputs to avoid overtraining. The maximum number of epochs needed for the training stop of the MLP was set to 1000. On the other hand, another criterion for the early stopping occurs when validation performance has increased by more than 200 times since the last time it decreased (a condition using maximum validation failures).

After training, a traditional cross-validation technique with ten folds was used to compare the results of the classification with the incremental training proposal [28]. In each case, the metric used to assess the performance accuracy was the percentage of correct classifications for all classes. This percentage was computed separately for each unit in the hidden layer and each run in the incremental training proposal in terms of re-training.

3   Results

Figure 3 shows the results for the traditional cross-validation technique when a run is developed in an independent way.

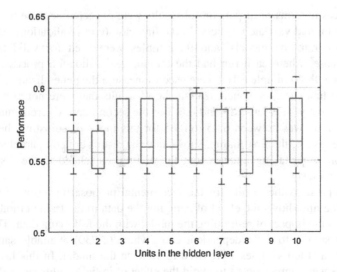

Fig. 3. Results for the validation set for cross-validation technique.

The results are shown in terms of the classification rate that is labeled as "performance." Box plots were employed to visualize the results for the ten folds validation where each run was taken as a set out-of-sample for the validation process. Each box represents the number of units used in the hidden layer.

Figure 4 shows the results for the current incremental proposal when the training and validation were developed in each run. As mentioned, the dataset, in this case, was composed of fewer examples, which led to the dispersion in the figure is wider in comparison to the exposed previous case.

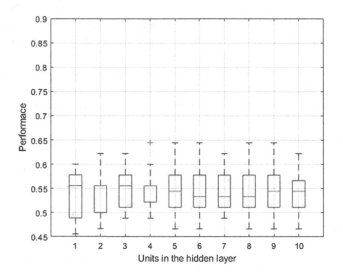

Fig. 4. Results for the validation set for the incremental training proposal.

Differences in terms of dispersion for results were expected because the number of samples to train and validate was less. In the first case (cross-validation), 810 samples were used to train the models, and 90 samples were used for validation. In the incremental case, where each run had the training and validation process, the model was adjusted with 20 samples. Ten were used to measure the generalization. According to this, in the first case, the generalization shows results that were not promising. The values were mostly between 0.55 and 0.6. For the second case (incremental training), the generalization was between 0.45 to 0.65 for this mode of adjusting the model. In light of these results, the first finding shows that the results are comparable in terms of generalization while taking into account the differences related to the dispersion, as explained previously.

Figure 5 presents the results for the incremental proposal in terms of re-training, where the blue line shows the effect of applying the data to the trained model. The red line visualizes the impact of re-training the model with the following data. The mode of training was similar to the independent case, where (20/30) available samples were used to train, and ten samples were used to validate the model. In this last case, 100 initializations were implemented to avoid the effect of initial conditions in the synaptic weights. From each collection of 100 training, the maximum values in terms of performance were selected to compute the figure.

It is possible to see that re-training has a positive effect - it increases the performance in terms of classification. This increment reaches at least 10% (see Fig. 5), changing from 0.55 to 0.68, and as a result, improves the classification. Another aspect is related to the dispersion of the results. For two first runs, there is overlapping for the standard deviation, showing similarities between the results. This event was decreasing for the last runs, showing a smaller quantity in the error bar.

One interesting fact to note is the possibility of observing what runs are more challenging to classify. In Fig. 5, the run number seven obtained a classification rate out of tendency for both cases. This result means that the present proposal also permits to evaluate and find the problematic runs, providing more information that can be useful to determine problems in the generalization of models.

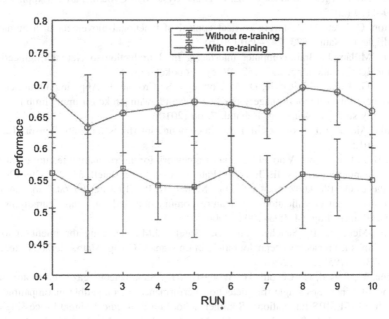

Fig. 5. Results comparison for the incremental proposal in terms of re-training. (Color figure online)

4   Conclusions

This paper introduced an incremental training proposal for BCI systems, where the re-training strategy improves the results of a classification model in at least 10%. Results show that there are no notable differences between a traditional cross-validation technique and an independent way in terms of runs, according to the generalization of the classification models. Considering these preliminary results, and as future work, more classifiers will be implemented to determine if artificial neural networks offer more advantages in comparison with other classification algorithms.

Acknowledgments. This research is supported by Doctorados Nacionales 2017 – Conv 785 funded by COLCIENCIAS and Universidad Antonio Nariño under grants 2018213.

References

1. Wolpaw, J.R., Birbaumer, N., McFarland, D.J., Pfurtscheller, G., Vaughan, T.M.: Brain-computer interfaces for communication and control. Clin. Neurophysiol. **113**, 767–791 (2002)
2. Hochberg, L.R., et al.: Neuronal ensemble control of prosthetic devices by a human with tetraplegia. Nature **442**, 164–171 (2006)
3. Daly, J.J., Wolpaw, J.R.: Brain–computer interfaces in neurological rehabilitation. Lancet Neurol. **7**, 1032–1043 (2008)
4. Frisoli, A., et al.: A new gaze-BCI-driven control of an upper limb exoskeleton for rehabilitation in real-world tasks. IEEE Trans. Syst. Man Cybern. Part C (Appl. Rev.) **42**, 1169–1179 (2012)
5. Bouton, C.E., et al.: Restoring cortical control of functional movement in a human with quadriplegia. Nature **533**, 247–250 (2016)
6. del R. Millán, J.: Brain-computer interfaces. In: Introduction to Neural Engineering for Motor Rehabilitation, pp. 237–252. Wiley, Hoboken (2013)
7. Prasad, G., Herman, P., Coyle, D., McDonough, S., Crosbie, J.: Applying a brain-computer interface to support motor imagery practice in people with stroke for upper limb recovery: a feasibility study. J. Neuroeng. Rehabil. **7**, 60 (2010)
8. Nicolas-Alonso, L.F., Gomez-Gil, J.: Brain computer interfaces, a review. Sensors **12**, 1211–1279 (2012)
9. Suk, H.-I., Lee, S.-W.: A novel bayesian framework for discriminative feature extraction in brain-computer interfaces. IEEE Trans. Pattern Anal. Mach. Intell. **35**, 286–299 (2013)
10. Dornhege, G., Blankertz, B., Curio, G., Muller, K.-R.: Boosting bit rates in noninvasive EEG single-trial classifications by feature combination and multiclass paradigms. IEEE Trans. Biomed. Eng. **51**, 993–1002 (2004)
11. Gudiño-Mendoza, B., Sánchez-Ante, G., Antelis, J.M.: Detecting the intention to move upper limbs from electroencephalographic brain signals. Comp. Math. Methods Med. **2016**, 3195373:1–3195373:11 (2016)
12. Meena, Y., Prasad, G., Cecotti, H., Wong-Lin, K.: Simultaneous gaze and motor imagery hybrid BCI increases single-trial detection performance: a compatible incompatible study. In: 9th IEEE-EMBS International Summer School on Biomedical Signal Processing (2015)
13. Hamedi, M., Salleh, S.-H., Noor, A.M., Mohammad-Rezazadeh, I.: Neural network-based three-class motor imagery classification using time-domain features for BCI applications. In: 2014 IEEE Region 10 Symposium, pp. 204–207 (2014)
14. Atkinson, J., Campos, D.: Improving BCI-based emotion recognition by combining EEG feature selection and kernel classifiers. Expert Syst. Appl. **47**, 35–41 (2016)
15. Lan, Y., Soh, Y.C., Huang, G.-B.: Ensemble of online sequential extreme learning machine. Neurocomputing **72**, 3391–3395 (2009)
16. Liang, N.-Y., Huang, G.-B., Saratchandran, P., Sundararajan, N.: A fast and accurate online sequential learning algorithm for feedforward networks. IEEE Trans. Neural Netw. **17**, 1411–1423 (2006)
17. Nastac, D.-I., Matei, R.: Fast retraining of artificial neural networks. In: Wang, G., Liu, Q., Yao, Y., Skowron, A. (eds.) RSFDGrC 2003. LNCS, vol. 2639, pp. 458–461. Springer, Heidelberg (2003). https://doi.org/10.1007/3-540-39205-X_77

18. Spüler, M., Rosenstiel, W., Bogdan, M.: Adaptive SVM-based classification increases performance of a MEG-based Brain-Computer Interface (BCI). In: Villa, A.E.P., Duch, W., Érdi, P., Masulli, F., Palm, G. (eds.) ICANN 2012. LNCS, vol. 7552, pp. 669–676. Springer, Heidelberg (2012). https://doi.org/10.1007/978-3-642-33269-2_84

19. Yokoi, T., Yoshikawa, T., Furuhashi, T.: Incremental learning to reduce the burden of machine learning for P300 speller. In: The 6th International Conference on Soft Computing and Intelligent Systems, and The 13th International Symposium on Advanced Intelligence Systems, pp. 167–170 (2012)

20. Woehrle, H., Krell, M.M., Straube, S., Kim, S.K., Kirchner, E.A., Kirchner, F.: An adaptive spatial filter for user-independent single trial detection of event-related potentials. IEEE Trans. Biomed. Eng. **62**, 1696–1705 (2015)

21. Lotte, F., et al.: A review of classification algorithms for EEG-based brain-computer interfaces. IOP Publishing (2007)

22. Orjuela-Cañón, A.D., Renteria-Meza, O., Hernández, L.G., Ruíz-Olaya, A.F., Cerquera, A., Antelis, J.M.: Self-organizing maps for motor tasks recognition from electrical brain signals. In: Mendoza, M., Velastín, S. (eds.) CIARP 2017. LNCS, vol. 10657, pp. 458–465. Springer, Cham (2018). https://doi.org/10.1007/978-3-319-75193-1_55

23. Pfurtscheller, G., Brunner, C., Schlögl, A., Lopes da Silva, F.H.: Mu rhythm (de) synchronization and EEG single-trial classification of different motor imagery tasks. Neuroimage **31**, 153–159 (2006)

24. Marquez-Chin, C., Marquis, A., Popovic, M.R.: EEG-triggered functional electrical stimulation therapy for restoring upper limb function in chronic stroke with severe hemiplegia. Case Rep. Neurol. Med. **2016**, 9146213 (2016)

25. Pfurtscheller, G., Linortner, P., Winkler, R., Korisek, G., Müller-Putz, G.: Discrimination of motor imagery-induced EEG patterns in patients with complete spinal cord injury. Comput. Intell. Neurosci. **2009**, 104180 (2009)

26. Haykin, S.S., Haykin, S.S.: Neural Networks and Learning Machines. Prentice-Hall/Pearson, Upper Saddle River (2009)

27. Riedmiller, M., Braun, H.: A direct adaptive method for faster backpropagation learning: the RPROP algorithm. In: IEEE International Conference on Neural Networks - Conference Proceedings, pp. 586–591. IEEE (1993)

28. Kohavi, R.: A study of cross-validation and bootstrap for accuracy estimation and model selection (1995)

Supervised Relevance Analysis for Multiple Stein Kernels for Spatio-Spectral Component Selection in BCI Discrimination Tasks

Camilo López-Montes[1], David Cárdenas-Peña[2(✉)],
and G. Castellanos-Dominguez[1]

[1] Signal Processing and Recognition Group, Universidad Nacional de Colombia,
Manizales, Colombia
{julopezm,cgcastellanosd}@unal.edu.co
[2] Automatic Research Group, Universidad Tecnológica de Pereira, Pereira, Colombia
dcardenasp@utp.edu.co

Abstract. Brain-Computer Interfaces (BCI) allow direct communication from the human brain to machines by analyzing sensorimotor activity. Within the most common paradigms of cognitive neuroscience for BCI, the Motor Imagery one relies on imaging movement of hands, feet, and tongue, usually. Standard BCI systems make use of electroencephalographic signals due to its high temporal resolution, portability, and easiness to implement. Recently, the filter-banked analysis of common spatial patterns has proved to be among the most discriminative approaches. Nonetheless, such an approach results in high dimensional spaces yielding to overtrained systems, or ill-posed spatial covariance matrices, when the number of EEG sensors increases. In this work, we propose a spatio-spectral relevance analysis, termed Multiple Stein Kernels (MSK), to simultaneously select channels and bands in a supervised criterion, so reducing the spatial and frequency dimension. We test our approach in three known BCI datasets with a varying number of subjects and channels. The attained classification results prove that MKS not only improves the classification accuracy and reduces the dimension of the representation space, but also select frequency bands that are physiologically interpretable.

1 Introduction

Nowadays, there are many challenges in rehabilitation in the development of systems that enhance communication between the brain and the exterior environment. Conventional Brain-Computer Interfaces (BCI) have been widely used to assist disabled people to re-establish their capabilities of environmental control decoding the brain activity [7], being the electroencephalography (EEG) signal the most commonly used acquisition technique due to its implementation easiness and noninvasive implementation [4].

© Springer Nature Switzerland AG 2019
I. Nyström et al. (Eds.): CIARP 2019, LNCS 11896, pp. 620–628, 2019.
https://doi.org/10.1007/978-3-030-33904-3_58

Among the BCI systems, the Motor Imagery (MI), corresponding to imagining a motor action without execution, is the most known paradigm. Feature extraction approaches, as common spatial patterns (CSP), are mainly devoted to MI classification tasks by reducing the characteristic noise, while highlighting mental imagination patterns. Further, the spectral CSP extensions try to find the frequency band with the highest performance focusing on the relationship between CSP features and the frequency. For instance, Sparse Filter Band Common Spatial Pattern (SFBCSP) performs a sparse regression to select the most discriminant bands from the full channel set [8]. Nonetheless, CSP-based approaches demand a linear projection of channels that hinders non-linear data distributions [6]. Recently, the Riemannian geometry has improved the performance of conventional BCI algorithms relying on the EEG covariance as the feature representation, so avoiding any linearity assumption. In fact, spatial covariance matrices carry useful information for MI, namely, their diagonal retrieves the sensor power, and the off-diagonal entries give extra information about channel correlations. However, in the case of EEG-based MI classification, the Riemannian geometry suffers from the curse of dimensionality. That is, the number of sensors hampers the classification performance, due to more samples are needed to build non-singular covariance matrices. When nearly singular covariance matrices are produced, their ill-posing cannot be efficiently handled via Riemannian geometry [5].

For overcoming the above issues, we propose a spatio-spectral relevance analysis simultaneously selecting channels and bands in a supervised scheme. The approach, termed Multiple Stein Kernel (MSK), reduces the spatial and frequency dimension by decreasing the chances of an EEG recording to result in an ill-posed covariance matrix. Then, the Stein kernel, that assesses the inner-product for covariance matrix spaces overcomes feature extraction stages. Since the Stein kernel is applied for each frequency band, a multiple kernel learning strategy provides a single similarity measure to feed classification machines. Attained results prove that the introduced relevance analysis improves the classification performance, and enhances the representation of the MI paradigm while reducing the spatio-spectral components.

2 Materials and Methods

2.1 EEG Filter-Bank Decomposition

Let $\boldsymbol{X} = \{\boldsymbol{x}_n^c \in \mathbb{R}^T : n \in N, c \in C\}$ be a set of N acquired EEG recordings of length T along C channels, where each recording is provided with a label $l_n \in \{-1, +1\}$. Each recording is further passed through a bank of B linearly distributed filters of elemental bandwidth $\Delta_\omega \in \mathbb{R}^+$ and overlap $\delta_\omega \in \mathbb{R}^+$. Thus, the set of bandpass-filtered EEG data is obtained, $\boldsymbol{X} = \{\boldsymbol{x}_{nb}^c \in \mathbb{R}^T : b \in B\}$. As recommended in the literature, we consider $B = 17$ band-pass filters with a bandwidth of $\Delta_\omega = 4\,\mathrm{Hz}$ aiming to cover the whole relevant EEG frequencies, ranging from 4 to 40 Hz and fixing the overlap between each other at $\delta_\omega = 2\,\mathrm{Hz}$.

2.2   Proposed Relevance Analysis

To account the relevance of a single channel at each frequency band, we build a relevance index based on the cross-correlation operator $R\{\cdot,\cdot\} \in [-1, 1]$ and the Centered-Kernel Alignment measure. We compute the cross-correlation function between all record pairs for each band and channel following:

$$R_\tau\{\boldsymbol{x}_{nb}^c, \boldsymbol{x}_{mb}^c\} = \frac{\mathbb{E}_t\{x_{nb}^c(t)x_{mb}^c(t-\tau)\}}{\sqrt{\mathbb{E}_t\{(x_{nb}^c(t))^2\}\mathbb{E}_t\{(x_{mb}^c(t-\tau))^2\}}}; \forall b \in [1, B], c \in [1, C] \quad (1)$$

where $\mathbb{E}_z\{v(z)\}$ stands for the expectation of the vector v across the z axis and $x(t)$ denotes the EEG measure at time instant t. From the cross-correlation function we build a symmetric matrix $\boldsymbol{\Gamma}_b^c \in [0, 1]^{N \times N}$ with elements $\gamma_b^c(n, m)$:

$$\gamma_b^c(n, m) = \max_\tau |R_\tau\{\boldsymbol{x}_{nb}^c, \boldsymbol{x}_{mb}^c\}| \in [0, 1]. \quad (2)$$

Relying on the properties of the cross-correlation operator, the symmetric matrix $\boldsymbol{\Gamma}_b^c$ is also positive definite, allowing to account for the matrix similarity using the kernel-devoted metrics. Particularly, we assess the centered-kernel alignment (CKA) $\rho_b^c \in [0, 1]$ between $\boldsymbol{\Gamma}_b^c$ and $\boldsymbol{L} \in \mathbb{R}^{N \times N}$, that holds elements $L_{nm} = \delta(l_n - l_m)$ and accounts for the supervised information:

$$\rho_b^c = \frac{\langle \bar{\boldsymbol{\Gamma}}_b^c, \bar{\boldsymbol{L}} \rangle_F}{\sqrt{\|\bar{\boldsymbol{\Gamma}}_b^c\|_F \|\bar{\boldsymbol{L}}\|_F}}; \forall b \in [1, B], c \in [1, C], \quad (3)$$

where $\bar{\boldsymbol{L}}$ stand for the centered version of $\boldsymbol{L}$. Notations $\|\cdot\|_F$ and $\langle \cdot, \cdot \rangle_F$ correspond to the Frobenius norm and dot product, respectively. As a result, ρ_b^c measures the similarity between EEG data and their labels at spatial and frequency levels, allowing to rank each channel and band according its discriminating capacity.

2.3   Classification Based on Stein Kernel Representation

The channels of each band are ranked according the score given by the CKA criterion, in this way the most relevant channels are selected for each band, so that the spatial covariance only holds the most discriminative information as $\boldsymbol{\Sigma}_{nb} = \tilde{\boldsymbol{X}}_{nb}\tilde{\boldsymbol{X}}_{nb}^\top$, being $\tilde{\boldsymbol{X}}_{nb} \in \mathbb{R}^{C' \times T}$ corresponds to the n-th EEG recording at band b with $C' \geq C$ selected channels. Covariance matrices belong to a special Riemannian manifold in the space of symmetric matrices, that can be endowed with dissimilarity metrics to estimate the matrix distances. Here, we consider the Stein divergence, also known as LogDet Divergence, that defines the not geodesic distance between two symmetric positive definite (SPD) matrices $\boldsymbol{\Sigma}_n$ and $\boldsymbol{\Sigma}_m$ as follows:

$$d_{SD}^2(\boldsymbol{\Sigma}_{nb}, \boldsymbol{\Sigma}_{mb}) = \log \det\left(\frac{\boldsymbol{\Sigma}_{nb} + \boldsymbol{\Sigma}_{mb}}{2}\right) - \log \det\left(\frac{\boldsymbol{\Sigma}_{nb}\boldsymbol{\Sigma}_{mb}}{2}\right) \quad (4)$$

In the proposed methodology, we feed the Stein divergence into a radial basis function to build a Gaussian Kernel on the Riemannian geometry as:

$$\kappa(\tilde{\boldsymbol{X}}_{nb}, \tilde{\boldsymbol{X}}_{mb}) = \exp(-d^2_{SD}(\boldsymbol{\Sigma}_{nb}, \boldsymbol{\Sigma}_{mb})/2\sigma^2) \tag{5}$$

where the kernel $\kappa(\cdot,\cdot) \in [0,1]$ corresponds to the Stein kernel and $\sigma \in \mathbb{R}^+$ to its bandwidth. Since each selected band b' yields its own similarity, we introduce the multiple kernel learning to gather all similarity functions:

$$\kappa(\boldsymbol{X}_n, \boldsymbol{X}_m) = \sum_{b'=1}^{B'} w_{b'}\kappa(\tilde{\boldsymbol{X}}_{nb'}, \tilde{\boldsymbol{X}}_{mb'}) \tag{6}$$

where $w'_b \geq 0$ holds the mixture weights that must be optimized to improve discrimination between classes based on the label set. By constraining the weights to sum the unit, $\sum_{b'} w_{b'} = 1$, yields a convex function that has a closed form solution that maximizes the CKA criterion. Therefore, we use a support vector machine equipped with the proposed Multiple Stein Kernel (MSK) for covariance matrices. As a result, the proposed approach reduces the spatio-spectral components based on the relevance analysis, with the additional benefit of simplifying the BCI system as the explicit feature extraction stage is avoided.

3 Experimental Set-Up

3.1 Dataset Description

Being the most used MI paradigm in the literature, this work considers the binary classification of left and right movement for three datasets:

BCICIII4a: The BCI Competition III dataset IVa[1] contains the EEG signals recorded from five subjects performing two different (right hand and left hand) MI tasks. 118 channels at the positions of the extended international 10/20 system were used for measuring the EEG signal, sampled at a rate of 1000 Hz. The dataset contains a total of 280 trials for each subject with an equal number of trials for each task. The data provided were interval filtered with a passband of 0.05–200 Hz the down-sampled data at 100 Hz is used. Visual cues, lasting 3.5 s indicated which of the following three motor imageries the subject should perform: (L) left hand, (R) right hand, (F) right foot. The presentation of target cues was intermitted by periods of random length, 1.75 to 2.25 s, in which the subject could relax.

BCICIV2a: The BCI competition IV dataset IIa[2] contains a collection of EEG signals recorded using a 22-electrode montage from nine subjects that compose the dataset. Each subject performs two sessions on different days with four motor

---

[1] http://www.bbci.de/competition/iii/.
[2] http://www.bbci.de/competition/iv/.

imagery tasks. Namely, left hand, right hand, both feet, and tongue on two sessions. The session includes six runs with twelve trials per task, obtaining 144 trials for each class. The signals were sampled with 250 Hz and bandpass-filtered between 0.5 Hz and 100 Hz. The subjects were sitting in a comfortable armchair in front of a computer screen. At the beginning of a trial ($t = 0$ s), a fixation cross appeared on the black screen. Also, a short acoustic warning tone was presented. After two seconds ($t = 2$ s), a cue in the form of an arrow pointing either to the left, right, down or up (corresponding to one of the four classes left hand, right hand, foot or tongue) appeared and stayed on the screen for 1.25 s. This fact prompted the subjects to perform the desired motor imagery task. No feedback was provided. The subjects were asked to carry out the motor imagery task until the fixation cross disappeared from the screen at $t = 6$ s.

GIGASCIENCE: The GIGASCIENCE[3] dataset contains an EEG signal recorded conducting a BCI experiment for the MI movement of the left and right hands with 52 subjects [1] of which we selected 44 subject for the current experiment because the remainder subject holds a large number of trials with the artifact. EEG data were collected using 64 Ag/AgCl active electrodes 64-channel montage based on the international 10-10 system of sampling rate 512 Hz. The dataset contains 100 or 120 trials for each MI class. Each trial starts with the monitor showed a black screen with a fixation cross for $t = 2$ s then one of two instructions ("left hand" or "right hand") appeared randomly on the screen for $t = 3$ s, and subjects were asked to move the appropriate hand depending on the instruction given. After the movement, when the blank screen reappeared, the subject was given a break for a random, ranging from 4.1 to 4.8 s. These procedures were repeated 20 times for one class (one run), and one run was performed.

3.2   Parameter Tuning

The proposed method MKS requires two procedures of parameter tuning, namely, the kernel bandwidth $\sigma \in \mathbb{R}^+$ and the threshold $\lambda \in (0, 1)$ controlling the amount of selected spatio-spectral components. The parameter σ and λ are tuned based on a cross-validation grid search for the best classification accuracy. The spatio-spectral components used for the classification are selected according to the relevance analysis ranking. Figure 1 illustrates the learning curve for each subject in the BCICIV IIa dataset, marking the best number of components. Note that the maximum accuracy for each subject is achieved at a different amount of components, evidencing the need for a subject-wise tuning of λ to properly encode the discriminant brain activity.

Further, observing the learning curves, we identify two kinds of learning, one for subjects with high performance (S1, S3, S5, S7, S8, S9) and another that embraces all subjects with lower performance (S2, S4, S6). In the first kind, the accuracy monotonously grows and stabilizes at a few components, so

---

[3] http://gigadb.org/dataset/100295.

revealing the goodness of our approach for the selection of relevant components. In the second kind, there are abrupt changes induced by noisy components that disturb the classifier. However, the proposed relevance allowed to select a subset of components outperforming the full component set.

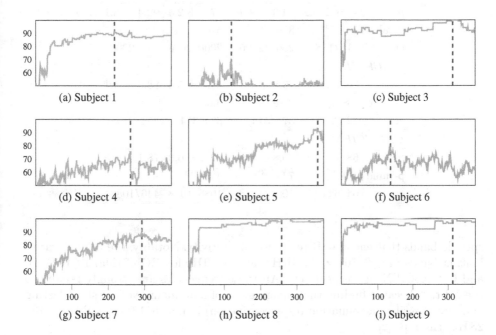

(a) Subject 1   (b) Subject 2   (c) Subject 3

(d) Subject 4   (e) Subject 5   (f) Subject 6

(g) Subject 7   (h) Subject 8   (i) Subject 9

Fig. 1. BCICIV2a learning curves. Accuracy versus the number of selected components. The best accuracy is marked by a red dashed line for each subject. (Color figure online)

3.3   Performance Results

We compare the MKS against the conventional CSP, Sparse Filter-bank (SFB) CSP [8], and the Stein kernel on a single band [3] for the considered datasets. Fivefold cross-validation is used to evaluate the performance of all experiments conducted using BCICIII4a, BCICIV2a, and GIGASCIENCE datasets. Table 1 summarizes the accuracy performed by each contrasted approach along with the standard deviation, the component reduction rate (crr), and the p-value of the significance test of the proposed approach against each competing one. Results evidence that MKS outperforms the baselines at all considered performance metrics for the three datasets. Particularly, our proposed approach obtains better accuracy, using fewer components that could be redundant or noisy for the MI task. Besides, obtaining a lower standard deviation proves that MKS also provides more confidence than the approaches of the state-of-the-art do.

Figure 2, showing the most selected bands per channel on each considered database, indicates that MKS suitably incorporates frequency information on

Table 1. Classification accuracy for the considered approaches on a five-fold cross-validation.

	CSP	Stein kernel	SFB	MKS
BCICIII IVa				
$\mu \pm \sigma$	87.6 ± 4.2	84.3 ± 4.4	87.7 ± 2.4	92.4 ± 3.1
p-value	$2.0e^{-3}$	$8.8e^{-6}$	$5.1e^{-3}$	–
crr	118/118	2006/2006	2006/2006	867/2006
BCICIV IIa				
$\mu \pm \sigma$	76.5 ± 5.6	79.8 ± 7.0	84.5 ± 5.4	88.6 ± 5.1
p-value	$4.8e^{-7}$	$2.1e^{-4}$	$6.6e^{-3}$	–
crr	22/22	374/374	374/374	252/374
GIGASCIENCE				
$\mu \pm \sigma$	68.9 ± 7.6	70.8 ± 6.6	72.2 ± 6.3	78.1 ± 6.5
p-value	$4.0e^{-7}$	$7.9e^{-28}$	$4.1e^{-13}$	–
crr	64/64	1088/1088	1088/1088	446/1088

specific bands that encode MI related brain activity. Notably, the most discriminating bands are 12–16 Hz, 14–16 Hz, and 8–10 Hz for BCICIII4a, BCICIV2a, and GIGASCIENCE, respectively. All these frequencies are strongly related to the MI task because finding that imagination of movement leads to short-lasting and circumscribed attenuation (or accentuation) in mu (8–12 Hz) and beta (13–28 Hz) rhythms [2].

4   Concluding Remarks and Future Work

This work proposes a spatio-spectral relevance analysis to select the most discriminating spatial bandpass filtered components of EEG recordings, termed MKS. As the accuracy results prove, the supervised relevance criterion information enhances the stability and the discriminative capability in the representation space. That is the case of the Stein kernel along with the Riemannian geometry that suffers from stability problems at large dimensionality, that in our application results from the number of sensors defining the size of covariance matrices. The proposed approach deals with such an issue by reducing the representation dimension. Also, MKS incorporates frequency information allowing the selection of bands that encode the information related to brain activity, as shown in Fig. 2. Attained results prove that the relevant information is found in bands 12–16 Hz, 14–16 Hz, and 8–10 Hz for BCIIII4a, BCIIV2a, and GIGASCIENCE, respectively. All those bands are strongly related to the MI task because the movement imagination leads to short-lasting and circumscribed attenuation (or accentuation) in μ and β rhythms. Including the above relevant frequency information in the feature selection stage improves the accuracy on subjects such as S5 and S2 of BCIIV2a. Therefore, our proposed approach obtains a better

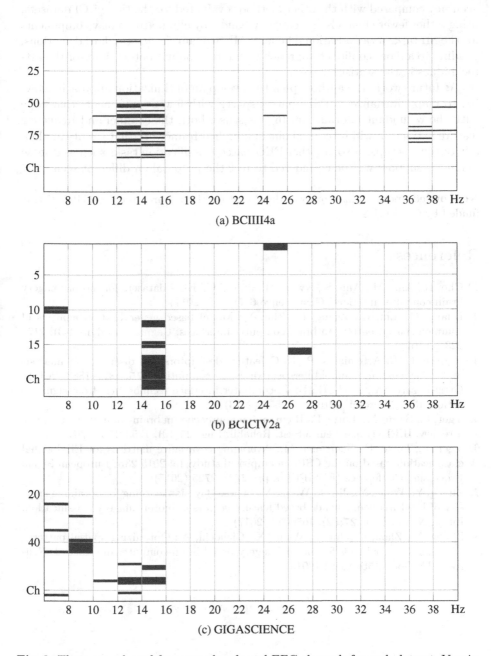

Fig. 2. The most selected frequency bands and EEG channels for each dataset. Y-axis represents the channels distributed from frontal (top) to occipital (bottom). X-axis indicates the center of the frequency band.

accuracy compared with the other methods evaluated for the three BCI datasets, using either fewer channels or frequency bands, by eliminating noisy components for the MI task. As an additional benefit, MKS attains lower standard deviations, yielding to a more confident approach in comparison with other state-of-the-art-techniques in the statistical sense.

As future work, the authors plan to develop an optimization approach allowing to map covariance to PSD matrix spaces with fewer components that cope with the redundant information. In this sense, both the channel and frequency feature extraction will be carried out in a unified framework that results in globally optimal setups. Also, another EEG-based classification tasks such as disease diagnosis support will be considered to test the proposal on different scenarios.

Acknowledgment. This work was developed for the research project 111974454238 funded by Colciencias.

References

1. Cho, H., Ahn, M., Ahn, S., Kwon, M., Jun, S.C.: EEG datasets for motor imagery brain-computer interface. GigaScience **6**(7), 1–8 (2017)
2. Huang, G., Liu, G., Zhang, D., Zhu, X.: Model based generalization analysis of common spatial pattern in brain computer interfaces. Cogn. Neurodyn. **4**(3), 217–223 (2010)
3. Nguyen, C.H., Artemiadis, P.: EEG feature descriptors and discriminant analysis under Riemannian manifold perspective. Neurocomputing **275**, 1871–1883 (2018)
4. Ortiz-Rosario, A., Adeli, H.: Brain-computer interface technologies: from signal to action. Rev. Neurosci. **24**(5), 537–552 (2013)
5. Yger, F., Berar, M., Lotte, F.: Riemannian approaches in brain-computer interfaces: a review. IEEE Trans. Neural Syst. Rehabil. Eng. **25**(10), 1753–1762 (2017)
6. Yger, F., Lotte, F., Sugiyama, M.: Averaging covariance matrices for EEG signal classification based on the CSP: an empirical study. In: 2015 23rd European Signal Processing Conference (EUSIPCO), pp. 2721–2725 (2015)
7. Zhang, Y., Wang, Y., Jin, J., Wang, X.: Sparse Bayesian learning for obtaining sparsity of EEG frequency bands based feature vectors in motor imagery classification. Int. J. Neural Syst. **27**(02), 1650032 (2017)
8. Zhang, Y., Zhou, G., Jin, J., Wang, X., Cichocki, A.: Optimizing spatial patterns with sparse filter bands for motor-imagery based brain-computer interface. J. Neurosci. Methods **255**, 85–91 (2015)

Performance of Different Average Methods for the Automatic Detection of Evoked Potentials

Idileisy Torres-Rodríguez[1] ⓘ, Carlos Ariel Ferrer-Riesgo[1(✉)] ⓘ,
Juan Carlos Oliva Pérez[2] ⓘ, and Alberto Taboada-Crispi[1] ⓘ

[1] Informatics Research Center, Universidad Central "Marta Abreu" de Las Villas, UCLV, 54830 Santa Clara, Cuba
{itrodriguez,cferrer,ataboada}@uclv.edu.cu
[2] Electronic and Telecommunications Department, Universidad Central "Marta Abreu" de Las Villas, UCLV, 54830 Santa Clara, Cuba
jpoliva@uclv.cu

Abstract. The evoked potentials can be auditory, visual or somatosensory. Noise reduction is the first step in most biomedical signal processing systems. The quality and accuracy of the rest of the operations carried out on the signal depend to a large extent on the quality of the noise reduction algorithms that have been used in the preprocessing of the signal. The method commonly used to enhance the signal of interest is the coherent average, however, this technique has some limitations that justify the search for alternatives to detect or extract the characteristics of these signals. The weighted average is a possible alternative; however, this is still not appropriate enough when the signal may have outliers within the epoch. Trimmed average techniques have a better solution when in the presence of impulsive noise. The modified trimmed average is adapted for use in auditory evoked potentials. In this work, we compare different techniques of trimmed average, with the weighted average and the coherent average; using quality measures in the frequency domain for detect Auditory Brainstem Evoked Potentials. The results showed that the proposed method and his variants are superior to the rest of the used ones. The Q-Sample Modified measure offers the best result.

Keywords: Auditory evoked potentials · Ensemble average · Weighted average · Trimmed average · Q-sample uniform · Q-Sample Modified · Watson Q-sample

1 Introduction

1.1 Evoked Potentials and Their Classification

The specific neural activity that arises from acoustic stimulation as a pattern of voltage fluctuations lasting approximately half a second is an auditory evoked potential [1]. Depending on the type and placement of the electrodes, the amplification of the signal, the selection of the filters and the post-stimulation period, it is possible to detect the

© Springer Nature Switzerland AG 2019
I. Nyström et al. (Eds.): CIARP 2019, LNCS 11896, pp. 629–636, 2019.
https://doi.org/10.1007/978-3-030-33904-3_59

neuronal activity that arises from different structures that span from the auditory nerve to the cerebral cortex [2, 3]. Noise reduction is the first step in most biomedical signal processing systems. The quality and accuracy of the rest of the operations carried out on the signal depend to a large extent on the quality of the noise reduction algorithms that have been used in the preprocessing of the signal. The coherent average (CA) or arithmetic mean, as it is also known, can be calculated from the ensemble matrix that is formed with the evoked responses. Where the response to the i-th stimulus is assumed as the sum of the deterministic component of the signal or response evoked s plus a random noise r_i (see Eq. 1), which is asynchronous with the stimulus. Where the noise in progress is assumed to be stationary, with zero mean. Consequently, the variance of the noise must be fixed and equal in all the potentials. The average is a simple and direct method. The estimated signal can be modeled as the sum of the deterministic component plus the attenuated noise by a factor of 1/M, as show in Eq. 2.

$$p_i = s + r_i \tag{1}$$

$$\hat{s} = \frac{1}{M} \sum_{i=1}^{M} p_i(n) \tag{2}$$

Considering the presence of several types of noise which cause degradation in the performance of the average, the development of new methods that correctly handle these problems is justified. One of these methods that have been proposed in the literature is the weighted average. Several criteria have been used in order to determine the vector of weights that best fits the problem. One of these criteria (minimization of the mean square error) is based on the noise variance of all the cycles. A potential with a high noise level is assigned a lower weight than one with a lower noise level [4–8].

The average ensemble and the weighted average represent linear techniques, and consequently they perform very well when the noise is of Gaussian type. However, in the case that out-of-range artifacts appear occasionally, with large amplitude values, these techniques are limited. The ensemble average and the ensemble median can be seen as special cases within a broad family of existing estimators known as trimmed means, within which are the trimmed average, the Winsorized average, the L-trimmed average, or mean TL and the average Tanh [9–12].

1.2   Automatic Detection of Evoked Potentials

Methods for the objective detection of Auditory Brainstem Evoked Potentials (ABR) can be characterized as template-based methods and non-template based methods, as reviewed by [6] and referenced in [13]. Most used methods for detection in the frequency domain. The original test of uniform sample scores q (Q-sample uniform) [14] is a nonparametric test that uses the phase ranges of the Fourier components of Q spectral bands to test whether the phases share the same distribution. The test only uses the phase angles in the form of their ranges and rejects the spectral amplitudes.

The most powerful test in the frequency domain according to [15] is Q-sample uniform. This test uses only the phase angles in the form of their ranges while the spectral information amplitude is rejected. In [16] a modification is introduced where the spectral amplitude is also considered, this test has come to occupy a better position than its predecessor in its use for the detection of auditory evoked potentials. Another test [17] that can be considered as a special type of Q-sample test is the Watson Q-sample. This test also uses both phase angles and spectral amplitudes.

2 Methods

Two versions of the Modified Trimmed Mean [18] are adapted for their use in Brainstem Auditory Evoked Potentials (BAEP). It also analyzes the characteristics of the database and how the comparison between the different methods will be carried out.

2.1 Modified Trimmed Mean Adapted to the BAEP

The Trimmed Mean Modified in [18], proposes to determine the cut-off factor t as $2 * \sigma_\Gamma$, where σ_Γ is the standard deviation of the average background noise, estimated from the isoelectric segment of the electrocardiographic signal as posted in [19–21]. In the case of evoked potentials, they do not have an isoelectric segment. However, in the literature consulted several ways of estimating the variance of background noise have been found. Following this approach, the trimming factor could be estimated as 3 or 2 standard deviations of noise, based on the fact that the variance is the square of the standard deviation. In several articles also reviewed, the estimation of background noise has been proposed as the average of the variances of several unique points, used in the estimation of the Fmp [22, 23], as a better approximation to the noise of the signal. In this paper we propose to determine the cut-off factor t as three standard deviations of the estimated background noise using the Fmp.

Another variant of modifying the cut average in order to determine the cut-off factor t was the use of the interquartile range (IQR), defined as the difference between the third quartile Q3 and the first quartile, Q1. The interquartile range is a robust measure, because it only takes into account 50% of the data.

$$[Q_1 - 1.5 * IQR, \quad Q_3 + 1.5 * IQR] \tag{3}$$

The range given in this equation depends on the factor 1.5, this is an arbitrary value, but it finds its justification according to a normal distribution. In this way, two versions of the Modified trimmed mean are obtained.

2.2 Data

The database used in this study consists of Transient Auditory Evoked Potentials registered in 39 neonatal patients between 1–3 months of age born in Hospital Materno

Ramón González Coro, in Havana, Cuba [27]. The signals were recorded with an AUDIX electroaudiometer. A click stimulus with duration 0.1 ms was provided at different intensities (100, 80, 70, 60, 30 dBnHL and 0 dBpSPL) via insert earphones (EarTone3A) [28, 29]. Ag/AgCl dry electrodes were used, which were fixed with electrolytic paste on the forehead (positive), ipsilateral mastoids (negative) and contralateral mastoids (earth). The impedance values were maintained below 5 kΩ. The sampling frequency used was 13.3 kHz, and the analysis windows to form the ensemble matrix P (Eq. 5) and calculate the coherent average were of approximately 15 ms, that is about 200 samples per window (N = 200). From this database, only records obtained at 100 dBnHL (78 signals) were used, where it was confirmed by specialists that a response was present. These signals were used in order to guarantee the maximum values of the quality measures for this database.

2.3 Description of the Experiment

In order to carry out the experiment, set matrices of each of the registers were formed using the time between stimuli of 15 ms as reference. With each record an ensemble matrix of approximately 2000 epochs was formed by 200 samples on average. Each ensemble matrix was averaged using the average methods described in previous sections. The automatic detection measures used to establish the comparison were measured in the frequency domain; they were the Q-Sample Uniform, the Modified Q-Sample and the Q-Sample of Watson U2.

To establish the comparison of the different methods using measures in the frequency domain, first, the records were transformed into set matrices. From each set matrix, sets of 250 epochs are randomly taken 100 times and averaged using each of the average methods that were described. A new matrix (100 × 200) composed of the average vectors obtained in the previous step was formed trying to simulate a Monte Carlo experiment. The matrices obtained for each of the average methods and for each of the different sizes of X are transformed to the frequency domain using the fft (Fast Fourier Transform, from its acronym in English), from which it is obtained an array of dimension 100 × 200 × 2, which contains the phase and amplitude values of the result of applying the transform. This arrangement is to which the different quality measures are calculated. The design of the experiment is based on the design of the experiment proposed in [32] for the comparison of detection measures in the frequency domain.

3 Results

To evaluate the results obtained, a Friedman test was performed. The multipair comparison returns an interactive plot that allows to visually determining which of the methods have differences between them as show the Fig. 1. In all cases, the test resulted in a value of $p < 0.05$, which suggests that there are significant differences between at least two methods. In order to identify the methods in which the differences existed, a post-hoc test was developed using the Bonferroni method.

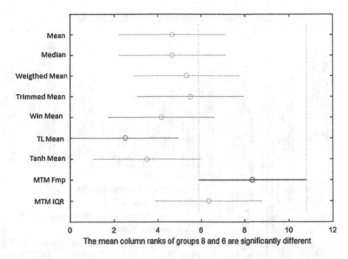

Fig. 1. Multiple comparison of methods using all quality measures

Table 1 gives the values of the average range returned by the Friedman test, giving a higher score to the method with the best performance. In this case, although the significant differences are only between the Average TL and the MTM Fmp, according to the Range the best method was the MTM Fmp, followed by the MTM IQR., Which constitute the proposals of this work.

Table 1. Values of the average ranges given by the Friedman test to each of the methods.

Mean	Median	Weighted average	Trimmed mean	Mean Winsorized	Mean TL	Mean Tanh	MTM Fmp	MTM IQR
4.667	4.667	5.333	5.500	4.167	2.500	3.500	8.333	6.333

Another analysis that is important to consider is to determine which method presented minor differences when it acted with the clean database and with the data without artifacts, values outside the range of ± 5 μV. Figure 2 is a total view of all the methods under analysis, in them it can be seen that for the coherent average there are visible differences in their behavior, while the proposed methods prove to be more robust in behaving in a similar way before the two variants of the data.

Next, the signals obtained from a random subject are shown using the different methods. In Fig. 3 it can be seen how the MTM Fmp method is the one that comes closest to the expected while other methods move away from the expected signal and present higher noise levels.

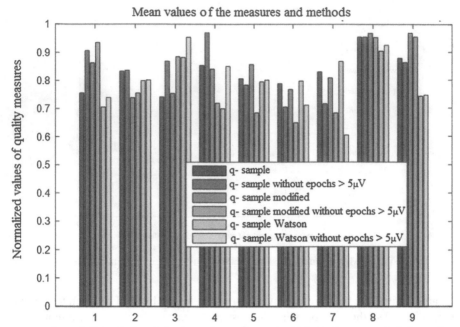

From left to rigth, 1 is Mean, 2 is Median, 3 is Weigthed mean, 4 is Trimmed mean, 5 is Win mean, 6 is TL mean, 7 is Tanh mean, 8 is MTM Fmp and 9 is MTM IQR

Fig. 2. Mean values of the measures and methods.

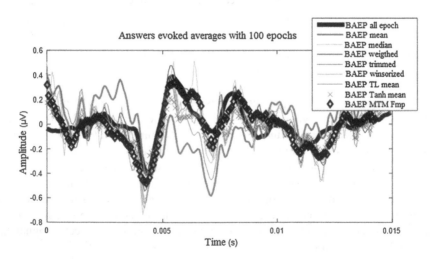

Fig. 3. Answers evoked averages with 100 epochs

4 Conclusions

Despite the variety of the proposed methods, this paper describes the proposals for modifications made to the Modified Trimmed Media in order to adapt it for the reduction of noise of auditory evoked potentials. The Modified Trimmed Media is a method that proposes to combine the solutions to the main drawbacks of the coherent average. The Friedman test used showed that the best method was the MTM using Fmp, however, the differences were significant only with respect to the Trimmed Media TL. From the review of the bibliography, the measurements in the frequency domain were selected to establish the comparison between the results obtained by each method. The measure that behaved better was modified Q-sample.

As a consequence of the analysis carried out, it is recommended that the experiment be developed with a data with higher noise levels, which is carried out in a more exhaustive way, checking the results in a Monte Carlo experiment for smaller steps in the formation of the set matrix trying to simulate a real recording environment. It is also recommended that the results obtained be compared with similar results, but in the time domain.

References

1. Sharma, A., Dorman, M.F., Kral, A.: The influence of a sensitive period on central auditory development in children with unilateral and bilateral cochlear implants. Hear. Res. **203**(1–2), 134–143 (2005)
2. Eggermont, J.J.: On the rate of maturation of sensory evoked potentials. Electroencephalogr. Clin. Neurophysiol. **70**(4), 293–305 (1988)
3. Kraus, N., Nicol, T.: Auditory evoked potentials. In: Binder, M.D., Hirokawa, N., Windhorst, U. (eds.) Encyclopedia of Neuroscience. Springer, Heidelberg (2008). https://doi.org/10.1007/978-3-540-29678-2
4. Sornmo, L., Laguna, P.: Bioelectrical Signal Processing in Cardiac and Neural Applications (2005)
5. Stapells, D.R.: Frequency-specific threshold assessment in young infants using the transient ABR and the brainstem ASSR. In: Comprehensive Handbook of Pediatric Audiology, pp. 409–448 (2011)
6. Suárez, A., De Armas, J.L., Aznielle, T., Charroó, L., Hernández, R.: SenseWitness: Un nuevo software para potenciales evocados (2003)
7. Nodarse, E.M., et al.: Cribado auditivo neonatal con potenciales evocados auditivos de estado estable a múltiples frecuencias. Acta Otorrinolaringol. Esp. **62**(2), 87–94 (2011)
8. Torres, I.: Estimadores Robustos de Promedio para la Detección de Potenciales Relacionados con Eventos. Universidad Central "Marta Abreu" de Las Villas (2014)
9. Elamir, E.A.H., Seheult, A.H.: Trimmed L-moments. Comput. Stat. Data Anal. **43**(3), 299–314 (2003)
10. Leonowicz, Z., Karvanen, J., Shishkin, S.L.: Trimmed estimators for robust averaging of event-related potentials. J. Neurosci. Methods **142**(1), 17–26 (2005)
11. Hampel, F., Ronchetti, E., Rousseeuw, P., Stahel, W.: Robust Statistics: The Approach Based on Influence Functions. Wiley, Hoboken (2011)
12. Valderrama, J.T., Alvarez, I., De La Torre, A., Segura, J.C., Sainz, M., Vargas, J.L.: A portable, modular, and low cost auditory brainstem response recording system including an algorithm for automatic identification of responses suitable for hearing screening. In: IEEE EMBS

Special Topics Conference on Point-of-Care Healthcare Technologies Synergy Towards Better Global Healthcare, PHT 2013, pp. 180–183 (2013)

13. Laciar, E., Jane, R.: An improved weighted signal averaging method for high-resolution ECG signals, pp. 69–72 (2001)

14. Chesnaye, M.A., Bell, S.L., Harte, J.M., Simpson, D.M.: Objective measures for detecting the auditory brainstem response: comparisons of specificity, sensitivity and detection time. Int. J. Audiol. **57**, 1–11 (2018)

15. Lins, O., Picton, P., Picton, T.: Auditory steady-state responses to tones amplitude-modulated at 80–110 Hz. J. Acoust. Soc. Am. **97**, 3051–3063 (1995)

16. Congedo, M., Korczowski, L., Delorme, A., Lopes da Silva, F.: Spatio-temporal common pattern: a companion method for ERP analysis in the time domain. J. Neurosci. Methods **267**, 74–88 (2016)

17. Stürzebecher, E., Cebulla, M., Wernecke, K.D.: Objective response detection in the frequency domain: comparison of several q-sample tests. Audiol. Neuro-Otology **4**(1), 2–11 (1999)

18. Taboada-Crispi, A., Lorenzo-Ginori, J.: Adaptive line enhancing plus modified signal averaging for ventricular late potential detection. Electronics **35**, 1293–1295 (1999)

19. Heckenlively, J., Arden, G.: Principles and Practice of Clinical Electrophysiology of Vision. MIT Press, Cambridge (2006)

20. Leski, J.M.: Robust weighted averaging. IEEE Trans. Biomed. Eng. **49**(8), 796–804 (2002)

21. The BC Early Hearing Programme: Audiology Assessment Protocol (BCEHP) (2012). http://www.phsa.ca/health-professionals/clinical-resources/hearing/ BCEarlyHearingProgram:Audiologyassessmentprotocols.pdf

22. Mijares, E., et al.: Sensitivity of the automated auditory brainstem response in neonatal hearing screening. Pediatrics **3**(3), 117–128 (2014)

23. Da Silva, I.G.A.: Objective estimation of loudness growth using tone burst evoked auditory responses. Diss. Abstr. Int. B Sci. Eng. **70**(8), 4987 (2010)

24. Paradis, J.: The development of English as a second language with and without specific language impairment: clinical implications. J. Speech Lang. Hear. Res. **59**(2), 171–182 (2016)

25. Pander, T.: A new approach to robust, weighted signal averaging. Biocybern. Biomed. Eng. **35**(4), 317–327 (2015)

26. Levit, Y., Himmelfarb, M., Dollberg, S.: Sensitivity of the automated auditory brainstem response in neonatal hearing screening. Pediatrics **136**(3), e641–e647 (2015)

27. Rodrigues, G.R.I., Ramos, N., Lewis, D.R.: Comparing auditory brainstem responses (ABRs) to toneburst and narrow band CE-chirp® in young infants. Int. J. Pediatr. Otorhinolaryngol. **77**(9), 1555–1560 (2013)

28. Kotas, M., Pander, T., Leski, J.M.: Averaging of nonlinearly aligned signal cycles for noise suppression. Biomed. Signal Process. Control **21**, 157–168 (2015)

29. Pander, T., Pietraszek, S., Przybyła, T.: A novel approach to robust weighted averaging of auditory evoked potentials. In: Piętka, E., Kawa, J., Wieclawek, W. (eds.) Information Technologies in Biomedicine, Volume 4. AISC, vol. 284, pp. 321–332. Springer, Cham (2014). https://doi.org/10.1007/978-3-319-06596-0_30

30. Valderrama, J.T., et al.: Automatic quality assessment and peak identification of auditory brainstem responses with fitted parametric peaks. Comput. Methods Programs Biomed. **114**(3), 262–275 (2014)

31. Berninger, E., Olofsson, A., Leijon, A.: Analysis of click-evoked auditory brainstem responses using time domain cross-correlations between interleaved responses. Ear Hear. **35**(3), 318–329 (2014)

32. Tsurukiri, J., Nagata, K., Kumasaka, K., Ueno, K., Ueno, M.: Middle latency auditory evoked potential index for prediction of post-resuscitation survival in elderly populations with out-of-hospital cardiac arrest. Signa Vitae **13**, 80–83 (2017)

Discrimination of Shoulder Flexion/Extension Motor Imagery Through EEG Spatial Features to Command an Upper Limb Robotic Exoskeleton

Ramón Amado Reinoso-Leblanch[1](✉), Yunier Prieur-Coloma[1],
Leondry Mayeta-Revilla[2], Roberto Sagaró-Zamora[3],
Denis Delisle-Rodriguez[4], Teodiano Bastos[4],
and Alberto López-Delis[1]

[1] Centre of Medical Biophysics, University of Oriente, Santiago de Cuba, Cuba
ramon.reinoso2308@gmail.com, yunier.prieur@gmail.com,
lopez.delis69@gmail.com
[2] Department of Biomedical Engineering, University of Oriente,
Santiago de Cuba, Cuba
lmayeta@uo.edu.cu
[3] Faculty of Mechanical Engineering, University of Oriente,
Santiago de Cuba, Cuba
sagaro@uo.edu.cu
[4] Postgraduate Program in Electrical Engineering,
Federal University of Espirito Santo, Vitoria, Brazil
delisle05@gmail.com, teodiano.bastos@ufes.br

Abstract. This work presents a comparison between two methods for spatial feature extraction applied on a system to recognize shoulder flexion/extension motor imagery (SMI) tasks to convey on-line control commands towards a 4 degrees-of-freedom (DoF) upper-limb robotic exoskeleton. Riemannian geometry and Common Spatial Pattern (CSP) are applied on the filtered EEG for spatial feature extraction, which later are used by the Linear Discriminant Analysis (LDA) classifier for motor imagery (MI) recognition. Three bipolar EEG channels were used on six healthy subjects to acquire our database, composed of two classes: rest state and shoulder flexion/extension MI. Our system achieved a mean accuracy (ACC) of 75.12% applying Riemannian, with the highest performance for Subject S01 (ACC = 89.68%, Kappa = 79.37%, true positive rate (TPR) = 87.50%, and FPR < 8.13%). In contrast, for CSP, a mean ACC of 66.29% was achieved. These findings suggest that unsupervised methods for feature extraction, such as Riemannian geometry, can be suitable for shoulder flexion/extension MI to command an upper-limb robotic exoskeleton.

Keywords: Riemannian geometry · Common Spatial Pattern ·
Brain-computer interface · Motor imagery · Upper limb · Robotic exoskeleton ·
Shoulder movement intention

© Springer Nature Switzerland AG 2019
I. Nyström et al. (Eds.): CIARP 2019, LNCS 11896, pp. 637–645, 2019.
https://doi.org/10.1007/978-3-030-33904-3_60

1 Introduction

A brain-computer interface (BCI) is a system that translates brain activity patterns of a user into messages or commands for an interactive application [1]. BCIs based on voluntary modulation of Sensorimotor Rhythm (SMR) have increased the interest of many researchers, due to evidence that real or imagery motor intention of upper and lower limbs induces changes the rhythmic activity recorded on the sensorimotor cortex [2]. Pfurtscheller *et al.* [3] indicate that an internal or external paced event results not only in generation of event-related potential (ERP), but also in a change in the ongoing EEG/MEG in form of an event-related desynchronization (ERD) or event-related synchronization (ERS). The ERP on the one side and the ERD/ERS on the other side are different responses of neuronal structures in the brain. Whereas the former is phase-locked, the latter is not phase-locked to the event. The most important difference between both phenomena is that the ERD/ERS is highly frequency band specific, whereby either the same or different locations on the scalp can display ERD and ERS simultaneously. Also, it was found that ERD/ERS are detectable from EEG in a majority of stroke patients while performing Kinesthetic Motor Imagery (KMI) and Visual Motor Imagery (VMI) [4–6]. A KMI can be described as the ability to imagine performing a movement without executing it, by imagining haptic sensations felt during the real movement (i.e., tactile, proprioceptive, and kinesthetic). In contrast, a VMI mainly relies on the visualization of the execution of that movement. VMI and KMI share common neural networks particularly in the primary motor cortex, motor cortex, supplementary motor areas, somatosensory cortex and cerebellum, but also involve different cortical structures, due to the intuitive nature of KMI tasks. More precisely, KMI produces a greater activation of the primary motor cortex and of the supplementary motor areas. The resulting synaptic plasticity phenomenon makes the use of KMI-based BCIs a promising instrument of acquisition and refinement of motor skills [7]. In fact, various studies have demonstrated that similar cortical locations are activated during the same execution of real or imagery movements [3, 8, 9], whose changes can be captured through electroencephalography (EEG) signals to be translated into commands for end applications, such as robotic exoskeletons. BCI is a useful technology in helping people who have suffered a nervous system injury by providing them with an alternative pathway of communication, mobility, and rehabilitation. Therefore, stroke patients suffering severe limb weakness, but still able to imagine movements of the paretic limb, can use a BCI to trigger a contingent feedback upon the detection of MI-related EEG signals, enhancing neuroplasticity and motor recovery [10].

The objective of this work is to propose a system based on spatial feature extraction to discriminate shoulder flexion/extension motor imagery (SMI) tasks to assist shoulder movements through a 4-DoF (degrees of freedom) upper-limb robotic exoskeleton applied to post-stroke patients that cannot voluntarily start a movement [11]. Both Riemannian geometry and Common Spatial Pattern (CSP) were used on six healthy subjects to obtain the best alternative to discriminate SMI tasks.

This work presents four sections ordered as follows: Sect. 2 describes the experimental protocol, the proposed system for SMI discrimination, and the methodology to evaluate the performance of our proposal. Section 3 presents the achieved results, and their discussion. Finally, Sect. 4 presents the conclusions and future works.

2   Materials and Methods

2.1   Experimental Protocol

An experimental protocol was conducted on six healthy participants (all males with age from 23 to 57 years) to evaluate the performance of the proposed system to discriminate SMI tasks. Four stages were carried out for the experiments: (a) completing the informed consent form; (b) protocol familiarization; (c) electrode preparation; (d) protocol execution.

During the experiments, the subjects were seated on a chair in front of a PC screen. They provided their written consent prior to the data collection. Later, each volunteer was familiarized with the protocol through the execution of kinesthetic motor imageries [7] related to shoulder flexion/extension. The electrode preparation was done using conductor gel to reduce the skin-electrode impedance, preserving it below 10 kΩ. Subsequently, the experiment was executed in two-phases: (1) Collecting an EEG dataset while the subject was executing the SMI tasks for the training phase of our BCI; (2) Data collection to validate offline the BCI after training phase. To record the database for both training and validation stages, a sequence of four visual cues (black screen, red circle, yellow circle, and green circle) was used with period of 5 s. Red, yellow and green cues indicate the rest state, ready (similar to rest state, but waiting for the next cue without mental activity) and SMI tasks, respectively. All volunteers were asked to avoid any voluntary during the red, yellow and green cues, but slight and unavoidable movements execution, such as eye blinks, were allowed on the black screen. This cue sequence was displayed on a PC screen 12 times per session, through a developed Graphical User Interface (GUI) to guide the subjects during the experiment. For each participant, six sessions were completed for both training and validation sets, giving 3 min for resting after each session.

2.2   Proposed System

Figure 1 shows the block diagram of the proposed system to discriminate the SMI by EEG signals acquired through the Nexus-10 Mark II equipment from MindMedia manufacturer to be used as control commands towards an upper-limb robotic exoskeleton [11], which was built to rehabilitate people with severe motor disabilities. Our recognition system is composed of the following three stages: raw EEG preprocessing by a 2nd order Butterworth band-pass filter with a frequency range from 8 to 30 Hz; feature extraction (here two approaches based on Riemannian covariance matrices and CSP were analyzed); and classification by LDA. Both training and testing feature datasets were normalized using the mean and the standard deviation.

Data Acquisition
To capture the brain response through SMI tasks, three pairs of bipolar electrodes (C3-FC1, C4-FC2, Cz-Fz) were located on the motor primary and supplementary motor areas, according the 10–20 international system. The ground electrode was attached between the eyebrows. The EEG signal was recorded with Nexus-10 Mark II using BioTrace+ application software from MindMedia manufacturer, in a frequency range from 0.1 to 100 Hz, and sampling rate at 256 Hz. Moreover, an Arduino board was used to synchronize both cues (visual and sound stimuli) and EEG acquisition system.

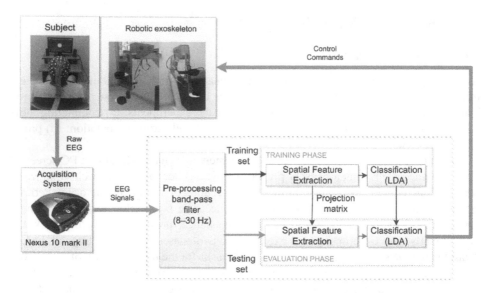

Fig. 1. Block diagram of the proposed system for the BCI based on SMI task.

2.3 Features Extraction

Common Spatial Pattern
Common Spatial Pattern (CSP) analysis was applied to analyze the multichannel data based on recordings from two classes (rest and imagery) from the band-pass filtered EEG data. This technique aims at using different spatial filters, such that the variance of the spatially filtered signals is maximized for one class and minimized for the other class. CSP yields a data-driven supervised decomposition of the signal parameterized by a matrix $\mathbf{W} \in \mathbb{R}^{C \times C}$ (C being the number of channels) that projects the signal $x(t) \in \mathbb{R}^C$ that belongs to trial t in the original sensor space to $x_{CSP}(t) \in \mathbb{R}^C$, which lives in the surrogate sensor space, as follows:

$$x_{CSP}(t) = \mathbf{W}^{\mathrm{T}} x(t) \tag{1}$$

where each column vector $w_j \in \mathbb{R}^C$ ($j = 1,\dots, C$) of $\mathbf{W}$, a spatial filter or simply a filter [12]. CSP maximizes the variance of the spatially filtered signal under one task while minimizing it for the other task, since the variance of a bandpass filtered signal is equal to band power, such as shown in [13]. This is done by a simultaneous diagonalization of the estimated covariance matrices $\sum_1 = X_1 X_1^{\mathrm{T}}$ and $\sum_2 = X_2 X_2^{\mathrm{T}}$ of the data for the two classes:

$$\mathbf{W}^{\mathrm{T}} \Sigma_1 \mathbf{W} = \Lambda_1, \tag{2}$$

$$\mathbf{W}^{\mathrm{T}} \Sigma_2 \mathbf{W} = \Lambda_2, \tag{3}$$

$$\text{s.t. } \Lambda_1 + \Lambda_2 = I \tag{4}$$

where Λ_1 and Λ_2 are diagonal matrices, and each λ on the diagonal corresponds to an eigenvector w^T. This way, the eigenvectors are the same for both decompositions and the same eigenvector, i.e. a spatial filter, corresponds to a large eigenvalue for one class and to a small eigenvalue for the other class. Since eigenvectors with large eigenvalues correspond to a large variance of the data, spatial filters with extreme eigenvalues maximize the difference in the variances for the two classes. Technically, this can be achieved by solving the generalized eigenvalue problem:

$$\Sigma_1 w = \lambda \Sigma_2 w \tag{5}$$

Choosing D filters corresponding to extreme eigenvalues (either close to 1 or close to 0), the filtered data $x_{CSP}(t) = W_D^T x(t)$ will have smaller dimensionality $D < N$ and the two classes will be maximally separated by their variance.

For our study, CSP was individually selected for each subject using the band-pass filtered signals from red (rest state) and green (imagery state) cues. The number of CSP components was automatically selected, and ranged between 2 and 3. Details about this technique can be found in [14].

Riemannian Geometry

Such as described by Yger et al. [15], this approach enables the direct manipulation of EEG signal covariance matrices and subspaces, with an appropriate and dedicated geometry. Also, they show the full Riemannian BCI pipeline, through signal preprocessing, representing them as covariance matrices, and classifying them using covariance matrix-based classifiers, which exploit the Riemannian geometry. These approaches have been used for EEG-based BCI, in particular for feature representation and learning, as well as classifier design and calibration time reduction [16–18]. In our study, we used Riemannian geometry for feature extraction using tangent space representation from covariance matrices of the EEG trials by mean of the standard covariance method and the average of Riemann, where the tangent space projection maps a set of covariance matrices to its tangent space, according to [16]. The tangent space projection can be seen as a kernel operation described by Barachant et al. [17]. After projection, each matrix is represented as a vector of size $\frac{N(N+1)}{2}$, where N is the dimension of the covariance matrices. The tangent space projection is a local approximation of the manifold, whose approximation will be bigger if the matrices in the set are scattered in the manifold, and lower if they are grouped in a small region of the manifold. Details about the Riemannian geometry can be found in [19].

2.4   Data Description

To evaluate our recognition system using Riemannian or CSP for spatial feature extraction, a total of 12 sessions for the training and validation sets were used. For every subject, the first 6 sessions were used for training, and the remaining 6 sessions were used for validation. Each session is formed by two classes (rest state and SMI execution), with each class performing 12 times for a period of 5 s. Then, labeled EEG segments of 1 s in length, overlapped of 0.5 s, were used throughout segments of 4 s, taking into account 0.5 s after each cue beginning.

2.5   Evaluation

Metrics such as Accuracy (ACC), Kappa, True Positive Rate (TPR) and False Positive Rate (FPR) were used to validate the performance of our system using one of the two spatial feature extraction methods (Riemannian geometry and CSP). This comparison looks for the best method of feature extraction, in order to use it in our on-line BCI for patients to trigger, by MI, the upper limb exoskeleton.

3   Results and Discussion

Table 1 presents a comparison about the performance of our system with both Riemannian geometry and CSP in addition to the features extracted during the SMI tasks, using three bipolar channels (C3-FC1, C4-FC2, Cz-Fz), and classifying them with LDA. The best performance was achieved with Subject S01 (ACC > 89.68%), but ACC was >68% for all subjects, with mean ACC > 75.12%, using Riemannian covariance matrices and tangent space on the band-pass filtered EEG data. Meanwhile, CSP reached mean ACC around 66% for all subjects. It is worth mentioning that Subject S03 began to feel mental fatigue in session 3 of the validation stage. Table 2 shows the performance achieved that subject for different training and validation sets. The best result for this subject was achieved for the case where the first 3 sessions were used (ACC > 70%), because after the volunteer felt mental exhaustion, he showed signs of sleep during the acquisition of the remaining sessions, not executing the protocol in the desired way. Thus, it can be appreciated as the subject involvement can influence the performance of the recognition system. Both methods show inter-subject variability in the EEG with respect to the brain signal characteristics mentioned in previous works [10].

Table 1. Performance of the proposed system during shoulder flexion/extension MI tasks.

Subjects	Riemannian geometry				Common spatial pattern				
	ACC (%)	Kappa (%)	TPR (%)	FPR (%)	ACC (%)	Kappa (%)	TPR (%)	FPR (%)	Components
S01	89.68	79.37	87.50	8.13	76.59	53.17	61.71	8.53	2
S02	68.75	37.5	52.78	15.28	65.97	31.94	55.36	23.41	3
S03	68.35	36.71	82.54	45.83	62.40	24.80	68.45	43.65	3
S04	69.25	38.49	58.33	19.84	63.99	27.98	59.52	31.55	2
S05	76.98	53.97	80.16	26.19	68.06	36.11	64.29	28.18	2
S06	77.68	55.36	80.36	25.00	60.71	21.43	51.19	29.76	3
Mean	75.12	50.23	73.61	23.38	66.29	32.57	60.09	27.51	
SD	8.28	16.56	14.34	12.85	5.67	11.34	6.20	11.48	

Table 2. Performance for subject S03, for each session of the training phase and validation phase.

Sessions total used for phase		Riemannian geometry			
Training	Validation	ACC (%)	κ (%)	TPR (%)	FPR (%)
3	3	73.81	47.61	76.19	28.57
4	4	69.79	39.58	81.25	41.66
6	6	68.35	36.71	82.54	45.83

4 Discussion

Bipolar channels have been widely used in several BCI studies, particularly in a set of reduced channels for upper-limb MI-based BCI [20–22]. In [23] the authors analyzed the largest weights of mu rhythms at C3 and C4 locations, finding no evident changes of this rhythm during one week. This conclusion is reasonable because the distribution of the mu rhythm is physiologically determined by the structure of the subject's sensorimotor cortex, which normally does not vary in a period of time. Although subjects may have varied intensities of ERD/ERS during on-line sessions, the selected electrodes have obtained always a promising result. A practical BCI system requires a stable electrode layout, but the variability of brain activity makes it difficult to satisfy this requirement. Other studies have presented different methodologies to recognize upper limb MI tasks in healthy subjects. For instance, in [24] the authors conducted an MI experiment of left and right hands with 52 healthy subjects. They used a 64-channel montage based on the international 10–10 system to record EEG signals at 512 Hz of sampling rate. For the training phase, each trial (from 0.5 to 2 s, after the cue onset) was bandpass filtered with a frequency range from 8 to 3 Hz and extracted temporally. They used both CSP and LDA for spatial feature extraction and classification, respectively. To calculate the performance over the MI data, they conducted a cross validation, achieving mean ACC of $67.46 \pm 13.17\%$ over 50 subjects. In [25] the authors showed a method based on the deep convolutional neural network (CNN) to perform feature extraction and EEG MI classification for two subjects. Similarly, they used a bandpass filter over the raw EEG from 8 to 30 Hz, with the EEG data recorded on 28 locations. For this purpose, the authors built a 5-layer CNN model based on the spatio-temporal characteristics of EEG to recognize the MI tasks linked to left- and right-hand movements, obtaining an average ACC of $86.41 \pm 0.77\%$. In [22] the authors used three bipolar channels, obtaining ACC of 77%, 84%, and 78% for three subjects, respectively.

5 Conclusions

This study showed the feasibility of using the Riemannian geometry to discriminate SMI tasks. The results obtained are encouraging for conveying control commands towards an upper-limb robotic exoskeleton by means of an online MI-based BCI. Our proposed system may be used in future works to increase the effectiveness of an upper-limb rehabilitation process for people with severe motor disabilities, providing them an alternative way to trigger, through EEG signals, the 4-DOF upper-limb robotic exoskeleton available in our lab. Also, the proposed system will be integrated to our robotic exoskeleton to be part of the motor rehabilitation system.

Acknowledgments. The authors would like to thank the Medical Biophysics Center (Centro de Biofísica Médica) of Cuba, UFES/Brazil, and the Belgian Development Cooperation, through VLIR-UO (Flemish Interuniversity Council-University Cooperation for Development), in the context of the Institutional University Cooperation program with the University of Oriente for supporting this research.

References

1. Lotte, F., et al.: A review of classification algorithms for EEG-based brain–computer interfaces: a 10 year update. J. Neural Eng. **15**(3), 031005 (2018)
2. Wolpaw, J., Wolpaw, E.W.: Brain-Computer Interfaces: Principles and Practice. OUP, New York (2012)
3. Pfurtscheller, G., Da Silva, F.L.: Event-related EEG/MEG synchronization and desynchronization: basic principles. Clin. Neurophysiol. **110**(11), 1842–1857 (1999)
4. Ang, K.K., et al.: A large clinical study on the ability of stroke patients to use an EEG-based motor imagery brain-computer interface. Clin. EEG Neurosci. **42**(4), 253–258 (2011)
5. Soekadar, S.R., et al.: ERD-based online brain–machine interfaces (BMI) in the context of neurorehabilitation: optimizing BMI learning and performance. IEEE Trans. Neural Syst. Rehabil. Eng. **19**(5), 542–549 (2011)
6. Ono, T., et al.: Brain-computer interface with somatosensory feedback improves functional recovery from severe hemiplegia due to chronic stroke. Front. Neuroeng. **7**, 19 (2014)
7. Rimbert, S., et al.: Can a subjective questionnaire be used as brain-computer interface performance predictor? Front. Hum. Neurosci. **12**, 529 (2018)
8. Neuper, C., et al.: Motor imagery and action observation: modulation of sensorimotor brain rhythms during mental control of a brain–computer interface. Clin. Neurophysiol. **120**(2), 239–247 (2009)
9. Tang, Z.-C., et al.: Classification of EEG-based single-trial motor imagery tasks using a B-CSP method for BCI. Front. Inf. Technol. Electron. Eng. **20**, 1087–1098 (2019)
10. Ang, K.K., Guan, C.: Brain–computer interface for neurorehabilitation of upper limb after stroke. Proc. IEEE **103**(6), 944–953 (2015)
11. Torres, M., et al.: Robotic system for upper limb rehabilitation. In: Braidot, A., Hadad, A. (eds.) VI Latin American Congress on Biomedical Engineering CLAIB 2014, vol. 49, pp. 948–951. Springer, Cham (2015). https://doi.org/10.1007/978-3-319-13117-7_240
12. Vidaurre, C., et al.: Time domain parameters as a feature for EEG-based brain–computer interfaces. Neural Netw. **22**(9), 1313–1319 (2009)
13. Sannelli, C., et al.: A large scale screening study with a SMR-based BCI: categorization of BCI users and differences in their SMR activity. PLoS ONE **14**(1), e0207351 (2019)
14. Blankertz, B., et al.: Optimizing spatial filters for robust EEG single-trial analysis. IEEE Sig. Process. Mag. **25**(1), 41–56 (2008)
15. Yger, F., Berar, M., Lotte, F.: Riemannian approaches in brain-computer interfaces: a review. IEEE Trans. Neural Syst. Rehabil. Eng. **25**(10), 1753–1762 (2017)

16. Barachant, A., et al.: Multiclass brain–computer interface classification by Riemannian geometry. IEEE Trans. Biomed. Eng. **59**(4), 920–928 (2012)
17. Barachant, A., et al.: Classification of covariance matrices using a Riemannian-based kernel for BCI applications. Neurocomputing **112**, 172–178 (2013)
18. Kalunga, E., Chevallier, S., Barthélemy, Q.: Data augmentation in Riemannian space for brain-computer interfaces. In: STAMLINS (2015)
19. Petersen, P.: Riemannian Geometry, vol. 171. Springer, Cham (2016). https://doi.org/10.1007/978-3-319-26654-1
20. Vidaurre, C., et al.: A fully on-line adaptive BCI. IEEE Trans. Biomed. Eng. **53**(6), 1214–1219 (2006)
21. Osuagwu, B.C., et al.: Rehabilitation of hand in subacute tetraplegic patients based on brain computer interface and functional electrical stimulation: a randomised pilot study. J. Neural Eng. **13**(6), 065002 (2016)
22. Scherer, R., et al.: Toward self-paced brain–computer communication: navigation through virtual worlds. IEEE Trans. Biomed. Eng. **55**(2), 675–682 (2008)
23. Lou, B., et al.: Bipolar electrode selection for a motor imagery based brain–computer interface. J. Neural Eng. **5**(3), 342 (2008)
24. Cho, H., et al.: A step-by-step tutorial for a motor imagery–based BCI. In: Brain–Computer Interfaces Handbook, pp. 445–460. CRC Press, Boca Raton (2018)
25. Tang, Z., Li, C., Sun, S.: Single-trial EEG classification of motor imagery using deep convolutional neural networks. Opt.-Int. J. Light. Electron Opt. **130**, 11–18 (2017)

Automatic Identification of Traditional Colombian Music Genres Based on Audio Content Analysis and Machine Learning Techniques

Diego A. Cruz, Cristian C. Cristancho, and Jorge E. Camargo[✉]

Universidad Nacional de Colombia, Bogota, Colombia
{diacruzmo,cccristanchoc,jecamargom}@unal.edu.co
http://www.unsecurelab.org

Abstract. Colombia has a diversity of genres in traditional music, which allows to express the richness of the Colombian culture according to the region. This musical diversity is the result of a mixture of African, native Indigenous, and European influences. Organizing large collections of songs is a time consuming task that requires that a human listens to fragments of audio to identify genre, singer, year, instruments and other relevant characteristics that allow to index the song dataset. This paper presents a method to automatically identify the genre of a Colombian song by means of its audio content. The method extracts audio features that are used to train a machine learning model that learns to classify the genre. The method was evaluated in a dataset of 180 musical pieces belonging to six folkloric Colombian music genres: Bambuco, Carranga, Cumbia, Joropo, Pasillo, and Vallenato. Results show that it is possible to automatically identify the music genre in spite of the complexity of Colombian rhythms reaching an average accuracy of 69%.

Keywords: Music genre classification · Audio feature extraction · Colombian music recognition

1 Introduction

Traditional Colombian music has a clear expression of the country's culture, achieving through its diffusion to share some characteristic of our society. Colombian musical diversity is the result of a mixture of African, native Indigenous, and European influences. The popularity of the Colombian music genres depends on the region. For instance, in the Andean region the most popular genres are Bambuco, Pasillo and Carranga; in the Orinoquia region genres such as the Joropo, Contrapunteo and Pajarillo are in the most popular; in the Caribbean region Cumbia, Vallenato and Porro are the most representative genres; in the Insular region genres such as Reggae, Pasillo Isleño, Vals Isleño are in the main genres in the Colombian islands; in the Pacific region Currulao, Patacorée and Mekerule are in the most popular.

© Springer Nature Switzerland AG 2019
I. Nyström et al. (Eds.): CIARP 2019, LNCS 11896, pp. 646–655, 2019.
https://doi.org/10.1007/978-3-030-33904-3_61

One of the main influences in the central region of Colombia (the Andean Region) was the Waltz genre, which typically sounds one chord per measure, and the accompaniment style particularly associated is to play the root of the chord on the first beat, the upper notes on the second and third beats.

Companies nowadays use music classification, by means of recommendation systems like "Spotify" or simply as a product like "Shazam". The traditional musical genres of Colombia are not widely recognized globally, and in some cases, they have lost popularity and remain exclusively in the regions to which they belong culturally. Identify the musical genres is the first step to exalt and make known more easily both internally and globally this traditional music. Indexing in music information retrieval systems is a very important task to allow search in a music dataset. Machine learning techniques have proved to be successful in the analysis of trends and patterns of music, which although over time and the study of them have increased research on their application in music, there are no enough research focused on the identification of musical genres.

This paper proposes an automatic method to identify the genre of Colombian music. Particularly, we focused on some popular Colombian genres: Bambuco, Carranga, Cumbia, Joropo, Pasillo, and Vallenato. Up to the best of our knowledge this is the first attempt to automatically classify these Colombian genres. The proposed model can be used in the construction of new music information retrieval systems and music recommendation systems to allow the access to Colombian music, which in many cases is inaccessible with current technology. We want to contribute also in the continuity of cultural heritage in these days when traditional Colombian music is increasingly forgotten in the new generations.

The rest of the paper is organized as follows: Sect. 2 presents related work; Sect. 3 presents the proposed model; in Sect. 4 results are presented; and Sect. 5 concludes the paper.

2   Related Work

The classification of musical genres is a field that has always been of great interest in the scientific community in the last times with the application of supervised machine learning techniques, such as Gaussian Mixture model [1] and k-nearest neighbour classifiers [2]. There have been several works that seek to refine more and more the methods to obtain better classifiers. In Bahuleyan [3], authors use conventional machine learning models like Logistic Regression, Random Forests and Gradient Boosting which are trained to classify the audio pieces. Feng [4] studies the pre-trained algorithms such as auto-encoders and restricted Boltzmann machine. Thiruvengatanadhan [5] applies a technique that uses support vector machines (SVM) to classify songs based on features using Mel Frequency Cepstral Coefficients (MFCC).

3   Proposed Method

This section presents the proposed method, which is composed of a song feature extraction process and the training of machine learning models that learn from the audio content.

3.1   Feature Extraction

Music information retrieval is a science that is responsible for the retrieval of the musical information, currently we can find applications like: the classification of musical genres, music transcription, speech recognition, among others. For this work we have used "Librosa" [6], a Python module specialized in performing this task. The features chosen in this investigation are the following ones:

Spectogram
It is a visual representation of the frequency spectrum in a signal, which varies with time (see Fig. 3). A common format is an image that indicates: on the vertical axis the frequency, on the horizontal axis time and a third dimension with the amplitude of a particular frequency at a given moment, represented by the intensity of the color (Fig. 1).

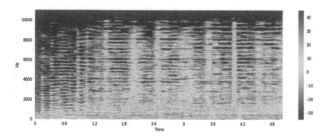

Fig. 1. Spectrogram of a Pasillo song

Spectral Centroid
This property tells us where the mass center is located in a spectrogram and is obtained with the weighted average of the frequencies (see Fig. 3). It helps to predict the "brightness" in a sound, so it is very useful when measuring the "timbre" in an audio (Fig. 2).

Chroma Features
It is strongly related to the 12 semitones in the music, it is a powerful tool for audio analysis that has tones that can be categorized in a significant way, one of its properties allows to capture the harmonic and melodic characteristics (see Fig. 3).

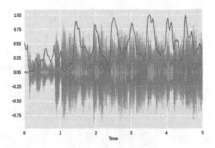

Fig. 2. Spectral centroid

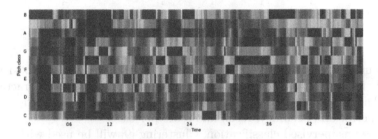

Fig. 3. Chroma features

Zero Crossing Rate

It is the amount of sign changes that a signal experiences, in other words, it is the number of times the signal changes its value, from positive to negative and vice versa. It is used to measure the amount of noise in a signal.

Mel-Frequency Cepstral Coefficients

It is small set of characteristics that concisely describe the general way of a spectral envelope. It is widely used in the retrieval of musical information, to obtain audio similarity measurements and in the classification of genres.

Spectral Rolloff

It is a measure of the shape of a signal, it indicates the frequency in Hz that is below a percentage of the total spectral energy.

3.2 Model Learning

The experimental section is realized through Co-laboratory, a free Jupyter Notebook environment that does not require configuration and that is completely executed in the cloud. Both supervised and unsupervised classification will be used.

We will use two methods of supervised classification, which are Random Forest [7] with the help of the Sci-kit Learn library (Sklearn) and the training of a neural network [8] of 4 layers with the help of the Keras library. Validation partitions will be created to verify that the training of the models does not

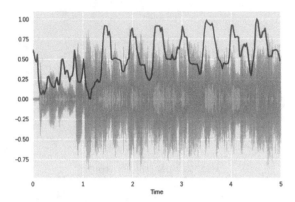

Fig. 4. Spectral rolloff

fall into underfitting or overfitting. Therefore, it will also be analyzed if it is necessary for our data set to make a reduction in the number of characteristics or not. The performance of the method will be evaluated using metrics such as accuracy, error and accuracy, recall and score per class.

For the unsupervised classification, Clustering [9] will be used with the help of centroid-based algorithms (k-means) to see how the groups will be distributed in 3 configurations; first with the original data, then with a reduction of dimensionality using PCA and finally applying t-SNE.

Supervised Classification with Random Forest
The first method that we use is Random Forest, it is a very accurate learning algorithm that consists of a combination of prediction trees, each tree depends on the values of a randomly tested vector independently and with the same distribution for each of these, building a long collection of uncorrelated trees and then averaging them. To apply this classifier, before training it, an analysis of its complexity is performed for different values of estimators, this with the objective of finding the model with the best relation between training error and generalization error (see Fig. 4). Choosing the value of 2^6.

Supervised Classification with Neural Network
The multilayer perceptron (MLP) is a special type of neural network [10] in which several layers of perceptrons are stacked. It is also called Feedforward neural network. The multilayer perceptron is motivated by the little ability of the simple perceptron to model nonlinear functions. For our neural network, we used the Keras library, which allows us to define the base model and add layers as required, in our case our network is built with 4 layers of 256, 128, 64 and 6 neurons respectively.

Clustering Analysis
The objective of Clustering is to group physical or abstract objects in classes of similar objects, it is an unsupervised task, because we do not know how to classify our objects, so the algorithm will only pass the data of the set and not its labels (Fig. 6).

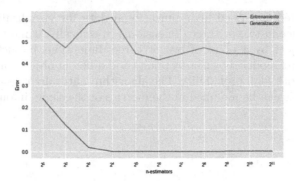

Fig. 5. Random forest

Dimensionality Reduction with PCA

Principal component analysis is a technique used to describe a set of data in terms of new uncorrelated variables called "components". The components are ordered by the amount of original variance, so the technique is useful to reduce the dimensionality of a set of data. This technique is used mainly in exploratory data analysis and in the construction of predictive models. For our particular case when evaluating our data set, which contained 26 different features, we were able to obtain the following accumulated variance graph (see Fig. 5).

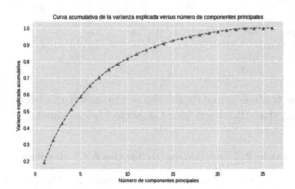

Fig. 6. Accumulated variance

Since our objective is to reduce these characteristics without losing valuable information, we decided to use the principal components analysis to transform our data set into one with 10 components, since its variance is around 82% and more than half of the characteristics are being eliminated.

Dimensionality Reduction with t-SNE

It is an automatic learning algorithm for visualization developed by Laurens Van Der Maaten and Geoffrey Hinton. It is a non-linear dimensionality reduction technique adapted to embed high-dimensional data for visualization in a low-dimensional space in two or three dimensions. Specifically, it models each high-dimensional object by a point of two or three dimensions in such a way that similar objects are modeled by nearby points and different objects are modeled by distant points with high probability. The algorithm naturally uses the Euclidean distance, but this can be modified in the metrics of the same, in our case we use Minkowski.

4   Results

This section presents the obtained results of the conducted classification experiments.

4.1   Classification Results

Table 1 presents the confusion matrix obtained using the Random Forest classifier. Note that the Cumbia genre is the most difficult for this classifier.

Table 1. Confusion matrix for the random forest classifier

	bambuco	carranga	cumbia	joropo	pasillo	vallenato
bambuco	4	0	1	0	1	0
carranga	0	5	1	0	0	0
cumbia	0	1	2	1	0	2
joropo	0	0	0	5	1	0
pasillo	1	0	0	0	5	0
vallenato	0	1	1	0	0	4

In Table 2 the confusion matrix for the neural network classifier is presented. It is worth noting that Bambuco genre is perfectly classified.

Table 2. Confusion matrix for the ANN classifier

	bambuco	carranga	cumbia	joropo	pasillo	vallenato
bambuco	6	0	0	0	0	0
carranga	0	3	0	2	0	1
cumbia	0	0	3	1	0	2
joropo	0	0	0	4	2	0
pasillo	3	0	0	0	3	0
vallenato	0	0	1	2	0	3

In Table 3 a comparison of random forest and ANN is performed. The performance metrics analyzed are accuracy, error, precision, recall and f1-score. The ANN obtained the highest classification performance in terms of accuracy, reaching 69%.

Table 3. Random forest vs neural network.

	RandomForest					
	bambuco	carranga	cumbia	joropo	pasillo	vallenato
Accuracy	0.6944444444					
Error	0.3055555556					
Precision	0.8	0.71428571	0.4	0.83333333	0.71428571	0.66666667
Recall	0.66666667	0.83333333	0.33333333	0.83333333	0.83333333	0.66666667
F1_score	0.72727273	0.76923077	0.36363636	0.83333333	0.76923077	0.66666667
	Red Neuronal					
	bambuco	carranga	cumbia	joropo	pasillo	vallenato
Accuracy	0.6111111111					
Error	0.3888888889					
Precision	0.6667	1	0.75	0.444	0.6	0.5
Recall	1	0.5	0.5	0.66666667	0.5	0.5
F1_score	0.8	0.66666667	0.6	0.53333333	0.54545455	0.5

4.2 2D Visualization

Figure 8 presents a 2D visualization of the songs using all the extracted features. The highlighted circles represent the centroids of the clusters found with k-means. Figure 9 shows a visualization using PCA and Fig. 9 using t-SNE. In this we found that the best number of clusters is 6 using the coefficient silhouette (sc) analysis. In each Figure the inertia score is reported.

It is important to note that reducing the dimensionality of the vector that represents a song produces a more compact feature representation. In this case, results show that t-SNE generates the best visualization of the complete song dataset.

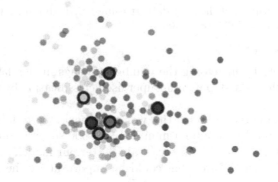

Fig. 7. 2D visualization of the song dataset using all the extracted features. Clusters = 6, inertia = 3227, sc = 0.096

On the unsupervised classification side, it is very clear to observe the improved in the visualization of the different groups or clusters formed after the application of dimensionality reduction methods such as PCA and t-SNE; being the latter one that better allows us to observe the separation between the

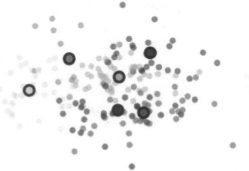

Fig. 8. 2D visualization of the song dataset using PCA to reduce the dimensionality. Clusters $= 6$, inertia $= 2384$, sc $= 0.127$

Fig. 9. 2D visualization of the song dataset using t-SNE to reduce the dimensionality. Clusters $= 6$, inertia $= 2477$, sc $= 0.387$

clusters. Possibly by increasing the number of songs in the data set, the use of t-SNE will be the next step if an unsupervised classification by means of clusters is desired.

After analyzing the results of this work, the next step to follow is the feeding of our data set until we obtain a minimum of 1000 songs in order to have more diversity of data, which will allow us train better our model, in addition to this, the analysis of other features that could be important for the classification and add them to the proposed model (Fig. 7).

In the other hand, if we see the Colombian''s traditional genres we could find that many of them have subgenres that also represent cultures of different regions, for this we think that this work can be applied to a specific genre to classify this sub-genres, for example in the Vallenato we could find the Son, Paseo, Merengue and Puya; in Joropo we could find Contrapunteo, Pasaje, Tonada, Golpe llanero and Copla; and in Carranga the subgenres Rumba and Merengue.

Using this we can improve our model to specify not only the genre but also the sub-genre of the song.

We also would like to obtain information about the instruments that are identified in each song, because also involve the culture of the regions and could be an important feature that helps in classification.

5 Conclusion and Future Work

This paper presented a method based on two classification techniques (supervised and unsupervised). The first technique was the Random Forest, which had the highest performance with 8% more success than the. This may be due to the previous evaluation of the relationship between training error and generalization error; with which we reduce the probability that when training this classifier is so complex as to remember the particularities of the training set (about adjustment/overfitting) or so flexible so as not to model the variability of the data (subfitting/underfitting); an aspect that we only manage with the validation partition in the training of the neural network.

Also we can highlight with the help of metrics and the confusion matrix that the Cumbia is the most difficult of identify musical genre for Random Forest and too one of the least accurate in the neural network, so we can conclude that it is the Colombian genre that is more difficult to classify using these methods.

References

1. Vanek, J., Machlica, L., Psutka, J.: Estimation of single-Gaussian and Gaussian mixture models for pattern recognition. In: Ruiz-Shulcloper, J., Sanniti di Baja, G. (eds.) CIARP 2013. LNCS, vol. 8258, pp. 49–56. Springer, Heidelberg (2013). https://doi.org/10.1007/978-3-642-41822-8_7
2. Taneja, S., Gupta, C., Goyal, K., Gureja, D.: An enhanced k-nearest neighbor algorithm using information gain and clustering (2014)
3. Bahuleyan, H.: Music genre classification using machine learning techniques (2018)
4. Feng, T.: Deep learning for music genre classification (2014)
5. Thiruvengatanadhan, R.: Music classification using MFCC and SVM (2018)
6. Raguraman, P., Mohan, R., Vijayan, M.: LibROSA based assessment tool for music information retrieval systems (2019)
7. Yekkala, I., Dixit, S.: Prediction of heart disease using random forest and rough set based feature selection (2018)
8. Ravi, N.D., Bhalke, D.: Musical instrument information retrieval using neural network (2016)
9. Nisha, Kaur, P.J.: Cluster quality based performance evaluation of hierarchical clustering method (2015)
10. Kukreja, H., Bharath, N., Siddesh, C.S., Kuldeep, S.: An introduction to artificial neural network (2016)

Dual Watermarking for Handwritten Document Image Authentication and Copyright Protection for JPEG Compression Attacks

Ernesto Avila-Domenech[1](✉) ⓘ, Anier Soria-Lorente[1] ⓘ,
and Alberto Taboada-Crispi[2] ⓘ

[1] Universidad de Granma,
Carretera Central vía Holguín Km 1/2, Bayamo, Granma, Cuba
eadomenech@gmail.com, asorial1983@gmail.com
[2] Universidad Central "Marta Abreu" de Las Villas, Santa Clara, Villa Clara, Cuba
ataboada@uclv.edu.cu

Abstract. For authentication and copyright protection of handwritten document images, a dual watermarking algorithm that connects the robust watermarking algorithm based on Krawtchouk moments with a fragile watermarking algorithm based on MD5 hash function is presented. Hence, the robust watermarking algorithm is used to guarantee robustness by modifying frequency coefficients in Krawtchouk moments. Thus, this study proposes a fragile watermarking algorithm, which can perceive in time when the protected image is tampered. Experimental results show that the proposed algorithm can be used for copyright protection for JPEG compression attacks and tampering detection of this images.

Keywords: Handwritten · Image · Watermarking

1 Introduction

The explosive growth of digital multimedia techniques, together with the rapid development of digital network communication has created a pressing demand for techniques that could be used for content authentication and copyright protection. Due to these needs, digital rights management (DRM) is gaining importance; it refers to a range of access control technologies used to limit or restrict the use of digital content. Digital watermarking is useful in DRM systems as it can hide information within the digital content like images, audio and video.

Watermarking technique is effectively applied to content authentication and copyright protection. In accordance with the desired robustness of the embedded watermark, digital watermarking techniques are divided into fragile watermarking and robust watermarking. The first one is designed to detect slight changes to the watermarked image with high probability and the second one is typically used for copyright protection, thus it is designed to resist attacks that attempt

© Springer Nature Switzerland AG 2019
I. Nyström et al. (Eds.): CIARP 2019, LNCS 11896, pp. 656–666, 2019.
https://doi.org/10.1007/978-3-030-33904-3_62

to remove or destroy the watermark without significantly degrading the visual quality of the watermarked image.

When users want to detect illegal tampering and protect the copyright at the same time, the single watermarking algorithm cannot meet the needs of users. Therefore, a dual watermarking algorithm is developed, as it can effectively combine the advantages and functions of the two watermarks [12].

Numerous dual watermarking algorithms have been proposed. In [7], a dual watermarking technique is presented which attempts to establish the owner's right to the image and detect the intentional and unintentional tampering of the image. However, this early research is simply a combination of visible and invisible watermarking algorithms. In [12], a dual watermarking algorithm that connects the robust watermarking algorithm based on singular value decomposition (SVD) with a fragile watermarking algorithm based on compressive sensing (CS) is presented. [11] uses cryptography and QR codes in combination with least significant bits (LSB) and discrete cosine transform (DCT), the authors combines the LSB and DCT approaches because LSB contains spatial domain property and DCT contains frequency domain property.

In [9], a gray scale logo used as copyright information of the owner is embedded imperceptibly into the singular values of the cover image in multiple locations, also performed pixel-wise authentication which fetched the advantage of accurate tamper localization in case of alterations. In [6], a blind dual watermarking mechanism for digital color images is presented. The first watermark is embedded by using the discrete wavelet transform (DWT) in YCbCr color space, and it can be extracted blindly without access to the host image. However, fragile watermarking is based on an improved LSB replacement approach in RGB components for image authentication. In [10], a lifting wavelet transform (LWT) and DCT based robust watermarking approach for tele-health applications is presented. They are based on LWT, which requires less memory, it has reduced aliasing effects and distortion, it is fast and is a good choice for low computational complexity than conventional DWT.

The aforementioned methods have been proposed for images in a general sense. Unlike these methods, our proposal is a dual watermark optimized for handwritten document images.

The rest of the paper is organized as follow; Sect. 2 describes the proposed method including robust watermarking and fragile watermarking. Experimental results are given in Sects. 3 and 4 concludes the paper.

2   Proposed Method

Dual watermarking implies embedding of fragile as well as robust watermarks into the same cover image. It facilitates integration of image authentication and copyright protection into the same scheme. First robust watermarking and then the fragile watermarking should be done because the fragile watermarking is sensitive to small changes. Unlike the fragile watermarking, the robust one resists changes caused by performing the fragile watermarking (Fig. 1).

Fig. 1. Dual watermarking

2.1   Robust Watermarking

The robust watermarking method proposed is similar to the one proposed in [2]. The difference consists of considering any binary image as a watermark. In the previous work only a QR code was considered as a watermark, so it was possible a restructuring of the extracted watermark making use of the characteristics related to the QR codes (Fig. 2).

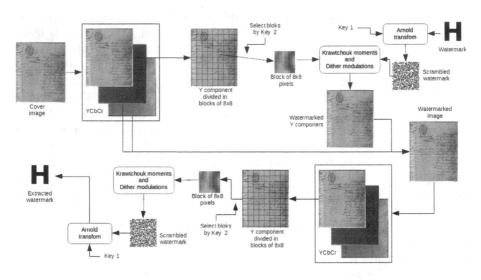

Fig. 2. Watermark embedding and extraction scheme. (Modified from [2])

The following steps are taken during the embedding process:

1. The binary watermark image is scrambled using Arnold transform [1].
2. The cover image is transformed from RGB to YCbCr color space, and the Y component, corresponding to the luminance information, is divided into small image blocks of 8×8 pixels.
3. A number of blocks equal to the number of bits to be inserted is selected from a given key.
4. The Krawtchouk moments [13] of the selected blocks are determined.

5. Watermark bit is embedded in the selected block moments using Dither modulation [3]. The values 19 and 128 are used as the coefficient and embedding strength values respectively. Watermarked blocks can be obtained.
6. The YCbCr to RGB color space is transformed to obtain RGB watermarked image.

For watermark extraction:

1. The watermarked image is transformed from the RGB to the YCbCr color space and the Y component is divided into 8×8 pixels blocks.
2. Some blocks are selected from which they will be extracted from the key used in the embedding process.
3. The Krawtchouk moments of the selected blocks are determined.
4. Scrambled watermark bits are obtained with the selected blocks moments using Dither modulations.
5. Finally, a watermark is constructed with the scrambled bits using Arnold transform.

2.2 Fragile Watermarking

As we know, a hash function, such as MD5 or SHA-256, can be utilized to authenticate the data. If the hash value of original message is exactly equal to the re-calculated hash value of the received message, the received data can be regarded as integrated, otherwise as false.

For the process of embedding the following steps are performed for each RGB component:

1. The component is divided into 32×32 non-overlapped blocks.
2. 128 pixels of each block are selected by a given key.
3. The least significant bit (LSB) of each selected pixel is assigned the value 0.
4. The MD5 hash value of the modified block is generated as a watermark.
5. The watermark is embedded into the LSB of the selected pixels and a watermarked block image is obtained.

Detecting a fragile watermark water is the reverse process of embedding watermark, which is used to detect whether the watermarked image has been tampered and what the precise position of the tampered parts is. For this:

1. The RGB image is divided into 32×32 non-overlapped blocks.
2. 128 pixels of each block are selected by a given key.
3. Three binary series are formed from the LSBs of the selected pixels.
4. The LSBs of each selected pixel are assigned the value 0.
5. The MD5 hash value of the modified block is generated and compared with obtained series.

3   Experiments and Results

The watermarking algorithm is evaluated through imperceptibility, tamper detection and robustness. Also, it is compared with the methods proposed in [9] and [6]. This last method has the variable k as a parameter, that corresponds to the strength of the watermark. A higher k can increase the strength of the embedded watermark, but it makes the watermarked image easier to perceive. For this reason and to make a better comparison, we have taken four different k values (0.2, 0.4, 0.8 and 1.0).

We used two handwritten document image databases: Saint Gall [4] and Parzival [5] database. The first one contains manuscripts from the 9th century using Carolingian scripts by a single writer, while the Parzival is compiled from 13th century Gothic scripts [8].

3.1   Imperceptibility

We calculated the larger peak signal-to-noise ratio (PSNR) which compares the similarity between the original image I, and the watermarked image I_w. A higher PSNR indicates that the watermarked image more closely resembles the original image meaning that the watermark is more imperceptible.

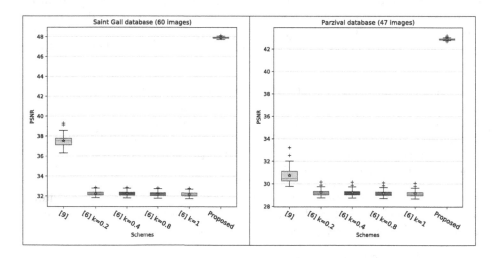

Fig. 3. PSNR values for Saint Gall and Parzival database watermarked images.

For both databases, the proposed method obtains higher values of PSNR compared to [9] and [6] in its four variants (see Fig. 3). Also, it can be noticed that [9] provides improvements with respect to [6]. In addition, it is observable that by varying the parameter k, similar values of PSNR are contained.

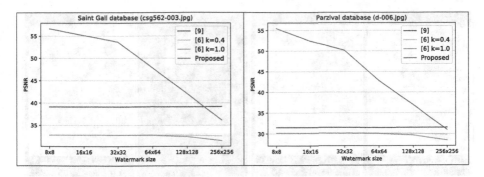

Fig. 4. PSNR behavior to mark the "csg562-003.jpg" image of Saint Gall database and "d-006.jpg" image of Parzival database with watermarks of different sizes.

As a preliminary experiment, two images were taken, one from each database, and tests were performed to obtain the corresponding PSNR by varying the size of the watermark. For this case, the dimensions 8×8, 16×16, 32×32, 64×64, 128×128 and 256×256 pixels were taken as a watermark. As shown in Fig. 4 the proposed method obtains better imperceptibility values for the first five dimensions, only in the sixth is it slightly exceeded by [9].

3.2   Tamper Detection

Tamper area detection capability is evaluated, by modifying the contents of images. We developed our proposed fragile watermarking particularly for integrity images and locating tampered areas. Figure 5 shows the modified water-marked image by text addition, word substitution, underline words, content removal, and their corresponding tamper detection results.

The results obtained by [9] are acceptable. Of the possible modifications, there is a 50% probability that $\left(\sum_{m=1}^{5} XOR\right) mod. 2$ is the same as the water-mark bit. Likewise, both the [6] method and the proposed one detect the modifications made in an acceptable way.

3.3   Robustness

The robustness is measured as the bit error rate (BER) corresponding to incorrectly formed binary values of the watermark image.

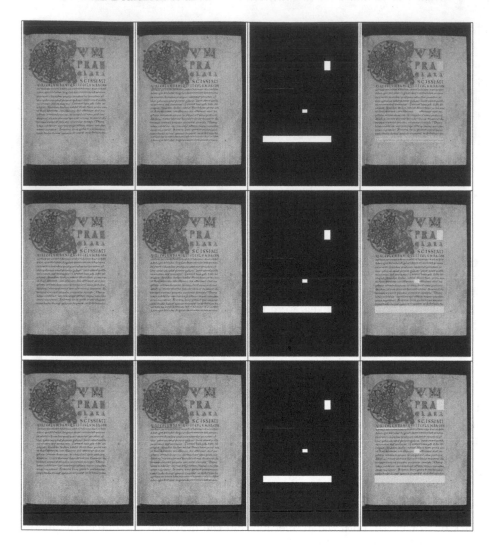

Fig. 5. Watermarked image, modified watermarked, tamper zone and tamper detection corresponding to [6,9] and proposed scheme.

The main contributions of this paper are twofold. First is the obtaining of better values of imperceptibility, and the second is the remarkable improvement in the strength when JPEG compression attacks at 75%, 50% and 25% are applied (see Figs. 6, 7 and 8). Similar to the PSNR, a preliminary experiment, using the same two images, was performed to obtain the corresponding BER by varying the size of the watermark (Figs. 9 and 10).

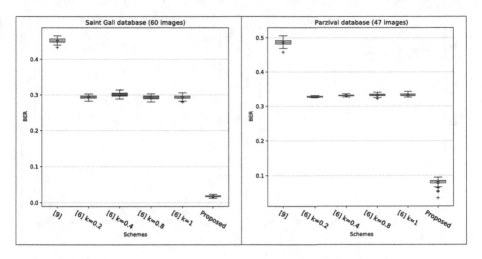

Fig. 6. BER values for watermarked images with JPEG compression (QF = 75%).

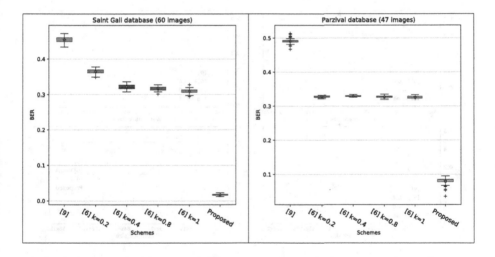

Fig. 7. BER values for watermarked images with JPEG compression (QF = 50%).

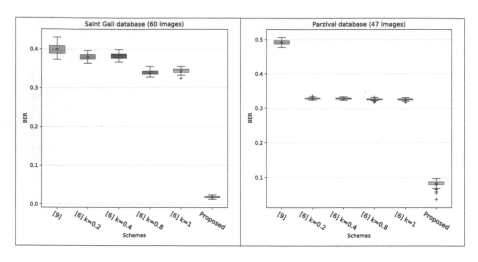

Fig. 8. BER values for watermarked images with JPEG compression (QF = 25%).

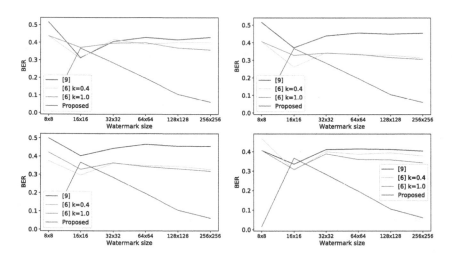

Fig. 9. BER behavior to mark the image "csg562-003.jpg" of Saint Gall database with watermarks of different sizes when no attack is applied, a JPEG compression is performed with QF = 75%, 50% and 25% respectively.

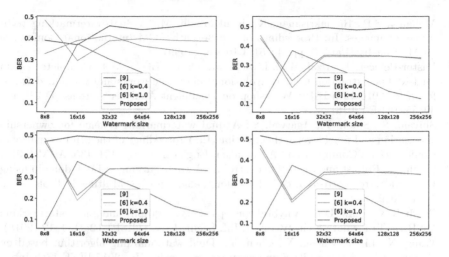

Fig. 10. BER behavior to mark the image "d-006.jpg" of the Parzival database with watermarks of different sizes when no attack is applied, a JPEG compression is performed with QF = 75%, 50% and 25% respectively.

4   Conclusions

In this paper, a dual watermarking technique based on Krawtchouk moments and MD5 hash function was implemented. The experimental results show that our scheme is robust to JPEG compression attacks. In addition, the values corresponding to the PSNR were improved compared to previously presented papers.

References

1. Arnol'd, V.I., Avez, A.: Ergodic problems of classical mechanics. In: The Mathematical Physics Monograph Series. W. A. Benjamin, New York (1968). http://cds.cern.ch/record/1987366
2. Avila-Domenech, E., Soria-Lorente, A.: Watermarking based on Krawtchouk moments for handwritten document images. In: Hernández Heredia, Y., Milián Núñez, V., Ruiz Shulcloper, J. (eds.) IWAIPR 2018. LNCS, vol. 11047, pp. 122–129. Springer, Cham (2018). https://doi.org/10.1007/978-3-030-01132-1_14
3. Chen, B., Wornell, G.W.: Quantization index modulation: a class of provably good methods for digital watermarking and information embedding. IEEE Trans. Inf. Theory **47**(4), 1423–1443 (2001)
4. Fischer, A., Frinken, V., Fornés, A., Bunke, H.: Transcription alignment of Latin manuscripts using hidden Markov models. In: Proceedings of the 2011 Workshop on Historical Document Imaging and Processing, pp. 29–36. ACM (2011)
5. Fischer, A., et al.: Automatic transcription of handwritten medieval documents. In: 2009 15th International Conference on Virtual Systems and Multimedia, pp. 137–142. IEEE (2009)
6. Liu, X.L., Lin, C.C., Yuan, S.M.: Blind dual watermarking for color images' authentication and copyright protection. IEEE Trans. Circ. Syst. Video Technol. **28**(5), 1047–1055 (2018)

7. Mohanty, S.P., Ramakrishnan, K., Kankanhalli, M.: A dual watermarking technique for images. In: Proceedings of the Seventh ACM International Conference on Multimedia (Part 2), pp. 49–51. Citeseer (1999)
8. Pastor-Pellicer, J., Afzal, M.Z., Liwicki, M., Castro-Bleda, M.J.: Complete system for text line extraction using convolutional neural networks and watershed transform. In: 2016 12th IAPR Workshop on Document Analysis Systems (DAS), pp. 30–35. IEEE (2016)
9. Shivani, S., Singh, P., Agarwal, S.: A dual watermarking scheme for ownership verification and pixel level authentication. In: Proceedings of the 9th International Conference on Computer and Automation Engineering, pp. 131–135. ACM (2017)
10. Singh, A.: Robust and distortion control dual watermarking in LWT domain using DCT and error correction code for color medical image. Multimed. Tools Appl. 1–11 (2019)
11. Singh, R.K., Shaw, D.K.: A hybrid concept of cryptography and dual watermarking (LSB_DCT) for data security. Int. J. Inf. Secur. Priv. (IJISP) **12**(1), 1–12 (2018)
12. Wang, N., Li, Z., Cheng, X., Chen, Y.: Dual watermarking algorithm based on singular value decomposition and compressive sensing. In: 2017 IEEE 17th International Conference on Communication Technology (ICCT), pp. 1763–1767. IEEE (2017)
13. Yap, P., Paramesran, R., Ong, S.H.: Image analysis by Krawtchouk moments. IEEE Trans. Image Process. **12**(11), 1367–1377 (2003)

Speech Recognition

A Survey of the Effects of Data Augmentation for Automatic Speech Recognition Systems

Jose Manuel Ramirez[✉], Ana Montalvo[✉], and Jose Ramon Calvo[✉]

Advanced Technologies Application Center, 7th A Street, #21406, Havana, Cuba
{jsanchez,amontalvo,jcalvo}@cenatav.co.cu
https://portal.cenatav.co.cu/

Abstract. Data augmentation has been proposed as a method to increase the quantity of training data. It is a common strategy adopted to avoid over-fitting, reduce mismatch and improve robustness of the models. But, would the system performance improve if we add data of any nature? This paper presents a survey about data augmentation techniques and its effect on Automatic Speech Recognition systems, some experiments were carried out to support the hypothesis that adding noise is not allways help.

Keywords: Data augmentation · Speech recognition

1 Introduction

Data augmentation is a popular technique for increasing the size of labeled training sets by applying class-preserving transformations to create copies of labeled data points [3]. Data augmentation in Automatic Speech Recognition (ASR) is an effective method to reduce mismatch between training and testing samples, improve robustness of the models and to avoid over-fitting.

Which are the most used techniques to get your data augmented? In which stage of an ASR system should be added more data? Are there any available tools or databases to get training data transformed and augmented? How much does ASR performance benefits from data augmentation?

Mostly motivated by deep learning revolution and its data greedy approach [11], data augmentation has played an important role in ASR, where the main focus of research has been on designing better network architectures that tend to over-fit easily and require large amounts of labeled training data [2].

Data augmentation strategy depends strongly on the ASR architecture. The simplicity of "end-to-end" models and their recent success in neural machine-translation have prompted considerable research into replacing conventional ASR architectures with a single "end-to-end" model, which trains the acoustic and language models jointly rather than separately. Recently, state-of-the-art results have been achieved [2] using an attention-based encoder-decoder model

© Springer Nature Switzerland AG 2019
I. Nyström et al. (Eds.): CIARP 2019, LNCS 11896, pp. 669–678, 2019.
https://doi.org/10.1007/978-3-030-33904-3_63

trained on over 12K h of speech data. However, on large publicly available corpora, such as "Librispeech" or "Fisher English", which are one order of magnitude smaller, performance still lags behind that of conventional systems.

The goal of data augmentation in "end-to-end" ASR systems is to leverage much larger text corpora alongside limited amounts of speech datasets to improve performance. Various methods of leveraging these text corpora have improved "end-to-end" ASR performance [21], for instance: composes recurrent neural network output lattices with a lexicon and word-level language model, while [1] simply re-scores beams with an external language model [13,20] incorporate a character-level language model during beam search, possibly disallowing character sequences absent from a dictionary, while [8] includes a full word level language model in decoding by simultaneously keeping track of word histories and word prefixes.

On the other hand there is conventional ASR architecture: hybrid Hidden Markov Model - Deep Neural Network (HMM-DNN), on which this paper is focused. In HMM-DNN based ASR systems, data augmentation can be done to improve the Acoustic Modeling (AM) or the Language Modeling (LM), and there are several ways to do so. The authors propose a taxonomy of the most used data augmentation techniques for HMM-DNN based ASR systems.

The remainder of the paper is organized as follows: Sect. 2 presents the taxonomy proposed by the authors relative to data augmentation methods for acoustic modeling in HMM-DNN based ASR systems. As part of the Sect. 3 data sets, tools and experimental setup are shown. Section 4 gives the experimental results and Sect. 5 is devoted to the conclusions of the paper.

2   Data Augmentation for HMM-DNN Based ASR Systems

Given augmenting data technique, an important question is how to best exploit the augmented data. The answer to this question will ultimately depend on the particular architecture the speech recognizer adopts and the nature and amount of augmented data used.

Concerning data augmentation methods to improve acoustic models, the following taxonomy is proposed:

- semi-supervised training,
- transformation of acoustic data,
- speech synthesis.

2.1   Semi-supervised Training

The semi-supervised training approach assumes the use of the text produced by an automatic speech recognition system to train acoustic models. In other words the unlabeled data may be adopted by recognizing it with an existing or boot-strapped system, filtering out those utterances that fail to pass confidence threshold [6,34] and re-training the system on supervised and filtered unlabeled training data.

The main advantage of this approach is that it is generally possible to collect vast amounts of such data, e.g., radio and television news broadcasts, covering all sorts of speaker and noise conditions [19].

The main disadvantage of this type of data is the lack of correct transcriptions. This limits possible gains from the approaches particularly sensitive to the accuracy of transcriptions supplied, such as discriminative training [31] and speaker adaptation based on discriminative criteria [32].

To date the majority of work has considered individual data augmentation schemes, with few consistent performance contrasts or examination of whether the schemes are complementary. In [29] two data augmentation schemes, semi-supervised training and vocal tract length perturbation, are examined and combined.

2.2 Transformation of Acoustic Data

The methods based on transformation of acoustic features include the variation of the Vocal Tract Length (VTL) on the stage of extracting the standard features [14] and its extended version to large vocabulary continuous speech recognition presented in [3]. In [3] instead of randomly choosing a warping factor for each utterance of a speaker, they deterministically perturb the estimated VTL warping factor of a speaker.

They also proposed a novel data augmentation approach based on Stochastic Feature Mapping (SFM) for utterance transformation. SFM estimates a maximum likelihood linear transformation in some feature space of the source speaker against the speaker dependent model of the target speaker. Different from vocal track and length perturbation (VTLP) which perturbs a speaker, SFM explicitly maps the features of a speaker to some target speaker based on a statistically estimated linear transformation.

In [16] they did research into an elastic spectral distortion method to artificially augment training samples to help HMM-DNNs acquire enough robustness even when there are a limited number of training samples. Three distortion methods were proposed: vocal tract length distortion, speech rate distortion, and frequency-axis random distortion.

The family of techniques based on acoustic data transformations includes methods such as audio signal speed alteration [18], applying noises, introduction of artificial reverberation into the records [24].

In [18] experiments are conducted with audio speed perturbation, which emulates a combination of pitch perturbation and VTLP, but it shows to perform better than either of those two methods. It is particularly recommended to change the speed of the audio signal, producing versions of the original signal with different speed factors.

In [23] the authors propose SpecAugment, an augmentation method that operates on the log mel spectrogram of the input audio, rather than the raw audio itself. This method is applied on Listen, Attend and Spell networks for end-to-end speech recognition tasks, which even when it is beyond this paper scope it is a very interesting proposal that could be evaluated over conventional ASR systems. SpecAugment consists of three kinds of deformations of the log mel spectrogram. The first is time warping, a deformation of the time-series in the time direction. The other two augmentations, proposed in computer vision [5], are time and frequency masking, where it is masked a block of consecutive time steps or mel frequency channels.

2.3 Speech Synthesis

The synthesized data may refer to existing but perturbed in a certain way data as well as new artificially generated data. One major advantage of synthesized data is that, similar to semi-supervised case, it is possible to collect vast amounts of such data. Another important advantage of synthesized datasets lies in the ability to approximate the required recognition conditions and get the necessary amount of training data. In addition, this method allows to obtain a precise alignment of noised data using known text transcriptions and the corresponding clean recordings.

Furthermore, different to semi-supervised case, the correctness of associated transcriptions is usually guaranteed. A major disadvantage of this type of data could be its quality.

Corrupting clean training speech with noise was found to improve the robustness of the speech recognizer against noisy speech. In [7,9], noisy audio has been synthesized via superimposing clean audio with a noisy audio signal.

The use of an acoustic room simulator has been explored in [17]. This paper describes a system that simulate millions of different utterances in millions of virtual rooms, and use the generated data to train deep-neural network models. This simulation based approach was employed on Google Home product and brought significant performance improvement.

3   Experimental Setup

In order to evaluate the impact of the augmented data on the effectiveness of an ASR task, we decided to measure the Word Error Rate (WER) obtained by several ASR systems facing the same decoding scenario. The five systems used in the experimentation: tri1, tri2, tri3, sgmm and dnn; were trained over the same training set, with the same LM and evaluated on the same set of test data; but in each system a new AM and a transformation in the feature space was tested. All systems share the same 3-gram LM and the acoustic feature. The acoustic feature selected was Mel Frequency Cepstral Coefficients (MFCC) [4] with a Mel Filter Bank of 40 filters (8 filter per octave), discarding the value of the energy of the frame for a total of 13 coefficients. This will allow knowing if the new noise data used in training enhance the performance of the ASR systems used in the experimentation.

The first three systems tri1, tri2 and tri3 have an Hidden Markov Model - Gaussian Mixture Model (HMM-GMM) triphone based architecture [33], this kind of statistical systems have the characteristic that after a certain amount of data the accuracy remained constant, unlike HMM-DNN that improve its accuracy when the amount of data increased. The feature extraction phase in system tri1 starts with the 13-dimensional MFCC feature, then Cepstral Mean and Variance Normalization (CMVN) [33] is applied over MFCC features and concatenated with their first and second order regression coefficients [33], to obtain a 39-dimensional vector. The feature extraction phase in system tri2 starts with 13-dimensional MFCC features that are spliced across ±4 frames to obtain 117-dimensional vectors. Linear Discriminant Analysis (LDA) [15] is applied to reduce the dimensionality to 40, using context-dependent HMM states as classes for the acoustic model estimation. Maximum Likelihood Linear Transform (MLLT) [28] is applied to the resulting features, making them more accurately modeled by diagonal-covariance Gaussians. The feature extraction phase in system tri3 is the same of tri2, but with one extra step before applying MLLT: a feature-space maximum likelihood linear regression (fMLLR) [27] is applied to normalize inter-speaker variability of the features. Systems tri1 and tri2 have similar AM, unlike tri3 that uses a speaker adaptive training (SAT) [33].

The fourth system (sgmm) have a triphonic Hidden Markov Model - Subspace Gaussian Mixture Model (HMM-SGMM) architecture [25]. The feature extraction phases of sgmm and tri3 are equals.

The last system (dnn) have a HMM-DNN architecture and its feature extraction phase is the same of tri3 and sgmm, but with three extra steps: before fMLLR comes a new spliced across ±4 frames, then another LDA over the spliced frames is applied and last another splice across ±4 frames for a 160-dimensional final vector.

3.1 Kaldi Toolkit

Kaldi is a set of free and open source tools, developed by Daniel Povey et al. [26] for research in ASR area. Kaldi allows to build ASR systems through a series of well-documented shell command routines. ASR systems in Kaldi are based on weighted finite state transducers that optimize training and decoding processes [22].

3.2 Speech Data

The TC-STAR project, funded by the European Commission, represents a long-term effort focused on advanced research in language technologies such as ASR, automatic speaker recognition, automatic speech translation and speech synthesis [10].

The TC-STAR recordings used in the experimentation, which we will call from now on TC_STAR_USED, correspond to sessions of the European Parliament or sessions of the Spanish Court where the announcers speak only Spanish. The recorded sessions of TC_STAR_USED come from different scenarios (two

Table 1. TC_STAR_USED corpus description.

	Training data	Test data
Sessions	17 163	1 908
Words	241 413	33 492
Unique words	15 722	4 178
Speakers	53 f 100 m	9 f 14 m
Hours	26:38:59	3:36:16

auditorium-like scenarios), it has multiple speakers of both genders (60 female and 112 male) and different ages, there are sessions of spontaneous speech and since some speakers talk at different times of the session or sometimes on different days, this adds variability between recorded sessions of the same speaker, which makes ASR task extremely complex and real. Criterion 90–10 was followed for the creation of training and test sets; where the training data consists of a set of recorded sessions corresponding to 90% of the total time (including silences) of the TC_STAR_USED data and the remaining 10% to the test data. The test data set shares 2 female and 2 male speakers with the training data set. Table 1 provides a summary of the characteristics of TC_STAR_USED where the term word refers to the spoken words, the interjections and sounds without linguistic information labelled in the database as noise or throat clearing. The format of the audio files is the standard RIFF (.wav) encoded PCM, 16 bits signed, at 16 kHz without compression.

3.3   Noise Database and Fant Tool-Kit

We used a noise database and a tool to simulate noisy conditions with different Signal-to-Noise Ratio (SNR) levels in the TC_STAR_USED train set, to evaluate the impact of augmenting training data in a ASR system over the WER in our clean test scenario. The noise database selected in this research has been DEMAND for more detail see description in [30]. All DEMAND noises can be classified into two groups, according to the nature of the noise type: in-door and out-door.

The selected tool to simulate the audio files of the test data set with the noises of the DEMAND database was FaNT - Filtering and Noised Adding Tool [12]. This tool allows us to add noise to speech sessions recorded with a desired SNR.

Using FaNT and the two groups of noises provided by DEMAND, we were able to augment the training set many times. This is possible by choosing a specific SNR value or range and one of the two groups of noises in DEMAND. For this research we used two range of SNR, the first one, 5 dB–15 dB, and the second one, 15 dB–25 dB. All this provided us with a four times bigger training data set, but just in duration because the amount of unique words and unique phrases said did not increase. Table 2 provides a summary of the characteristics of the new training data set, called TC_STAR_AUGMENTED.

Table 2. TC_STAR_AUGMENTED corpus description.

	Training data	Test data
Sessions	85 815	1 908
Words	1 207 065	33 492
Unique words	15 722	4 178
Speakers	53 f 100 m	9 f 14 m
Hours	133:12:25	3:36:16

4 Results

In this section we present the result of decoding the test data set with and without data augmentation. Tables 3 shows the difference between the WER of the decoding processes for all models tested, training with TC_STAR_USED and TC_START_AUGMENTED. Because the WER is an error metric, lower values are more desired than higher ones, positive differences mean that the efficiency of the recognition using data augmentation is better than when it is not used.

Table 3. Differences of WER (TC_STAR_USED - TC_START_AUGMENTED)

Systems	WER	INS	DEL	SUB
tri1	0.30%	0.20%	−0.58%	1.36%
tri2	0.26%	−0.48%	0.75%	−0.27%
tri3	0.32%	1.38%	−2.01%	0.64%
sgmm	0.11%	1.47%	−0.34%	−1.13%
dnn	**1.10%**	−0.92%	0.75%	0.17%

In data presented in Table 3 the most important difference occurs in the dnn model, where the impact of data augmentation was better. This behavior is consistent with what was described in the previous sections. The interesting thing about this experiment is that the increase in the accuracy of DNN-based models is small (and only due to the large amount of data involved, statistically relevant). Our hypothesis about the cause of this behavior is that the augmentation of the data using different noises did not contribute to having new phonetic realizations in the data, but rather redundancy. This would explain why the data augmentation method using noise is usually used to improve robustness of the recognition, because the network can "learn" about the nature of noise processing the augmented data; but this doesn't entail improving the recognition of clean signals.

5   Conclusions

Data augmentation is intended as a procedure that starting from your available data, multiplies the amount of it by producing versions of the original. Data augmentation strategy depends on the goal of the classification. For ASR all different ways of doing data augmentation could be put in one of the three categories proposed.

From the experiments carried out, it can be concluded that noise-based data augmentation methods are not suitable for raising the recognition rate in ASR systems, because from the data augmented the network learns more about the nature of the noise than about the phonetic combinations present in the utterances; this knowledge allows a system to deal better with different acoustic conditions but does not help to obtain better recognition rates.

From the results obtained it can be inferred that data augmentation procedures based on the simulation of acoustic conditions different from those present in the training set, for example simulating different acoustic channels or increasing the reverberation in the new signals, will not have an impact on the recognition rate of ASR systems. Future studies, with other data augmentation methods, might allow us to generalize our hypothesis.

References

1. Chan, W., Jaitly, N., Le, Q.V., Vinyals, O.: Listen, attend and spell: a neural network for large vocabulary conversational speech recognition. In: ICASSP (2016). http://williamchan.ca/papers/wchan-icassp-2016.pdf
2. Chiu, C.C., et al.: State-of-the-art speech recognition with sequence-to-sequence models. In: 2018 IEEE International Conference on Acoustics, Speech and Signal Processing (ICASSP), April 2018. https://doi.org/10.1109/icassp.2018.8462105
3. Cui, X., Goel, V., Kingsbury, B.: Data augmentation for deep neural network acoustic modeling. In: 2014 IEEE International Conference on Acoustics, Speech and Signal Processing, ICASSP 2014, vol. 23, pp. 5582–5586, May 2014. https://doi.org/10.1109/ICASSP.2014.6854671
4. Davis, S.B., Mermelstein, P.: Comparison of parametric representations for monosyllabic word recognition in continuously spoken sentences. In: Readings in Speech Recognition, pp. 65–74. Morgan Kaufmann Publishers Inc., San Francisco (1990). http://dl.acm.org/citation.cfm?id=108235.108239
5. DeVries, T., Taylor, G.W.: Improved regularization of convolutional neural networks with cutout (2017)
6. Evermann, G., Woodland, P.: Large vocabulary decoding and confidence estimation using word posterior probabilities (2000)
7. Gales, M.J.F., Ragni, A., AlDamarki, H., Gautier, C.: Support vector machines for noise robust ASR. In: 2009 IEEE Workshop on Automatic Speech Recognition Understanding, pp. 205–210, November 2009. https://doi.org/10.1109/ASRU.2009.5372913

8. Graves, A., Jaitly, N.: Towards end-to-end speech recognition with recurrent neural networks. In: Proceedings of the 31st International Conference on International Conference on Machine Learning, ICML 2014, vol. 32, pp. II-1764–II-1772. JMLR.org (2014). http://dl.acm.org/citation.cfm?id=3044805.3045089
9. Hannun, A., et al.: Deep speech: scaling up end-to-end speech recognition (2014)
10. Van den Heuvel, H., Choukri, K., Gollan, C., Moreno, A., Mostefa, D.: TC-STAR: new language resources for ASR and SLT purposes. In: LREC, pp. 2570–2573 (2006)
11. Hinton, G., et al.: Deep neural networks for acoustic modeling in speech recognition. Sig. Process. Mag. **29**, 82–97 (2012)
12. Hirsch, H.G.: Fant-filtering and noise adding tool. Niederrhein Univ. Appl. Sci. (2005). http://dnt.kr.hsnr.de/download.html
13. Hori, T., Watanabe, S., Zhang, Y., Chan, W.: Advances in joint CTC-attention based end-to-end speech recognition with a deep CNN encoder and RNN-LM. CoRR abs/1706.02737 (2017). http://arxiv.org/abs/1706.02737
14. Jaitly, N., Hinton, E.: Vocal tract length perturbation (VTLP) improves speech recognition. In: Proceedings of the 30th International Conference on Machine Learning (2013)
15. Kajarekar, S.S., Yegnanarayana, B., Hermansky, H.: A study of two dimensional linear discriminants for ASR. In: Proceedings of the 2001 IEEE International Conference on Acoustics, Speech, and Signal Processing (Cat. No. 01CH37221), vol. 1, pp. 137–140, May 2001. https://doi.org/10.1109/ICASSP.2001.940786
16. Kanda, N., Takeda, R., Obuchi, Y.: Elastic spectral distortion for low resource speech recognition with deep neural networks. In: 2013 IEEE Workshop on Automatic Speech Recognition and Understanding, Olomouc, Czech Republic, 8–12 December 2013, pp. 309–314 (2013). https://doi.org/10.1109/ASRU.2013.6707748
17. Kim, C., et al.: Generation of large-scale simulated utterances in virtual rooms to train deep-neural networks for far-field speech recognition in Google home, pp. 379–383 (2017). http://www.isca-speech.org/archive/Interspeech_2017/pdfs/1510.PDF
18. Ko, T., Peddinti, V., Povey, D., Khudanpur, S.: Audio augmentation for speech recognition. In: INTERSPEECH (2015)
19. Lamel, L., Gauvain, J.L., Adda, G.: Lightly supervised and unsupervised acoustic model training. Comput. Speech Lang. **16**, 115–129 (2002). https://doi.org/10.1006/csla.2001.0186
20. Maas, A.L., Xie, Z., Jurafsky, D., Ng, A.Y.: Lexicon-free conversational speech recognition with neural networks. In: Proceedings of the North American Chapter of the Association for Computational Linguistics (NAACL) (2015)
21. Miao, Y., Gowayyed, M., Metze, F.: EESEN: end-to-end speech recognition using deep RNN models and WFST-based decoding. CoRR abs/1507.08240 (2015). http://arxiv.org/abs/1507.08240
22. Mohri, M., Pereira, F., Riley, M.: Speech recognition with weighted finite-state transducers. In: Benesty, J., Sondhi, M.M., Huang, Y.A. (eds.) Springer Handbook of Speech Processing. SH, pp. 559–584. Springer, Heidelberg (2008). https://doi.org/10.1007/978-3-540-49127-9_28
23. Park, D.S., et al.: SpecAugment: a simple data augmentation method for automatic speech recognition. CoRR abs/1904.08779 (2019). http://arxiv.org/abs/1904.08779
24. Peddinti, V., Chen, G., Povey, D., Khudanpur, S.: Reverberation robust acoustic modeling using i-vectors with time delay neural networks. In: INTERSPEECH (2015)

25. Povey, D.: A tutorial-style introduction to subspace Gaussian mixture models for speech recognition. Microsoft Research, Redmond, WA (2009)
26. Povey, D., et al.: The Kaldi speech recognition toolkit. Technical report, IEEE Signal Processing Society (2011)
27. Povey, D., Yao, K.: A basis representation of constrained MLLR transforms for robust adaptation. Comput. Speech Lang. **26**(1), 35–51 (2012)
28. Psutka, J.V.: Benefit of maximum likelihood linear transform (MLLT) used at different levels of covariance matrices clustering in ASR systems. In: Matoušek, V., Mautner, P. (eds.) TSD 2007. LNCS (LNAI), vol. 4629, pp. 431–438. Springer, Heidelberg (2007). https://doi.org/10.1007/978-3-540-74628-7_56
29. Ragni, A., Knill, K.M., Rath, S.P., Gales, M.J.: Data augmentation for low resource languages (2014)
30. Thiemann, J., Ito, N., Vincent, E.: The diverse environments multi-channel acoustic noise database (DEMAND): a database of multichannel environmental noise recordings. In: Proceedings of Meetings on Acoustics, ICA 2013, vol. 19, p. 035081. ASA (2013)
31. Wang, L., Gales, M.J.F., Woodland, P.C.: Unsupervised training for mandarin broadcast news and conversation transcription. In: 2007 IEEE International Conference on Acoustics, Speech and Signal Processing, ICASSP 2007, vol. 4, pp. IV-353–IV-356, April 2007. https://doi.org/10.1109/ICASSP.2007.366922
32. Wang, L., Woodland, P.C.: Discriminative adaptive training using the MPE criterion. In: 2003 IEEE Workshop on Automatic Speech Recognition and Understanding (IEEE Cat. No. 03EX721), pp. 279–284, November 2003. https://doi.org/10.1109/ASRU.2003.1318454
33. Young, S.: HMMs and related speech recognition technologies. In: Benesty, J., Sondhi, M.M., Huang, Y.A. (eds.) Springer Handbook of Speech Processing. SH, pp. 539–558. Springer, Heidelberg (2008). https://doi.org/10.1007/978-3-540-49127-9_27
34. Zavaliagkos, G., Colthurst, T.: Utilizing untranscribed training data to improve performance. In: DARPA Broadcast News Transcription and Understanding Workshop, Landsdowne, pp. 301–305 (1998)

Multi-channel Convolutional Neural Networks for Automatic Detection of Speech Deficits in Cochlear Implant Users

Tomas Arias-Vergara[1,2,3(✉)], Juan Camilo Vasquez-Correa[1,2],
Sandra Gollwitzer[3], Juan Rafael Orozco-Arroyave[1,2], Maria Schuster[3],
and Elmar Nöth[2]

[1] Faculty of engineering, Universidad de Antioquia UdeA,
Calle 70 No. 52-21, Medellín, Colombia
[2] Pattern Recognition Lab, Friedrich-Alexander University,
Erlangen-Nürnberg, Germany
tomas.ariasvergara@lmu.de
[3] Department of Otorhinolaryngology, Head and Neck Surgery,
Ludwig-Maximilians University, Munich, Germany

Abstract. This paper proposes a methodology for automatic detection of speech disorders in Cochlear Implant users by implementing a multi-channel Convolutional Neural Network. The model is fed with a 2-channel input which consists of two spectrograms computed from the speech signals using Mel-scaled and Gammatone filter banks. Speech recordings of 107 cochlear implant users (aged between 18 and 89 years old) and 94 healthy controls (aged between 20 and 64 years old) are considered for the tests. According to the results, using 2-channel spectrograms improves the performance of the classifier for automatic detection of speech impairments in Cochlear Implant users.

Keywords: Speech processing · Time-frequency analysis ·
Multi-channel CNN · Deep learning · Cochlear Implants

1 Introduction

Speech disorders affect the communication ability of people affected by certain medical conditions such as hearing loss, laryngeal and oral cancer, neurodegenerative diseases such as Parkinson's disease, and others. For the case of hearing loss, there are different treatments available depending on the degree and type of deafness. Cochlear Implants (CIs) are the most suitable devices when hearing aids no longer provide sufficient auditory feedback. However, CI users often experience alteration in speech even after rehabilitation, such as decreased intelligibility and changes in terms of articulation [1]. Thus, the development of computer aided systems will contribute to support the diagnosis and monitoring of speech. In the literature, few studies have addressed acoustic analysis

© Springer Nature Switzerland AG 2019
I. Nyström et al. (Eds.): CIARP 2019, LNCS 11896, pp. 679–687, 2019.
https://doi.org/10.1007/978-3-030-33904-3_64

of speech of CI users by implementing machine learning methods. In [2] speech intelligibility of 50 CI users is evaluated using an automatic speech recognition system and compared with 50 Healthy Controls (HC). Recently in [3] automatic classification using Support Vector Machines (SVM) between 20 CI users and 20 healthy speakers was performed in order to evaluate articulation disorders considering acoustic features. For the case of pathological speech detection, CNNs have outperformed classical machine learning methods [4–6]. In these studies, the conventional method is to perform time-frequency analysis by computing spectrograms over the speech signals to feed the CNNs with single channel inputs. However, using one channel as input may limit the potential of the model to learn more complex representations of speech signals.

In this study we propose a deep learning-based approach for the automatic detection of disordered speech in postlingually deafened CI users, i.e, when hearing loss occurs after speech acquisition. The method consists of 2-channel spectrograms as input to a CNN. Time-frequency analysis is performed considering Mel-scaled and Gammatone spectrograms, which are computed from short-time segments extracted from the recordings. These segments are defined as the transitions from voiceless to voiced sounds (onset) and voiced to voiceless sounds (offset). Our main hypothesis is that using the spectrograms as a 2-channel input will allow the CNN to complement the information from the two time-frequency representations. On the one hand, Mel-based features have been established as the standard feature set for different speech and audio processing applications. On the other hand, previous studies have shown that Gammatone-based features are more robust to noise compared with Mel features [7].

The rest of the paper is organized as follows: Sect. 2 includes details of the data and methods. Section 3 describes the experiments and results. Section 4 provides conclusions derived from this work.

2   Materials and Methods

2.1   Data

Standardized speech recordings of 107 CI users (56 male, 51 female) and 94 HC (46 male, 48 female) German native speakers are considered for the experiments. All of the CI users and 31 of the 94 healthy speakers were recorded at the Clinic of the Ludwig-Maximilians University in Munich (LMU), with a sampling frequency of 44.1 kHz and a 16 bit resolution. The recordings of the remaining 63 HC speakers were extracted from the PhonDat 1 (PD1) corpus from the Bavarian Archive For Speech Signals (BAS), which is freely available for European academic users[1]. For this corpus, the subjects were labeled as "old" and "young", however, the age of the speakers is not included in the description of the dataset. Speech recordings from the BAS corpus have a sampling frequency of 16 kHz. The mismatch in the acoustic conditions of the BAS corpus and our recordings is addressed in Sect. 2.2. The speech recordings include the

---

[1] http://hdl.handle.net/11858/00-1779-0000-000C-DAAF-B.

reading of *Der Nordwind und die Sonne* (*The North Wind and the Sun*) text. Information about the subjects considered in this study is presented in Table 1.

Table 1. Information of the speakers. HC-clinic: healthy speakers recorded in the clinic. HC-BAS: healthy speakers extracted from the BAS repository. μ: mean. σ: standard deviation

	CI		HC-clinic		HC-BAS	
	Male	Female	Male	Female	Male	Female
N. speakers	56	51	11	20	35	28
Range of age	18–89	28–84	26–53	20–64	-	-
Age ($\mu \pm \sigma$)	65 ± 16	62 ± 15	35 ± 9	37 ± 13	-	-

2.2 Preprocessing

The first step is to remove any possible DC offset induced by the microphone and to re-scale the amplitude of the speech signals between -1 and 1. Then, a noise reduction method and a compression technique are applied to normalize the acoustic conditions of the recordings from the clinic and BAS. Then, onset and offset transitions are extracted to model speech disorders in CI users. The details of the methods implemented are as follows:

Noise Reduction. Background noise is removed using the SoX codec[2]. The noise reduction algorithm is based on spectral gating, which consists in obtaining a profile of the background noise to enhance the quality of the audio. In order to get the profile, the Short-Time Fourier Transform (STFT) is computed over short-time frames extracted from a noisy signal (silence region from the recording to be denoise). Then, the mean power is computed over each point of the STFT in order to get thresholds per each frequency band. The STFT of the complete signal is calculated and the sounds with energies lower than the thresholds are attenuated for noise reduction. For more details regarding the implementation, please refer to the official SoX website.

Compression. The GSM full-rate compression technique is considered to normalize the acoustic conditions of the recordings from the clinic and the BAS repository [8]. The denoised speech signals are down-sampled to 8 kHz and the resolution is lowered to 13 bits, with a compression factor of 8. Additionally, a bandpass filter between 200 Hz and 3.4 kHz is applied in order to meet the specifications of a GSM transmission network. Figure 1 shows the STFT spectrograms of a speech recording before and after applying noise reduction and compression. The figures correspond to a speech segment of 600 ms extracted from the speech signal of one of the healthy speakers recorded in the clinic.

---

[2] http://sox.sourceforge.net/.

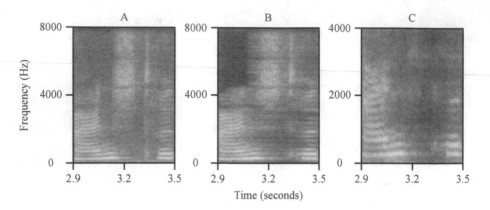

Fig. 1. Time-frequency representation of a segment from a speech signal. The figure shows (A) the original signal, (B) the signal after noise reduction, and (C) the signal after compression.

Segmentation. Speech signals are analyzed based on the automatic detection of onset and offset transitions, which are considered to model the difficulties of the patients to start/stop the movement of the vocal folds. The method used to identify the transitions is based on the presence of the fundamental frequency of speech (pitch) in short-time frames as it was shown in [9]. The transition is detected, and 80 ms of the signal are taken to the left and to the right of each border, forming segments with 160 ms length (Fig. 2).

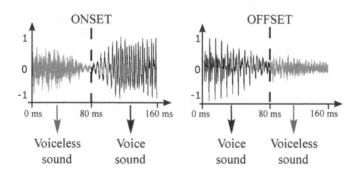

Fig. 2. ONSET and OFFSET transition frames.

2.3 Acoustic Analysis

Acoustic features are extracted from the onset/offset transitions based on two different auditory filter banks. First, acoustic features are extracted by applying triangular filters on the Mel scale. Frequencies in Hz can be converted to Mel scale as:

$$M(f_{Hz}) = 1125 \ln(1 + f_{Hz}/700) \tag{1}$$

In the second approach, features are extracted using Gammatone filter banks, which are based on the cochlear model proposed in [10]. The model consists of an array of bandpass filters organized from high frequency at the base of the cochlea, to low frequencies at the apex (innermost part of the cochlea). The Gammatone filter bank is defined in the time domain by Eq. 2 as:

$$g(t) = at^{n-1} \exp(-2\pi bt) \cos(2\pi f_c t + \phi) \tag{2}$$

Where f_c is the filter's center frequency in Hz, ϕ is the phase of the carrier in radians, a is the amplitude, n is the order of the filter, b is the bandwidth in Hz, and t is the time. The number of filters used for both Mel-scale and Gammatone based features is $n = 64$. The Gammatone filters are implemented following the procedure described in [11]. Features are extracted from the transitions using Hanning windows of 20 ms length with a time step of 5 ms.

2.4 Baseline Model

Mel-Frequency Cepstral Coefficients (MFCCs) and Gammatone-Frequency Cepstral Coefficients (GFCC) are extracted by dividing the transitions into short-time segments $X = \{x_1, \ldots, x_n\}$. Then, the Mel/Gammatone filter bank is applied and the discrete cosine transform is calculated upon the logarithm of the energy bands using Eq. 3.

$$coef[k] = 2 \sum_{i=0}^{n-1} x_f[i] \cos(\pi k(2i+1)/2n) \tag{3}$$

Where k are the coefficients and x_f is the resulting signal after applying the filter banks. In this work, 13 MFCCs (including the energy of the signal) and 12 GFCCs are considered. The mean, standard deviation, kurtosis, and skewness are computed from the descriptors. The automatic classification between CI users and HC speakers is performed with a radial basis SVM with margin parameter C and a Gaussian kernel with parameter γ. C and γ are optimized through a grid-search up to powers of ten with $10^{-4} < C < 10^4$ and $10^{-6} < \gamma < 10^3$. The selection criterion is based on the performance obtained in the training stage. The SVM is implemented with scikit-learn [12].

2.5 Proposed Model

Mel-scaled and the Gammatone spectrograms (Cochleagram) are computed from the onset/offset transitions by applying the filter banks described before. Then, the spectrograms are combined into a 2-channel tensor to fed the CNN, which is implemented using PyTorch [13]. From the documentation, it can be observed that the output of the convolutional layer for an input signal (Bs, C_{in}, H, W) is described as:

$$h(Bs_i, C_{out_j}) = \text{bias}(C_{out_j}) + \sum_{k=0}^{C_{in}-1} \omega(C_{out}, k) * \text{input}(Bs_i, k) \qquad (4)$$

Where Bs is the batch size ($Bs = 100$), ω are the weights of the network, C is the number of channels ($C = 2$) of the input tensor, H is the height of the input signal ($H = 64$, number of filter banks), and W is the width of the input signal ($W = 28$, number of frames in the onset/offset transitions). The architecture of the CNN implemented in this study is summarized in Fig. 3. It consists of two convolutional layers, two max-pooling layers, dropout to regularize the weights, and two fully connected hidden layer followed by the output layer to make the final decision using a softmax activation function. The CNN is trained using the Adam optimization algorithm [14] with a learning rate of $\eta = 10^{-4}$. The cross–entropy between the training labels y and the model predictions $\hat{y}$ is used as the loss function. The size of the kernel in the convolutional layers is $k_c = 3 \times 3$. For the pooling layers the kernel's size is $k_p = 2 \times 2$. None of the network hyper-parameters are optimized in order to have comparable models across experiments.

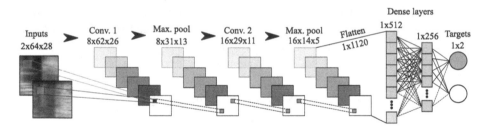

Fig. 3. Architecture of the CNN implemented in this study. The size of the kernel in the convolutional (Conv. i) and pooling layers (Max. pool) is 3×3 and 2×2, respectively.

3   Experiments and Results

The SVMs and CNNs are tested following a 10-Fold Cross-validation strategy. The performance of the system is evaluated by means of the accuracy (Acc), sensitivity (Sen), and the specificity (Spe). The SVM and the multi-channel CNN are trained with features/spectrograms extracted from onset and offset transitions, individually. Table 2 shows the results obtained for the baseline model. It can be observed that the accuracies are higher for the offset transitions, when MFCCs and GFCCs are considered to train the SVMs individually, however, the best performance is achieved when the two feature sets are combined (Onset-Acc = 82.4%; Offset-Acc = 83.4%). Additionally, note that the sensitivity values

Table 2. Classification results for the SVM trained with MFCCs and GFCCs features. **Acc:** accuracy. **Sen:** sensitivity. **Spe:** specificity. Fusion: combination of MFCCs and GFCCs features.

Segment	Features	Acc (%)	Sen (%)	Spe (%)
Onset	MFCC	80.6	85.2	76.0
	GFCC	75.3	78.7	72.0
	Fusion	82.4	86.0	78.7
Offset	MFCC	83.4	93.5	73.5
	GFCC	82.8	88.8	76.9
	Fusion	83.4	91.7	75.2

are higher in all of the experiments. This can be explained considering that the speech of some of the CI users may be not affected, thus, it is closer to the speech of healthy speakers. Table 3 shows the results obtained with the proposed approach. In general, the accuracies of the CNNs are higher than those of the baseline model. This is mainly because the CNNs are able to classify more CI users, which is not the case for the HC. As explained before not every CI users may present speech deficits, thus, it is not expected that the classifiers discriminate all of the speakers correctly. Nevertheless, it can be observed that the combination of the Mel-spectrogram and Cochleagram into a 2-channel tensor is suitable for the automatic detection of speech deficits. Additionally, note that this methodology is not restricted only to analyze speech of CI users, but it can be adapted to study other pathologies or to recognize other paralinguistic aspects from speech signals such as emotions.

Table 3. Classification results for the CNNs trained with Mel-spectrograms and Cochleagrams. **Acc:** accuracy. **Sen:** sensitivity. **Spe:** specificity. **Fusion:** 2-channel spectrograms (Mel-spectrogram–Cochleagram).

Segment	Inputs	Acc (%)	Sen (%)	Spe (%)
Onset	Mel-spectrogram	83.5	98.1	68.8
	Cochleagram	85.4	95.4	75.4
	Fusion	86.8	98.2	75.4
Offset	Mel-spectrogram	85.9	96.4	75.5
	Cochleagram	86.8	96.3	77.3
	Fusion	86.8	97.3	76.3

4   Conclusions

In this paper we presented a methodology for automatic detection of speech deficits in CI users using multi-channel CNNs. The method consists in combining two types of time-frequency representations into a 2-channel tensor which is used to fed a CNN. In order to do this, Mel-spectrograms and Cochleagrams are computed from onset and offset transitions extracted from the recordings of CI users and healthy speakers. Cepstral coefficients and SVM classifiers were considered for comparison. According to the results, it is possible to differentiate between CI users and HC with accuracies of up to 86.8% when the multi-channel CNN is considered. We are aware of a mismatch regarding the age of the CI users and HC. Currently, we are collecting more HC, however, we don't expect the outcome of the experiments to change. Additionally, note that the multi-channel CNN may be suitable for other speech processing tasks such as emotion detection or as the feature stage for speech recognition. Future work should include more time-frequency analysis such as Perceptual Linear Prediction and the analysis of other pathologies in other to validate the proposed approach.

Acknowledgments. The authors acknowledge to the Training Network on Automatic Processing of PAthological Speech (TAPAS) funded by the Horizon 2020 programme of the European Commission. Tomás Arias-Vergara is under grants of Convocatoria Doctorado Nacional-785 financed by COLCIENCIAS. The authors also thanks to CODI from University of Antioquia (grant number 2018-23541).

References

1. Hudgins, C.V., Numbers, F.C.: An investigation of the intelligibility of the speech of the deaf. Genet. Psychol. Monogr. **25**, 289–392 (1942)
2. Ruff, S., Bocklet, T., Nöth, E., Müller, J., Hoster, E., Schuster, M.: Speech production quality of cochlear implant users with respect to duration and onset of hearing loss. ORL **79**(5), 282–294 (2017)
3. Arias-Vergara, T., Orozco-Arroyave, J.R., Gollwitzer, S., Schuster, M., Nöth, E.: Consonant-to-vowel/vowel-to-consonant transitions to analyze the speech of cochlear implant users. In: Ekštein, K. (ed.) TSD 2019. LNCS (LNAI), vol. 11697, pp. 299–306. Springer, Cham (2019). https://doi.org/10.1007/978-3-030-27947-9_25
4. Nakashika, T., Yoshioka, T., Takiguchi, T., Ariki, Y., Duffner, S., Garcia, C.: Dysarthric speech recognition using a convolutive bottleneck network. In: 2014 12th International Conference on Signal Processing (ICSP), pp. 505–509. IEEE (2014)
5. Takashima, Y., et al.: Audio-visual speech recognition using bimodal-trained bottleneck features for a person with severe hearing loss. In: Proceedings of the Seventeenth Annual Conference of the International Speech Communication Association, pp. 277–281 (2016)
6. Vásquez-Correa, J.C., Orozco-Arroyave, J.R., Nöth, E.: Convolutional neural network to model articulation impairments in patients with Parkinson's disease. In: Proceedings of the Eighteenth Annual Conference of the International Speech Communication Association, pp. 314–318 (2017)
7. Zhao, X., Wang, D.: Analyzing noise robustness of MFCC and GFCC features in speaker identification. In: 2013 IEEE International Conference on Acoustics, Speech and Signal Processing, pp. 7204–7208. IEEE (2013)

8. Huerta, J.M., Stern, R.M.: Speech recognition from GSM codec parameters. In: Fifth International Conference on Spoken Language Processing (1998)
9. Orozco-Arroyave, J.R.: Analysis of Speech of People with Parkinson's Disease. Logos Verlag, Berlin (2016)
10. Patterson, R.D., Robinson, K., Holdsworth, J., McKeown, D., Zhang, C., Allerhand, M.: Complex sounds and auditory images. In: Auditory Physiology and Perception, pp. 429–446. Elsevier (1992)
11. Slaney, M., et al.: An efficient implementation of the Patterson-Holdsworth auditory filter bank. Technical report 35(8). Apple Computer, Perception Group (1993)
12. Pedregosa, F., et al.: Scikit-learn: machine learning in python. J. Mach. Learn. Res. **12**, 2825–2830 (2011)
13. Paszke, A., et al.: Automatic differentiation in PyTorch (2017)
14. Kingma, D.P., Ba, J.: Adam: a method for stochastic optimization. In: International Conference on Learning Representation (ICLR) (2015)

Articulation and Empirical Mode Decomposition Features in Diadochokinetic Exercises for the Speech Assessment of Parkinson's Disease Patients

Juan Camilo Vásquez-Correa[1,2](✉), Cristian D. Rios-Urrego[2], Alice Rueda[3],
Juan Rafael Orozco-Arroyave[1,2], Sri Krishnan[3], and Elmar Nöth[1]

[1] Pattern Recognition Lab, Friedrich -Alexander Universität,
Erlangen-Nürnberg, Germany
juan.vasquez@fau.de
[2] Faculty of Engineering. Universidad de Antioquia UdeA,
Calle 70 No. 52-21, Medellín, Colombia
[3] Department of Electrical and Computer Engineering, Ryerson University,
Toronto, Canada

Abstract. Speech impairments are one of the earliest manifestations in patients with Parkinson's disease. Particularly, articulation impairments related to the capability of the speaker to move the limbs and muscles of the vocal tract have been observed in the patients. Articulation deficits have been evaluated in the patients mainly using diadochokinetic exercises, which consist in the rapid repetition of syllables like /pa-ta-ka/. This study considered different features to model several aspects of the diadochokinetic exercises, including the capacity to start/stop the vocal fold vibration, the speech rate, and the regularity of the diadochokinetic task. Articulation features are combined with others that result from an empirical mode decomposition procedure, which have been recently used to model dysphonia in Parkinson's patients. The features are used to classify Parkinson's patients and healthy speakers, and to predict the dysarthria severity of the participants according to a clinical scale. According to the results, articulation features are able to classify the presence of the disease with an accuracy up to 76%, and to predict the dysarthria level of the speakers with a Spearman's correlation of up to 0.68.

Keywords: Parkinson's disease · Articulation ·
Empirical mode decomposition · Dysarthria · Diadochokinetic exercises

1 Introduction

Parkinson's disease (PD) is a neurological disorder that alters the function of the basal ganglia in the midbrain, producing motor and non–motor deficits in

© Springer Nature Switzerland AG 2019
I. Nyström et al. (Eds.): CIARP 2019, LNCS 11896, pp. 688–696, 2019.
https://doi.org/10.1007/978-3-030-33904-3_65

the patients [1]. Motor symptoms include among others, bradykinesia, rigidity, resting tremor, and different speech impairments. Non–motor symptoms include depression, sleep disorders, impaired language, and others. Speech impairments are an early and prominent manifestation that can contribute primarily to the diagnosis of PD [2]. The main symptoms of the speech of PD patients are grouped and called hypokinetic dysarthria. They include monopitch, reduced stress, imprecise consonants, and reduced loudness. Several studies in the literature have described the speech impairments of PD patients in terms of phonation, articulation, and prosody [3,4]. Particularly, articulation impairments are related to the modification of position, stress, and shape of several limbs and muscles to produce speech. One of the first observed articulation symptoms was the imprecise production of stop consonants such as /p/, /t/, /k/, /b/, /d/, and /g/ [5,6]. Other symptoms include reduced duration of voiced segments and transitions, and increased voice onset time (VOT) [6,7]. These symptoms have been considered among the most important for the assessment of PD from speech [4–6].

Articulation analysis have been evaluated mainly using diadochokinetic exercises (DDK), which consist in the rapid repetition of syllables like /pa-ta-ka/. The execution of these exercises requires the continuous movement of different articulators such as lips, tongue and velum. Several researchers have addressed the task of modeling articulatory deficits considering DDK tasks. In [8], the authors modeled six different articulatory deficits in PD patients using DDK exercises: vowel quality, coordination of laryngeal and supra-laryngeal activity, precision of consonant articulation, tongue movement, occlusion weakening, and speech timing. The authors reported an accuracy of 88% discriminating between PD patients and HC speakers, using a support vector machine (SVM) classifier. Another articulation model was proposed in [9], where the energy content in the transitions from unvoiced to voiced (onset) and from voiced to unvoiced (offset) segments was considered. The authors classified PD patients and HC speakers with speech recordings in three different languages (Spanish, German, and Czech) and reported accuracies in a range between 80% and 94% depending on the language. In [10] the authors proposed an articulation model based on temporal and spectral features extracted from the VOT segments from DDK exercises. The temporal features included the VOT duration, the VOT ratio, the vowel variability quotient, and the articulation rate. The spectral features considered 13 Mel frequency cepstral coefficients (MFCCs) extracted from the VOT segments. The authors considered a SVM classifier, and reported an accuracy of up to 92.2%. Recently, in [11], the authors proposed an articulatory model based on forced Gaussian mixture models (GMMs) to get time-stamps for the different phonetic structures that appear in an utterance. The different phonemes were segmented and grouped to train separate GMMs for each phoneme unit. The proposed method allowed to compare the features of each phonetic unit between PD and HC subjects independently. The classification was performed based on a threshold of the difference between the posterior probabilities from the models created for HC subjects and PD patients. The authors reported accuracies in a range from 81.0% to 94%.

This paper aims to evaluate the performance of three different articulation feature sets extracted from DDK exercises to model the speech deficits of PD patients. The first set is based on the proposed in [9], which aims to model the difficulty of the patients to start/stop the vocal fold vibration. These features are based on modeling the transition between voiced and unvoiced segments in the speech signal. The second group of articulation features aims to model the speech rate and the regularity of the DDK exercises. Finally, the third feature set considers descriptors extracted from the empirical mode decomposition (EMD), which have been recently used to model dysphonia in PD patients [12,13]. The combination of the articulation features to model the start/stop movement of the vocal folds with the features to model the regularity of the DDK exercises showed to be highly accurate to classify PD patients and HC subjects, and to predict the dysarthria level of the participants, according to a modified version of the Frenchay dysarthria assessment scale (m-FDA), which was proposed recently in [14].

2   Methods

2.1   Transition Features

The first group of features aims to model the difficulties exhibited by PD patients to start/stop the movement of the vocal folds [9], and are based on the energy content in onset and offset segments. The border between voiced and unvoiced segments is detected based on the presence of the F_0. Once the borders are detected, 40 ms of the signal are taken to the left and to the right, forming a segment with 80 ms length. The spectrum of the transitions is distributed into the first 17 critical bands according to the Bark scale, and the Bark band energies (BBE) are computed. 13 MFCCs and their first two derivatives are also computed in the transitions to complete the feature vector. Then, the mean, standard deviation, skewness, and kurtosis are computed for the features of consecutive transitions. Finally, the features obtained for onset and offset are concatenated, forming the final feature vector per utterance. Figure 1 shows the process to extract the articulation features.

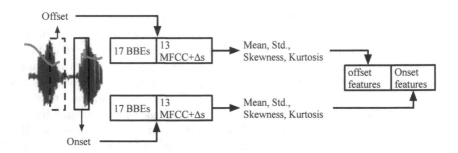

Fig. 1. Scheme for the extraction of articulation features

2.2   DDK Regularity Features

The second group of articulation features aims to model the regularity of the DDK exercises in terms of regularity, tone, rhythm, and duration. The feature set is formed with 55 features per utterance based on the speech rate, duration, and the F_0 contour when the subjects pronounce the DDK exercises. A detailed description of the features included in this set is shown in Table 1.

Table 1. Description of the features to model the regularity of the DDK exercises. **P:** pause duration, **V:** voiced duration, **U:** unvoiced duration

Num.	Feature	Description
1–4	F0-contour	Average, standard deviation, maximum, minimum
5	Voiced rate	Number of voiced segments per second
6–9	Duration of voiced	Average, standard deviation, maximum, minimum
10–11	Duration of unvoiced	Maximum, minimum
12	Pause rate	Number of pauses per second
13–14	Duration of pauses	Average, standard deviation
15–18	F0 in first voiced segment	Average, standard deviation, skewness, kurtosis
19–22	F0 in last voiced segment	Average, standard deviation, skewness, kurtosis
23–24	Linear estimation of F0	Tilt, mean square error (MSE)
25–30	Duration ratios	P/(V+U), P/U, U/(V+U), V/(V+U), V/P, U/P
31	#voiced/#unvoiced	
32–55	Estimated F0 with a 5-degree Lagrange polynomial	Average, standard deviation, skewness, and kurtosis

2.3   Empirical Mode Decomposition Features

The third group of articulation features is based on the EMD computed on the DDK exercises. The EMD is a time-domain decomposition method commonly used in signal denoising. The method is based on the estimation of intrinsic mode functions (IMFs) through a sifting process in time-domain [15]. The original signal $s[n]$ can be reconstructed by adding all IMFs together, as shown in Eq. 1. EMD-based features have been recently considered for the analysis of dysphonia in PD patients [12,13].

$$s[n] = \sum_{i=1}^{N} \text{IMF}_i[n] \qquad (1)$$

The process to extract the EMD-based features considered in this study is as follows: the vocalic and plosive segments from the speech signal are segmented using the envelope of the Hilbert transform of the signal. The inflection points of the envelope are used to segment the plosive sounds from the DDK exercises and the vowels (see Fig. 2). Once the plosives and the vowels are segmented, 12 descriptors are computed for each segment based on the IMFs decomposition.

The computed descriptors include the number of IMFs obtained from the decomposition, the occupied bandwidth (OBW) of the first 10 IMFs, and the adaptive SNR estimated based on the cross-correlation between the IMFs and the original speech signal [12,13]. Four statistical functionals (mean, standard deviation, skewness, and kurtosis) are computed for the 12 descriptors extracted from the plosives and vocalic sounds from an utterance, forming two 48-dimensional feature vectors (one for plosives and one for vowels). Finally, the features obtained for plosives and vowels are concatenated, forming the final feature vector per utterance.

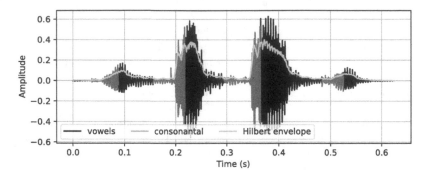

Fig. 2. Segmentation of the DDK exercises using the envelope of the Hilbert transform

2.4   Evaluation

The capability of the articulation and EMD-based feature sets to classify PD patients and HC subjects was evaluated with a SVM classifier with a Gaussian kernel. The complexity hyperparameter C and the bandwidth of the kernel γ were optimized in a grid-search with selection criterion based on the accuracy obtained in the train set, where $C \in \{10^{-5}, 10^{-4}0, \ldots 10^3\}$ and $\gamma \in \{10^{-5}, 10^{-4}, \ldots 10^3\}$. In addition, the feature sets were used to predict the dysarthria level of the participants according to the m-FDA scale. The prediction was performed with a support vector regression (SVR) with a Gaussian kernel and an ε-insensitive loss function. The hyperparameter C, γ, and ε of the regressor were optimized in a grid-search with selection criterion based on the Spearman's correlation ρ obtained in the train set, where $C \in \{10^{-5}, 10^{-4}, \ldots 10^3\}$, $\gamma \in \{10^{-5}, 10^{-4}, \ldots 10^3\}$, and $\varepsilon \in \{10^{-4}, 10^{-3}, \ldots 10^1\}$. A Leave-one-out cross-validation strategy was performed for all classification and regression experiments.

3   Data

3.1   m-FDA Scale

The evaluation of PD patients according to the MDS-UPDRS-III scale is suitable to assess general motor impairments of PD patients; however, the deterioration of the communication skills is not properly evaluated because such a scale only considers speech impairments in one of its items. We recently introduced the m-FDA scale, which is focused on speech impairments showed by PD patients and can be administered based on speech recordings [14]. The scale includes several aspects of speech: respiration, lips movement, palate/velum movement, larynx, tongue, monotonicity, and intelligibility. It has a total of 13 items and each of them ranges from 0 (normal or completely healthy) to 4 (very impaired), thus the range of the total score is from 0 to 52. The labeling process of the recordings was performed by three phoniatricians who agreed in the first ten speakers. Afterwards, each phoniatrician evaluated the remaining recordings independently. The inter-rater reliability among the labelers is 0.75, which was computed as the average Spearman's correlation between all pairs of labelers.

3.2 Participants

Recordings of the PC-GITA database [16] are considered in this study. The data contain speech utterances from 50 PD and 50 HC Colombian Spanish native speakers balanced in age and gender. All patients were recorded in ON state, i.e., no more than three ours after their daily medication. The DDK exercises pronounced by the subjects included the rapid repetition of the syllables /pa-ta-ka/, /pa-ka-ta/, /pe-ta-ka/, /pa/, /ta/, and /ka/. Additional information from the participants is shown in Table 2. The results from the statistical tests show that the data are gender, age, and education level balanced, and that there is a significant difference between the m-FDA scores assigned for PD patients and HC subjects.

Table 2. General information of the subjects. Time since diagnosis, age and education are given in years. [a]p–value calculated through chi–square test. [b]p–value calculated through t-test.

	PD patients	Healthy controls	Patients vs. controls
Gender [F/M]	25/25	25/25	$p = 1.00$[a]
Age [F/M]	60.7(7.3)/61.3(11.7)	61.4(7.1)/60.5(11.6)	$p = 0.98$[b]
Education level [F/M]	11.5(4.1)/10.9(4.5)	11.5(5.2)/10.6(4.4)	$p = 0.88$[b]
Total m-FDA score [F/M]	29.8(8.6)/28.2(9)	7.6(9.2)/5.1(7.3)	$p \ll 0.005$[b]
Time since diagnosis [F/M]	12.6(11.5)/8.7(5.8)		
MDS–UPDRS–III [F/M]	37.6(14.0)/37.8(22.1)		

4   Experiments and Results

Two experiments are performed: (1) the classification of PD patients and HC subjects using the three different feature sets and their combination, and (2) the prediction of the dysarthria severity following the m-FDA scale. Table 3 shows the results for the first experiment.

The transition, regularity, and the EMD feature sets were considered separately and also their combination using an early fusion strategy. The classification was performed with the features computed from the six DDK exercises, and also with their combination. The results suggest that the combination of transition and regularity features are the most accurate to classify PD patients and HC subjects. A detailed evaluation of the importance of the different features have been performed in related research [17]. In addition, note that the combination of the six exercises slightly improved the accuracy w.r.t. the obtained with the individual exercises per feature set. When all of the exercises and feature sets are merged the classification accuracies decrease w.r.t. those obtained with the combination of transition and regularity features only. This result indicates that EMD features are not complementary to the other two feature sets. For the separate exercises, note that the repetition of /pa-ta-ka/, /pa-ka-ta/, and /pe-ta-ka/ produce the highest accuracies, which can be explained because the difficulties for the patients to perform those exercises, i.e., the patients have to move the lips, the tongue and the velum continuously, while in the other exercises the patients only have to move one articulator.

Table 3. Classification of PD patients and HC subjects using the transition (**Trans.**), regularity (**Reg.**), and EMD features computed upon DDK exercises. **Acc**: accuracy (%), **Prec.** Precision (%), **Rec.** Recall (%), **F1.** F1-score. **Fusion.** Fusion of the features of the six DDK exercises

Exercise	Trans.				Reg.				EMD				Trans.+Reg.				Trans.+Reg.+EMD			
	Acc	Prec	Rec	F1	Acc	Prec	Rec	F1	Acc	Prec	Rec	F1	Acc	Prec	Rec	F1	Acc	Prec	Rec	F1
/pa-ta-ka/	74	81	64	0.74	69	72	62	0.69	70	72	66	0.70	75	82	64	0.75	71	72	68	0.71
/pa-ka-ta/	67	68	64	0.67	58	59	54	0.58	53	51	57	0.54	73	76	68	0.73	67	68	64	0.67
/pe-ta-ka/	70	70	70	0.71	64	65	60	0.64	53	53	54	0.53	72	72	72	0.72	66	69	58	0.66
/pa/	61	63	52	0.61	54	54	50	0.50	52	52	56	0.52	67	7	58	0.67	66	69	58	0.66
/ta/	60	61	59	0.60	70	75	60	0.69	67	68	64	0.67	58	59	54	0.58	67	68	64	0.67
/ka/	70	79	60	0.71	64	71	68	0.73	63	70	54	0.62	72	81	58	0.71	72	79	60	0.72
Fusion	75	77	73	0.75	63	67	52	0.63	71	77	60	0.71	**76**	**77**	**74**	**0.76**	68	73	58	0.68

The features from the fusion of the six DDK exercises are used to predict the dysarthria severity of the patients according to the m-FDA scale. The results are shown in Table 4 for the combination of transition and regularity features, and for the EMD features. Similarly to the classification experiments, the highest correlation is obtained with the combination of transition and regularity features ($\rho = 0.6782$). The "strong" correlation obtained is statistically significant, and it is comparable to the obtained in related studies, where the same problem was addressed [14,18].

Table 4. Prediction of the m-FDA score of PD patients and HC subjects using transition (**Trans.**), regularity (**Reg.**) and EMD-based features computed on DDK exercises. ρ Spearman's correlation coefficient.

Exercise	Trans.+Reg.		EMD		Trans.+Reg.+EMD	
	ρ	p-val	ρ	p-val	ρ	p-val
Fusion of exercises	0.678	$\ll$0.005	0.500	$\ll$0.005	0.629	$\ll$0.005

5   Conclusion

This study evaluated and compared the performance of three different articulation features for the assessment of the speech of PD patients when they perform different DDK exercises. The first set of features are used to model the capabilities of the patients to start/stop the vibration of the vocal folds. The second feature set aimed to model the regularity of the movement of the articulators when they perform the exercises. Finally, the third feature set based on EMD features was included to evaluate the harmonic structure and the noise presence in the speech signal.

The results indicated that the combination of the first two feature sets are the most accurate to classify the presence of the disease, and also to predict the dysarthria level of the participants. The combination of different exercises seems to be more appropriate and accurate than considering DDK exercises separately. Additionally, the most complex exercises like /pa-ta-ka/ or /pa-ka-ta/, where the patients have to move more articulators showed to be more accurate than the exercises where the patients have to move only one articulator. Further research should consider an exhaustive analysis about the importance of individual articulation features for the assessment of the dysarthria severity of the patients.

Acknowledgments. The work reported here was financed by CODI from University of Antioquia by grant Number 2017–15530. This project has received funding from the European Union's Horizon 2020 research and innovation programme under the Marie Sklodowska-Curie Grant Agreement No. 766287.

References

1. Hornykiewicz, O.: Biochemical aspects of Parkinson's disease. Neurology **51**(2), S2–S9 (1998)
2. Rusz, J., et al.: Quantitative acoustic measurements for characterization of speech and voice disorders in early untreated Parkinson's disease. J. Acoust. Soc. Am. **129**(1), 350–367 (2011)
3. Orozco-Arroyave, J.R., Vásquez-Correa, J.C., et al.: Neurospeech: an open-source software for Parkinson's speech analysis. Digit. Signal Proc. **77**, 207–221 (2018)
4. Hlavnicka, J., Cmejla, R., Tykalova, T., Sonka, K., Ruzicka, E., Rusz, J.: Automated analysis of connected speech reveals early biomarkers of Parkinson's disease

in patients with rapid eye movement sleep behaviour disorder. Nat. Sci. Rep. **7**(12), 1–13 (2017)

5. Ackermann, H., Ziegler, W.: Articulatory deficits in Parkinsonian dysarthria: an acoustic analysis. J. Neurol. Neurosurg. Psychiatry **54**(12), 1093–1098 (1991)
6. Tykalova, T., Rusz, J., Klempir, J., Cmejla, R., Ruzicka, E.: Distinct patterns of imprecise consonant articulation among Parkinson's disease, progressive supranuclear palsy and multiple system atrophy. Brain Lang. **165**, 1–9 (2017)
7. Forrest, K., Weismer, G., Turner, G.S.: Kinematic, acoustic, and perceptual analyses of connected speech produced by Parkinsonian and normal geriatric adults. J. Acoust. Soc. Am. **85**(6), 2608–2622 (1989)
8. Novotný, M., Rusz, J., et al.: Automatic evaluation of articulatory disorders in Parkinson's disease. IEEE/ACM Trans. Audio Speech Lang. Process. **22**(9), 1366–1378 (2014)
9. Orozco-Arroyave, J.R.: Analysis of Speech of People with Parkinson's Disease, vol. 41. Logos Verlag, Berlin (2016). GmbH
10. Montaña, D., Campos-Roca, Y., Pérez, C.J.: A diadochokinesis-based expert system considering articulatory features of plosive consonants for early detection of Parkinson's disease. Comput. Methods Programs Biomed. **154**, 89–97 (2018)
11. Moro-Velazquez, L., et al.: A forced Gaussians based methodology for the differential evaluation of Parkinson's disease by means of speech processing. Biomed. Sig. Process. Control **48**, 205–220 (2019)
12. Rueda, A., Krishnan, S.: Feature analysis of dysphonia speech for monitoring Parkinson's disease. In: 39th Annual International Conference of the IEEE Engineering in Medicine and Biology Society (EMBC), pp. 2308–2311. IEEE (2017)
13. Rueda, A., Krishnan, S.: Clustering Parkinson's and age-related voice impairment signal features for unsupervised learning. Adv. Data Sci. Adapt. Anal. **10**(02), 1840007 (2018)
14. Vásquez-Correa, J.C., Orozco-Arroyave, J.R., Bocklet, T., Nöth, E.: Towards an automatic evaluation of the dysarthria level of patients with Parkinson's disease. J. Commun. Disord. **76**, 21–36 (2018)
15. Huang, N.E., et al.: The empirical mode decomposition and the Hilbert spectrum for nonlinear and non-stationary time series analysis. Proc. R. Soc. London. Ser. A: Math. Phys. Eng. Sci. **454**(1971), 903–995 (1998)
16. Orozco-Arroyave, J.R., et al.: New Spanish speech corpus database for the analysis of people suffering from Parkinson's disease. In: Language Resources and Evaluation Conference, (LREC), pp. 342–347 (2014)
17. Rueda, A., Vasquez-Correa, J.C., Rios-Urego, C.D., Orozco-Arroyave, J.R., Krishnan, S., Nöth, E.: Feature representation of pathophysiology of Parkinsonian dysarthria. In: Proceedings of INTERSPEECH, pp. 1–5 (2019)
18. Cernak, M., et al.: Characterisation of voice quality of Parkinson's disease using differential phonological posterior features. Comput. Speech Lang. **46**, 196–208 (2017)

Convolutional Neural Networks and a Transfer Learning Strategy to Classify Parkinson's Disease from Speech in Three Different Languages

Juan Camilo Vásquez-Correa[1,2(✉)], Tomas Arias-Vergara[1,2,3],
Cristian D. Rios-Urrego[2], Maria Schuster[3], Jan Rusz[4],
Juan Rafael Orozco-Arroyave[1,2], and Elmar Nöth[1]

[1] Pattern Recognition Lab, Friedrich-Alexander Universität,
Erlangen-Nürnberg, Germany
`juan.vasquez@fau.de`
[2] Faculty of Engineering, Universidad de Antioquia UdeA,
Calle 70 No. 52-21, Medellín, Colombia
[3] Department of Otorhinolaryngology, Head and Neck Surgery,
Ludwig-Maximilians Universität, Munich, Germany
[4] Department of Circuit Theory, Faculty of Electrical Engineering,
Czech Technical University in Prague, Prague, Czech Republic

Abstract. Parkinson's disease patients develop different speech impairments that affect their communication capabilities. The automatic assessment of the speech of the patients allows the development of computer aided tools to support the diagnosis and the evaluation of the disease severity. This paper introduces a methodology to classify Parkinson's disease from speech in three different languages: Spanish, German, and Czech. The proposed approach considers convolutional neural networks trained with time frequency representations and a transfer learning strategy among the three languages. The transfer learning scheme aims to improve the accuracy of the models when the weights of the neural network are initialized with utterances from a different language than the used for the test set. The results suggest that the proposed strategy improves the accuracy of the models in up to 8% when the base model used to initialize the weights of the classifier is robust enough. In addition, the results obtained after the transfer learning are in most cases more balanced in terms of specificity-sensitivity than those trained without the transfer learning strategy.

Keywords: Parkinson's disease · Speech processing · Convolutional neural networks · Transfer learning

1 Introduction

Parkinson's disease (PD) is a neurodegenerative disorder characterized by the progressive loss of dopaminergic neurons in the mid-brain producing several

© Springer Nature Switzerland AG 2019
I. Nyström et al. (Eds.): CIARP 2019, LNCS 11896, pp. 697–706, 2019.
https://doi.org/10.1007/978-3-030-33904-3_66

motor and non-motor impairments in the patients [1]. Motor symptoms include among others, bradykinesia, rigidity, resting tremor, micrographia, and different speech impairments. The speech impairments observed in PD patients are typically grouped as hypokinetic dysarthria, and include symptoms such as vocal folds rigidity, bradykinesia, and reduced control of muscles and limbs involved in the speech production. The effects of dysarthria in the speech of PD patients include increased acoustic noise, reduced intensity, harsh and breathy voice quality, increased voice nasality, monopitch, monoludness, speech rate disturbances, imprecise articulation of consonants [2], and involuntary introduction of pauses [3]. Clinical observations in the speech of patients can be objectively and automatically measured by using computer aided methods supported in signal processing and pattern recognition with the aim to address two main aspects: (1) to support the diagnosis of the disease by classifying healthy control (HC) subjects and patients, and (2) to predict the level of degradation of the speech of the patients according to a specific clinical scale.

Most of the studies in the literature to classify PD from speech are based on computing hand-crafted features and using classifiers such as support vector machines (SVMs) or K-nearest neighbors (KNN). For instance, in [4], the authors computed features related to perturbations of the fundamental frequency and amplitude of the speech signal to classify utterances from 20 PD patients and 20 HC subjects, Turkish speakers. Classifiers based on KNN and SVMs were considered, and accuracies of up to 75% were reported. Later, in [5] the authors proposed a phonation analysis based on several time frequency representations to assess tremor in the speech of PD patients. The extracted features were based on energy and entropy computed from time frequency representations. Several classifiers were used, including Gaussian mixture models (GMMs) and SVMs. Accuracies of up to 77% were reported in utterances of the PC-GITA database [6], formed with utterances from 50 PD patients and 50 HC subjects, Colombian Spanish native speakers. The authors from [7] computed features to model different articulation deficits in PD such as vowel quality, coordination of laryngeal and supra-laryngeal activity, precision of consonant articulation, tongue movement, occlusion weakening, and speech timing. The authors studied the rapid repetition of the syllables /pa-ta-ka/ pronounced by 24 Czech native speakers, and reported an accuracy of 88% discriminating between PD patients and HC speakers, using an SVM classifier. Additional articulation features were proposed in [8], where the authors modeled the difficulty of PD patients to start/stop the vocal fold vibration in continuous speech. The model was based on the energy content in the transitions between unvoiced and voiced segments. The authors classified PD patients and HC speakers with speech recordings in three different languages (Spanish, German, and Czech), and reported accuracies ranging from 80% to 94% depending on the language; however, the results were optimistic, since the hyper-parameters of the classifier were optimized based on the accuracy on the test set. Another articulation model was proposed in [9]. The authors considered a forced alignment strategy to segment the different phonetic units in the speech utterances. The phonemes were segmented and grouped to train different GMMs. The classification was performed based on a threshold of the difference between the posterior probabilities from the models created for HC subjects and PD patients. The model was tested with Colombian Spanish utterances from the PC-GITA database [6] and with the Czech data from [10]. The authors reported accuracies of up to 81% for the Spanish data, and of up to 94% for the Czech data.

In addition to the hand-crafted feature extraction models, there is a growing interest in the research community to consider deep learning models in the assessment of the speech of PD patients [11–13]. Deep learning methods have the potential to extract more abstract and robust features than those manually computed. These features could help to improve the accuracy of different models to classify pathological speech, such as PD [14]. A deep learning based articulation model was proposed in [12] to model the difficulties of the patients to stop/start the vibration of the vocal folds. Transitions between voiced and unvoiced segments were modeled with time-frequency representations and convolutional neural networks (CNNs). The authors considered speech recordings of PD patients and HC speakers in three languages: Spanish, German, and Czech, and reported accuracies ranging from 70% to 89%, depending on the language. However, in a language independent scenario, i.e., training the CNN with utterances from one language and testing with the remaining two, the results were not satisfactory (accuracy <60%).

The classification of PD from speech in different languages has to be carefully conducted to avoid bias towards the linguistic content present in each language. For instance, Czech and German languages are richer than Spanish language in terms of consonant production, which may cause that it is easier to produce consonant sounds by Czech PD patients than by Spanish PD patients. Despite these language dependent issues, the results in the classification of PD in different languages could be improved using a transfer learning strategy among languages, i.e., to train a base model with utterances from one language, and then, to perform a fine-tuning of the weights with utterances from the target language [15]. Similar approaches based on transfer learning have been recently considered to classify PD using handwriting [16]. In the present study, we propose a methodology to classify PD via a transfer learning strategy with the aim to improve the accuracy in different languages. CNNs trained with utterances from one language are used to initialize a model to classify speech utterances from PD patients in a different language. The models are evaluated with speech utterances in Spanish, German, and Czech languages. The results suggest that the use of a transfer learning strategy improved the accuracy of the models over 8% with respect to those obtained when the model is trained only with utterance from the target language.

2   Materials and Methods

2.1   Data

Speech recordings of patients in three different languages are considered: Spanish, German, and Czech. All of the recordings were captured in noise controlled conditions. The speech signals were down-sampled to 16 kHz. The patients in the three datasets were evaluated by a neurologist expert according to the third section of the movement disorder society, unified Parkinson's disease rating scale (MDS-UPDRS-III) [17]. Table 1 summarizes the information about the patients and healthy speakers.

Spanish. The Spanish data consider the PC-GITA corpus [6], which contains utterances from 50 PD patients and 50 HC, Colombian Spanish native speakers. The participants were asked to pronounce a total of 10 sentences, the rapid repetition of /pa-ta-ka/, /pe-ta-ka/, /pa-ka-ta/, /pa/, /ta/, and /ka/, one text with 36 words, and a monologue. All patients were in ON state at the time of the recording, i.e., under the effect of their daily medication.

German. Speech recordings of 88 PD patients and 88 HC speakers from Germany are considered [18]. The participants performed four speech task: the rapid repetition of /pa-ta-ka/, 5 sentences, one text with 81 words, and a monologue.

Czech. A total of 100 native Czech speakers (50 PD, 50 HC) were considered [19]. The speech tasks performed by the participants include the rapid repetition of the syllables /pa-ta-ka/, a read text with 80 words, and a monologue.

Table 1. Information of the speakers in the three datasets. **Subjects:** number of speakers. **G:** gender (**M.** male or **F.** female). **T:** time after diagnosis in years.

	G	Spanish		German		Czech	
		PD	HC	PD	HC	PD	HC
Subjects	M	25	25	47	44	30	30
	F	25	25	41	44	20	20
Range of age	M	33–81	31–86	44–82	26–83	43–82	41–77
	F	49–75	49–76	42–84	28–85	41–72	40–79
Age	M	61.3 (11.4)	60.5 (11.6)	66.7 (8.7)	63.8 (12.7)	65.3 (9.6)	60.3 (11.5)
	F	60.7 (7.3)	61.4 (7.0)	66.2 (9.7)	62.6 (15.2)	60.1 (8.7)	63.5 (11.1)
T	M	8.7 (5.9)	–	7.0 (5.5)	–	6.7 (4.5)	–
	F	12.6 (11.6)	–	7.1 (6.2)	–	6.8 (5.2)	–
MDS-UPDRS-III	M	37.8 (22.1)	–	22.1 (9.9)	–	21.4 (11.5)	–
	F	37.6 (14.1)	–	23.3 (12.0)	–	18.1 (9.7)	–

2.2 Segmentation

Speech signals are analyzed based on the automatic detection of onset and offset transitions, which model the difficulties of the patients to start/stop the movement of the vocal folds. The detection of the transitions is based on the presence of the fundamental frequency of speech in short-time frames, as it was shown in [8]. The border between voiced and unvoiced frames is detected, and 80 ms of the signal are taken to the left and to the right, forming segments with 160 ms length. The transition segments are modeled with two different approaches: (1) a baseline model based on hand-crafted features, which are classified using an SVM, and (2) a model based on time-frequency representations used as input to train a CNN, which then will be used for the transfer learning strategy. Further details are given in the following subsections.

2.3 Baseline Model

The features extracted from the transitions include 12 Mel-Frequency Cepstral Coefficients (MFCCs) with their first and second derivatives, and the log energy of the signal distributed into 22 Bark bands. The total number of descriptors corresponds to 58. Four statistical functionals (mean, standard deviation, skewness, and kurtosis) are computed for each descriptor, obtaining a 232-dimensional feature-vector per utterance. The classification of PD patients and HC speakers is performed with a radial basis SVM with margin parameter $C = 10$ and a Gaussian kernel with parameter $\gamma = 0.0001$. The SVM is tested following a 10-fold Cross-Validation strategy, speaker independent.

2.4 CNN Model

Time frequency representations based on the short-time Fourier transform (STFT) are used as input to a CNN, which extract the most suitable features to discriminate between PD patients and HC subjects. The STFT with 256 frequency bins is computed for each segmented transition, for a window length of 16 ms and a step-size of 4 ms, forming 41 time frames per transition. The obtained spectrogram is transformed into the Mel-scale using 80 filters, forming an spectrogram with a size of 80 × 41, which is used to train the CNNs. The architecture of the implemented CNN is summarized in Table 2. It consists of four convolutional and max-pooling layers, dropout to regularize the weights, and two fully connected layers followed by the output layer to make the final decision using a softmax activation function. The number of feature maps on each convolutional layer is twice the previous one in order to get more detailed representations of the input space in the deeper layers. The CNN is trained using the cross-entropy as the loss function, using an Adam optimizer [20].

2.5 Transfer Learning

Transfer learning allows to use a neural network trained for one task to be used in another domain. We use transfer learning to classify patients and healthy

Table 2. Architecture of the CNN implemented in this study.

CNN architecture	Input size	Output size
Conv $(1 \times 4 \times 3, 1)$+dropout	$1 \times 80 \times 41$	$4 \times 80 \times 41$
Max pool $(2, 2)$	$4 \times 80 \times 41$	$4 \times 40 \times 20$
Conv $(4 \times 8 \times 3, 1)$+dropout	$4 \times 40 \times 20$	$8 \times 40 \times 20$
Max pool $(2, 2)$	$8 \times 40 \times 20$	$8 \times 20 \times 10$
Conv $(8 \times 16 \times 3, 1)$+dropout	$8 \times 20 \times 10$	$16 \times 20 \times 10$
Max pool $(2, 2)$	$16 \times 20 \times 10$	$16 \times 10 \times 5$
Conv $(16 \times 32 \times 3, 1)$+dropout	$16 \times 10 \times 5$	$32 \times 10 \times 5$
Max pool $(2, 2)$	$32 \times 10 \times 5$	$32 \times 5 \times 2$
Lineal $(320, 128)$+dropout	$32 \times 5 \times 2$	1×128
Lineal $(128, 64)$+dropout	1×128	1×64
Lineal $(64, 2)$	1×64	1×2

speakers in three different languages. The CNN architecture described before is used to train a CNN with utterances from one language. Then, the pre-trained model is used as a base to initialize two different models with the remaining languages. Figure 1 summarizes this procedure.

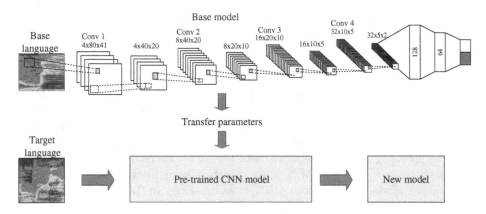

Fig. 1. Transfer learning strategy proposed in this study to classify PD from speech with utterances from different languages

3   Experiments and Results

The experiments are divided as follows: First, the baseline and the CNN models are trained considering each language individually. Then, the trained CNNs for each language are used as a base model in the transfer learning strategy in order to improve the accuracy in the other two languages. All speech exercises performed by the participants were considered for the classification strategy. The final decision for each speaker was obtained by a majority voting strategy among the different speech exercises.

3.1 Baseline and Individual CNN Models

Table 3 shows the results obtained for the baseline and the CNNs trained for each language individually. Similar accuracies are obtained between the baseline and the CNN model for Spanish language, which also exhibit the highest accuracy among the three languages. Note that the highest accuracy for German language was obtained with the baseline model. Conversely, for Czech language the CNN produces the highest accuracy. Note also that for the three languages, the results are unbalanced towards one of the two classes according to the specificity and sensitivity values. The difference in the results obtained among the three languages can be explained considering the information provided in Table 1. For the patients in the Spanish language, the average MDS-UPDRS-III score is higher compared with the German and Czech patients, i.e, there are patients with higher disease severity in the Spanish data compared to German and Czech patients.

Table 3. Classification results for the baseline and CNN models trained in three different languages. **Acc:** accuracy. **Sen:** sensitivity. **Spe:** specificity. **MCC:** matthews correlation coefficient.

Language	Baseline				CNN			
	Acc (%)	Sen (%)	Spe (%)	MCC	Acc (%)	Sen (%)	Spe (%)	MCC
Spanish	73.7 (13.0)	74.5 (16.7)	77.1 (16.2)	0.50	71.0 (15.9)	74.0 (25.0)	68.0 (28.6)	0.42
German	69.3 (9.9)	71.8 (12.4)	68.7 (10.0)	0.39	63.1 (11.7)	43.1 (38.0)	83.1 (17.7)	0.30
Czech	61.0 (12.5)	64.5 (19.5)	60.2 (11.9)	0.27	68.5 (14.1)	94.0 (13.5)	42.0 (33.2)	0.43

3.2 Transfer Language Among Languages

The results with the transfer learning strategy among languages are shown in Table 4. A CNN trained with utterances from the base language is fine-tuned with utterances from the target language. Note that the accuracy improved considerably when the target languages are German and Czech, with respect to the results observed for baseline and the CNN in Table 3. The accuracy improved over 8% for German (from 69.3% in the baseline to 77.3% when the model is fine-tuned from Spanish), and over 4.1% for Czech language (from 68.5% with the initial CNN to 72.6% when the model is fine-tuned from Spanish). Particularly, the highest accuracy for German and Czech languages is obtained when the base language is Spanish. This can be explained considering that Spanish speakers have the best initial separability, thus, the other two languages benefit from the best initial model. The results obtained with the transfer learning strategy among languages are also more balanced in terms of the specificity and sensitivity than the observed in the baseline and with the initial CNNs. The standard deviation of the transfered CNNs is also lower, which leads to an improvement in the generalization of the models.

Table 4. Classification results for the transfer learning among languages using CNNs. **Acc:** accuracy. **Sen:** sensitivity. **Spe:** specificity. **MCC:** matthews correlation coefficient. **Base lang:** language used to pre-train the CNN model. **Target lang:** language used for transfer learning from the base model.

Base lang.	Target lang.	Acc (%)	Sen (%)	Spe (%)	MCC
German	Spanish	70.0 (12.5)	62.0 (19.9)	78.0 (23.9)	0.41
Czech		72.0 (13.1)	67.0 (11.6)	78.0 (23.9)	0.46
Spanish	German	77.3 (11.3)	86.2 (13.8)	68.3 (14.3)	0.57
Czech		76.7 (7.9)	87.5 (11.0)	66.0 (15.6)	0.55
Spanish	Czech	72.6 (13.9)	82.0 (14.8)	62.0 (28.9)	0.46
German		70.7 (14.5)	80.0 (16.3)	62.5 (26.3)	0.38

The receiver operating characteristic (ROC) curves from Fig. 2 show with more detail the effect of the transfer learning strategy in the performance of the CNNs to classify PD speakers in different languages. The area under the ROC curve (AUC) when the target language is Spanish (Fig. 2A) is slightly higher when the base language is Czech. When the target languages are German and Czech (Fig. 2B and C) the highest AUC is obtained when the base model is trained with Spanish utterances.

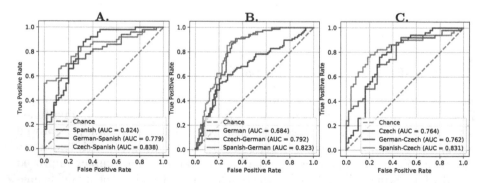

Fig. 2. ROC curves for the transfer learning among languages, when the target language is **A.** Spanish, **B.** German, and **C.** Czech.

4   Conclusion

This study proposed the use of a transfer learning strategy based on fine-tuning to classify PD from speech in three different languages: Spanish, German, and Czech. The transfer learning among languages aimed to improve the accuracy when the models are initialized with utterances from a different language than the one used for the test set. Mel-scale spectrograms extracted from the transitions between voiced and unvoiced segments are used to train a CNN for each language. Then, the trained models are used to fine-tune a model to classify utterances in the remaining two languages.

The results indicate that the transfer learning among languages improved the accuracy of the models in up to 8% when a base model trained with Spanish utterances is used to fine-tune a model to classify PD German utterances. The results obtained after the transfer learning are also more balanced in terms of specificity-sensitivity and have a lower variance. In addition, the transfer learning among languages scheme was accurate to improve the accuracy in the target language only when the base model was robust enough. This was observed when the model trained with Spanish utterances was used to initialize the models for German and Czech languages.

Further experiments will include the development of more robust base models using hyper-parameter optimization strategies like those based on Bayesian optimization. In addition, the base models will be trained considering two of the languages instead of only one of them. The trained models will also be evaluated to classify the speech of PD patients in several stages of the disease based on the MDS-UPDRS-III score, or based on their dysarthria severity [21]. Further experiments will also include transfer learning among diseases, for instance training a base model with utterances to classify PD, and use such a model to initialize another one to classify other neurological diseases such as Huntington's disease.

Acknowledgments. The work reported here was financed by CODI from University of Antioquia by grant Numbers 2017–15530 and PRG2018–23541. This project has received funding from the European Union's Horizon 2020 research and innovation programme under the Marie Sklodowska-Curie Grant Agreement No. 766287. T. Arias-Vergara is also under grants of Convocatoria Doctorado Nacional-785 financed by COLCIENCIAS.

References

1. Hornykiewicz, O.: Biochemical aspects of Parkinson's disease. Neurology **51**(2 Suppl 2), S2–S9 (1998)
2. Tykalova, T., Rusz, J., Klempir, J., Cmejla, R., Ruzicka, E.: Distinct patterns of imprecise consonant articulation among Parkinson's disease, progressive supranuclear palsy and multiple system atrophy. Brain Lang. **165**, 1–9 (2017)
3. Moretti, R., et al.: Speech initiation hesitation following subthalamic nucleus stimulation in a patient with Parkinson's disease. Eur. Neurol. **49**(4), 251–253 (2003)

4. Sakar, B.E., et al.: Collection and analysis of a Parkinson speech dataset with multiple types of sound recordings. IEEE J. Biomed. Health Inform. **17**(4), 828–834 (2013)
5. Villa-Cañas, T., Arias-Londoño, J.D., Orozco-Arroyave, J.R., Vargas-Bonilla, J.F., Nöth, E.: Low-frequency components analysis in running speech for the automatic detection of Parkinson's disease. In: Proceedings of the Sixteenth Annual Conference of the International Speech Communication Association, pp. 100–104 (2015)
6. Orozco-Arroyave, J.R., et al.: New Spanish speech corpus database for the analysis of people suffering from Parkinson's disease. In: Proceedings of the Ninth International Conference on Language Resources and Evaluation, pp. 342–347 (2014)
7. Novotný, M., Rusz, J., et al.: Automatic evaluation of articulatory disorders in Parkinson's disease. IEEE/ACM Trans. Audio Speech Lang. Process. **22**(9), 1366–1378 (2014)
8. Orozco-Arroyave, J.R.: Analysis of Speech of People with Parkinson's Disease. Logos Verlag, Berlin (2016)
9. Moro-Velazquez, L., et al.: A forced gaussians based methodology for the differential evaluation of Parkinson's disease by means of speech processing. Biomed. Signal Process. Control **48**, 205–220 (2019)
10. Rusz, J., Cmejla, R., et al.: Imprecise vowel articulation as a potential early marker of Parkinson's disease: effect of speaking task. J. Acoust. Soc. Am. **134**(3), 2171–2181 (2013)
11. Grósz, T., Busa-Fekete, R., Gosztolya, G., Tóth, L.: Assessing the degree of nativeness and Parkinson's condition using Gaussian processes and deep rectifier neural networks. In: Proceedings of the Sixteenth Annual Conference of the International Speech Communication Association, pp. 919–923 (2015)
12. Vásquez-Correa, J.C., Orozco-Arroyave, J.R., Nöth, E.: Convolutional neural network to model articulation impairments in patients with Parkinson's disease. In: Proceedings of the Eighteenth Annual Conference of the International Speech Communication Association, pp. 314–318 (2017)
13. Tu, M., Berisha, V., Liss, J.: Interpretable objective assessment of dysarthric speech based on deep neural networks. In: Proceedings of the Eighteenth Annual Conference of the International Speech Communication Association, pp. 1849–1853 (2017)
14. Cummins, N., Baird, A., Schuller, B.: Speech analysis for health: current state-of-the-art and the increasing impact of deep learning. Methods **151**, 41–54 (2018)
15. Wang, D., Zheng, T.F.: Transfer learning for speech and language processing. In: Proceedings of the Asia-Pacific Signal and Information Processing Association Annual Summit and Conference (APSIPA), pp. 1225–1237. IEEE (2015)
16. Naseer, A., et al.: Refining Parkinson's neurological disorder identification through deep transfer learning. Neural Comput. Appl. 1–16 (2019)
17. Goetz, C.G., et al.: Movement disorder society-sponsored revision of the unified Parkinson's disease rating scale (MDS-UPDRS): scale presentation and clinimetric testing results. Mov. Disord. **23**(15), 2129–2170 (2008)
18. Skodda, S., Visser, W., Schlegel, U.: Vowel articulation in Parkinson's disease. J. Voice **25**(4), 467–472 (2011)
19. Rusz, J.: Detecting speech disorders in early Parkinson's disease by acoustic analysis. Habilitation thesis, Czech Technical University in Prague (2018)
20. Kingma, D.P., Ba, J.: Adam: a method for stochastic optimization. In: Proceedings of International Conference on Learning Representations (ICLR), pp. 1–15 (2015)
21. Vásquez-Correa, J.C., Orozco-Arroyave, J.R., Bocklet, T., Nöth, E.: Towards an automatic evaluation of the dysarthria level of patients with Parkinson's disease. J. Commun. Disord. **76**, 21–36 (2018)

Bidirectional Alignment of Glottal Pulse Length Sequences for the Evaluation of Pitch Detection Algorithms

Carlos A. Ferrer[1,2]([⊠]) [iD], Reinier Rodríguez Guillén[1] [iD],
and Elmar Nöth[2] [iD]

[1] Informatics Research Center, Central University "Marta Abreu" de Las Villas,
Santa Clara, Cuba
cferrer@uclv.edu.cu, rrguillen@uclv.cu
[2] Pattern Recognition Lab, Friedrich Alexander University Erlangen-Nuemberg,
Erlangen, Germany
elmar.noeth@fau.de

Abstract. This paper describes a problem in a reported Dynamic Time Warping (DTW) alignment procedure to compare the reference and detected glottal pulse length sequences, oriented to compare the evaluation of Pitch Detection Algorithms (PDAs) in pathological voices. The problem in the existing alignment method tends to overestimate the failure of the PDA, by aligning only the detected to the reference sequence. A solution is presented, which performs a bidirectional alignment reducing the differences present in the definitive comparison. The proposal is evaluated in both synthetic and real voice signals, by running three well-known PDAs, and the magnitude of the error reduction along with comments on the possible factors influencing its value, are given. The alignment variant introduced in this paper allows to perform a fairer comparison of the PDAs performances.

Keywords: Dynamic Time Warping · Pitch Detection Algorithms · Jitter · Alignment

1 Introduction

In human oral communication, the sounds where the vocal cords vibrate show a quasi-periodic pattern in their acoustic waveforms. The origin of this periodicity is found in the alternating opening and closing of the vocal folds at the glottis, allowing a pulse-like flow of air coming from the lungs to travel forward and create sound [1]. The determination of glottal pulse boundaries is a common problem in several speech processing tasks, either oriented for healthy (e.g. singing [2], fluent speaking [3, 4] and similar uses) or pathologic speech [5]. The methods used for determining the pulse location are specific types of Pitch Detection Algorithms (PDAs), working on a cycle-by-cycle basis [6].

When facing pathological voices, a degradation in the PDAs performance appears [7, 8], due to the higher levels of periodicity perturbations present. There are dozens of

© Springer Nature Switzerland AG 2019
I. Nyström et al. (Eds.): CIARP 2019, LNCS 11896, pp. 707–716, 2019.
https://doi.org/10.1007/978-3-030-33904-3_67

PDAs to choose from, and selecting a particular approach should be based on thorough performance comparisons [9–12]. Key in this evaluation is the selection of the measures of PDAs performance extracted.

The most commonly used measure of PDA performance is to compare the pulse's durations variability in the detected sequence $T_d(n)$ with the variability of the reference, known sequence $T_r(n)$. Variability of the pulse duration contour is a concept denoted as *jitter* in the Voice Measurement literature [13], and there are several expressions proposed to measure it [14], among which a representative ones is:

$$\alpha = \frac{1}{N-1} \sum_{n=1}^{N-1} \frac{|T(n+1) - T(n)|}{0.5 * (T(n+1) + T(n))} * 100 \qquad (1)$$

The difference $|\alpha_r - \alpha_d|$, i.e. the values obtained when using $T_r(n)$ and $T_d(n)$ on Eq. (1), is widely reported as a measure of PDA accuracy [8–12]. However, this practice has been heavily criticized [7, 15, 16], since equal variability of two sequences does not imply sequences equality. An alternative approach suggested since [7] is to use an inter-sequence variability measure, termed β as a successor to α in Eq. (1). The expression for β used in [7] was slightly modified to closely resemble α in [15], as:

$$\beta = \frac{1}{N} \sum_{n=1}^{N} \frac{|T_d - T_r(n)|}{T_r(n)} * 100 \qquad (2)$$

However, the internal functioning of the PDAs frequently produces T_ds which cannot be related to T_r by simply iterating through n, the pulse index. A Dynamic Time Warping (DTW) of both glottal pulse lengths sequences, $T_r(n)$ and $T_d(n)$, was proposed [15] in order to correct the misalignments described, for a better evaluation of the performance of the PDAs. The DTW by itself produced several new performance measures by counting the amount of the different types of misalignment. The DTW procedure described in [15] actually detected the different types of misalignment between both sequences, but, as will be shown in the next section, the value of β produced failed to represent the one corresponding to the aligned sequences.

In this paper we describe the problem present in the reported DTW, we introduce the required modifications to solve it, and perform some experiments showing the relevance of the correction.

2   Existing DTW Procedure: Problem and Solution Proposed

A flowchart depicting the DTW procedure proposed in is shown in the left panel of Fig. 1. The algorithm departs from the sequences of reference and detected pulse positions, $P_r(n)$ and $P_d(n)$, from which the sequences of pulse lengths $T_r(n)$ and $T_d(n)$ are obtained by a difference (discrete derivative) operation. The additional measures of performance produced by the DTW algorithm are the number of significant/gross errors (GE) between both aligned sequences, the number of pulse insertions (PI) and deletions (PD) and the number of contour shifts to the left (SL) or to the right (SR) of the detected pulse boundaries $P_d(n)$ as compared to the reference, hand-marked contour $P_r(n)$. A copy of the detected pulse length contour is made to store the dynamically aligned contour, labeled T_dAl.

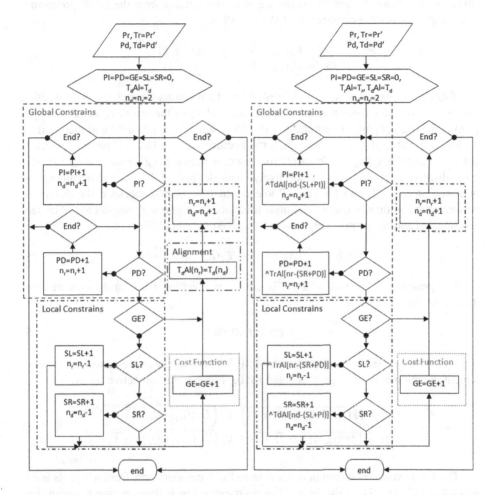

Fig. 1. Flowchart used for calculating the measures of performance as reported in [15] (left panel) and here (right panel). Basic inputs are the reference and detected pulse boundary positions $P_r(n)$ and $P_d(n)$. Bounding labeled squares indicate the relevant stages on a DTW procedure. Positive outcomes in the conditional blocks have been signaled by a black circle in the corresponding corner of the rhombus. Notice that the "Alignment" block has been replaced by suppression of pointed-by ($\wedge$) elements in the processing of PI, PD, SR & SL conditions.

The DTW procedure works by incrementally moving through the contours by using separate vector indexes for both, namely n_r and n_d. Both indexes compose a bi-dimensional grid of points (n_r, n_d) where the best path for alignment is searched for using some heuristic constraints.

Regarding the Global Constraints for the DTW, the alignment of length values (Ts) can only be performed if position values (Ps) are actually related in time. If for a particular position this is not met, then a pitch mark deletion (PD) or insertion (PI) occurred. Both PD and PI counts are metrics available from the DTW procedure. The conditions for the occurrence of PD and PI are:

$$PD? = P_d(n_d - 1) \ > \ P_r(n_r + 1)$$
$$PI? = P_d(n_d + 1) \ < \ P_r(n_r - 1) \tag{3}$$

If Global Constraints are not violated, Local Constraints were checked on whether the length contours require alignments, either a shift to the left (SL) or to the right (SR) in the (n_r, n_d) path. A first check is that the contours are not similar at their current positions in the grid (n_r, n_d), for which the occurrence of a Gross Error (GE) is evaluated. An auxiliary large-difference/error function between $T_r(n_r)$ and $T_d(n_d)$ for arbitrary displacements a and b, respectively, was defined in [15] for determining the presence of GE,, SL and SR. This Boolean function $D(a,b)$ checks if the difference between the contours at certain positions (displaced a and b with respect to the current positions n_r and n_d, respectively) exceeds a given threshold Thr:

$$D(a, b)? = |T_r(n_r + a) - T_d(n_d + b)| \ > \ Thr \tag{4}$$

The presence or absence of a GE is given by checking for D at the current coordinates of the search grid:

$$GE? = D(0, 0)? \tag{5}$$

And then, the need to perform an SL or SR is given by the expressions in (6), where logical complements have been denoted by bars above the respective terms:

$$SL? = \left(\overline{D?(-1, 0)} \ \& \ \overline{D(0, 1)} \right) \ \& \ \left(\overline{D?(0, -1) \& \overline{D(1, 0)}} \right)$$
$$SR? = \left(\overline{D?(0, -1)} \ \& \ \overline{D(1, 0)} \right) \ \& \ \left(\overline{D?(-1, 0) \& \overline{D(0, 1)}} \right) \tag{6}$$

The GE measure is common in short-term PDA comparisons, from where its name was taken [17, 18]. The value of Thr has been expressed in the literature either in time units [17, 19, 20] or in percentage of the average T_r [18, 21, 22]. The latter approach was used in [15], choosing a value of Thr equal to 3% of the average T_r.

After all Global and Local Constraints have been checked, and the indexes n_r and n_d properly modified, T_dAl is correspondingly modified (if needed) to better match T_r, and the algorithm proceeds to check the contour at the next values of n_r and n_d. The DTW ends whenever any of the contours contour reaches its end as indexed by the current values of n_r and n_d. The alignment performed on T_dAl consists in replacing its value in index n_r with the value of $T_d(n_d)$.

2.1 Modification to Make Alignment Bidirectional

The alignment performed in [15] could be considered, however, incomplete, due to this unilateral alignment: only T_d is modified to correspond to T_r. In this way, there will be indexes in T_dAl with no corresponding aligned values in T_r. In this paper we slightly modify the DTW procedure so that the alignment occurs bilaterally: from Tr indexes to T_dAl (as the original) but also from T_d indexes to an aligned version of T_r, namely T_rAl. In this way, T_dAl and T_rAl are better suited as aligned detected and reference contours, to evaluate an inter-contour variability measure, than the previously used pair T_dAl and T_r. The modified flow diagram of the DTW procedure used here is shown in the right panel of Fig. 1. There are two main differences between the alignments performed on both methods, represented in left and right panels:

- First, the alignment of the contours occurs in this case not after all constraints checking were performed ("Alignment" section in left panel), but within the actions taken for a particular condition producing a modification of either index n_r or n_d (i.e. the occurrence of a PI, PD, SL or SR in right panel).
- Second: The alignment is not performed here by assignment, but by suppression of the value without a corresponding element in the companion contour. The element to suppress in each case is pointed to by the ^ sign within the square encasing the actions corresponding to the particular condition (right panel).

In short, the element $[n_d - (SL + PI)]$ in T_dAl is suppressed every time a PI or SL occurs, while the element $[n_r - (SR + PD)]$ in T_rAl is suppressed every time a PD or SR occurs. These actions are in line with the classical 'insertion' and 'deletion' operations defined for string sequences in [23]. The third operation ('change') is still present when a GE occurs without an SL or SR condition met.

3 Experiments Performed

Both alignment procedures were evaluated by applying them to reference and detected contours corresponding to synthetic and real signals. Synthetic signals allow introducing periodicity perturbations of any magnitude desired, while real signals serve as validation, in case the synthesis procedure were to be contested.

Three well known PDAs were used, so that different performances are to be expected and tested. Among the PDAs are the ones included by default in the freely available Praat system [24], namely the Peak Picking (P-P) and the Cross Correlation (C-C) based methods. Praat is used as a reference in the evaluation of other systems and methods in [9, 12, 25–29], among many other studies. A third PDA evaluated is the Super-Resolution (S-R) method described in [30], which has shown very good results elsewhere [7, 15, 18]. The internal implementation of these PDAs is out of the scope of this paper, however, they are a representative sample of the best performing and more frequently used PDAs, as the previous references may prove.

3.1   Synthetic Signals

Synthetic signals were obtained according to the source filter model [31]. The standard pulse duration T_E was chosen to correspond to a frequency of 150 Hz, with sampling frequency of 22050 Hz. A single glottal pulse waveform was generated according to the polynomial model type C in [32], with rising time of 0.33 T_E and falling time of 0.09 T_E, which reportedly produced the most natural-sounding synthesis.

A vocal tract configuration corresponding to a vowel "*a*" with resonances defined by the frequencies, amplitudes and bandwidths already used in [7, 15, 30, 33–35]. Periodicity perturbations were introduced in seven levels of perturbation, ranging from quasi-periodic to highly aperiodic signals, incorporating random *jitter* and *shimmer* (pulse amplitude variability) values per individual pulse up to a maximum % as given in Table 1. Additive noise is added such that the signal-to-noise ratio (SNR) expressed in dB is also given in the table.

Table 1. Values of the individual perturbations per level, combined in the synthetic signals.

Level	1	2	3	4	5	6	7
Jitter (%)	3.4	6.8	10.2	13.6	17.0	20.4	23.8
Shimmer (%)	6.8	13.6	20.4	27.2	34.0	40.8	47.6
Signal-to-Noise (dB)	22	18	15	12	8	5	2

A total of 300 contiguous pulses, corresponding to roughly two seconds of signal, are synthesized for each perturbation level, with the reference pulse boundaries $P_r(n)$ available from the synthesis procedure. More details in the synthesis procedure can be found in the references provided.

3.2   Real Signals

The hand-marked pulse positions from the 29 pathological signals used in [15] were available for this paper as $P_r(n)$. The acoustic waveforms are available from the Massachusetts Eye and Ear Infirmary Database [36]. The hand-marking includes 4695 pulse markers, with average F0 of 161 Hz, close to the value in the synthetic signals.

3.3   Evaluation

Both alignment procedures, as depicted in left and right panels of Fig. 1, were applied to the available reference $P_r(n)$ and detected $P_d(n)$ resulting from the different PDAs. In the case of the synthetic signals, 100 realizations were obtained for each level, to report averaged measures.

All methods were programmed in MatLab R2017, and a script calling Praat's PDAs was executed within the MatLab environment. Three results are to be reported for β values: the first being value obtained for non aligned contours, denoted simply as β. The second is the value obtained from the pair of sequences T_dAl and T_r, following the procedure in [15] and shown in the left panel of Fig. 1, and denoted β_O. Finally, the third value is obtained from applying Eq. (2) to the sequences T_dAl and T_rAl, obtained according to the corrections described here and represented in the right panel of Fig. 1 and denoted as β_C.

4   Results and Discussion

4.1   Synthetic Signals

As mentioned in description of the synthetic signals, 100 realizations of the two second synthesis were obtained. The pulse sequences were aligned to the reference ones, and the 100 Beta values corresponding to each level per PDA were averaged. These values are shown in Table 2.

Table 2. Values of the three alternatives for β obtained for the three PDAs on synthetic signals

Level		1	2	3	4	5	6	7
β	P-P	0.04	8.96	29.27	56.60	118.30	415.01	NaN
	C-C	0.00	4.84	10.23	15.01	18.07	22.33	28.42
	S-R	0.00	0.10	0.18	0.21	0.35	0.75	3.60
β_O	P-P	0.04	5.28	23.23	47.85	105.14	371.29	NaN
	C-C	0.00	2.10	5.07	8.68	11.39	16.21	23.03
	S-R	0.00	0.10	0.18	0.21	0.35	0.67	1.47
β_C	P-P	0.04	0.33	1.46	4.78	10.64	12.90	NaN
	C-C	0.00	0.24	0.79	2.57	6.12	11.79	19.37
	S-R	0.00	0.10	0.18	0.21	0.35	0.65	1.41

A first conspicuous result is that the S-R PDA greatly outperforms both Praat based variants. In fact, in the most degraded level the P-P variant did not report an average value, since some sequences returned empty from the Praat system, and for the 5th and 6th perturbation level the variability was higher than 100%. But PDAs performance is not the main subject of analysis here, but the behavior of our proposed modifications in the value of β_C.

With respect to this issue, it is noticeable the manifest tendency for reduction from the un-aligned β, to the aligned β_O, to our alignment proposal β_C. This tendency is of course larger for the worst PDAs, where the occurrence of misalignments is expected to be higher. But, even for the almost-unaffected S-R PDA, the values of β_C are always equal or lower than β_O.

4.2   Real Signals

The average values for the three variants of β obtained for the 29 signals are shown in Table 3, with rows corresponding to each PDA considered.

Table 3. Averaged values of the three alternatives for β obtained for the three PDAs on the 29 hand-marked signals

PDA	β	β_O	β_C
C-C	21.69	16.89	8.84
P-P	17.58	15.37	11.74
S-R	4.22	2.10	1.35

The tendency to obtain reduced inter-contour variability as we move from the non aligned variant to the original alignment described in [15] and then to the corrections of the alignment described here is manifest for the three PDAs. Here again the Praat based PDAs show the poorest performance, reinforcing the need to check for tweaking the default settings. However, the PDA performance is not our objective in this paper. More interesting results the fact that the reduction is more noticeable in the C-C PDA. It seems that previous β and β_O values were more affected by insertion and deletions of pulses in C-C than by the magnitudes of the GE errors, the opposite of the P-P variant, which shows a smaller reduction towards β_C.

5   Conclusions

A correction to the DTW procedure proposed in [15] has been described, which allows to perform a more realistic comparison of the PDAs performances in terms of β, the variability between the reference and detected pulse lengths contours which can be credited completely to the functioning of the PDA.

The other measures of performance produced by the DTW algorithm (GE, PI, PD, SR & SL) are unaffected by the corrections made, so any previous result obtained with the uncorrected DTW method hold for all these measures. The ability of the correction to reduce the described inflation of β was shown in both real and synthetic signals. The modification introduced is simple to implement, and the authors will make public the code in the near future.

Acknowledgements. This work was partially supported by an Alexander von Humboldt Foundation Fellowship granted to one of the authors (Ref 3.2-1164728-CUB-GF-E).

References

1. Carding, P.N., Mathieson, L.: Voice and speech production. In: Gleeson, M. (ed.) Scott Brown's Otorhinolaryngology, Head & Neck Surgery, vol. II, 7th edn, pp. 2164–2169. Hodder Education, London (2008)
2. Babacan, O., Drugman, T., D'Alessandro, N., Henrich, N., Dutoit, T.: A quantitative comparison of glottal closure instant estimation algorithms on a large variety of singing sounds. In: Proceedings of the Annual Conference of the International Speech Communication Association, INTERSPEECH 2013, pp. 1702–1706 (2013)
3. Rao, K.S., Vuppala, A.K.: Non-uniform time scale modification using instants of significant excitation and vowel onset points. Speech Commun. **55**(6), 745–756 (2013)

4. Rao, K.S., Maity, S., Reddy, V.R.: Pitch synchronous and glottal closure based speech analysis for language recognition. Int. J. Speech Technol. **16**(4), 413–430 (2013)
5. Deshpande, P.S., Manikandan, M.S.: Effective glottal instant detection and electroglotto-graphic parameter extraction for automated voice pathology assessment. IEEE J. Biomed. Heal. Inform. **22**(2), 398–408 (2018)
6. Linder, R., Albers, A.E., Hess, M., Pöppl, S.J., Schönweiler, R.: Artificial neural network-based classification to screen for dysphonia using psychoacoustic scaling of acoustic voice features. J. Voice **22**(2), 155–163 (2008)
7. Parsa, V., Jamieson, D.G.: A comparison of high precision F0 extraction algorithms for sustained vowels. J. Speech Lang. Hear. Res. **42**(1), 112–126 (1999)
8. Veprek, P., Scordilis, M.S.: Analysis, enhancement and evaluation of five pitch determi-nation techniques. Speech Commun. **37**(3–4), 249–270 (2002)
9. Dejonckere, P.H., Schoentgen, J., Giordano, A., Fraj, S., Bocchi, L., Manfredi, C.: Validity of jitter measures in non-quasi-periodic voices. Part I: perceptual and computer performances in cycle pattern recognition. Logop. Phoniatr. Vocology **36**(March), 70–77 (2011)
10. Manfredi, C., Giordano, A., Schoentgen, J., Fraj, S., Bocchi, L., Dejonckere, P.H.: Validity of jitter measures in non-quasi-periodic voices. Part II: the effect of noise. Logop. Phoniatr. Vocology **36**(2), 78–89 (2011)
11. Dejonckere, P.H., Giordano, A., Schoentgen, J., Fraj, S., Bocchi, L., Manfredi, C.: To what degree of voice perturbation are jitter measurements valid? a novel approach with synthesized vowels and visuo-perceptual pattern recognition. Biomed. Sig. Process. Control **7**(1), 37–42 (2012)
12. Manfredi, C., Giordano, A., Schoentgen, J., Fraj, S., Bocchi, L., Dejonckere, P.H.: Perturbation measurements in highly irregular voice signals: Performances/validity of analysis software tools. Biomed. Sig. Process. Control **7**(4), 409–416 (2012)
13. Baken, R.J., Orlikoff, R.F.: Clinical Measurement of Speech and Voice, 2nd edn. Cengage Learning, Boston (2000)
14. Buder, E.H.: Acoustic analysis of voice quality: a tabulation of algorithms 1902–1990. In: Kent, R.D., Ball, M.J. (eds.) Voice Quality Measurement, pp. 119–244. Singular, San Diego (2000)
15. Ferrer, C., Torres, D., Hernández-Díaz, M.E.: Using dynamic time warping of T0 contours in the evaluation of cycle-to-cycle pitch detection algorithms. Pattern Recogn. Lett. **31**(6), 517–522 (2010)
16. Tsanas, A., Zañartu, M., Little, M.A., Fox, C.M., Ramig, L.O., Clifford, G.D.: Robust fundamental frequency estimation in sustained vowels: detailed algorithmic comparisons and information fusion with adaptive Kalman filtering. J. Acoust. Soc. Am. **135**(5), 2885–2901 (2014)
17. Rabiner, L.R., Cheng, M.J., Rosenberg, A.E., McGonegal, C.A.: A comparative perfor-mance study of several pitch detection algorithms. Acoust. Speech Sig. Process. IEEE Trans. **24**(5), 399–418 (1976)
18. Bagshaw, P.C., Miller, S.M., Jack, M.A.: Enhanced pitch tracking and the processing of F0 contours for computer aided intonation teaching. In: 3rd European Conference on Speech Communication and Technology EUROSPEECH 1993, pp. 1003–1006 (1993)
19. Wise, J.D., Caprio, J.R., Parks, T.W.: Maximum likelihood pitch estimation. IEEE Trans. Acoust. **24**(5), 418–423 (1976)
20. Shahnaz, C., Zhu, W.P., Ahmad, M.O.: Robust pitch estimation at very low SNR exploiting time and frequency domain cues. In: ICASSP, IEEE International Conference Acoustics, Speech, and Signal Processing – Proceedings, vol. I, no. February 2005 (2005)
21. Nakatani, T., Irino, T.: Robust and accurate fundamental frequency estimation based on dominant harmonic components. J. Acoust. Soc. Am. **116**(6), 3690–3700 (2004)

22. de Cheveigné, A., Kawahara, H.: YIN, a fundamental frequency estimator for speech and music. J. Acoust. Soc. Am. **111**(4), 1917–1930 (2002)
23. Wagner, R.A., Fischer, M.J.: The string-to-string correction problem. J. ACM **21**(1), 168–173 (1974)
24. Boersma, P.: Praat, a system for doing phonetics by computer. Glot Int. **5**(9/10), 5 (2002)
25. Amir, O., Wolf, M., Amir, N.: A clinical comparison between two acoustic analysis softwares: MDVP and Praat. Biomed. Sig. Process. Control **4**(3), 202–205 (2009)
26. Maryn, Y., Corthals, P., De Bodt, M.S., Van Cauwenberge, P., Deliyski, D.D.: Perturbation measures of voice: a comparative study between multi-dimensional voice program and praat. Folia Phoniatr. Logop. **61**(4), 217–226 (2009)
27. Hanschmann, H., Gärtner, S., Berger, R.: Comparability of computer-supported concurrent voice analysis. Folia Phoniatr. Logop. **67**(1), 8–14 (2015)
28. Burris, C., Vorperian, H.K., Fourakis, M., Kent, R.D., Bolt, D.M.: Quantitative and descriptive comparison of four acoustic analysis systems: vowel measurements. J. Speech Lang. Hear. Res. **57**(1), 26–45 (2014)
29. Hagmüller, M., Kubin, G.: Poincaré pitch marks. Speech Commun. **48**(12), 1650–1665 (2006)
30. Medan, Y., Yair, E., Chazan, D.: Super resolution pitch determination of speech signals. IEEE Trans. Signal Process. **39**(1), 40–48 (1991)
31. Fant, G.: Acoustic Theory of Speech Production, 1st edn. Mouton, The Hage (1960)
32. Rosenberg, A.E.: Effect of glottal pulse shape on the quality of natural vowels. J. Acoust. Soc. Am. **49**(2B), 583–590 (1971)
33. Ferrer, C.A., González, E., Hernández-Díaz, M.E.: Evaluation of time and frequency domain-based methods for the estimation of harmonics-to-noise-ratios in voice signals. In: Martínez-Trinidad, J.F., Carrasco Ochoa, J.A., Kittler, J. (eds.) CIARP 2006. LNCS, vol. 4225, pp. 406–415. Springer, Heidelberg (2006). https://doi.org/10.1007/11892755_42
34. Ferrer, C., González, E., Hernández-Díaz, M.E., Torres, D., Del Toro, A.: Removing the influence of shimmer in the calculation of harmonics-to-noise ratios using ensemble-averages in voice signals. EURASIP J. Adv. Sig. Process. **2009**(1), 784379 (2009). https://doi.org/10.1155/2009/784379
35. Ferrer, C., Hernández-Díaz, M.E., González, E.: Using waveform matching techniques in the measurement of shimmer in voiced signals. In: Interspeech 2007: 8th Annual Conference of the International Speech Communication Association, pp. 2436–2439 (2007)
36. Disordered Voice Database v1.03. Kay Elemetrics Corp. (1994)

Video Analysis

Gait Recognition Using Pose Estimation and Signal Processing

Vítor Cézar de Lima and William Robson Schwartz[✉]

Smart Sense Laboratory, Department of Computer Science,
Universidade Federal de Minas Gerais, Belo Horizonte, Brazil
{vitorcezar,william}@dcc.ufmg.br

Abstract. Gait is a biometry that differentiates individuals by the way they walk. Research on this topic has gained evidence since it is unobtrusive and can be collected at distance, which is desirable in surveillance scenarios. Most of the previous works have focused on human silhouette as representation. However, they suffer from many factors such as movement on scene, clothing and carrying conditions. To avoid such problems, this work employs pose estimation to retrieve the coordinates of body parts, which are transformed into signals and movement histograms to be used as feature descriptors. While the former descriptors are used with the Subsequence Dynamic Time Warping that compares signals from probe and gallery, the Euclidean distance is used on the latter to find the person on gallery that is closest to probe. Finally, the outputs of both are fused. This work was evaluated on all views of CASIA Dataset A and compared to existing ones, demonstrating its efficacy.

Keywords: Gait recognition · Biometry · Computer vision

1 Introduction

Gait is a biometry employed to identify individuals by the way they walk [7]. It has gained evidence in recent years due to its advantages comparing with other biometrics (e.g., iris, face and fingerprint), since it can be collected at distance and it does not need the subject cooperation.

Works on gait recognition can be divided in two main approaches: model-based and model-free. While the former models characteristics of human motion pattern, the latter employs other features to represent gait, such as silhouette [22], symmetry of human motion [5] and Fourier descriptors [14].

An example of a model-based approach is the work of Wang et al. [20], that uses as dynamic feature a tracking operation that calculates joint-angle trajectories of the main lower limbs. Another work is Wagg et al. [19], where anatomical data is used to generate shape models consistent with regular human body proportions and create prototype gait motion models adapted to fit each subject. The problem is that model-based methods present a high computational cost since they need to model human motion patterns.

© Springer Nature Switzerland AG 2019
I. Nyström et al. (Eds.): CIARP 2019, LNCS 11896, pp. 719–728, 2019.
https://doi.org/10.1007/978-3-030-33904-3_68

The methods of model-free approach are simple and present lower computational requirements. They usually use the silhouette to represent a subject, such as the gait features [10,12,16], where the gait energy image [13] is the most influential, in which silhouettes from a gait cycle are normalized and aligned and then, have their mean values calculated and used as template. However, this approach is severely effected by clothing, carried objects and the camera view of the silhouette.

There are works that address the problems related to camera view, clothing and carrying conditions. Liu and Zheng [1] developed a cross-view approach that does not require prior information regarding the probe angle. It uses Gaussian process for view angle estimation and correlation strength from canonical correlation analysis to perform gait recognition. The algorithm was tested on CASIA Dataset B [23] and performed better than other existing works. However, the results still need to be improved, specially when other conditions are tested, such as carrying and clothing. Isaac et al. [6] proposed a genetic template segmentation to select silhouette parts to be used for classification. For each angle, the genetic algorithm automates the boundary positions and selects the parts to use, from which a view-estimator is able to determine the probe angle and selects the suitable view-specific classifier to recognize.

Despite the advances on gait recognition with methods based on silhouettes, there are intrinsic limitations on such representation. Silhouettes are not robust on outdoor scenes, due to movements of people and objects. In addition, as Sarkar et al. [18] pointed out, they change on different surface types and silhouettes from the same person obtained at different days are hard to be recognized. Therefore, due to these limitations, we employ pose estimation to obtain the coordinates of the body parts [2]. This information is processed generating signals and movement histograms, which are used as features, robust to the aforementioned aspects. Two methods are employed to perform the recognition. While the first uses Subsequence Dynamic Time Warping to compare signals from probe and gallery, ranking based on minimum matching distance cost, the second computes the Euclidean distance of the movement histograms to define the person from gallery that is closest to the probe. Finally, a score fusion is used on the result of these two methods to provide the identity of a subject.

Our main contributions are the following. The creation of a gait representation that is robust to clothing and carrying conditions and can be used on all views and the development of two signal processing methods based on the pose estimation that are efficient on gait discrimination among individuals.

We evaluate the proposed approach on all views of the CASIA Dataset A [22]. Even though our proposed approach neither uses machine learning techniques nor deep learning-based approaches, we are able to achieve recognition accuracy above 92.5% on all views. The results are compared to state-of-art works, showing that our approach achieves similar accuracy to the best works on lateral and oblique views and the same accuracy of the best work on the frontal view.

2   Feature Extraction

To extract the body coordinates, we use the pose estimator proposed by Cao et al. [2]. Their method returns $P_{b,t}^i$ for each frame t on gait sequence i and body part indexed by b. $1 \leq t \leq T^i$, where T^i is the total number of frames on sequence i. $P_{b,t}^i$ is a coordinate (x, y) and x and y values are referenced by $P_{b,t}^i.x$ and $P_{b,t}^i.y$, respectively. The coordinates have non-negative values, except when the body part is not found due to occlusion, when they have the invalid values $(-1, -1)$.

The coordinates returned by the pose estimator are from 18 body parts: neck, nose and both ears, wrists, elbows, hips, knees, ankles, shoulders and eyes. While ears, eyes and nose are not considered because their positions do not help on the gait recognition, the remaining 13 body parts, indexed by b ranging from 1 to 13, are used. For instance, the index b for the neck is 1 and $P_{1,t}^i$ is the neck coordinate at the t-th frame and in the i-th sequence.

The invalid coordinates generated by the pose estimator due to occlusion may interfere with the results. So, to avoid interference, a tracking of invalid body parts is performed, creating the noise indexing N^i for sequence i that saves the indexes of all body parts whose percentage of invalid coordinates is higher than a defined threshold. It is used on classification to eliminate "noisy" body parts on the Subsequence Dynamic Time Warping and the Euclidean distance calculation (see Sects. 3.1 and 3.2). N^i is defined as

$$N^i = \left\{ b : \frac{\#\{P_{b,t}^i.x = -1 \; \forall t \in [1, 2, ..., T^i]\}}{T^i} > \gamma \right\}, \tag{1}$$

where γ is a parameter that represents the noise tolerance. Therefore, it should be defined in a way that increases recognition accuracy, and this is done on Sect. 4.

2.1   Feature Based on Body Part Signals

Body part signals are used to get dynamic information on gait sequence, differentiating individuals by the way their body locations vary on time relative to neck position.

Signals to represent gait movement from sequence i are created, represented by S^i. Each b from 2 to 13 (the neck is used on the formula, but signals for it are not created because they would have only zeros) will generate two lines on S^i: one for its x coordinate value and the other for y. S^i has 24 lines and T^i columns and it is obtained using

$$S_{2b-3,t}^i = \begin{cases} -1, & \text{if } P_{b,t}^i.x = -1 \\ \frac{P_{b,t}^i.x - P_{1,t}^i.x}{max_j P_{j,t}^i.y - P_{1,t}^i.y}, & \text{otherwise} \end{cases} \tag{2}$$

$$S_{2b-2,t}^i = \begin{cases} -1, & \text{if } P_{b,t}^i.y = -1 \\ \frac{P_{b,t}^i.y - P_{1,t}^i.y}{max_j P_{j,t}^i.y - P_{1,t}^i.y}, & \text{otherwise} \end{cases} \tag{3}$$

According to the equations, the person position on frame is not considered because the distances are relative to neck position. The denominator of S^i is the vertical distance of the neck to one of the feet (one foot is always on the floor, so maximum y is from it), making the signals invariant to the person distance on the video.

After the signal creation, each line of S^i has its invalid values removed using linear interpolation and median filter is also applied. Figure 1 shows examples of signals from all views, where some periodicity can be observed.

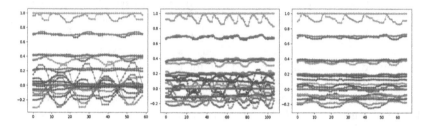

Fig. 1. Examples of signals from lateral, oblique and frontal view, respectively. The lines correspond to x and y distance of body parts to the neck position on coordinates. The distances vary with time, changing the amplitude.

2.2   Feature Based on Movement Histograms

The movement histograms capture static gait information, dividing the values of body part signals on intervals to represent a gait sequence.

After the signals creation and processing as described in the previous section, the movement histogram H^i is created for sequence i. S^i values are divided on intervals in $(-1, 1]$, and each occurrence of a value in S^i increments its corresponding position on H^i. Then, each value on H^i is divided by T^i, turning the histogram values invariant to sequence size. This is represented by

$$H_{l,j}^i = \frac{\#\left\{ \left\lceil \frac{nVals(S_{l,t}^i+1)}{2} \right\rceil = j \ \forall t \in [1, 2, ..., T^i] \right\}}{T^i}, \tag{4}$$

for $1 \le l \le 24$ and $1 \le j \le nVals$, where $nVals$ is the number of intervals in $(-1, 1]$.

Increasing $nVals$ make easier to differentiate individuals. The problem is that if $nVals$ is extremely high, the division of values of S^i on H^i will be sparse and the recognition will be affected. Therefore, it is necessary to find an optimum value of $nVals$ that can differentiate individuals without decreasing recognition (see experimental results section). Figure 2 shows movement histograms for different values of $nVals$.

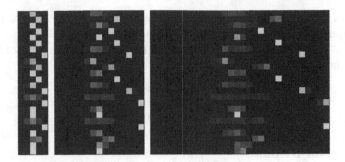

Fig. 2. Histograms with $nVals$ equal to 5, 15 and 30, respectively. Each line is related to one body part and each column corresponds to an interval on histogram. The images show that increasing $nVals$ increases the distribution of signal values on histogram, but causes sparsity on some intervals.

3   Gait Recognition

This section present two methods used for recognition and their fusion. They use the features of body part signals and movement histograms presented on the last section.

In the recognition stage, each person g from gallery is compared to probe p. The goal is to find the person g that minimizes the cost function $D^{p,g}_{sdtw}$, $D^{p,g}_{dist}$ or $D^{p,g}_{fusion}$.

The union operation is applied on the indexes of body parts on N^p and N^g, creating $N^{p,g}$. S^{i*}, S^{p*}, H^{i*} and H^{p*} are created from S^i, S^p, H^i and H^p, removing lines corresponding to body parts indexed on $N^{p,g}$.

The following subsection presents how $D^{p,g}_{sdtw}$ is calculated using Subsequence Dynamic Time Warping. Then, Sect. 3.2 describes the use of the Euclidean distance with the movement histograms to calculate $D^{p,g}_{dist}$. Finally, the last section discusses the fusion $D^{p,g}_{fusion}$ of the results.

3.1   Subsequence Dynamic Time Warping

Dynamic Time Warping (DTW) [17] is a nonlinear dynamic programming time normalization technique, where a warping function is computed to map the time axis from the probe to gallery samples. Its advantages are that signals which are compared do not need to have the same number of samples and this technique is robust to changes in walking speed [8]. It is used on gait applications for signal normalization [4,20] and matching [8]. In the latter case, gait sequences that have low minimum matching cost are more likely to be from the same person.

On this work, signal matching is applied, using a variation of DTW called Subsequence Dynamic Time Warping (SDTW) to find S^{g*} from the gallery where a subsequence within S^{p*} is optimally fitted using squared distance. This subsequence is composed of 26 consecutive columns, which is the size of a gait cycle.

The result of SDTW is also normalized by the number of lines of S^{g*}. This operation is important, giving the variation on number of lines of S^{g*} for different persons from gallery, because of N^g information. The cost function of SDTW is defined in the equation below:

$$D^{p,g}_{sdtw} = \frac{SDTW(S^{p*}_{:,1:26}, S^{g*})}{\sqrt{num\_lines(S^{g*})}} \qquad (5)$$

3.2   Euclidean Distance

To recognize gait using movement histograms, Euclidean distance is applied comparing probe H^{p*} with H^{g*} from gallery. These features have two dimensions, so they are vectorized for comparison. The Euclidean distance used on movement histograms is defined in the equation below:

$$D^{p,g}_{dist} = \frac{\|vec(H^{p*}) - vec(H^{g*})\|}{\sqrt{num\_lines(H^{g*})}} \qquad (6)$$

The distance is used to rank the individuals from gallery based on their distance from probe. The result is normalized according to the the number of lines on H^{g*} for the same reason discussed in the previous subsection.

3.3   Score Fusion

Fusion of features is a common operation used to improve recognition accuracy. In this work it is done using information from SDTW and Euclidean distance. Score fusion is applied, combining results of the two methods as

$$D^{p,g}_{fusion} = \alpha D^{p,g}_{dist} + (1 - \alpha) D^{p,g}_{sdtw}, \qquad (7)$$
$$\text{with } 0 \leq \alpha \leq 1.$$

In this equation, increasing α favors the results of Euclidean distance and decreasing α favors SDTW. An experiment described in the next section estimates the best value of α in which the fusion leads to better gait recognition.

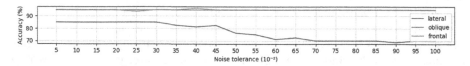

Fig. 3. Accuracy on D_{sdtw} varying the noise tolerance γ. The best results are found when γ is between 0.05 to 0.2.

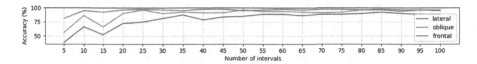

Fig. 4. Accuracy on D_{dist} varying the number of intervals $nVals$. The value 85 gives the best results.

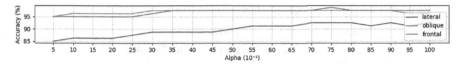

Fig. 5. Accuracy on D_{fusion} varying the score fusion weight α. The value 0.75 gives the best results.

4   Experimental Results

Experiments are conducted on the CASIA Dataset A [22], in which data were captured at 25 FPS and has frame resolution of 352×240 pixels. This dataset has sequences from lateral, oblique and frontal views. There are 20 subjects and they have four gait sequences for each view angle. Two of these sequences have the same walking direction and on the other half the walking direction is reversed. For each pair of videos with the same configuration of person, angle and movement direction, one is used as gallery and the other as probe.

The next paragraphs describe and analyze the feature, classification parameters and the results achieved are compared to other methods in the literature.

Noise tolerance. The noise tolerance γ defines which body parts will not be used with SDTW and Euclidean distance because of occlusion. This parameter is tested on the SDTW method and was varied from 0.05 to 1. According to the results showed in Fig. 3, occlusion has a great impact on recognition and the best results were found when γ is between 0.05 and 0.2. Giving that lower values remove more body parts on calculations and make the computation simpler as consequence, 0.05 has been chosen to be used in the remaining experiments.

Number of intervals. This experiment evaluates the number of intervals $nVals$ on the movement histogram. It is responsible for the size of intervals, which are used to distinguish gait on sequences. $nVals$ was varied from 5 to 100. According to Fig. 4, the best results are achieved by $nVals$ equals to 85, which maximizes accuracy on D_{dist}. This value is used in the remaining experiments.

Score fusion weight. The score weight α is used to fuse results from SDTW and Euclidean distance. It ranges from 0 to 1 and, when α presents higher values, it favors Euclidean distance and when it has lower values, SDTW is favored. In our experiment, we varied α from 0.05 to 1. According to Fig. 5, the best results were achieved with α equals to 0.75 (i.e., the Euclidean distance should receive more importance than SDTW).

Table 1. Rank 1 recognition on CASIA Dataset A.

Work/view	Lateral	Oblique	Frontal
Zhang et al. [24]	55%	80%	85%
Wang et al. [21]	88.75%	87.5%	90%
Liu et al. [11]	85%	87.5%	95%
Chen et al. [3]	90%	75%	–
Wang et al. [20]	97.5%	–	–
Nizami et al. [15]	**100%**	–	–
Kusakunniran et al. [9]	**100%**	**100%**	**98.75%**
SDTW method	85%	95%	95%
Euclidean method	92.5%	96.25%	97.5%
Fusion of methods	92.5%	97.5%	**98.75%**

Comparisons. The best results of the proposed methods was achieved when γ is 0.05, $nVals$ is 85 and α is 0.75, achieving accuracy of 92.5%, 97.5% and 98.75% on lateral, oblique and frontal views, respectively. We believe the lower accuracy on lateral view is caused by occlusion, which is more common on that view. Euclidean distance method is more efficient than SDTW, showing that static information is more efficient than dynamic on gait recognition. However, the accuracy is increased when dynamic and static gait information from SDTW and Euclidean distance are fused.

The work is compared to others that use the same dataset. According to the results showed in Table 1, our approach has some of the best results on lateral, the second best on oblique and the best on frontal view (together with Kusakunniran et al. [9]). It is also interesting to note that our results are better than Liu et al. [11], which is a recent model-based method that uses deep learning. These results prove the efficiency of our method, despite the simplicity of the developed algorithms.

5   Conclusions

This work used pose estimation on gait recognition to retrieve body parts coordinates and transform them to signals and movements histograms, that are then used as features. Two methods were used for recognition. The first employs Subsequence Dynamic Time Warping to compare signals from the probe and gallery and rank them based on minimum matching cost; and the other uses Euclidean distance on the movement histograms to define the gallery sample closest to probe. A score fusion is then used on the methods results.

Experiments were performed on all views of CASIA Dataset A and we conclude that occlusion has a great effect on gait recognition. Therefore, noisy body parts should not be used on calculations. Euclidean distance was proved to be better than Subsequence Dynamic Time Warping and the recognition is improved when the two methods are fused.

The accuracy of 92.5%, 97.5% and 98.75% was found for lateral, oblique and frontal views, respectively, which is compared with state-of-art works. This work is one of the best on lateral, the second on oblique, and the best on frontal (together with Kusakunniran et al. [9]), proving our work efficiency on gait recognition applications.

Acknowledgments. The authors would like to thank the National Council for Scientific and Technological Development – CNPq (Grants 311053/2016-5 and 438629/2018-3), the Minas Gerais Research Foundation – FAPEMIG (Grants APQ-00567-14 and PPM-00540-17), the Coordination for the Improvement of Higher Education Personnel – CAPES (DeepEyes Project) and Petrobras (Grant 2017/00643-0).

References

1. Bashir, K., Xiang, T., Gong, S.: Cross view gait recognition using correlation strength. In: Bmvc, pp. 1–11 (2010)
2. Cao, Z., Simon, T., Wei, S.E., Sheikh, Y.: Realtime multi-person 2d pose estimation using part affinity fields. In: CVPR, vol. 1, p. 7 (2017)
3. Chen, C., Liang, J., Zhao, H., Hu, H., Tian, J.: Factorial HMM and parallel HMM for gait recognition. IEEE Trans. Syst. Man Cybern. Part C (Appl. Rev.) **39**(1), 114–123 (2009)
4. Derawi, M.O., Nickel, C., Bours, P., Busch, C.: Unobtrusive user-authentication on mobile phones using biometric gait recognition. In: 2010 Sixth International Conference on Intelligent Information Hiding and Multimedia Signal Processing (IIH-MSP), pp. 306–311. IEEE (2010)
5. Hayfron-Acquah, J.B., Nixon, M.S., Carter, J.N.: Automatic gait recognition by symmetry analysis. Pattern Recogn. Lett. **24**(13), 2175–2183 (2003)
6. Isaac, E.R., Elias, S., Rajagopalan, S., Easwarakumar, K.: View-invariant gait recognition through genetic template segmentation. IEEE Sig. Process. Lett. **24**(8), 1188–1192 (2017)
7. Jain, A.K., Ross, A., Prabhakar, S.: An introduction to biometric recognition. IEEE Trans. Circ. Syst. video Technol. **14**(1), 4–20 (2004)
8. Kale, A., Cuntoor, N., Yegnanarayana, B., Rajagopalan, A.N., Chellappa, R.: Gait Analysis for Human Identification. In: Kittler, J., Nixon, M.S. (eds.) AVBPA 2003. LNCS, vol. 2688, pp. 706–714. Springer, Heidelberg (2003). https://doi.org/10.1007/3-540-44887-X_82
9. Kusakunniran, W., Wu, Q., Li, H., Zhang, J.: Automatic gait recognition using weighted binary pattern on video. In: Sixth IEEE International Conference on Advanced Video and Signal Based Surveillance, 2009. AVSS 2009, pp. 49–54. IEEE (2009)
10. Lee, H., Hong, S., Nizami, I.F., Kim, E.: A noise robust gait representation: motion energy image. Int. J. Control Autom. Syst. **7**(4), 638–643 (2009)
11. Liu, D., Ye, M., Li, X., Zhang, F., Lin, L.: Memory-based gait recognition. In: BMVC (2016)
12. Liu, J., Zheng, N.: Gait history image: a novel temporal template for gait recognition. In: 2007 IEEE International Conference on Multimedia and Expo, pp. 663–666. IEEE (2007)

13. Man, J., Bhanu, B.: Individual recognition using gait energy image. IEEE Trans. Pattern Anal. Mach. Intell. **28**(2), 316–322 (2006)
14. Mowbray, S.D., Nixon, M.S.: Automatic Gait Recognition via Fourier Descriptors of Deformable Objects. In: Kittler, J., Nixon, M.S. (eds.) AVBPA 2003. LNCS, vol. 2688, pp. 566–573. Springer, Heidelberg (2003). https://doi.org/10.1007/3-540-44887-X_67
15. Nizami, I.F., Hong, S., Lee, H., Lee, B., Kim, E.: Automatic gait recognition based on probabilistic approach. Int. J. Imaging Syst. Technol. **20**(4), 400–408 (2010)
16. Ogata, T., Tan, J.K., Ishikawa, S.: High-speed human motion recognition based on a motion history image and an eigenspace. IEICE Trans. Inf. Syst. **89**(1), 281–289 (2006)
17. Sakoe, H., Chiba, S.: Dynamic programming algorithm optimization for spoken word recognition. IEEE Trans. Acoust. Speech Sig. Process. **26**(1), 43–49 (1978)
18. Sarkar, S., Phillips, P.J., Liu, Z., Vega, I.R., Grother, P., Bowyer, K.W.: The humanid gait challenge problem: data sets, performance, and analysis. IEEE Trans. Pattern Anal. Mach. Intell. **27**(2), 162–177 (2005)
19. Wagg, D.K., Nixon, M.S.: On automated model-based extraction and analysis of gait. In: 2004 Sixth IEEE International Conference on Automatic Face and Gesture Recognition, Proceedings, pp. 11–16. IEEE (2004)
20. Wang, L., Ning, H., Tan, T., Hu, W.: Fusion of static and dynamic body biometrics for gait recognition. IEEE Trans. Circ. Syst. video Technol. **14**(2), 149–158 (2004)
21. Wang, L., Tan, T., Hu, W., Ning, H.: Automatic gait recognition based on statistical shape analysis. IEEE Trans. Image Process. **12**(9), 1120–1131 (2003)
22. Wang, L., Tan, T., Ning, H., Hu, W.: Silhouette analysis-based gait recognition for human identification. IEEE Trans. Pattern Anal. Mach. Intell. **25**(12), 1505–1518 (2003)
23. Yu, S., Tan, D., Tan, T.: A framework for evaluating the effect of view angle, clothing and carrying condition on gait recognition. In: 18th International Conference on Pattern Recognition, 2006. ICPR 2006, vol. 4, pp. 441–444. IEEE (2006)
24. Zhang, E.H., Ma, H.B., Lu, J.W., Chen, Y.J.: Gait recognition using dynamic gait energy and pca+lpp method. In: 2009 International Conference on Machine Learning and Cybernetics, vol. 1, pp. 50–53. IEEE (2009)

A Novel Scheme for Training Two-Stream CNNs for Action Recognition

Reinier Oves García$^{(\boxtimes)}$, Eduardo F. Morales, and L. Enrique Sucar

Instituto Nacional de Astrofísica, Óptica y Electrónica, San Andrés Cholula, Mexico
ovesreinier@gmail.com, {emorales,esucar}@inaoep.mx

Abstract. Human actions recognition from realistic video data constitutes a challenging and relevant research area. Leading the *state-of-the-art* we can find those methods based on Convolutional Neural Networks (CNNs) and specially two-stream CNNs (appearance and motion). In this paper we present a novel scheme for training two-stream CNNs that increases the accuracy of the fusion (when one of the channels does not perform as well as the other one) and reduces the total time used for training the entire architecture. In addition, we introduce a new descriptor for motion representation that improves the *state-of-the-art*. Based on this more efficient scheme, we developed an early recognition system. The proposed approach is evaluated on the UCF101 data set with competitive results.

Keywords: Human Action Recognition · Convolutional Neural Networks

1 Introduction

The automatic recognition of human actions from video data is an important topic withing the Computer Vision area due to its strength in providing a personalized support for several real-world applications, such as medicine, human-computer interaction and robotics, among others. Action recognition deals with the problem of assigning a predefined label to an input video and constitutes a challenging task for both areas, Computer Vision and Machine Learning. This research topic has been developed constantly in the last two decades [13] with a considerable improvement when using non-handcrafted features [18].

The best performance so far been achieved by multi-stream approaches [3], specifically by two-stream CNNs [21], turning obsolete those approaches based on handcrafted features [27]. During the last two years several works have been focused on 3D CNNs and transfer learning [2,5]. 3D CNNs are able to capture spatio-temporal relations and constitute the more powerful tool for action recognition so far. In 2017 [2] established a new mark in the *state-of-the-art* by transferring the knowledge from a 2D pre-trained model to 3D models, creating the most powerful two-stream CNN for action recognition (I3D).

© Springer Nature Switzerland AG 2019
I. Nyström et al. (Eds.): CIARP 2019, LNCS 11896, pp. 729–739, 2019.
https://doi.org/10.1007/978-3-030-33904-3_69

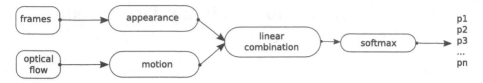

Fig. 1. General overview of two-stream CNNs. **"Appearance"** is a CNN that receives multiple RGB frames at once for producing the appearance outcome. **"Motion"** is another CNN that receives multiple optical-flows at once for producing the motion outcome. Both outcomes: appearance and motion usually contribute equally to a linear combination. Finally, such a combination is transformed into a probabilistic distribution throughout a softmax layer.

Two-stream architectures are based on the late fusion of two CNNs which are trained independently over two different domains; appearance and motion. The appearance-stream learns features from the RGB frames while the motion-stream learns the features from the optical flow [30]. For the final fusion, a linear combination of the last layer of each stream is applied and transformed into a probabilistic distribution by a softmax layer [2] (See Fig. 1).

The main drawback of two-stream CNNs is given by the fusion layer. When all streams perform reasonably well, a sophisticated combination is not needed since the contribution will be minimal. However, when one of the streams has a weak performance the final results can be severely affected, producing in that way a result worse than the produced by the best of the streams. On the other hand, the quality of video data, the sampling rate and interpolation methods used while sampling, also affect the performance.

In this paper, we present a novel strategy that outperforms the results achieved by the *state-of-the-art* when using two-stream CNNs [2] over the UCF101-1, UCF101-2 and UCF101-3 datasets [23]. In addition, we included a new motion descriptor, easy to understand and fast to compute. This descriptor (*CurlDiv*) is composed by the curl and divergence of the optical flow and is used in order to replace the motion stream. Although this descriptor itself does not perform as well as the optical flow, when it is combined with the appearance channel under our training scheme, it outperforms the classical fusion. Given that the computation of the *CurlDiv* descriptor is faster than the optical flow, we report a modification of [2] for early action recognition, based on a Bayesian approach.

The rest of the paper is organized as follow. In Sect. 2 the related work is presented in action recognition using deep CNNs. Section 3 explains how our *CurlDiv* motion representation is computed. The proposed scheme for training two-stream CNNs is presented in Sect. 4 while Sect. 5 describes the approach proposed for early action recognition. In Sect. 6 a set of experiments are presented in order to evaluate the performance of our method as well as the comparison with the *state-of-the-art* methods. Finally, in Sect. 7 the conclusions and future work are given.

2   Related Work

Human Action Recognition task has presented a significant improvement in accuracy during the last decade by applying CNNs. The most prominent approaches have adapted classical architectures from image recognition [12,22] to represent spatio-temporal relations [6,11,14,15]. These approaches are differentiated by the way they process data: spatio-temporal convolutions (3D CNNs) [15,26], Recurrent Neural Networks [7,29] or multi-streams CNNs [2,28]. This last approach has been used for action localization and video segmentation [19,20].

Spatio-temporal Convolutions or 3D CNNs model the video as a three-dimensional volume and apply a set of 3D convolutions at different levels of depth [10,15]. On the other hand, Recurrent Neural Networks approaches model video data as a sequence of frames [1,16]. In this approach exists two common ways for data representation: (1) using a classical image classification CNN for feature extraction [17], where frames are preprocessed and transformed into a sequence of non-handcrafted features and (2) by extracting skeleton information [8,16], videos are transformed into sequences of human poses. The main drawback of (1) is given by the action localization and motion in the background while (2) is fully affected by occlusions, video resolution and the target-camera distance; resulting (2) in a more accurate approach than (1).

Multi-streams approaches constitute the most revolutionary approach for action recognition. As is mentioned before, these approaches count with multiple CNNs trained independently and merged in the testing phase. Within these approaches, the best results have been achieved by those that transfer the knowledge from image processing domains [2,3].

Two-streams architectures transform the video into appearance (purely RGB frames) and motion (optical flow) throughout a set of offline algorithms [30]. The RGB channel is usually sampled at 25 fps, using a bilinear interpolation method. In order to get a best performance it is recommended to generate optical flow after video sampling [2]. Recently, I3D architecture [2] emerges as a two-stream architecture with spatio-temporal convolutions (3D CNNs) and pooling operators, inflated from an image classification CNN with spatial convolutions and pooling layers [25]. Contributions of [2] rely on the transfer learning from 2D convolutions (from image classification) to 3D convolutions (video classification). In 2018 the authors of [3] encode the skeleton information into a set of RGB images and utilize a shallow CNN for classification. Although this architecture does not show the best performance, when it is fused with I3D the authors report an improvement.

All these architectures based on deep CNNs can be characterized by the large amount of time for training multiple channels as well as for preprocessing. Although our approach is based on I3D architecture, it considerably reduces the training times for one of the streams without sacrificing the accuracy. In addition, our motion representation is about 6 times faster to compute than optical flow [30] and is more accurate than the skeleton based approach proposed by [3].

3 *CurlDiv* Motion Representation

The motion channel constitutes the more important channel for two-stream architectures. Usually, optical flow [30] is used in order to estimate such a motion. In the OpenCV library we can find implementations of these algorithms for both architectures: GPU and CPU. The problem with both implementations is the time they take to process data. In order to reduce preprocessing times, we introduce a new representation for the motion channel based on the curl and divergence of the optical flow [24]. The curl is a scalar property of vector fields that represents how the field rotates around a point. On the other hand, the divergence measures the density of the outward flux of a vector field from an infinitesimal boundary around a given point.

Let $\vec{V}$ be a 2-dimensional vector field computed with the Gunner Farneback algorithm [9] with scalar components $P(x, y)$ and $Q(x, y)$.

$$\vec{V}(x, y) = \begin{bmatrix} P(x, y) \\ Q(x, y) \end{bmatrix} \tag{1}$$

The Curl of $\vec{V}(x, y)$ can be defined as:

$$Curl(x, y) = \left\| \frac{\partial Q}{\partial x} - \frac{\partial P}{\partial y} \right\| \tag{2}$$

and the Divergence as follow:

$$Div(x, y) = \frac{\partial P}{\partial x} + \frac{\partial Q}{\partial y} \tag{3}$$

Second order partial derivatives of $\vec{V}$ are computed by using a finite difference strategy for reducing computational times. In this way, our *CurlDiv* motion descriptor takes the following formulation:

$$CurlDiv(x, y) = \begin{bmatrix} Curl(x, y) \\ Div(x, y) \end{bmatrix} \tag{4}$$

Following the schema proposed in [2] for motion data manipulation we also rescaled the *CurlDiv* descriptor in the range $[-1, 1]$.

4   Training Strategy and Fusion

Our training strategy follows the assumption that one of the streams has a weak performance. Based on this, we train first the strongest stream (appearance); detecting the set of samples in which this stream does not perform well. Over these samples the motion CNN is trained. In order to train the motion channel over the *CurlDiv* representation we reutilize the motion CNN from I3D. Such training is conducted in two ways: (1) by using all training data and (2) using the proposed strategy.

Table 1. This table shows the top2 accuracy for the appearance channel on I3D two-stream architecture as well as the accuracy achieved by our motion channel and the fusion of the two streams (RGB+$CurlDiv$).

Dataset	top2 accuracy	$CurlDiv$ accuracy	RGB+$CurlDiv$
UCF101-1	99.10	75.81	89.26
UCF101-2	99.03	74.50	88.99
UCF101-3	99.24	77.08	90.99
UCF101	99.12	75.79	89.74

Through the former strategy (1), the $CurlDiv$ channel is trained over the entire dataset, without taking into consideration the quality of the other channel; the appearance. For such reason, the training time is the same as that consumed by the appearance channel. On the other hand, our motion representation does not improve the optical flow although we used the same CNN. This is because we start training with a model pretrained over the optical flow in order to accelerate the convergence. During experimentation we realized that the top_2 accuracy of the appearance channel is above 99% in all the UCF101 databases (See Table 1).

Our second strategy (2) for training two-stream architectures is based on the assumption that one of the channels is better than the other one. Taking this into account we utilize a simple, but effective heuristic proposed in [4] that helps us to improve the training phase (Eq. 5).

$$max(softmax) - 2nd(softmax) \geq 2 * 2nd(softmax) \tag{5}$$

where $max(softmax)$ is the top of the softmax layer and $2nd(softmax)$ corresponds with the second most probable element in the softmax layer. Following this criteria, the training data and training times can be considerably reduced for the second channel. In other words, the motion channel will only be fine tuned over those samples that not fulfill the condition proposed in Eq. 5.

5   Bayesian Strategy for Early Action Recognition

An early response of the action that is taking place can be useful for real world applications (*e.g*, video surveillance). In this section we show how retraining the I3D (RGB + $CurlDiv$) over few frames can be successfully used as input to a Bayesian classifier in order to determine the action of a video in an early way. The core idea of this method is based on accumulating evidence about the observed activity while sliding windows are processed and making a decision when the highest probability fulfills the condition in Eq. 5.

Taking the vector of probabilities P obtained from the $softmax$ layer we can apply a Bayesian approach in order to make predictions, observing a few segments of the video.

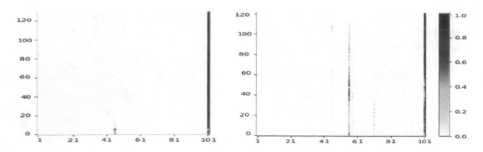

Fig. 2. This image shows the sliding windows classification over two different videos of class 101. The y axis enumerates the sliding windows (32 frames) and the x axis enumerates the 101 classes within the dataset. The outcome of the $softmax$ layer is visualized horizontally, according to the y - index. Note how the higher probability only tends to vary between two classes. The black color represents the higher value while light blue represents the smaller value. (Best seen in color). (Color figure online)

$$P(C_j|f) = \frac{P(C_i) * P(f|C_i)}{P(f)} \tag{6}$$

In Eq. 6 the idea is to estimate the maximum a posteriori probability (MAP) of the segment C_j given the video features. Notice that, as the denominator does not depends on the class we can go without it and rewrite the equation as follow.

$$P(C_j|f) \sim P(C_i) * P(f|C_i) \tag{7}$$

where f is the feature set extracted by the CNN, $P(C_i)$ is the probability of the previous segment and $P(f|C_i)$ is the probability estimated for the current segment. As video data can be seen as a continuous sequence of sliding windows, we can use Eq. 7 for predicting the action label. Following the condition proposed in Eq. 5 we can compute the most probable action for a given video and stop the processing. Figure 2 presents how the higher probability only varies between two classes across time.

6   Experiments and Results

In this paper experiments were conducted over three datasets in which two-stream architectures have been tested: **UCF101-1**, **UCF101-2** and **UCF101-3** [23]. Each of this dataset counts with 101 classes and 13,320 videos recorded from internet, divided into training and testing sets. The experiments presented in this section are focused in measuring how the training times can be reduced following our scheme, to what extent the RGB+$CurlDiv$ outperforms previous results and how good can a Bayesian strategy be for early action recognition.

Table 2. Hours (h) used by our architecture for training the I3D two-stream model following both schemes: classical and our (*CurlDiv*). Hardware: Core i7 32GB RAM and NVidia GeForce GTX 1080 with 11 GB of VRAM.

Dataset	RGB	FLOW	*CurlDiv*
UCF101-1	72 h	72 h	22 h
UCF101-2	72 h	72 h	22 h
UCF101-3	72 h	72 h	21 h

6.1 Preprocessing and Training Times

This experiment relates the times during preprocessing and training phases. These two phases are crucial factors when deep learning architectures are used. Our hardware architecture took almost 1 h for extracting all the frames of 13,320 videos and 18 h for optical flow computation. The *CurlDiv* only took less than 3 h for preprocessing the same amount of data, which represents a considerable reduction in time. To train the I3D architecture following the classical scheme took about 6 days for both channels: appearance and motion (3 days for each channel, over each dataset). By using our approach, we only expend 4 days for training the two channels (3 days for RGB and 1 day for *CurlDiv*) (See table 2). The considerable reduction in training times is given by applying the condition presented in Eq. 5 while the appearance channel is being trained.

6.2 Results Achieved by Our RGB+*CurlDiv* Scheme

We detect visually that noise and distortion are widely presented in the datasets. For such a reason we apply over all the frames a smoothing algorithm for image blurring with a 3×3 kernel for border and distortion reduction. Table 3 shows how applying a smoothing strategy before preprocessing data outperforms the *state-of-the-art* results.

Taking into consideration the improvement produced by image smoothing, we compute the *CurlDiv* motion representation and retrain the model following

Table 3. **RGB**, **FLOW** and **RGB+FLOW** columns show the accuracy achieved with the I3D without applying image smoothing. **RGB_s**, **FLOW_s** and **RGB_s+ FLOW_s** columns show the accuracy achieved by the I3D after applying image smoothing.

Dataset	RGB	FLOW	RGB+FLOW	RGB_s	FLOW_s	RGB_s+FLOW_s
UCF101-1	94.55	96.3	97.6	**94.90**	**96.4**	**98.01**
UCF101-2	95.66	97.1	97.99	**95.98**	**97.99**	**98.04**
UCF101-3	94.34	95.99	97.72	**94.56**	**96.18**	**97.99**
UCF101	94.85	96.46	97.77	**95.14**	**96.85**	**98.01**

Table 4. Comparison with the *state-of-the-art* methods: I3D [2] and PoTion [3]. The UCF101* represents the mean accuracy achieved by three datasets. The metric used is accuracy.

Dataset	I3D (RGB+FLOW)	I3D+PoTion	I3D RGB$_{smooth}$ + CurlDiv
UCF101-1	97.6	–	**98.62**
UCF101-2	97.99	–	**98.39**
UCF101-3	97.72	–	**97.97**
UFC101*	97.77 (our) | 98.0 [2]	98.2	**98.32**

the proposed in Sect. 4. For testing we follow the same strategy used for training the appearance stream (guided by the condition presented in Eq. 5). First we let the appearance channel to decide. If this channel fulfills the condition we take the selected class. In case the condition is not satisfied, we apply the condition over the motion channel. If the motion channel satisfies the condition, then we take the predicted class. In case neither the appearance channel nor the motion channel fulfills the condition, the classical fusion of both channels is carried out. Results achieved by our framework outperform the *state-of-the-art* results in accuracy and are shown in Table 4.

6.3 Impact of I3D in Early Action Recognition

Each dataset presented in these experiments have a total of 7,352,018 frames distributed into 13,320 videos. The aim of these experiments is to quantify how many frames can be saved using the strategy proposed in Sect. 5 without sacrificing the performance of the global method (See Table 5). For that, we retrain the I3D with clips of 32 frames (usually clip-size is 64) randomly selected within the input videos during 40 epochs. In this experiment we take into consideration three different early conditions for stopping the algorithm: [**First**] taking only the first 32 frames, [**Bayes**] applying our early recognition framework and [**Total**] spanning entire videos.

Table 5 shows how spanning entire videos [**Total**] is the worst case. This happen because activities occur in a low number of frames and videos are extremely large. This situation causes that the evidence of real classes is less than the rest of the classes. On the other hand, the [**First**] strategy behaves decently for short videos or videos where the action performed starts from the beginning. Finally, a more accurate and safe way is to let a high level classifier [**Bayes**] to decide the true class while the video is spanned. In general, can be seen how following increase the number of non analyzed frames without sacrificing the performance.

Table 5. [**First**]: This condition utilizes only the first 32 frames of each video. [**Total**]: With this condition the video is spanned entirely. [**Bayes**]: Employs the proposed condition for stopping video spanning. **f-saved** is the number of non analyzed frames for each strategy.

Dataset	First		Total		Bayes	
	Accuracy	f-saved	Accuracy	f-saved	Accuracy	f-saved
UCF101-1	95.40	95 %	93.18	0 %	97.27	93.52 %
UCF101-2	96.30	95 %	94.10	0 %	97.99	94.0 %
UCF101-3	95.18	95 %	93.80	0 %	96.48	92.23 %

7 Conclusions and Future Work

In this paper was presented a novel strategy for training two-stream CNNs that considerably reduces the training times. We showed how by reducing the noise and distortions from input frames we can improve the performance of the classical two-stream CNNs; especially the I3D architecture. Our new representation for the motion channel results in a weak classifier by itself, but when it is combined with the appearance channel, the *state-of-the-art* methods are outperformed. Presented results shows that the *CurlDiv* representation is easier and faster to compute than the commonly used optical flow [30]. On the other hand, despite the reduction in training and preprocessing times, we reported the best results achieved so far over the UCF101 dataset. At the same time we presented an approach based on the I3D architecture for early action recognition that avoids the need of spanning entire videos without sacrificing the accuracy of the method. In future work we will introduce another scalar properties derived from the optical flow in order to encode the motion channel and we will extend the architecture for multiple CNNs streams (streams > 2).

References

1. Baccouche, M., Mamalet, F., Wolf, C., Garcia, C., Baskurt, A.: Sequential deep learning for human action recognition. In: Salah, A.A., Lepri, B. (eds.) HBU 2011. LNCS, vol. 7065, pp. 29–39. Springer, Heidelberg (2011). https://doi.org/10.1007/978-3-642-25446-8_4
2. Carreira, J., Zisserman, A.: Quo vadis, action recognition? a new model and the kinetics dataset. In: 2017 IEEE Conference on CVPR, pp. 4724–4733. IEEE (2017)
3. Choutas, V., Weinzaepfel, P., Revaud, J., Schmid, C.: Potion: pose motion representation for action recognition. In: CVPR. pp. 7024–7033 (2018)
4. Cruz, C., Sucar, L.E., Morales, E.F.: Real-time face recognition for human-robot interaction. In: 2008 8th IEEE International Conference on Automatic Face & Gesture Recognition, pp. 1–6. IEEE (2008)
5. Diba, A., et al.: Temporal 3d convnets: new architecture and transfer learning for video classification. arXiv preprint arXiv:1711.08200 (2017)

6. Diba, A., Pazandeh, A.M., Van Gool, L.: Efficient two-stream motion and appearance 3d cnns for video classification. arXiv preprint arXiv:1608.08851 (2016)
7. Donahue, J., et al.: Long-term recurrent convolutional networks for visual recognition and description. In: CVPR, pp. 2625–2634 (2015)
8. Du, Y., Wang, W., Wang, L.: Hierarchical recurrent neural network for skeleton based action recognition. In: CVPR, pp. 1110–1118 (2015)
9. Farnebäck, G.: Two-frame motion estimation based on polynomial expansion. In: Bigun, J., Gustavsson, T. (eds.) SCIA 2003. LNCS, vol. 2749, pp. 363–370. Springer, Heidelberg (2003). https://doi.org/10.1007/3-540-45103-X_50
10. Hara, K., Kataoka, H., Satoh, Y.: Can spatiotemporal 3d cnns retrace the history of 2d cnns and imagenet? In: CVPR. pp. 6546–6555 (2018)
11. He, K., Zhang, X., Ren, S., Sun, J.: Deep residual learning for image recognition. In: CVPR, pp. 770–778 (2016)
12. He, K., Zhang, X., Ren, S., Sun, J.: Identity mappings in deep residual networks. In: Leibe, B., Matas, J., Sebe, N., Welling, M. (eds.) ECCV 2016. LNCS, vol. 9908, pp. 630–645. Springer, Cham (2016). https://doi.org/10.1007/978-3-319-46493-0_38
13. Herath, S., Harandi, M., Porikli, F.: Going deeper into action recognition: a survey. Image Vis. Comput. **60**, 4–21 (2017)
14. Ioffe, S., Szegedy, C.: Batch normalization: Accelerating deep network training by reducing internal covariate shift. arXiv preprint arXiv:1502.03167 (2015)
15. Ji, S., Xu, W., Yang, M., Yu, K.: 3d convolutional neural networks for human action recognition. IEEE Trans. Pattern Anal. Mach. Intell. **35**(1), 221–231 (2013)
16. Liu, J., Shahroudy, A., Xu, D., Wang, G.: Spatio-temporal LSTM with trust gates for 3D human action recognition. In: Leibe, B., Matas, J., Sebe, N., Welling, M. (eds.) ECCV 2016. LNCS, vol. 9907, pp. 816–833. Springer, Cham (2016). https://doi.org/10.1007/978-3-319-46487-9_50
17. Ma, S., Sigal, L., Sclaroff, S.: Learning activity progression in lstms for activity detection and early detection. In: CVPR, pp. 1942–1950 (2016)
18. Nanni, L., Ghidoni, S., Brahnam, S.: Handcrafted vs. non-handcrafted features for computer vision classification. Pattern Recogn. **71**, 158–172 (2017)
19. Peng, X., Schmid, C.: Multi-region two-stream R-CNN for action detection. In: Leibe, B., Matas, J., Sebe, N., Welling, M. (eds.) ECCV 2016. LNCS, vol. 9908, pp. 744–759. Springer, Cham (2016). https://doi.org/10.1007/978-3-319-46493-0_45
20. Saha, S., Singh, G., Sapienza, M., Torr, P.H., Cuzzolin, F.: Deep learning for detecting multiple space-time action tubes in videos. arXiv preprint arXiv:1608.01529 (2016)
21. Simonyan, K., Zisserman, A.: Two-stream convolutional networks for action recognition in videos. In: Advances in Neural Information Processing Systems, pp. 568–576 (2014)
22. Simonyan, K., Zisserman, A.: Very deep convolutional networks for large-scale image recognition. arXiv preprint arXiv:1409.1556 (2014)
23. Soomro, K., Zamir, A.R., Shah, M.: Ucf101: A dataset of 101 human actions classes from videos in the wild. arXiv preprint arXiv:1212.0402 (2012)
24. Suter, D.: Motion estimation and vector splines. In: CVPR, vol. 94, pp. 939–942 (1994)
25. Szegedy, C., Vanhoucke, V., Ioffe, S., Shlens, J., Wojna, Z.: Rethinking the inception architecture for computer vision. In: CVPR, pp. 2818–2826 (2016)
26. Tran, D., Ray, J., Shou, Z., Chang, S.F., Paluri, M.: Convnet architecture search for spatiotemporal feature learning. arXiv preprint arXiv:1708.05038 (2017)
27. Wang, H., Schmid, C.: Action recognition with improved trajectories. In: ICCV, pp. 3551–3558 (2013)

28. Wang, Y., Song, J., Wang, L., Van Gool, L., Hilliges, O.: Two-stream sr-cnns for action recognition in videos. In: BMVC (2016)
29. Yue-Hei Ng, J., Hausknecht, M., Vijayanarasimhan, S., Vinyals, O., Monga, R., Toderici, G.: Beyond short snippets: deep networks for video classification. In: CVPR, pp. 4694–4702 (2015)
30. Zach, C., Pock, T., Bischof, H.: A duality based approach for realtime TV-L^1 optical flow. In: Hamprecht, F.A., Schnörr, C., Jähne, B. (eds.) DAGM 2007. LNCS, vol. 4713, pp. 214–223. Springer, Heidelberg (2007). https://doi.org/10.1007/978-3-540-74936-3_22

Parkinsonian Ocular Fixation Patterns from Magnified Videos and CNN Features

Isail Salazar[1,3], Said Pertuz[2,3], William Contreras[4], and Fabio Martínez[1,3(✉)]

[1] Biomedical Imaging, Vision and Learning Laboratory (BIVL2ab),
Bucaramanga, Colombia
famarcar@saber.uis.edu.co
[2] Grupo de Investigación en Conectividad y Procesamiento de Señales (CPS),
Bucaramanga, Colombia
[3] Universidad Industrial de Santander., Bucaramanga, Colombia
[4] Departamento de Neurocirugía, Clínica FOSCAL.,
Bucaramanga, Colombia

Abstract. Recent neurological studies suggest that oculomotor alterations are one of the most important biomarkers to detect and characterize Parkinson's disease (PD), even on asymptomatic stages. Nevertheless, only global and simplified gaze trajectories, obtained from tracking devices, are generally used to represent the complex eye dynamics. Besides, such acquisition procedures often require sophisticated calibration and invasive configuration schemes. This work introduces a novel approach that models very subtle ocular fixational movements, recorded with conventional cameras, as an imaging biomarker for PD assessment. For this purpose, a video acceleration magnification is performed to enhance small fixational patterns on standard gaze video recordings of test subjects. Subsequently, feature maps are derived from spatio-temporal video slices by means of convolutional layer responses of known pre-trained CNN architectures, allowing to describe the depicted oculomotor cues. The set of extracted CNN features are then efficiently coded by means of covariance matrices in order to train a support vector machine and perform an automated disease classification. Promising results were obtained through a leave-one-patient-out cross-validation scheme, showing a proper PD characterization from fixational eye motion patterns in ordinary sequences.

Keywords: Parkinson's disease · Ocular fixation · Oculomotor patterns · Motion magnification · Cnn features

1 Introduction

Parkinson's disease (PD) is the second most common neurodegenerative disorder, with a prevalence exceeding more than 6 million people worldwide [6]. This disease is directly related to deficiencies in the production of the dopamine neurotransmitter, which plays a key role in sending and receiving messages from the

© Springer Nature Switzerland AG 2019
I. Nyström et al. (Eds.): CIARP 2019, LNCS 11896, pp. 740–750, 2019.
https://doi.org/10.1007/978-3-030-33904-3_70

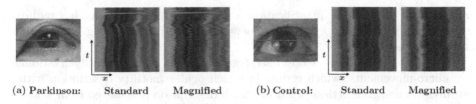

(a) Parkinson: Standard   Magnified   (b) Control: Standard   Magnified

Fig. 1. Ocular fixation comparison of PD and control subjects. Magnified xt-slices improve the visual difference between both classes. Video magnification was applied around a frequency of 5.7 Hz, previously identified in PD fixational behavior [8]. (a) PD sequence where the magnification enhances an oscillatory fixational instability. (b) Control sequence where no particular oculomotor pattern can be detected at a glance.

brain to the body. Such affectation results on progressive motor alterations, e.g., tremor, bradykinesia, stiffness, and postural instability [14].

A main critical issue nowadays is the poor understanding of dopamine loss, resulting in diagnostic procedures based mostly on conspicuous motor alterations. In clinical practice, specific macro and observational tests that coarsely relate such physical signs with PD are often used as the main diagnosis tool [11]. This evaluation is however prone to bias and subjectivity due to the high interpersonal motion variability and the particular perception dependency [16]. To tackle these problems, researchers have proposed a variety of quantitative and computerized approaches including the use of body-worn motion sensors [10], computer vision schemes [4] or machine and deep learning algorithms to recognize PD behaviors during handwriting, speech and gait tasks [15]. These approaches rely on the quantification of upper and lower limb movements, which generally appear in advanced stages [3] and therefore are not suitable for an early disease detection.

Recently, eye movement has been pointed out as a potential and very promising PD biomarker [5]. Several works report a strong association between subtle oculomotor abnormalities and dopamine concentration changes, even at the disease onset [1,12]. This fact suggests a highly sensitive predictor of neurodegeneration before it reaches critical levels. For instance, the representative study of [8] evidenced the presence of ocular tremor responses while fixating a static target in a large cohort of PD patients at different stages. Interestingly enough, one of the control subjects, who turned out to exhibit fixational tremor, eventually developed the disease over a period of 2 years. Such studies rely on classical oculomotor monitoring protocols like electro-oculography (EOG) and video-oculography (VOG) [7]. The main disadvantage of EOG setups is their high susceptibility to electronic noise. Alternatively, modern VOG devices offer more reliable and noise-free recordings, but in both settings obtained signals only depict global and simplified trajectories of the whole eye motion field and its intrinsic deformations. Furthermore, they require individual calibration steps and invasive configurations, usually making contact with the entire area around the eyes and thus affecting the natural visual gesture. Other approaches [19] include the Poculomotor quantification from velocity field patterns, but require experiments with strong eye displacements to differentiate among control and disease classes.

This paper introduces a novel strategy to enhance and quantify fixational eye micro-movements, which results in a rich ocular motility description without invasive and comfortless protocols. Such description is achieved from spatio-temporal video slices, extracted from standard videos, instead of relying on the acquisition of global gaze trajectories through sophisticated devices. Firstly, tiny fixation patterns are highlighted by means of a video acceleration magnification. As shown in Fig. 1, magnified slices can better depict the oscillatory fixational patterns that visually differentiate between control and PD eye motion. Then, a dense profile description is obtained by means of a bank of filter responses computed from early layers of deep learning architectures. These features are compactly codified as the covariance matrix of filter channels with major information energy according to an eigenvalue decomposition. The resulting covariance is then mapped to a previously trained support vector machine to classify the disease given a particular slice. A preliminary evaluation with 6 PD patients and 6 control subjects by a leave-one-patient-out cross-validation scheme shows high effectiveness to codify fixation motility and differentiate between both classes. The proposed approach could potentially provide support and assistance in the PD diagnosis and follow-up.

2   Video Acceleration Magnification

A particular limitation in the analysis of fixational PD motion lies in the quantification of involuntary eye micro-movements [9]. Additionally, it should be considered that eye displacement can be masked by comparatively larger head motion. This work hence starts by performing an optical spatio-temporal amplification over fixation sequences. A remarkable fact of PD fixations is their oscillatory behavior, as shown in [8,9]. Therefore, we can amplify specific frequency bands to characterize PD. For doing so, we use the acceleration magnification approach of [20], which allows to amplify subtle motion (fixational) even in the presence of large motion (head). The acceleration magnification works by analyzing the variation of pixel signals over time. In such case, any amplified pixel signal $\hat{I}(\mathbf{x}, t)$ can be represented by a Taylor decomposition:

$$\hat{I}(\mathbf{x}, t) = f(\mathbf{x} + \beta\,\delta(t)) \approx f(\mathbf{x}) + \underbrace{\beta\,\delta(t)\frac{\partial f(\mathbf{x})}{\partial \mathbf{x}}}_{A} + \underbrace{\alpha\,\delta(t)^2\frac{\partial^2 f(\mathbf{x})}{\partial^2 \mathbf{x}}}_{B}, \qquad (1)$$

where $\delta(t)$ is the small displacement function of the original pixel signal $I(\mathbf{x}, t) = f(\mathbf{x} + \delta(t))$, and $\alpha = \beta^2$ is the magnification factor.

Equation 1 part A, is related to first-order linear changes (velocity) and will not be considered. Part B represents second order changes, i.e., acceleration, which is amplified by an α factor that allows to accentuate every small deviation of linear motion. For measuring motion information, local phase

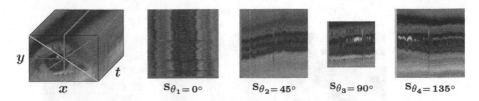

Fig. 2. Spatio-temporal video slices. At different slice directions, relevant cues in fixation recordings can be captured.

information is computed through a complex-valued steerable pyramid $\Psi_{w,\theta}$: $\mathbf{I}(\mathbf{x},t) * \Psi_{w,\theta} = A_{w,\theta}(\mathbf{x},t)\, e^{i\,\Phi_{w,\theta}(\mathbf{x},t)}$, where $\mathbf{I}(\mathbf{x},t)$ is a video frame and w, θ are spatial frequency bands and orientations. Then, phases are amplified by using a Laplacian temporal filtering:

$$\hat{\Phi}_{w,\theta}(\mathbf{x},t) = \Phi_{w,\theta}(\mathbf{x},t) \ + \ \alpha\ \Phi_{w,\theta}(\mathbf{x},t) * \frac{\partial^2 G_\sigma(\mathbf{x},t)}{\partial t^2}\,, \tag{2}$$

where σ is the Laplacian filter scale. This computation allows to take the second-order derivative of the smoothed phase signals and thus amplify them by α at a given temporal frequency $f = \frac{\text{frame rate}}{8\pi\sqrt{2}\,\sigma}$.

3 Learning a CNN Fixation Representation

The resulting amplified video $\hat{I}(\mathbf{x},t)$ is split-up into a set of spatio-temporal slices $\mathbf{S}_\theta = \{\mathbf{s}_{\theta_1}, \mathbf{s}_{\theta_2}, \dots, \mathbf{s}_{\theta_N}\}$ at N orientations. For doing so, different radial directions on the spatial xy-plane were used as reference along time. Typical slice configurations are illustrated in Fig. 2. These slices capture small eye iris displacements and herein constitute an ideal source of information to analyze small ocular movements.

Each slice $\mathbf{s}_{\theta_i}$ is represented as a bank of separated band responses of high and low frequency filters, with some coverage of mid frequencies. This was achieved by mapping $\mathbf{S}_\theta$ on the first layers of known and pre-trained deep convolutional frameworks, which have been implemented for the classification of natural scenes and trained on the ImageNet dataset (around 1.2 million samples). In brief, such architectures progressively compute linear transformations, followed by contractive nonlinearities, projecting information on a set of C learned filters $\Psi^j = \{\psi_1^j, \psi_2^j, \dots, \psi_C^j\}$ at a given layer j. Hence, each eye slice $\mathbf{s}_{\theta_i}$ is filtered by a particular Ψ^j set, obtaining a feature representation $\Phi^j = \sum_{c=1\dots C} \mathbf{s}_{\theta_i} * \psi_c^j$, with $\phi_c^j = \mathbf{s}_{\theta_i} * \psi_c^j$ as each independent feature channel. In this work, three different pre-trained architectures were independently used and evaluated for eye slice dense representation. A succinct description of the studied deep architectures is presented below.

- **VGG-19** [17]: is a classical CNN architecture with a total of 19 layers. For low-level representation purposes, in this work we considered the first block

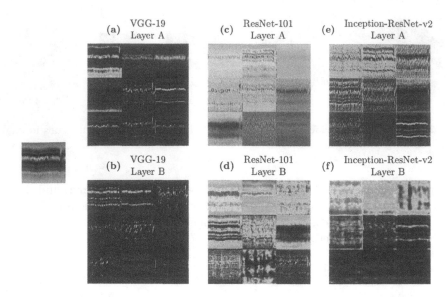

Fig. 3. Sample filter responses from each CNN architecture. In general, selected layers exhibit a high response rate to the small local slice patterns, such as lines, edges, and corners. This representation can thus provide an adequate feature map for the depicted fixational cues.

pooling layer with a total of $C = 64$ filter channels, and responses of size $W = 112 \times H = 112$. Hereafter named VGG-19 Layer A. Also, a second representation was evaluated from the second block pooling layer ($C = 128$, $W = 56 \times H = 56$), referred to as VGG-19 Layer B.

- **ResNet-101** [13]**:** is a deep net that includes residual maps serving as recursive inputs on superior layers throughout shortcut connections. A primary representation was herein obtained from the first block pooling layer ($C = 64$, $W = 112 \times H = 112$), hereafter named ResNet-101 Layer A. Additionally, the first residual-added ReLu layer was computed as second representation ($C = 256$, $W = 56 \times H = 56$), designated as ResNet-101 Layer B.
- **Inception-ResNet-v2** [18]**:** is one of the most recent approaches that combines the inception blocks, i.e., multiple sub-networks that learn independently, with residual connections that allow optimal learning rates. The third block ReLu layer of $C = 64$ filters and responses of $W = 147 \times H = 147$ was used as primary representation, referred to as Inception-ResNet-v2 Layer A. The first residual-concatenated layer ($C = 320$, $W = 35 \times H = 35$) was also considered, named Inception-ResNet-v2 Layer B.

For illustration, Fig. 3 shows sample responses of the utilized deep architectures for a given input video slice.

3.1 Recognizing Parkinsonian Patterns: A Compact Fixational Descriptor

The feature representation $\mathbf{\Phi} \in \mathbb{R}^{H \times W \times C}$ for each slice $\mathbf{s}_\theta$ is composed by C filter responses $\boldsymbol{\phi}_c$ with dimensions $H \times W$. We vectorize each $\boldsymbol{\phi}_c$, reshaping $\mathbf{\Phi}$ to $HW \times C$. This information is nevertheless redundant on common background and could lead to a wrong motion description. Hence, a very compact analysis is herein carried out by computing a feature covariance matrix $\mathbf{\Sigma} = \frac{1}{HW}\left[(\mathbf{\Phi} - \mu(\mathbf{\Phi}))(\mathbf{\Phi} - \mu(\mathbf{\Phi}))^T\right]$, with μ as the mean $1 \times C$ feature vector repeated HW times. The covariance $\mathbf{\Sigma} \in \mathbb{R}^{C \times C}$ describes a second statistical moment on the whole feature space, that compactly summarizes the fixational motion representation. From a spectral matrix analysis, we found that the energy of $\mathbf{\Sigma}$ is fully concentrated on only a few eigenvalues, which is why we only keep information related to the k major eigenvalues. In this way, a new compact reduced covariance $\mathbf{\Sigma_r}$ that captures the most variability of the C feature channels is computed as $\mathbf{\Sigma_r} = \mathbf{W^T \Sigma W}$, where $\mathbf{W} \in \mathbb{R}^{C \times k}$ is the reduced eigenvector matrix of $\mathbf{\Sigma}$ with $k < C$.

Due to the semi-definite and positive properties of covariance matrices, they exist on a semi-spherical Ricmannian space. This fact limits the application of classic machine learning approaches that assume Euclidean structured data. We therefore project $\mathbf{\Sigma_r}$ onto the Euclidean space by the matrix logarithm of $\mathbf{\Sigma_r}$. That is, $\log(\mathbf{\Sigma_r}) = \mathbf{V_r}\log(\mathbf{\Lambda_r})\mathbf{V_r}^T$, where $\mathbf{V_r}$ are the eigenvectors of $\mathbf{\Sigma_r}$ and $\log(\mathbf{\Lambda_r})$ the corresponding logarithmic eigenvalues. This reduced covariance represents the fixational motion descriptor to be fed into a machine learning algorithm in order to obtain a prediction of Parkinson's disease, under a supervised learning scheme. In this work, we selected a support vector machine (SVM) as supervised model due to its demonstrated capability at defining non-linear boundaries between classes. Also, SMVs have widely reported proper performance on high dimensional data with low computational complexity. Finally, as non-linear SVM kernel classifier, we utilized the classical yet powerful Radial Basis Function (RBF) kernel $K = \exp\left(-\gamma \|\log(\mathbf{\Sigma_r})_i - \log(\mathbf{\Sigma_r})_j\|^2\right)$ [2].

3.2 Imaging Data

We implemented a protocol to record fixational motion on PD-diagnosed and control patients. Participants were invited to observe a fixed spotlight projected on a screen with a dark background. A Nikon D3200 camera with spatial resolution of 1920 × 1080 pixels and a temporal resolution of 30 fps was fixed in front of the subjects to capture the whole face. The eye region was manually cropped (180 × 120 pixels) to obtain the sequences of interest. A total of 6 PD patients (average age of 71.8 ± 12.2) and 6 control subjects (average age of 66.2 ± 6.6) were captured and analyzed for validation of the proposed approach. PD patients were diagnosed in second (2 patients), third (3 patients) and fourth (1 patient) stage of the disease by a physician using standard protocols of the Hoehn-Yahr scale [11]. A total of 24 sequences were recorded, i.e., 2 samples per

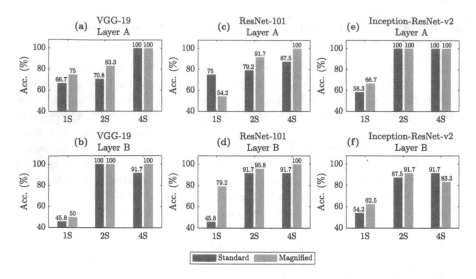

Fig. 4. Obtained accuracy results by using different deep architectures and different number of slices on standard and magnified videos. In general, the proposed approach achieves outstanding results regarding PD prediction.

patient, with duration of 5 s. This study was approved by the Ethics Committee of the Universidad Industrial de Santander and a written informed consent was obtained. The recorded dataset was possible thanks to the support of the local Parkinson foundation FAMPAS *(Fundación del Adulto Mayor y Parkinson Santander)* and the local elderly institution *Centro Vida Años Maravillosos*.

4   Evaluation and Results

Experiments were performed on the 24 recorded sequences through a leave-one-patient-out cross-validation, in which at each iteration one patient is left out to test and the remaining ones are used for training the model. For comparison purposes, we considered both standard and magnified videos. Figure 1 illustrates standard and magnified ocular fixational motion for sample PD and control subjects. Parkinsonian slices show well-defined amplitudes of slight oscillatory motility at specific frequencies (later discussed). A total of 4 slices that recover eye motion cues were extracted from each video (see Fig. 2), and thereafter mapped to the selected CNN architectures in order to obtain a deep representation. Then, a very compact covariance descriptor per slice was computed by using the minimal number of eigenvalues that concentrate 95% of information.

Figure 4 shows a first quantitative analysis of average accuracies over the patient-fold scheme. Initially, it should be noted that an increasing number of eye slices yields a performance improvement. This is valid for all of the considered architectures. Secondly, video magnification consistently contributes to improve the disease prediction, being a major contribution when fewer slices are utilized

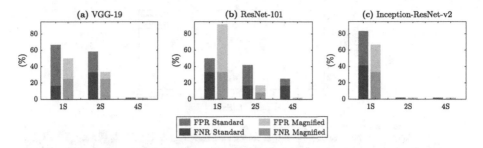

Fig. 5. FPR and FNR indices for Layer A of selected architectures.

in the analysis. Interestingly enough, outstanding performance is achieved on layer A of all CNNs by using a 4-slice representation. In this case, standard slices are able to obtain complete predictions on the VGG-19 and Inception-ResNet-v2, and magnified ones on all the three nets. Such fact could be related to a more comprehensive description of eye slices in the earliest representation levels, which suggest that low-level primitives were able to properly represent the depicted fixational patterns even on standard slices. Also, it is remarkable the Inception-ResNet-v2 performance that achieves complete predictions even with only two slices. Although Layer A in such scheme is not directly related to residual inputs or inception blocks, the overall network minimization allows a better description of relevant slice features. Layer B of VGG-19 achieves the same results for two slices, but in general, this net layer requires more forward steps and denser representations.

Thereafter, a deeper analysis of layer A was performed in Fig. 5 by quantifying the False Negative Rate (FNR) and the False Positive Rate (FPR). The FNR is related to the percentage of PD patterns incorrectly classified as control, while the FPR represents the percentage of control patterns miss-classified as PD. In all cases, the use of four slices in magnified sequences yields negligible prediction of false conditions. In general, an exponential error decay can be observed as there are more slices involved on the descriptor. For the Inception-ResNet-v2 architecture, this trend is more accentuated, achieving the best condition classification with a very compact descriptor, i.e., by using fewer slices. On the other side, the ResNet-101 architecture presents the slowest decay, requiring more information to achieve proper performance.

Finally, video magnification parameters were evaluated to correctly enhance fixational motion patterns. In the proposed approach, the temporal frequency to be magnified was fixed w.r.t. the fundamental PD ocular tremor frequency. In the literature, such frequency during ocular fixation has been quantified as $f = 5.7\,\text{Hz}$ [8]. As observed in Fig. 6-(a), the proposed approach achieves equally outstanding performance on a range of characteristic PD tremor frequencies, in terms of improved accuracies per layer. Regarding the magnification factor, a visually reasonable value to emphasize this subtle oscillatory pattern was found to be $\alpha = 15$, as shown in the spatio-temporal comparison of Fig. 6-(b).

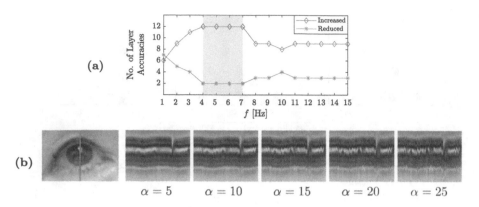

Fig. 6. Video magnification variables. (a) Influence of the magnification frequency choice in the disease classification task. Reported values were obtained by counting the number of layer representations (layer A and layer B at the three considered slice quantities) with increased performance due to the magnification step. (b) Visual effect of different magnification factors.

5   Conclusions

In this work, we proposed a quantitative strategy to characterize ocular fixational motion as an imaging biomarker for Parkinson's disease (PD). This approach achieved a robust eye motion modeling over conventional video sequences. For so doing, we recorded eye fixation experiments in order to capture the oculomotor activity of test subjects. Acquired videos were then magnified using an optical acceleration-based framework that allowed to enhance small motility patterns. Video slice features based on primary CNN layer responses were used to classify control and PD-diagnosed patients under a supervised machine learning framework. Preliminary experiments on a pilot case-control set of 12 subjects yielded promising results in terms of high classification accuracies and low false-positive and false-negative rates. Despite the relatively limited sample size, a common problem in PD research [9,19], obtained results demonstrated a feasible alternative for PD assessment upon ordinary and magnified videos, avoiding complex and sophisticated acquisition setups like EOG and VOG. The proposed strategy therefore represents a potential approach to understand and quantify the association between PD and eye motility, aiming to support diagnosis and follow-up of the disease. In order to validate our findings, further evaluation with a larger population sample is warranted. Future work also includes the distinction of PD sensibility regarding different stages of the disease.

Acknowledgements. The authors thank the *Vicerrectoría de Investigación y Extensión* of the Universidad Industrial de Santander for supporting this research work by the project *"Reconocimiento continuo de expresiones cortas del lenguaje de señas registrado en secuencias de video"*, with SIVIE code 2430.

References

1. Anderson, T., Macaskill, M.R.: Eye movements in patients with neurodegenerative disorders. Nat. Rev. Neurol. **9**, 74–85 (2013)
2. Chang, C.C., Lin, C.J.: LIBSVM: a library for support vector machines. ACM Trans. Intell. Syst. Technol. **2**, 27:1–27:27 (2011)
3. Cheng, H.C., Ulane, C.M., Burke, R.E.: Clinical progression in parkinson disease and the neurobiology of axons. Ann. Neurol. **67**(6), 715–725 (2010)
4. Contreras, S., Salazar, I., Martínez, F.: Parkinsonian hand tremor characterization from magnified video sequences. In: 14th International Symposium on Medical Information Processing and Analysis, SPIE, vol. 10975, p. 1097503 (2018)
5. Ekker, M.S., Janssen, S., Seppi, K., et al.: Ocular and visual disorders in parkinson's disease: common but frequently overlooked. Parkinsonism Related Disord. **40**, 1–10 (2017)
6. Feigin, V.L., Abajobir, A.A., Abate, K.H., et al.: Global, regional, and national burden of neurological disorders during 1990–2015: a systematic analysis for the global burden of disease study 2015. LANCET Neurol. **16**(11), 877–897 (2017)
7. Furman, J.M., Wuyts, F.L.: vestibular laboratory testing. In: Aminoff, M.J. (ed.) Aminoff's Electrodiagnosis in Clinical Neurology, pp. 699–723, sixth edn. W.B. Saunders, London (2012)
8. Gitchel, G.T., Wetzel, P.A., Baron, M.S.: Pervasive ocular tremor in patients with parkinson disease. Arch. Neurol. **69**(8), 1011–1017 (2012)
9. Gitchel, G.T., Wetzel, P.A., Qutubuddin, A., Baron, M.S.: Experimental support that ocular tremor in parkinson's disease does not originate from head movement. Parkinsonism Related Disord. **20**(7), 743–747 (2014)
10. Godinho, C., Domingos, J., Cunha, G., et al.: A systematic review of the characteristics and validity of monitoring technologies to assess parkinson's disease. J. Neuroeng. Rehabil. **13**(1), 24 (2016)
11. Goetz, C.G., Poewe, W., Rascol, O., et al.: Movement disorder society task force report on the hoehn and yahr staging scale: status and recommendations the movement disorder society task force on rating scales for parkinson's disease. Mov. Disord. **19**(9), 1020–1028 (2004)
12. Gorges, M., Müller, H.P., Lulé, D., et al.: The association between alterations of eye movement control and cerebral intrinsic functional connectivity in parkinson's disease. Brain Imaging Behav. **10**(1), 79–91 (2016)
13. He, K., Zhang, X., Ren, S., Sun, J.: Deep residual learning for image recognition. In: Proceedings of the IEEE Conference on Computer Vision and Pattern Recognition, pp. 770–778 (2016)
14. Jankovic, J.: Parkinson's disease: clinical features and diagnosis. J. Neurol. Neurosurg. Psychiatry **79**(4), 368–376 (2008)
15. Pereira, C.R., Pereira, D.R., Weber, S.A., et al.: A survey on computer-assisted parkinson's disease diagnosis. Artif. Intell. Med. **95**, 48–63 (2018)
16. Rizzo, G., Copetti, M., Arcuti, S., et al.: Accuracy of clinical diagnosis of parkinson disease a systematic review and meta-analysis. Neurology **86**(6), 566–576 (2016)
17. Simonyan, K., Zisserman, A.: Very deep convolutional networks for large-scale image recognition. arXiv preprint arXiv:1409.1556 (2014)
18. Szegedy, C., Ioffe, S., Vanhoucke, V., Alemi, A.A.: Inception-v4, inception-resnet and the impact of residual connections on learning. In: Thirty-First AAAI Conference on Artificial Intelligence (2017)

19. Trujillo, D., Martínez, F., Atehortúa, A., et al.: A characterization of parkinson's disease by describing the visual field motion during gait. In: 11th International Symposium on Medical Information Processing and Analysis, SPIE, vol. 9681 (2015)
20. Zhang, Y., Pintea, S.L., Van Gemert, J.C.: Video acceleration magnification. In: Computer Vision and Pattern Recognition (2017)

Does Pooling Really Matter? An Evaluation on Gait Recognition

Claudio Filipi Goncalves dos Santos[1]([X]), Thierry Pinheiro Moreira[2],
Danilo Colombo[3]([X]), and João Paulo Papa[2]([X])

[1] Federal University of São Carlos - UFSCar, São Carlos, Brazil
cfsantos@ufscar.br
[2] State University of Sao Paulo - UNESP, Sao Paulo, Brazil
thierrypin@gmail.com, joao.papa@unesp.br
[3] Cenpes, Petróleo Brasileiro S.A. – Petrobras, Rio de Janeiro - RJ, Brazil
colombo.danilo@petrobras.com.br

Abstract. Most Convolutional Neural Networks make use of subsampling layers to reduce dimensionality and keep only the most essential information, besides turning the model more robust to rotation and translation variations. One of the most common sampling methods is the one who keeps only the maximum value in a given region, known as max-pooling. In this study, we provide pieces of evidence that, by removing this subsampling layer and changing the stride of the convolution layer, one can obtain comparable results but much faster. Results on the gait recognition task show the robustness of the proposed approach, as well as its statistical similarity to other pooling methods.

Keywords: Convolutional Neural Networks · Deep learning · Gait recognition

1 Introduction

Sub-sampling layers, known as pooling, perform two essential tasks on Convolutional Neural Networks (CNN): (i) to reduce the number of hyperparameters, thus decreasing the computational cost for training and inference; and (ii) to hold a certain degree of space invariance by keeping the most relevant information. Deep learning techniques have achieved state-of-the-art results on image processing tasks since 2010. Image classification and localization competitions, such as ImageNET Large Scale Visual Recognition Challenge (ILSVRC) [22] and COCO (Common Objects in Context) [15], comprise such neural models in their top results mostly. Inception-V4 [24] and ResNET [6], for instance, achieved outstanding results in image classification tasks. Their basic structure has been used in several other works by adopting transfer learning techniques [3,4,11,20].

João Paulo Papa–The authors acknowledge FAPESP grants 2013/07375-0, 2014/12236-1, 2016/06441-7, and 2017/25908-6, CNPq grants 429003/2018-8, 427968/2018-6 and 307066/2017-7, as well as Petrobras research grant 2017/00285-6.

I. Nyström et al. (Eds.): CIARP 2019, LNCS 11896, pp. 751–760, 2019.
https://doi.org/10.1007/978-3-030-33904-3_71

However, a considerable drawback of these networks concerns the computational cost for both training and inference, taking several days (or even weeks) to achieve the desired results. Therefore, any gain on speed is always welcomed in such models. This work aimed to introduce a more efficient way to reduce the number of parameters and still to keep the spatial invariance expected in CNN-based models. The idea is to replace pooling layers by 2D convolutions with stride as of two. Such modification keeps the average accuracy in different networks, with the boost in both training and inference time.

The remainder of this work is organized as follows: Sect. 2 describes several types of sub-sampling approaches, and Sect. 3 presents the proposed approach. Sections 4 and 5 discuss the methodology and the experiments, respectively. Finally, Sect. 6 states conclusions and future works[1].

2   Related Works

Convolutional Neural Networks were designed based on human visual cortex [13]. In short, such a brain region has two main types of cells: (i) simple cells, which are computationally emulated by the CNN kernels; and (ii) complex cells, that can be found either in the primary visual cortex [7], secondary visual cortex, and the Broadman area 19 of the human brain [9]. The former cells are allocated in the primary visual cortex, and such structures respond mainly to edges and bars [8]. The former cells respond both to edges and gradings, like a simple cell, but also to spatial invariance. It means that such cells react to light patterns in a large receptive field on a given orientation.

Based on this biological information, LeCun et al. [13] developed the first successful CNN model. Its structure consists of a total of seven layers: two pairs of convolutions followed by an average pooling, two multi-layer perceptrons layer, and a final layer responsible for classification. Roughly speaking, a CNN uses pooling since its beginning.

Max-pooling was first proposed in 2011 [17] as a solution for gesture recognition problems. Since then, several works claim that such operation is the best sub-sampling rule for a CNN. However, some other rules, such as Global Averaging Pooling [14], may also be applied in other circumstances: in this specific case, it was designed to replace a multi-layered perceptron network in the final layers of a CNN since it tries to impose correspondences between feature maps and categories. Another sub-sampling approach is a forced concatenation of information from *MaxPooling* combined with the convolution of stride two. The work of Romera et al. [21], for instance, aimed at performing real-time pixel-level segmentation using such paradigm, achieving near state-of-the-art segmentation results.

Sometimes, data sub-sampling is not desired because spatial information is quite important, and any loss could affect the results. DeepMind claims, on its reinforcement learning work [16], that any kind of pooling could remove relevant

---

[1] The source code is available at https://github.com/thierrypin/gei-pool.

spatial information in several games so that the CNN used in their work consists only on convolutional and perceptron layers. Therefore, such arguments suggest it may be necessary to develop new pooling techniques in order to improve results on several problems.

In this work, we proposed GEINet, a deep network for the problem of gait recognition that does not contain any pooling layer. Besides, we also showed that the lack of such a layer could provide satisfactory results, but pretty much faster.

3   Proposed Approach

The main goal of this work is to find out the best neural structure in order to perform gait recognition successfully. Proposed by Han and Bhanu [5], the Gait Energy Image (GEI) approach can be used to classify or identify a given individual. Such technique consists of an average of pictures from a person in a given activity, such as walking or jogging. Roughly speaking, it can be understood as a heatmap indicating what the most frequent positions assumed by a person are. Figure 1 depicts some examples of images generated by the GEI approach.

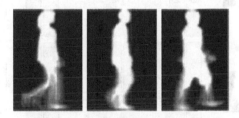

Fig. 1. Example of a GEI image for three different people. Image extracted from the "OU-ISIR Gait Database, Large Population Dataset (OULP)" [10].

State-of-the-art GEI classification results were achieved by Shiraga et al. [23], which proposed three other architectures to identify people from their gait images. The original network is straightforward, consisting of two blocks with a convolutional step (18 7 × 7 and 45 5 × 5 kernels), a 2 × 2 max-pooling, as well as local response normalization [12]. Following the convolutions, are two fully-connected layers of size 1,024 and 956 (number of classes). All layer outputs are activated with ReLU, except for the last one, which is activated with the well-known *softmax* function.

In this paper, we proposed three other architectures for comparison purposes:

1. A re-trained GEINet structure composed of two sets of layers of convolution, pooling, and Local Response Normalization (LRN) [12]. Such layers are then followed by two multilayer perceptrons and finally by a *softmax* for baseline purposes;

2. A similar model, but removing the pooling layer, and changing the convolution stride from one to two (GEINet no-pool); and
3. A third model based on the first one, but replacing the pooling layer for a convolution layer of stride two, acting as a dimensionality reducer. This model doubles the number of convolution layers in comparison to the other two (Double-conv).

Figure 2 depicts the architectures of the neural networks proposed in this work.

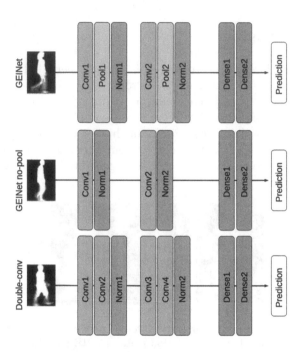

Fig. 2. Architecture of the neural networks proposed in this work.

We followed the protocol described by Shiraga et al. [23] to construct the energy images, which consists of taking four consecutive video silhouette masks to further obtaining their pixel-wise averages.

4   Methodology

In this section, we described the methodology employed to validate the robustness of the proposed approach. The equipment used in the paper was an Intel Xeon Bronze® 3104 CPU with 6 cores (12 threads), 1.70 GHz, 96 GB RAM 2666 Mhz, and GPU Nvidia Tesla P4 8 GB. The framework MXNet [1] was used for the neural network architecture implementation. We provided a better description of data sets used, models, and the evaluation protocol in the following subsections.

4.1 Data Set

We considered the "OU-ISIR Gait Database, Large Population Dataset (OULP)" [10], which consists of silhouettes from 3,961 people from several ages, size, and gender, walking on a controlled environment. Data have been collected since March 2009 through outreach activity events in Japan and recorded at 30 frames per second, from four different angles: 55, 65, 75 and 85 degrees. The original images have a resolution of 640 × 480 pixels, but the silhouettes were further cropped originating another set of image with a resolution of 88 × 128 pixels. In this work, we resized the images to a resolution of 44 × 64 pixels for the sake of computational load.

4.2 Evaluation Protocol

We performed the cross-validation protocol described by Iwama et al. [10]. The dataset is divided into five subgroups of 1,912 people each, and each subset i is further divided into two equal parts of 956 individuals, hereinafter called g_{i1} and g_{i2}, respectively, $\forall i = 1, 2, \ldots, 5$. The former group (g_{i1}) is used for feature extraction purposes using the proposed approaches and baseline, and the latter set (g_{i2}) is employed for the classification step. Each subset is further divided in half, i.e., $g_{i1} = g_{i1}^T \cup g_{i1}^V$ and $g_{i2} = g_{i2}^T \cup g_{i2}^V$, where g_{ij}^T and g_{ij}^V stand for training and validating sets, respectively, $\forall j = 1, 2$. In this work, we opted to use two fast and parameterless techniques for the classification step: the well-known nearest neighbor (NN) [2] and the Optimum-Path Forest (OPF) [18,19][2]. Figure 3 depicts the aforementioned protocol.

As mentioned earlier, the dataset provides four camera angles: 55°, 65°, 75°, and 85°. Therefore, we opted to use a cross-angle methodology, i.e., we used a given angle for training purposes and all angles to evaluate the models. Each video contains between 15 and 45 frames, but we used only 4 to build the gait energy images[3]. To train the neural networks, in each batch iteration, we selected four random contiguous frames. For evaluation purposes, we divided the videos into consecutive non-overlapping clips and further classified each. The final prediction is the mode of all predictions in the sequence.

Since the networks are trained with a single video from each subject, we employed data augmentation to improve training diversity. For this purpose, we employed four image transformations, each with 50% chance of occurring independently: horizontal flip, Gaussian noise with zero mean and standard deviation as of 0.02, as well as random vertical and horizontal black stripes of width 3. Additionally, the random temporal cropping step functions as augmentation. Lastly, due to the low number of videos and the high number of possible variations in the augmentation step, we trained the networks on 12,500 epochs. In addition, we considered three measurements: (i) training and (ii) classification

---

[2] We used the Python OPF implementation available at https://github.com/marcoscleison/PyOPF.

[3] We observed that only four images were enough to obtain a reasonable energy image.

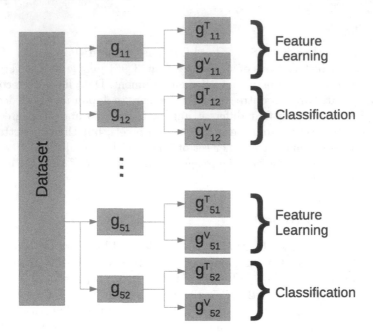

Fig. 3. Protocol adopted in the work, as described by Iwama et al. [10]. The dataset is divided into 5, and each part is further subdivided twice: one is used for feature learning and the other for classification. Then the parts are switched, so that there are a total of 10 evaluation steps.

times, and (iii) accuracy. Notice we used the Wilcoxon signed-rank test [25] for the statistical analysis of each measurement.

5 Results

In this section, we presented the experimental results and discussion. We showed that replacing the pooling layer by a larger convolutional stride is sufficient to obtain a good trade-off between computational load and accuracy. As aforementioned, in this paper we evaluated three models and compared their performance.

5.1 Accuracy

We evaluated how the models perform when predicting with different camera angles. All the training step was performed on a single camera angle and the same for the classifier. Therefore, the idea is to predict gaits from all four viewpoints. Tables 1 and 2 depict the accuracy results using NN and OPF, respectively. The results concern the average from all five folds, as described in Sect. 4.2. It is worth noticing that, the closer the test angle is to 90°, the better the overall accuracy is, i.e., when the camera records the actor from the side view. As expected, the accuracies tend to be higher when the train and test angles are the same.

Table 1. Mean accuracies using NN classifier.

Train angle	Method	Test angle			
		55	65	75	85
55	GEINet	88.77	88.56	87.51	88.35
	GEINet no-pool	87.82	87.26	86.05	87.34
	Double-conv	88.28	88.18	86.86	86.51
65	GEINet	85.61	89.52	90.27	90.88
	GEINet no-pool	83.56	88.41	88.87	89.77
	Double-conv	84.21	89.23	89.06	90.00
75	GEINet	79.71	86.80	90.29	91.59
	GEINet no-pool	79.54	87.13	90.15	91.78
	Double-conv	80.00	87.49	89.83	91.11
85	GEINet	72.57	78.85	87.07	91.42
	GEINet no-pool	73.37	79.48	87.15	91.17
	Double-conv	75.71	80.48	87.59	91.44

Table 2. Mean accuracies using OPF classifier.

Train angle	Method	Test angle			
		55	65	75	85
55	GEINet	88.14	88.01	87.02	88.14
	GEINet no-pool	87.36	86.99	85.67	87.07
	Double-conv	87.55	87.89	86.34	86.19
65	GEINet	85.12	89.08	89.81	90.48
	GEINet no-pool	82.80	88.08	88.26	89.41
	Double-conv	83.62	88.70	88.85	89.52
75	GEINet	79.27	86.09	89.79	91.28
	GEINet no-pool	79.16	86.67	89.67	91.42
	Double-conv	79.27	86.57	89.54	90.61
85	GEINet	71.99	78.10	86.36	90.86
	GEINet no-pool	72.87	78.81	86.97	90.86
	Double-conv	75.17	79.95	87.05	91.05

When replacing the pooling layer from GEINet by a stride in its convolution layer, the accuracy results go down marginally – around 1%. Besides, Wilcoxon test returned a p-value around to 10^{-7}, indicating they probably do not diverge. Trading the pooling step by a new convolutional layer with stride as of 2 results in slightly better results, but still not quite better than the original model. The Wilcoxon test outputted a p-value as of 0.102, indicating that their distribution might be similar as well.

5.2   Execution Time

Since the protocol employed in this paper establishes ten runs, and the models were trained once for each angle, each model has 40-time measurements. Therefore, all results presented in this section correspond to the average of all runs.

Table 3 presents the network training and inference times. Although the non-pooling model achieved slightly smaller accuracies than the original one, its training time is considerably lower. The reduction from 3,753 seconds to 3,322 corresponds to a gain of 11.5%, while such gain was 8.3% for inference purposes.

Table 3. Training and inference times: replacing the pooling layer by a convolutional stride resulted in considerably faster training time.

Training time			
Model	Per epoch (s)	Total (s)	Inference Time (s)
GEINet	0.300	3,753.71	0.108
GEINet no pool	0.266	3,322.18	0.099
Double-conv	0.320	4,004.93	0.109

6   Conclusion and Future Works

In this work, we introduced two variants of a simple but efficient model for gait recognition purposes (*GEINet*): one replaces the pooling layers by a convolutional stride (*GEINet no-pool*), and the other replaces the pooling layers by a convolutional layer with stride (*double-conv*). We showed the non-pooling version achieved slightly smaller accuracies than GEINet, but with a considerable speed-up (11.5%). On the other hand, the double-conv model ran 6.3% slower without any perceptible gain in accuracy. Regarding future works, we intend to use GEI to identify people directly from the video streams. Besides, different activation functions shall be investigated too.

References

1. Chen, T., et al.: Mxnet: A flexible and efficient machine learning library for heterogeneous distributed systems. CoRR abs/1512.01274 (2015). http://arxiv.org/abs/1512.01274
2. Cover, T.M., Hart, P.E., et al.: Nearest neighbor pattern classification. IEEE Trans. Inf. Theory **13**(1), 21–27 (1967)
3. Esteva, A., et al.: Dermatologist-level classification of skin cancer with deep neural networks. Nature **542**(7639), 115 (2017)

4. Habibzadeh, M., Jannesari, M., Rezaei, Z., Baharvand, H., Totonchi, M.: Automatic white blood cell classification using pre-trained deep learning models: Resnet and inception. In: Tenth International Conference on Machine Vision (ICMV 2017), International Society for Optics and Photonics, vol. 10696, p. 1069612 (2018)

5. Han, J., Bhanu, B.: Individual recognition using gait energy image. IEEE Trans. Pattern Anal. Mach. Intell. **28**(2), 316–322 (2006)

6. He, K., Zhang, X., Ren, S., Sun, J.: Deep residual learning for image recognition. CoRR abs/1512.03385 (2015). http://arxiv.org/abs/1512.03385

7. Hubel, D., Wiesel, T.: Receptive fields, binocular interaction, and functional architecture in the cat's visual cortex. J. Physiol. **160**, 106–154 (1962)

8. Hubel, D.H., Wiesel, T.N.: Receptive fields of single neurons in the cat's striate cortex. J. Physiol. **148**, 574–591 (1959)

9. Hubel, D.H., Wiesel, T.N.: Receptive fields and functional architecture in two nonstriate visual areas (18 and 19) of the cat. J. Neurophysiol. **28**(2), 229–289 (1965)

10. Iwama, H., Okumura, M., Makihara, Y., Yagi, Y.: The ou-isir gait database comprising the large population dataset and performance evaluation of gait recognition. IEEE Trans. Inf. Forensics Secur. **7**(5), 1511–1521 (2012)

11. Kong, B., Wang, X., Li, Z., Song, Q., Zhang, S.: Cancer metastasis detection via spatially structured deep network. In: Niethammer, M., Styner, M., Aylward, S., Zhu, H., Oguz, I., Yap, P.-T., Shen, D. (eds.) IPMI 2017. LNCS, vol. 10265, pp. 236–248. Springer, Cham (2017). https://doi.org/10.1007/978-3-319-59050-9_19

12. Krizhevsky, A., Sutskever, I., Hinton, G.E.: Imagenet classification with deep convolutional neural networks. In: Advances in Neural Information Processing Systems, pp. 1097–1105 (2012)

13. LeCun, Y., Bottou, L., Bengio, Y., Haffner, P.: Gradient-based learning applied to document recognition. Proc. IEEE **86**(11), 2278–2324 (1998)

14. Lin, M., Chen, Q., Yan, S.: Network in network. CoRR abs/1312.4400 (2013). http://arxiv.org/abs/1312.4400

15. Lin, T., et al.: Microsoft COCO: common objects in context. CoRR abs/1405.0312 (2014). http://arxiv.org/abs/1405.0312

16. Mnih, V., et al.: Human-level control through deep reinforcement learning. Nature **518**(7540), 529–533 (2015). https://doi.org/10.1038/nature14236

17. Nagi, J., et al.: Max-pooling convolutional neural networks for vision-based hand gesture recognition. In: 2011 IEEE International Conference on Signal and Image Processing Applications (ICSIPA), pp. 342–347. IEEE (2011)

18. Papa, J.P., Falcão, A.X., Suzuki, C.T.N.: Supervised pattern classification based on optimum-path forest. Int. J. Imaging Syst. Technol. **19**(2), 120–131 (2009). https://doi.org/10.1002/ima.v19:2

19. Papa, J.P., Falcão, A.X., Albuquerque, V.H.C., Tavares, J.M.R.S.: Efficient supervised optimum-path forest classification for large datasets. Pattern Recogn. **45**(1), 512–520 (2012)

20. Rakhlin, A., Shvets, A., Iglovikov, V., Kalinin, A.A.: Deep convolutional neural networks for breast cancer histology image analysis. In: Campilho, A., Karray, F., ter Haar Romeny, B. (eds.) ICIAR 2018. LNCS, vol. 10882, pp. 737–744. Springer, Cham (2018). https://doi.org/10.1007/978-3-319-93000-8_83

21. Romera, E., Alvarez, J.M., Bergasa, L.M., Arroyo, R.: Efficient convnet for real-time semantic segmentation. In: IEEE Intelligent Vehicles Symposium (IV), pp. 1789–1794 (2017)

22. Russakovsky, O., et al.: ImageNet large scale visual recognition challenge. Int. J. Comput. Vis. (IJCV) **115**(3), 211–252 (2015). https://doi.org/10.1007/s11263-015-0816-y
23. Shiraga, K., Makihara, Y., Muramatsu, D., Echigo, T., Yagi, Y.: Geinet: view-invariant gait recognition using a convolutional neural network. In: 2016 International Conference on Biometrics (ICB), pp. 1–8. IEEE (2016)
24. Szegedy, C., Ioffe, S., Vanhoucke, V.: Inception-v4, inception-resnet and the impact of residual connections on learning. CoRR abs/1602.07261 (2016). http://arxiv.org/abs/1602.07261
25. Wilcoxon, F.: Individual comparisons by ranking methods. Biom. Bull. **1**(6), 80–83 (1945)

Virtual Hand Training Platform Controlled Through Online Recognition of Motion Intention

César Quinayás[1]([✉]), Fernando Barrera[1]([✉]), Andrés Ruiz[1]([✉]),
and Alberto Delis[2]([✉])

[1] Antonio Nariño University, Popayán 190001, Colombia
{cquinayas,fernando.barrera,andresru}@uan.edu.co
[2] Center of Medical Biophysics, University of Oriente,
901000 Santiago de Cuba, Cuba
lopez.delis69@gmail.com

Abstract. Patients with amputation or defects in their limbs use prosthetic devices that require a great cognitive and physical effort to control them, especially during rehabilitation and training phases, being this one of the most frequent reasons of why the patients gave up their prosthesis. This paper presents a platform for the training of patients to control prostheses using a virtual robotic hand. The virtual environment combines a graphics engine as Unity with a homemade bracelet that acquires surface Electromyography signals (sEMG) from a patient's arm. The virtual training platform computes two features from the patient's sEMG signals. Firstly, the Absolute value of the Summation of Square root (ASS), and then the Mean value of the Square Root (MSR). Once this is done, they are concatenated and passed through a multi-layer neural network, which has been trained to detect 4 different movement intentions generated by the patient (rest, open hand, power and precision grips). Finally, classifier outcomes are used to control the position of joints of a virtual robotic hand that is simulated in Unity. Experimental results have shown a classification accuracy of 86.6% on a patient with a congenital amputation of its left arm.

Keywords: Virtual robotic hand · Surface electromyography signals bracelet · Motion intention · Training platform

1 Introduction

Actions carried out by people in daily life require the movements of their hands and fingers, which are controlled by muscles of the forearm, biceps, and triceps.

This work was supported by Colombian state funds managed by the Universidad Antonio Nariño within the Innovacción Cauca program, under reference ID-3848 (Innovacción Cauca). Project name: Desarrollo de una prótesis de mano robótica controlada a través de señales EMG.

© Springer Nature Switzerland AG 2019
I. Nyström et al. (Eds.): CIARP 2019, LNCS 11896, pp. 761–768, 2019.
https://doi.org/10.1007/978-3-030-33904-3_72

People with amputation or deficiencies in their hands cannot perform many of these activities. Because of this, most of research efforts have focused on restoring the functions that patients can do with their hands and fingers through prosthetic devices [4,5,7]. Among these devices, it could be highlight active commercial prostheses such as the i-limb [2], bebionic [1], and Michelangelo [3].

Using hand prosthesis requires an extensive training process with the patient to have optimal control of the prosthesis before starting to use it. In the training process, patients experience a great mental and physical effort to control hand prosthesis with many degrees of freedom, with a reduced number of sEMG signals. Some studies show that many amputees do not use their prosthesis because they are not able to control them efficiently or because the prosthesis does not offer an adequate human-machine interface to be controlled.

In this work, we propose to develop a human-machine interface in real-time based on augmented reality and embedded systems for the acquisition, conditioning, and classification of sEMG signals. The proposed system uses a pattern recognition algorithm based on a multi-layer neural network that classifies motion intentions generated by patients, then these are reproduced by a virtual robotic hand. The virtual prosthesis is configured to perform four types of actions: rest, open hand, power grip, and tripod grip, these correspond to the number of classes recognized by the neural network.

2   Virtual Training Platform

A schematic diagram of the proposed system is presented in Fig. 1. It consists of an inertial measurement unit (IMU), which sense the rotation and acceleration of the whole system in x, y, and z axis. This information is used for position controlling of the prosthesis in the virtual environment and like feature for classification purposes. Also, the system employs 4 independent channels for acquiring and conditioning of sEMG signals. Both IMO and sEMG data are processed by an Arduino mega. This device is responsible for the processing of all the data passed through Analog-to-Digital Converter (ADC) as well as signal sampling, feature extraction, and patient's movement intention (features classification).

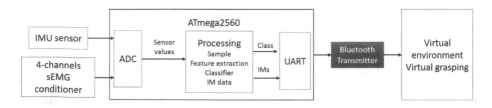

Fig. 1. Schematic diagram of the virtual hand training platform.

As can be seen in Fig. 1, the system establish wireless communication between the Arduino and virtual environment by means the Bluetooth protocol. This communication channel is used to transmit both classification results and inertial measurements from the patient arm toward the virtual environment. The Universal Asynchronous Receiver-Transmitter or UART is the entity that manages the data exchange between Arduino and desktop computer, being the latter who executes the virtual environment (Unity application).

The proposed system is shown in Fig. 2. Patient muscle activity is acquired using four sEMG electrodes, which are attached to a velcro strap (sEMG bracelet). During the experiments, the bracelet is placed on the forearm as it shows Fig. 2. The figure also shows the IMO, Arduino, and sEMG signal conditioning circuit.

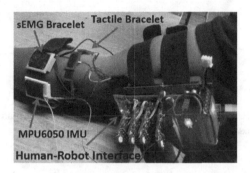

Fig. 2. Proposed system of acquisition, conditioning and processing.

3 EMG Conditioner

All sEMG signals are passed through an instrumentation amplifier with a gain of 10, then the resulting signal is band-pass filtered, with upper and lower cutoff frequencies of 500 Hz and 10 Hz, respectively. To reduce the electromagnetic noise induced by the electrical grid, the sEMG signals are passed through a notch filter whose cutoff frequency is 60 Hz. Finally, signals are amplified six times, and its offset level compensated by a combination of amplifiers and voltage followers. Then, signals are input to Arduino through their 10-bits ADCs. Following the Nyquist Theorem, the EMG signal is sampled at 1 KHz to register frequency components up to 500 Hz.

3.1 Data Processing

The feature extraction methods include a combination of parameters in the time domain (Absolute value of the summation of Square Root-ASS and Mean value of Square Root-MSR) from a given analysis windows k, where x_n denote the data within the corresponding analysis window. The ASS computation consists of three principal steps using the data in the analysis window: the square root

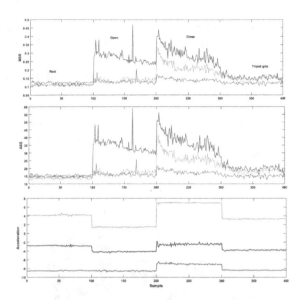

Fig. 3. Traces of MSR and ASS features and triaxial acceleration associated with three movements (open, close, and tripod grip).

of all of the values in the analysis window is first computed, the summation of the resultant values is determined, and lastly, the absolute value is computed [6], see Eq. 1. MSR feature is determined from the square root of all the values in a given analysis window, followed by the mean of the resultant values [6], see Eq. 2.

$$ASS = \sum_{n=1}^{k-1} | (x_n)^{1/2} | \tag{1}$$

$$MSR = \frac{1}{k} \sum_{n=1}^{k-1} (x_n)^{1/2} \tag{2}$$

The performance of the ASS and MSR features was examined in others work from sEMG recordings of eight (four transradial and four transhumeral) amputees by using four different metrics [6]. The results obtained in this study suggest that the proposed time-domain features would potentially improve the overall performance of sEMG pattern recognition control strategy for multifunctional myoelectric prosthesis.

The features vectors are extracted using a sliding analysis windows of 125 ms in length, spaced 50 ms for both training and testing process, besides, the mean of acceleration data is synchronized with EMG signals in time windows. Each sEMG channel and its features vectors (ASS and MSR) are concatenated with average acceleration data, resulting in 11 coefficients (4 channels x 2 features vectors + 3 data of acceleration/channel). In Fig. 3 is shown the features extracted from the EMG signals.

After computing the signal features according to the previous section, we apply them to a LM multi-layer perceptron neural network. The extracted features correspond to the inputs to a three-layer LM neural network, with 10 nodes in the input layer, 5 nodes in the hidden layer, and 1 node in the output layer (this output represents the estimated joint angle). We chose the network's architecture and size empirically, aiming at the maximum possible reduction of the final mean squared error (MSE).

For neural network training, we used the same initial weight values for all three network layers (null weight for all neurons). The maximum number of iterations was set to 200 and the stop criterion was MSE of $10\,e^{-10}$. For the training process 100 features vector was used and the same number for the testing process. These features represent the rest, hand open, power grip and tripod grip for each channel.

4 Virtual Environment

This module is responsible for recreating in the virtual environment an articulated hand prosthesis that performs a set of preset actions, each of them under user's demand through sEMG signals. In the virtual environment, the hand prosthesis can animate movements such as open hand, close hand, tripod grip and rest, as well as orienting the hand on the x, y and z axes to reach the object. In Fig. 4 the virtual environment developed in Unity 3D software is shown.

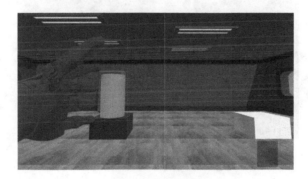

Fig. 4. Virtual environment for patient training.

5 Experimental Protocol

The patient was instrumented with a bracelet of 4-sEMG channel located on her residual arm. The patient gave written informed consent prior to participation. The patient was asked to perform voluntary contractions for identifying muscle with myoelectric activity. During the exercises channel 1 and 4 were the most active.

For identification the channels in the previous phase, the subject is cued to training sessions for modulating the intensity of contraction. In a Labview program, the sEMG signals of voluntary contraction were visualized. These images act as feedback so that the subject check and qualifies the information. The aim of this training phase is for the subject becomes familiar with the adequate muscles to generate the valid sEMG signals.

Once the viable muscles have been identified and the person is able to contract them, emulating a normal action as if the upper limb were intact, it is time for the sEMG signal to be used as input to the processing system. The subject sitting comfortably on a chair was asked to imagine grasping and releasing the object that appears in the virtual environment. In this way, data of rest, open, close and tripod grip are recorded for 3 s. With the collected data, the feature vectors of training and test are obtained to train the neural network off-line. The experimental setup is shown in Fig. 5.

Fig. 5. Experimental setup.

In this study, two types of metrics were adopted to evaluate the performance of the proposed of virtual platform of training. The first metric that evaluates the performance of the use of the ASS and MSR features and inertial data in the classification of gestures using the testing dataset is based on the classification accuracy of classifier as the relationship between the number of correct classifications and the total number of testing samples.

The second metric evaluates the success percentage by developing gestures to grasp and leave the object in the virtual environment.

6   Results

The performance of use the ASS and MSR features and mean value of the acceleration in x, y and z-axes in terms of classification accuracy is shown in Table 1.

Table 1. Success rate for each movement.

	Rest	Open hand	Close hand	Tripod grip
Rest	100	0	0	0
Open hand	0	100	0	0
Close hand	0	0	100	0
Tripod grip	10	4	3	83

Gestures identified from the patient allowed controlling the movements of open, close and tripod grip in the virtual hand. These movements were simulated separately as well as simultaneously. Each movement was tested for 20 trials. Table 2 summarizes the success rate for each movement for the 20 trails. The overall success rate was found to be 86.6 %.

Table 2. Success rate for each movement.

	Number of trials	Numbers of success	Percentage of success
Rest	–	–	–
Open hand	20	19	95
Close hand	20	18	90
Tripod grip	20	15	75

7   Conclusion

In this work, a virtual hand-training platform was presented that allows reproducing gestures obtained from a neural classifier. The platform also allows orienting a virtual hand in a virtual environment developed in Unity to locate and grasp an object.

This platform can be used for patients as a myoelectric prosthesis trainer, thus facilitating the learning process and the study on the suitability of the patient to the management of their prosthetic device.

Average accuracy of 86.6% was obtained for the classification of 3 movements, taking into account that the results are related to one patient. The hypothesis is maintained that the system allows training with few variations in the percentage of success, if the capture and processing conditions of the sEMG signal are maintained and that in the future it can become a tool for training people with amputation.

References

1. Bebionic hand. https://www.ottobockus.com/prosthetics/upper-limb-prosthetics/solution-overview/bebionic-hand. Accessed 27 Oct 2019
2. How the i-limb works. http://www.touchbionics.com/products/how-i-limb-works. Accessed 27 Oct 2019
3. Michelangelo prosthetic hand. https://www.ottobockus.com/prosthetics/upper-limb-prosthetics/solution-overview/michelangelo-prosthetic-hand. Accessed 27 Oct 2019
4. Chadwell, A., Kenney, L., Thies, S., Galpin, A., Head, J.: The reality of myoelectric prostheses: understanding what makes these devices difficult for some users to control. Front. Neurorobotics **10**, 7 (2016). https://doi.org/10.3389/fnbot.2016.00007. https://www.frontiersin.org/article/10.3389/fnbot.2016.00007
5. Roche, A.D., Rehbaum, H., Farina, D., Aszmann, O.C.: Prosthetic myoelectric control strategies: a clinical perspective. Curr. Surg. Rep. **2**(3), 44 (2014). https://doi.org/10.1007/s40137-013-0044-8
6. Samuel, O.W., et al.: Pattern recognition of electromyography signals based on novel time domain features for amputees' limb motion classification. Comput. Electr. Eng. **67**, 646–655 (2018). https://doi.org/10.1016/j.compeleceng.2017.04.003. http://www.sciencedirect.com/science/article/pii/S0045790617303932
7. Vujaklija, I., Farina, D., Aszmann, O.C.: New developments in prosthetic arm systems. Orthop. Res. Rev. **8**, 31–39 (2016). https://doi.org/10.2147/ORR.S71468. https://www.ncbi.nlm.nih.gov/pubmed/30774468

Video Popularity Forecasting to Improve Cache Miss Rate in Content Delivery Networks

Jairo Rojas-Delgado[1]([✉]) [iD], Rafael Trujillo-Rasúa[2] [iD], Rafael Bello[3], and Gerdys E. Jiménez Moya[1] [iD]

[1] Universidad de las Ciencias Informáticas, Havana, Cuba
{jrdelgado,gejimenez}@uci.cu
[2] IN Switch Solutions S.A., Montevideo, Uruguay
rafael.trujillo@inswitch.com
[3] Universidad Central de las Villas, Santa Clara, Cuba
rbellop@uclv.edu.cu

Abstract. Video transmission is of critical interest in several practical applications. Recent studies show that video content is highly cacheable in content delivery networks. Proactive and hybrid reactive-proactive caching policies, with the use of media popularity forecasting, are being developed as a better approach to conventional reactive cache strategies. Neural networks have been extensively used in popularity forecasting, however, training a neural network is a challenging NP-hard optimization problem. In this paper, we propose to train neural networks for video popularity forecasting with a novel continuation approach and Particle Swarm Optimization algorithm to improve forecasting accuracy. We create a dataset from an online video transmission platform and develop a cache simulation to find the relationship between forecasting accuracy and cache efficiency. Our findings support that higher accuracy have a significant effect in cache efficiency. Further results show that our neural network training approach is able to improve forecasting accuracy respect gradient based algorithms and therefore improve cache efficiency.

Keywords: Neural-network · Popularity-forecasting · Video · Continuation

1 Introduction

Video transmission is of critical interest in several practical applications. In 2017 video accounted for 75% of internet traffic and in 2022 will raise up to 82% [29]. Distributed transmission platforms, such as Content Delivery Networks (CDNs), have shown to be very reliable and efficient solutions for video-on-demand. CDNs implement highly efficient caches to reduce the effect of delays, jitters and traffic congestion [17]. Caching strategies provide an effective mechanism for mitigating the massive bandwidth requirements by replicating the most popular content closer to the network edge and not storing it in a central site [6].

© Springer Nature Switzerland AG 2019
I. Nyström et al. (Eds.): CIARP 2019, LNCS 11896, pp. 769–779, 2019.
https://doi.org/10.1007/978-3-030-33904-3_73

Recent studies show that video content is highly cacheable due to the skew distribution of video requests, the strong dominance of popular content and the ephemeral nature of content popularity [22,23]. For some video publishing platforms, the 70% of the videos are requested only once from the edge network and 10% of the most popular videos account for 80% of video requests [7]. Several cache replacement and insertion policies, with use of media popularity forecasting, are being developed as a better approach to conventional pure reactive cache strategies [2,10,14,21]. Automated demand forecast and performance prediction can help with capacity planning and quality control so that sufficient server bandwidth can always be supplied to each video channel without wastage.

The best cache strategy is the oracle policy [3], which uses exact knowledge of all future requests of the resources in the cache. Even if it is not possible to use oracle policy in practical applications, we can use it as a higher bond of efficiency. Several related papers approximates content popularity by means of a power law distribution known as Zipf distribution [11] or a Mandelbrot-Zipf variation [22]. Such approximation of content popularity allows substantial room for improvement if content popularity models with higher accuracy are proposed as noticed by [7,23].

However, content popularity forecasting depends on several factors such as the viewers culture, the editorial publishing policies, scope of the video platform and the content itself. This is why, some authors are proposing domain-specific solutions for popularity forecasting [21]. In this area, Artificial Neural Networks (ANNs) have been extensively used to predict media content popularity [12, 18,30]. Among the most popular algorithms for training ANNs is Stochastic Gradient Descent (SGD) which is a gradient based first order method that has been proven to converge into local minima in the parameter space.

Recently, several meta-heuristic algorithms for optimization have been actively used in ANN training. Meta-heuristic algorithms provide features to escape local minima and increase the probability of global convergence. In the literature there are reports of nature-inspired meta-heuristics such as Cuckoo Search [8], Wolf Search Optimization [1], Particle Swarm Optimization (PSO) [9] and others. For current applications, the rule of thumb to obtain higher accuracy have been to increase the number of parameters of ANNs and the amount of instances mainly because over-fitting [13]. In this scenario, the use of meta-heuristic algorithms is limited by the increased computational complexity of the fitness function evaluation. Hence, methods such as continuation [15] have been introduced to reduce execution time of training.

In this paper, we propose a domain-specific model based in ANNs to predict content popularity and reduce cache miss rate in a video transmission platform. We use data from an online video transmission platform deployed in a closed university campus. We perform a simple simulation to clarify the relation between forecasting accuracy and cache efficiency for this particular transmission platform. Our findings support the need for higher accuracy in popularity forecasting. That way, by using ANNs trained with PSO and continuation, we improve model accuracy when compared with SGD and ADAM.

The paper is structured as follows: Sect. 2 describes feature extraction procedure conducted for creating the training dataset. Section 3 presents some insights on continuation methods and introduce an algorithm to train ANNs based on optimization by continuation and PSO. In Sect. 4 we present simulation analysis and the results for the accuracy and execution time of training. Finally, some conclusions and recommendations are given. Throughout this article we use x for scalars, $\boldsymbol{x}$ for vectors and X for sets.

2   Dataset and Feature Extraction

To predict media popularity, we collect meta-data and traffic information[1] from an online video transmission platform named Internos currently deployed in the Universidad de las Ciencias Informáticas. Internos is online in a closed university campus via network intranet and serves about 7.000–10.000 users. The dataset is build in a time interval of two years of activity between 04/16 and 04/18.

The core concept in the platform is the media resource. Media resources have a section. One media resource belongs to only one section and a section has several media resources. There are several types of media resource: Video, Movie, Serial, Audio and Other. Users can post comments and vote for specific media resources. Finally, a user may choose to reproduce a media resource requesting the video frames from the streaming server.

In total, there are 4394 media resources, 2504763 video requests events, 1866 comments and 14695 votes for the media resources. For each media resource, we create several instances considering the number of video requests in each day after the media publication up to 20 days. Hence, the dataset contains 75782 instances after we removed duplicated records.

In a recent study [12], authors suggest the use of different video features for video popularity forecasting. Social attributes, such as the number of comments and the number votes, are more important than visual attributes extracted from video frames [28]. In addition, temporal features such as publication date and the age of a video relates to an observed ephemeral property of video popularity [7]. Moreover, video category is an important feature with a significant relation with video popularity [20]. Considering that for Internos dataset no video frames are available, we extracted the attributes presented in Table 1 after conducting an exploratory data analysis.

We encode several ordinal and nominal attributes. The ordinal attributes, such as day of the week or day of the year, show a cyclic nature. We encode these attributes in two new attributes $x'_{i,j} = \sin(2\pi x_{i,j}/|\Omega_j|)$ and $x''_{i,j} = \cos(2\pi x_{i,j}/|\Omega_j|)$, where $|\Omega_j|$ is the size of the domain of possible values for the j-th attribute. In addition, attribute video type is nominal. Although there are several encoding strategies for nominal variables, a recent comparative study suggests that for neural networks the Backward Difference Encoding gets promising results [27]. After variable encoding, we normalize all attributes using Min-Max feature scaling.

---

[1] Dataset available at: https://data.mendeley.com/datasets/69g55s83mn/1.

Table 1. Features extracted for media popularity forecasting in Internos video transmission platform.

No.	Name	Type	Description
1	Requests one week	Numerical	Number of requests one week ago
2	Requests two week	Numerical	Number of requests two weeks ago
3	Requests one day	Numerical	Number of requests one day ago
4	Requests two days	Numerical	Number of requests two days ago
5	Day of week	Ordinal	Day of the week: Monday, Tuesday, etc.
6	Day of month	Ordinal	Day of the month: first, second, etc.
7	Day of year	Ordinal	Day of the year
8	Month	Ordinal	Number of the month
9	Comments	Numerical	Number of comments in the publication
10	Comments in section	Numerical	Number of comments in section for the last week
11	Publications	Numerical	Number of publications for the last week
12	Publications in section	Numerical	Number of publications in section for the last week
13	Votes	Numerical	Number of votes in the publication
14	Votes in section	Numerical	Number of votes in section for the last week
15	Days published	Numerical	Number of days since the publication was published
16	Holidays left	Numerical	Number of days of remaining holidays
17	Next holidays	Numerical	Number of days until the next holidays
18	Video type	Nominal	Type of video: video, audio, movie, etc.
19	Requests	Numerical	Number of video requests in the next 24 h

3   Neural Networks and Optimization by Continuation

In practice the fitness function of an ANN is a sort of loss function such as the MSE in Eq. 1, where $X = \{x_1, ..., x_p\}, x_i \in \mathbb{R}^d$ is the set of instances, $w \in \mathbb{R}^n$ contains the neural network parameters and $\hat{y}_i$ and y_i are the expected and real output of the network for the x_i instance respectively.

$$f(w, X) = \frac{\sum_{i=1}^{p} (\hat{y}_i - y_i)^2}{p} \tag{1}$$

In Eq. 1 the computational cost comes from calculating the real output y_i that requires propagating an instance through each layer of the neural network. Actually, according to [24], minimizing $f(w, X)$ is an NP-hard optimization problem defined in Eq. 2.

$$w = \operatorname*{argmin}_{w \in \mathbb{R}^n} f(w, X) \tag{2}$$

Having an ANN formed by two weights, its parameter space would look like a landscape with valleys and hills. As our goal is to find a point in such parameter space near of the global optimum, it is possible to use a coarse-grained representation of the error surface generated by the instances. The former principle is known in optimization as continuation method. The general idea behind the so-called continuation method, is to start solving an easy problem and progressively change it to the actual complex task [26].

PSO [5] is a population based meta-heuristic algorithm. Standard PSO has several hyper-parameters such as: population size (r), global contribution (α), local contribution (β), maximum speed (v_{max}) and minimum speed (v_{min}). In ANN training by PSO, the i-th particle in the population represents the ANN parameters vector $\boldsymbol{w}_i$. Let $\boldsymbol{w}^*$ be the position of the particle with the lowest fitness function value in the population and $\boldsymbol{w}_i^*$ be the position where $\boldsymbol{w}_i$ achieved its lowest fitness function value. Each particle $\boldsymbol{w}_i$ moves towards a direction determined by $\boldsymbol{w}^*$, $\boldsymbol{w}_i^*$ and a velocity factor. Particle $\boldsymbol{w}^*$ is also known as best-global particle and $\boldsymbol{w}_i^*$ as best-so-far particle of $\boldsymbol{w}_i$.

Algorithm 1 describes how to use Continuation-Particle-Swarm-Optimization (CPSO) for ANN training[2]. The algorithm input is a set of instances X and a fitness function in the form of $f(\boldsymbol{w}, X)$. In the first steep of the algorithm, a sequence of subsets of instances $\hat{X}_1, \hat{X}_2, ..., \hat{X}_k$ is created. The last subset of the sequence, denoted by $\hat{X}_k$, is the entire set of instances $(\hat{X}_k = X)$ and each subset $\hat{X}_i, 1 \leq i \leq k-1$ is built by randomly removing a $\beta^\dagger \cdot p$ number of training patterns from the subset $\hat{X}_{i+1}$ with $\beta^\dagger \in (0,1)$. Hence, the sequence holds that $\hat{X}_1 \subset \hat{X}_2 \subset ... \subset \hat{X}_k$. For each iteration, the optimization continues the search using the population from the previous iteration instead of restarting the population randomly. Hence, the name of *continuation*. Finally, we return the *best-global* particle of the population in line 8.

Algorithm 1. Neural network training by PSO and continuation.

Input: Set of instances $X = \{\boldsymbol{x}_1, \boldsymbol{x}_2, ..., \boldsymbol{x}_p\}$
Input: Fitness function $f(\boldsymbol{w}, X)$, $\boldsymbol{w} \in \mathbb{R}^n$
Output: Best particle $\boldsymbol{w}^*$
1: Build a sequence of subsets of instances $\hat{X}_1 \subset \hat{X}_2 \subset ... \subset \hat{X}_k$
2: Generate initial population $\boldsymbol{w}_1, \boldsymbol{w}_2, ..., \boldsymbol{w}_r$ with $\boldsymbol{w}_j \in \mathbb{R}^n$
3: **for** i = 1 : k **do**
4:     Update fitness value of $\boldsymbol{w}^*$ by evaluating $f(\boldsymbol{w}^*, \hat{X}_i)$
5:     Update fitness value of $\boldsymbol{w}_j^*$ by evaluating $f(\boldsymbol{w}_j^*, \hat{X}_i)$ for $1 \leq j \leq r$
6:     Perform a number of optimization iterations that accounts for η/k fitness function evaluations using $f(\boldsymbol{w}_i, \hat{X}_i)$ with PSO
7: **end for**
8: **return** best particle $\boldsymbol{w}^*$ in the population

---

4   Results and Discussion

In this section we present an analysis of cache miss rate of different reactive and proactive cache replacement strategies under different error scenarios. Further

---

[2] Source code available at: https://github.com/jairodelgado/dnn_opt.

we show the accuracy and execution time of training neural networks when considering SGD, ADAM, PSO and CPSO.

4.1   Analysis of Cache Miss Rate

We compare reactive, proactive and hybrid reactive-proactive cache replacement strategies using exact information of the future number of video requests for the proactive cases. As in practical applications future information about the number of video requests is not available, we further study the effect of accuracy in the proactive and hybrids cache replacement strategies. The goal with this analysis is to establish a target accuracy for the latter predictive models and establish a relation between forecasting accuracy and cache efficiency.

Cache miss rate is an indicator of the number of video requests performed to an edge server that refers to a video that is not previously stored in the server's cache. We define cache miss rate as $r = m/t$, where m is the number of requests when the video is not in cache and t is the total number of requests.

In order to compare different cache replacement strategies, we develop a cache simulation model. We consider a transmission platform with a single edge server that receives video requests. The edge server has a cache capable of storing a number s of videos. When the edge server receives a request, the system first checks if the requested video is in the cache. If the requested video is not in the cache, the edge server retrieves the video from a unique central server.

After the central server provides the video, the edge server adds the new video in the cache. If the cache is full, that is, the number of videos in the cache is equals to s, the new video replaces an old video according to some replacement strategy. Here, we will compare six cache replacement strategies: Least Recently Used (LRU), Simple Proactive (PRO, replace less popular content), Hybrid Least Recently Used and Simple Proactive (LRU-PRO) [21], Oracle proactive (OPT) [3], Segmented Least Recently Used (SLRU) [19] and Adaptive Replacement Caching (ARC) [25].

For the simulation, we use traffic data of Internos transmission platform with video requests of real users. We use different cache sizes: from $s = 50$ to $s = 500$. Figure 1a shows the comparison between the cache replacement strategies. The cache replacement policy PRO has the biggest cache miss rate followed by LRU. In addition, the LRU-PRO cache replacement strategy shows the best results, specifically for $\delta = 0.1$, which means that 10% of the cache is managed by LRU and the other 90% is managed by PRO. In the case of other popularity aware replacement policies, SLRU and ARC, the performance is similar to LRU.

In practical applications exact information about resources popularity is not available, hence, we study the effect of accuracy in the cache miss rate of the LRU-PRO cache replacement policy with $\delta = 0.1$. We denote $\bar{f}$ to the MSE between real popularity and predicted popularity. In this case, we calculate predicted accuracy by adding random noise to the real popularity.

Figure 1b shows the cache miss rate when using the LRU-PRO cache replacement strategy and different cache sizes under different accuracy scenarios. The accuracy of the media popularity forecasting has a direct impact in the cache miss rate. For $\bar{f}$ = 2.0E-3 the cache miss rate of the LRU-PRO policy is less than the LRU policy, however, there is room for improvement when considering forecasting accuracy. The results presented in Fig. 1b supports the need of further developments in forecasting methods for video popularity.

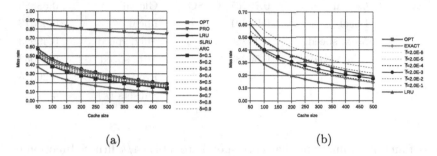

(a)   (b)

Fig. 1. (a) Cache miss rate comparison between different cache replacement strategies. (b) Cache miss rate for the LRU-PRO cache replacement strategy using an estimation of resource popularity.

4.2   Analysis of the Accuracy and Execution Time of Training

This section describes the experimental setup and results of the proposed neural network model for media popularity forecasting. We compare the accuracy of a linear model and a neural network. Moreover, we compare the accuracy different training algorithms for the neural network: SGD, ADAM, PSO and CPSO. In addition, we conduct an analysis of the execution time of training when using PSO and CPSO.

In order to calculate generalization error, we split the original dataset in training data, test data and validation data (60% - 20% - 20%). The performance measure is defined as the MSE between actual prediction and expected prediction (ground-truth) as in Eq. 1. We use test data for hyper-parameters optimization. We optimize hyper-parameters related to training algorithms with Sequential Model-based Algorithm Configuration (SMAC) [16]. Table 2 shows hyper-parameters of SGD, ADAM PSO and CPSO after 100 SMAC evaluations.

For model hyper-parameters, we used a grid search, increasing the model capacity until test error stops dropping [4]. Our final neural network model has logistic neurons with three fully connected feed-forward hidden layers of $26 \times 45 \times 1$ neurons. The allowed search domain for the neural network weights is $w_{i,j} \in [-5.0, 5.0]$ and the number of particles of PSO is $r = 200$.

We optimize a linear model with the Least Squared method. The generalization error for this model is 3.01E-3. For training the neural network, we used a total of 1.0E+4 iterations for SGD and ADAM with a mini-batch size of 40 instances. We used a fixed budget of 4.0E+4 fitness function evaluations for PSO algorithm. In the case of continuation versions, we used a fixed budget of 4.0E+4

Table 2. Selected hyper-parameters for each training algorithm.

Algorithm	Hyper-parameter	Value	Algorithm	Hyper-parameter	Value
SGD	Learning rate	0.25439	ADAM	First moment	0.49482
	Momentum	0.99407		Second moment	0.66610
PSO	Global	0.70267	CPSO	Global	0.58889
	Local	0.93391		Local	0.93578
	Max. speed	0.44886		Max. speed	0.49993
	Min. speed	0.56283		Min. speed	0.64033
				Proportion $(\beta^\dagger)$	0.20318
				Subsets (k)	4

fitness function evaluations that corresponds to $4.0E+4/k$ fitness function evaluations for each subset of instances.

Figure 2 shows the generalization error and training error. The bottom and top lines in the box plots represents the first and third quartile, the line in the middle represents the median of the measurements and the whiskers represents standard deviation. The small square represents the average of 50 measurements and the (x) marks represent the maximum and minimum value.

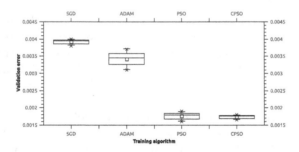

Fig. 2. Comparison between different neural network training algorithms for media popularity forecasting.

In order to find statistical significant differences between the training algorithms we performed Friedman test to find out whether the means of the generalization error of different training algorithm are statistically significant. We obtained a p-value < 0.01 and an F of 8.93 rejecting the null hypothesis of equal population means. Furthermore, Nemenyi post-hoc test revealed statistically significant differences between SGD-PSO (p-value < 0.01), SGD-CPSO (p-value < 0.01), ADAM-PSO (p-value $= 0.02$) and ADAM-CPSO (p-value < 0.01) for a 0.05 level of confidence.

We calculated execution time of training in a Core i3 microprocessor at 2.8 GHz with 4 GB of RAM running in an Ubuntu 18.04 operating system.

For PSO, we see an average execution time of 5.45E+7 s with a standard deviation of 3.18 s in 50 measurements. For CPSO we see an average execution time of 3.27E+7 s with a standard deviation of 12.54 s that represents a reduction of 28.56% of the execution time respect PSO.

5   Conclusions and Recommendations

In this paper we presented an approach to improve cache miss rate of video CDN based in a hybrid reactive-proactive cache replacement strategy (LRU-PRO). We reduced cache miss rate by improving video popularity forecasting accuracy based on neural networks trained by Particle Swarm Optimization and continuation. Continuation allows to reduce execution time of training without statistical significant loss of accuracy in about 28.56%. We see a relation between video popularity forecasting accuracy and cache miss rate reduction.

Although, we see an improvement in cache miss rate when using LRU-PRO replacement policy, there is substantial room for improvement when considering the optimal cache replacement policy. Further research should investigate robust cache replacement policies that can manage inexact popularity information. In addition, further reduction of popularity forecasting generalization error is desirable to reduce cache miss rate.

References

1. Ahmed, H.M., Youssef, B.A., Elkorany, A.S., Saleeb, A.A., El-Samie, F.A.: Hybrid gray wolf optimizer–artificial neural network classification approach for magnetic resonance brain images. Appl. Opt. **57**(7), B25–B31 (2018)
2. Alghamdi, F., Mahfoudh, S., Barnawi, A.: A novel fog computing based architecture to improve the performance in content delivery networks. Wirel. Commun. Mob. Comput. **2019** (2019)
3. Applegate, D., Archer, A., Gopalakrishnan, V., Lee, S., Ramakrishnan, K.: Optimal content placement for a large-scale VoD system. IEEE/ACM Trans. Netw. **24**(4), 2114–2127 (2016)
4. Bengio, Y.: Practical recommendations for gradient-based training of deep architectures. In: Montavon, G., Orr, G.B., Müller, K.-R. (eds.) Neural Networks: Tricks of the Trade. LNCS, vol. 7700, pp. 437–478. Springer, Heidelberg (2012). https://doi.org/10.1007/978-3-642-35289-8_26
5. Bonyadi, M.R., Michalewicz, Z.: Particle swarm optimization for single objective continuous space problems: a review (2017)
6. Borst, S., Gupta, V., Walid, A.: Distributed caching algorithms for content distribution networks. In: 2010 Proceedings IEEE INFOCOM, pp. 1–9. Citeseer (2010)
7. Carlsson, N., Eager, D.: Ephemeral content popularity at the edge and implications for on-demand caching. IEEE Trans. Parallel Distrib. Syst. **28**(6), 1621–1634 (2017)
8. Chatterjee, S., Dey, N., Ashour, A.S., Drugarin, C.V.A.: Electrical energy output prediction using cuckoo search based artificial neural network. In: Yang, X.-S., Nagar, A.K., Joshi, A. (eds.) Smart Trends in Systems, Security and Sustainability. LNNS, vol. 18, pp. 277–285. Springer, Singapore (2018). https://doi.org/10.1007/978-981-10-6916-1_26

9. Chen, W., Wang, X.A., Zhang, W., Xu, C.: Phishing detection research based on PSO-BP neural network. In: Barolli, L., Xhafa, F., Javaid, N., Spaho, E., Kolici, V. (eds.) EIDWT 2018. LNDECT, vol. 17, pp. 990–998. Springer, Cham (2018). https://doi.org/10.1007/978-3-319-75928-9_91

10. Feng, G., Qin, S., Yum, T.S.P., Cao, G., et al.: Multi-agent reinforcement learning for efficient content caching in mobile D2D networks. IEEE Trans. Wirel. Commun. **18**, 1610–1622 (2019)

11. Friedlander, E., Aggarwal, V.: Generalization of LRU cache replacement policy with applications to video streaming. arXiv preprint arXiv:1806.10853 (2018)

12. Goian, H.S., Al-Jarrah, O.Y., Muhaidat, S., Al-Hammadi, Y., Yoo, P., Dianati, M.: Popularity-based video caching techniques for cache-enabled networks: a survey. IEEE Access (2019)

13. Goodfellow, I., Bengio, Y., Courville, A.: Deep Learning. MIT Press, Cambridge (2016)

14. Hassine, N.B., Minet, P., Marinca, D., Barth, D.: Popularity prediction-based caching in CDN. Ann. Telecommun. **74**, 1–14 (2019)

15. Heredia, Y.H., Núñez, V.M., Shulcloper, J.R.: IWAIPR 2018. LNCS, vol. 11047. Springer, Cham (2018). https://doi.org/10.1007/978-3-030-01132-1

16. Hutter, F., Hoos, H.H., Leyton-Brown, K.: Sequential model-based optimization for general algorithm configuration. In: Coello, C.A.C. (ed.) LION 2011. LNCS, vol. 6683, pp. 507–523. Springer, Heidelberg (2011). https://doi.org/10.1007/978-3-642-25566-3_40

17. Izvorski, A., Deven, R., Dharmapurikar, M.: Method and apparatus for controlling source transmission rate for video streaming based on queuing delay. US Patent 9,860,605, 2 January 2018

18. Jiang, Y.G., Wu, Z., Wang, J., Xue, X., Chang, S.F.: Exploiting feature and class relationships in video categorization with regularized deep neural networks. IEEE Trans. Pattern Anal. Mach. Intell. **40**(2), 352–364 (2018)

19. Karedla, R., Love, J.S., Wherry, B.G.: Caching strategies to improve disk system performance. Computer **27**(3), 38–46 (1994)

20. Koch, C., Pfannmüller, J., Rizk, A., Hausheer, D., Steinmetz, R.: Category-aware hierarchical caching for video-on-demand content on YouTube. In: Proceedings of the 9th ACM Multimedia Systems Conference, pp. 89–100. ACM (2018)

21. Koch, C., Werner, S., Rizk, A., Steinmetz, R.: Mira: proactive music video caching using convnet-based classification and multivariate popularity prediction. In: 2018 IEEE 26th International Symposium on Modeling, Analysis, and Simulation of Computer and Telecommunication Systems, pp. 109–115. IEEE (2018)

22. Lee, M.C., Ji, M., Molisch, A.F., Sastry, N.: Performance of caching-based D2D video distribution with measured popularity distributions. arXiv preprint arXiv:1806.05380 (2018)

23. Liu, Z., Dong, M., Gu, B., Zhang, C., Ji, Y., Tanaka, Y.: Fast-start video delivery in future internet architectures with intra-domain caching. Mob. Netw. Appl. **22**(1), 98–112 (2017)

24. Livni, R., Shalev-Shwartz, S., Shamir, O.: On the computational efficiency of training neural networks. In: Advances in Neural Information Processing Systems, pp. 855–863 (2014)

25. Megiddo, N., Modha, D.S.: Outperforming LRU with an adaptive replacement cache algorithm. Computer **37**(4), 58–65 (2004)

26. Mobahi, H., Fisher III, J.W.: A theoretical analysis of optimization by Gaussian continuation. In: AAAI, pp. 1205–1211 (2015)

27. Potdar, K., Pardawala, T., Pai, C.: A comparative study of categorical variable encoding techniques for neural network classifiers. Int. J. Comput. Appl. **175**(4), 7–9 (2017)
28. Trzciński, T., Rokita, P.: Predicting popularity of online videos using support vector regression. IEEE Trans. Multimed. **19**(11), 2561–2570 (2017)
29. Cisco Visual: Cisco visual networking index: forecast and trends, 2017–2022. Technical report, Cisco Visual (2017)
30. Zhang, Y., et al.: Proactive video push for optimizing bandwidth consumption in hybrid CDN-P2P VoD systems. In: IEEE INFOCOM 2018-IEEE Conference on Computer Communications, pp. 2555–2563. IEEE (2018)

Piggybacking Detection Based on Coupled Body-Feet Recognition at Entrance Control

Dirk Siegmund[1(✉)], Vinh Phuc Tran[1], Julian von Wilmsdorff[1], Florian Kirchbuchner[1], and Arjan Kuijper[2]

[1] Fraunhofer Institute for Computer Graphics Research (IGD), Fraunhoferstrasse 5, 64283 Darmstadt, Germany
{dirk.siegmund,vinh.phuc.tran,julian-von.wilmsdorff, florian.kirchbuchner}@igd.fraunhofer.de
[2] Technische Universität Darmstadt, Hochschulstr. 10, 64289 Darmstadt, Germany

Abstract. A major risk of an automated high-security entrance control is that an authorized person takes an unauthorized person into the secured area. This practice is called "piggybacking". Known systems try to prevent it by using physical barriers combined with sensory or camera based algorithms. In this paper we present a multi-sensor solution for verifying the number of persons that stand within a defined transit area. We use sensors that are installed in the floor to detect feet as well as camera shots taken from above. We propose an image-based approach that uses change detection to extract motion from a sequence of images and classify it by using a convolutional neural network. Our sensor-based approach shows how user interactions can be used to facilitate safe separation. Both methods are computationally efficient so they can be used in embedded systems. In the evaluation, we were able to achieve state-of-the-art results for both approaches individually. Merging both methods sustainably prevents piggybacking, at a BPCER of 7.1%, where bona fide presentations are incorrectly classified as presentation attacks.

1 Introduction

The main goal of an automated access control system is the prevention of unrestricted access of an unauthorized person. Such un-staffed control systems can be found at many different places like office buildings, prisons or airports. Usually, access is granted providing a physical item (e.g. key) or a biometric property. A general problem is that an authorized person can take an unauthorized person into the secured area. This practice is entitled as "tailgating" or "piggybacking". Many barriers like (drop-arm-) turnstiles are therefore equipped with sensors like infrared break-beams in order to eliminate this vulnerability. However, these available systems are designed to achieve high flow rates and can easily be defeated. In places where higher security is needed, mantrap portals or video surveillance systems are being used. These systems regulate the access

© Springer Nature Switzerland AG 2019
I. Nyström et al. (Eds.): CIARP 2019, LNCS 11896, pp. 780–789, 2019.
https://doi.org/10.1007/978-3-030-33904-3_74

of only a single person through a transit space. Permitted subjects enter and close the portal, so that a software can verify the number of people present in the transit space. After a successful verification, the system unlocks a second door to give access to the secured area. Previous research in the field of tailgating/piggybacking prevention shows that none of the existing systems is completely safe. Computer vision approaches using video captures can detect intrusion analyzing movements and distinguish it from certain behavior. Often, optical flow is used in such application to detect motion. In contrast, we present an approach based on an change detection that uses an adaptive background model. Compared to optical flow this method is more computational efficient and allows real-time calculation as we will show late in the paper. Another advantage over optical flow is that the data can be augmented more easily because the original feature space is preserved. The fundamental problem of all imaging approaches, however, is that in practice they are limited to the field of view of the camera. Persons who want to overcome these systems can do so in the simplest case by hiding behind a permitted person (see Sect. 2). Sensors mounted in the floor can here be of great help, but are limited to the fact that all feet must be on the floor (Fig. 1).

Fig. 1. Bona fide authentification of mantrap portal (left) piggybacking (right)

In this paper we combine both approaches and establish a system that can not be overcome easily. We are the first to show how to couple the use of ground-based position data of the foot with an image-based algorithm from the top-view. In this multi-model approach we verify the number of feet by using an active user interaction scheme and couple it with an image-based verification from the top-view perspective. The used dataset includes multiple humans in a scene, close to each other and causing occlusion and illumination changes. It consists of 21 2D image frames from a regular camera, taken during 3–4 s of recording [1]. We want to point out three important benefits of our method: (1) Our system is designed to work on low-cost hardware like a Raspberry Pi in real time. This is important, as images of these cameras should not be transmitted to be processed elsewhere than the place where they are mounted. (2) An interactive capacitive sensing approach ensures a limited amount of conductive material on the floor (see Sect. 3). (3) Our new feature descriptor maximize movement detection by using foreground segmentation and convolutional neural networks (see Sect. 3.3). A discussion of related work follows in Sect. 2 while we present our results obtained in Sect. 4.

2   Related Work

In research, several computer-vision methods have been developed in order to overcome the weaknesses of non-smart entrance gates. Most methods use a top-view perspective and pattern recognition methods to distinguish between one and more than one person in an observed area. An imaging approach using thermal sensors was introduced by Siegmund et al. [2] but shows disadvantages especially, in cases where the intruder uses equipment to hide himself. RGB-D images have been used by the same authors to create models of different verification attempts [3] and evaluated them in attempts with and without identity claim. Their method consists of change- and blob detection and uses an AdaBoost machine-learning classifier where they achieved an EER of 5% for scenarios with id. claim and of 11% without id. claim. When analyzing sequences of 2D images, change detection using a gaussian mixture model was used in order to detect and count contours [4]. Rauter introduced a motion based head-shoulder detector in order to detect intrusion [5]. Optical flow is a strong feature descriptor as it extracts motion and direction besides its position. A latter method using that descriptor achieved an EER of 5.17% by creating histograms of image sequences and classifying them via machine learning [1]. The disadvantage of optical flow is the high computational cost. A drawback of all these methods is the camera angle, which allows people to hide on the floor or between the legs of a permitted person. A recently published study makes use of capacitive sensors on the ground to detect and count feet [6] on the floor. It is build upon a floor-based indoor positioning system in grid layout [7]. Its active capacitive measuring system [8] is efficient for remote sensing and can reliably recognize a foot in up to 10 cm height. An obvious limitation, is when two people are standing with only one foot each on the sensor surface.

3   Methodology

We assume that people have vested interest in coming through the access control point with as little hassle as possible. Previous studies have shown that users "optimize" their behavior in a way that it is convenient to them [9]. Nevertheless, a system that can be used in practice must find a compromise of usability and safety. But there are other things that need to be considered when developing such a system:

1. The detection method must be flexible with regard to light and clothing as well as different stature of the users.
2. The proposed solution should verify a subject, without the need to claim their identity.
3. Authorized subjects sometimes need to pass carrying different objects, which can be of and kind and appearance.
4. The proposed method should be reasonably fast, in order to be used in an embedded computer close by.

Camera based method do not seem sufficient to guarantee reliable piggybacking detection as they suffer from the limited viewing range of the camera. For this reason, we monitor the floor area by additional verification, verifying that there is nobody hiding on the floor. In the next Section, we therefore present an approach, which interactively controls the that area by means of a capacitive grid of sensors. In Sect. 3.3, we present a new image-based method that combines time-based change detection and convolutional neural networks (CNN) in compliance with the conditions mentioned here.

3.1   Dataset

In the image-based method we use the dataset introduce by Siegmund et al. [1] which includes 60 bona fide verification attempts and 216 piggybacking attacks by 12 different participants (see Fig. 2). The participants cover a wide range of physical characteristics, like different height, weight, body shape. The attack schemes were shot with two subjects present at a transit area. Six different scenarios are carried out where the attackers showed different approaches to spoof the system and/or hide. Each recording consists of a total of 21 RGB images, recorded over a period of 3–4 s. In case of the capacitive sensing grid a test-group, consisting of 12 people with different shoe size (between 37 and 48) was acquired. Each subject was recorded at least once alone and several times with another subject. When evaluating the combined approach, a total of 87 assaults was carried out in different compositions of the test group. The test group was also explained the function of the system and direct feedback of their success was provided.

Fig. 2. Some of the attack attempts included in the database.

3.2   Capacitive Sensing Grid

In an earlier study, an approach using capacitive active feet detection sensors was presented [6]. Capacitive sensors are proximity sensors that detect nearby conductive objects by creating an electric field [10] (Fig 3).

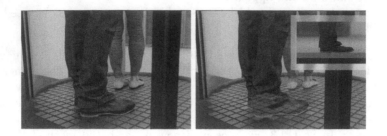

Fig. 3. Schematic representation of capacitive verification.

Since the range of these sensors depend on the size of the electric field, it is possible to detect feet even away from the ground. Therefore, this technology is particularly suitable for the application described. We use the same sensors as the authors of that paper which provide a continuous signal for analysis at a frequency of 4 Hz. In our prototype, we have selected a monitored area of 800×800 mm, which acts as the transit area. We mounted 7×7 sensors in the floor, located in a grid used for the alignment. The sensors are mounted in the middle of each cell at a horizontal distance of 100 mm between each sensor. We use a copper plate as electrode because it shows the best ratio of range and sensibility compared to other material. The initial capacitance value of each sensor acts as a baseline value.

Active Feet Verification. Previous studies have shown that although capacitive sensors are able to reliably detect feet, but they can not always ensure that they are only a single person. We think that access to high-security areas can be expected to include following interaction of the person entitled to access. First, we propose to ask the to put both feet on the ground. In a next step, the user is asked to lift one foot. So if more than two people are in the area, the intruder would now have to prevent all his feet from touching the ground. We evaluated this procedure in a first scenario by using marked positions on the floor and in a second in which the user is able to freely choose the position of their feet on the floor. Since the measured capacitance changes even when approaching a sensor, we define a threshold ϵ above which a sensor is considered activated. For this we asked our test group to stand on certain sensors areas without touching the surrounding ones. We then calculated the difference between activated and surrounded sensors for all sensors. We determined ϵ based on the minimum difference between activated and surrounding sensors plus 20% of the delta. We interpret the sensor grid output as image sg with x rows and

y columns. Equation 1 applies fixed-level thresholding to the nth single channel matrix $sg^n(x, y)$, $n = 0.0, \ldots, 1.0$ using ϵ as threshold.

$$dst^n(x, y) = \begin{cases} sg^n(x, y) \ if \ sg^n(x, y) > \epsilon \\ 0 \qquad\qquad otherwise \end{cases} \tag{1}$$

We get the activated sensors in the resulting image dst where $dst^n(x, y)$ is not 0. In order to detect the lift of a foot, we evaluate for a period of 8 frames whether the previously defined sensors have been activated. If this is the case, the user is asked to lift one leg. Successful validation is achieved when the number of activated sensors has halved in at least three out of eight dst images. We determined the number of only three validation images through experiments that revealed that users need some reaction time. In the second scenario, where users were not given a marked position, successful validation also takes place in two steps. First, the number of activated sensors is counted over 8 frames. The number must not exceed a defined number of sensors. Then it is validated whether the number of activated sensors has halved in the second step.

3.3 Image Based Approach

Our method is based on extracting motion features from image sequences using change detection. The reason for this is that a learning algorithm based on the very complex and limited data that we use can be difficult to generalize. Therefore, the complexity must be reduced without losing the information necessary for the detection. Since the background model dynamically adapts to any background, the proposed method is applicable to every background. This is done by finding the difference between the current and previous frames (background model subtraction). For the first three background frames from each scene we create a model of background pixels by K Gaussian's and then check the weight of the mixture representing color proportions. After calculating the background we take the next frames as the foreground and calculate the difference from the background frame.

Motion Detection via Background Subtraction. Our method presented here aims learning models for the cases of bona fide and attack. We assume that the amount, intensity and location of movements in a room differ according to whether one or two people are in it. In doing so, we interpret pixels in the foreground mask as movements. For this reason, objects that are carried by people are not getting included into the feature vector, if they do not move. In our dataset, for each shot situation there are image sequences consisting of 21 pictures collected over 3–4 s, we denote them as $i_0, i_1, \ldots i_{20}$. The first three frames in a sequence are not getting used for feature extraction as they are needed for training the change detection algorithm. For calculating the movement models, we use the frame instances $i_3, i_4, \ldots i_{20}$ over time. Thereby, we are able to determine the changes detected in image CD from the background model to the next time instance denoted as $CD_{i_3:}, CD_{i_4}, \ldots CD_{i_{20}}$. We use a Gaussian mixture model based approach [11] where the decision that a pixel belongs to the background is made if:

$$p(x^{\rightarrow(i)}|BG > c_{thr}(= p(x^{\rightarrow(i)}|CD)i(CD)|p(BG)), \qquad (2)$$

where c_{thr} is a threshold value and the value of a pixel at time i in RGB is denoted by $x^{\rightarrow(t)}$. We will refer $p(x^{\rightarrow}|BG)$ as the background model. The background model is estimated from a training set i_0, i_1, i_2. The estimated model is denoted by $p(x^{\rightarrow}|X, BG)$ and depends on the training set as denoted explicitly. So we calculate the background model using the first three frames and set the learning rate to 0.001 afterward. By doing so, the model is getting updated every 1000 frames which is slow enough to detect all changes background and foreground in the following 18 frames and fast enough to capture changes in e.g. the illumination conditions. We apply change detection frame by frame calculating individual grayscale foreground masks (see Fig. 4) for each instance. Another property that we want to depict is the amount of movement. We do so by accumulating the individual foreground masks to a single result image dst. Equation 3 scales each foreground mask dividing the each pixel of the single-channel mask $CD_i(x, y)$, ranging from $0, \dots, 255$ by 255 and weighting it with a factor of σ. In our experiments we achieved the best results with a σ of 30.

$$dst^n(x, y) = (CD_i^n(x, y)/255) * \sigma, \qquad (3)$$

By doing so, pixel that got recognized as foreground multiple times get a higher value than pixel that have been foreground only for a short time. Therefore, micro movements get visible in the resulting image dst (see Fig. 4).

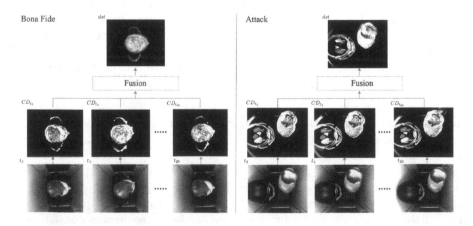

Fig. 4. Generation of feature vectors for both scenarios.

For the classification task we decided to train a convolutional neural network classifier. As we expect that our relatively small dataset could cause the network to show overfitting, we decided to augment the data as follows. First we mirror each *dst* image horizontally and vertically, then we rotate them clockwise 179 times by 2°. On 10% of the data we add additionally Gaussian noise in order to improve generalization of the network. We balance both classes of our trainings data by skipping some rotation steps for the attack image classes. After data augmentation we received 33.124 images of the bona fide class and 41.041 images in the attack class.

Learning a Binary Classifier. Following method is architectural inspired of the proposed Google-LeNet [12] but uses an architecture that is quite different from a traditional CNN design like LeNet-5 model. We used inception modules, which perform multiple convolution operations and max pooling in parallel. Therefore its not obligatory to choose a certain convolution kernel size for a certain layer. This approach is not just efficient in classification results but also in computational efficiency. The reason for the computational gain is 1×1 convolution operation which is applied before every 3×3 or 5×5 convolution of the inception module, it results in dimensionality reduction.

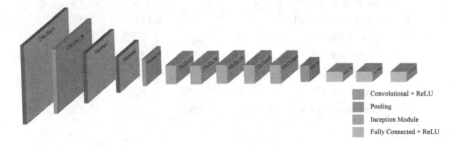

Convolutional + ReLU
Pooling
Inception Module
Fully Connected + ReLU

Fig. 5. CNN architecture based on GoogLeNet model.

Proposed architecture (see Fig. 5) takes a grayscale image of 256×256 pixel as input. To avoid internal covariate shift batch normalization is used for all convolutional and fully connected layer [13]. All weights are initialized with a normal distribution, using a standard deviation of 0.1 and zero mean. A batch size of 300 is used with the equal representation of both classes. The Training process ran for 200 epoch for each fold of the data set and each epoch contained 3 batches. As both classes are mutually exclusive, a softmax classifier is used with cross-entropy loss function. For all fully-connected layers a dropout of 0.5 is implemented. For stochastic optimization, an Adam optimizer is used with a learning rate of 0.01. Except the final layer, all layers including those inside the inception modules use ReLU activation. Sigmoid activation is used for the final layer.

4 Experiments and Results

We performed individual experiments in order to ensure that our assumptions about both approaches are correct. The experiments are evaluated based on

APCER[1] and BPCER[2]. As the imaging dataset is collected under conditions where the users did not follow any constrains regarding their position, we can not give any separate results. However, it can be assumed that the results on marked position would rather improve here. Due to the small amount of data available, the evaluation is performed using a leave-one-out approach. To make sure that there is no augmented data in the training set, we omitted the complete shot. Using only the imaging approach #4 we achieved an APCER of 1.93% at BPCER of 3.80%. We observed false positives especially in cases where a second person was hiding on the floor. However, it must be said that the attackers had no knowledge about the algorithm used and therefore could not respond specifically.

Table 1. Results in comparison with competitive methods.

No	Scenario	Marked Pos.		Random Pos.	
		APCER	BPCER	APCER	BPCER
1	Thermal Variance [2]	-	-	20.2%	20.2%
2	Morph. on RGB-D [3]	-	-	11.0%	11.0%
3	Optical Flow [1]	-	-	5.2%	5.7%
4	BG Substr NN	-	-	2.5%	6.1
5	Capacitive [6]	5.4%	5.4%	7.1%	7.1%
6	Capacitive LF	0	5.3%	6.3%	18.1%
7	Combining #4 and #6	0%	7.8%	4.3%	15.2%

Experimented with the number of sensors activated in the sensing grid we found our, that a number of 2×4 sensors represent a good compromise between flexibility against great feet and safety. In the case of the marked position on the ground we could not detect any successful attacks in these experiments. However, there were cases in which unintentionally surrounding sensors were activated, which led to an increased BPCER. The approach without marking the position on the floor got circumvented in particular when feet were arranged diagonally to the grid. Since we did not use machine learning in comparison to the comparative study, a single threshold for the activation sensors proved to be too weak. Due to the good results in the case of detection of attacks, we have conducted a fusion on decision level to combine both approaches. Through this procedure, we were able to successfully detect all attacks, but a BPCER of 7.1% must be accepted (Table 1).

---

[1] APCER: Proportion of attack presentations using the same PAI species incorrectly classified as bona fide presentations in a specific scenario.

[2] BPCER: Proportion of bona fide presentations incorrectly classified as presentation attacks in a specific scenario.

5   Conclusion

We presented a novel approach for identifying attacks in an autonomous access control system. We identified piggybacking attacks in which attackers try to pass through the system separately by using image and floor-mounted sensors. Our evaluation proved the layout and performance of the proposed interactive sensor-grid and recognized all piggybacking attacks when combined with the imaged based method. A limitation of the system is the requirement of the user to place himself on a position marked on the ground, since the performance otherwise decreases strongly. Both presented methods do not require high computing power and can therefore be used on single-board computers.

References

1. Siegmund, D., Fu, B., Samartzidis, T., Wainakh, A., Kuijper, A., Braun, A.: Attack detection in an autonomous entrance system using optical flow. In: 7th International Conference on Crime Detection and Prevention (ICDP 2016), pp. 1–6. IET (2016)
2. Siegmund, D., Handtke, D., Kaehm, O.: Verifying isolation in a mantrap portal via thermal imaging. In: 2016 International Conference on Systems, Signals and Image Processing (IWSSIP), pp. 1–4, May 2016
3. Siegmund, D., Wainakh, A., Braun, A.: Verification of single-person access in a mantrap portal using RGB-D images. In: XII Workshop de Visao Computacional (WVC), November 2016
4. Chan, T.W., Yap, V.V., Soh, C.S.: Embedded based tailgating/piggybacking detection security system. In: 2012 IEEE Colloquium on Humanities, Science and Engineering (CHUSER), pp. 277–282. IEEE (2012)
5. Rauter, M.: Reliable human detection and tracking in top-view depth images. In: Proceedings of the IEEE Conference on Computer Vision and Pattern Recognition Workshops, pp. 529–534 (2013)
6. Siegmund, D., Dev, S., Fu, B., Scheller, D., Braun, A.: A look at feet: recognizing tailgating via capacitive sensing. In: Streitz, N., Konomi, S. (eds.) DAPI 2018. LNCS, vol. 10922, pp. 139–151. Springer, Cham (2018). https://doi.org/10.1007/978-3-319-91131-1_11
7. Braun, A., Heggen, H., Wichert, R.: CapFloor – a flexible capacitive indoor localization system. In: Chessa, S., Knauth, S. (eds.) EvAAL 2011. CCIS, vol. 309, pp. 26–35. Springer, Heidelberg (2012). https://doi.org/10.1007/978-3-642-33533-4_3
8. Braun, A., Wichert, R., Kuijper, A., Fellner, D.W.: Capacitive proximity sensing in smart environments. J. Ambient Intell. Smart Environ. 7(4), 483–510 (2015)
9. Perš, J., Sulić, V., Kristan, M., Perše, M., Polanec, K., Kovačič, S.: Histograms of optical flow for efficient representation of body motion. Pattern Recogn. Lett. 31(11), 1369–1376 (2010)
10. Baxter, L.K.: Capacitive Sensors: Design and Applications. Wiley, Hoboken (1996)
11. Zivkovic, Z., et al.: Improved adaptive Gaussian mixture model for background subtraction. In: ICPR (2), pp. 28–31. Citeseer (2004)
12. Szegedy, C., et al.: Going deeper with convolutions. In: Computer Vision and Pattern Recognition (CVPR) (2015)
13. Ioffe, S., Szegedy, C.: Batch normalization: accelerating deep network training by reducing internal covariate shift. arXiv preprint arXiv:1502.03167 (2015)

Author Index

Printed in the United States
By Bookmasters